Lecture Notes in Computer Science 16972

Founding Editors

Gerhard Goos
Juris Hartmanis

The series Lecture Notes in Computer Science (LNCS), including its subseries Lecture Notes in Artificial Intelligence (LNAI) and Lecture Notes in Bioinformatics (LNBI), has established itself as a medium for the publication of new developments in computer science and information technology research, teaching, and education.

LNCS enjoys close cooperation with the computer science R & D community, the series counts many renowned academics among its volume editors and paper authors, and collaborates with prestigious societies. Its mission is to serve this international community by providing an invaluable service, mainly focused on the publication of conference and workshop proceedings and postproceedings. LNCS commenced publication in 1973.

Gernot A. Fink · Alicia Fornés · Koichi Kise ·
Daniel Lopresti
Editors

Document Analysis and Recognition – ICDAR 2026

20th International Conference
Vienna, Austria, August 31 – September 2, 2026
Proceedings, Part I

Springer

Editors
Gernot A. Fink
TU Dortmund University
Dortmund, Germany

Alicia Fornés
Universitat Autònoma de Barcelona
Barcelona, Spain

Koichi Kise
Osaka Metropolitan University
Osaka, Japan

Daniel Lopresti
Lehigh University
Bethlehem, PA, USA

ISSN 0302-9743 ISSN 1611-3349 (electronic)
Lecture Notes in Computer Science
ISBN 978-3-032-36022-9 ISBN 978-3-032-36023-6 (eBook)
https://doi.org/10.1007/978-3-032-36023-6

This Springer imprint is published by the registered company Springer Nature Switzerland AG
The registered company address is: Gewerbestrasse 11, 6330 Cham, Switzerland

Foreword

Welcome to the 20th IAPR International Conference on Document Analysis and Recognition (ICDAR 2026), which took place from August 30 to September 4, 2026, at TU Wien in Vienna, Austria. Located in the heart of Vienna's city center, the venue places participants within walking distance of many of the city's world-famous historical and cultural attractions.

ICDAR 2026 continues a long-standing tradition of bringing together researchers, academics, and professionals from around the world in the field of Document Analysis and Recognition. Since the first ICDAR in Saint-Malo, France, in 1991, the conference has been held in Tsukuba, Japan (1993); Montreal, Canada (1995); Ulm, Germany (1997); Bangalore, India (1999); Seattle, USA (2001); Edinburgh, UK (2003); Seoul, Korea (2005); Curitiba, Brazil (2007); Barcelona, Spain (2009); Beijing, China (2011); Washington, DC, USA (2013); Nancy, France (2015); Kyoto, Japan (2017); Sydney, Australia (2019); Lausanne, Switzerland (2021); San Jose, USA (2023); Athens, Greece (2024); and most recently Wuhan, China (2025). ICDAR serves as a leading international forum for researchers and practitioners working on document analysis and recognition, covering topics such as optical character recognition, handwriting and text recognition, layout analysis, document understanding, graphics recognition, historical documents, digital libraries, document forensics, and scene-text analysis.

The choice of TU Wien as the conference venue reflects the academic character of ICDAR and its close connection to university-based research. With a history spanning more than 200 years, TU Wien provides an excellent environment for scholarly exchange. As a university of technology, it covers a broad spectrum of scientific inquiry, ranging from fundamental research to applied technological research and industry collaboration. For more than two centuries, TU Wien has been a place of research, teaching, and learning in the service of progress.

The scientific program will feature three keynote talks. The first keynote will be delivered by C. V. Jawahar, Professor at IIIT Hyderabad, India, and Co-Founder of the Center for Visual Information Technology (CVIT), who will discuss the changing landscape of Document Understanding. The second keynote, by Christoph Rass, Professor at Osnabrück University, Germany, will address what historians need from Document Understanding, considering the topic from archives to algorithms and back. Finally, Tong Sun, Senior Director at Adobe, will share her perspective on the next frontier: from Document Understanding to Document Agency.

In keeping with previous ICDAR conferences, ICDAR 2026 is a rigorously peer-reviewed, single-track conference focusing on document analysis and recognition. All submissions underwent a double-blind review process with nearly three reviews, evaluating originality, research quality, or analysis of experience, relevance to document analysis systems, and clarity of presentation. The technical program will include 40 oral presentations and 108 poster presentations over three days. In addition, the conference will host both traditional and new workshops and tutorials on focused topics, offering opportunities for in-depth study of emerging problems in the field. Participants will also be able to learn about new algorithms and results through the competitions that continue to drive research interest in the community.

A conference such as ICDAR depends on the commitment of volunteers who contribute tirelessly to the program, infrastructure, and organization. The General Chairs would first like to thank the Program Co-Chairs, Gernot A. Fink, TU Dortmund University, Germany; Alicia Fornés, Computer Vision Center, Spain; Koichi Kise, Osaka Metropolitan University, Japan; and Dan Lopresti, Lehigh University, USA, for their work in assembling a first-class technical program.

Special thanks are due to the Publication Chairs, Vincent Christlein, FAU Erlangen-Nürnberg, Germany, and Foteini Liwicki, Luleå University of Technology, Sweden; the Workshop Chairs, Jihad El-Sana, Ben-Gurion University of the Negev, Israel, Elisa Barney Smith, Luleå University of Technology, Sweden, and George Retsinas, NVIDIA, USA; and the Competition Chairs, Dimosthenis Karatzas, Computer Vision Center, Spain, and Srirangaraj Setlur, University at Buffalo, USA.

Gratitude is also extended to the Industry Chairs, Josep Lladós, Computer Vision Center, Spain, and Stefan Fiel, Anyline GmbH; the Tutorial Chairs, Anna Scius-Bertrand, University of Applied Sciences and Arts Western Switzerland, Switzerland, Marcus Liwicki, Luleå University of Technology, Sweden, and Anna Zhu, Wuhan University of Technology, China; the Publicity Chairs, Jean-Marc Ogier, University of La Rochelle, France, and Umapada Pal, Indian Statistical Institute, India; the Doctoral Consortium Chairs, Anna Scius-Bertrand, University of Applied Sciences and Arts Western Switzerland, Switzerland, and Momina Moetesum, National University of Sciences and Technology (NUST), Pakistan; the Posters and Demo Chairs, Eyad Elyan, Robert Gordon University Aberdeen, UK, and Liangrui Peng, Tsinghua University, Beijing, China; the Awards Chair, Richard Zanibbi, Rochester Institute of Technology, USA; and the Sponsorship Chairs, Basilis Gatos, NCSR Demokritos, Greece, and Apostolos Antonacopoulos, University of Salford, Manchester, UK, for their valuable contributions to the conference.

The conference gratefully acknowledges the financial support of its sponsors, which helps to reduce costs and support various awards, including student travel awards.

We would also like to thank the local organizing team, in particular Marco Peer, University of Fribourg, Switzerland, as well as Marvin Burges, Katharina Pois, and Food and Beverage Chair Rafael Sterzinger, all from TU Wien.

Finally, the conference would not be possible without the excellent papers contributed by the authors. The Chairs thank all authors for their contributions to ICDAR 2026 and all external reviewers for their diligent work in reviewing submissions. This program is expected to further stimulate research and provide practitioners with improved techniques, algorithms, and deployment tools. It is an honor and a privilege to share the latest developments in Document Analysis and Recognition through this exciting program.

We hope that all participants enjoy the conference, the city of Vienna, its food and culture, and many stimulating exchanges with colleagues.

June 2026

Robert Sablatnig
Florian Kleber
ICDAR 2026 General Chairs

Preface

Welcome to the proceedings of the 20th International Conference on Document Analysis and Recognition (ICDAR) 2026 – the premier international event for scientists and practitioners involved in document analysis and recognition.

This year, we received 295 valid submissions. In order to create a high-quality scientific program for the conference, we recruited 148 regular and 41 senior program committee (PC) members. Regular PC members provided a total of 817 reviews for the submitted papers (2.8 on average per paper). Senior PC members who were overlooking the review phase for typically 8 to 9 submissions took care of consolidating the evaluations expressed in the reviews and suggested paper decisions in their meta-reviews. Based on the information provided in both the reviews and the prepared meta-reviews, we the PC Chairs, then selected 140 submissions (47.5%) for inclusion in the scientific program of ICDAR 2026. From the accepted papers, 40 were selected for oral presentation, and 100 for poster presentation.

The final conference program was complemented by three more groups of presentations. First of all, we invited three renowned researchers to give keynote presentations at the conference and share with the audience their thoughts and insights on current trends and future challenges in document analysis and recognition. C. V. Jawahar, IIT Hyderabad, India, presented "The Changing Landscape of Document Understanding". Christoph Rass, Osnabrück University, Germany, presented "From Archives to Algorithms and Back: What Historians Need from Document Understanding". Finally, Tong Sun, Adobe, USA, presented "The Next Frontier: From Document Understanding to Document Agency".

In addition to the papers submitted directly to ICDAR 2026, again a special journal track was organized in collaboration with the International Journal on Document Analysis and Recognition (IJDAR). In this journal track submissions underwent the same rigorous review process as regular IJDAR submissions. PC Chairs served as Guest Editors for handling the review processes. From the 49 manuscripts submitted to the journal track, 12 were accepted (24.5%) and were published in a Special Issue of IJDAR. In addition, all papers accepted in the journal track were also included as oral presentations in the conference program.

Last, but not least, as with every ICDAR, a number of competitions were organized in conjunction with the event. After reviewing competition reports by the Competitions Chairs, eight competition reports were included as additional papers in the conference proceedings and were presented in a special Competitions session during the conference.

Finally, we would like to thank all the people who contributed to the preparation of the scientific program of ICDAR 2026, namely the authors of the scientific papers submitted to the journal track and directly to the conference, the reviewers for journal-track papers, and both our regular and senior PC members.

July 2026

Gernot A. Fink
Alicia Fornés
Koichi Kise
Daniel Lopresti
ICDAR 2026 Program Committee
Co-Chairs

Organization

General Chairs

Robert Sablatnig	TU Wien, Austria
Florian Kleber	TU Wien, Austria

Program Chairs

Gernot A. Fink	TU Dortmund University, Germany
Alicia Fornés	Universitat Autonòma de Barcelona, Spain
Koichi Kise	Osaka Metropolitan University, Japan
Daniel Lopresti	Lehigh University, USA

Publication Chairs

Vincent Christlein	FAU Erlangen-Nürnberg, Germany
Foteini Liwicki	Luleå University of Technology, Sweden

Industry Chairs

Josep Llados	Universitat Autònoma de Barcelona, Spain
Stefan Fiel	Austria

Tutorial Chairs

Anna Scius-Bertrand	University of Applied Sciences and Arts Western Switzerland, Switzerland
Marcus Liwicki	Luleå University of Technology, Sweden
Anna Zhu	Wuhan University of Technology, China

Workshop Chairs

Jihad El-Sana	Ben-Gurion University of the Negev, Israel
Elisa Barney Smith	Luleå University of Technology, Sweden
George Retsinas	NVIDIA, USA

Publicity Chairs

Jean-Marc Ogier	University of La Rochelle, France
Umapada Pal	Indian Statistical Institute, Kolkata, India

Doctoral Consortium Chairs

Anna Scius-Bertrand	University of Applied Sciences and Arts Western Switzerland, Switzerland
Momina Moetesum	National University of Sciences and Technology, Pakistan

Posters and Demo Chairs

Eyad Elyan	Robert Gordon University, UK
Liangrui Peng	Tsinghua University, China

Competition Chairs

Dimosthenis Karatzas	Universitat Autónoma de Barcelona, Spain
Srirangaraj Setlur	University at Buffalo, USA

Awards Chair

Richard Zanibbi	Rochester University of Technology, USA

Sponsorship Chairs

Basilis Gatos	NCSR Demokritos, Greece
Apostolos Antonacopoulos	University of Salford, UK

Local Organization Chairs

Marvin Burges	TU Wien, Austria
Marco Peer	University of Fribourg, Switzerland
Katharina Pois	TU Wien, Austria

Food and Beverage Chair

Rafael Sterzinger	TU Wien, Austria

Senior Program Committee

Apostolos Antonacopoulos	University of Salford, UK
Xiang Bai	Huazhong University of Science and Technology, China
Michael Blumenstein	University of Technology Sydney, Australia
Łukasz Borchmann	Snowflake, Poland
Jean-Christophe Burie	Université de La Rochelle, France
Silvia Cascianelli	University of Modena and Reggio Emilia, Italy
Joseph Chazalon	EPITA, France
Vincent Christlein	FAU Erlangen-Nürnberg, Germany
Bertrand Coüasnon	University of Rennes, IRISA, France
Mickaël Coustaty	University of La Rochelle, France
Kenny Davila	DePaul University, USA
David Doermann	University at Buffalo, USA
Véronique Eglin	INSA Lyon, France
Andreas Fischer	University of Applied Sciences and Arts Western Switzerland, Switzerland
Liangcai Gao	Peking University, China
Daichi Haraguchi	NEC, Japan
Rajiv Jain	Adobe, USA
C. V. Jawahar	IIT Hyderabad, India
Lianwen Jin	South China University of Technology, China
Dimosthenis Karatzas	Universitat Autónoma de Barcelona, Spain
Christopher Kermorvant	Teklia, France

Bart Lamiroy	Université de Reims Champagne-Ardenne, France
Cheng-Lin Liu	Jiangnan University, China
Lu Liu	Alibaba, China
Josep Lladós	Universitat Autònoma de Barcelona, Spain
Muhammad Imran Malik	National University of Sciences and Technology, Pakistan
Simone Marinai	Universitá degli Studi di Firenze, Italy
Ronaldo Messina	Mitek, France
Harold Mouchère	Nantes Université, France
Wataru Ohyama	Tokyo Denki University, Japan
Umapada Pal	Indian Statistical Institute, Kolkata, India
Thierry Paquet	University of Rouen Normandy, France
Antonio Parziale	University of Salerno, Italy
Liangrui Peng	Tsinghua University, China
Faisal Shafait	National University of Sciences and Technology, Pakistan
Elisa H. Barney Smith	Luleå University of Technology, Sweden
Jun Sun	Fujitsu, Singapore
Seiichi Uchida	Kyushu University, Japan
Jerod J. Weinman	Grinnell College, USA
Richard Zanibbi	Rochester Institute of Technology
Anna Zhu	Wuhan University of Technology, China

Program Committee

Irfan Ahmad
Alireza Alaei
Vlad Atanasiu
Ahmad Montaser Awal
Raphael Baena
Souhail Bakkali
Killian Barrere
Byron L.D. Bezerra
Ujjwal Bhattacharya
Sanket Biswas
Mélodie Boillet
Carlos Boned Riera
Alceu Britto
Jorge Calvo-Zaragoza
Edwin Carlinet
Sukalpa Chanda
Clement Chatelain
Jin Chen
Shanxiong Chen
Zhineng Chen
Christian Clausner
Mark Clement
Florence Cloppet
Denis Coquenet
Pengwen Dai
Axel De Nardin
Claudio De Stefano
Qi Dong
Antoine Doucet
Jonathan Fabrizio
Francesco Fontanella
Akio Fujiyoshi
Andrea Gemelli
Petra Gomez-Krämer
Carlos-Emiliano González-Gallardo
Sheng He

Nina S.T. Hirata
Kai Hu
Florent Imbert
Donato Impedovo
Brian K. Iwana
Maham Jahangir
Mohammed Javed
Ehsanollah Kabir
Lei Kang
Aurélie Lemaitre
Lzhenjiang Li
Xiao-Hui Li
Minghui Liao
Laurence Likforman
Yijun Lin
Chang Liu
Yuliang Liu
Artemis Llabrés
Muhammad Muzzamil Luqman
Nam Tuan Ly
Jiahao Lyu
Carlos D. Martínez-Hinarejos
Carlos Mello
Tomo Miyazaki
Momina Moetesum
Hussein Mohammed
Adrià Molina Rodríguez
Ajoy Mondal
Carlos F. Moreno-Garcia
Nibal Nayef
Clemens Neudecker
Shinichiro Omachi
Shivakumara Palaiahnakote
Mohammad T. Parvez
Marco Peer
Julien Perret
Vincent Poulain d'Andecy
Ioannis Pratikakis
Xugong Qin
Irina Rabaev
Jean-Yves Ramel
Oriol Ramos Terrades
Frédéric Rayar
Christophe Rigaud
Verónica Romero
Henry A. Rowley
Joan Andreu Sánchez
Ravi Kiran Sarvadevabhatla
Martin Schall
Ayush Kumar Shah
Anuj Sharma
Baoguang Shi
Steven J. Simske
Sukhdeep Dr. Singh
Mohamed Ali Souibgui
Yann Soullard
Daniel Stoekl
Suresh Sundaram
Solène Tarride
Kengo Terasawa
Iuliia Tkachenko
Pau Torras
Pierrick Tranouez
Oliver Tüselmann
Kurban Ubul
Stephan M. Unter
Michal Uricar
Szilard Vajda
Ernest Valveny
Emanuele Vivoli
Da-Han Wang
Qiufeng Wang
Runmin Wang
Zhihui Wang
Hongxi Wei
Jin Wei
Fei Wu
Yejing Xie
Fei Xu
Pengyu Yan
Chuang Yang
Chun Yang
Fei Yin
Gangyan Zeng
Chongsheng Zhang
Heng Zhang
Yanming Zhang
Yaping Zhang
Yuchen Zheng
Guoqiang Zhong
Yuanping Zhu

Contents

Document Image Processing

Benchmarks and Datasets

Graphics Representation

Layout Analysis

Recognition of Tables and Formulas

Text and Symbol Recognition

Document Analysis Systems 2

Spatially-Grounded Gaussian-Prior Attention for Handwritten Mathematical Expression Recognition

Ibtissem Haj Ali(✉) and Harold Mouchère

Laboratoire des Sciences du Numérique de Nantes (LS2N), Nantes Université, École Centrale Nantes, CNRS, LS2N, UMR 6004, 44000 Nantes, France
{ibtissem.hajali,harold.mouchere}@univ-nantes.fr

Abstract. Transformer-based architectures have significantly advanced Handwritten Mathematical Expression Recognition (HMER). However, standard attention mechanisms often suffer from spatial drift and lack explicit geometric grounding, especially in dense hierarchical layouts. In this paper, we propose a spatially-grounded Transformer framework that introduces a continuous geometric inductive bias into the decoding process. Our core contribution is a **Latent Gaussian Prior (LGP)** module that acts as a foveal anchor, dynamically constraining cross-attention to predicted symbol regions. This is complemented by a **Geometry-Aware Encoder** that preserves 2D structural topology through grid-augmented self-attention. Furthermore, we introduce an **adaptive gating mechanism** to decouple the spatial grounding of entity tokens from the global contextual modeling of structural markers. Evaluated on CROHME 2016/2019/2023 benchmarks, our approach consistently improves upon strong Transformer-based baselines across multiple CROHME benchmarks, demonstrating significant improvements in recognition accuracy and symbol localization robustness. Notably, the LGP effectively regularizes attention even in weakly-supervised settings, providing a more interpretable and stable decoding process.

Keywords: Handwritten Mathematical Expression Recognition · Transformer · Spatial Attention · Latent Gaussian Prior · Geometry-Aware Encoding

1 Introduction

Handwritten Mathematical Expression Recognition (HMER) is not merely an image-to-text translation task; it is fundamentally a problem of structured spatial reasoning. Unlike linear text, mathematical notation encodes semantics in two dimensions, where spatial configurations—such as superscripts, subscripts, fractions, and nested radicals—convey intrinsic structural meaning. Recent encoder-decoder architectures based on the Transformer [9] have significantly improved recognition accuracy by leveraging self-attention to capture long-range

G. A. Fink et al. (Eds.): ICDAR 2026, LNCS 16972, pp. 3–19, 2027.
https://doi.org/10.1007/978-3-032-36023-6_1

dependencies. However, these models typically process visual features as flattened 1D sequences, relying primarily on content similarity and absolute positional encodings. This transition from 2D manifolds to 1D sequences often results in a "structural gap," where the model discards the inherent geometric topology of the expression. Consequently, such mechanisms provide only weak structural constraints, which may lead to unstable attention drift when decoding deeply nested or spatially dense mathematical structures.

Existing attempts to address this limitation often rely on auxiliary structural objectives (e.g., symbol counting) [17], structural decoding (e.g., tree-based transformers) [14], or position-enhanced architectures [21]. Nevertheless, these approaches face a fundamental trade-off: while unconstrained attention offers flexibility, it is prone to spatial instability; conversely, hard geometric supervision improves localization but often requires costly annotations and reduces the model's ability to generalize across diverse handwriting styles. An ideal solution should introduce a spatial inductive bias that encourages structural consistency without enforcing rigid, hand-crafted constraints.

Inspired by human visual perception, which processes complex mathematical notation through localized foveal fixations rather than global sequential scanning [4], we propose a **Latent Gaussian Spatial Prior (LGP)** for cross-attention modeling. Instead of enforcing rigid coordinate alignment, our approach introduces a learnable 2D Gaussian distribution that dynamically guides the decoder's attention toward coherent spatial regions. This prior acts as a soft spatial constraint, anchoring the attention around physically consistent areas while preserving the flexibility of the Transformer to capture global context. Importantly, our framework is designed to be annotation-agnostic: the Gaussian prior can be trained in a weakly-supervised manner using only LaTeX transcriptions. When spatial annotations are available, such as in the CROHME23 datasets, we introduce a **Spatial Grounding Loss** to optionally align the predicted Gaussian centers with ground-truth symbol locations, further enhancing localization accuracy and interpretability. To the best of our knowledge, this is one of the first HMER approaches to integrate a continuous 2D probabilistic spatial prior directly into Transformer cross-attention, enabling intrinsic geometric grounding in an end-to-end manner without relying on discrete sampling strategies or external detection modules.

Precise spatial anchoring in the decoder, however, requires inherently structured visual representations. To this end, we propose a **Relative Geometry-Aware Encoder** that explicitly models pairwise geometric relationships between grid cells of the feature map before the flattening operation. By embedding structured spatial interactions into the latent space, the encoder provides a topologically organized representation that facilitates the decoder's Gaussian grounding.

The contributions of this work are threefold: (i) We introduce the **Latent Gaussian Spatial Prior (LGP)**, a learnable continuous spatial bias integrated into the attention mechanism that provides foveal grounding for entity symbols; (ii) we propose a **Relative Geometry-Aware Encoder** that preserves the 2D topology of mathematical expressions, mitigating the information loss caused by feature flattening; (iii) we demonstrate through comprehensive evaluations on the CROHME benchmarks that incorporating geometric inductive bias consistently improves performance over state-of-the-art Transformer baselines, particularly in complex structural scenarios, with further gains achieved through optional spatial supervision.

2 Related Work

Handwritten Mathematical Expression Recognition (HMER) is a long-standing problem in document analysis, aiming to convert two-dimensional handwritten expressions into structured markup languages such as LaTeX [1]. Compared with conventional handwritten text recognition, HMER is considerably more challenging due to nested hierarchical layouts, variable symbol scales (e.g., superscripts and fractions), and ambiguities introduced by diverse handwriting styles.

Early approaches relied on explicit symbol segmentation followed by grammar-based structural analysis, often using context-free grammars or tree transformations to generate parse trees [2,3,5]. While these syntactic and sequential systems provided interpretable results, they required handcrafted rules and struggled to generalize across notation variations.

With the advent of deep learning, HMER shifted toward end-to-end encoder–decoder architectures that implicitly learn symbol recognition and structural relationships through attention mechanisms.

2.1 Sequence Modeling for HMER

Early neural approaches such as WAP [7] and DenseWAP [8] introduced CNN–RNN architectures with attention-based decoding [6]. A convolutional encoder extracts visual features, and a recurrent decoder predicts LaTeX tokens sequentially. PAL and PAL-v2 [10] further enhanced performance via adversarial learning on handwritten–printed pairs, improving style invariance. However, recurrent decoders struggle to model long-range dependencies and the complex two-dimensional structure of mathematical notation.

Transformer-based approaches, including BTTR [11] and CoMER [12], replaced recurrent decoding with multi-head self-attention, enabling stronger global context modeling. BTTR introduced bidirectional training to leverage forward and backward language information, while CoMER incorporated coverage modeling to mitigate repetitive attention and improve alignment consistency. Despite these advances, attention in these models remains primarily content-driven, with spatial alignment implicitly learned.

2.2 Structural Modeling and Auxiliary Supervision

Recognizing the hierarchical and relational nature of mathematical expressions, several works incorporate explicit structural priors. Tree-based models such as TAMER [13] and TST [14] integrate syntactic structure prediction alongside sequence decoding, improving structural validity. Graph-based approaches, including LGAP [15] and edge-weighted graph attention networks [16], formulate HMER as a graph construction problem to enhance interpretability and symbol relationship modeling.

Other methods introduce auxiliary supervision to regularize decoding. The Counting-Aware Network (CAN) [17] supervises symbol counting to mitigate over- and under-parsing. General Category Networks [18] incorporate coarse-grained recognition as a supplementary task. Weakly supervised learning strategies and ICAL [19] enhance decoder robustness via additional semantic signals. Syntax-Aware Network (SAN) [20] models decoding as a grammar-guided tree traversal process.

While effective, these methods operate at the sequence or structural level, relying on auxiliary objectives or discrete modeling rather than directly constraining spatial attention distributions within the Transformer framework.

2.3 Spatial Awareness and Geometry-Constrained Attention

Recent works emphasize the importance of spatial information in HMER. PosFormer [21] introduces a position forest representation to encode hierarchical spatial relationships, while PACM [22] integrates visual density maps with linguistic modeling to alleviate semantic bias.

Beyond HMER, spatial bias has been incorporated into attention mechanisms in broader sequence modeling tasks. Deformable attention [24] introduces sparse offset-based sampling for object detection. Gaussian-based attention mechanisms have been explored in neural machine translation [25] and automatic speech recognition [26] to encourage locality and monotonic alignment. However, most existing approaches rely on discrete sampling strategies or one-dimensional locality assumptions. Continuous two-dimensional probabilistic spatial modeling for structured layout recognition remains underexplored.

2.4 Our Position

In contrast to structural supervision, auxiliary correction tasks, or discrete sampling mechanisms, we introduce a continuous geometric inductive bias directly into Transformer cross-attention. Our method factorizes attention into content similarity and spatial likelihood through a dynamically predicted Gaussian prior, while enriching encoder representations with relative geometric features. This formulation preserves end-to-end differentiability and models spatial grounding intrinsically within the attention mechanism.

3 Methodology

3.1 Overview

Handwritten Mathematical Expression Recognition (HMER) requires modeling complex two-dimensional spatial layouts while generating a linear LaTeX sequence. Unlike natural text recognition, mathematical notation encodes semantic meaning through hierarchical and spatial relationships such as superscripts, subscripts, fractions, and radicals.

Given an input handwritten expression image $X \in \mathbb{R}^{H \times W}$, our objective is to predict the corresponding LaTeX sequence

$$Y = \{y_1, y_2, \ldots, y_T\}.$$

Our architecture is built upon the CoMER framework [12], following an encoder–decoder Transformer paradigm. While CoMER effectively integrates coverage modeling via an Attention Refinement Module (ARM), it remains fundamentally content-driven and lacks explicit spatial grounding during decoding.

To address this limitation, we introduce a spatially grounded decoding mechanism based on a **Latent Gaussian Prior (LGP)**. Additionally, inspired by PosFormer [21], we explicitly distinguish between entity and structural tokens to prevent attention corruption caused by syntactic symbols.

Our model consists of three main components, as illustrated in Fig. 1:

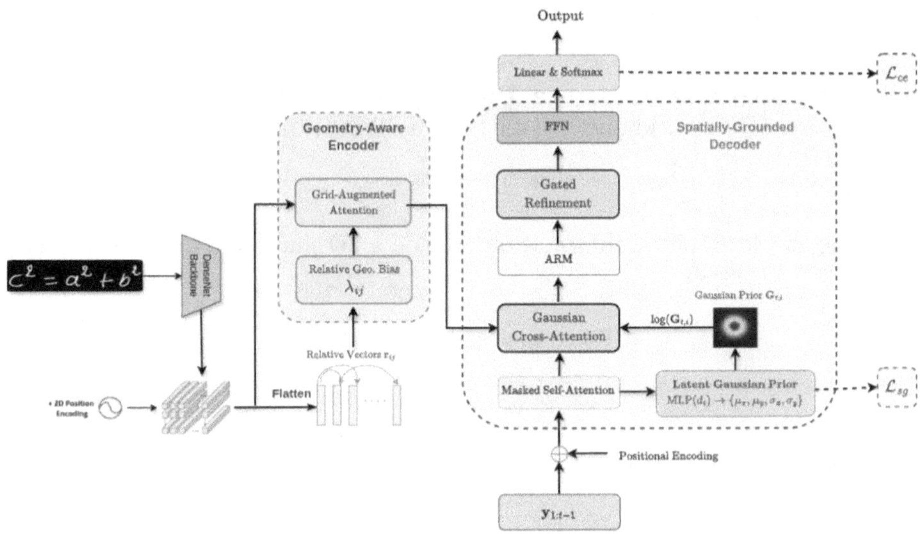

Fig. 1. Overall architecture of the proposed model. The framework consists of three main components: (1) a Geometry-Aware Encoder with grid-augmented self-attention and relative geometry bias, (2) a Gaussian Cross-Attention module integrated into the Transformer decoder, and (3) an Attention Refinement Module (ARM) with gated refinement for enhanced feature modeling.

- **DenseNet Backbone**: extracts low-level visual features.
- **Geometry-Aware Encoder**: augments self-attention with relative geometric bias through a Grid-Augmented mechanism.

- **Spatially-Grounded Transformer Decoder**: integrates a Latent Gaussian Prior into cross-attention and applies a gated refinement strategy for structural tokens.

3.2 Geometry-Aware Encoder

Mathematical expressions are inherently defined by precise two-dimensional spatial arrangements. In standard Transformer-based encoders, the extracted 2D feature map is flattened into a 1D sequence before being processed by self-attention layers. Although effective for sequence modeling, this flattening operation inevitably discards the intrinsic geometric topology of mathematical notation.

To mitigate this limitation, we propose a **Geometry-Aware Encoder** that incorporates a Grid-Augmented (GA) module. The proposed design preserves the structural topology of the expression by combining absolute positional encoding with a learnable relative geometric bias. As a result, the self-attention mechanism remains anchored to the original 2D spatial manifold rather than operating purely in a flattened sequence space.

Visual Feature and Absolute Positioning. Given an input image $X \in \mathbb{R}^{H\times W}$, a DenseNet backbone [27] extracts a 2D feature map:

$$V \in \mathbb{R}^{h\times w\times d_{model}}.$$

To encode absolute spatial information, we add a 2D sinusoidal positional encoding $P \in \mathbb{R}^{h\times w\times d_{model}}$:

$$\tilde{V} = \text{Flatten}(V + P) \in \mathbb{R}^{N\times d_{model}}, \tag{1}$$

where $N = h \times w$.

Relative Geometry Bias. For two grid cells i and j located at normalized coordinates (x_i, y_i) and (x_j, y_j), we define:

$$r_{ij} = [\log(|x_i - x_j| + \epsilon), \log(|y_i - y_j| + \epsilon)]^T. \tag{2}$$

The geometric bias is computed as:

$$\lambda_{ij}^{g(h)} = w_g^{(h)T}\text{ReLU}\left(W_g^{(h)} r_{ij}\right), \tag{3}$$

where $W_g^{(h)} \in \mathbb{R}^{d_g\times 2}$ and $w_g^{(h)} \in \mathbb{R}^{d_g}$ are head-specific learnable parameters.

Grid-Augmented Multi-head Self-attention. The attention energy for head h is defined as:

$$\text{Energy}_{ij}^{(h)} = \frac{Q_i^{(h)}(K_j^{(h)})^T}{\sqrt{d_k}} + \lambda_{ij}^{g(h)}. \tag{4}$$

The attention weights are computed as:

$$A_{ij}^{(h)} = \frac{\exp(\text{Energy}_{ij}^{(h)})}{\sum_{k=1}^{N} \exp(\text{Energy}_{ik}^{(h)})}. \tag{5}$$

The output feature is:

$$Z_i^{(h)} = \sum_{j=1}^{N} A_{ij}^{(h)} V_j^{(h)}. \tag{6}$$

3.3 Spatially-Grounded Transformer Decoder

The decoder follows the three-layer Transformer structure of CoMER [12]. Each layer consists of: Masked Self-Attention, Cross-Attention, and a Feed-Forward Network.

Additionally, CoMER introduces an Attention Refinement Module (ARM) to model coverage and alleviate over-parsing and under-parsing.

At decoding step t, the decoder hidden state d_t predicts the next token:

$$p(y_t|y_{<t}, X) = \text{softmax}(W_o d_t). \tag{7}$$

Standard Cross-Attention. Given decoder query Q_t and encoder keys K_i, the energy score is:

$$E_{t,i} = \frac{Q_t K_i^T}{\sqrt{d_k}}. \tag{8}$$

The attention weights are computed via softmax:

$$A_{t,i} = \frac{\exp(E_{t,i})}{\sum_{j=1}^{N} \exp(E_{t,j})}. \tag{9}$$

The context vector is:

$$C_t = \sum_{i=1}^{N} A_{t,i} V_i. \tag{10}$$

CoMER refines attention using coverage accumulation:

$$\hat{E}_{t,i} = E_{t,i} - \phi(A_{t-1}), \tag{11}$$

where $\phi(\cdot)$ extracts coverage features.

Spatial Prior for Grounded Decoding. Although coverage modeling reduces repetition errors, cross-attention remains purely similarity-driven. In dense expressions containing repeated symbols, attention may drift toward incorrect spatial regions. We introduce the LGP as a *foveal anchor*. Inspired by the human visual system, the model learns to predict a 2D "glance" area for each step. This provides a geometric constraint that guides the attention to remain focused on a specific physical entity, effectively regularizing the search space of the Multi-Head Attention.

Given decoder state d_t, a lightweight MLP predicts spatial parameters:

$$(\hat{\mu}_x, \hat{\mu}_y, \hat{\rho}_x, \hat{\rho}_y) = \mathrm{MLP}(d_t). \tag{12}$$

They are constrained as:

$$\mu_x, \mu_y = \mathrm{Sigmoid}(\hat{\mu}_x, \hat{\mu}_y), \tag{13}$$

$$\sigma_x, \sigma_y = \mathrm{Softplus}(\hat{\rho}_{x,y}) + \epsilon. \tag{14}$$

Note that Softplus, defined as $f(x) = \ln(1 + \exp(x))$, provides a smooth, non-zero approximation of the ReLU function, ensuring that the Gaussian variance remains strictly positive and numerically stable during backpropagation. The spatial prior over encoder grid locations (x_i, y_i) is defined as:

$$G_{t,i} = \exp\left(-\frac{(x_i - \mu_x)^2}{2\sigma_x^2} - \frac{(y_i - \mu_y)^2}{2\sigma_y^2}\right). \tag{15}$$

After normalization:

$$\tilde{G}_{t,i} = \frac{G_{t,i}}{\sum_j G_{t,j} + \eta}. \tag{16}$$

The prior is injected into the cross-attention energy in log-space:

$$E^*_{t,i} = \frac{Q_t K_i^T}{\sqrt{d_k}} + \log(\tilde{G}_{t,i} + \eta) - \phi(A_{t-1}), \tag{17}$$

where $\tilde{G}_{t,i}$ is the normalized Gaussian map and $\phi(A_{t-1})$ represents the ARM coverage refinement. This formulation can be interpreted as a probabilistic factorization where attention is the product of content similarity, spatial likelihood, and coverage regularization.

We apply the LGP only in the first decoder layer. According to the analysis in "Attention Is All You Need" [9], lower layers capture local alignment patterns while higher layers encode more abstract semantic dependencies. Constraining early layers ensures stable spatial grounding without limiting high-level syntactic reasoning.

Gated Refinement for Structural Tokens. Mathematical LaTeX vocabulary contains two fundamentally different token types:

- **Entity tokens**: visible symbols with physical correspondence in the image (digits, variables, operators).
- **Structural tokens**: syntax markers describing hierarchical relations (e.g., {, ^, _).

Structural tokens do not correspond to localized visual entities. For such tokens, enforcing spatial grounding is unnecessary and may corrupt coverage modeling.

Following the observation of PosFormer [21], we prevent structural tokens from influencing spatial refinement.

We define the following indicator function:

$$I_t = \begin{cases} 1 & y_t \in \Omega_{entity} \\ 0 & y_t \in \Omega_{struct} \end{cases} \tag{18}$$

The cross-attention energy is then computed as:

$$E_{t,i}^{final} = \frac{Q_t K_i^T}{\sqrt{d_k}} + I_t \cdot \log(\tilde{G}_{t,i} + \eta) - I_t \cdot \phi(A_{t-1}). \tag{19}$$

Thus:

- For entity tokens ($I_t = 1$), spatial grounding and coverage refinement are applied.
- For structural tokens ($I_t = 0$), the model falls back to standard content-based attention without spatial constraint.

Additionally, attention maps corresponding to structural tokens are excluded from coverage accumulation to avoid phantom coverage effects.

3.4 Training Objective

To anchor the LGP, we supplement the standard Cross-Entropy loss ($\mathcal{L}_{ce}$) with a task-specific supervision for spatial grounding.

Spatial Grounding Loss ($\mathcal{L}_{sg}$). When symbol-level coordinates are available, we supervise the predicted Gaussian centers. Unlike standard bounding box regression, our **Spatial Grounding Loss** only supervises the "foveal center" of entity symbols:

$$\mathcal{L}_{sg} = \frac{1}{T} \sum_{t=1}^{T} I_t \cdot \text{SmoothL1}\left((\mu_t^x, \mu_t^y), (\bar{x}_t, \bar{y}_t)\right) \tag{20}$$

where $(\bar{x}_t, \bar{y}_t)$ are the ground-truth centroids. The indicator I_t ensures that the model is not penalized for the lack of localization on purely syntactic tokens.

Total Objective. The final optimization objective is a weighted combination of the sequence generation and spatial localization tasks:

$$\mathcal{L}_{total} = \mathcal{L}_{ce} + \lambda\mathcal{L}_{sg} \tag{21}$$

We set $\lambda = 0.1$ to maintain the primacy of the LaTeX prediction while providing sufficient signal for the LGP modules to converge. This dual supervision forces the model to develop an internal 2D coordinate system, significantly improving performance on complex, and nested expressions.

4 Experiments

4.1 Datasets and Metrics

Datasets. We evaluate our proposed framework on the most widely recognized benchmarks for HMER: the **CROHME** series.

- **CROHME 2023**: We utilize Task 2 (Off-line) of the latest ICDAR 2023 competition [31]. The training set consists of 20,979 images (combining rendered and real handwritten formulas). The testing set contains 2,300 samples. Unlike previous versions, we exploit the symbol-level Bounding Box (BBOX) annotations provided in the 2023 metadata to supervise our Gaussian center prediction $(\bar{x}_t, \bar{y}_t)$ via the Spatial Grounding Loss.
- **CROHME 2016/2019**: To ensure comparability with existing state-of-the-art models [12], we also evaluate on earlier versions. We follow the standard protocol using 8,836 expressions for training. The testing sets consist 1,147 for CROHME 2016 [29], and 1,199 for CROHME 2019 [30].

Evaluation Metrics. The primary metric is the **Expression Recognition Rate (ExpRate)**, denoting the percentage of expressions recognized with zero errors. We also report accuracy for ≤ 1 and ≤ 2 symbol errors.

4.2 Implementation Details

Network Architecture. Our encoder utilizes a DenseNet backbone with three blocks, each containing $D = 16$ bottleneck layers, and a transition layer with a reduction factor $\theta = 0.5$. In the **Geometry-Aware Encoder**, we set the geometric head dimension $d_g = 64$ and utilize $h = 8$ heads for the Grid-Augmented Multi-Head Self-Attention. The 2D feature map is enriched with sinusoidal positional encodings before applying the relative geometric bias.

For the decoder, we employ a 3-layer Transformer with $d_{model} = 256$, $h = 8$, and $d_{ff} = 1024$. The Latent Gaussian Prior (LGP) module consists of a 2-layer MLP with a hidden dimension of 128. The Attention Refinement Module (ARM) [12] is implemented with a 5×5 kernel size and an intermediate dimension $d_c = 32$, starting from the second decoder layer.

Training Strategy. We use the PyTorch framework [33] with SGD optimizer (momentum 0.9, weight decay 10^{-4}). The initial learning rate is 0.08. To stabilize the integration of the Geometry-Aware modules and the LGP, we implement a **Learning Rate Warm-up** for the first 5 epochs, followed by a standard cosine annealing schedule. This warm-up phase is critical to allow the spatial anchoring mechanism to converge before the model optimizes the linguistic sequence generation.

Data augmentation [32] includes scale factors $s \in [0.7, 1.4]$. All experiments are conducted on NVIDIA GPUs with a batch size of 8. For inference, we perform approximate joint search as used in CoMER [12]. Symbol-level bounding box annotations provided in CROHME 2023 are used only during training to supervise Gaussian center prediction. No spatial annotations are required during inference, ensuring that the proposed method remains fully end-to-end and does not depend on detection modules.

4.3 Comparison with State-of-the-Art

Fair Comparison Without Spatial Supervision. To ensure a fair comparison with previous work [12], we first evaluate our model **without Spatial Grounding Loss**, i.e., without using any bounding box supervision.

We report the original results of CoMER [12] as published with scale augmentation, following the standard evaluation protocol.

Table 1 reports the comparison with representative state-of-the-art methods on CROHME 2016/2019. Our method achieves ExpRates of **62.87%** (2016), and **63.56%** (2019), consistently outperforming prior Transformer-based and structure-aware models.

Compared to CoMER [12], which achieves 59.81%, and 62.97% respectively, our method yields improvements of +3.06%, and +0.59%. We also surpass recent structural modeling approaches such as TAMER [13] and SSAN [23], demonstrating that introducing a continuous geometric inductive bias is more effective than discrete tree supervision or auxiliary correction tasks.

Results on CROHME 2023. We further evaluate our method on CROHME 2023. The results are reported in Table 2. To ensure fairness, we reproduce CoMER [12] on the same training split without spatial supervision.

The reproduced CoMER baseline achieves 60.12% ExpRate, 78.00% for $\leq$1 error, and 79.87% for $\leq$2 errors.

Our method without spatial supervision already improves performance significantly, achieving 63.28% ExpRate, 80.54% for $\leq$1, and 82.78% for $\leq$2 errors, corresponding to a +3.16% gain in ExpRate over CoMER.

Effect of Spatial Supervision. CROHME 2023 provides symbol-level bounding box annotations. By enabling Spatial Grounding Loss during training, the ExpRate further improves from 63.28% to **64.69%**, while the $\leq$1 and $\leq$2 accuracies increase to **82.85%** and **84.85%**, respectively.

Table 1. Comparison with state-of-the-art methods on CROHME2016/2019 (ExpRate %).

Method	2016	2019
DenseWAP [8]	47.50	-
CAN-ABM [17]	56.15	55.96
SAN [20]	53.60	53.50
GCN [18]	58.94	61.63
BTTR [11]	52.31	52.96
CoMER [12]	59.81	62.97
ICAL [19]	58.79	60.51
SSAN [23]	60.07	62.14
TAMER [13]	60.26	61.97
Ours (no spatial supervision)	**62.87**	**63.56**

Importantly, spatial annotations are used *only during training* and are not required at inference time. The moderate yet consistent improvement demonstrates that the proposed Latent Gaussian Prior already captures spatial alignment effectively in a **weakly-supervised** manner, while explicit centroid supervision further stabilizes localization for complex and densely nested expressions.

Table 2. Results on CROHME 2023.

Method	ExpRate	≤ 1	≤ 2
CoMER (reproduced) [12]	60.12	78.00	79.87
Ours (no spatial supervision)	63.28	80.54	82.78
Ours (+ Spatial Supervision)	**64.69**	**82.85**	**84.85**

4.4 Ablation Study

To validate the effectiveness of each component in our framework, we conduct ablation experiments on CROHME 2023, all results are reported in Table 3. All variants are trained under identical settings. We progressively introduce the proposed modules on top of the CoMER baseline [12].

Effect of Geometry-Aware Encoder. Introducing the Geometry-Aware (GA) encoder improves ExpRate from 60.12% to 61.32%. This demonstrates that explicitly modeling relative spatial relationships between grid cells enhances the structural representation before decoding.

Table 3. Ablation study on CROHME 2023.

Model Variant	ExpRate	≤ 1	≤ 2
Baseline (CoMER re-trained)	60.12	78.00	79.87
+ Geometry-Aware Encoder (GA)	61.32	–	–
+ Latent Gaussian Prior (LGP)	61.89	–	–
+ LGP (Supervised)	62.70	–	–
+ GA + LGP	63.92	–	–
+ GA + LGP + Gated Refinement (Full Model)	**64.69**	**82.85**	**84.85**

Effect of Latent Gaussian Prior. Adding the Latent Gaussian Prior (LGP) without supervision further increases performance to 61.89%, confirming that spatially grounded cross-attention reduces alignment drift. When Gaussian centers are supervised using symbol-level centroids, performance rises to 62.70%, indicating that explicit spatial anchoring stabilizes the localization mechanism.

Synergy Between GA and LGP. Combining GA and LGP leads to a substantial gain (63.92%), suggesting that spatially enriched encoder features and probabilistic spatial grounding in the decoder are complementary.

Full Model with Hybrid Structural Gating. Finally, integrating the gated refinement mechanism yields the best performance (64.69%), achieving a +4.57% absolute improvement over the baseline. This confirms that separating entity tokens from structural tokens prevents spatial corruption and improves alignment consistency.

Layer-Wise Placement of LGP. An important design choice is where to inject the Gaussian spatial prior within the 3-layer Transformer decoder. We conduct experiments by applying the LGP at different levels, as summarized in Table 4.

- **First Layer Only**: This configuration achieves the best results. We hypothesize that the first layer acts as the primary spatial alignment stage. Establishing a strong foveal anchor early on allows subsequent layers to focus on abstract semantic refinement and structural hierarchy modeling.
- **Last Layer Only**: Performance is lower, suggesting that if the initial alignment is diffuse, the final layer struggles to "correct" the spatial drift while simultaneously predicting the next token.
- **All Layers**: While intuitive, enforcing the Gaussian prior across all layers leads to a slight performance degradation compared to the first-layer-only setup. This likely imposes an overly rigid constraint, limiting the Transformer's ability to utilize global context for complex syntactic reasoning in deeper layers.

Table 4. Impact of LGP placement across decoder layers on CROHME 2023.

Placement	ExpRate (%)
First Layer (Ours)	**63.28**
Last Layer	61.85
All Layers	62.74

4.5 Qualitative Analysis

To evaluate the effectiveness of the Latent Gaussian Prior (LGP) and the gating mechanism, we visualize the step-wise cross-attention maps for a complex expression, $r_2^2 = (\alpha - \beta)^2$, in Fig. 2.

As illustrated, the model exhibits a dynamic, token-dependent attention behavior. For **entity tokens** (e.g., r, α, β), the attention maps show sharply localized, Gaussian-shaped activations centered on the character strokes. This confirms that the LGP acts as a spatial foveal anchor, constraining the search space and mitigating the attention drift common in purely content-based Transformers. Consequently, spatial consistency is preserved even across visually similar symbols.

Conversely, for **structural tokens** (e.g., $\{, \wedge, \},)$), the gating mechanism adaptively relaxes spatial constraints. The attention maps expand to a global pattern, capturing the hierarchical context necessary for tokens that lack a physical image correspondence. This transition between localized foveal focus and global aggregation demonstrates that the model successfully decouples visual recognition from syntactic structuring, ensuring a geometry-aware and interpretable decoding process.

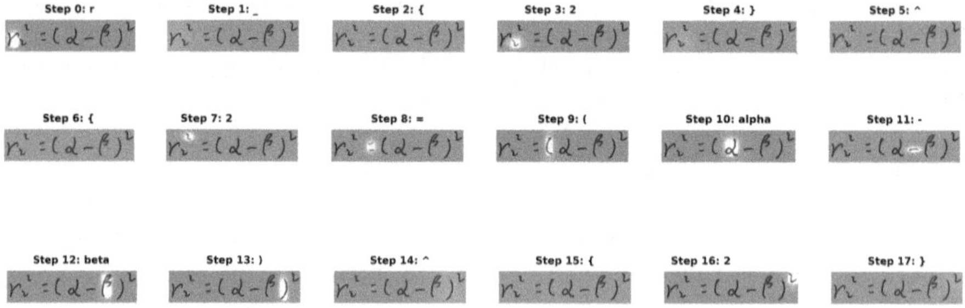

Fig. 2. Cross-attention maps at each decoding step for the expression $r_2^2 = (\alpha - \beta)^2$. Entity tokens exhibit sharp foveal localization via LGP, while structural tokens trigger global attention through the gating mechanism.

5 Conclusion

This paper presented a framework for HMER that integrates probabilistic spatial modeling within a Transformer paradigm. By factorizing attention into content similarity and a **Latent Gaussian Prior**, we established a foveal anchoring mechanism that significantly reduces attention drift. The synergy between our **Geometry-Aware Encoder** and the **gated refinement strategy** allows the model to effectively distinguish between localized visual recognition and global syntactic structuring.

Our experiments on the CROHME series demonstrate that the proposed method achieves **highly competitive results**, particularly in mitigating alignment errors in complex expressions. The qualitative analysis further confirms that the LGP successfully stabilizes foveal fixations on character strokes. While this work focuses on mathematical notation, the principle of latent geometric grounding offers a promising direction for other structured document analysis tasks where spatial-semantic alignment is critical.

References

1. Mouchère, H., Zanibbi, R., Garain, U., Viard-Gaudin, C.: Advancing the state of the art for handwritten math recognition: the CROHME competitions, 2011–2014. Int. J. Document Anal. Recognit. (IJDAR) **19**(2), 173–189 (2016)
2. Álvaro, F., Sánchez, J.A., Benedí, J.M.: Recognition of on-line handwritten mathematical expressions using 2D stochastic context-free grammars and hidden Markov models. Pattern Recognit. Lett. **35**, 58–67 (2014)
3. Álvaro, F., Sánchez, J.A., Benedí, J.M.: An integrated grammar-based approach for mathematical expression recognition. Pattern Recognit. **51**, 135–147 (2016)
4. Kohlhase, A., Fürsich, M.: Understanding mathematical expressions: an eye-tracking study. In: FM4M/MathUI/ThEdu/DP/WIP@CIKM, pp. 42–50 (2016)
5. Zanibbi, R., Blostein, D., Cordy, J.R.: Recognizing mathematical expressions using tree transformation. IEEE Trans. Pattern Anal. Mach. Intell. **24**(11), 1455–1467 (2002)
6. Truong, T.N., Nguyen, C.T., Zanibbi, R., Mouchère, H., Nakagawa, M.: A survey on handwritten mathematical expression recognition: the rise of encoder-decoder and GNN models. Pattern Recognit. **153**, 110531 (2024)
7. Zhang, J., et al.: Watch, attend and parse: an end-to-end neural network based approach to handwritten mathematical expression recognition. Pattern Recognit. **71**, 196–206 (2017)
8. Zhang, J., Du, J., Dai, L.: Multi-scale attention with dense encoder for handwritten mathematical expression recognition. In: Proceedings of ICPR, pp. 2245–2250 (2018)
9. Vaswani, A., et al.: Attention is all you need. In: Advances in Neural Information Processing Systems (NeurIPS), vol. 30 (2017)
10. Wu, J.-W., Yin, F., Zhang, Y.-M., Zhang, X.-Y., Liu, C.-L.: Handwritten mathematical expression recognition via paired adversarial learning. Int. J. Comput. Vis. **128**, 2386–2401 (2020)
11. Zhao, W., Gao, L., Yan, Z., Peng, S., Du, L., Zhang, Z.: Handwritten mathematical expression recognition with bidirectionally trained transformer. In: Proceedings of ICDAR, pp. 570–584 (2021)

12. Zhao, W., Gao, L.: CoMER: Modeling coverage for transformer-based handwritten mathematical expression recognition. In: Proceedings of ECCV. LNCS, vol. 13688, pp. 392–408. Springer, Cham (2022)
13. Zhu, J., Zhao, W., Li, Y., Hu, X., Gao, L.: TAMER: tree-aware transformer for handwritten mathematical expression recognition. In: Proceedings of AAAI, pp. 10950–10958 (2025)
14. Xie, Y., Mouchère, H.: TST: Tree structured transformer for handwritten mathematical expression recognition. In: Proceedings of ICDAR, Springer, Cham (2025)
15. Shah, A.K., Zanibbi, R.: Line-of-sight with graph attention parser for math formulas. In: Proceedings of ICDAR, pp. 104–119 (2023)
16. Xie, Y., Zanibbi, R., Mouchère, H.: Local and global graph modeling with edge-weighted graph attention network for handwritten mathematical expression recognition. arXiv preprintarXiv:2410.18555 (2024)
17. Li, B., et al.: When counting meets HMER: Counting-aware network for handwritten mathematical expression recognition. In: Proceedings of ECCV, LNCS, pp. 197–214. Springer, Cham (2022)
18. Zhang, X., Ying, H., Tao, Y., Xing, Y., Feng, G.: General category network: handwritten mathematical expression recognition with coarse-grained recognition task. In: Proceedings of ICASSP, pp. 1–5 (2023)
19. Zhu, J., Gao, L., Zhao, W.: ICAL: implicit character-aided learning for enhanced handwritten mathematical expression recognition. arXiv preprint arXiv:2405.09032 (2024)
20. Yuan, Y., et al.: Syntax-aware network for handwritten mathematical expression recognition. In: Proceedings of AAAI (2023)
21. Guan, Y., Lin, C., Shen, W., Yang, X.: PosFormer: recognizing complex handwritten mathematical expression with position forest transformer. In: Proceedings of ECCV (2024)
22. Li, Z., Wei, J., Shen, Z., Ma, C., Wu, Y., Zhou, Y.: PACM: position-aware cross-modality decoder for handwritten mathematical expression recognition. In: Proceedings of ICDAR, pp. 96–114. Springer, Cham (2025)
23. Zhang, H., Su, X., Zhou, X., Gao, G.: SSAN: a symbol spatial-aware network for handwritten mathematical expression recognition. In: Proceedings of the AAAI Conference on Artificial Intelligence, vol. 39, no. 21, pp. 22398–22406 (2025)
24. Zhu, X., Su, W., Lu, L., Li, B., Wang, X., Dai, J.: Deformable DETR: Deformable transformers for end-to-end object detection. arXiv preprint arXiv:2010.04159 (2020)
25. Zhang, X.: Modeling concentrated cross-attention for neural machine translation with Gaussian mixture model. In: Proceedings of ACL (2020)
26. Kim, S., et al.: T-GSA: transformer with Gaussian-weighted self-attention for speech enhancement. In: Proceedings of ICASSP (2021)
27. Huang, G., Liu, Z., van der Maaten, L., Weinberger, K.Q.: Densely connected convolutional networks. In: Proceedings of the IEEE Conference on Computer Vision and Pattern Recognition (CVPR), pp. 4700–4708 (2017)
28. Mouchere, H., Viard-Gaudin, C., Zanibbi, R., Garain, U.: ICFHR 2014 competition on recognition of on-line handwritten mathematical expressions (CROHME 2014). In: 14th ICFHR, pp. 791–796 (2014)
29. Mouchere, H., Viard-Gaudin, C., Zanibbi, R., Garain, U.: ICFHR 2016 CROHME: competition on recognition of online handwritten mathematical expressions. In: 15th ICFHR, pp. 607–612 (2016)

30. Mahdavi, M., Zanibbi, R., Mouchere, H., Viard-Gaudin, C., Garain, U.: ICDAR 2019 CROHME + TFD: competition on recognition of handwritten mathematical expressions and typeset formula detection. In: ICDAR 2019, pp. 1533–1538
31. Xie, Y., Mouchère, H., et al.: ICDAR 2023 CROHME: competition on recognition of handwritten mathematical expressions. In: ICDAR 2023, pp. 553–565
32. Li, Z., Jin, L., Lai, S., Zhu, Y.: Improving attention-based handwritten mathematical expression recognition with scale augmentation and drop attention. In: 17th International Conference on Frontiers in Handwriting Recognition (ICFHR), pp. 175–180 (2020)
33. Paszke, A., et al.: PyTorch: an imperative style, high-performance deep learning library. In: Advances in Neural Information Processing Systems (NeurIPS), pp. 8024–8035 (2019)

Bar-JEPA: Extracting Values from Bar Chart with Joint-Embedding Predictive Architecture

Poonam Poonam(✉), Alexander Epple, and Timo Ropinski

Institute of Media Informatics, Visual Computing Group, Ulm University, Ulm, Germany
{poonam.poonam,alexander.epple,timo.ropinski}@uni-ulm.de

Abstract. Bar charts are commonly used in data visualization, and while they are easily understood by humans, it is non-trivial to extract the underlying data computationally. For a machine-learning-based approach, training chart de-rendering models usually requires labeled, real-world data. Labeling data is a time consuming task, which is why annotated data is scarce. Models can learn more efficiently when provided with features of high semantic quality, which a joint-embedding predictive architecture (JEPA) is designed to learn in a self-supervised manner. We present a per-bar, numerical value recovery pipeline for bar charts, where a JEPA encoder is used to produce semantically rich latent features. The decoder model consuming these features is simple and quick to train and outputs the coordinates of ticks and bars, which can be used to recover bar values. The effectiveness of self-supervised finetuning and quality of the extracted features is evident when comparing our model to end-to-end supervised baselines. Code, datasets and checkpoints are available on GitHub.

Keywords: Document analysis · Bar chart value recovery · Document understanding

1 Introduction

Charts are a popular tool in data visualization, as they can effectively leverage human vision to communicate the underlying data. The raw data is often lost or unavailable when charts are published, as it is easier for humans to read charts than tables [29]. For machines on the other hand, having access to raw data is not only preferable, but necessary for tasks such as data analysis or for accessibility [17]. When extracting data from, or de-rendering charts, it is important for the recovered data to be both numerically and factually accurate, as well as true to the source. This is a challenging task, as there are many graphical and textual components, which need careful, task-specific handling and vary visually [15].

Many ways have been proposed previously to recover data from charts, ranging from semi-automatic [31], to relying on mostly handcrafted features and

G. A. Fink et al. (Eds.): ICDAR 2026, LNCS 16972, pp. 20–37, 2027.
https://doi.org/10.1007/978-3-032-36023-6_2

methods [30, 34] and machine-learning based approaches [6, 22, 25]. Deep learning methods have achieved state-of-the-art (SoTA) performance on tasks such as chart question answering, chart-to-table, and more, but require vast amounts of labeled data for training [27]. Synthetic data can readily be generated or accumulated [25], but the availability of real-world data, such as the CHART-Info 2024 dataset [8], is much more limited. The problem with synthetic data is, however, that it often does not contain artifacts and struggles to capture the full diversity of real-world data.

In this paper, we propose an approach to deal with real-world, labeled data scarcity by leveraging the self-supervised JEPA [20], which is trained on unlabeled data. We finetune I-JEPA [1], which has been pretrained on ImageNet-1K, on bar charts to produce semantically rich feature maps. For all training tasks, we generate a synthetic bar chart dataset. Then, to test the effectiveness of JEPA for chart de-rendering, we train a set of decoders to extract per-bar numerical values. The decoders first upsample the feature maps and then regress heatmaps for the bars, ticks, and coordinate system origin. They are intentionally designed to be simple and quick to train, as it is not the main focus of the paper. Another problem we address is that aspect ratios of charts can vary significantly. For example, the bar charts in the UB PMC training dataset [9], which consists of manually annotated charts, range all the way from 4.45 to 0.35. Vision transformers (ViT) [10], which I-JEPA uses, require fixed-resolution inputs, so images have to either be resized or padded. To this end, we modify JEPA to accept variable-resolution inputs as proposed in Pix2Struct [21].

To summarize the contributions of this work, we (1) propose a chart data extraction pipeline that uses I-JEPA as a feature extractor, (2) extend I-JEPA to accept variable-resolution inputs and (3) provide a synthetic, 100k bar chart dataset and accompanying generator.

2 Related Work

Our work touches two distinct topics: Using JEPA as a feature extractor for some downstream task and chart de-rendering. Our focus, however, is only on extracting values from bar charts.

2.1 Joint-Embedding Predictive Architectures

Since JEPA was proposed by LeCun in 2022 [20], it has been used for various tasks across many modalities to predict representations rather than reconstructing data. Both Fei *et al.* [11] and more recently Tuncay *et al.* [38] use JEPA for audio classification. Fei *et al.* employ a novel curriculum masking strategy, which they find to be crucial for downstream performance and note that the correct strategy depends on the input modality. Tuncay *et al.* on the other hand, note that their training regime required far less pre-training as well as less training data. Riou *et al.* [32] use JEPA for musical stem compatibility estimation, which indicates to what degree an audio file of a single instrument matches some musical context. They do fall behind their baselines in some tasks, but note the significantly lower amount of training data they had available ($\approx 1\%$ of what the baselines used). These findings support the claim that JEPA is a more data-efficient learning framework than other self-supervised methods.

Point-JEPA by Saito *et al.* [33] learns to predict point clouds, achieves SoTA performance, and the pre-training converges much faster than comparable methods. They point out that JEPA excels in environments where the available data heavily skews towards being unlabeled. Similar to charts, geometry and structure are very important in point clouds, which can be well represented by JEPA.

In the realm of vision models, I-JEPA [1] manages to improve upon Masked Autoencoders (MAE) [14] and Context Autoencoders (CAE) [5] in linear probing at a fraction of the compute. They highlight how I-JEPA converges faster than pixel reconstruction methods such as MAE and CAE, while learning features of a high semantic level. V-JEPA [3] and the more recent V-JEPA 2 [2], which build upon I-JEPA, can solve tasks such as spatio-temporal action detection and image classification without model parameter adaptation. Notably, V-JEPA 2 is particularly effective at tasks that require fine-grained motion understanding, such as pick-and-place on real-world robots. Overall, this shows that pixel-perfect reconstructions are not necessary to achieve good performance on many downstream tasks.

JEPA has also been used in textual contexts, for example TI-JEPA by Vo *et al.* [39] achieves SoTA performance on multimodal sentiment analysis, with the pre-training goal being multimodal alignment of images and text. They manage to bridge the semantic gap between text and image, and this is possible due to the model learning robust and generalizable features. Finally, T-JEPA by Thimonier *et al.* [37] uses JEPA as a pre-training technique for tabular data classification and regression. They manage to match or outperform gradient-boosted decision trees, which are considered to be the go-to method for tabular data.

Clearly, JEPA has been shown to be effective across a wide array of modalities, while being data efficient, learning semantic abstractions and converging fast. To the best of our knowledge it has, however, not yet been applied to charts.

2.2 Chart De-rendering

Many prior works have applied neural networks to tasks such as chart derendering, data recovery and plot question-and-answering (QA). The approaches range from employing convolutional neural networks (CNNs) to transformer architectures and hybrid approaches.

CNNs are a natural fit for charts, as they can effectively exploit the spatial relations of images. For bar charts specifically, object detection-based methods are particularly popular. Liu *et al.* use Faster-RCNN to extract text and values from bar and pie charts [24], while Ma *et al.* rely on ResNet-50 for feature extraction and Cascade R-CNN for element detection [26]. Both networks have multi-stage pipelines, chart specific heads and thus strong inductive biases. The performance suffers when evaluating real-world data compared to synthetic charts, which Liu *et al.* attribute to the unavailability of labeled data. Shahira *et al.* propose using Mask R-CNN for bar chart de-rendering, but also note that their approach requires a lot of training data, as well as being slow to train [35]. Several works have instead relied on point-based methods [13,36], where key points are extracted and further processed for value recovery. ChartOCR uses Hourglass Net to classify the chart type and extract key points from which data can be extracted, where the method depends on the chart type [25]. At the time, ChartOCR achieved SoTA performance and it is easy to include additional chart types, which indicates that key point based methods are a good choice for charts. In Zhou *et al.*'s work, bar charts are de-rendered by a CNN in combination with an attention mechanism and long short-term memory (LSTM) [42]. Although the attention mechanism learns to focus on key points, the outputs are vectors containing the coordinates of bar centers and normalized heights. They also note, that their method is unlikely to generalize well to unseen data due to the lack of labeled, real-world data and the diversity thereof.

With the ever growing popularity of transformers, they have also been used in chart de-rendering, being especially popular for chart QA. Pix2Struct [21], MatCha [23] and DePlot [22] all build upon each other and use ViT [10] encoders and text decoders. The models achieve SoTA results on plot QA tasks and significantly outperform the CNN-based ChartOCR [25]. Pix2Struct introduces variable-resolution inputs, as maintaining the original aspect ratio helps to improve model performance. There are many more examples of transformer-based chart-to-table models [6,27,28], which also show promising results for chart de-rendering. Key point detection is also possible with transformers as shown by Xue *et al.* in ChartDETR [41], building upon DETR [4], which is an end-to-end object detector. Their method is straightforward and, while being conceptually similar, improves upon ChartOCR in terms of robustness and performance. These findings indicate that transformers can often outperform CNN-based approaches and therefore, are a good choice for the task of chart de-rendering.

It is worth pointing out, however, that even though the training objectives are diverse, all prior work, as opposed to JEPA [1], fall into the field of supervised learning. To our knowledge, little to no work has been done on self-supervised pretraining in chart de-rendering. Existing methods focus on optimizing end-to-end performance and representation quality is rarely taken into account.

3 Method

This section covers the dataset used for training and the simplifying assumptions we make to scope the research area. As our goal is to study representation quality, we limit complexity and ensure a controlled setting. An overview of I-JEPA follows, as well as an explanation of our modifications to the framework and the self-supervised pretraining. The decoders and their supervised training regime are outlined next, before finally the numerical value extraction is detailed.

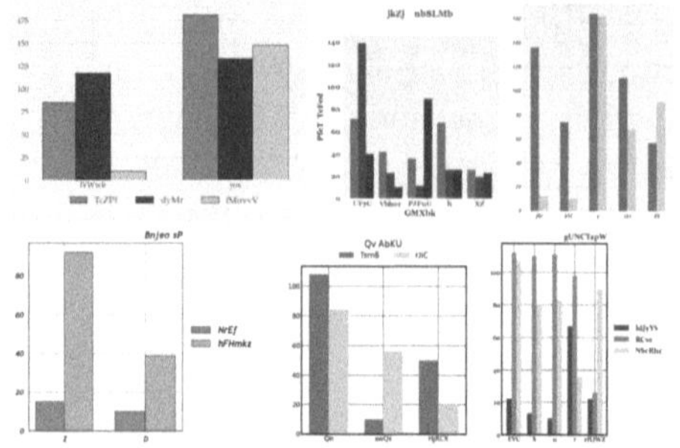

Fig. 1. Examples of generated bar charts as used for encoder finetuning and decoder training. Charts are generated according to the parameters in Table 5.

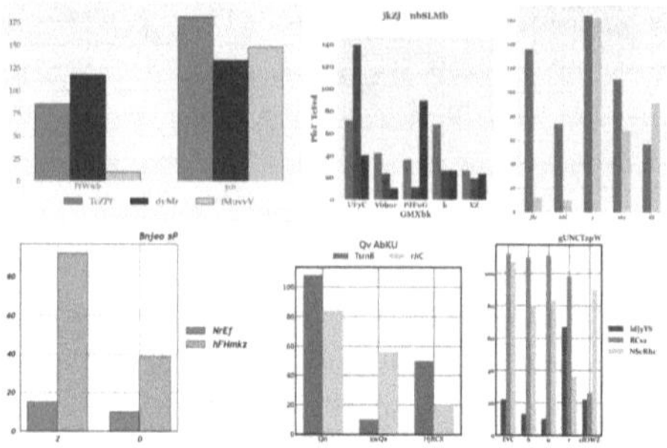

Fig. 2. Overview of the I-JEPA training procedure. The context encoder is used after training concludes, predictor and target encoder are discarded.

3.1 Synthetic Data Generator and Datasets

Bar charts come in considerable visual variety and can contain a lot of visual clutter, so in order to narrow the amount of variables, we make some assumptions for the dataset we generate for training. We only consider vertical bar charts and the bars may not be stacked, contain error bars or 3D effects. This ensures the charts have visual diversity but are kept simple. The parameters used to generate our datasets are listed in Table 5 and we use matplotlib [16] to generate charts. We extend the work of Zhou *et al.* [42] by, for example, varying which random distribution is used to generate values and by outputting a more exhaustive annotation file. The annotations contain precise bounding boxes and values of all chart elements. Some examples of the charts we generate can be seen in Fig. 1. For the I-JEPA finetuning objective, we generate $100k$ charts, all of which are used for training, as there is no validation or test step. For one experiment, we include an additional finetuning step on $15k$ real-world charts from Chart-to-Text [18]. The dataset for the decoder pretraining consists of $17k$ synthetic charts. Additionally, we use the UB PMC dataset [8] (specifically, ICPR CHART-Infographics 2022) for decoder finetuning. We can only use 1316 of the vertical bar charts, as the remaining data does not have the required labels. Both datasets are split into 80 : 20% training: validation sets.

3.2 I-JEPA Encoder

Our encoding pipeline starts with variable-resolution patch extraction, following the implementation from Pix2Struct [21]. Given a maximum number N of patches, for each input image of size (w, h) the amount of feasible rows and columns (R, C) are computed according to Eq. 1 and Eq. 2. Then, the image is resized to (W, H) and a sequence of $M \leq N$ non-overlapping patches of size p are extracted. An overview of the process can be seen in Fig. 3.

$$R = \max\left(1,\ \min\left(N,\ \left\lfloor \sqrt{N}\sqrt{\frac{h}{w}} \right\rfloor\right)\right), \qquad H = Rp \tag{1}$$

$$C = \max\left(1,\ \min\left(N,\ \left\lfloor \sqrt{N}\sqrt{\frac{w}{h}} \right\rfloor\right)\right), \qquad W = Cp \tag{2}$$

The training objective of I-JEPA is similar to that of MAEs [14], but the key differences are that it is (1) non-generative and (2) predicts in latent as opposed to pixel space. I-JEPA consists of three ViTs [10]: The context encoder f_e, target encoder $\bar{f}_e$ and predictor g_p. The context and target encoders are structurally identical, while the predictor is narrower and shallower. Given a latent context block from an encoded image and a target mask, the predictor learns to reproduce the latent patches of the target block in the same image. The target and context blocks are obtained using multi-block masking [1]. For the exact parameters, see Table 6 and refer to Fig. 2 for a visual overview of the model.

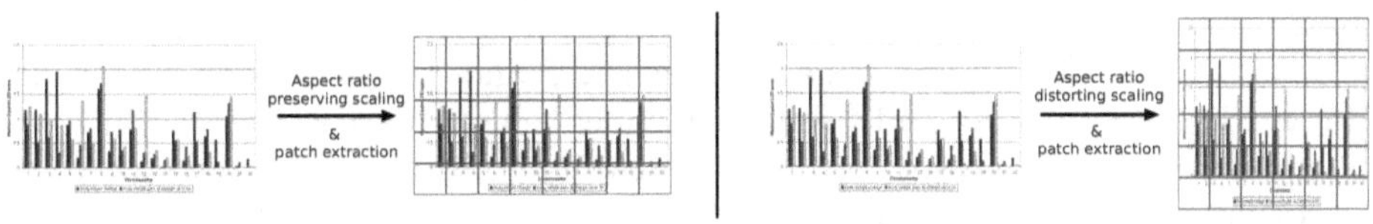

Fig. 3. Aspect-ratio preserving patch extraction ensures at most N patches are obtained after resizing an image. The operation guarantees the new size to be close to the original aspect-ratio as much as possible. The chart is an example from [9].

Targets. For each image, target blocks are obtained by first sampling i blocks B_i from the M input patches. The size and aspect-ratio of each B_i is random and the blocks can overlap. They are encoded by $\bar{f}_e$ to yield sequences $\bar{s}_i$, where each sequence $\bar{s}_i$ contains up to j latent patches. The target encoder is frozen and receives no gradient flow during backpropagation, which prevents the encoders from collapsing. Without this, the encoders' output could simply always be a constant value, which makes predictions trivial and the representations meaningless.

Context. There is usually only one context block B_x of random size and aspect-ratio, where any patch regions that overlap with the target blocks B_i are removed. The remaining patches in B_x are encoded by f_e into a sequence s_x of latent patches. The context encoder receives gradients during backpropagation and updates the target encoder $\bar{f}_e$ via exponential moving average (EMA).

Predictions. Instead of predicting all target blocks at once, the predictor g_p is invoked for each target B_i individually. It receives as input (1) the context patches s_x and (2) learned mask tokens t_i with added positional embeddings. The mask tokens t_i tell the predictor where the target block B_i is and enables it to generate the sequence $\hat{s_i}$ of latent patch predictions.

Loss. After the target and prediction sequences have been obtained, the loss $\mathcal{L}$ is calculated according to Eq. 3 by accumulating the smooth L1 loss (Eq. 4) of the targets $\bar{s_i}$ and predictions $\hat{s_i}$ for each block B_i.

$$\mathcal{L} = \sum_i \text{smoothL1}(\bar{s}_i, \hat{s}_i) \tag{3}$$

$$\text{smoothL1}(x, y) = \begin{cases} \frac{1}{2}(x-y)^2 & \text{if } |x-y| < 1 \\ |x-y| - \frac{1}{2} & \text{otherwise} \end{cases} \tag{4}$$

Training. We use the ViT-H ImageNet-1K checkpoint from the original I-JEPA paper [1] and finetune for 50 epochs. The model is trained using the AdamW optimizer, with cosine weight decay between 0.02 and 0.04 and a cosine learning rate schedule from $5.0e{-}5$ to $1.0e{-}6$ with a 6-epoch warm-up starting at $5.0e{-}6$. We train both variable-resolution and fixed-resolution models on the aforementioned dataset of $100k$ generated charts using an effective batch size of 40 on two NVIDIA RTX A6000 GPUs. For one experiment, we finetune the variable-resolution checkpoint for an additional 25 epochs on $15k$ real-world charts [18], with all hyperparameters halved. Similarly to Thimonier *et al.* [37], we observe that the loss starts at a collapsed equilibrium before rising and converging. This is also the case for I-JEPA pretraining [1] and expected behavior.

3.3 Key Point Extractor

Our extraction pipeline, which can be seen in Fig. 4, is intentionally simple and starts with the features extracted by the frozen encoder. These features with latent channel dimension size d and patch size p are of shape $F_{jepa} \in \mathbb{R}^{d \times \frac{H}{p} \times \frac{W}{p}}$. The decoder consumes these features directly and is the only part that is trained.

Decoders. To keep the decoders lightweight, we use the simple and classic decoders described in ViTPose [40] but with $N_k = 32$ heatmap channels. Conceptually, this is supposed to encourage the model to put one key point into each channel during training. We enforce that the first channel contains the coordinate system origin, channels $1-15$ ticks, and the rest is reserved for bars. This is accomplished by the ground truth heatmaps being in the given order. Ticks are to be output from the bottom to the top and bars from left to right.

The simple decoder consists of a ReLU activation, followed by $4\times$ bilinear upsampling and a 3×3 convolution. For the classic decoder, there are two blocks

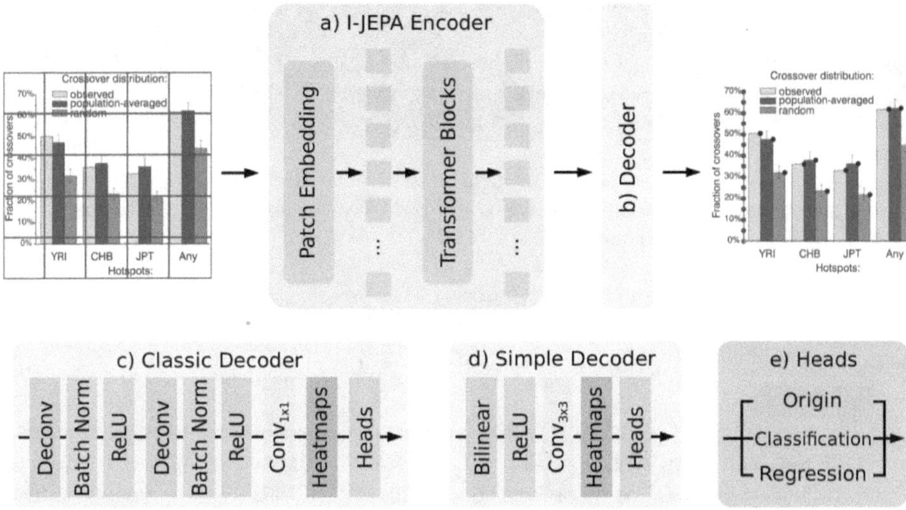

Fig. 4. The decoder pipeline consists of the frozen, pretrained I-JEPA encoder (a) and a decoder (b), which produces key point heatmaps. The heatmaps are consumed by the output heads (e), which in turn generate a combined heatmap. There are two possible configurations, a simple (d) and a slightly more sophisticated (c) decoder model. The chart is an example from [9].

each consisting of a deconvolution, batch normalization and ReLU activation. The two blocks are followed by a single 1×1 convolution. The final convolutions shape the feature maps to $F_{kp} \in \mathbb{R}^{N_k \times \frac{H}{4} \times \frac{W}{4}}$ by reducing the channels down to N_k key point maps.

Heatmap Heads. The key point maps are consumed by three heads, which produce an origin, classification and regression map and mirror the method of Soto *et al.* [36]. The origin map $H_{org} \in \mathbb{R}^{1 \times \frac{H}{4} \times \frac{W}{4}}$ helps guide learning early on and predicts the location of the coordinate system origin. The channels of the classification map $H_{cls} \in \mathbb{R}^{3 \times \frac{H}{4} \times \frac{W}{4}}$ correspond to the likelihood of a pixel being background, bar or tick. The regression map $H_{reg} \in \mathbb{R}^{2 \times \frac{H}{4} \times \frac{W}{4}}$ accounts for point offsets accrued by the low resolution. An example of the outputs can be seen in Fig. 5.

Heatmap Loss. We use a composite loss function, consisting of separate losses for each of the three output maps and an extra loss for the key point channels. Since the decoders are trained with supervision, ground truth maps are used to compute the loss (Eq. 5). We find through experimentation that the model is not able to output individual key point heatmaps into the key point channels directly, but the additional loss does significantly help the classification head to separate ticks from bars.

Fig. 5. An example of the heatmaps produced by the decoder. From left to right: Input image, (inverted) background map, top right bar corners, ticks and input image overlaid with all three heatmaps.

$$\mathcal{L}_{total} = \lambda_1 \mathcal{L}_{org} + \lambda_2 \mathcal{L}_{cls} + \lambda_3 \mathcal{L}_{reg} + \lambda_4 \mathcal{L}_{kp} \quad (5)$$

$\mathcal{L}_{org}$: The origin loss is computed as the MSE between a Gaussian target heatmap ($\sigma = 1.0$) and the output H_{org}. The origin loss is weighted equally to the classification loss ($\lambda_1 = 1.0$) and, on average, contributes $\approx 25\%$ in relation to $\mathcal{L}_{cls}$ during training.

$\mathcal{L}_{cls}$: For classification we use cross entropy on H_{cls}, with the background class being weighted 0.05, whereas bars and ticks are weighted equally at 1.0 due to class imbalance. Classification being most important for value recovery is weighted $\lambda_2 = 1.0$.

$\mathcal{L}_{reg}$: Regression uses mean squared error (MSE) and only tick and bar pixels in H_{reg} receive gradients; all other pixels are masked out. This loss is more sparse, so we set $\lambda_3 = 4.0$, but during training $\mathcal{L}_{reg}$ is, on average, only $\approx 33\%$ of $\mathcal{L}_{cls}$.

$\mathcal{L}_{kp}$: The heatmap point loss operates on the key point maps F_{kp} and is supposed to encourage the decoder to output key points earlier, so the heads can combine them into the final outputs. As described earlier, each channel corresponds to one specific key point and its loss contribution is calculated the same as $\mathcal{L}_{org}$. Only channels with an associated ground truth key point receive gradients and the loss is normalized by the number of active channels. We weight $\lambda_4 = 4.0$, although $\mathcal{L}_{kp}$ and $\mathcal{L}_{cls}$ contribute equally to the overall loss.

Training. The decoders are trained for 50 epochs on our generated dataset consisting of $17k$ charts. We use a cosine learning rate schedule from $1.0e-3$ to $1.0e-5$ and a 3 epoch warm-up starting at $2.0e-4$. Optimization is handled by AdamW with a cosine weight decay schedule between 0.04 and 0.1. The training batch size is 1360 on a single NVIDIA RTX A6000 GPU. After pretraining concludes, we finetune for 30 more epochs on the UB PMC training dataset [8]. Decoder finetuning follows the same regime as in Sect. 3.2, with a batch size of 263 on a single NVIDIA RTX A6000 GPU, the batch size being tailored to divide the small number of samples evenly.

3.4 Value Recovery

To translate the heatmaps into useful data, we follow the method of Soto *et al.* [36] with some modifications: First, the bar and tick maps from H_{cls} are background masked at a threshold of 0.75 and surviving candidates from each map with confidence $c \geq 0.75$ are transformed to normalized image coordinates. The offsets from H_{reg} are added to allow for sub-pixel accuracy. Next, confidence-based non-maximum suppression (NMS) is applied to the candidates, where the radius r_{nms} can account for variable-resolution inputs and is the equivalent distance of 1.5 pixels in heatmap size (Eq. 6). Afterwards, the remaining coordinates are the predicted bar and tick positions in normalized image space.

$$r_{\text{nms}} = \frac{1.5 \cdot p}{4 \cdot \sqrt{HW}} \tag{6}$$

The next step consists of converting tick labels to numerical data. As optical character recognition (OCR) is not the focus of this work, we use PaddleOCR [7] for this task, specifically the latin_PP-OCRv5_mobile_rec text recognition model with the language set to English. For all detections containing numerical text (determined using regular expressions), the extracted values and bounding box centers are matched with the predicted tick positions using the Hungarian algorithm [19]. A match is only considered to be valid if the distance d between label and tick is $d \leq 5.0 \times r_{\text{nms}}$ to avoid numerical text that does not stem from labels being matched with ticks. Similar to Zhou *et al.* [42], we fit a line for matched value and tick coordinates with RANSAC regression [12]. The resulting function can then be used to predict values from bar coordinates.

4 Experiments

This section outlines the metrics used for evaluation, shows the effectiveness of finetuning the encoder on domain-specific data, and explains the influence of variable-resolution inputs, as well as the decoder choice has on performance. Finally, a short comparison to prior work highlights the strengths and shortcomings of our approach.

4.1 Metrics and Test Datasets

All results are evaluated on two datasets, namely 100 synthetic charts that we generate specifically for testing and split 4 of the UB PMC [8] test set (ICPR CHART-Infographics 2022). There are only 139 vertical bar charts in this split that have the labels required for evaluation. The results we report are the F1 score for bar and tick detections, as well as the value recovery accuracy.

To calculate the F1 score, we use the Hungarian algorithm [19] with the precision threshold $r_{\text{f1}} = r_{\text{nms}}$ (Eq. 6) to match predicted and ground truth coordinates. When evaluating fixed-resolution inputs, the heatmaps are 64×64 pixels, so the error of a correct prediction is at most $\varepsilon \approx 2.3\%$. For consistency,

Table 1. Quantitative results of six models evaluated on two datasets. The variable- and fixed-resolution models, as well as the vanilla model, use the *classic* decoder from Fig. 4 and the *simple* decoder receives variable-resolution inputs. *Real-world FT* and *Oracle OCR* are finetuned versions of the variable-resolution model.

Dataset	Model	Bar F1	Tick F1	Acc. ($\varepsilon = 0.05$)	Acc. ($\varepsilon = 0.02$)
UB PMC [8]	Oracle OCR	0.785	0.842	0.512	0.389
	Real-world FT	0.785	0.842	0.499	0.365
	Variable Resolution	0.740	0.827	0.450	0.341
	Fixed Resolution	0.716	0.812	0.422	0.292
	Vanilla	0.161	0.024	0.000	0.000
	Simple Decoder	0.085	0.178	0.009	0.005
Synthetic Charts	Oracle OCR	0.961	0.951	0.783	0.652
	Real-world FT	0.961	0.951	0.792	0.657
	Variable Resolution	0.956	0.940	0.778	0.629
	Fixed Resolution	0.882	0.892	0.654	0.498
	Vanilla	0.389	0.182	0.056	0.051
	Simple Decoder	0.124	0.253	0.014	0.006

we use the same threshold for both variable and fixed-resolution inputs, even though this could, in some cases, mean a fixed-resolution detection would not be considered valid in a variable-resolution setup and vice-versa.

Value recovery is measured using accuracy. We specifically use the same criterion as Zhou *et al.* [42] as described in Eq. 7.

$$\frac{|h_g - h_p|}{h_g} \leq \varepsilon \tag{7}$$

Here, h_g is the ground truth value, h_p the predicted value and ε the strictness. For our results, we use the same relaxed ($\varepsilon = 0.05$) and hard ($\varepsilon = 0.02$) accuracy thresholds. We also account for missing and additional predictions, but for simplicity no OCR correction is applied. In an oracle OCR experiment, where detected labels were replaced with ground truths, we verified that OCR is not detrimental to prediction accuracy (see Table 1).

4.2 Ablation Study and Analysis

In order to evaluate the influence of our contributions and implementation choices, we train four different models as described in Table 2 according to the schedule outlined in Sect. 3.3.

The Influence of Encoder Finetuning. Arguably, the most important question was whether or not self-supervised encoder finetuning on charts has an impact on downstream performance. As is evident from the results in Table 1,

Table 2. The models used for the ablation studies. All encoders originate from the ViT-H ImageNet-1K checkpoint. For aspect ratio preserving (ARP) models the encoder was finetuned with ARP enabled. The decoder types are described in Fig. 4.

Model	ARP	Decoder	Encoder
Real-world FT & Oracle OCR	✓	Classic	Real-world Finetuned
Variable Resolution	✓	Classic	Finetuned
Fixed Resolution	×	Classic	Finetuned
Vanilla	×	Classic	ViT-H ImageNet-1K
Simple Decoder	✓	Simple	Finetuned

this very much appears to be the case: On synthetic charts, the F1 scores of the fixed-resolution, finetuned encoder are more than double that of the vanilla encoder. In case of value recovery, the difference is even more pronounced. This likely is the result of compounding errors, as it depends on both bars and ticks being accurately detected. When evaluating the models on real-world data, the vanilla encoder completely fails at recovering any values, whereas the finetuned encoder has a relaxed accuracy of 42%. With the only difference between these two models being the additional training step, self-supervised finetuning on charts clearly is able to yield a significantly more capable encoder for downstream tasks. Additionally, the method continues to scale with more data, as finetuning the encoder further on real-world data yields even better results (e.g., 5% better relaxed accuracy on the real-world test set).

The Influence of Variable-Resolution Inputs. The second question we want to answer concerns the impact of variable-resolution inputs. In Pix2Struct [21], this simple change resulted in $\approx 5\%$ accuracy uplift in their warm-up stage. We see a similar benefit in value recovery accuracy, the difference being $\approx 5\%$ for real world data and $\approx 13\%$ for synthetic charts at the hard accuracy threshold. Considering this modification comes at negligible cost during training and inference time, the integration is well worth it.

The Influence of Decoder Choice. Finally, we find that the results of ViTPose [40] do not transfer to our method. Whereas they see almost no performance difference between the two decoders for human pose estimation, the simple decoder performs significantly worse for our task. It fails to recover values from both synthetic and real-world charts, with the accuracy hovering at $\approx 1\%$. When evaluating the heatmaps directly, the outputs are diffuse and poorly localized compared to the classic decoder, which produces sharp, focused activations (Fig. 7). Our hypothesis is that the decoder is over-constrained and does not have enough capacity for the task at hand, which is supported by the fact that instead of confusing bars and ticks, detections are simply missed (Fig. 6). Our network is not trained end-to-end, which could possibly explain the different findings: If the encoder in ViTPose does a lot of the heavy-lifting, the decoder choice would matter a lot less.

Table 3. The mean F1 score of our model and Soto *et al.*'s [36] trained on synthetic data. The test datasets, while not being the same, are very similar, and our precision threshold is also slightly stricter (2.3% vs. 2.7%).

Method	Data	F1 Score
Ours	Synthetic	0.956
Soto *et al.* [36]	Synthetic	**0.981**
Ours	Real World	**0.814**
Soto *et al.* [36]	Real World	0.555

Table 4. Value recovery accuracy at the relaxed threshold ($\varepsilon = 0.05$) of our model and Zhou *et al.*'s [42] without OCR correction. The test datasets differ. Our real-world data is more varied, whereas the synthetic data is nearly identical.

Method	Dataset	Accuracy
Ours	Synthetic	79%
Zhou *et al.* [42]	Synthetic	**82%**
Ours	Real World	50%
Zhou *et al.* [42]	Real World	**71%**

4.3 Comparison to Other Papers

It is somewhat difficult to compare our results to prior work directly, as our outputs are limited to bar and tick coordinates. Nevertheless, we can draw a comparison to the works of Soto *et al.* [36] and Zhou *et al.* [42]. We can only compare results conceptually on qualitatively similar data due to the unavailability of the datasets they used in testing. As can be seen in Table 4, our method performs similarly well on synthetic data at the relaxed threshold ($\varepsilon = 0.05$). Since our chart generator builds on theirs, the comparison on synthetic data is fair. While our method does fall behind on real-world data, most of the charts in our test set do not fulfill the criteria Zhou *et al.* assume (e.g. include error bars/stacked bars). To compare our work to Soto *et al.*, we use the mean tick and bar F1 scores over the entire test set. Their synthetic data is, again, very comparable to ours and the real-world data they curated, just like UB PMC [8], consists of charts from PubMedCentral. The results in Table 3 show that, while our model performs slightly worse on synthetic data (≈ 0.03 difference), we can improve upon their results when it comes to real-world data by ≈ 0.26. It should be noted, however, that we do not have many trainable parameters in the value recovery pipeline, as the bulk of the parameters live in the frozen encoder. Additionally, Zhou *et al.* [42] train on $30k$ charts for 300 epochs and Soto *et al.* 1750 epochs on $5k$ charts, which is significantly more than it takes for our model to converge (see Sect. 3.3). This suggests that self-supervised finetuning, as well as the semantic quality of the extracted features are the driving factor behind our results.

5 Discussion and Conclusion

In this study, we propose utilizing I-JEPA, a self-supervised learning model, as a feature extractor for chart de-rendering. To our knowledge, this is the first time that JEPA has been used in the context of chart understanding, a traditionally supervision-heavy task. We are able to show that finetuning the feature extractor is highly beneficial for downstream performance and our results suggest that the features are indeed of high semantic quality. This is supported by the fact that our lightweight value recovery model converges quickly during training and does not require a large corpus to achieve good results. We also introduce variable resolution inputs to I-JEPA, which yields modest performance improvements with negligible overhead.

Our work does, however, have some limitations: (1) Our model is rather simplistic and can only recover values of one type of chart, (2) it does not achieve SoTA results at this task and (3) it may prove difficult to integrate our feature extractor into multimodal language models (e.g. for chart QA). Although we are not trying to compete with SoTA models, it would be beneficial to build a more competent decoder in future work to show the full potential of our method. Our encoder is frozen during training and I-JEPA operates exclusively in latent space, so aligning a text-based model could be challenging. Considering many SoTA models are to some extent based on large language models [6,22,27,28], the ability to train end-to-end might be beneficial.

There are several possible avenues for future work. Implementing a more powerful decoder, such as the transformer-based ChartDETR [41] could result in better de-rendering capabilities. Including a large language model in the pipeline or replacing the decoder with a large language model would allow testing if our method is also suited for chart QA. Taking ChartGemma [28] as an example, it may be possible to use I-JEPA as the vision encoder and align it with the language model by training an embedding layer. Finally, scaling the model to more chart types and finetuning on a more diverse corpus would be a logical next step, as it is unclear how much data is required for finetuning. Judging the quality of the extracted features directly is a hard task by itself and it is entirely possible that more data and longer training would yield even stronger features.

Acknowledgments. We acknowledge the EuroHPC Joint Undertaking for awarding this project access to the EuroHPC supercomputer LEONARDO, hosted by CINECA (Italy) and the LEONARDO consortium through a EuroHPC Development Access call. The authors acknowledge support by the state of Baden-Württemberg through bwHPC. We also thank Laurna Epple and Hannah Kniesel for proofreading and valuable discussions.

A Parameters

Table 5. Parameters used to generate our datasets. The colors are selected to be rich in contrast, to avoid white-on-white charts. Fonts have to support the roman alphabet and we filter out specialty fonts (e.g. math, pictographs).

Parameter	Details
Chart size	600–1600 px (width & height)
Bar series	2–5
Bars per series	1–3
Bar values	10–200
Bar width & spacing	Random in accordance with size
Bar colors	Randomly sampled from color maps
Font & font size	Randomly selected from roman fonts
Axis label length	5–15 characters
Ticks label length	1–5 characters
Title length	5–15 characters
Title location	left, center, right
Legend length	3–6 characters
Legend location	top, bottom, right

Table 6. I-JEPA finetuning parameters. We use the ViT-H preset for training and the default multiblock sampling strategy, with the same settings as used during pre-training. We generally allow at most 256 patches of size 14 px.

Context Encoder	
Max Patch Count	256
Block Count	1
Block Size	85–100%
Predictor	
Block Aspect Ratio	0.75–1.5
Block Count	4
Block Size	15–20%

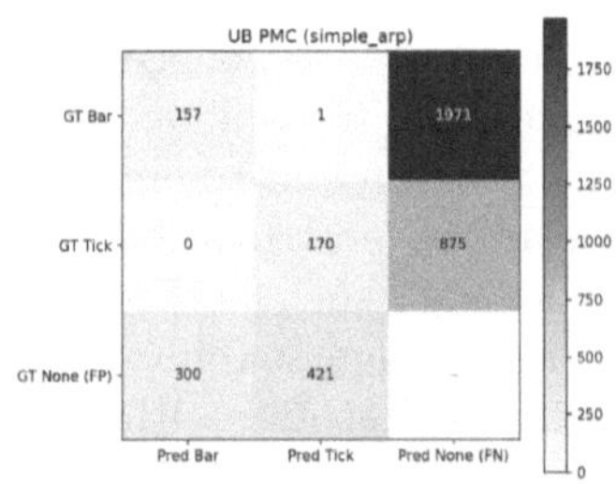

Fig. 6. The confusion matrix of the simple decoder evaluated on [9], showing that most bars and ticks are not detected.

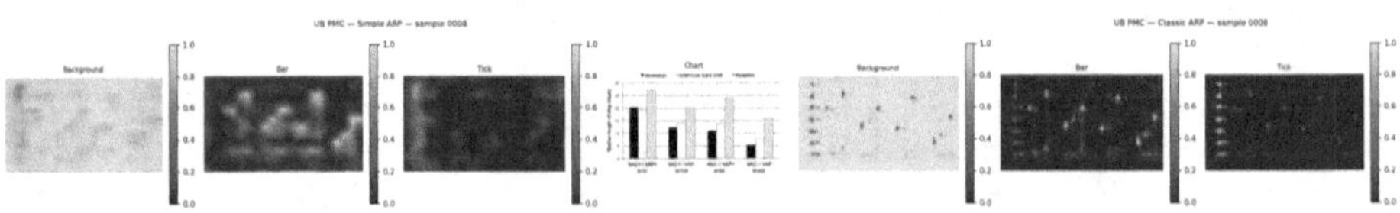

Fig. 7. Simple vs. classic decoder on a chart from [9]. The activations of the simple decoder are blurry, whereas the classic decoder produces pronounced, sharp peaks.

References

1. Assran, M., et al.: Self-supervised learning from images with a joint-embedding predictive architecture. In: Proceedings of the IEEE/CVF CVPR (2023)

2. Assran, M., et al.: V-jepa 2: self-supervised video models enable understanding, prediction and planning. arXiv preprint arXiv:2506.09985 (2025)
3. Bardes, A., et al.: V-JEPA: latent video prediction for visual representation learning (2024). https://openreview.net/forum?id=WFYbBOEOtv. Accessed 13 Feb 2026
4. Carion, N., Massa, F., Synnaeve, G., Usunier, N., Kirillov, A., Zagoruyko, S.: End-to-end object detection with transformers. In: ECCV. Springer, Cham (2020)
5. Chen, X., et al.: Context autoencoder for self-supervised representation learning. Int. J. Comput. Vis. **132**(1) (2024)
6. Cheng, Z.Q., Dai, Q., Hauptmann, A.G.: Chartreader: a unified framework for chart derendering and comprehension without heuristic rules. In: Proceedings of the IEEE/CVF ICCV (2023)
7. Cui, C., et al.: Paddleocr 3.0 technical report. arXiv preprint arXiv:2507.05595 (2025)
8. Davila, K., et al.: Chart-info 2024: A dataset for chart analysis and recognition. In: ICPR. Springer, Cham (2024)
9. Davila, K., Xu, F., Ahmed, S., Mendoza, D.A., Setlur, S., Govindaraju, V.: Icpr 2022: challenge on harvesting raw tables from infographics (chart-infographics). In: 2022 26th ICPR, IEEE (2022)
10. Dosovitskiy, A.: An image is worth 16x16 words: Transformers for image recognition at scale. arXiv preprint arXiv:2010.11929 (2020)
11. Fei, Z., Fan, M., Huang, J.: A-jepa: Joint-embedding predictive architecture can listen. arXiv preprint arXiv:2311.15830 (2023)
12. Fischler, M.A., Bolles, R.C.: Random sample consensus: a paradigm for model fitting with applications to image analysis and automated cartography. Commun. ACM **24**(6) (1981)
13. Hassan, M.Y., Singh, M.: Lineex: data extraction from scientific line charts. In: Proceedings of the IEEE/CVF WACV (2023)
14. He, K., Chen, X., Xie, S., Li, Y., Dollár, P., Girshick, R.: Masked autoencoders are scalable vision learners. In: Proceedings of the IEEE/CVF CVPR (2022)
15. Huang, W., Tan, C.L.: A system for understanding imaged infographics and its applications. In: Proceedings of the 2007 ACM Symposium on Document Engineering (2007)
16. Hunter, J.D.: Matplotlib: A 2d graphics environment. Comput. Sci. Eng. **9**(3) (2007). https://doi.org/10.1109/MCSE.2007.55. Accessed 07 June 2026
17. Joshi, I.A.: Unblind the charts: Towards making interactive charts accessible in android applications. arXiv preprint arXiv:2109.12442 (2021)
18. Kantharaj, S., et al.: Chart-to-text: a large-scale benchmark for chart summarization. In: Proceedings of the 60th Annual Meeting of the ACL (Volume 1: Long Papers) (2022)
19. Kuhn, H.W.: The Hungarian method for the assignment problem. Naval Res. Logistics Q. **2**(1-2) (1955)
20. LeCun, Y.: A path towards autonomous machine intelligence version 0.9. 2, 2022-06-27. Open Rev. **62**(1) (2022)
21. Lee, K., et al.: Pix2struct: screenshot parsing as pretraining for visual language understanding. In: ICML. PMLR (2023)
22. Liu, F., et al.: Deplot: one-shot visual language reasoning by plot-to-table translation. In: ACL 2023 (2023)
23. Liu, F., et al.: Matcha: enhancing visual language pretraining with math reasoning and chart derendering. In: Proceedings of the 61st Annual Meeting of the ACL (Volume 1: Long Papers) (2023)

24. Liu, X., Klabjan, D., NBless, P.: Data extraction from charts via single deep neural network. arXiv preprint arXiv:1906.11906 (2019)
25. Luo, J., Li, Z., Wang, J., Lin, C.Y.: Chartocr: data extraction from charts images via a deep hybrid framework. In: Proceedings of the IEEE/CVF WACV (2021)
26. Ma, W., et al.: Towards an efficient framework for data extraction from chart images. In: ICDAR. Springer, Cham (2021)
27. Masry, A., Kavehzadeh, P., Do, X.L., Hoque, E., Joty, S.: Unichart: a universal vision-language pretrained model for chart comprehension and reasoning. arXiv preprintarXiv:2305.14761 (2023)
28. Masry, A., Thakkar, M., Bajaj, A., Kartha, A., Hoque, E., Joty, S.: Chartgemma: visual instruction-tuning for chart reasoning in the wild. In: Proceedings of the 31st International Conference on Computational Linguistics: Industry Track (2025)
29. Meyer, J., Shamo, M., Gopher, D.: Information structure and the relative efficacy of tables and graphs. Hum. Factors: J. Hum. Factors Ergon. Soc. **41** (1999). https://doi.org/10.1518/001872099779656707. Accessed 13 Feb 2026
30. Mishchenko, A., Vassilieva, N.: Chart image understanding and numerical data extraction. In: 2011 Sixth International Conference on Digital Information Management. IEEE (2011)
31. PlotDigitizer: PlotDigitizer: Extract Data from Graph Image Online — plotdigitizer.com. https://plotdigitizer.com. Accessed 19 Jan 2026
32. Riou, A., Lattner, S., Hadjeres, G., Anslow, M., Peeters, G.: Stem-jepa: a joint-embedding predictive architecture for musical stem compatibility estimation. In: International Society for Music Information Retrieval Conference (2024)
33. Saito, A., Kudeshia, P., Poovvancheri, J.: Point-jepa: a joint embedding predictive architecture for self-supervised learning on point cloud. In: 2025 IEEE/CVF WACV, pp. 7348–7357. IEEE (2025)
34. Savva, M., Kong, N., Chhajta, A., Fei-Fei, L., Agrawala, M., Heer, J.: Revision: automated classification, analysis and redesign of chart images. In: Proceedings of the 24th Annual ACM Symposium on User Interface Software and Technology (2011)
35. Shahira, K., Joshi, P., Lijiya, A.: Data extraction and question answering on chart images towards accessibility and data interpretation. IEEE Open J. Comput. Soc. **4** (2023)
36. Soto, C., Yoo, S.: An extensible point-based method for data chart value detection. arXiv preprint arXiv:2308.11788 (2023)
37. Thimonier, H., Costa, J.L.D.M., Popineau, F., Rimmel, A., Doan, B.L.: T-jepa: augmentation-free self-supervised learning for tabular data. arXiv preprint arXiv:2410.05016 (2024)
38. Tuncay, L., Labbé, E., Benetos, E., Pellegrini, T.: Audio-jepa: joint-embedding predictive architecture for audio representation learning. arXiv preprint arXiv:2507.02915 (2025)
39. Vo, K.H., Nguyen, D.P., Nguyen, T.T., Quan, T.T.: Ti-jepa: an innovative energy-based joint embedding strategy for text-image multimodal systems. In: International Symposium on Information and Communication Technology. Springer, Cham (2024)
40. Xu, Y., Zhang, J., Zhang, Q., Tao, D.: Vitpose: simple vision transformer baselines for human pose estimation. In: Koyejo, S., Mohamed, S., Agarwal, A., Belgrave, D., Cho, K., Oh, A. (eds.) Advances in Neural Information Processing Systems, vol. 35. Curran Associates, Inc. (2022)

41. Xue, W., Chen, D., Yu, B., Chen, Y., Zhou, S., Peng, W.: Chartdetr: a multi-shape detection network for visual chart recognition. arXiv preprint arXiv:2308.07743 (2023)
42. Zhou, F., et al.: Reverse-engineering bar charts using neural networks. J. Visual. **24**(2) (2021)

GeoLogVQA: A Borehole Log Documents Dataset for Explicit and Implicit Spatial Reasoning

Stanislas Bagnol[1,2](✉), Killian Barrere[1], Véronique Eglin[1], Előd Egyed-Zsigmond[1], David Pitaval[2], and Jean-Marie Côme[2]

[1] INSA Lyon, CNRS, Universite Claude Bernard Lyon 1, LIRIS, UMR5205, 69621 Villeurbanne, France
killian.barrere@insa-lyon.fr
[2] Ginger BURGEAP, 69003 Lyon, France
s.bagnol@groupeginger.com

Abstract. Document information extraction remains a significant challenge for automated document analysis systems. While Vision-Language Models (VLMs) show strong performance on document information extraction, they often struggle with spatial and numerical inference and counting. Borehole log documents provide vertically structured records about subsoil layers. Extracting depth values from borehole logs represents a major difficulty for state-of-the-art models as some values must be inferred from depth scale, requiring joint visual grounding and numerical reasoning. We introduce GeoLogVQA, a dataset comprising 2,554 images of authentic borehole log documents from BRGM (https://www.brgm.fr/en/identity/brgm-glance), the French public Earth Science institute. We generated 92k question-answer pairs, from human verified JSON annotations, targeting counting, spatial reasoning, depth inference, and general information extraction. We evaluated the performance of both fine-tuned open-source VLMs and zero-shot closed-source VLMs on VQA tasks, as well as full-structure information extraction. Our results show that, despite strong performance on text-explicit fields, current VLMs still struggle significantly when depth values must be inferred, or when counting tasks is involved, highlighting a critical gap in document reasoning capabilities. The GeoLogVQA dataset is publicly available for download here.

Keywords: Document Information Extraction · VQA Dataset · Implicit Spatial Reasoning Dataset

1 Introduction

Document Information Extraction is a core task in document analysis and recognition, aiming to transform unstructured or semi-structured documents into structured, machine-readable representations. Recent advances in Vision Language Models (VLMs) have demonstrated remarkable capabilities in Document

G. A. Fink et al. (Eds.): ICDAR 2026, LNCS 16972, pp. 38–54, 2027.
https://doi.org/10.1007/978-3-032-36023-6_3

Information Extraction tasks such as Visual Question Answering (VQA) or Optical Character Recognition (OCR). These models demonstrate strong performance in extracting explicitly stated information, but still exhibit limitations in tasks involving spatial and multimodal reasoning, numerical inference, or counting.

Existing document analysis benchmarks largely focus on text recognition, structural layout analysis, and information extraction from text-dominant documents such as invoices, receipts, and forms. However, these benchmarks rarely evaluate models ability to understand implicit, spatially-relative information, which often requires multimodal reasoning, integrating textual cues with visual structures, graphical scales, or proportional relationships. In particular, the ability to map visual spatial coordinates to quantitative values remains largely untested, despite being central to many real-world technical documents (e.g., engineering reports, geological surveys, and scientific plots), where key information is encoded geometrically rather than textually. Consequently, models may achieve high scores on conventional benchmarks, they often fail in real-world technical documents that require joint visual grounding and numerical reasoning.

Borehole log documents, as illustrated in Fig. 1, are challenging and under explored case of document involving implicit reasoning. Widely used in geotechnical engineering and environmental studies, these documents present vertically structured records of subsurface investigations. Lithological layers are associated with depth intervals, which may be either explicitly stated or implicitly defined relative to a depth scale. Extracting implicit depths requires interpreting graphical scales, converting proportional pixel distances into metric values, and performing interpolation or counting operations without any direct textual supervision. This form of reasoning grounded in spatial interpretation and proportional inference goes beyond standard text extraction and poses significant challenges for current multimodal systems.

To address this gap, we introduce GeoLogVQA, a publicly available dataset for document implicit reasoning evaluation, demonstrated through a borehole log information extraction case study. The dataset comprises 2,554 authentic borehole log documents. From human-verified structured annotations, we generated 92k question–answer pairs targeting counting, spatial reasoning, depth inference, and general information extraction.

GeoLogVQA enables evaluation along two complementary axes: (1) Visual Question Answering, with emphasis on implicit reasoning and counting, and (2) full-data extraction via JSON reconstruction. We benchmark both fine-tuned open-source VLMs and zero-shot closed-source models, showing that while models perform strongly on explicitly written textual fields, they struggle significantly when depth values must be inferred or when counting tasks are involved.

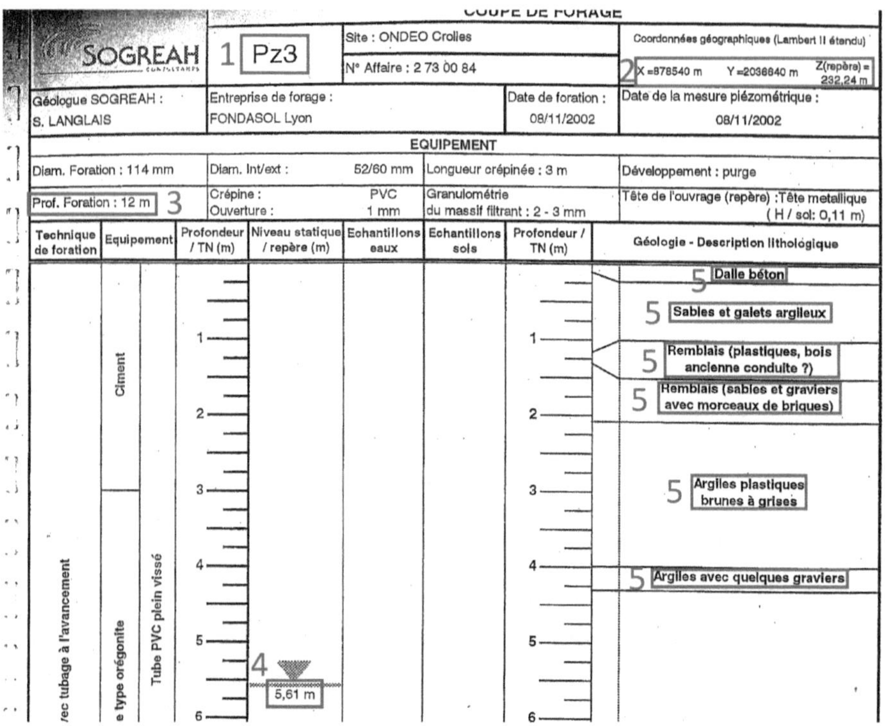

Fig. 1. Borehole Log information example: (1) ID: Pz3; (2) LONGITUDE: 878540, LATITUDE: 2036640, ALTITUDE: 232,24; (3) FINAL_DEPTH: 12; (4) WATER: 5,61; (5) LITHOLOGIES: ("Dalle béton"; 0.10), ("Sables et galets argileux"; 1.2), ..., ("Argiles avec quelques graviers", 4.3).

2 Related Work

2.1 Vision-Language Models for Document Understanding

Recent advances in Vision-Language Models (VLMs) have significantly improved document understanding and recognition, enabling tasks such as Visual Question Answering (VQA), Optical Character Recognition (OCR), and structured information extraction. Open-source models like Qwen3-VL [1], InternVL3 [22], LLaVA [8], and Gemma [19] demonstrate strong performance on standard benchmarks, while smaller alternatives like SmolVLM [12] achieves competitive accuracy with reduced computational cost. DeepSeek-VL2 [20] further optimizes performance via Mixture-of-Experts (MoE) architectures [2]. These models can be fine-tuned for domain-specific applications, enhancing task-specific performance.

Closed-source VLMs, such as GPT-5-2, Gemini 3 or Claude Sonnet 4.6 lead state-of-the-art benchmarks due to their massive scale (hundreds of billions of parameters) and extensive pre training. However, the restricted access to model weights and training details limits controlled fine tuning and fully reproducible experimentation.

Additionally, document-specific multimodal models like Donut [6], DocKylin [21] or LayoutLLM [9] have emerged, leveraging specialized architectures for document understanding and parsing.

Despite these advancements, VLMs still struggle with tasks requiring spatial reasoning, numerical inference or implicit information extraction.

2.2 Document Information Extraction Datasets

Document Information Extraction benchmarks have progressed from text-centric tasks to multimodal reasoning challenges. Early datasets such as FUNSD [5] and SROIE [3] focused on key-value pair extraction from forms and receipts, primarily evaluating text recognition and basic layout analysis. TableBank [7] and PubTables-1M [18] focus on table detection and table structure recognition. Although foundational, these benchmarks lacked spatial reasoning or implicit value inference, as answers were typically explicitly stated in the text. Recent benchmarks have introduced Visual Question Answering (VQA) for documents, enabling more comprehensive evaluation of multimodal understanding. DocVQA [15] and InfographicVQA [14] expanded the scope to diverse document types, but their questions predominantly test direct text extraction or simple layout comprehension rather than complex spatial or numerical reasoning.

2.3 Spatial and Implicit Reasoning in Documents

Few datasets address spatial reasoning or implicit value inference in documents. Methani et al. introduce PlotQA [17], a dataset comprising 28.9M question-answer pairs over 224k plots. Notably, 23.3M questions involve values that are not explicitly present in images, thus requiring reasoning. Masry et al. proposed the ChartQA dataset [13], which include 33k pairs involving spatial and numerical reasoning. However, this reasoning relies on explicitly stated in the images. Islam et al. extend this work with ChartQA* [4], a variant of ChartQA without value labels on chart. Their study shows that VLMs struggle significantly when labels are absent, relying heavily on explicitly written values rather than spatial cues.

Meanwhile, technical documents exhibit greater complexities beyond simple plots or charts. They often contain dense information within a single page and use diverse formats. To the best of our knowledge, no existing benchmark evaluates implicit spatial to numerical reasoning in real world technical documents.

3 GeoLogVQA

In this section, we present the GeologVQA dataset[1], which consists of 2,554 authentic borehole log documents from the *Banque du Sous-Sol*[2] (BSS) database. The BSS is provided by the *Bureau de Recherches Géologiques et Minières*

[1] https://github.com/stanouz/GeoLogVQA.

[2] https://infoterre.brgm.fr/page/banque-sol-bss.

(BRGM), the French public institute for applied Earth sciences. The dataset comprises 92k question–answer pairs designed to evaluate counting, spatial reasoning, depth inference, and general information extraction capabilities. Notably, 24% of the depth values present a significant challenge for state-of-the-art models due to the absence of explicit textual labels, requiring numerical inference from visual and structural cues within the documents. We refer to these cases as *implicit* depths, which involve linear interpolation between the lithology cell bottom line and closest depth ticks.

3.1 Borehole Logs

Borehole logs are geotechnical documents used in engineering and environmental studies to report the results of subsurface investigations carried out through mechanical drilling. They describe the sequence, thickness, physical condition, and engineering properties of soil and rock layers encountered at depth, along with groundwater conditions, sampling intervals, and in-situ test results. The documents are typically structured into two parts: a header section and a table containing the results.

The header includes various metadata fields: borehole ID, date, client name, borehole final depth, geospatial data (i.e. longitude, latitude and elevation), drilling tool or method, etc.

The body is the main part of the document, reporting the results of the investigation, in a vertically structured tabular format. The core information is a geological description of soil layers (i.e. lithological facies or lithologies), where each layer is associated with a depth interval. A depth scale is generally provided, with vertical proportionality preserved across the document body (e.g. 2 cm high equals 1 m depth). Layers depth values can be either explicitly reported (i.e. explicit depth Fig. 2a) or must be inferred from the depth scale (i.e. implicit depth Fig. 2b). Depth values may be expressed either as ground relative depth or as absolute elevation value, or both.

Extracting implicit depth values from borehole logs is particularly challenging for current models as values must be inferred from the depth scale through visual grounding and numerical interpolation or counting (i.e. converting vertical distances into metric values). This requires reasoning beyond simple alignment detection, as models must interpret scales. Such multimodal reasoning and interpolation are rarely evaluated in existing document benchmarks.

3.2 Dataset Description

The dataset comprises 2,554 annotated documents, divided into two subsets: 1,936 borehole logs with explicit layer depths and 618 logs with implicit layer depths. Additionally, 87 documents contain two borehole logs on a single page, showing explicit depths. Details are provided in Table 1a.

For each document image, an associated JSON label file specifies the information to be extracted, as illustrated in Fig. 3. Each record corresponds to a single borehole and generally includes a unique borehole identifier (`ID`). Geospatial

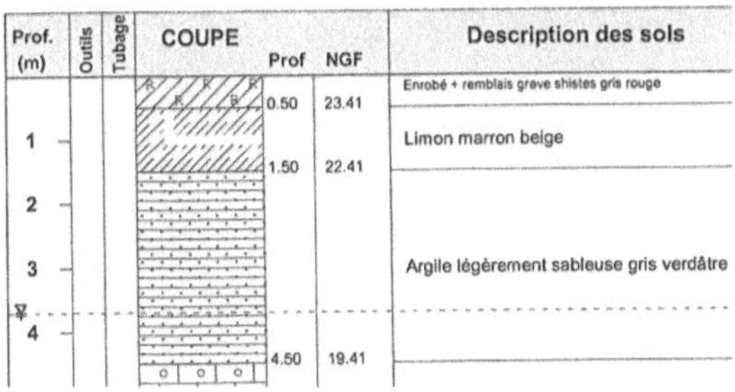

(a) Explicit depth representation; for example, "Limon marron beige" layer depth (0.50–1.50 m) and corresponding elevation (23.41–22.41 m) are explicitly provided.

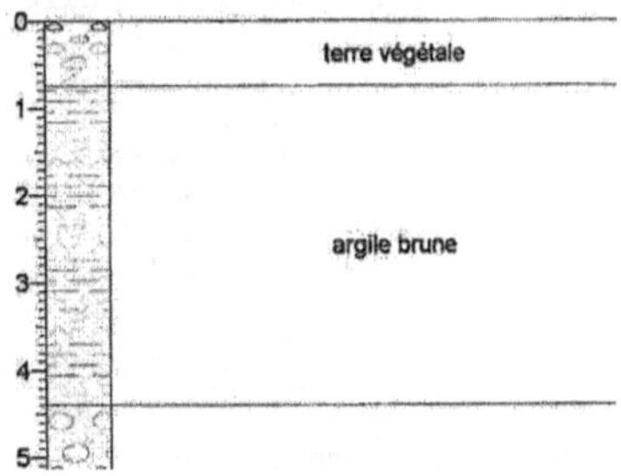

(b) Implicit depth representation; for example, "terre végétale" layer end-depth (0.75 m) must be inferred from the depth scale.

Fig. 2. Explicit (a) and implicit (b) representations of the depth of lithology.

Table 1. Dataset statistics.

(a) Document type distribution

Explicit	Implicit	Multi logs (explicit)	Total
1,936 (75.8%)	618 (24.2%)	87 (3.4%)	2,554 (100%)

(b) Field occurrence distribution

Altitude	Longitude	Latitude	Water Level
1,340 (52.4%)	1,227 (48.0%)	1,226 (47.9%)	1,062 (41.5%)

attributes (LONGITUDE, LATITUDE, ALTITUDE) specify the borehole location and elevation, although these fields are not systematically present (see in Table 1b). The FINAL_DEPTH corresponds to the borehole final depth, either explicitly stated in the header or inferred from the maximum recorded lithological depth, and WATER indicates the groundwater depth in meters, when reported. Finally, LITHOLOGIES contains the ordered sequence of encountered lithological layers, each defined by a textual description (LITHO_DESC) and its associated end-depth (LITHO_DEPTH). All LITHO_DESC are in french and LITHO_DEPTH are expressed in meters and increase positively downward from ground level, while elevation values are expressed relative to sea level when available. The number of lithological layers varies among borehole logs; across the 2,554 documents, a total of 13,319 lithological units are recorded.

We generated 92k question-answer pairs from JSON data structure, targeting counting, spatial reasoning, depth inference, and general information extraction. Further details are provided in Sect. 3.4.

```
"image": "BSS000AAAA.png"
"depth_type": "explicit",
"data": [{
  "ID": "TA3",
  "LONGITUDE": "498218.45",
  "LATITUDE": "6623796.09",
  "ALTITUDE": "109.68",
  "FINAL_DEPTH": "4.20",
  "WATER": "3.0",
  "LITHOLOGIES": [
    {"LITHO_DESC": "Terre végétale", "LITHO_DEPTH": "0.30"},
    {"LITHO_DESC": "Argile limoneuse brune ocre", "LITHO_DEPTH": "4.20"}
  ]
}]
```


Fig. 3. Example of a Borehole Log JSON Format.

3.3 Data Annotation Process

To ensure an efficient annotation process, we trained a Qwen3-VL-2B model on a synthetic, temporary, dataset to generate the JSON data structure shown in Fig. 3 as a pre-annotation step. The synthetic dataset comprises 200 documents automatically generated using dedicated borehole log authoring software[3].

The previously mentioned model was subsequently used to produce pre-annotations for the final dataset. This approach significantly accelerated the annotation process, as annotators only needed to review and correct model predictions rather than annotate each document from scratch. The pre-annotation review and correction process was performed manually twice for each document, with validation by domain experts, to improve annotation quality and consistency.

Lithological implicit end-depths were computed by mapping the lower boundary line of the lithology cell to the depth scale and interpolating pixel-to-depth distances based on the two nearest depth scale ticks. Depths were reported with two decimal precision. To deal with cases where the bottom line is slightly displaced to accommodate the lithological description of thin layers (Fig. 1), depth were determined at the intersection of the bottom line and the column boundary.

Finally, the documents were categorized into two groups: (i) borehole logs with **explicit** layer depths (written, see Fig. 2a) and (ii) borehole logs with **implicit** layer depths (i.e., that must be inferred from depth scale, see Fig. 2b).

3.4 Visual Question Answering

Using the JSON annotations, we generated 92k closed-ended question–answer pairs covering extractive, counting, indexing, and spatial or numerical reasoning tasks. The generation script is available. Questions are grouped into four categories: Metadata Extraction, Lithology Extraction, Counting and Indexing, and Spatial and Numerical reasoning. All questions are summarized in Table 2, with examples provided in Fig. 4.

[3] Jean Lutz – EXGTE software http://www.jeanlutzsa.fr/geotechnique/.

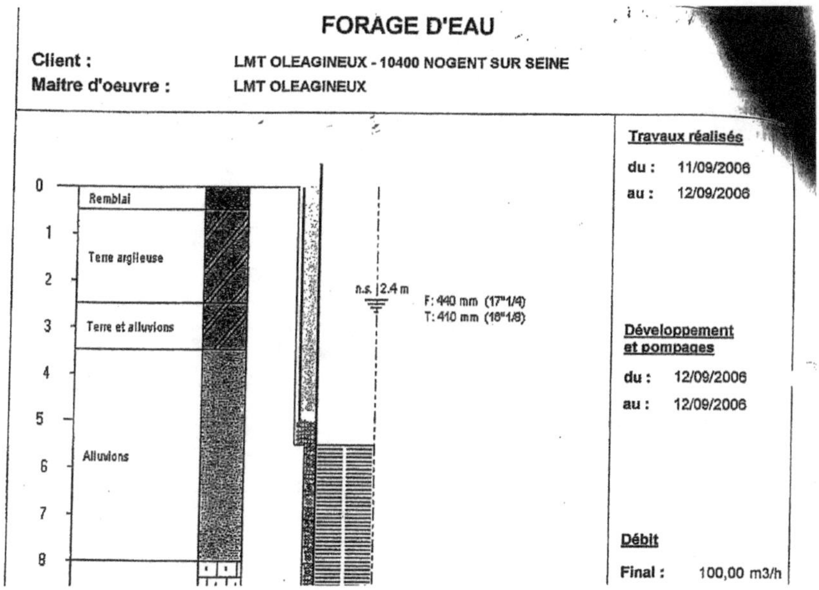

Fig. 4. Visual Question Answering examples: LITHO_INDEX_TO_DEPTH **Q**: What is the 2nd lithogical layer end-depth? **A**: 2.5 m; LITHO_DESC_TO_THICKNESS **Q**: What is the thickness of the "Alluvions" layer? **A**: 4.5 m; WATER **Q**: What is the water level on this borehole log? **A**: 2.4 m.

Table 2. Question types. Total of 92,060 question-answer pairs, 53,347 involving reasoning. Reasoning is sorted into 4 categories: indexing, counting, numerical reasoning, and spatial reasoning.

Question ID	Description	Index	Count.	Num. R.	Spat. R.	#
Metadata Extraction						
ID	Get ID value					2,467
FINAL_DEPTH	Get final depth value					2,467
LONGITUDE	Get longitude value					2,467
LATITUDE	Get latitude value					2,467
ALTITUDE	Get altitude value					2,467
WATER	Get water level value					2,467
Lithology Extraction						
LITHO_DEPTH_TO_DESC	Lithological description from a lithological depth					12,796
LITHO_DESC_TO_DEPTH expl	Lithogical end-depth from a lithological description					8,648
LITHO_LIST	List all lithological descriptions					2,467
Counting and Indexing						
LITHO_COUNT	Get number of lithologies		√			2,467
LITHO_INDEX_TO_DESC	Lithological description from an index	√				12,796
LITHO_INDEX_TO_DEPTH expl	Lithogical end-depth from an index	√				9,417
Spatial and Num. Reasoning						
INTERNAL_DEPTH_TO_DESC	Lithological description from a lithological inner depth			√		8,727
LITHO_DESC_TO_DEPTH impl	Lithogical end-depth from a lithological description			√	√	3,078
LITHO_INDEX_TO_DEPTH impl	Lithogical end-depth from an index	√		√	√	3,379
LITHO_DESC_TO_ELEVATION	Lithological elevation value from a description			√		5,913
LITHO_DESC_TO_THICKNESS	Lithological layer thickness from a description			√	√	7,570

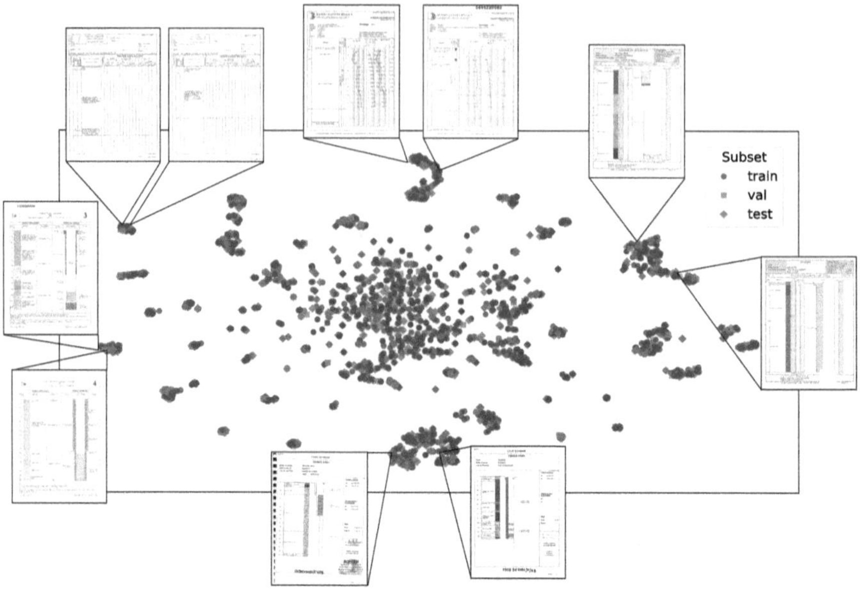

Fig. 5. 2D t-SNE visualization of the document embeddings and the proposed training, validation and testing sets.

3.5 Dataset Partitioning Protocol

To partition the dataset into representative training, validation and testing sets, we aimed to preserve the global data distribution using visual similarity. To this end, embeddings from the vision encoder of a pre-trained Qwen3-VL (Qwen3-VL-2B-Instruct) [1] were used as a perceptual surrogate to approximate visual similarity between documents. All images were resized to a fixed resolution of 560×800 pixels to ensure a consistent number of visual tokens across samples.

The resulting visual tokens were then flattened and concatenated, producing a vector of length 921,600 for each image. We then reduced the embedding dimensionality to 50 using Principal Component Analysis (PCA) [16], followed by K-means clustering [11] in the reduced space to obtain 30 clusters. This value was chosen empirically based on visualization inspection. Each cluster was split into training, validation and testing sets using stratified sampling to preserve the proportion of documents with explicit and implicit depths. Additionally, to assess the effect of training set size, we define three training scenarios: 7%, 35% and 100% of the training set. A summary of the resulting five subsets is available in Table 3.

Lastly, we empirically validated the representativeness of the proposed subsets using a t-SNE [10] visualization on PCA embeddings. As illustrated in Fig. 5, similar documents appear close to each other while being well distributed across subsets.

Table 3. Dataset statistics for training, validation, and test subsets.

	Train (7%)	Train (35%)	Train (100%)	Val	Test
Documents	137 (5.4%)	635 (25%)	1,758 (68.8%)	145 (5.7%)	653 (25.5%)
QA pairs	4,801	22,594	63,458	4,767	23,834
Implicit Depth	24.09%	23.82%	24.29%	24.14%	24.2%

4 Experiments

Our dataset supports evaluation of two complementary tasks. The first focuses on VQA task, as detailed in Sect. 3.4, emphasizing implicit reasoning, including visual grounding, numerical inference, and counting or indexing tasks. The second task involves full-structure data extraction, aiming to reconstruct the JSON structure shown in Fig. 3 and directly evaluate the ability to extract and reproduce structured information from the input documents. For all experiments, including both training and evaluation, we used original image sizes of documents.

4.1 Models

Open-Source Models. We conducted an evaluation of several open-source models: Qwen3-VL-2B [1], InternVL3-2B [22], SmolVLM2 2.2B [12], and DeepSeekVL-2-tiny [20]. Models are evaluated in fine-tuned settings using the complete training set and reduced training subsets.

Each model was fine-tuned with the vision encoder frozen, while training the remaining layers. Fine-tuning was conducted on a single NVIDIA A100 Tensor Core GPU with a batch size of 8 for 5 epochs (extended to 10 epochs when training on limited training data scenarios) using validation set as a best model criterion. Other hyperparameters (e.g., learning rate, weight decay, and optimizer) were set according to the author's recommendations. Models were fine-tuned separately on VQA and JSON generation tasks.

Closed-Source Models. For the sake of comparison, we conduct minimal experiments using closed source models GPT-5.2 and Claude-Sonnet-4.6 in a strict zero-shot setting, since fine-tuning is not accessible. Prompting method consisted of providing the document image, the pre-prompt "For the following question, just write the answer." and the question itself. This ensure the model generates a direct answer, facilitating evaluation.

We only evaluated closed-source models on VQA tasks, as zero-shot methods do not support reliable JSON structure generation. Common failures include incorrect or inconsistent object names, missing fields, and incorrect formatting. This motivates our observation that prompt-based approaches alone are not sufficient for reliable structured generation without fine-tuning or additional constraints.

4.2 Evaluation Metrics

Metadata Extraction. For both VQA and JSON generation tasks, metadata fields (`ID`, `LONGITUDE`, `LATITUDE`, `FINAL_DEPTH`, `ALTITUDE` and `WATER`) are evaluated using exact match.

Since `LONGITUDE`, `LATITUDE`, `ALTITUDE` and `WATER` values are not systematically present (see Table 1b) we additionally assess the model's ability to predict the presence of each field using the F1-score.

Numerical Values. Explicitly stated depths are evaluated using exact match accuracy, whereas implicit depths are evaluated using relaxed accuracy with a given tolerance ε defined as:

$$\varepsilon = \max\left(0.1,\ 0.01 \times D_{\max}\right) \tag{1}$$

where $D_{\max}$ denotes the final borehole depth. This threshold enforce a minimum tolerance of 0.1 m, while relaxing the bound for deep boreholes, therefore accounting for uncertainty introduced by pixel-to-depth interpolation.

Since `LITHO_DESC_TO_ELEVATION` and `LITHO_DESC_TO_THICKNESS` involve intermediate numerical calculations, we adopted a relaxed accuracy using a $\varepsilon = 0.1$ m threshold.

Finally, we evaluated the number of extracted borehole logs (accounting for cases where multiple borehole logs are present in the JSON generation task) and `LITHO_COUNT` using exact match accuracy.

Lithological Description. Lithological description extraction is evaluated using accuracy based on Levenshtein similarity ratio. A prediction is considered correct if the similarity ratio exceeds 0.9.

Lithologies List Extraction. Tasks involving lithologies list prediction (JSON generation task and `LITHO_LIST` questions in the VQA task) are evaluated using a best ordered alignment metric. We compute the optimal ordered alignment using a Levenshtein similarity ratio (with a given threshold of 0.9) between the predicted lithologies and the ground truth.

This strategy penalizes inserted or missed lithologies by assigning a zero score to predicted lithologies that have no similar entry in ground truth and to unmatched ground-truth lithologies. The metric therefore captures both the correctness and the completeness of the predicted lithology sequence.

For the JSON generation task, matched lithologies are further evaluated on depth, using the previously defined metrics. Lithologies that do not satisfy correctness criteria for either description or depth are considered incorrect.

4.3 Results

Table 4. VQA results with different open-source models (Qwen3-VL-2B, InternVL3-2B, SmolVLM2 2.2B and DeepSeekVL2-tiny) fine-tuned on 7%, 35% and 100% of the training set.

	Qwen3-VL			InternVL3			SmolVLM2			DeepSeekVL2		
	7%	35%	100%	7%	35%	100%	7%	35%	100%	7%	35%	100%
Metadata Extraction												
ID	90.2	92.5	94.0	85.0	89.6	90.2	74.7	77.8	79.6	67.1	71.8	68.7
FINAL_DEPTH	82.8	81.8	85.0	76.3	81.7	82.9	63.3	67.9	69.1	59.8	66.0	67.4
LONGITUDE (F1)	98.2	99.3	99.4	98.5	99.2	99.2	84.8	90.6	91.1	78.7	82.3	84.9
LONGITUDE (value)	67.6	67.7	71.3	65.0	69.3	66.7	39.2	50.2	52.9	37.9	42.6	41.0
LATITUDE (F1)	97.9	98.8	99.0	98.01	99.0	98.2	87.2	90.8	91.7	78.2	84.3	81.4
LATITUDE (value)	65.4	62.6	69.7	60.9	67.9	61.5	36.8	45.6	45.2	33.3	40.5	34.4
ALTITUDE (F1)	92.4	93.4	95.4	90.2	94.4	94.2	76.1	82.2	85.5	76.4	80.7	80.3
ALTITUDE (value)	83.9	86.9	91.5	81.0	89.6	89.3	44.1	57.7	63.8	31.2	39.9	38.6
WATER (F1)	74.3	85.5	88.3	70.1	80.6	81.6	26.5	27.7	42.0	38.3	62.7	67.3
WATER (value)	50.8	64.6	68.7	39.7	53.2	55.2	7.4	9.2	13.3	9.0	11.1	14.2
Lithology Extraction												
LITHO_DEPTH_TO_DESC	82.7	86.1	87.9	70.6	78.1	80.1	49.5	55.5	58.3	47.9	56.1	59.5
LITHO_DESC_TO_DEPTH (expl)	83.7	88.1	90.8	77.5	84.7	85.8	60.7	67.6	70.4	58.9	65.0	67.2
LITHO_LIST	89.9	90.3	90.9	72.1	79.5	81.3	30.0	41.3	42.6	53.0	58.0	62.3
Counting and Indexing												
LITHO_COUNT	60.8	71.3	70.7	58.4	69.5	68.4	33.9	41.0	43.7	45.6	49.4	49.7
LITHO_INDEX_TO_DESC	68.3	75.1	78.2	59.1	70.3	72.6	39.4	42.7	44.0	43.6	44.8	45.0
LITHO_INDEX_TO_DEPTH (expl)	60.7	68.8	73.8	57.0	67.6	69.3	33.6	39.1	40.7	36.2	35.5	35.6
Spatial and Num. Reasoning												
LITHO_DESC_TO_DEPTH (impl)	28.0	32.0	38.8	20.6	26.5	27.5	14.6	17.6	19.7	18.6	22.6	23.0
LITHO_INDEX_TO_DEPTH (impl)	18.9	26.9	31.1	16.3	23.1	22.0	7.4	8.9	10.7	15.1	15.5	15.4
LITHO_DESC_TO_ELEVATION	34.6	47.4	53.8	36.4	51.2	56.9	15.4	17.9	18.1	16.0	12.4	10.6
INTERNAL_DEPTH_TO_DESC	66.3	70.5	72.1	53.9	62.0	63.3	34.2	41.8	44.0	38.8	41.9	45.2
LITHO_DESC_TO_THICKNESS	34.6	48.5	57.5	34.8	51.9	58.5	9.3	11.3	12.1	8.9	12.8	11.9

Evaluation of Fine-Tuned Open-Source Models. Table 4 presents the performance of open-source models fine-tuned on the VQA task. Each model is evaluated under three training scenarios (as explained in Sect. 3.5). Qwen3-VL-2B consistently achieves the best results, closely followed by InternVL3-2B, while SmolVLM2 and DeepSeekVL-2-tiny underperform across most metrics. Across all models, performance on implicit information extraction is significantly lower. On lithological end-depth extraction the relative performance drop ranges from 57.3% (Qwen3-VL) to 72.0% (SmolVLM2). This drop is more pronounced under limited data scenarios (i.e., 66.5% with 7% of training data compared to 57.3% with full training), indicating that additional training data is beneficial for implicit tasks.

Table 5. VQA results with closed-source VLMs in a zero-shot setting.

	GPT-5.2	Claude Sonnet 4–6
Metadata Extraction		
`ID`	64.2	64.6
`FINAL_DEPTH`	87.7	84.7
`LONGITUDE (F1)`	86.7	93.6
`LONGITUDE (Value)`	46.5	36.9
`LATITUDE (F1)`	49.3	56.2
`LATITUDE (Value)`	20.0	24.9
`ALTITUDE (F1)`	88.6	94.2
`ALTITUDE (Value)`	71.4	71.8
`WATER (F1)`	75.9	75.7
`WATER (Value)`	45.6	48.5
Lithology Extraction		
`LITHO_DEPTH_TO_DESC`	85.2	68.9
`LITHO_DESC_TO_DEPTH (expl)`	88.1	83.3
`LITHO_LIST`	87.1	79.1
Counting and Indexing		
`LITHO_COUNT`	75.8	66.3
`LITHO_INDEX_TO_DESC`	71.2	64.0
`LITHO_INDEX_TO_DEPTH (expl)`	66.9	58.0
Spatial and Num. Reasoning		
`LITHO_DESC_TO_DEPTH (impl)`	50.9	35.8
`LITHO_INDEX_TO_DEPTH (impl)`	35.5	25.5
`LITHO_DESC_TO_ELEVATION`	60.7	43.8
`INTERNAL_DEPTH_TO_DESC`	71.8	65.4
`LITHO_DESC_TO_THICKNESS`	76.3	58.0

Performance of Zero-Shot Closed-Source Models. In Table 5, we report zero-shot performance of closed-source models on the VQA task. On metadata extraction tasks, Claude Sonnet 4.6 outperforms GPT-5.2, but achieves only 74.0% of the performance obtained by the fine-tuned Qwen3-VL model. This highlights the benefits of task-specific fine-tuning for structured document formats.

On tasks involving reasoning, counting or indexing, GPT-5.2 shows significantly better results, even surpassing Qwen3-VL, with an average relative improvement of 10.3% on these tasks. Notably, the largest gains are observed on `LITHO_DESC_TO_DEPTH (impl)` and `INTERNAL_DEPTH_TO_DESC` tasks, showcasing GPT-5.2's visual grounding abilities. These differences likely reflect the advantage of ultra-large model scale and extensive pre-training in closed-source models.

To confirm our findings, we conducted further experiments with Gemini 3.1 Pro. To comply with API rate limits, evaluations were focused on `LITHO_DESC_-TO_DEPTH (impl)` questions, one of the most challenging tasks, involving spatial and numerical reasoning. Gemini 3.1 Pro achieves an accuracy of 37.3%, significantly lower than GPT-5.2 (50.9%), while still outperforming fine-tuned open-source models.

Table 6. JSON generation results with different open-source models (Qwen3-VL-2B, InternVL3-2B, SmolVLM2 2.2B and DeepSeekVL2-tiny) fine-tuned on 7%, 35% and 100% of the training set.

	Qwen3-VL			InternVL3			SmolVLM2			DeepSeekVL2		
	7%	35%	100%	7%	35%	100%	7%	35%	100%	7%	35%	100%
Metadata Extraction												
`ID`	88.2	93.0	93.8	85.4	89.1	91.8	56.7	74.5	83.8	84.0	90.0	92.0
`FINAL_DEPTH`	81.0	81.7	85.8	74.5	81.1	82.0	55.5	67.5	72.8	65.0	67.0	68.6
`LONGITUDE (F1)`	97.9	99.0	99.0	94.9	97.8	98.2	75.3	90.0	93.3	54.1	77.1	90.9
`LONGITUDE (values)`	67.0	67.9	67.5	59.5	67.2	64.9	27.5	48.1	54.7	23.0	39.1	52.5
`LATITUDE (F1)`	98.2	99.0	99.0	95.9	98.1	99.0	74.2	90.2	93.4	54.1	77.1	90.9
`LATITUDE(values)`	61.1	65.9	67.2	60.5	67.3	66.0	29.1	46.1	52.9	21.2	37.3	51.5
`ALTITUDE (F1)`	91.9	95.0	94.4	88.0	92.0	92.8	75.7	86.8	89.9	61.2	74.4	83.1
`ALTITUDE(values)`	82.5	88.6	90.2	78.4	84.9	89.6	46.8	63.4	71.3	42.5	50.3	68.1
`WATER (F1)`	69.4	82.9	87.1	71.5	90.3	81.9	47.3	55.5	62.1	60.9	43.6	56.2
`WATER (values)`	42.7	60.5	65.8	38.0	51.8	52.7	8.6	13.8	23.5	24.7	19.9	25.5
Lithology Extraction												
`LITHO_COUNT`	84.5	88.7	90.4	72.3	82.3	84.7	48.2	54.5	64.0	63.8	64.4	60.4
`LITHOLOGIES (expl)`	74.7	79.8	83.3	53.3	63.5	69.2	33.3	41.7	51.7	51.8	55.2	57.0
`LITHOLOGIES (impl)`	16.8	19.8	22.4	10.1	12.0	16.2	4.4	6.9	9.7	11.0	11.6	11.6
Multi Borehole												
`BOREHOLE_COUNT`	100	100	100	100	99.8	99.8	96.8	97.0	99.1	98.4	98.4	99.1

Evaluations on JSON Generation Task. Finally, Table 6 shows that open-source models fine-tuned on their respective tasks achieve comparable results between VQA and full data extraction via JSON generation.

Key Findings on Implicit Multimodal Reasoning. Across all evaluated models, performance on implicit information extraction remains significantly lower. For example, on the `LITHO_INDEX_TO_DEPTH (impl)` task, Qwen3-VL attains an accuracy of 31.1% while GPT-5.2 reaches 35.5%, highlighting the challenge of reasoning over implicit multimodal content.

5 Conclusion

In this work, we introduced GeoLogVQA, a large-scale Visual Question Answering dataset comprising 92k question-answer pairs from 2,554 authentic borehole log documents. These complex technical documents exhibit diverse formats and densely packed information within a single page. Notably, 53k question-answer pairs specifically address reasoning tasks (i.e., indexing, counting, numerical reasoning and spatial reasoning), with 24% of the documents requiring implicit inference where depth values are not explicitly written. These challenges remain largely underexplored in existing document VQA benchmarks.

Through experiments with both fine-tuned open-source and zero-shot closed-source Vision Language Models, we demonstrate a clear performance gap between explicit information extraction and tasks involving numerical and spatially grounded reasoning. Models achieve strong results on explicitly stated information, for example Qwen3-VL reaches 90.8% accuracy on direct lithology depth extraction. However, their performances decrease significantly on tasks requiring implicit spatial and numerical reasoning, such as inferring lithology depths from visual depth scale. The best-performing open-source model achieves only 38.8% accuracy (Qwen3-VL), while the strongest closed-source model (GPT-5.2) reaches 50.9%, highlighting the substantial difficulty of these visual reasoning tasks.

GeoLogVQA therefore provides a challenging diagnostic benchmark for evaluating spatially grounded multimodal reasoning in technical documents. Indeed, the results emphasize the necessity for improved spatially aware multimodal architectures.

Acknowledgment. This project was provided with computing AI and storage resources by GENCI at IDRIS thanks to the grant 2025-AD011016469 on the supercomputer Jean Zay's A100 partition.

References

1. Bai, S., Cai, Y., Chen, R., Chen, K., Chen, X., et al.: Qwen3-VL Technical Report (2025). https://arxiv.org/abs/2511.21631
2. Dai, D., Deng, C., Zhao, C., Xu, R., Gao, H., et al.: Deepseekmoe: towards ultimate expert specialization in mixture-of-experts language models. In: Proceedings of the 62nd Annual Meeting of the Association for Computational Linguistics, vol. 1: Long Papers, pp. 1280–1297 (2024)
3. Huang, Z., et al.: ICDAR2019 competition on scanned receipt OCR and information extraction. In: 2019 International Conference on Document Analysis and Recognition (ICDAR), pp. 1516–1520 (2019). https://doi.org/10.1109/ICDAR.2019.00244
4. Islam, M.S., Rahman, R., Masry, A., Laskar, M.T.R., Nayeem, M.T., Hoque, E.: Are large vision language models up to the challenge of chart comprehension and reasoning? An extensive investigation into the capabilities and limitations of LVLMs (2024). http://arxiv.org/abs/2406.00257

5. Jaume, G., Kemal Ekenel, H., Thiran, J.P.: FUNSD: a dataset for form understanding in noisy scanned documents. In: 2019 International Conference on Document Analysis and Recognition Workshops (ICDARW), vol. 2, pp. 1–6 (2019). https://doi.org/10.1109/ICDARW.2019.10029
6. Kim, G., et al.: OCR-free document understanding transformer. In: European Conference on Computer Vision, pp. 498–517. Springer, Cham (2022). https://doi.org/10.1007/978-3-031-19815-1_29
7. Li, M., Cui, L., Huang, S., Wei, F., Zhou, M., Li, Z.: TableBank: table benchmark for image-based table detection and recognition. In: Proceedings of the Twelfth Language Resources and Evaluation Conference, pp. 1918–1925. European Language Resources Association, Marseille (2020). https://aclanthology.org/2020.lrec-1.236/
8. Liu, H., Li, C., Wu, Q., Lee, Y.J.: Visual instruction tuning. Adv. Neural Inf. Process. Syst. **36**, 34892–34916 (2023). https://papers.nips.cc/paper_files/paper/2023/hash/6dcf277ea32ce3288914faf369fe6de0-Abstract-Conference.html
9. Luo, C., et al.: LayoutLLM: layout instruction tuning with large language models for document understanding. In: Proceedings of the IEEE/CVF conference on computer vision and pattern recognition, pp. 15630–15640 (2024). https://openaccess.thecvf.com/content/CVPR2024/html/LuovLayoutLLM_Layout_Instruction_Tuning_with_Large_Language_Models_for_Document_CVPR_2024_paper.html
10. Maaten, L.v.d., Hinton, G.: Visualizing data using t-SNE. J. Mach. Learn. Res. **9**(86), 2579–2605 (2008). http://jmlr.org/papers/v9/vandermaaten08a.html
11. MacQueen, J.: Some methods for classification and analysis of multivariate observations. In: Proceedings of the Fifth Berkeley Symposium on Mathematical Statistics and Probability, vol. 1: Statistics, vol. 5.1, pp. 281–298. University of California Press (1967)
12. Marafioti, A., Zohar, O., Farré, M., Noyan, M., Bakouch, E., et al.: SmolVLM: redefining small and efficient multimodal models (2025). http://arxiv.org/abs/2504.05299
13. Masry, A., Long, D.X., Tan, J.Q., Joty, S., Hoque, E.: ChartQA: a benchmark for question answering about charts with visual and logical reasoning. In: Findings of the Association for Computational Linguistics: ACL 2022, pp. 2263–2279. Association for Computational Linguistics, Dublin (2022). https://doi.org/10.18653/v1/2022.findings-acl.177
14. Mathew, M., Bagal, V., Tito, R., Karatzas, D., Valveny, E., Jawahar, C.V.: InfographicVQA. In: Proceedings of the IEEE/CVF Winter Conference on Applications of Computer Vision, pp. 1697–1706 (2022). https://openaccess.thecvf.com/content/WACV2022/html/Mathew_InfographicVQA_WACV_2022_paper.html
15. Mathew, M., Karatzas, D., Jawahar, C.V.: DocVQA: a dataset for VQA on document images. In: 2021 IEEE Winter Conference on Applications of Computer Vision (WACV), pp. 2199–2208. IEEE, Waikoloa (2021). https://doi.org/10.1109/WACV48630.2021.00225
16. Maćkiewicz, A., Ratajczak, W.: Principal components analysis (PCA). Comput. Geosci. **19**(3), 303–342 (1993). https://doi.org/10.1016/0098-3004(93)90090-R
17. Methani, N., Ganguly, P., Khapra, M.M., Kumar, P.: PlotQA: reasoning over scientific plots. In: Proceedings of the IEEE/CVF Winter Conference on Applications of Computer Vision, pp. 1527–1536 (2020). https://openaccess.thecvf.com/content_WACV_2020/html/Methani_PlotQA_Reasoning_over_Scientific_Plots_WACV_2020_paper.html

18. Smock, B., Pesala, R., Abraham, R.: PubTables-1M: towards comprehensive table extraction from unstructured documents. In: Proceedings of the IEEE/CVF Conference on Computer Vision and Pattern Recognition, pp. 4634–4642 (2022). https://openaccess.thecvf.com/content/CVPR2022/html/Smock_PubTables-1M_Towards_Comprehensive_Table_Extraction_From_Unstructured_Documents_CVPR_2022_paper.html
19. Team Gemma: Gemma 3 Technical Report (2025). https://arxiv.org/abs/2503.19786
20. Wu, Z., Chen, X., Pan, Z., Liu, X., Liu, W., et al.: DeepSeek-VL2: mixture-of-experts vision-language models for advanced multimodal understanding (2024). http://arxiv.org/abs/2412.10302
21. Zhang, J., Yang, W., Lai, S., Xie, Z., Jin, L.: DocKylin: a large multimodal model for visual document understanding with efficient visual slimming. In: Proceedings of the AAAI Conference on Artificial Intelligence, vol. 39, no. 9, pp. 9923–9932 (2025). https://doi.org/10.1609/aaai.v39i9.33076
22. Zhu, J., Wang, W., Chen, Z., Liu, Z., Ye, S., et al.: InternVL3: exploring advanced training and test-time recipes for open-source multimodal models (2025). https://arxiv.org/abs/2504.10479

AdaNav: Query-Adaptive Multi-granularity Navigation for Long Document Understanding

Yiming Xu(✉), Eric López, Artemis Llabrés, Maximiliano Hormazábal, Ernest Valveny, and Dimosthenis Karatzas

Computer Vision Center, Universitat Autònoma de Barcelona, Barcelona, Spain
yiming@cvc.uab.es

Abstract. Long document understanding remains challenging for Vision-Language Models (VLMs) due to the need to locate relevant information across dozens of pages with diverse content types like text, tables, and figures. Existing embedding-based retrieval methods can miss relevant pages when the query–evidence relation is compositional. Embedding-free approaches avoid this by feeding all pages to VLMs, but incur high inference cost. Neither exploits the inherent hierarchical structure of documents for efficient navigation. In this paper, we propose AdaNav, an adaptive tree-based navigation framework for visual question answering in long documents. AdaNav constructs a multimodal document tree that hierarchically organizes sections, pages, and visual elements, enabling structured navigation without embedding retrieval models. We design the navigator to adapt the retrieval granularity to the question ensuring that content queries navigate from sections to pages in a coarse-to-fine manner. To support this, we build a multi-agent system with a question router and specialized navigators that differ in their action spaces, enabling traversal of the multimodal tree at the appropriate granularity. Experiments on MMLongBench-Doc demonstrate that AdaNav outperforms existing open-source VLM-based agent systems by over 5% accuracy without requiring embedding models. Compared to embedding-free approaches, AdaNav retrieves fewer pages through tree-structured navigation. Our results show that heterogeneous tree representations with query-adaptive navigation offer an effective alternative to both paradigms for multimodal long document understanding.

Keywords: Long Document Understanding · Visual Question Answering · Adaptive Navigation · Multimodal Document Tree · Retrieval-Augmented Generation

1 Introduction

Visually rich long documents, such as financial reports, technical manuals, and scientific papers, typically span dozens of pages, with key evidence distributed

G. A. Fink et al. (Eds.): ICDAR 2026, LNCS 16972, pp. 55–72, 2027.
https://doi.org/10.1007/978-3-032-36023-6_4

across running text, tables, figures, and layout structure. Existing long-document QA pipelines [10,12,15] often fall into two extremes, making it difficult to achieve both high accuracy and efficiency. On the one hand, embedding-based retrieval [2,8,9] uses page or segment embeddings for scalable search, but can miss relevant pages when the query–evidence relation is compositional (e.g., requiring multi-hop, cross-page, or cross-modal aggregation), causing downstream reasoning to fail due to missing evidence. On the other hand, embedding-free approaches [1,15,17] that feed all pages to a VLM avoid retrieval bias, but require processing the full document, leading to large token budgets and high inference cost; they also do not scale well to longer documents.

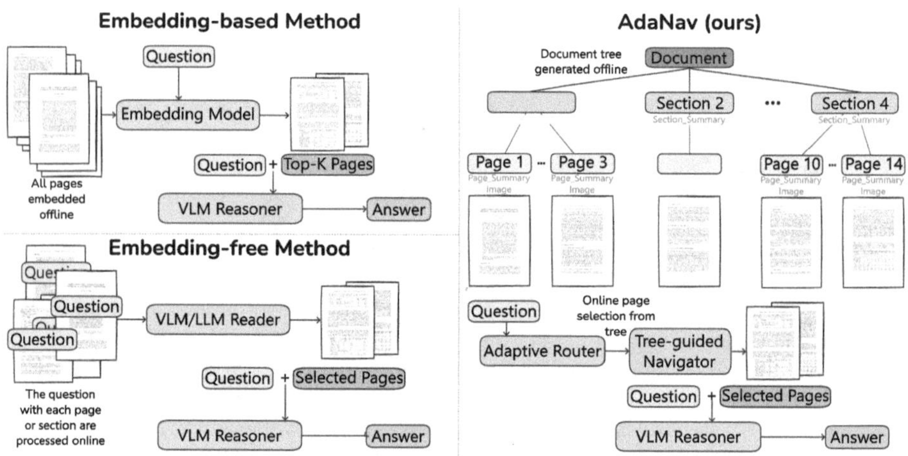

Fig. 1. Comparison between embedding-based methods (top-left), embedding-free methods (bottom-left) and AdaNav (right).

We observe that documents are inherently hierarchical: sections group related content, pages contain localized information, and visual elements (tables, figures, charts) carry structured evidence. This structure naturally supports coarse-to-fine navigation, allowing a system to traverse from sections to pages to elements while filtering irrelevant branches early. Such navigation is particularly beneficial for cross-page and multi-hop questions, where evidence must be aggregated from multiple locations. This motivates a navigation-centric alternative that does not rely on extra embedding retrieval models, yet can structurally traverse a document and focus on only the necessary granularity for a given question.

Addressing this goal poses several challenges. First, long documents exhibit a natural hierarchy (Section → Page → Element), but choosing the right retrieval granularity is non-trivial: retrieving too coarsely introduces excessive irrelevant context, while retrieving too finely exponentially increases the search space. Building a multi-granularity, navigable representation and adapting the navigation granularity to the question remains an open challenge. Second, critical evidence often resides in table cells, chart annotations, legends, or localized visual regions. OCR-only representations lose layout and visual relations, while using page-level vision alone makes it difficult to align reliably the content of the page to question semantics; we need consistent grounding across section/page summaries and visual elements. Third, multi-page VLM inference is expensive, and errors can compound in long-document settings due to OCR noise, parsing errors, and cross-page dependencies. A practical system must reduce the number of pages sent to the VLM through structured filtering and query-type-aware routing.

In this paper, we propose AdaNav, Fig. 1, a query-adaptive multi-granularity navigation framework for long-document visual question answering. AdaNav is based on building a hierarchical multimodal document tree based on sections, pages and visual elements. Using this tree, a router agent derives the question to specialized navigation agents that traverse the tree at different levels of granularity depending on the question type, retrieving a set of candidate pages to answer the question. Finally, the selected pages are passed through the reasoner agent to obtain the final answer. Our contributions are as follows. (1) We construct a multimodal document tree (Document–Section–Page–Element) that hierarchically organizes sections, pages, and visual elements (e.g., cropped tables and figures), preserving both logical and visual structure for downstream navigation. (2) We introduce an adaptive tree-based navigation framework that enables structured traversal of the tree without embedding retrieval models and adapts navigation granularity in a coarse-to-fine manner. (3) We build a routing-based multi-agent system comprising a question router and specialized navigators with different action spaces, enabling query-type-aware paths (Specific/General/Normal) and traversal of the multimodal tree at the appropriate granularity. We validate the proposed method through comprehensive evaluation on MMLongBench-Doc and multiple datasets spanning diverse document types and question categories, demonstrating that AdaNav consistently improves end-to-end accuracy and retrieval efficiency using both Qwen3-VL-30B-A3B and Qwen2.5-VL-32B as backbone VLMs. Code and doctree data are available at https://github.com/ym-xu/AdaNav.

2 Related Work

2.1 Long Document Understanding

Long document understanding studies how to answer questions that require locating and aggregating evidence across many pages with heterogeneous content types (running text, tables, figures, and layout cues). Compared to short-form document QA [11,12], the key difficulty is not only reasoning, but also where to look: evidence can be sparse, distributed, and often grounded in visual structures rather than plain OCR text. Recent benchmarks have made this challenge explicit. MMLongBench-Doc [10] focuses on long-context document understanding with visualizations and provides evidence page annotations to evaluate retrieval and end-to-end QA. LongDocURL [3] further broadens the scope by integrating understanding, reasoning, and locating in multimodal long documents. These benchmarks motivate systems that jointly address evidence localization and multimodal reasoning under strict context and cost constraints.

2.2 Embedding-Based Retrieval for Document QA

One common way to address long document understanding is to use some form of Retrieval Augmented Generation (RAG). These approaches commonly use dense embeddings to index pages, regions, or chunks, reducing the amount of context passed to a downstream reader. Recent systems such as SimpleDoc [8] combine multi-modal cues for page retrieval and design careful prompting to improve document QA performance. Meanwhile, M3DOCRAG [2] argues that multi-modal retrieval is crucial for multi-page/multi-document understanding and highlights the limitations of text-only signals for evidence localization. Agentic frameworks like MDocAgent [6] employ a multi-agent team—consisting of general, critical, and modality-specific agents—to refine retrieved evidences. Despite these advances, embedding-based pipelines typically require an additional retrieval model and may still miss evidence under complex reasoning relationships, like when the query–evidence relation is compositional (e.g., multi-hop or cross-page).

2.3 Embedding-Free Document QA

An alternative line of work avoids explicit embedding retrieval by directly applying long-context VLMs (or agentic orchestration) to large portions of a document. Such embedding-free paradigms can reduce retrieval bias but often "read" excessive irrelevant content, incurring high token cost and making inference inefficient for long documents; they may also increase the risk of hallucination when the model cannot reliably ground its answers. Hi-VT5 [15] and GRAM [1] utilize hierarchical or local-global strategies to process all pages in a single model. DocLens [17] exemplifies tool-augmented, multi-agent frameworks that attempt to improve long visual document understanding through orchestration and tool use.

2.4 Tree-Structured Document Navigation

Hierarchical representations provide a principled way to navigate long inputs by exploiting document structure. Structural document maps, for instance, aim to preserve global organization and support controlled traversal; DMAP [5] studies human-aligned structural maps for multimodal document understanding. On the tooling side, document parsing systems such as MinerU2.5 [14] expose structural metadata and detected elements that can serve as building blocks for hierarchical navigation. In parallel, reasoning-based "vectorless" retrieval narratives like PageIndex [16] emphasize structured search actions over embeddings. Nevertheless, many hierarchical approaches are still text-centric or fail to preserve fine-grained visual semantics, limiting their effectiveness on visual-rich documents.

In contrast to embedding-based retrieval and embedding-free full-document prompting, AdaNav combines the strengths of both paradigms: it leverages tree-structured navigation to avoid exhaustive page reading, while eliminating the need for additional embedding retrieval models through query-adaptive, multi-granularity traversal over a multimodal document tree.

3 Methodology

3.1 Overview

AdaNav is a query-adaptive navigation framework for long-document visual question answering. The key insight is that different questions require different navigation granularities: some can be resolved by direct reference lookup (e.g., "What does Fig. 3 show?"), others need document-wide scanning (e.g., "How many tables are there?"), and most benefit from coarse-to-fine section-to-page navigation (e.g., "What is the proposed method?").

Based on this observation, AdaNav operates in two stages: (1) an offline stage that parses the PDF and constructs a multimodal document tree capturing the hierarchy from sections to pages to visual elements; (2) an online stage that routes each question to a specialized navigator, retrieves relevant pages via tree traversal, and generates an answer with a VLM-based reasoner. We describe the tree construction in Sect. 3.2, then detail the navigation pipeline in Sect. 3.3.

3.2 Multimodal Document Tree Construction

As illustrated in Fig. 2, we represent a document as a rooted tree $\mathcal{T} = (D, \{S_i\}, \{P_j\}, \{E_k\})$ with four levels of nodes: a single Document node D, a set of Section nodes $\{S_1, \ldots, S_M\}$, a set of Page nodes $\{P_1, \ldots, P_N\}$, and a set of Element nodes $\{E_1, \ldots, E_L\}$. The parent–child relations are $D \to S_i \to P_j \to E_k$, forming a strict hierarchy. Each node carries a heterogeneous feature vector described below.

Section Construction. We first partition the document into M sections. Each Section node is defined as $S_i = (t_i, [a_i, b_i], \mathbf{s}_i)$, where t_i is the section title, $[a_i, b_i]$ is the page range (0-indexed), and $\mathbf{s}_i$ is a section-level summary (populated later). When the PDF provides an explicit table of contents (ToC), we directly extract level-1 TOC entries as sections, each associated with a title and a start page. When no reliable TOC is available, we derive section boundaries from MinerU-predicted headings and group semantically related titles to form coherent sections. We only materialize a single layer of Section nodes. If a section spans more than τ pages, we further split it using its subheadings to keep the downstream navigation budgeted while preserving local topical coherence.

Page and Element Construction. Given the section structure, we assign each physical page to its parent Section according to page ranges, yielding N Page nodes. Each Page node is defined as $P_j = (I_j, x_j, \mathbf{p}_j, \{E_k\}_{k \in \mathcal{E}_j})$, where I_j is the rendered page image, x_j is the page-level textual content comprising the

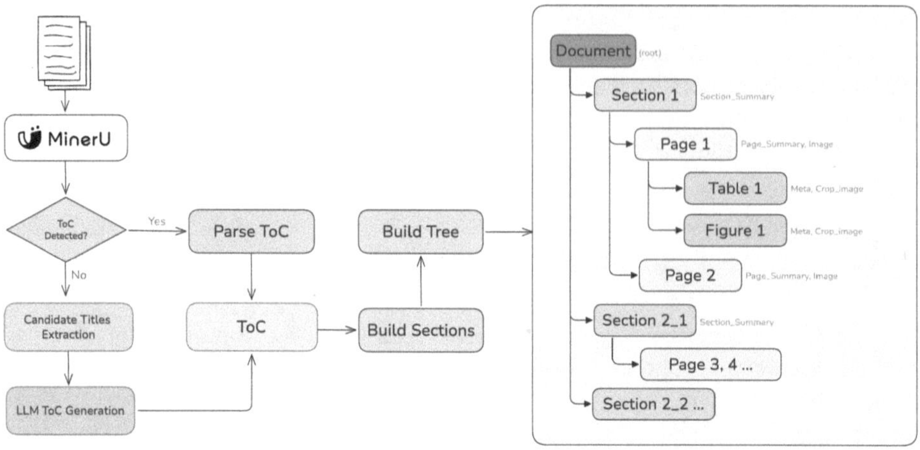

Fig. 2. Multimodal document tree construction. The tree organizes documents into four levels: Document (root), Section (with text summaries), Page (with screenshots), and Element (with cropped visual regions). This hierarchy enables coarse-to-fine navigation.

concatenated OCR text of all layout blocks extracted by MinerU, with each block annotated with its semantic type and bounding-box coordinates $[x_0, y_0, x_1, y_1]$, $\mathbf{p}_j$ is a page-level summary, and $\mathcal{E}_j$ indexes the element children of P_j.

Each Element node is defined as $E_k = (\text{type}_k, \text{bbox}_k, t_k, c_k, I_k^{\text{crop}})$, where $\text{type}_k \in \{$`table`, `figure`, `chart`, `image`, `diagram`, `graph`, `text`, `page_number`$\}$ is the MinerU-assigned layout category, bbox_k is the bounding box on the source page, t_k is the block-level text content, c_k is the associated caption for figure and table elements, and I_k^{crop} is a cropped image present only for visual-category elements whose region exceeds a minimum size threshold.

Hierarchical Summary Generation. Beyond the structural hierarchy, we attach retrieval-oriented features at two granularities. For each Page node, we generate $\mathbf{p}_j$ by prompting a VLM with the page image I_j and OCR text x_j, using a structured template that elicits three components: (1) a free-text description of the page content; (2) textual and visual evidence highlights, with separate fields for tables, figures, and charts; and (3) a categorised entity list covering numbers, names, dates, terms, and cross-page references. For each Section node, we generate $\mathbf{s}_i$ by prompting a Large Language Model (LLM) with the concatenated page summaries $\{\mathbf{p}_j\}_{j \in [a_i, b_i]}$, producing a structured index card that includes topic tags, evidence lists with page-level provenance, visual element inventories, a consolidated entity list. Crucially, the section card is intentionally designed to be queryable rather than abstractive, preserving specific values, figure/table identifiers, and cross-references verbatim to support precise retrieval matching. Prompt templates are provided in the code repository.

3.3 Query-Adaptive Navigation

As illustrated in Fig. 3, given a question and the prebuilt document tree, AdaNav answers through three stages: routing, navigation, and reasoning. Algorithm 1 summarizes the complete procedure.

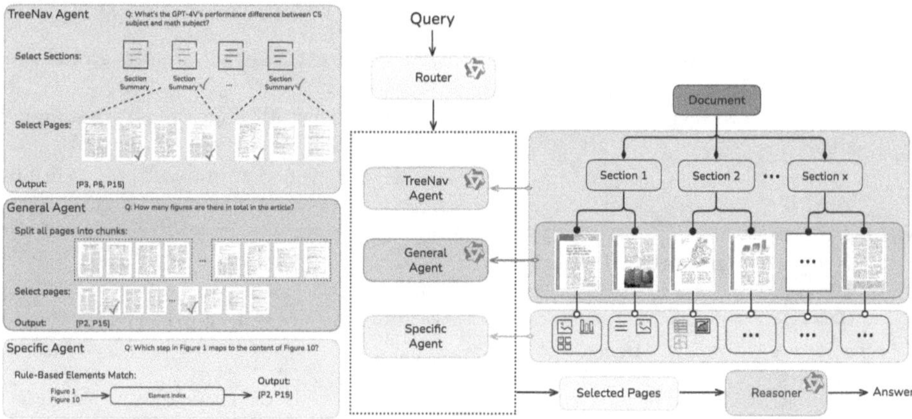

Fig. 3. AdaNav framework overview. **Left panel:** Workflows of three specialized agents. TreeNav selects top-K sections then navigates to relevant pages; General scans all pages in chunks for aggregation queries; Specific resolves element references via index lookup. **Right panel:** Query-adaptive routing dispatches each question to the appropriate agent based on its type, retrieving pages from the document tree at the optimal granularity before VLM reasoning.

Router. The router classifies each question into one of three categories:

- **SPECIFIC**: the question contains explicit references such as "page X", "figure Y", or "table Z" (e.g., "What does Table 3 show?"), which can be resolved by direct index lookup.
- **GENERAL**: the question requires document-wide coverage (e.g., "How many figures are in this document?"), handled by scanning all pages.
- **NORMAL**: the question is a standard content query (e.g., "What method does the paper propose?"), handled by coarse-to-fine navigation from candidate sections to pages.

Routing proceeds in two stages. In the first stage, a rule-based detector $f_{\text{rule}}(q)$ pattern-matches explicit page, figure, and table references in the question; if any are found, the router assigns $c = $ SPECIFIC and terminates without invoking an LLM. In the second stage, an LLM-based classifier $f_{\text{llm}}(q,\ \{(t_i, \mathbf{s}_i)\}_{i=1}^{M})$ reads all M section titles and summaries jointly to determine whether the question can be localised to a subset of sections or requires

Algorithm 1. AdaNav Online Inference

Require: question q; document tree $\mathcal{T} = (D, \{S_i\}_{i=1}^{M}, \{P_j\}_{j=0}^{N-1}, \{E_k\})$
Ensure: answer $\hat{a}$

// **Stage 1: Question Router**
$(refs, \mathcal{P}^*) \leftarrow f_{\text{rule}}(q, \mathcal{T})$ ▷ rule-based reference detection
if $refs \neq \emptyset$ **then**
  $c \leftarrow$ SPECIFIC ▷ direct lookup, zero LLM cost
else
  $(\mathcal{S}^*, g) \leftarrow f_{\text{llm}}(q, \{(t_i, \mathbf{s}_i)\}_{i=1}^{M})$ ▷ top-K sections + global flag
  if $g = 1$ **then**
    $c \leftarrow$ GENERAL ▷ global query; all pages must be scanned
  else
    $c \leftarrow$ NORMAL; $\hat{\mathcal{S}} \leftarrow \mathcal{S}^*$ ▷ router-selected sections
  end if
end if

// **Stage 2: Specialized Navigation**
if $c =$ SPECIFIC **then**
  $\hat{\mathcal{P}} \leftarrow \mathcal{P}^*$ ▷ resolved by index lookup, zero LLM cost
else if $c =$ GENERAL **then**
  partition $\{P_0, \dots, P_{N-1}\}$ into overlapping chunks $\{\mathcal{C}_1, \dots, \mathcal{C}_L\}$
  $\hat{\mathcal{P}} \leftarrow \emptyset$
  for each chunk $\mathcal{C}_l$ **do** ▷ all N pages, token-budget chunking
    $\hat{\mathcal{P}} \leftarrow \hat{\mathcal{P}} \cup \pi_{\text{page}}(q, \mathcal{C}_l, B_{\text{gen}})$
  end for
else ▷ $c =$ NORMAL
  $\hat{\mathcal{P}} \leftarrow \emptyset$
  for each $S_i \in \hat{\mathcal{S}}$ **do** ▷ K candidate sections
    $\hat{\mathcal{P}} \leftarrow \hat{\mathcal{P}} \cup \pi_{\text{page}}(q, S_i, B_{\text{nav}})$
  end for
end if

// **Stage 3: Multimodal Reasoning**
$\hat{a} \leftarrow \text{VLM}(q, \{(I_j, x_j)\}_{j \in \hat{\mathcal{P}}})$ ▷ scratchpad reasoning over raw page content
return $\hat{a}$

full-document coverage. In the former case, it assigns $c =$ NORMAL and returns the top-K candidate sections $\mathcal{S}^*$; in the latter, it assigns $c =$ GENERAL, indicating that targeted section selection is insufficient and all pages must be scanned. The routing decision c is passed to the downstream navigation agent; the candidate sections $\mathcal{S}^*$ are additionally forwarded when $c =$ NORMAL.

Specialized Navigation Agents. Based on the routing decision c, the system dispatches the query to one of three specialized navigation agents. Each agent traverses the document tree at an appropriate granularity and returns an evidence page set $\hat{\mathcal{P}}$ for the downstream reasoner.

TreeNav Agent. When $c =$ NORMAL, the Router has already narrowed the search space to K candidate sections $\mathcal{S}^*$. The TreeNav Agent performs visual page navigation within each of these sections: for every section $S_i \in \mathcal{S}^*$, it constructs an interleaved multimodal sequence from the page images I_j and structured summaries $\mathbf{p}_j$ of all pages $P_j \in S_i$, and presents this sequence to a VLM to identify the top-B_{nav} most relevant pages. For sections exceeding a length threshold τ_{pre}, a two-stage procedure is employed: an LLM first performs a coarse text-based pre-filter to reduce the candidate set, after which the VLM conducts fine-grained visual retrieval over the pruned pages. Results across all K sections are merged and deduplicated to form $\hat{\mathcal{P}}$.

General Agent. When $c = \text{GENERAL}$, the question demands document-wide coverage such as counting or aggregation across the entire document. The General Agent operates directly at the page level of the document tree, scanning all N page nodes without section-level filtering. As the VLM context is constrained by a token budget, the full page sequence is partitioned into overlapping chunks of size C, where the overlap prevents evidence loss at chunk boundaries. For each chunk, the interleaved sequence of page images I_j and summaries $\mathbf{p}_j$ is presented to the VLM for visual retrieval. The results from all chunks are merged via union to produce $\hat{\mathcal{P}}$.

Specific Agent. When $c = \text{SPECIFIC}$, the question contains explicit references that can be resolved through direct index lookup. The Specific Agent first applies regular-expression patterns to extract reference targets from the question, such as page numbers, figure identifiers, and table identifiers. It then resolves each reference through a cascaded lookup over three offline-built indices: a logical-to-physical page number map, an element identifier index constructed from captions and block text matching "Figure X" or "Table X", and a page-level entity index for broader reference matching. The entire procedure incurs zero LLM cost, and the resolved page indices directly constitute $\hat{\mathcal{P}}$.

Reasoner. Once navigation produces the evidence set $\hat{\mathcal{P}}$, a VLM generates the final answer:

$$\hat{a} = \text{VLM}\big(q,\ \{(I_j, x_j)\}_{j\in\hat{\mathcal{P}}}\big)$$

where I_j is the rendered page image and x_j is the OCR text of page P_j. Notably, the structured retrieval indices built during the offline stage, namely page summaries $\mathbf{p}_j$, section index cards $\mathbf{s}_i$, and element-level crops I_k^{crop}, are used exclusively for navigation and are not forwarded to the reasoner. The reasoner instead receives the raw page content directly, allowing the VLM to ground its answer in the original visual layout without interference from pre-computed abstractions. The prompt adopts a scratchpad structure, instructing the model to first identify key evidence from both modalities across the retrieved pages before synthesising the final answer.

4 Experiments

4.1 Experimental Setup

Datasets. We evaluate on three publicly available DocVQA benchmarks that span diverse document types and question categories. (1) MMLongBench-Doc [10] contains 1,073 questions over 135 long PDF documents with an average of 47.5 pages per document; questions require reasoning over text, tables, charts, figures, and layouts, and are annotated with evidence page indices. (2) PaperTab [7] focuses on tabular data extracted from academic papers, providing 393 questions over 307 documents. (3) FetaTab [13] contains 1,023 free-form questions over 878 documents built from Wikipedia tables. Together these benchmarks cover complementary difficulty axes: long multi-modal documents (MMLongBench-Doc) and structured table understanding (PaperTab, FetaTab).

Baselines. We compare against three representative open-source VLM-based retrieval systems, all using ColQwen-2.5 [4] for visual embedding retrieval and Qwen2.5-VL-32B-Instruct for generation, following the evaluation protocol of SimpleDoc [8]. (1) M3DocRAG [2] retrieves the top-k pages via image embedding similarity and feeds them directly to a VLM. (2) MDocAgent [6] employs parallel text and image retrieval agents together with a dedicated critical-information extractor in a multi-agent pipeline. (3) SimpleDoc [8] augments embedding retrieval with LLM-based summary re-ranking and an iterative reasoning loop. We also include a Direct VLM baseline (Qwen2.5-VL-32B without any retrieval) and an Oracle condition in which the VLM receives the ground-truth evidence pages directly, serving as an upper bound on the $n = 847$ questions that carry evidence annotations.

Metrics. For end-to-end QA we report binary accuracy: each response is judged correct or incorrect by GPT-4.1 used as an automatic evaluator at temperature 0, following [8]. For retrieval quality on MMLongBench-Doc we additionally report page-level **Recall** (fraction of ground-truth evidence pages retrieved), **All-Hit Rate** (fraction of questions for which every evidence page is retrieved), and page-level **F1** (harmonic mean of per-question precision and recall).

Implementation Details. Our primary experiments use Qwen3-VL-30B-A3B as the VLM backbone for all stages of AdaNav. To enable a direct comparison with the baselines under identical generation conditions, we also report AdaNav results with Qwen2.5-VL-32B-Instruct as the backbone. The tree index is constructed offline once per document; at inference time we set the per-section page budget $B_{\text{nav}} = 3$ and the general-agent budget $B_{\text{gen}} = 2$ pages per section.

4.2 Main Results

Table 1 reports end-to-end accuracy on three DocVQA benchmarks. AdaNav with Qwen3-VL-30B-A3B achieves the highest average accuracy, surpassing all open-source RAG baselines. On MMLongBench-Doc, AdaNav attains 65.73%, a gain of +6.2 points over SimpleDoc and +10.4 points over MDocAgent, while retrieving significantly fewer pages. When evaluated on the answerable subset ($n = 847$), AdaNav achieves 63.79%, outperforming SimpleDoc by +4.2 points and MDocAgent by +8.5 points on this subset.

Table 1. Accuracy (%) on three DocVQA benchmarks: MMLongBench-Doc (MMLB), PaperTab, and FetaTab. *Pg. Ret.* is the average number of pages fed to the VLM per question. †Numbers are taken from SimpleDoc [8]; all those methods use ColQwen-2.5 for visual embedding retrieval and Qwen2.5-VL-32B-Instruct for generation. ‡Oracle is evaluated only on questions with annotated evidence pages ($n = 847$ of 1,073); the full pipeline covers all questions including unanswerable ones. Best results are **bold**; second-best are <u>underlined</u>.

Method	Pg. Ret.	MMLB	PaperTab	FetaTab	Avg.
Direct VLM (no retrieval)					
Qwen2.5-VL-32B†	–	22.18	7.12	16.14	15.15
Oracle upper bound (GT evidence pages)‡					
Qwen2.5-VL-32B + GT pages†	–	67.94	–	–	–
Qwen3-VL-30B-A3B + GT pages	–	67.65	–	–	–
RAG Methods (Qwen2.5-VL-32B)†					
M3DocRAG [2]	6	41.80	60.10	79.80	60.57
MDocAgent [6]	12	55.30	**64.90**	**84.50**	<u>68.23</u>
SimpleDoc [8]	3.2	<u>59.55</u>	<u>64.38</u>	80.31	68.08
AdaNav (Ours)					
AdaNav (Qwen3-VL-30B-A3B, answerable only)‡	–	63.79	–	–	–
AdaNav (Qwen2.5-VL-32B)	4.17	59.46	61.06	82.89	67.80
AdaNav (Qwen3-VL-30B-A3B)	4.17	**65.73**	63.61	<u>83.75</u>	**71.02**

Across the three benchmarks, AdaNav shows complementary strengths and limitations relative to the baselines. On FetaTab, AdaNav (Qwen3) achieves 83.75%, the second highest score, trailing MDocAgent (84.50%) by less than one point; this small gap is consistent with observations in prior work [8] that MDocAgent's dedicated image retrieval agent confers a specific advantage on table-centric datasets. On PaperTab, AdaNav (Qwen3) achieves 63.61%, slightly below SimpleDoc (64.38%) and MDocAgent (64.90%), indicating that structured tabular retrieval from academic papers remains a challenging domain where embedding-based retrieval complements tree navigation.

To enable a fair comparison under identical generation conditions, we also evaluate AdaNav with Qwen2.5-VL-32B-Instruct, the backbone used by all baselines. AdaNav (Qwen2.5) achieves 59.46% on MMLongBench-Doc, which is on par with SimpleDoc (59.55%) while operating without any pre-computed visual embeddings. On FetaTab, AdaNav (Qwen2.5) outperforms SimpleDoc by +2.6

points (82.89% vs. 80.31%), confirming that the tree-structured navigation provides a reliable retrieval signal independent of the backbone. The improvement from Qwen2.5 to Qwen3 (+6.3 points on MMLongBench-Doc, +2.5 on FetaTab) reflects the stronger visual reasoning capability of Qwen3-VL-30B-A3B, particularly for multi-modal evidence types.

4.3 Ablation Study

Table 2 presents the ablation results. The base variant uses TreeNav for all queries, achieving 61.4% accuracy. Adding the General Agent (+2.0%) improves recall from 79.6% to 83.9% by scanning all pages in chunks for GENERAL queries. Adding the Specific Agent (+1.4%) enables direct element lookup with zero LLM cost for SPECIFIC queries. The full AdaNav combines all three agents, achieving 65.7% accuracy (+4.3%), validating the effectiveness of query-adaptive routing.

Table 2. Ablation study on MMLongBench-Doc with Qwen3-VL-30B-A3B (evaluated by GPT-4.1). Each row shows which agent handles each query category; "TreeNav" means the default tree-navigation path is used as a fallback. Recall and F1 are page-level retrieval metrics (computed on $n = 847$ questions with annotated evidence pages). ACC is overall accuracy on all 1,073 questions.

Variant	Query Routing			Recall	F1	ACC
	Spec.	Gen.	Norm.			
AdaNav (Base)	TreeNav	TreeNav	TreeNav	79.6	48.9	61.4
w/o SpecificAgent	TreeNav	General	TreeNav	83.9	56.2	63.4
w/o GeneralAgent	Specific	TreeNav	TreeNav	83.2	53.7	62.8
AdaNav (Agentic)	Specific	General	TreeNav	**85.2**	**57.1**	**65.7**

4.4 Analysis

Query Routing Distribution. Table 3 reports the routing decisions on MMLongBench-Doc, providing the interpretive context for all findings that follow. The vast majority of queries (76.7%) follow the Normal path, reflecting that most document questions require structured coarse-to-fine navigation via the TreeNavAgent. The Specific path accounts for 10.5% of queries (113 questions) involving explicit references to named pages, figures, or tables; these are resolved by the SpecificAgent through programmatic index lookup without any LLM call. The General path covers 12.8% of queries (137 questions) requiring document-wide aggregation, handled by the GeneralAgent through a full-document scan. This distribution validates the multi-path design: each path serves a structurally distinct query type, and the SpecificAgent alone eliminates LLM overhead for over one in ten questions.

Table 3. Distribution of query routing decisions made by the Router on MMLongBench-Doc (1,073 questions). The SPECIFIC path incurs zero LLM cost; GENERAL and NORMAL paths invoke the text-based PageNavAgent.

Path	Handler	Count	Proportion
SPECIFIC	SpecificAgent (rule-based, zero LLM)	113	10.5%
GENERAL	GeneralAgent (full-document scan)	137	12.8%
NORMAL	TreeNavAgent (coarse-to-fine)	823	76.7%
Total		1,073	100%

Retrieval Efficiency. The routing design translates into distinctive retrieval characteristics, as shown in Table 4. Pure embedding-based retrieval (ColQwen-2.5) exhibits a clear coverage–precision tension: increasing k from 2 to 10 raises All-Hit from 64.1% to 83.6% but causes F1 to collapse from 38.75% to 18.38%, as many irrelevant pages accumulate. SimpleDoc mitigates this with summary re-ranking, achieving F1 $\approx$62% at 3.5 pages, but All-Hit remains capped at 67.4%, over 12 points below AdaNav. AdaNav (Base) achieves All-Hit of 77.5% and Recall of 83.2%, matching ColQwen at $k = 10$ in coverage but requiring only 5.94 pages. AdaNav (Agentic) further improves to All-Hit of 80.2%, Recall of 85.5%, and F1 of 57.1% at only 4.17 pages, achieving the best precision–coverage balance among all compared systems. This demonstrates that the offline tree index enables targeted navigation that avoids the indiscriminate page accumulation inherent in top-k embedding retrieval.

Table 4. Retrieval performance on MMLongBench-Doc ($n = 847$ questions with annotated evidence pages). *Avg. Pages* is the mean number of pages fed to the VLM. *All-Hit* is the fraction of questions for which every gold evidence page is retrieved. *Recall* is the mean fraction of gold pages retrieved per question. *F1* is the macro-averaged page-level F1 score. †Numbers are from SimpleDoc [8]; Recall is not reported for those methods.

Method	Avg. Pages	All-Hit (%)	Recall (%)	F1 (%)
Embedding-based retrieval (ColQwen-2.5)†				
ColQwen-2.5 ($k = 2$)	2	64.12	–	38.75
ColQwen-2.5 ($k = 6$)	6	76.42	–	24.36
ColQwen-2.5 ($k = 10$)	10	83.60	–	18.38
RAG with re-ranking†				
SimpleDoc (top-10)	3.19	65.72	–	61.42
SimpleDoc (top-30)	3.46	67.37	–	62.22
AdaNav (Ours, Qwen3-VL-30B-A3B)				
AdaNav (Base)	5.94	77.5	79.6	48.9
AdaNav (Agentic)	4.17	80.2	85.5	57.1

Evidence-Type Breakdown. The retrieval gains from the two specialized agents manifest clearly at the answer level. Table 5 decomposes accuracy by evidence source and page-span category. AdaNav (Agentic, Qwen3) achieves the best score in eight of nine columns, with two particularly striking findings that directly reflect the routing distribution above. First, on table-centric questions (TAB), AdaNav attains 68.7%, outperforming SimpleDoc (51.2%) by +17.5 points and MDocAgent (56.1%) by +12.6 points: the 10.5% of SPECIFIC queries routed to the SpecificAgent are dominated by "Table X" references, resolved via exact caption-index lookup at zero LLM cost. The same mechanism drives the large gain on figure-centric questions (FIG: 47.3→57.7, +10.4 pts from Base to Agentic). Second, on multi-page questions (MUL), AdaNav Agentic reaches 54.9%, outperforming SimpleDoc (39.2%) by +15.7 points, with the largest absolute gain over Base (+11.4 pts). This is driven by the GeneralAgent: the 12.8% of GENERAL queries requiring document-wide aggregation are answered by a full-document scan that uniform tree navigation cannot match. The one exception is the unanswerable (UNA) category, where AdaNav Base (76.0%) outperforms Agentic (71.4%): committing to a specific retrieval path reduces the model's tendency to abstain when evidence is absent. On single-page questions (SIN), AdaNav Agentic (71.0%) narrowly surpasses SimpleDoc (70.8%), confirming that tree-based coarse-to-fine navigation is competitive even on straightforward localization tasks.

Table 5. Per-category accuracy (%) on MMLongBench-Doc. Evidence sources: text (TXT), layout (LAY), chart (CHA), table (TAB), figure (FIG); page count: single-page (SIN), multi-page (MUL), unanswerable (UNA). †Baseline numbers are from SimpleDoc [8] using Qwen2.5-VL-32B-Instruct. Our results are evaluated by GPT-4.1. Oracle provides GT evidence pages directly and has no UNA subset. **Bold**: best per column (non-Oracle); underlined: second best.

Method	TXT	LAY	CHA	TAB	FIG	SIN	MUL	UNA	ACC
Oracle upper bound (Qwen3-VL-30B-A3B, GT pages)									
Oracle	64.2	63.3	66.3	65.7	63.6	73.3	59.9	–	67.7
Baselines (Qwen2.5-VL-32B)†									
M3DocRAG [2]	46.7	41.5	45.4	39.2	43.8	58.6	30.6	22.5	41.8
MDocAgent [6]	57.5	50.0	54.7	56.1	52.8	68.7	42.9	45.4	55.3
SimpleDoc [8]	59.9	51.3	54.9	51.2	51.2	70.8	39.2	67.4	59.6
AdaNav (Ours, Qwen2.5-VL-32B)									
AdaNav (Agentic)	61.6	54.6	50.6	62.7	51.8	66.7	46.5	64.2	59.5
AdaNav (Ours, Qwen3-VL-30B-A3B)									
AdaNav (Base)	59.8	52.1	57.1	59.4	47.3	67.7	43.5	**76.0**	61.4
AdaNav (Agentic)	**66.7**	**57.1**	**60.9**	**68.7**	**57.7**	**71.0**	**54.9**	71.4	**65.7**

4.5 Case Study

Figure 4 compares AdaNav with SimpleDoc on three representative question types. For the Normal question (Top), TreeNav selects relevant sections then navigates to pages containing the target table, while SimpleDoc's embedding retrieval fails to locate structural evidence. For the General question (Middle), the General Agent scans all pages in chunks to aggregate counts, whereas top-k

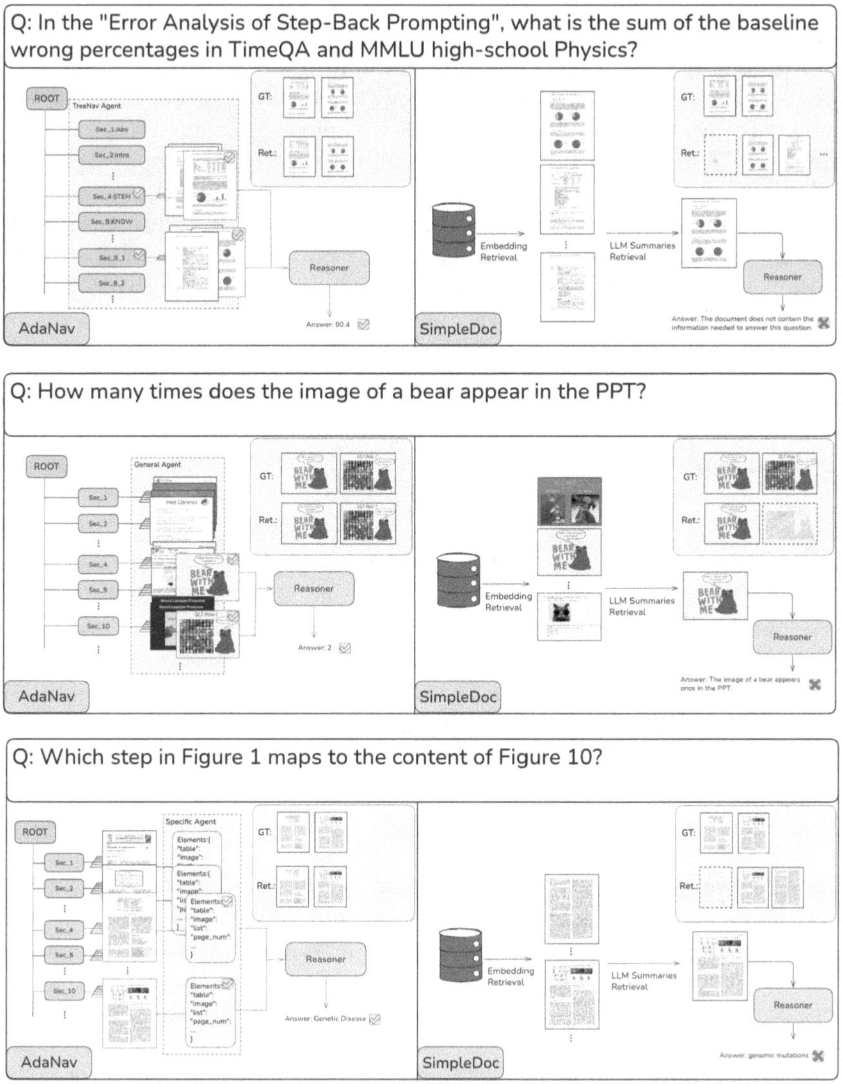

Fig. 4. Qualitative examples. A Normal question case where the TreeNav Agent is used (Top), a General question case where the General Agent is used (Middle), and a Specific Question where the Specific Agent is used (Bottom).

retrieval misses occurrences. For the Specific question (Bottom), direct element index lookup retrieves the exact pages for Fig. 1 and Fig. 10 at zero LLM cost, while embedding similarity cannot reliably match figure references. These cases validate that different question types benefit from specialized navigation strategies.

5 Limitations

While AdaNav achieves strong performance on the evaluated benchmarks, several limitations remain. The quality of the multimodal document tree depends directly on the reliability of the underlying parser. OCR errors in dense tables, incorrect heading detection in documents without a machine-readable table of contents, and missed figure captions introduce noise that propagates through every level of the tree; confidence-aware tree construction that can flag and recover from low-quality subtrees is a promising direction. The current implementation constructs a tree per document. For book-length documents or cross-document settings, the Section level may need further decomposition, and inter-document linking is not yet supported. Finally, the two-stage router is evaluated only on English documents in the MMLongBench-Doc domain; its generalization to domain-specific terminology or non-English text remains to be studied.

6 Conclusion

We have presented AdaNav, a query-adaptive multi-granularity navigation framework for long-document visual question answering. AdaNav offline constructs a multimodal document tree that hierarchically organizes sections, pages, and cropped visual elements, enabling structured navigation without embedding retrieval models. At inference time, a two-stage router classifies each question into one of three paths, and dispatches it to a specialized agent: the SpecificAgent resolves explicit element references through programmatic index lookup at zero LLM cost, the GeneralAgent performs a document-wide scan for global aggregation queries, and the TreeNavAgent executes coarse-to-fine section-to-page navigation for the remaining content questions. A shared GlobalReasonerAgent generates the final answer from the retrieved pages and visual elements.

Evaluated on MMLongBench-Doc, AdaNav (Agentic, Qwen3-VL-30B-A3B) achieves 65.7% end-to-end accuracy, outperforming all open-source VLM-based agent baselines by a substantial margin while retrieving only 4.17 pages per query on average. A per-category breakdown confirms that each specialized agent yields gains on its intended evidence type, validating the multi-path design. These results suggest that structured offline indexing combined with query-adaptive online navigation constitutes an effective and practical alternative to both embedding-based and embedding-free paradigms for multimodal long document understanding.

Acknowledgments. This publication has been supported by the Consolidated Research Group 2021 SGR 01559 from the Research and University Department of the Catalan Government, and by project PID2023-146426NB-I00 funded by MCIU /AEI/10.13039/501100011033 and FEDER, UE. This work was also supported by the FI-STEP predoctoral grant program from the Department of Research and Universities of the Generalitat de Catalunya and co-financing by the European Social Fund Plus (2025STEP00060).

This work was also supported by the European Union's Horizon Europe research and innovation programme under grant agreements No 101214398 (ELLIOT) and No 101070617 (ELSA).

Funded by the European Union. Views and opinions expressed are however those of the author(s) only and do not necessarily reflect those of the European Union or the European Commission. Neither the European Union nor the European Commission can be held responsible for them.

References

1. Blau, T., et al.: Gram: global reasoning for multi-page VQA. In: Proceedings of the IEEE/CVF Conference on Computer Vision and Pattern Recognition, pp. 15598–15607 (2024)
2. Cho, J., Mahata, D., Irsoy, O., He, Y., Bansal, M.: M3DocVQA: multi-modal multi-page multi-document understanding. In: Proceedings of the IEEE/CVF International Conference on Computer Vision, pp. 6178–6188 (2025)
3. Deng, C., et al.: LongDocURL: a comprehensive multimodal long document benchmark integrating understanding, reasoning, and locating. In: Proceedings of the 63rd Annual Meeting of the Association for Computational Linguistics (Volume 1: Long Papers), pp. 1135–1159 (2025)
4. Faysse, M., et al.: ColPali: efficient document retrieval with vision language models. In: International Conference on Learning Representations (ICLR) (2025)
5. Fu, S., Zhang, Y., Xiang, Y., Du, X., Tang, J.: DMAP: human-aligned structural document map for multimodal document understanding. In: The Web Conference. WWW (2026)
6. Han, S., et al.: MDocAgent: a multi-modal multi-agent framework for document understanding, arXiv preprint arXiv:2503.13964 (2025)
7. Hui, Y., Lu, Y., Zhang, H.: UDA: a benchmark suite for retrieval augmented generation in real-world document analysis. Adv. Neural. Inf. Process. Syst. **37**, 67200–67217 (2024)
8. Jain, C., et al.: SimpleDoc: multi-modal document understanding with dual-cue page retrieval and iterative refinement. In: Proceedings of the 2025 Conference on Empirical Methods in Natural Language Processing, pp. 28398–28415 (2025)
9. López, E., Llabrés, A., Valveny, E.: Enhancing document VQA models via retrieval-augmented generation. In: International Conference on Document Analysis and Recognition, pp. 92–107. Springer (2025)
10. Ma, Y., et al.: MMLongBench-DOC: benchmarking long-context document understanding with visualizations. Adv. Neural. Inf. Process. Syst. **37**, 95963–96010 (2024)
11. Masry, A., Do, X.L., Tan, J.Q., Joty, S., Hoque, E.: ChartQA: a benchmark for question answering about charts with visual and logical reasoning. In: Findings of the Association for Computational Linguistics, ACL 2022, pp. 2263–2279 (2022)

12. Mathew, M., Karatzas, D., Jawahar, C.: DocVQA: a dataset for VQA on document images. In: Proceedings of the IEEE/CVF Winter Conference on Applications of Computer Vision, pp. 2200–2209 (2021)
13. Nan, L., et al.: FeTaQA: free-form table question answering. Trans. Assoc. Comput. Linguist. **10**, 35–49 (2022)
14. Niu, J., et al.: MinerU2. 5: a decoupled vision-language model for efficient high-resolution document parsing, arXiv preprint arXiv:2509.22186 (2025)
15. Tito, R., Karatzas, D., Valveny, E.: Hierarchical multimodal transformers for multipage DocVQA. Pattern Recogn. **144**, 109834 (2023)
16. Zhang, M., Tang, Y., Team, P.: Pageindex: next-generation vectorless, reasoning-based rag. PageIndex Blog (2025). https://pageindex.ai/blog/pageindex-intro. Accessed 14 June 2026
17. Zhu, D., Meng, R., Chen, J., Li, S., Pfister, T., Yoon, J.: DocLens: a tool-augmented multi-agent framework for long visual document understanding, arXiv preprint arXiv:2511.11552 (2025)

Document Forensics and Provenance

Doc-Protector: A Self-healing Approach for Digital Documents

Sudev Kumar Padhi(✉), Archana Tiwari, Umesh Kashyap, and Sk. Subidh Ali

Department of Computer Science and Engineering, Indian Institute of Technology Bhilai, Bhilai, Chhattisgarh, India
{sudevp,archanat,umeshk,subidh}@iitbhilai.ac.in

Abstract. Tamper localization and document recovery are critical challenges in digital document forensics, where ensuring the authenticity and integrity of documents such as contracts, certificates, invoices, and identification records is essential. With the rapid advancement of document editing tools and AI-driven manipulation techniques, including content splicing, copy-move forgery, text alteration, signature replacement, and synthetic document generation, safeguarding digital documents has become increasingly complex. We propose *Doc-Protector*, a deep learning–based self-healing framework that enables both tamper localization and semantic-consistent document reconstruction. The method extracts hierarchical latent representations using a ResNet-34 backbone with a Feature Pyramid Network (FPN), capturing stroke-level, text-structure, and layout-level information. These compressed latents are embedded into the document via a mask-guided U-Net–based residual embedding network to ensure imperceptibility in text-sensitive regions. Given a tampered document, an extraction network recovers the embedded payload, which is decoded using a reverse-FPN hierarchical decoder to reconstruct the document in a coarse-to-fine manner. Tamper localization is obtained through discrepancy analysis between the received and reconstructed documents. Experiments under diverse manipulation scenarios demonstrate strong robustness in both in-domain and cross-domain settings, while reconstructed documents preserve structural and textual consistency under severe attacks.

1 Introduction

In today's world of digitization, the sharing of digital images over various online platforms has become an integral part of daily life, enabling the rapid dissemination of visual information across social media, messaging apps, and professional networks. However, this convenience also brings about concerns regarding unauthorized tampering and data security. The issue regarding digital images further escalates when we consider the case of a digital document images. In this work, the term digital document refers specifically to rasterized or image-based representations of documents (e.g., scanned documents, screenshots, or document images) rather than editable document formats such as PDF, Word, or XML

G. A. Fink et al. (Eds.): ICDAR 2026, LNCS 16972, pp. 75–91, 2027.
https://doi.org/10.1007/978-3-032-36023-6_5

files. Digital documents play a pivotal role across sectors such as academia, finance, healthcare, and law, where they are trusted as credible sources of information. Unlike natural images, where tampering can sometimes be identified through inconsistencies in texture or appearance, digital document are composed of structured text and uniform layouts that lack such visible tampering indicators [1,5,10,15,23]. Even minor manipulations, such as altering a single character, number, or symbol, can significantly change the semantic meaning of a document while remaining visually imperceptible to human observers. This underscores the need for reliable techniques that can verify the authenticity of digital documents by self-healing them from various tampering techniques, thereby preserving their integrity and trustworthiness.

There are techniques for detecting tampering in images [14,18,19]. Certain advanced techniques even allow for the perfect recovery of the original image if tampering is detected [7,11,12,16,17,20,21]. In this line, tamper localization in digital document have gained significant attention in research to combat forgery and associated security risks [3,10,15]. However, these techniques struggle with the vast array of potential tampering techniques and tend to underperform on compressed or low-resolution images. This limits their applicability to digital documents, such as academic transcripts, digital licenses, and other confidential documents. Furthermore, the complexity increases if we want to recover the original document from the tampered one, as identifying tampered regions alone doesn't provide enough context to infer the intent behind the tampering. Despite their ability to recover original content, existing techniques suffer from low tamper localization accuracy, limited effectiveness in handling digital documents, poor reconstruction performance under multi-region tampering, and vulnerability to sophisticated attacks such as image inpainting. A significant gap remains in developing a unified solution that not only detects tampering but also facilitates the high-fidelity reconstruction of digital documents.

In light of these challenges, we propose *Doc-Protector*, a novel deep learning-based framework for self-healing (tamper-aware reconstruction) of tampered documents through tamper localization and document reconstruction. Our technique extracts a latent representation of a digital document using a Feature Pyramid Network (FPN) that captures high-level semantic features in a compressed form. This latent representation is then seamlessly embedded into the same digital document in a visually imperceptible way, creating a protected document. In the event of tampering, the embedded latent representation can be extracted from the protected document and transformed back into the original document, ideally reconstructing the input with accuracy. Now, we can compare the reconstructed document to the tampered protected document to accurately localize tampered regions. Thus, the protected document can heal itself and avoid tampering. This technique provides a comprehensive solution for protecting against document forgery. The primary contributions of this paper are:

1. We introduce *Doc-Protector*, a hierarchical latent embedding framework for digital document self-healing that enables tamper localization and semantic-

consistent reconstruction by embedding multi-scale document representations.
2. We design a mask-guided residual embedding network and a reverse-FPN hierarchical decoder that together ensure imperceptible watermark insertion in text-sensitive regions while maintaining robustness against diverse content and signal-level manipulations, including copy-move, splicing, inpainting, compression, and geometric distortions.
3. We conduct extensive in-domain and cross-domain evaluations under multiple attack scenarios, demonstrating improved localization accuracy and reconstruction stability compared to existing approaches.

2 Background

With increasing digitization, digital documents are becoming widely accepted, raising concerns about unauthorized tampering. Even simple tampering like copy-move or splicing for altering a single character can compromise document integrity [5,10]. Tamper localization and recovery involve detecting tampered regions in the digital document (tamper localization) and restoring the original document (recovery). This ensures the authenticity and integrity of digital documents. Over the years, techniques for tamper localization and self-recovery primarily developed for digital images have evolved significantly [9,11–13,18,19,21]. Usually, techniques use irregularity in the pixels and edge distribution to localize the tampering. Once the tamper is localized, different techniques are used to recover the image. Early self-recovery methods embed most significant bits (MSB) information in least significant bits (LSB) or distribute redundant image copies to enable partial recovery under tampering [2]. Transform-domain features such as Discrete Cosine Transform (DCT) coefficients, Polar Harmonic Transform (PHT) moments, and Zernike Moments (ZMs), along with block-based and source–channel coding schemes, have further improved robustness against tampering [9,11–13,18,19]. Still, these techniques often rely on handcrafted features, struggle with adaptability, and are not easily generalizable. Furthermore, recovering heavily tampered regions, particularly in cases of large-scale or multi-region modifications that can drastically alter document semantics, remains a significant challenge. *Doc-Protector* is used for the self-healing of tampered documents through tamper localization and document reconstruction, ensuring integrity in the digital document.

3 Proposed Methodology

In this section, we propose *Doc-Protector*, a novel deep learning-based technique designed to improve the authenticity and integrity of digital documents by reconstructing the original digital document from its tampered version while achieving robustness against various attacks. *Doc-Protector* consists of three major modules. The first module consists of a Feature Pyramid Network (FPN) with ResNet-34 backbone, which is used to create a latent representation. The second module consists of the embedding network, which takes a digital document, its latent representation and mask as input to produce a protected document. Lastly, the third module consists of the extractor network, which extracts the embedded latent representation from the protected document. This extracted latent representation is used to reconstruct the original document for self-healing and tamper localization.

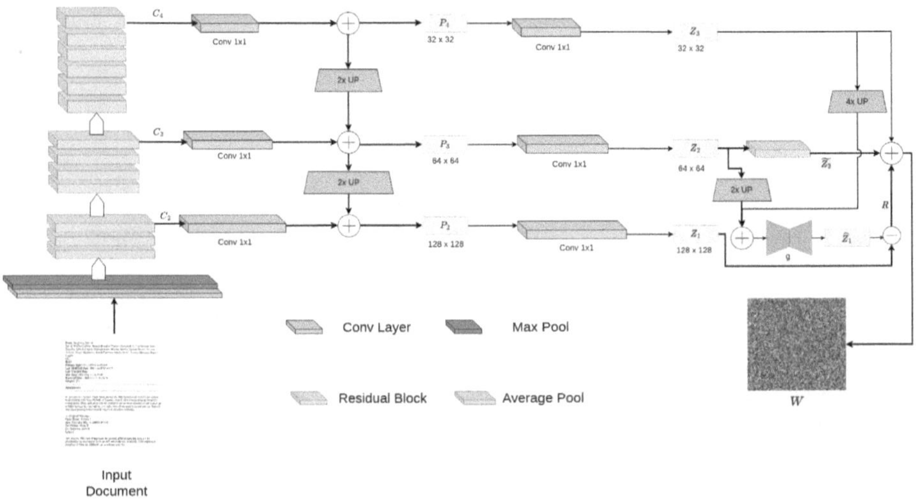

Fig. 1. Hierarchical latent representation and watermark payload formation. The input digital document is processed using a ResNet-34 backbone with a Feature Pyramid Network (FPN) to extract multi-scale features. These features are projected into hierarchical latent components corresponding to stroke-level (Z_1), text-structure-level (Z_2), and layout-level (Z_3) representations. After spatial pooling and residual computation, the compressed watermark payload W is formed by concatenating Z_3, $\tilde{Z}_2$, and the stroke residual R.

3.1 Creating the Latent Representation

Given an input digital document $I \in \mathbb{R}^{3\times H\times W}$, hierarchical latent representations are generated using a ResNet-34 backbone augmented with a Feature Pyramid Network (FPN) [8]. Feature maps are extracted from three residual stages of res2, res3, and res4, which correspond to output strides of 4, 8, and 16, respectively. The choice of these blocks is motivated by their progressively increasing receptive fields and semantic abstraction. Early-stage features of res2 preserve fine spatial detail and encode character strokes and glyph boundaries, while res3 features capture word- and line-level text structure. In contrast, res4 features represent global page layout, including columns, margins, and paragraph blocks. To enhance semantic consistency across scales, a top-down FPN fusion is applied:

$$P_4 = \psi_4(C_4), \quad P_3 = \psi_3(C_3) + \mathrm{Up}(P_4), \quad P_2 = \psi_2(C_2) + \mathrm{Up}(P_3),$$

where ($\psi_i(\cdot)$) denotes a (1×1) projection and ($\mathrm{Up}(\cdot)$) is a $2\times$ upsampling operation. Each fused feature map is then projected into a fixed-dimensional latent space to form the hierarchical watermark components:

$$Z_1 = \phi_1(P_2), \quad Z_2 = \phi_2(P_3), \quad Z_3 = \phi_3(P_4).$$

where (ϕ_i) are (1×1) conv projections followed by normalization. Here, (Z_1), (Z_2), and (Z_3) correspond to stroke-level, text-structure-level, and layout-level latent representations, respectively. These hierarchical latents collectively form the basis of the watermark payload, which is subsequently compressed and embedded into the digital document. The raw latents are too large to embed directly, so the watermark creation stage includes payload formation, i.e., transforming (Z_1, Z_2, Z_3) into a compact (W) suitable for embedding using the step below:

1. The layout-level latent Z_3 captures global structural information of the document, including margins, column organization, paragraph alignment, and large-scale text grouping. Owing to the Feature Pyramid Network design, $Z_3 \in \mathbb{R}^{c_3 \times h_3 \times w_3}$ is already spatially compact (e.g., $h_3 = H/16$, $w_3 = W/16$) and channel-reduced through the projection ϕ_3. Consequently, it forms a low-dimensional yet semantically rich representation of the overall document layout. Preserving Z_3 in its entirety ensures that the global document geometry remains recoverable even when higher-resolution components are partially corrupted. During hierarchical decoding, Z_3 acts as a structural scaffold, enabling coarse layout reconstruction before finer text-structure and stroke-level refinements are applied. This design promotes graceful degradation, ensuring semantic readability under severe tampering.
2. Mid-level features (word/line structure) do not require the same spatial resolution as stroke-level features. Therefore, after channel reduction, the mid-level latent ($\bar{Z}_2$) is spatially pooled:

$$\tilde{Z}_2 = \mathrm{Pool}(\bar{Z}_2)$$

   to match the spatial resolution of the layout latent (Z_3). This step reduces the payload quadratically with spatial size while retaining text organization information that is invariant to small spatial shifts.
3. The stroke-level latent (Z_1) is the largest contributor to payload size. Instead of embedding it directly, a autoencoder g is trained to predict an approximation ($\hat{Z}_1$) from coarser latents $\hat{Z}_1 = g(Z_2, Z_3)$. Only the residual $R = Z_1 - \hat{Z}_1$ is encoded. Since most stroke information is already explained by higher-level structure, the residual is sparse.

Finally, the compressed components are concatenated into a structured watermark payload as shown in Fig. 1:

$$W = \text{Concat}\big((Z_3), (\tilde{Z}_2), (R)\big)$$

3.2 Embedding Network

Document are structurally heterogeneous, i.e. they contain text regions that are highly sensitive to distortion, while background regions tolerate stronger modifications. Uniform watermark embedding across the digital document often leads to visible artifacts in text or insufficient robustness in background areas. Therefore, prior to embedding, we perform document region analysis to guide the spatially adaptive insertion of watermarks. A lightweight DBNet-style network [22] is used to identify semantically distinct regions, primarily text regions (characters, words, lines) and background regions (blank or textured areas). The network produces a dense probability map indicating the likelihood of text presence at each pixel. This map is binarized and optionally smoothed to obtain a region mask M where 1 denotes text-sensitive regions and 0 denotes background regions. This region mask is used only to control embedding strength.

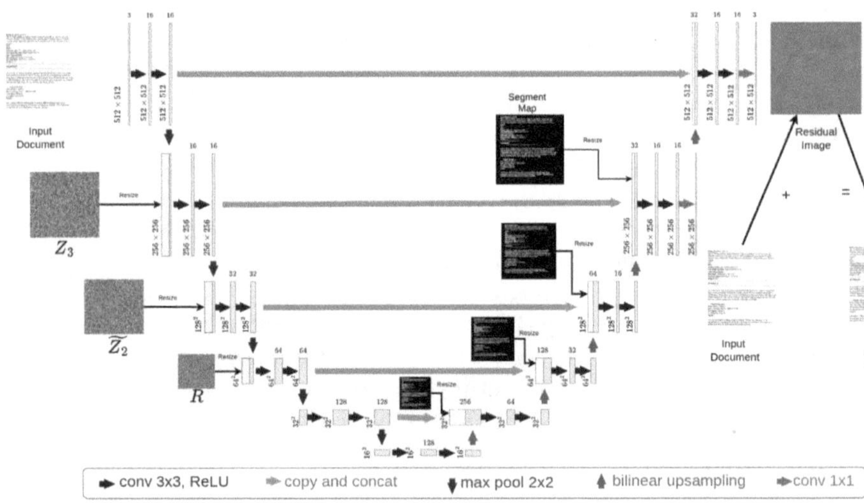

Fig. 2. Watermark embedding network architecture. The original digital document I, compressed watermark payload W, and region mask M are provided as inputs to a U-Net-based embedding network. The payload is spatially aligned and fused with the features of the digital document at multiple encoder stages and mask-guided weighting is used to control embedding strength in text-sensitive regions. The network predicts a residual watermark signal ΔI, which is added to the original digital document to produce the protected document $I_p = I + \Delta I$.

Watermark embedding is implemented using a U-Net–style encoder–decoder network as shown in Fig. 2, which is well-suited for dense, pixel-level residual prediction while preserving spatial detail through skip connections. The embedding network takes as input the original digital document (I), the compressed watermark payload (W), and the document region mask (M), and outputs a visually imperceptible residual that is added to the host digital document. The U-Net consists of an encoder path that extracts multi-scale feature representations from the input digital document (I). A decoder path that progressively upsamples features to reconstruct a full-resolution residual image and, at last, Skip connections that propagate fine spatial details from encoder to decoder. The network predicts a residual watermark signal (ΔI), and the final Protected Document (I_p) is obtained as:

$$I_p = I + \Delta I.$$

This residual formulation ensures minimal visual distortion while enabling robust watermark insertion. The compressed watermark payload (W) is a low-dimensional representation and must be spatially aligned with the digital document's features before fusion. To achieve this, (W) is upsampled to match the resolution of intermediate U-Net feature maps. This produces a spatial watermark feature map $W_s \in \mathbb{R}^{c_w \times H \times W}$, which can be fused with digital document features at multiple scales in the encoder part. Spatial adaptivity is enforced by incorporating the region mask (M) directly into the fusion process. The mask is resized to match the spatial resolution of the feature maps at each decoder stage and used to modulate watermark strength. At a given decoder level (l), watermark fusion is performed as:

$$F_l = F_l^{\text{doc}} + \alpha_l(M) \cdot F_l^{\text{wm}},$$

where (F_l^{doc}) are the features of the digital document from the U-Net decoder, (F_l^{wm}) are the watermark features derived from (W_s), and ($\alpha_l(M)$) is a spatial weighting map computed from the region mask. The weighting function ($\alpha_l(M)$) assigns lower weights to text regions and higher weights to background regions, ensuring that watermark energy is suppressed in distortion-sensitive areas while being reinforced in areas where perceptual tolerance is higher. Because the watermark payload is hierarchical, different components of (W) are fused at different decoder depths:

1. Layout-level information is fused at coarse decoder stages, where global context dominates.
2. Text-structure information is fused at intermediate stages to align with line-level features.
3. Sparse stroke residuals are fused only at late decoder stages with low embedding strength.

This multi-scale fusion ensures that each component of the watermark is embedded at a scale consistent with its semantic role. After watermark fusion at all relevant decoder stages, the U-Net outputs a residual image (ΔI). The final protected document is obtained as I_p where the residual magnitude is implicitly constrained by training losses to remain imperceptible.

3.3 Extraction Network and Document Recovery

At the receiver side, the watermarked document may have undergone unknown and potentially severe manipulations such as cropping, inpainting, text replacement, diffusion-based editing, compression, or geometric distortions. Let the received (possibly tampered) document be denoted as Attacked document $I_a \in \mathbb{R}^{3\times H'\times W'}$. The objective of watermark extraction is to robustly recover the embedded watermark payload (W) from (I_a), even when parts of the document content are missing or altered. Watermark extraction is performed using an extraction network designed to be robust to both geometric and semantic document attacks. The extractor follows a CNN-based spatial transformer [6] architecture, mirroring the embedding network in complexity but not symmetry. The extraction network consists of an feature encoder that processes (I_a), multi-scale convolutional blocks to handle distortions at different resolutions, and a payload reconstruction head that estimates the embedded watermark as shown in Fig. 3. Unlike classical watermarking, no explicit inverse transform is applied. Instead, the extractor directly learns the mapping:

$$\hat{W} = \mathcal{X}(I_a),$$

where $(\mathcal{X}(\cdot))$ is trained end-to-end to recover the compressed payload. To improve robustness against non-linear edits (e.g., inpainting or diffusion), the extractor is trained with heavy attack augmentation, forcing it to rely on redundant and spatially distributed watermark cues rather than local pixel correspondences.

The extracted watermark $(\hat{W})$ is a packed representation containing compressed latent components and associated metadata. Payload decoding proceeds deterministically as the payload is parsed to recover individual components: $\hat{Z}_3$, $\hat{Z}_2$, $\hat{R}$ corresponding to layout-level latents, mid-level text-structure latents, and sparse stroke residuals, respectively. At this stage, the recovered latents may be incomplete or corrupted; however, the decoding process is explicitly designed to tolerate such imperfections. Document recovery is performed using a hierarchical decoder that mirrors the encoder's feature pyramid structure in reverse. Rather than reconstructing the document directly from all latents simultaneously, the decoder follows a coarse-to-fine reconstruction strategy, which significantly improves robustness. The decoder begins from the lowest-resolution latent $\hat{Z}_3$ which encodes global page layout information. This latent serves as a structural scaffold for reconstruction. The decoder first processes $(\hat{Z}_3)$ using convolutional and residual blocks to reconstruct global document structure such as margins, columns, and paragraph blocks. The resulting feature map X is upsampled to an intermediate resolution. The mid-level latent $(\hat{Z}_2)$ is projected and additively fused with the upsampled features X:

$$X \leftarrow X + \phi(\hat{Z}_2).$$

This stage refines word grouping, line spacing, and text density without disrupting the global layout established earlier. Fine-grained character details are

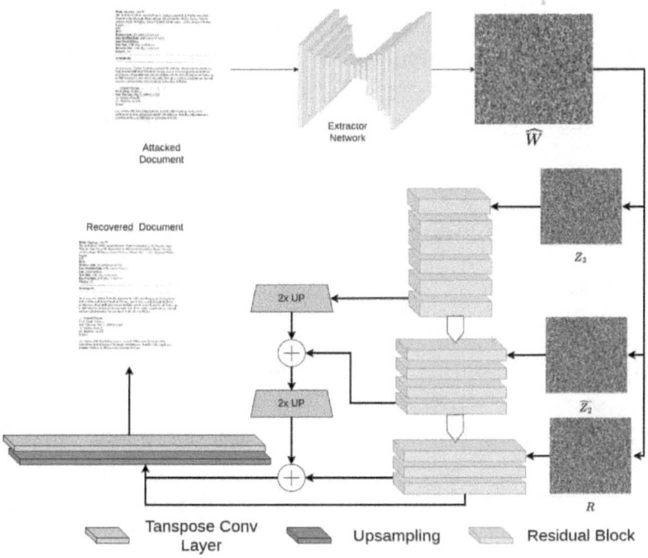

Fig. 3. Watermark extraction and hierarchical document recovery framework. The attacked document I_a is processed by a CNN-based extraction network to recover the embedded watermark payload $\hat{W}$. The decoded hierarchical latent components $(\hat{Z}_3, \hat{Z}_2, \hat{R}_1)$ are fed into a reverse-FPN hierarchical decoder, which reconstructs the original digital document $\hat{I}$ in a coarse-to-fine manner. The reconstructed document enables both self-healing and tamper localization.

incorporated through sparse residual correction rather than direct decoding of (Z_1). The decoder learns to infer a base approximation of stroke-level features from $(\hat{Z}_2)$ and $(\hat{Z}_3)$, while the recovered residual $(\hat{R}_1)$ corrects local distortions such as broken strokes or blurred glyphs. This final refinement is applied at higher resolutions before reconstructing the full-resolution document:

$$\hat{I} = \mathcal{D}(\hat{Z}_3, \hat{Z}_2, \hat{R}).$$

A key advantage of hierarchical decoding is graceful degradation. If fine-scale residuals are missing or heavily corrupted, the decoder still reconstructs a readable document using layout and structure cues alone. This behaviour is learned during training by randomly dropping or corrupting latent components, forcing the decoder to rely on semantic priors rather than exact feature recovery. Recovered documents are also used to localize tampered regions. A tamper map is obtained by computing the discrepancy between the received document and the reconstructed document:

$$T(x, y) = |I_a(x, y) - \hat{I}(x, y)|.$$

Regions with high discrepancy indicate content that is inconsistent with the embedded latent watermark, enabling accurate tamper detection and localization. It is important to clarify that *Doc-Protector* does not perform lossless pixel-exact recovery in the strict information-theoretic sense. Instead, recovery is defined as high-fidelity reconstruction of the original digital document from the embedded hierarchical latent representation. Thus, the term *self-healing* in this work refers to semantic-consistent reconstruction rather than strict reversible watermarking.

4 Experimental Setup

4.1 Data Preparation

For data preparation, we exclusively use the RVL-CDIP dataset [4] as the source of authentic digital documents and synthesize tampered samples using the selective tampering synthesis strategy introduced in DocTamper [10]. RVL-CDIP consists of 400,000 digital documents across 16 document categories, including letters, forms, invoices, scientific publications, and legal documents. From this corpus, we randomly select 200,000 digital documents and generate realistic tampered documents by applying selective copy-move, selective splicing, and selective generation-based text replacement. Specifically, text regions are first localized using OCR tools, and their visual attributes such as font size, foreground color, background color, and spatial layout are analyzed. Tampering is then performed by copying text between visually similar regions within the same document (copy-move), inserting text from compatible documents (splicing), or erasing and regenerating text with matched appearance statistics (generation), followed by post-processing smoothing and blending operations to reduce boundary artifacts. All tampered samples are accompanied by pixel-level tamper masks. Digital documents are converted to RGB format, resized to 512×512 with aspect-ratio-preserving padding, and normalized to ([0, 1]). The resulting dataset is divided into an in-domain training set of 160,000 digital documents, an in-domain validation set of 20,000 images, and an in-domain test set of 20,000 digital documents, with tampering types uniformly distributed. To assess cross-domain generalization, we construct cross-domain test sets of 20,000 digital documents using RVL-CDIP categories excluded from training, specifically Scientific Publications, Laws & Regulations, Questionnaires, and Forms, which differ substantially in layout density, text distribution, and structural regularity from the training categories. This setup enables a comprehensive evaluation of both in-domain performance and robustness under domain shift for document self-healing and tamper localization.

4.2 Network Architecture

The proposed framework consists of four main components: a hierarchical feature encoder, a watermark embedding network, a watermark extraction network, and a hierarchical decoder. The encoder adopts a ResNet-34 backbone augmented with a Feature Pyramid Network (FPN) to extract multi-scale document representations. Feature maps from intermediate backbone stages are fused in a top-down manner to generate three hierarchical latent representations corresponding to stroke-level details, text-structure information, and global document layout. The embedding network follows a U-Net–style architecture and takes the original digital document and the compressed payload as input, predicting a residual watermark signal that is added to the original digital document to produce a visually imperceptible protected document. The extraction network consists of a spatial transformer network [6]. This spatial transformer is composed of 7 CNN layers in both the encoder and decoder, and is designed to provide invariance against various transformations, including translation, scaling, rotation, cropping, deformations, and changes in brightness and contrast. Finally, document reconstruction is performed using a reverse-FPN hierarchical decoder, which progressively upsamples and refines features in a coarse-to-fine manner by first reconstructing global layout, then injecting text-structure latents, and finally applying sparse stroke-level corrections to recover the digital document.

4.3 Loss Function and Scheme Objective

Embedding a high-capacity watermark into digital document often leads to visible artifacts, particularly in text-sensitive regions. The proposed Doc-Protector framework mitigates this issue by embedding a hierarchical latent representation of the document and optimizing the entire pipeline using a carefully designed composite loss function. The training objective is threefold i.e., preserve perceptual fidelity between the original digital documents (I) and the protected document (I_p), enable accurate reconstruction of the original digital document (I) from the decoded latent representation, and ensure robust recovery of the embedded hierarchical latent watermark under severe tampering. To achieve these goals, we employ a combination of pixel-level reconstruction loss and perceptual similarity loss, jointly applied across the encoder, embedding, extraction, and decoding stages. The L_1 loss is used to enforce pixel-wise consistency between an input digital documents X and its reconstructed counterpart X'. It is defined as:

$$L_1(X, X') = \|X - X'\|_1 \tag{1}$$

This loss penalizes absolute intensity differences and helps preserve fine-grained spatial details such as character edges and stroke continuity, which are critical in digital document. While the L_1 loss ensures local accuracy, it does not fully capture perceptual and structural similarity. Therefore, we additionally employ the Learned Perceptual Image Patch Similarity (LPIPS) loss, which measures perceptual distance using deep features extracted from a pre-trained VGG-19 network. For two digital images A and B, the LPIPS loss is defined as:

$$L_P(A, B) = \sum_k \tau^k \left(\text{VGG}^k(A) - \text{VGG}^k(B)\right) \tag{2}$$

where features are extracted from (k) intermediate layers of the VGG-19 network. In our implementation, we use five layers (k = 5). The function τ projects the feature-wise differences into a scalar perceptual distance by averaging across layers. This loss encourages preservation of high-level semantic structures such as text alignment, spacing, and layout consistency. The Feature Pyramid Network (FPN)–based encoder–decoder is trained to produce a compact yet semantically rich hierarchical latent representation of the digital document. The objective is to reconstruct the original digital document I as accurately as possible from its decoded latent representation (W). The autoencoder loss is defined as:

$$L_{AE} = L_1(I, I') + \alpha L_P(I, I') \tag{3}$$

here (I') denotes the reconstructed document. This loss encourages the encoder to distribute document information across multiple latent scales (Z_1, Z_2, Z_3) in a manner that supports stable reconstruction and redundancy, which is crucial for robustness against partial latent corruption. The embedding network inserts the compressed hierarchical watermark W into the original digital document in the form of a residual signal. To ensure imperceptibility, the protected document (I_p) is constrained to remain visually similar to the original document (I). The embedding loss is defined as:

$$L_{Emb} = L_1(I, I_p) + \alpha L_P(I, I_p) \tag{4}$$

This loss directly controls the magnitude and spatial distribution of the embedding residual (ΔI), ensuring that watermark energy is suppressed in text-sensitive regions while remaining sufficiently strong in background regions for robust extraction. The extraction network is trained to recover the embedded hierarchical latent watermark from a potentially tampered document (I_a). To enforce accurate latent recovery, we apply a binary cross-entropy loss between the original embedded watermark W and the extracted watermark $\hat{W}$. In addition, the recovered latents must enable faithful reconstruction of the original digital document through the hierarchical decoder. The extraction and reconstruction loss is defined as:

$$L_{Ext} = L_{BCE}(W, \hat{W}) + L_1(\mathcal{D}(\hat{W}), I) \tag{5}$$

where ($\mathcal{D}(\cdot)$) denotes the reverse-FPN hierarchical decoder and $\mathcal{D}(\hat{W}) = \hat{I}$ is the reconstructed document. This formulation explicitly couples latent recovery with digital documents reconstruction, ensuring that extracted latents are not only numerically accurate but also semantically meaningful. The complete Doc-Protector framework is trained end-to-end by jointly optimizing the embedding and extraction objectives. The total loss function is defined as:

$$L_{\text{total}} = \lambda_1 L_{Emb} + \lambda_2 L_{Ext} \tag{6}$$

where λ_1 and λ_2 balance imperceptibility and reconstruction robustness. This unified objective encourages the network to embed a compact yet resilient hierarchical watermark, recover it reliably under diverse attacks, and reconstruct a high-quality digital document suitable for accurate tamper localization.

4.4 Training of Network

All components of the framework are trained jointly in an end-to-end manner, except for the encoder backbone, which is initialized with ImageNet-pretrained weights. Training is performed using the Adam optimizer with an initial learning rate of 1×10^{-4}, which is decayed progressively during training. A batch size of 8 is used, and the network is trained for approximately 250,000 iterations. To improve robustness against partial watermark loss and severe tampering, dropout is applied in the latent space during training, encouraging distributed and redundant latent representations. In addition, extensive post-embedding attack simulation, including JPEG compression, noise, blurring, cropping, geometric distortions, and brightness variations, is applied to the protected documents during training, forcing the extraction and decoder networks to learn stable recovery under diverse degradation conditions. Early stopping based on validation performance is employed to prevent overfitting.

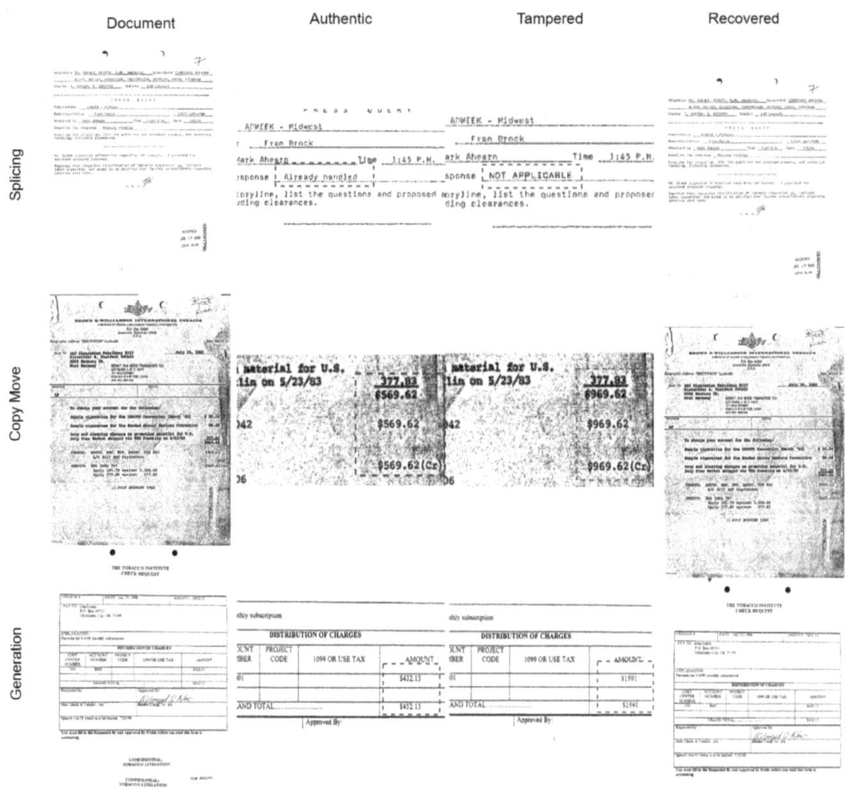

Fig. 4. Visual quality of the original digital documents and recovered document along with the tampered region. The first row shows splicing attacks, the second row copy-move attacks, and the third row generation-based tampering.

Table 1. Robustness under Noise, JPEG Compression, Gaussian, and Motion Blur

Data	Noise		JPEG			G-Blur			M-Blur		
	S&P	G	50	60	70	3	5	7	3	5	7
In-Domain	0.92	0.92	0.89	0.92	0.92	0.92	0.92	0.92	0.92	0.92	0.92
Cross-Domain	0.86	0.86	0.81	0.86	0.86	0.86	0.86	0.86	0.86	0.86	0.86

5 Results and Discussion

Figure 4 illustrates the visual quality of the original digital document alongside the protected document after the embedding process. Additionally, it presents the recovery (self-healing) and localization of the original digital documents under various tampering scenarios, including splicing (first row), copy-move (second row), inpainting and cross-modality image editing (third row). The results reveal that the visual quality of the digital documents remains high, indicating successful self-healing capabilities.

5.1 Evaluation Metrics

To verify the quality of the original digital document, protected document and reconstructed document, we used structural similarity index ($SSIM$) and peak signal-to-noise rate ($PSNR$) matrices. $SSIM$ compares local patterns of pixel intensities, which represent the attributes that represent the structure of objects in the digital documents, along with luminance and contrast. A higher $SSIM$ value represents that the two digital documents are structurally the same. On the other hand, $PSNR$ measures the distortion between the cover and the protected documents/cropped digital documents and also validates the reconstructed document's quality. $PSNR$ value greater than 30 dB is considered to be of good visual quality of the digital document. To identify the accuracy of localization, we have used Intersection over Union (IoU), which measures the ratio of the intersection area to the union area of the predicted and ground truth masks. The higher value of IoU indicates high accuracy in the localization of the predicted mask relative to the ground truth. Since digital documents are ultimately consumed as textual information, we evaluate semantic fidelity using OCR-based consistency metrics. We apply Tesseract OCR to original, tampered, and reconstructed documents and compute Character Error Rate (CER). Let T denote the ground-truth text extracted from the original document and $\hat{T}$ the OCR output from the reconstructed document. CER is computed as:

$$CER = \frac{S + D + I}{N} \tag{7}$$

where S, D, and I denote substitutions, deletions, and insertions, and N is the total number of characters.

5.2 Real-World Performances

In real-world scenarios, a single type of tampering technique is typically applied. However, we evaluate robustness under a wide range of simulated signal-level and content-level attacks, as summarized in Table 1 and Table 2. We randomly select 5000 digital documents from both the in-domain and cross-domain test sets to assess localization performance using IoU. The results indicate that the proposed framework maintains strong localization accuracy across diverse attack settings, supporting practical applicability. To assess perceptual and semantic fidelity, we evaluate document quality between the original and protected documents, and between the original and reconstructed documents using SSIM, PSNR, and OCR-based CER. In the in-domain setting, protected documents achieve near-identical similarity to the originals (SSIM = 0.99, PSNR = 41 dB), confirming that hierarchical latent embedding introduces negligible perceptual distortion. Reconstructed documents maintain strong visual quality (SSIM = 0.97, PSNR = 32 dB) while significantly restoring textual consistency, achieving a CER of 4.7% compared to 21.3% for tampered documents. In the cross-domain setting, protected documents preserve high fidelity (SSIM = 0.99, PSNR = 39 dB). Reconstructed documents remain perceptually close to the originals (SSIM = 0.95, PSNR = 36 dB) and substantially reduce OCR degradation, achieving a CER of 6.3% compared to 24.8% under tampering. These results demonstrate that *Doc-Protector* achieves imperceptible embedding with reliable structural and semantic document reconstruction across in- and cross-domain settings.

Table 2. Robustness of Doc-Protector under content-level manipulations, including image cropping and various tampering operations (edge modification, copy-move, splicing, and inpainting), evaluated on both in-domain and cross-domain datasets.

Data	Crop		Tampering			
	S	M	Edge	CM	Splice	Inpaint
In-Domain	0.92	0.89	0.92	0.89	0.92	0.92
Cross-Domain	0.86	0.82	0.86	0.81	0.84	0.86

5.3 Recovery Rate and Hiding Capacity

To evaluate recovery robustness, we simulate large-area content removal by randomly cropping, copy-moving, or splicing up to 50% of the protected document area. Across 1000 randomly selected test samples, *Doc-Protector* successfully reconstructs semantically readable documents in 92.8% of cases. Recovered documents achieve SSIM of 0.97, PSNR: 39 dB and OCR-CER: 5.2%. Importantly, recovery remains stable even when high-frequency stroke-level information is partially corrupted, due to the hierarchical decoding strategy that first reconstructs global layout and text structure before applying fine-grained residual refinement. This demonstrates graceful degradation behavior rather than catastrophic failure under extreme tampering.

5.4 Comparison with State-of-the-Art

We compare *Doc-Protector* against different document tampering detection methods, including DocTamper [10], Mantra-Net [14], and the self-recovery method of the reversible watermarking framework [13]. For fair comparison, all methods are evaluated on the same in-domain and cross-domain test sets under identical attack configurations. Doc-Protector achieves superior localization accuracy while uniquely enabling high-fidelity reconstruction. Unlike detection-only models, our framework reduces OCR-CER by a substantial margin, demonstrating that recovered documents retain semantic readability with structural similarity (Table 3).

Table 3. Comparison of Doc-Protector with state-of-the-art document methods. Higher IoU, F1-score, and reconstruction PSNR indicate better performance, while lower OCR character error rate (CER) reflects superior document recoverability.

Method	IoU ↑	F1 ↑	PSNR (Rec.) ↑	OCR-CER ↓
Mantra-Net [14]	0.74	0.79	–	–%
DocTamper [10]	0.83	0.87	–	–%
Reversible WM [13]	0.78	0.82	27.3 dB	15.9%
Doc-Protector (Ours)	**0.92**	**0.94**	**32.0 dB**	**4.7%**

6 Conclusions and Future Scope

In this paper, we introduced *Doc-Protector*, an innovative deep learning technique designed for the self-healing of the digital document from the tampered document. *Doc-Protector* effectively embeds and extracts latent representation, ensuring the integrity and authenticity of digital documents. Our extensive experiments demonstrate the technique's ability to accurately recover the original document and localize tampered regions, even in complex scenarios involving diverse types of documents such as invoices, forms, memos, letters, etc. This functionality of protecting against tampering while enabling high-fidelity reconstruction, enables *Doc-Protector* as a comprehensive solution for safeguarding digital documents against various forms of forgery and tampering. Future work will explore adaptive payload allocation for large-scale document systems.

References

1. Abramova, S., Böhme, R.: Detecting copy-move forgeries in scanned text documents. Electronic Imaging **28**, 1–9 (2016)
2. Bhalerao, S., Ansari, I.A., Kumar, A.: Analysis of DNN based image watermarking data generation for self-recovery. In: IEEE CAPS, pp. 1–6 (2021)
3. Dong, L., Liang, W., Wang, R.: Robust text image tampering localization via forgery traces enhancement and multiscale attention. IEEE TCE (2024)

4. Harley, A.W., Ufkes, A., Derpanis, K.G.: Evaluation of deep convolutional nets for document image classification and retrieval. In: ICDAR (2015)
5. Hayder, M.A., Alhaidery, M.M.A.: Robust passive image authentication scheme based on serial companied approach. Multimedia Tools Appl. (2024)
6. Jaderberg, M., Simonyan, K., Zisserman, A., et al.: Spatial transformer networks. Adv. Neural Inf. Process. Syst. **28** (2015)
7. Jin, Z., Iqbal, M.Z., Bobkov, D., Zou, W., Li, X., Steinbach, E.: A flexible deep CNN framework for image restoration. IEEE Trans. Multimedia **22**(4) (2019)
8. Lin, T.Y., Dollár, P., Girshick, R., He, K., Hariharan, B., Belongie, S.: Feature pyramid networks for object detection. In: IEEE/CVF CVPR, pp. 2117–2125 (2017)
9. Molina-Garcia, J., Garcia-Salgado, B.P., Ponomaryov, V., Reyes-Reyes, R., Sadovnychiy, S., Cruz-Ramos, C.: An effective fragile watermarking scheme for color image tampering detection and self-recovery. Signal Process. Image Commun. **81**, 115725 (2020)
10. Qu, C., et al.: Towards robust tampered text detection in document image: new dataset and new solution. In: IEEE/CVF CVPR, pp. 5937–5946. IEEE, Berlin (2023)
11. Sarreshtedari, S., Akhaee, M.A.: A source-channel coding approach to digital image protection and self-recovery. IEEE Trans. Image Process. **24**(7) (2015)
12. Tang, Y., Wang, C., Xiang, S., Cheung, Y.M.: A robust reversible watermarking scheme using attack-simulation-based adaptive normalization and embedding. IEEE Trans. Inf. Forensics Secur. (2024)
13. Wang, H., Yao, H., Qin, C., Zhang, X.: When robust reversible watermarking meets cropping attacks. IEEE TCSVT (2024)
14. Wu, Y., AbdAlmageed, W., Natarajan, P.: Mantra-net: manipulation tracing network for detection and localization of image forgeries with anomalous features. In: IEEE/CVF CVPR, pp. 9543–9552 (2019)
15. Yang, P., Fang, W., Zhang, F., Bai, L., Gao, Y.: Document image forgery detection based on deep learning models. In: 2022 International Symposium on Electrical, Electronics and Information Engineering (ISEEIE), pp. 36–41. IEEE (2022)
16. Ying, Q., Qian, Z., Zhou, H., Xu, H., Zhang, X., Li, S.: From image to imuge: immunized image generation. In: Proceedings of the 29th ACM International Conference on Multimedia, pp. 3565–3573 (2021)
17. Ying, Q., Zhou, H., Qian, Z., Li, S., Zhang, X.: Learning to immunize images for tamper localization and self-recovery. IEEE TPAMI (2023)
18. Zhang, X., Qian, Z., Ren, Y., Feng, G.: Watermarking with flexible self-recovery quality based on compressive sensing and compositive reconstruction. IEEE Trans. Inf. Forensics Secur. **6**(4), 1223–1232 (2011)
19. Zhang, X., Wang, S.: Fragile watermarking with error-free restoration capability. IEEE Trans. Multimedia **10**(8), 1490–1499 (2008)
20. Zhang, X., Wang, S., Qian, Z., Feng, G.: Reference sharing mechanism for watermark self-embedding. IEEE Trans. Image Process. **20**(2) (2010)
21. Zhang, X., Wang, S., Qian, Z., Feng, G.: Self-embedding watermark with flexible restoration quality. Multimedia Tools Appl. **54**, 385–395 (2011)
22. Zhang, Y., Yuan, L., Guo, Y., He, Z., Huang, I.A., Lee, H.: Discriminative bimodal networks for visual localization and detection with natural language queries. In: IEEE/CVF CVPR (2017)
23. Zhao, L., Chen, C., Huang, J.: Deep learning-based forgery attack on document images. IEEE Trans. Image Process. **30**, 7964–7979 (2021)

Temporal Modeling of Optically Variable Devices in Identity Documents

Glen Pouliquen[1,2](✉), Joseph Chazalon[2], Guillaume Chiron[1], Oriol Ramos Terrades[3], Thierry Géraud[2], and Ahmad Montaser Awal[1]

[1] IDnow Research Center, Cesson-Sévigné, France
{glen.pouliquen,guillaume.chiron,ahmad.awal}@idnow.io
[2] EPITA Research Lab. (LRE), Le Kremlin-Bicêtre, France
{glen.pouliquen,joseph.chazalon,thierry.geraud}@epita.fr
[3] Computer Vision Center (CVC), Barcelona, Spain
oriolrt@cvc.uab.cat

Abstract. Robust remote verification of identity documents relies on analyzing faint, transparent security features like Optically Variable Devices (OVDs), or *"holograms"*, within user-captured videos under uncontrolled conditions. Current systems, however, face critical limitations: existing methods often treat video frames in isolation, neglecting the intrinsic dynamic nature of OVDs and leaving systems vulnerable to swapping attacks, or focus on general holographic presence and lack the ability to verify specific OVD types. Moreover, the economic infeasibility of frame-by-frame video annotation makes supervised training impractical. In this work, we introduce two novel approaches for verifying the dynamic behavior of transparent OVDs protecting the holder's portrait, specifically designed for open-set scenarios where attack types are unknown during training. We demonstrate that these approaches can be trained without any attack samples in a self-supervised setting, surpassing previous state-of-the-art methods on public datasets while adhering strictly to industrial constraints. Our results confirm that modeling temporal dynamics is essential for defeating sophisticated attacks under realistic conditions, and underscores the promise of sequence modeling and anomaly detection for OVD verification. Code is available at https://github.com/EPITAResearchLab/pouliquen.26.icdar.

Keywords: Identity Documents · OVD Verification · Fraud Detection · Sequence Modeling · Anomaly Detection

1 Introduction

This paper addresses the automated verification of Optically Variable Devices (OVDs) on identity documents captured with commodity smartphones in remote Know Your Customer (KYC) scenarios. Such scenarios are increasingly common in various applications, such as opening a bank account. The overall process typically involves the user capturing a short video of their identity document,

G. A. Fink et al. (Eds.): ICDAR 2026, LNCS 16972, pp. 92–108, 2027.
https://doi.org/10.1007/978-3-032-36023-6_6

which is then analyzed by an automated system to verify its authenticity. Other security measures are integrated into the process, such as liveness checks to ensure the user is present and prevents the use of static images, as well as safeguards against digital content injection, but are beyond the scope of this work. Ultimately, this global KYC process consists in binding an online identity to a physical document, which serves as the primary source of truth for the holder's identity.

OVDs, often referred to as *holograms*, are critical visible security features, particularly when they overlap and protect the photo portrait linking the document to its holder. While current industrial pipelines rely heavily on attack detection (e.g., identifying screen recaptures or photocopies), they provide limited guarantees that the expected OVD is actually present and behaves as specified. Furthermore, attack scenarios are inherently unpredictable, and training samples for such attacks are scarce. This data scarcity severely restricts the performance of existing approaches in realistic settings, as demonstrated by the recent MIDV-DynAttack dataset study [15], which highlights that dynamic attacks should primarily be used for evaluation rather than training.

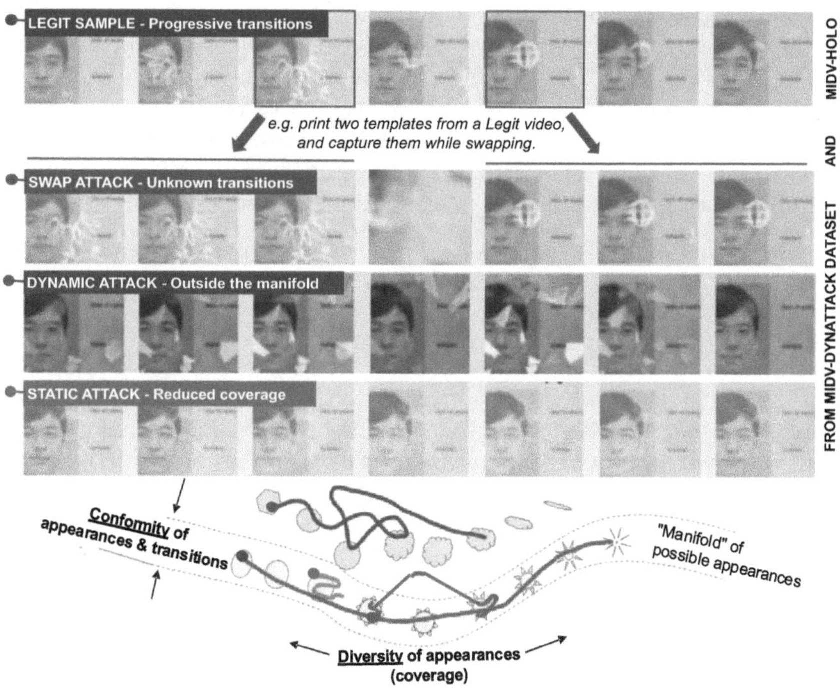

Fig. 1. Importance of sequence-level modeling, and complementary directions for OVD verification to detect advanced attacks from the MIDV-DynAttack dataset [15].

In this work, we adopt a complementary perspective based on model verification rather than specific attack detection. We aim to verify that the OVD in the portrait region is consistent with its expected visual behavior under realistic capture constraints—such as short, low-frame-rate clips from heterogeneous devices—and in the absence of fraud samples during training, which is critical for industrial applications. As detailed in Sect. 2, the few existing approaches capable of handling an open attack space still rely on attack samples for training. Moreover, they typically focus on frame-level analysis or global video features at the pixel level, failing to model the *dynamics* of the OVD, which is a fundamental aspect of its security design, and thus leaving the system vulnerable to several attacks.

This work addresses these limitations with the following contributions: 1. We introduce two approaches—a discriminative span-based classifier that learns from synthetically corrupted sequences, and a generative masked sequence model that uses reconstruction error in an embedding space as an anomaly score—for verifying the dynamic behavior of transparent OVDs; 2. We demonstrate through extensive evaluation on the MIDV-Holo [12] and MIDV-DynAttack [15] datasets that these methods surpass the previous state-of-the-art while being trained in a self-supervised manner without attack samples, strictly adhering to industrial constraints. Ultimately, the *Masked Sequence Modeling* (MSM) we introduce aims to learn a proper manifold of some OVD's dynamics and to verify the two essential properties: *conformity* to the expected visual behavior (i.e., correct color or shape changes) and *variety* of appearances (i.e., sufficient change) to ensure the target is not a static counterfeit, as illustrated in Fig. 1.

The remainder of this paper is organized as follows: Sect. 2 reviews related work in document fraud detection, Sect. 3 details our proposed architectures and training methodologies, Sect. 4 presents the experimental setup and results, followed by conclusions in Sect. 5.

2 Related Work

This section reviews existing approaches for Optically Variable Device (OVD) verification. As mentioned in the introduction, we assume that the acquisition channel is secured, so that digital content injection (such as deep fakes or computer-generated graphics) cannot be directly sent for verification. This does not prevent screens or other substitutes from being used as attack vectors, but we do not address such cases specifically as 1. Presentation Attack Detection (PAD) is a mature field with established solutions for detecting such digital replays [5]; and 2. we believe that such approaches complement a precise verification of the expected security features of the document, in an open-set setting where exact attack implementation details are not known in advance. Consequently, we focus on the subsequent challenge: detecting physical counterfeits—manipulations of the document itself and tampering with acquisition conditions—which bypass standard liveness/PAD checks.

Evolution of Datasets and Benchmarks. Public benchmarks for identity document verification have evolved from static tasks like text alteration detection [3,6] to presentation attack detection (PAD) involving printed copies or screen replays [9,13]. However, verification of holographic security features remains under-explored compared to face morphing or splicing. To date, the only public datasets explicitly targeting OVD verification are *MIDV-Holo* [12] and its extension *MIDV-DynAttack* [15], which include static and dynamic holographic attacks, respectively. Consequently, OVD verification is a nascent research topic that is still consolidating.

Early Approaches Relying on Extensive Annotation. Early methods for hologram verification relied on handcrafted features, such as color histograms and gradient statistics, to detect expected patterns [1]. More recent deep learning approaches employ direct classifiers trained on frame-level annotations [11]. While effective on isolated, non-transparent OVDs found on some parts of banknotes or passports, such systems cannot be generalized to the transparent OVDs overlapping the portrait. These are more challenging to analyze and more critical for security, unless extensive frame-level annotations are available for training. As a result, such supervised approaches were largely superseded in subsequent works.

Detection of Holographic Content Against Static Attacks. Shifting from modeling the collection of expected frame appearances and processing frames in isolation, two approaches proposed recovering the mask of the OVD layer of a document from a video. Kada *et al.* [7] first proposed a technique computing per-pixel color statistics over accurately registered video frames, producing a partial OVD mask for each video and enabling control over the resulting shape. Koliaskina *et al.* [12] applied a similar technique to design a hologram detector, aiming to defeat static document copies featured in the MIDV-Holo dataset they released. Pouliquen *et al.* [14] proposed a weakly-supervised contrastive framework that learns a dispersive representation of genuine OVD representations. While providing a first, practical answer to such attacks, these approaches do not model the specific visual appearance of the target OVD nor its temporal transitions, making them susceptible to dynamic attacks such as a simple holographic layer applied to a printed copy.

Facing Dynamic Attacks. Dynamic attacks, as defined by Pouliquen *et al.* [15], encompass a wide range of manipulations that can be applied to the document or the acquisition conditions, such that the light reflection of the document may exhibit some degree of resemblance to the expected OVD behavior, without actually being the target OVD. Because such attacks defeat holographic detectors, these authors proposed assessing both the *conformity* of OVD appearances to the expected visual behavior and the *variety* of appearances to ensure that the target is not a static counterfeit. They proposed *HoloVerif* [15], a method based on a pipeline comprised of background subtraction—which improves the focus on the OVD signal—, pseudo-labeling of frames likely to contain *Valid* OVD content vs *Non-Valid* frames which do not, and finally estimating the average of *Valid* frames as a proxy for classifying the video as genuine or attack. Despite its ability to counter a wider range of attacks, *HoloVerif* treats video inputs primarily as bags of frames, neglecting the fine-grained temporal dynamics that constitute the essence of OVD security. As a result, it remains vulnerable to swapping attacks, where several static counterfeits (with different OVD appearances) are printed and swapped during video capture, and achieves limited performance on other dynamic attacks. Furthermore, it still relies on attack samples for training and calibration, which limits its applicability in real-world scenarios. Our work addresses this gap by modeling the *transition* between OVD appearances over time, trained exclusively on genuine samples.

3 Proposed Methods

This section introduces two methods for video-level OVD verification designed around three shared principles: **transition awareness**, and **attack-free, self-supervised** training. Unlike prior frame-level approaches, both methods process temporal frame spans (groups of 5–10 consecutive frames) to model the dynamics of holographic transitions. Crucially, they are trained in a self-supervised manner exclusively on legitimate videos, avoiding the reliance on curated attack samples which are scarce and unpredictable in real-world scenarios.

We explore two distinct paradigms to achieve this: 1. **HoloVerif-Span**, a discriminative baseline extending HoloVerif [15] to temporal spans (5 frames) using a VideoMAE [10] backbone. It learns to distinguish genuine transitions from anomalies by training against synthetically corrupted versions of legitimate sequences. 2. **Masked Sequence Modeling (MSM)**, a generative approach inspired by BERT [4] which leverages WSL [14] as a frame encoder. It frames verification as an Out-of-Distribution (OOD) detection task, learning to reconstruct masked portions of valid OVD sequences and flagging reconstruction errors as anomalies, thus eliminating the need for (synthetic) negative samples during training.

3.1 HoloVerif-Span—Discriminative Approach

HoloVerif-Span (Fig. 2) is a sequence-level extension of HoloVerif [15] which maintains its core principle: a binary classifier is trained on sub-video elements (*spans*) that are either left intact or corrupted, and assigned pseudo-labels (*Valid*/*Non-Valid*) accordingly. At inference, the final video-level verdict (*Legit*/*Non-Legit*) is obtained through aggregation of span-level classifier outputs to produce a single video-level score. We preserve the overall pipeline structure, including background subtraction and asymmetric augmentation of *Valid* and *Non-Valid* samples, while redesigning each component to operate on temporal spans rather than isolated frames. Crucially, these modifications enable the model to train on legitimate samples only.

Backbone. The frame-level MobileViT backbone is replaced by a VideoMAE [10] encoder capable of capturing OVD transition dynamics directly from short clips.

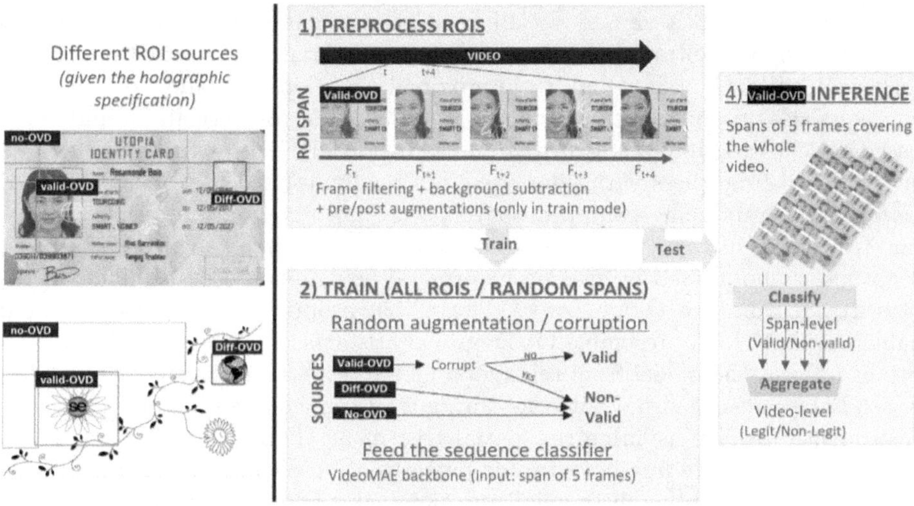

Fig. 2. HoloVerif-Span overview. The OVD ROI is extracted and preprocessed via background subtraction. During training, temporal spans are pseudo-labeled as Valid or Non-Valid using synthetic corruptions, and a VideoMAE classifier is trained on these spans. At inference, a sliding window produces per-span scores aggregated into a video-level verdict.

Span Sources, Pseudo-labeling and Pre-augmentations. Spans of five consecutive ROI frames are extracted from rectified videos sampled at 15 fps. *Valid* spans are sampled from ROI regions covering the target OVD. *Non-Valid* spans originate from two complementary sources:

- **ROI-based negatives:** regions that do not contain the target OVD—either because no hologram is present, or because the region covers a different OVD region.
- **Temporal corruptions of Valid spans:** synthesized by disrupting the natural temporal structure of genuine holographic transitions via (i) *frame repetition*—frames are selected and repeated to simulate static or buffered content; (ii) *temporal shuffling*—frames are randomly permuted, destroying transition order; (iii) *freeze simulation*—a single frame is repeated across the full span, mimicking a static replay. Each strategy preserves per-frame OVD appearance while destroying temporal coherence, forcing the classifier to learn sequence-level rather than frame-level cues.

Unlike HoloVerif, where pseudo-labels are assigned at the frame level by thresholding luminance variance, pseudo-labels here are assigned at the span level based on the ROI source and corruption type.

Post-augmentations. Background subtraction is adapted to operate on spans by computing the temporal median across the entire span, effectively isolating dynamic holographic content while suppressing the static background. All spans then undergo a shared set of photometric augmentations designed to emulate realistic capture conditions: color jitter (brightness, contrast, saturation, hue), mild Gaussian blur to approximate motion blur and defocus, random equalization, autocontrast, sharpness adjustment, posterization, and small erased patches to simulate sensor or compression artifacts. Synthetic shadows and localized brightness variations are additionally applied using plasma-based operators [8], producing complex, spatially-varying illumination effects. A custom transform further simulates holographic reflections in *Non-Valid* spans: brightness is modulated by plasma-like fractal fields masked by random silhouettes drawn from the *Quick!Draw!* dataset [2], producing spatially-structured highlights that closely resemble OVD optical artifacts. Together, these augmentations prevent the model from relying on global brightness cues while enriching it with hologram-like visual patterns during training. *Valid* and *Non-Valid* spans subsequently receive asymmetric spatial treatment. *Valid* spans are resized to 224×224 to preserve fine holographic patterns. *Non-Valid* spans are subjected to random resized crops, flips, small rotations, and additional photometric perturbations, increasing their diversity without altering their temporal invalidity.

Training. The VideoMAE encoder (22M parameters) is fine-tuned with binary cross-entropy to classify spans as *Valid* or *Non-Valid.* Training uses AdamW with a learning rate of 10^{-4}, cosine decay with a 10% linear warm-up, and a batch size of 64. Convergence is reached in three epochs per fold.

Inference. At inference time, a sliding window of width 5 frames is applied over the input video sequence, yielding a per-span legitimacy score from the classifier. These scores are then aggregated via averaging to produce a single video-level legitimacy score, which can be thresholded for a binary decision or used to compute an ROC curve.

Limitations. While this approach effectively leverages temporal information, provides a practical solution to the absence of real attacks in the training set, and serves as a comparison baseline for previous state-of-the-art, it suffers from two limitations which motivate our second, generative approach: 1. *HoloVerif-Span*'s attack coverage is inherently limited by the quality and diversity of the hardcoded augmentations, weakly calibrating the boundaries around the OVD representation manifold; and 2. its learned representation, optimized to discriminate between *Valid* and *Non-Valid* spans, does not ensure a diverse representation of genuine OVD appearances. This limits the use of embedding diversity as a proxy for *dynamic appearance variety*, potentially restricting its ability to detect simpler, static attacks.

3.2 Masked Sequence Modeling—Generative Approach

Masked Sequence Modeling (MSM) is a generative approach that frames OVD verification as an Out-of-Distribution (OOD) detection problem (Fig. 3). Rather than training a discriminative classifier with synthetic negatives, MSM

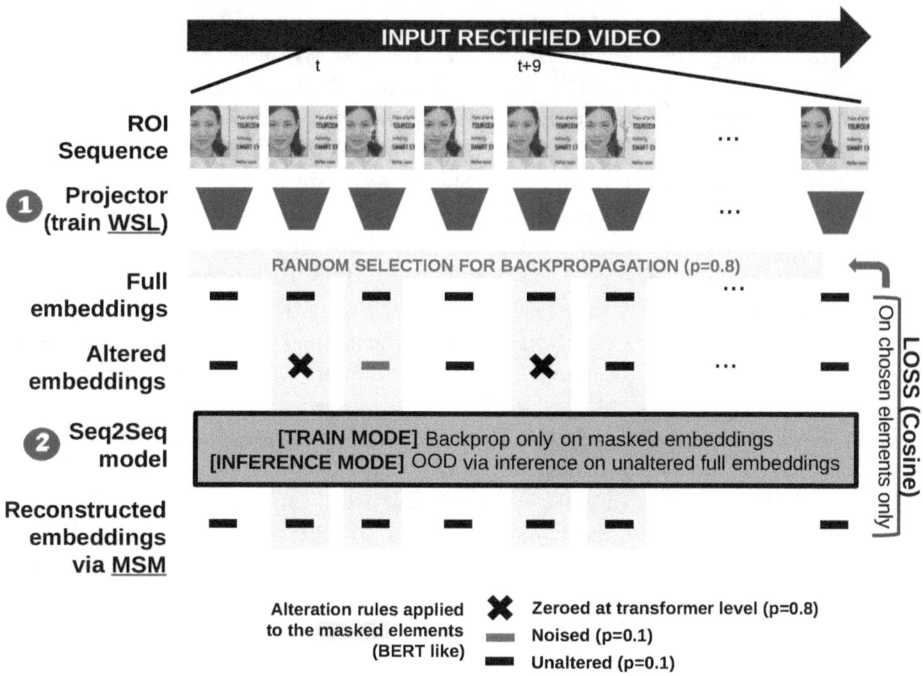

Fig. 3. Masked Sequence Modeling (MSM) pipeline. A frozen projector encodes ROI frames into OVD-focused embeddings. During training, embeddings are randomly masked and the Seq2Seq model learns to reconstruct them from legitimate sequences only. At inference, reconstruction error serves as the anomaly score.

learns to model the manifold of legitimate OVD transitions and detects attacks as deviations from this learned distribution. This is achieved through a two-stage architecture: a frame-level projector optimized for OVD content representation extracts embeddings, and a transformer encoder is trained to reconstruct masked embedding spans extracted from legitimate video sequences only. The transformer encoder provides a contextualized representation of OVD dynamics, effectively learning the underlying physical laws governing its behavior.

At inference, the method computes two complementary criteria at the span level (sequences of 10 frames), which are later aggregated at the video level: **Reconstructability** measures how well the model (trained exclusively on spans from legitimate samples) can reconstruct a given span's representation. It assumes that different OVD dynamics will deviate from the learned manifold and yield high reconstruction errors, hence serving as a *dynamic appearance conformity* check to detect attacks with incorrect dynamics. **Similarity** assesses the temporal diversity among frames in the learned embedding space. It serves as a proxy for *dynamic appearance variety* to detect static template attacks that might otherwise achieve low reconstruction error due to their stable nature.

Stage 1: Training the Projector. The projector serves as a frame-level encoder that captures hologram-specific visual features and produces per-frame embeddings. To remain consistent with the "Legit Only Training" constraint, the projector is expected to be pretrained on a large dataset (e.g., ImageNet, CLIP) regardless of the chosen backbone (e.g., MobileViT, DINOv2), and then fine-tuned on holographic content following the Weakly Supervised Learning (WSL) approach [14]. This approach samples neighboring pairs of frames from the same video and optimizes a contrastive loss to pull their embeddings together while pushing apart embeddings from distant frames. This encourages the model to learn a representation that captures the underlying structure of OVD appearances and their temporal evolution, while being robust to noise and irrelevant background features. Once trained, the projector part of the pipeline is frozen.

Stage 2: Training the Seq2Seq Model. The sequence-to-sequence (Seq2Seq) model is trained using a masked reconstruction objective that forces the model to predict masked frame embeddings from the visible temporal context. This training procedure naturally encourages the model to learn temporal dependencies between holographic states. The key insight is that valid transitions within genuine OVDs are predictable from their context, whereas attack-induced transitions will not be, as they do not follow the temporal dynamics observed during training.

The masking strategy follows a VideoMAE-like [10] high masking ratio combined with a BERT-inspired [4] corruption protocol adapted to the continuous embedding domain. Approximately 80% of the embeddings in a sequence are randomly selected for alteration, while the remaining elements are left untouched and excluded from the loss computation. Among those selected elements, 80% are zeroed at the transformer level, 10% are corrupted with Gaussian noise (standard deviation of 0.1), and 10% are left unaltered. This distribution mirrors the original BERT masking strategy adapted to the continuous embedding domain.

During training, data augmentation includes random temporal flipping ($p = 0.5$), which reverses the sequence order and teaches the model that transitions are valid in both temporal directions. The forward pass consists of applying the masking operation to produce altered embeddings and a binary mask, passing the altered sequence through the model to obtain reconstructed embeddings, and computing the reconstruction loss between the reconstructed and original embeddings on masked positions only.

The total loss combines two complementary reconstruction objectives for each frame embedding: a Mean Squared Error (MSE) loss that penalizes absolute reconstruction differences on masked positions, and a cosine distance loss that enforces directional similarity in the embedding space. The total loss is computed as a weighted sum of both terms with default weights of 1.0 each. We found that combining these two losses facilitated learning convergence.

Inference. At inference, videos are processed in spans of 10 frames, and the same projector is applied to extract frame embeddings. The trained Seq2Seq model is then used to reconstruct embeddings for each span (without masking) and compute the two criteria: **Reconstructability**, measured as the cosine similarity between the reconstructed and original embeddings across the span, later averaged over spans to produce a video-level score; and **Similarity**, computed as the mean pairwise cosine similarity between frame embeddings within the span, averaged across spans to produce a video-level score.

The final decision is made by fitting an envelope model (we use a Gaussian Model in the experiments reported here for illustrative purposes) over the joint 2D space of *Reconstructability* and *Similarity* on a validation set composed of legitimate samples only. A query video is classified (*Legit*/*Non-Legit*) based on its likelihood of belonging to this envelope, as illustrated in Fig. 5. An ROC curve can be computed by varying the decision threshold on this likelihood score.

Limitations. A key limitation of the MSM approach lies in its cascaded architecture, which stacks three distinct models: the projector (frame encoder), the sequence-to-sequence masked reconstruction model, and the final decision model. Each stage depends on the output quality of the previous one, meaning that errors can propagate and compound through the pipeline. In particular, if the projector fails to adequately isolate holographic content from the background or document template, the downstream embeddings will carry irrelevant or noisy information. This degrades both the reconstruction model's ability to learn meaningful transition patterns and the decision model's capacity to discriminate legitimate from anomalous sequences. The overall system therefore relies on the strong assumption that the projector produces embeddings that faithfully capture OVD-specific visual features which are both discriminative and robust to noise. This is a non-trivial requirement given the weak OVD signal, the diversity of OVD designs, and the limited availability of training data for this specific domain.

4 Experiments

Methods. We evaluate prior state-of-the-art methods under both *Legit-only* and *Legit/Non-Legit* training regimes. We then focus on our two proposed methods, which are specifically designed for the *Legit-only* setting—a key objective of this work, as it aligns with industrial requirements where genuine data is abundant but attack samples are scarce.

Existing methods (trainable from both Legits and Non-Legits):

- **MIDV-Holo-Baseline** [12] computes per-pixel chromatic variance in HSV space and applies handcrafted thresholds to detect holographic activity, serving as a signal-processing baseline. No real training is performed, only calibration.
- **WSL** [14] (MobileViT backbone) learns discriminative frame-level holographic representations via contrastive triplet sampling from video-level labels only.

- **HoloVerif** [15] (MobileViT backbone) combines temporal background subtraction with pseudo-labeling and targeted asymmetric augmentation. It currently represents the state of the art on MIDV-DynAttack.

Our proposed methods (designed for training on Legits only):

- **HoloVerif-Span** (VideoMAE backbone) is a sequence-level discriminative method classifying temporal spans of five frames, with *Non-Valid* spans synthesized from legitimate recordings via temporal corruptions (Sect. 3.1).
- **MSM** is a generative anomaly detection method combining the WSL frame encoder (MobileViT backbone) with a Masked Sequence Modeling reconstruction head and a single Gaussian model fitted exclusively on legitimate validation embeddings (Sect. 3.2).

Datasets. The evaluation is performed on two complementary public benchmarks:

- **MIDV-Holo** [12]: A standard benchmark containing genuine identity documents with OVDs captured under varied conditions. It contains 700 videos—300 legitimate and 400 non-legitimate (mostly static template attacks). We use the identity-disjoint splits proposed in [14,15].
- **MIDV-DynAttack** [15]: An extension featuring over 1,200 attack videos spanning three main categories:
  - *Static attacks* (550 videos): Printed documents, plastic copies with or without reflections, laser and LED lighting attacks.
  - *Static-swap attacks* (200 videos): Alternation between multiple static templates to simulate holographic behavior.
  - *Dynamic attacks* (450 videos): Real holograms overlaid on fraudulent documents, including plain hologram foils, leaf patterns, double stickers, masks, and decorative holograms.

Calibration. Decision thresholds are determined from the validation set which contains only *Legit* samples. We use the 95th and 99th percentiles of the score distribution on the validation set as candidate operating points, corresponding to target False Positive Rates (FPR) of 5% and 1%, respectively. When *non-Legit* samples are available at validation time, a ROC curve can be constructed and an operational point selected in order to balance the True Acceptance Rate (TAR) against the False Rejection Rate (FRR). All reported AUC values use the threshold-free metric, ensuring comparability regardless of the calibration strategy.

Metrics. We report the Area Under the ROC Curve (AUC) as the primary metric, enabling threshold-independent comparison across methods and attack types. All experiments are repeated over 5-fold cross-validation, reporting mean $\pm$ standard deviation. AUC values are computed per attack type and aggregated as a weighted mean over all attack videos ("Mix" column).

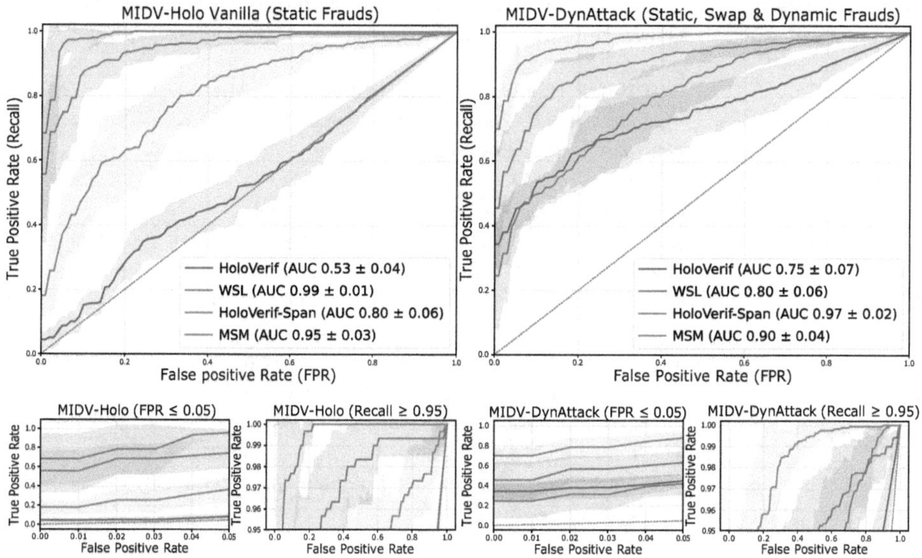

Fig. 4. ROC curves for all methods trained exclusively on legitimate videos, on MIDV-Holo and MIDV-DynAttack. Top row: full ROC curves on MIDV-Holo static template frauds (left) and MIDV-DynAttack covering static template, swap, and dynamic attacks (right). Since MIDV-DynAttack contains only non-Legit samples, Legits from MIDV-Holo are used as the negative class for AUC computation. Bottom row: zoomed operating-point view at FPR $\leq$ 0.05 (left) and TPR $\geq$ 0.95 (right). Shaded bands denote ± 1 std over 5 folds.

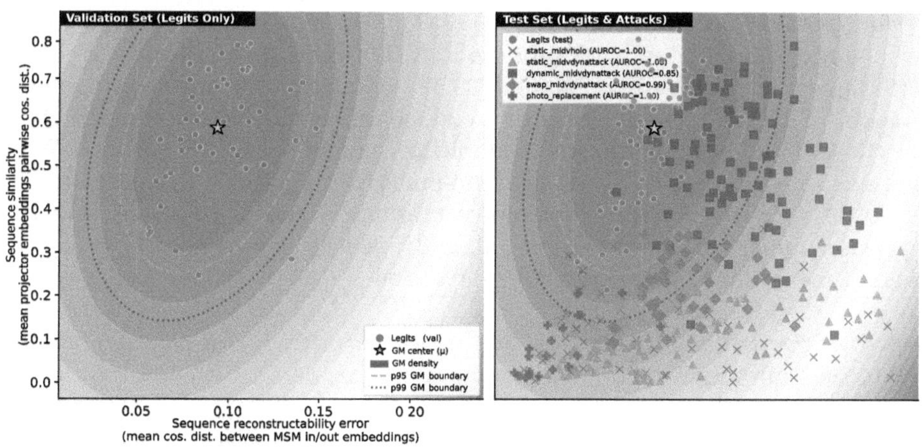

Fig. 5. MSM decision space (first fold). Each point represents a video projected onto the (Reconstructability, Similarity) plane. Left: GMM fitted (k = 1) on legitimate validation samples only—materialized by its density and boundaries. Right: Attack videos (from the test set) are overlaid—static attacks cluster in the low-similarity bottom region, while dynamic attacks show similar similarity but higher reconstruction error (right).

Table 1. Uncalibrated results (mean AUC, 5 folds) for different training regimes (with only Legits or not) on MIDV-Holo and MIDV-DynAttack datasets (test split).

	MIDV-Holo (L vs A)		MIDV-Holo (L) vs MIDV-DynAttack (A)			
	Vanilla	Photo Rep.	Static	Static-swap	Dynamic	Mix
#Vids. Legit/Attack →	60L+60A	60L+20A	60L+110A	60L+40A	60L+60A	60L+380A
Methods ↓	AUC	AUC	AUC	AUC	AUC	AUC
SOTA methods - Trained on ***Legits + Non-Legits.***						
MIDV-Holo-Baseline [12]	93.4 ± 1.9	74.8 ± 2.3	77.2 ± 2.2	48.5 ± 3.4	8.8 ± 2.4	57.3 ± 1.2
WSL [14]	**99.2 ± 0.9**	**98.4 ± 1.1**	95.7 ± 2.1	74.8 ± 8.2	60.7 ± 4.7	84.1 ± 2.9
HoloVerif [15]	**99.3 ± 0.6**	94.3 ± 2.3	**98.6 ± 0.9**	79.2 ± 3.7	92.7 ± 1.9	**94.4 ± 1.2**
SOTA methods - Trained only on ***Legits***						
WSL [14]	98.7 ± 1.0	97.8 ± 1.3	95.2 ± 2.2	80.6 ± 7.0	61.6 ± 9.1	84.7 ± 4.4
HoloVerif [15]	52.9 ± 4.3	42.7 ± 4.1	80.9 ± 4.5	62.2 ± 10.9	73.1 ± 7.6	68.7 ± 5.9
Proposed methods - Trained only on ***Legits***						
MSM (ours)	94.8 ± 3.1	89.7 ± 6.1	93.5 ± 4.2	92.3 ± 4.5	85.6 ± 6.7	91.1 ± 3.8
HoloVerif Span (ours)	80.4 ± 6.4	84.5 ± 4.6	96.5 ± 2.4	**96.9 ± 1.7**	**98.6 ± 1.3**	93.4 ± 2.6

Results. Table 1 presents uncalibrated results in order to compare the raw potentials of our methods versus the state-of-the-art. Since MIDV-DynAttack contains only non-Legit samples (negative class), Legits from MIDV-Holo are used as the positive class for AUC computation. Figure 4 shows the corresponding ROC curves. Table 2 show calibrated results at given operational points in order to illustrate explicit performances (true/false rejections), aligning with the evaluation protocol defined in [15].

Performances of SOTA Methods (Original Configuration). When including non-Legits during training (and/or calibration), MIDV-Holo-Baseline achieves strong performance on MIDV-Holo Vanilla (93.4% AUC) but collapses on dynamic attacks (8.8% AUC): chromatic variance alone cannot distinguish genuine OVD reflections from holographic foils exhibiting similar optical signatures. WSL improves robustness (84.1% overall) but similarly struggles on dynamic attacks (60.7% AUC), where frame-level representations saturate. HoloVerif sets the supervised upper bound at 94.4% overall AUC, benefiting from explicit sequence-level augmentation and background subtraction.

Impact of the Training Regime on SOTA Methods. WSL transfers remarkably well to the legit-only regime (84.7% overall AUC, vs. 84.1% with non-legit data), confirming that its contrastive training objective generalizes without requiring explicit fraud supervision. This makes WSL the most practical legit-only baseline: it requires no additional design effort and degrades gracefully. In contrast, HoloVerif degrades substantially (68.7% overall AUC) when deprived of non-legit data, as its discriminative pseudo-labeling was designed to contrast legitimate and fraudulent patterns explicitly.

Table 2. Calibrated results (mean Fscore, Specificity and Recall reported over 5 folds) reported on MIDV-Holo and MIDV-DynAttack datasets (test split). Numbers for the SOTA methods (calibrated by Fscore maximization on the validation set) slightly differs from [15] due to improved augmentation parameters. Our proposed methods are calibrated according to OOD thresholds of 95% and 99% (Legit correctly classified from the validation set). Specificities close to their respective OOD thresholds show a good generalization between splits.

	MIDV-Holo (L+A)				MIDV-DynAttack (A)		
		Vanilla		Photo Rep.	Static	Static-swap	Dynamic
#Vids. Legit/Attack →	120 L+A	60 L	60 A	20 A	110 A	40 A	90 A
Methods ↓	Fscore	Specificity	Recall	Recall	Recall	Recall	Recall
*SOTA methods - Trained/Calibrated on **Legits** + **Non-Legits*** - Maximizing Fscore							
MIDV-Holo-Baseline [12]	84.8 ± 3.2	84.9 ± 9.6	85.0 ± 11.6	38.0 ± 20.8	72.4 ± 3.1	19.5 ± 7.6	3.1 ± 3.1
WSL [14]	95.8 ± 2.7	95.3 ± 2.8	96.3 ± 5.8	88.0 ± 12.5	79.3 ± 11.2	16.5 ± 13.5	4.2 ± 4.4
HoloVerif [15]	95.9 ± 1.5	94.2 ± 3.3	97.3 ± 3.0	67.0 ± 21.7	92.5 ± 3.7	35.0 ± 7.5	68.9 ± 7.9
*Proposed methods—Trained/Calibrated on **Legits only** - OOD@99/OOD@95*							
MSM@99 (ours)		96.2 ± 3.4	72.0 ± 9.7	34.0 ± 25.2	60.7 ± 18.7	54.0 ± 10.1	58.4 ± 17.1
MSM@95 (ours)		92.6 ± 1.8	83.3 ± 8.8	56.0 ± 24.0	73.6 ± 13.1	71.0 ± 13.6	67.1 ± 12.0
HoloVerif-Span@99 (ours)		99.0 ± 1.5	19.0 ± 8.1	24.0 ± 2.2	67.3 ± 13.2	49.5 ± 16.5	87.3 ± 6.6
HoloVerif-Span@95 (ours)		94.6 ± 1.8	35.3 ± 10.0	44.0 ± 10.2	85.1 ± 7.5	81.5 ± 8.0	94.4 ± 3.0

Competitiveness of Our Methods Trained Exclusively on Legits. Both proposed methods, trained without any fraudulent sample, achieve competitive performance against fully supervised methods. HoloVerif-Span reaches 93.4% overall AUC and MSM 91.1%, both substantially above the strongest legit-only baseline WSL (84.7%), and closing to within one point of the supervised upper bound HoloVerif (94.4%).

Table 2 confirms these trends under real operational conditions. The specificities measured on the test set closely match the targeted OOD thresholds derived from the validation set (OOD@99: ∼96.2–99%; OOD@95: ∼92.6–94.6%), demonstrating good generalization of the calibration protocol across splits—with a slightly larger gap for MSM. On dynamic attacks, HoloVerif-Span@95 achieves 94.4% Recall, substantially outperforming the supervised HoloVerif (68.9%), while both proposed methods significantly outperform all baselines on swap attacks. However, recall on static template attacks and Photo Replacement remains lower, particularly at the strict OOD@99 threshold, showing limited discriminability of temporal features in this setting.

Figure 5 illustrates the GM decision space of the MSM: static attacks cluster in the low-similarity region (near-zero intra-span variation), while dynamic attacks exhibit higher overlap with legitimate sequences, consistent with the harder detection rates in this category.

Capture Condition Sensitivity. The performance gap of HoloVerif-Span on MIDV-Holo Vanilla, with an AUC of 80% (19 point bellow HoloVerif) surfaces a broader challenge: methods trained on MIDV-Holo legitimate videos may be sensitive to the capture conditions of that specific campaign. Future benchmarks should systematically vary acquisition devices, lighting rigs, and document nationalities to disentangle method robustness from dataset-specific biases.

Transition Modeling as a Learnable Signal. The central finding of this evaluation is that genuine OVD temporal dynamics constitute a learnable and discriminative signal, accessible from legitimate data alone. Notably, learning temporal dynamics finally provides a way to detect swapping attack, while strongly reinforcing the detection rate of dynamic attacks. Frame-level methods fail on swap and dynamic attacks by design: WSL stagnates at 74.8% and 60.7% AUC even with non-legit supervision. Transition modeling closes this gap from legitimate data alone—HoloVerif-Span reaches 96.9% (+22.1 pts) and 98.6% (+37.9 pts), with MSM independently confirming the trend (92.3%, 85.6%).

Complementarity of Generic Holographic Detectors and Temporal Models. The HoloVerif-Span outperforms all methods on transition-based attacks—Static-swap (96.9% AUC, +17.7 pts vs. supervised SOTA) and Dynamic (98.6% AUC, +5.9 pts)—but underperforms on static attacks (MIDV-Holo Vanilla: 80.4% AUC), where no temporal discontinuity is available. MSM, on the other hand, is consistent across all attack types (Static: 93.5%, Swap: 92.3%, Dynamic: 85.6% AUC), demonstrating that a model trained exclusively on legitimate sequences can generalize to unseen fraud categories—any deviation from the learned legitimate distribution is flagged, regardless of the attack nature.

5 Conclusion

We addressed Optically Variable Device (OVD) verification for remote KYC using short smartphone videos under an open-set threat model. In this challenging scenario, fraud samples are scarce, capture conditions are uncontrolled, and the OVD signal is weak and noisy compared to the static background (suffering from transparency, spatial/temporal sparsity, and light reflections). Previous methods either relied on heavily supervised training with attack samples, which are unavailable in production, or focused on static frame-level properties, leaving them vulnerable to swapping attacks and unable to verify the dynamic nature of OVDs.

We proposed two novel approaches for modeling OVD dynamics: HoloVerif-Span, a discriminative span classifier trained with synthetic temporal corruptions, and MSM, a generative masked sequence model that learns the manifold of genuine OVD dynamics via reconstruction in an embedding space. Crucially, both methods are trained exclusively on legitimate samples without any attack supervision nor manual frame-level labeling. They demonstrate strong performance on the MIDV-Holo and MIDV-DynAttack datasets, approaching previous state-of-the-art methods that rely on attack samples for training, while finally defeating swapping attacks and exhibiting robustness across all attack types. Overall, this confirms the importance of sequence-level temporal modeling for OVD verification and the viability of *Legit-only* training.

Although the performance of proposed methods on static attacks is slightly lower than the previous WSL approach—acting as a generic OVD detector—combining both kind of approachs is a straightforward way to improve robustness across the attack spectrum, as they are complementary in their detection capabilities. Furthermore, the proposed MSM approach provides a unique theoretical advantage: by defining verification through *reconstructability* and *similarity*, it offers a unified framework that enforces both the conformity of visual patterns and the necessity of dynamic change. This allows it to reject static counterfeits and dynamic anomalies based on principled deviations from the genuine manifold, rather than relying on discriminative boundaries learned from synthetic corruptions.

Both HoloVerif-Span and MSM are highly promising approaches for practical deployment since they do not require anticipating specific attack implementations and can be trained on abundant production data without costly annotation. However, the performance of both methods is still constrained by the quality of frame-level representations, which must be sufficiently focused on the OVD signal and robust to noise. Currently, the diversity of training data is limited to a single OVD model. Future work should focus on improving the frame-level projector and acquiring more diverse OVD data—either genuine or synthetic—to enhance representation learning, which is the main bottleneck of all approaches.

This project has received funding from the European Union's Horizon Europe research and innovation program under Grant Agreement No 101189689.

References

1. Hartl, A., Grubert, J., Schmalstieg, D., Reitmayr, G.: Mobile interactive hologram verification. In: IEEE International Symposium on Mixed and Augmented Reality (ISMAR), pp. 75–82 (2013). https://doi.org/10.1109/ISMAR.2013.6671766
2. Jongejan, J., Rowley, H., Kawashima, T., Kim, J., Fox-Gieg, N.: The Quick, Draw! Dataset (2017). https://github.com/googlecreativelab/quickdrawdataset
3. Arlazarov, V.V., Bulatov, K., Chernov, T., Arlazarov, V.L.: MIDV- 500: a dataset for identity documents analysis and recognition on mobile devices in video stream. Comput. Opt. **43** (2019). https://doi.org/10.18287/2412-6179-2019-43-5-818-824
4. Devlin, J., Chang, M.-W., Lee, K., Toutanova, K.: Bert: pre-training of deep bidirectional transformers for language understanding. In: Proceedings of NAACL-HLT, pp. 4171–4186 (2019)
5. Marcel, S., Nixon, M.S., Fierrez, J., Evans, N. (eds.) Handbook of Biometric Anti-Spoofing: Presentation Attack Detection. Advances in Computer Vision and Pattern Recognition. Springer (2019). https://doi.org/10.1007/978-3-319-92627-8
6. Soares, A., Junior, R.N., Bezerra, B.: BID dataset: a challenge dataset for document processing tasks. In: Proceedings of Conference on Graphics, Patterns and Images (SIBGRAPI), pp. 143–146 (2020). https://doi.org/10.5753/sibgrapi.est.2020.12997

7. Kada, O., Kurtz, C., van Kieu, C., Vincent, N.: Hologram detection for identity document authentication. In: Proceedings of International Conference on Pattern Recognition and Artificial Intelligence (ICPRAI), pp. 346–357 (2022). https://doi.org/10.1007/978-3-031-09037-0_29
8. Nicolaou, A., Christlein, V., Riba, E., Shi, J., Vogeler, G., Seuret, M.: TorMentor: deterministic dynamic-path, data augmentations with fractals. In: Proceedings of CVPR, pp. 2707–2711 (2022)
9. Polevoy, D.V., et al.: Document liveness challenge dataset (DLC-2021). J. Imaging **8**, 181 (2022). https://doi.org/10.3390/jimaging8070181
10. Tong, Z., Song, Y., Wang, J., Wang, L.: Videomae: masked autoencoders are data-efficient learners for self-supervised video pre-training. In: NeurIPS, vol. 35, pp. 10078–10093 (2022)
11. Chapel, M.-N., Al-Ghadi, M., Burie, J.-C.: Authentication of holograms with mixed patterns by direct LBP comparison. In: IEEE International Workshop on Multimedia Signal Processing (MMSP), pp. 1–6 (2023). https://doi.org/10.1109/MMSP59012.2023.10337669
12. Koliaskina, L.I., et al.: MIDV-Holo: a dataset for ID document hologram detection in a video stream. In: Proceedings of ICDAR, pp. 486–503 (2023). https://doi.org/10.1007/978-3-031-41682-8_30
13. Park, E.-J., Back, S.-Y., Kim, J., Woo, S.S.: KID34K: a dataset for online identity card fraud detection. In: Proceedings of ACM International Conference on Information and Knowledge Management (CIKM), pp. 5381–5385 (2023). https://doi.org/10.1145/3583780.3615122
14. Pouliquen, G., Chiron, G., Chazalon, J., Géraud, T., Awal, A.M.: Weakly supervised training for hologram verification in identity documents. In: Proceedings of ICDAR, pp. 17–33 (2024). https://doi.org/10.1007/978-3-031-70533-5_2
15. Pouliquen, G., Chazalon, J., Chiron, G., Géraud, T., Awal, A.M.: Verification of dynamic holographic behavior in identity documents. In: Proceedings of ICDAR, pp. 323–339 (2025). https://doi.org/10.1007/978-3-032-04624-6_19

DocQT: Improving Document Forgery Localization Robustness via Diverse JPEG Quantization Tables

Kylian Ronfleux–Corail[1,2](✉), Guillaume Bernard[1], Mickaël Coustaty[2], and Nicolas Sidère[2]

[1] MAIF, Niort, France
kylian.ronfleux-corail@univ-lr.fr
[2] L3i Laboratory, La Rochelle University, La Rochelle, France

Abstract. Document manipulation localization models achieve strong performance on public benchmarks yet fail to generalize to operational document workflows. We identify a critical and overlooked source of this gap: the mismatch between the narrow distribution of JPEG quantization tables used during training—restricted to standard libjpeg quality factors—and the heterogeneous compression profiles encountered in real-world insurance document pipelines. To isolate this factor, we conduct a controlled factorial study comparing two architectures with contrasting levels of quantization table awareness—FFDN [2] and Mesorch [20]—each trained under either standard quality factor augmentation (*Standard-QT*) or operationally calibrated quantization tables sampled from DocQT, a quantization-table bank derived from a MAIF operational image corpus (*Real-QT*), and evaluated under three recompression conditions. Training under *DocQT* yields substantial localization gains on DocTamper [15] and significantly reduces the pixel-level false positive rate on authentic operational documents, but only for architectures that explicitly ingest the quantization table as input. The released DocQT quantization-table dataset and compression-reproduction material are directly available at https://github.com/Kyliroco/Improving-Document-Forgery-Localization-Robustness-via-Diverse-JPEG-Quantization-Tables These results demonstrate that standard quality factor augmentation does not adequately proxy operational compression diversity, and that architectural choices explicitly conditioning on the quantization table provide a meaningful robustness advantage for real-world deployment.

Keywords: Document Manipulation Localization · JPEG Quantization Tables · Image Forensics · Compression Robustness

G. A. Fink et al. (Eds.): ICDAR 2026, LNCS 16972, pp. 109–126, 2027.
https://doi.org/10.1007/978-3-032-36023-6_7

1 Introduction

1.1 Document Manipulation Localization in Insurance Operational Workflows

Organizations such as insurance companies, public agencies, and financial institutions handle thousands of documents every day—ranging from claims and contracts to invoices and supporting paperwork. These documents constitute a primary vector for fraud and money laundering attempts [1,3,13]. As an illustrative example of operational scale, internal observations at MAIF, a French Mutual insurance company, indicate that over 100,000 documents are received per month for claim management alone. At such volumes, manual inspection is impractical, and automated localization systems are required [1]. In this context, false positives—that is, authentic documents incorrectly flagged as tampered—trigger additional manual reviews, increasing both processing time and operational costs. As argued by Guillaro et al. [5], in a realistic deployment where manipulated images are rare, a high false alarm rate can render a system counterproductive, with false positives drastically outnumbering true positives.

In this work, we focus on *Document Manipulation Localization* (DML), defined as the task of producing a pixel-level segmentation mask identifying tampered regions, independently of any intent classification. Importantly, the presence of tampering does not imply fraudulent intent: many legitimate documents undergo modifications such as highlighting, digital annotations, or watermarking that do not constitute forgery [13]. Fraud determination remains a business-level decision [1]. Throughout this paper, the terms *tampering* and *manipulation* are used interchangeably to denote pixel-level alteration; *forgery* is reserved for cases in which the intent to deceive is established.

The majority of manipulation localization models are developed and evaluated on natural images, while document-oriented methods remain comparatively scarce [4,15]. We argue that the performance reported on public benchmarks such as DocTamper [15] overstates the practical robustness of these models when deployed in operational workflows. This paper investigates the primary source of this generalization gap: the mismatch between the narrow distribution of JPEG quantization tables used during model training and the heterogeneous compression profiles present in real-world document pipelines.

1.2 Problem Statement: Quantization Table Mismatch

JPEG compression artifacts constitute a central forensic signal exploited by state-of-the-art localization models [2,9,15]. However, the forensic cues these models learn—in particular double-quantization residuals observed in the DCT domain—depend directly on the specific identity of the quantization table rather than merely a nominal quality factor [9,17]. Most existing training pipelines restrict JPEG augmentation to standard quality factors derived from *libjpeg*, which correspond to a small, discrete subset of all possible quantization tables [8,17].

This assumption breaks down in operational settings. A MAIF operational image corpus reveals a broad diversity of acquisition devices, scanning software, and document conversion chains encountered in insurance workflows [14]; from this corpus, we derive DocQT, a dataset of **859 distinct luminance quantization tables**. In comparison, public benchmarks such as DocTamper [15] and T-SROIE [18] expose only a single quantization table, while SROIE [6] contains at most 6 distinct configurations. This mismatch between training-time and inference-time quantization table distributions is expected to degrade localization accuracy and increase false positives on intact documents [9,17].

To address this gap, we conduct a comparative experimental study that systematically controls the quantization table distribution during both training and evaluation, following the evaluation framework of ForensicHub [4] and the benchmarking methodology of IMDL-BenCo [12]. Our contributions are twofold: first, we quantify the domain shift induced by the mismatch between training-time quantization tables and those encountered in operational workflows; second, we demonstrate that retraining with operational insurance quantization tables significantly reduces false positives on authentic documents while preserving localization performance on tampered ones.

2 Related Works

2.1 Document Manipulation Localization

Document manipulation localization is an active research field in image manipulation detection and localization, first studied extensively on natural scene images before being adapted to the specific challenges of documents [4,12]. In this section, we review both lines of work, focusing on methods related to the JPEG compression artifacts that are central to our problem.

Alteration Detection and Localization in Natural Images. Natural images constitute the primary domain in which manipulation localization methods have been developed [2,4,10]. A recurring principle across this literature is the exploitation of low-level statistical inconsistencies—noise patterns, compression artifacts, or boundary discontinuities—that betray the presence of a manipulation even when it is visually convincing [2,9,20].

ManTra-Net [19] operationalizes this principle through self-supervised training: a manipulation-trace feature extractor is trained to classify 385 distinct manipulation types, and forgery localization is then cast as a local anomaly detection problem using a ConvLSTM-based module that computes Z-score deviations across multiple window scales. While powerful in its generality, this approach does not explicitly model compression artifacts, which limits its sensitivity to the JPEG-specific traces prevalent in document images. CAT-Net [9] directly addresses this limitation by processing DCT coefficients through a dedicated stream that learns the statistical distribution of double-quantization residuals, combined with a standard RGB stream; the method relies on the observation that a region compressed twice leaves characteristic periodic patterns in the DCT coefficient histograms, whose structure depends on the ratio between the two quantization steps. PSCC-Net [10] pursues a different strategy, focusing on multi-scale spatial consistency rather than compression traces: a top-down path built on a lightweight HRNet backbone extracts multi-scale features, while a bottom-up path progressively refines manipulation masks from coarse to fine scales through a Spatio-Channel Correlation Module (SCCM) that captures both spatial and channel-wise correlations.

TruFor [5] extends this line of work by introducing a multimodal fusion framework that combines the RGB image with Noiseprint++, a learned noise-sensitive fingerprint trained via self-supervised contrastive learning on real images subjected to 512 distinct editing histories; both modalities are fused through a SegFormer-based transformer encoder, and the framework additionally produces a reliability map identifying regions where localization predictions are uncertain, along with a global integrity score. This reliability mechanism is particularly relevant in operational settings where false alarms carry a significant cost [5]. Building on the limitations of purely microscopic approaches, Mesorch [20] argues that manipulation operates simultaneously at the microscopic level through low-level forensic traces and at the macroscopic level through semantic object-level alterations, and addresses both by running a CNN branch and a Transformer branch in parallel: the CNN processes high-frequency DCT-enhanced features to capture local textural anomalies, while the Transformer processes low-frequency DCT-enhanced features to model global semantic context, with an adaptive weighting module dynamically adjusting the contribution of each scale.

These models have become standard references in image manipulation localization [4,12]. However, they are designed and evaluated under assumptions—rich visual content, diverse textures, standard compression profiles—that do not hold for administrative document images, motivating the development of document-specific approaches.

Specific Case of Documents. Document images differ fundamentally from natural images in their forensic properties. The presence of text, regular layouts, and uniform backgrounds suppresses the boundary artifacts that natural image methods rely upon, while artifacts introduced by scanning, printing, or document conversion processes create additional confounding signals [3,11,15]. Even a single substituted character can substantially alter the semantic content of a document [3,11,15], which distinguishes document forensics from natural image forensics where manipulations typically affect larger semantic regions. Accordingly, document-focused approaches have predominantly targeted pixel-level localization rather than whole-image detection [4,12,15].

As illustrated in Fig. 1, the manipulation operations encountered in document forensics—copy-move, splicing, generation, and coverage—are predominantly applied at the character or word level [1,11,15]. These operations are typically confined to small areas and forensic traces they leave are subtle and easily masked by the uniform backgrounds features [2,3].

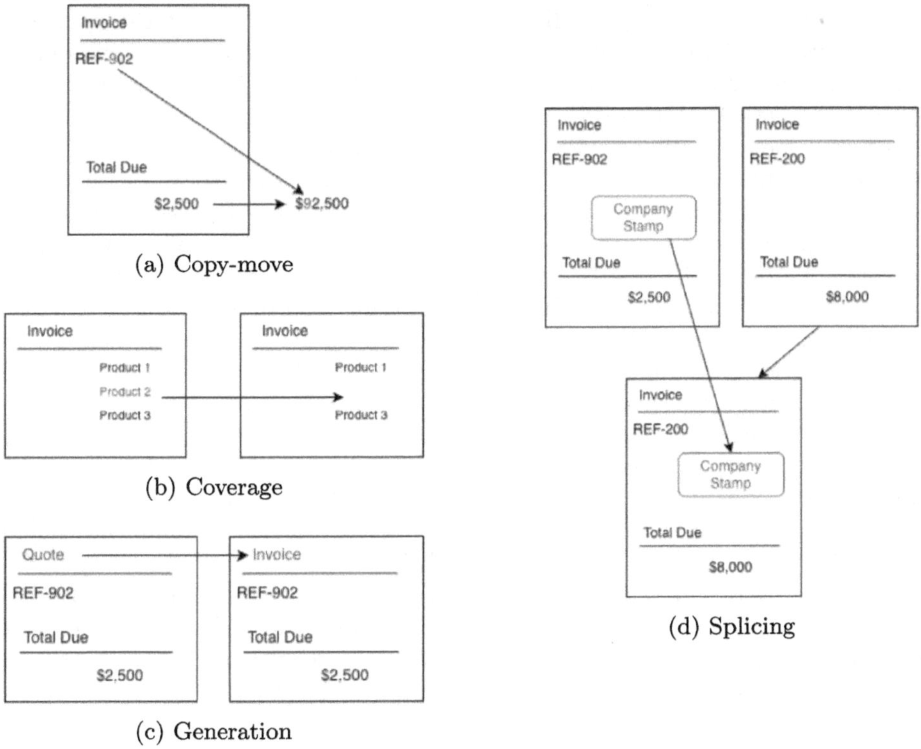

(a) Copy-move

(b) Coverage

(c) Generation

(d) Splicing

Fig. 1. Illustration of the 4 manipulation types considered in document forensics - Red elements denote the tampered regions. **(a)** Copy-move: a region is duplicated within the same document. **(b)** Coverage: a patch conceals existing content. **(c)** Generation: new pixel content is synthesized to replace a region. **(d)** Splicing: a region extracted from a different source document is inserted into the target. (Color figure online)

DTD [15] establishes the foundational framework for this domain by combining a Visual Perception Head with a Frequency Perception Head operating on raw DCT coefficients, fused through a Swin-Transformer encoder and a Multi-view Iterative Decoder (MID) that progressively aggregates features at multiple scales; to improve robustness against JPEG compression, the authors further propose Curriculum Learning for Tampering Detection (CLTD), a training paradigm that progressively increases compression difficulty with quality factors drawn from [75, 100]. FFDN [2] extends this dual-branch design by identifying two residual weaknesses: the incomplete integration of frequency information into the RGB feature space, and the loss of fine-grained high-frequency traces during downsampling. It addresses these through a Visual Enhancement Module (VEM) that injects frequency information into the RGB stream using zero-initialized convolutions, and a Wavelet-like Frequency Enhancement (WFE) module that explicitly decomposes features into high- and low-frequency components. Both DTD and FFDN share a common architectural property that is central to this study: **they explicitly ingest the JPEG quantization table as an input to the network**, making their learned representations inherently dependent on the distribution of quantization tables seen during training.

In contrast, CAFTB [16] operates in the noise domain: two parallel branches—one extracting spatial features from the RGB image, the other applying an SRM filter to expose global noise inconsistencies—are fused through a cross-attention module to integrate local and global forgery cues. TIFDM [3] similarly enhances forgery traces from multiple domains through a dedicated deep module before feeding them into an encoder-decoder network with a multiscale attention module, with a particular emphasis on robustness to real-world distortions. Neither CAFTB nor TIFDM explicitly processes quantization tables, relying instead on implicit pixel-domain or noise-domain artifacts.

Recent benchmarking efforts have begun to unify the evaluation of these heterogeneous approaches. ForensicHub [4] provides a modular and configuration-driven architecture that decomposes forensic pipelines into interchangeable components across datasets, transforms, models, and evaluators, enabling cross-domain comparisons across the four main forensic tasks including document manipulation localization. Similarly, IMDL-BenCo [12] standardizes the training and evaluation protocols for image manipulation detection and localization. Both frameworks highlight a persistent gap: the robustness of document forensic models to compression variability remains insufficiently characterized, as evaluations are typically restricted to standard quality factor r anges.

2.2 JPEG Quantization as an Alteration Detector

JPEG compression artifacts have long constituted a central forensic signal in image authenticity analysis [8,9,17]. One of the earliest practical methods exploiting these artifacts is Error Level Analysis (ELA) [8], which detects inconsistencies in residual compression error across image regions: when a tampered image is re-saved at a fixed quality, authentic regions and manipulated regions exhibit different error levels because they carry distinct compression histories. Since these artifacts originate from block-based DCT coding followed by coefficient quantization, forensic cues are commonly analyzed through compression inconsistencies observed in the DCT domain [2,9,15], whose manifestation is governed by the underlying quantization tables [9,17].

JPEG compression partitions the image into non-overlapping 8×8 blocks, transforms each block into the frequency domain via the Discrete Cosine Transform (DCT), and then divides the resulting 64 coefficients component-wise by a 8×8 quantization matrix before rounding to the nearest integer [9,17]. This quantization step is the primary source of information loss: large matrix entries aggressively suppress high-frequency coefficients, while small entries preserve detail at the cost of larger file size [8,17]. As illustrated in Fig. 2, the conventional quality factor $q \in [1, 100]$ scales a reference matrix according to a fixed formula [17], producing quantization tables with markedly different coefficient magnitudes—and thus different compression artifacts—across quality levels. When a JPEG image undergoes a second compression cycle, the DCT coefficient histograms exhibit characteristic periodic patterns whose structure depends on the ratio between the two quantization matrices [9,17]. It is precisely these double-quantization residuals that methods such as CAT-Net [9] and the Frequency Perception Head of DTD and FFDN [2,15] are designed to detect, which explains why the identity of the quantization matrix—rather than merely the quality level—is critical to their correct operation. The JPEG standard defines separate quantization tables for luminance (Y) and chrominance (Cb, Cr) channels [17]; since the forensic models considered in this study operate exclusively on the luminance table [2,9,15], all subsequent references to *quantization table* denote the luminance table.

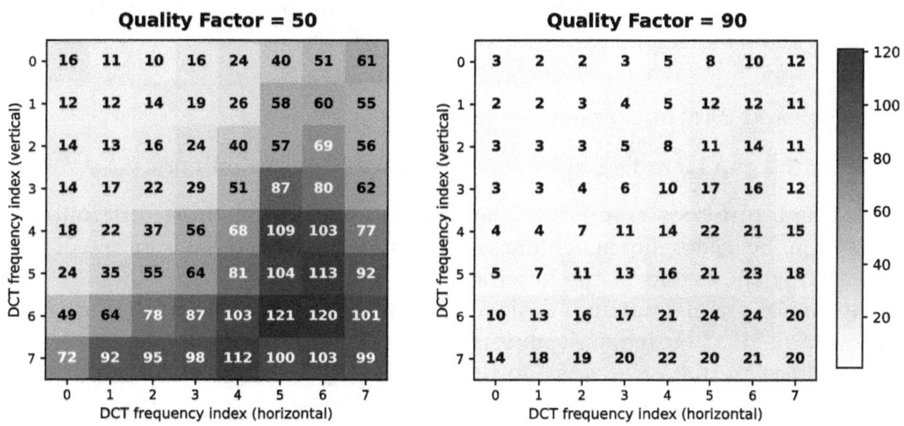

Fig. 2. Standard *libjpeg* luminance quantization tables at two quality factors. Each cell value represents the quantization step for the corresponding DCT frequency coefficient. At $q = 50$ (left), high-frequency coefficients are aggressively quantized (steps up to 121), discarding fine detail. At $q = 90$ (right), steps are much smaller (maximum 24), preserving more information. Both tables are derived from the same base matrix scaled by q [17], and represent only two points in the discrete family of tables reachable through standard quality factors.

A key aspect of the JPEG standard is that quantization tables are not universally fixed: virtually all graphical applications and digital cameras rely on hard-coded, manufacturer-specific tables optimized for their own hardware and color pipelines [8]. As a consequence, a given nominal quality level does not correspond to a unique quantization table across encoders. For example, a JPEG saved at quality 80% in Adobe Photoshop uses quantization tables equivalent to quality 91% in libjpeg [8]. In practice, however, many existing forensic works implicitly assume that compression follows standard libjpeg-based encoding conventions: they parameterize compression exclusively through integer quality factors, which map to quantization tables via a fixed formula [17] and thus cover only a small, discrete subset of the tables encountered in practice.

Among models that incorporate explicit compression augmentation, the quality factor ranges remain narrow: CAT-Net [9] uses factors in [60, 100], DTD [15] and FFDN [2] apply random recompression restricted to [75, 100], and TruFor [5], the broadest of these, uses [30, 100]. Several other models, including ManTra-Net [19], PSCC-Net [10], and Mesorch [20], include no explicit JPEG augmentation at all. All these configurations are derived from integer quality factors, which correspond to a small discrete subset of possible quantization tables and implicitly assume that compression follows standard tool-based encoding conventions [17]. **Despite the known diversity of quantization tables in real-world imaging pipelines** [8], **no existing work has systematically analyzed the impact of this diversity on the performance of document manipulation localization models**, nor proposed training strategies to mitigate the resulting distribution shift.

2.3 Proposed Methodology

To address the gap identified in the preceding sections, we propose:

1. A **dataset analysis** comparing the JPEG quantization table distributions across public scientific benchmarks and an operational insurance corpus, quantifying the extent of the heterogeneity gap (Sect. 3).
2. An **experimental protocol** designed to isolate the impact of JPEG quantization table distribution shift on forgery localization performance, through a factorial design that crosses two recompression pipelines with two structurally contrasting models (Sect. 4).

3 Dataset Analysis

3.1 Datasets

The datasets considered in this study were selected to cover the majority of publicly available document image datasets containing both altered and non-altered samples, spanning a wide range of document types, manipulation methods, and acquisition conditions. This selection includes DocTamper [15], RTM [11], T-SROIE [18], SROIE [6], FUNSD [7], and Find-it/Find-it Again [1,13] as public benchmarks, alongside the MAIF operational image corpus used to evaluate model behavior on real operational data exhibiting heterogeneous acquisition and compression pipelines. Due to privacy and confidentiality constraints, the underlying MAIF document images cannot be publicly released. From this image corpus, however, we derive DocQT, a public dataset of header-extracted luminance quantization tables used to define the *Real-QT* condition; these tables do not contain document content and are released, together with the material required to reproduce the compression protocol, on Softawre Heritage or on github. This combination enables a comprehensive evaluation of model robustness across both controlled and operational environments.

We establish a terminological distinction between *document images* and *natural images*. Document images refer to administrative or insurance-related documents processed in operational contexts. In such images, character-level tampering can be visually imperceptible: the absence of texture suppresses the boundary artifacts that detection methods rely upon, making localization significantly more challenging [2,3,11,15]. Documents processed in operational contexts exhibit substantial diversity both in content type—ranging from administrative forms to photographic evidence captured on smartphones [1,11,15]—and in image format, as they are commonly submitted as native PDFs, scanned PDFs, or raster images (JPEG or PNG) [14]. Raster photographs are predominantly stored in JPEG format, which introduces compression artifacts whose characteristics depend directly on the quantization table used during encoding [8,9].

3.2 Quantization Table Diversity Across Corpora

A central motivation of this study is the discrepancy between the JPEG compression diversity observed in public benchmarks and that encountered in operational settings. As reported in Table 1, SROIE [6] and Find-it [1] are the only public datasets exhibiting more than a single luminance quantization table, with 6 distinct configurations each arising from the heterogeneity of scanning equipment; all other public datasets—DocTamper [15], RTM [11], and T-SROIE [18]—expose a single uniform quantization table throughout, consistent with systematic generation using a standard encoder at a fixed quality factor.

In contrast, the MAIF dataset exhibits 859 distinct luminance quantization tables, reflecting the diversity of acquisition devices, scanning software, and document conversion chains encountered in insurance workflows [3,14]. As illustrated in Fig. 3, the distribution follows a pronounced long-tail pattern: even SROIE, the most diverse public benchmark, concentrates the vast majority of its images within 6 configurations, whereas the operational corpus distributes images across hundreds of non-standard tables with no dominant configuration. Although this frequency profile comes from a single French insurance workflow, it aggregates customer-submitted documents acquired through diverse devices and software chains, so overlap in quantization-table support with other administrative settings is plausible even if mixture weights differ. These observations confirm that public benchmarks do not capture the quantization table variability present in operational document environments [1,3,11], and directly motivate the *Real-QT* experimental condition described in Sect. 4.1.

Table 1. Datasets considered in this work. CM: copy-move; SP: splicing; GN: generation; IP: inpainting; CV: coverage. The symbol ⋆ denotes unaltered reference sets used for false positive evaluation only.

Dataset	Subset	# Images	Mod. Type	# Altered	# QT
DocTamper [15]	Train	120 000	CM/GN/SP	120 000	1
	Test	30 000	CM/GN/SP	30 000	1
	FCD	2 000	CM/GN/SP	2 000	1
	SCD	18 000	CM/GN/SP	18 000	1
RTM [11]	Train	5 803	CM/GN/SP/IP/CV	4 000	1
	Test	3 197	CM/GN/SP/IP/CV	2 000	1
T-SROIE [18]	Train	626	GN	626	1
	Test	360	GN	360	1
SROIE⋆ [6]	–	973	–	–	6
Find-it [1]	–	1 180	CM/GN/SP/IP	240	6
FUNSD⋆ [7]	–	199	–	–	–
Find-it Again [13]	–	988	CM/GN/SP	163	–
MAIF operational corpus⋆ (real documents, France)	–	13 455	–	–	**859**

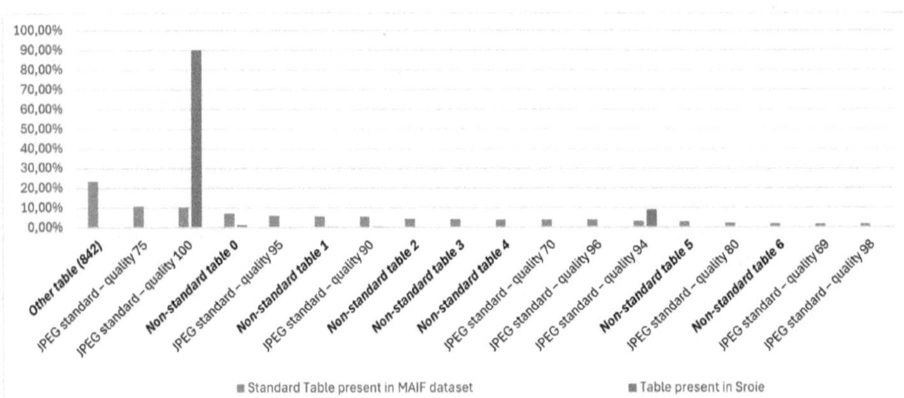

Fig. 3. Distribution of the most frequent luminance JPEG quantization tables in the MAIF operational dataset (orange) and SROIE [6] (blue). Each bar represents a distinct quantization table, ranked by frequency of occurrence in the operational corpus. The first bar aggregates the 851 remaining tables, each individually representing less than 25% of the operational dataset. While a single table accounts for over 90% of SROIE images, no dominant configuration emerges in the operational corpus, confirming its substantially higher compression diversity. (Color figure online)

4 Benchmark Protocol

This section describes the experimental protocol designed to isolate the effect of JPEG quantization table distribution shift on document manipulation localization. Two training pipelines are tested: *Standard-QT*, which applies classical quality factor augmentation, and *Real-QT*, which uses quantization tables extracted from real operational document headers. These pipelines are crossed with two structurally contrasting models to form a factorial design.

4.1 JPEG Recompression Pipelines

The central experimental variable of this study is the JPEG quantization table distribution encountered during both training and evaluation. Two recompression pipelines are defined and applied independently, yielding a factorial design that combines training pipeline with evaluation recompression condition.

The first pipeline, referred to as *Standard-QT*, recompresses all images with quality factors drawn uniformly at random from $[30, 100]$ using OpenCV, covering the broadest quality factor range reported in the literature (see Sect. 2.2 for a detailed comparison across methods).

The second pipeline, referred to as *Real-QT*, recompresses all images using quantization tables sampled from DocQT, a dataset extracted from the JPEG headers of the MAIF operational image corpus following the fingerprinting approach described in [8]. This dataset yields over 800 distinct luminance quantization tables reflecting heterogeneous real-world acquisition and processing pipelines.

Each model variant is evaluated under three conditions: no forced recompression (*Orig.*), *Standard-QT* recompression (*Std*), and *Real-QT* recompression (*Real*). This factorial design enables us to disentangle the effect of the training distribution from that of the evaluation distribution on localization and false positive performance.

4.2 Training Pipeline

All images are normalized using ImageNet statistics to match the expected input distribution of the pretrained backbones used by both models. Data augmentation is applied during training only: horizontal and vertical flips with probability 0.5, random rotation by multiples of 90° with probability 0.5, random brightness and contrast adjustment, and Gaussian blur with probability 0.2. These transformations are chosen to improve generalization without degrading the low-level forensic artifacts that both models rely upon [9,19].

4.3 Methods Used

All experiments are conducted within the ForensicHub framework [4], which provides a unified training and evaluation infrastructure across forensic tasks and natively implements both FFDN [2] and Mesorch [20]. These two models are selected because they represent opposite ends of a structural spectrum directly relevant to the central question of this paper—the impact of JPEG quantization table distribution shift on localization performance.

As described in Sect. 2.1, FFDN inherits from DTD [15] a Frequency Perception Head that ingests the quantization table as an explicit input, making it by design sensitive to any mismatch between training-time and inference-time quantization tables. Mesorch, as described in Sect. 2.1, does not exploit the quantization table explicitly: it relies on implicit pixel-domain and DCT-domain artifacts captured through parallel CNN and Transformer branches, which makes it a particularly informative counterpoint for assessing the role of explicit quantization table awareness under distribution shift.

The remaining models available in ForensicHub—such as ManTra-Net [19], CAT-Net [9], or PSCC-Net [10]—do not exhibit this structural contrast with respect to quantization table awareness, and their inclusion would not add discriminative information for the specific hypothesis under investigation.

Two model variants are trained, one under each pipeline: **FFDN-*Std*** and **Mesorch-*Std*** are trained under *Standard-QT*; **FFDN-*Real*** and **Mesorch-*Real*** are trained under *Real-QT*.

4.4 Data

Dataset Splits. Training data is drawn from four sources: the full DocTamper training set [15] (120,000 images), and 5,000 crops each from the training portions of Find-it [1], Find-it Again [13], and T-SROIE [18]. Additionally, 400 crops from FUNSD [7]—approximately half the dataset—and a subset of unaltered documents from the MAIF operational image corpus are included as negative examples to expose the models to authentic document statistics during training. RTM [11] is deliberately excluded from training in order to evaluate the capacity of models trained exclusively on synthetically generated manipulations to generalize to real manual forgeries. All remaining data are reserved for evaluation.

4.5 Evaluation Criteria

All models produce a pixel-level probability map $P \in [0,1]^{H \times W}$, from which a binary prediction mask $\hat{M}$ is derived by thresholding at $\tau = 0.5$:

$$\hat{M}(i,j) = \mathbb{I}\big(P(i,j) \geq \tau\big). \tag{1}$$

Given the ground-truth binary mask M_{gt} provided by each dataset, the following pixel-level quantities are defined:

$$TP_{\text{pix}} = \textstyle\sum_{i,j} \mathbb{I}\big(\hat{M}(i,j) = 1 \wedge M_{gt}(i,j) = 1\big), \tag{2}$$

$$FP_{\text{pix}} = \textstyle\sum_{i,j} \mathbb{I}\big(\hat{M}(i,j) = 1 \wedge M_{gt}(i,j) = 0\big), \tag{3}$$

$$FN_{\text{pix}} = \textstyle\sum_{i,j} \mathbb{I}\big(\hat{M}(i,j) = 0 \wedge M_{gt}(i,j) = 1\big). \tag{4}$$

Localization on Tampered Images. On images containing at least one altered pixel, localization performance is jointly measured by the pixel-level **F1 score** and **Intersection over Union** (IoU) [4,12]:

$$F_1^{\text{pix}} = \frac{2TP_{\text{pix}}}{2TP_{\text{pix}} + FP_{\text{pix}} + FN_{\text{pix}}}, \qquad \text{IoU} = \frac{TP_{\text{pix}}}{TP_{\text{pix}} + FP_{\text{pix}} + FN_{\text{pix}}}. \tag{5}$$

Both metrics are computed per image and averaged across the evaluation set.

False Positive Rate on Unaltered Images. For unaltered reference sets, the ground-truth mask is identically zero ($M_{gt} = \mathbf{0}$), so every predicted positive pixel is by definition a false positive. The **pixel-level false positive rate** is defined as:

$$\text{FPR}_{\text{pix}} = \frac{FP_{\text{pix}}}{H \times W}, \tag{6}$$

where $H \times W$ is the total number of pixels. This quantity is computed per image and reported as a mean across the reference set [5]. A low FPR_{pix} combined with high F1 and IoU on tampered images constitutes the primary robustness criterion of this study: it captures whether a model can accurately localize tampered regions without being misled by unfamiliar JPEG compression profiles into flagging authentic documents.

5 Results

Quantization Table Distribution as a Primary Source of Generalization Failure. Across all DocTamper subsets, evaluation without recompression (*Orig.*) consistently yields the highest scores, with substantial drops under both recompression conditions. The benefit of operational training is, however, strongly conditioned on the model's capacity to exploit this distributional alignment explicitly. On DT-Test under *Real-QT* evaluation, **FFDN-Real gains 14.5 F1 points over FFDN-Std** (0.853 vs. 0.708), with consistent improvements across FCD and SCD. By contrast, Mesorch-*Real* and Mesorch-*Std* achieve nearly identical scores under the same condition (0.704 vs. 0.703). This contrast directly reflects the architectural distinction identified in Sect. 2.1: FFDN's Frequency Perception Head can exploit the distributional match through its explicit quantization table input, whereas Mesorch has no mechanism to bind its representations to a specific quantization matrix [2,20]. **Architectures that explicitly condition on the quantization table therefore provide a meaningful robustness advantage in operational compression pipelines** [9,17].

Standard Quality Factor Augmentation Is Not a Sufficient Proxy for Operational Conditions. Real-QT evaluation consistently outperforms *Standard-QT*, showing that broad quality-factor augmentation is not a reliable proxy for operational compression variability. Quality-factor sampling only covers libjpeg-compatible matrices and misses many application-specific tables that dominate the operational distribution (Sect. 3.2) [3].

Table 2. Pixel-level F1 score on tampered evaluation sets. Columns correspond to model variants trained under *Standard-QT* (suffix *-Std*) or *Real-QT* (suffix *-Real*). Rows report three evaluation conditions per dataset subset: *Orig.* (no recompression), *Std* (*Standard-QT* recompression), *Real* (*Real-QT* recompression). **Bold**: best result per row within each architecture family (FFDN and Mesorch separately).

			FFDN-*Std* [2]	FFDN-*Real* [2]	Mesorch-*Std* [20]	Mesorch-*Real* [20]
DocTamper [15]	Test	*Orig.*	0.927	**0.954**	0.751	**0.818**
		Std	0.633	**0.647**	**0.676**	0.578
		Real	0.708	**0.853**	0.703	**0.704**
	FCD	*Orig.*	0.942	**0.951**	**0.555**	0.509
		Std	0.536	**0.537**	**0.556**	0.476
		Real	0.630	**0.832**	**0.540**	0.530
	SCD	*Orig.*	0.879	**0.900**	0.591	**0.709**
		Std	0.516	**0.542**	**0.514**	0.472
		Real	0.604	**0.776**	0.537	**0.587**
RTM [11]	–	*Orig.*	**0.048**	0.040	0.058	**0.074**
		Std	**0.034**	0.031	**0.048**	0.042
		Real	**0.030**	0.023	0.051	**0.057**
T-SROIE [18]	–	*Orig.*	**0.900**	0.893	**0.873**	0.865
		Std	0.746	**0.798**	**0.833**	0.754
		Real	0.727	**0.827**	**0.838**	0.798
Find-it [1]	–	*Orig.*	0.296	**0.418**	0.309	**0.364**
		Std	0.168	**0.216**	**0.223**	0.189
		Real	0.186	**0.303**	0.252	**0.253**
Find-it Again [13]	Test	*Orig.*	0.130	**0.158**	0.068	**0.212**
		Std	0.052	**0.058**	0.059	**0.077**
		Real	**0.073**	0.067	0.075	**0.125**
	Val	*Orig.*	**0.204**	0.201	0.063	**0.259**
		Std	0.071	**0.078**	**0.069**	0.064
		Real	**0.114**	0.076	0.043	**0.183**

The Synthetic-to-Real Gap Remains an Open Problem. On RTM [11], excluded from training, all variants remain near zero (best F1: 0.074 for Mesorch-*Real*), indicating a limitation beyond compression shift. This motivates either larger manually tampered datasets or architectures less dependent on low-level artifact cues [5,20].

Benchmark Diversity Does Not Reflect Operational Diversity. Models achieve substantially higher scores on T-SROIE (F1 up to 0.900) than on Find-it (best F1: 0.418) and Find-it Again (best F1: 0.259). T-SROIE's uniform JPEG configuration aligns with the DocTamper training signal, whereas Find-it's multi-tool edits and Find-it Again's lossless format expose failure modes that single-configuration benchmarks obscure [4,12].

False Positive Rates. On the operational corpus, **FFDN-Real under Real-QT reaches a** FPR_{pix} **of** 8.69×10^{-5}, nearly an order of magnitude lower than FFDN-*Std* at *Orig.* (5.64×10^{-4}). This reduction supports the hypothesis of Sect. 1.2: **quantization table mismatch is a major source of false alarms in deployment** (Tables 3 and 4).

Table 3. Mean pixel-level false positive rate (FPR_{pix}, lower is better) on unaltered reference sets under the same protocol as Table 2. **Bold**: lowest result per row within each architecture family.

		FFDN-*Std*	FFDN-*Real*	Mesorch-*Std*	Mesorch-*Real*
SROIE [6]	*Orig.*	3.50×10^{-5}	$\mathbf{2.82 \times 10^{-5}}$	3.21×10^{-5}	$\mathbf{1.60 \times 10^{-5}}$
	Std	$\mathbf{2.77 \times 10^{-5}}$	2.36×10^{-5}	4.12×10^{-5}	$\mathbf{2.88 \times 10^{-5}}$
	Real	4.65×10^{-5}	$\mathbf{1.69 \times 10^{-5}}$	4.06×10^{-5}	$\mathbf{4.00 \times 10^{-5}}$
FUNSD [7]	*Orig.*	2.49×10^{-4}	$\mathbf{6.01 \times 10^{-5}}$	1.48×10^{-4}	$\mathbf{5.98 \times 10^{-5}}$
	Std	$\mathbf{7.63 \times 10^{-5}}$	1.42×10^{-4}	$\mathbf{1.76 \times 10^{-4}}$	1.78×10^{-4}
	Real	8.53×10^{-4}	$\mathbf{7.79 \times 10^{-5}}$	1.87×10^{-4}	$\mathbf{1.47 \times 10^{-4}}$
MAIF	*Orig.*	5.64×10^{-4}	$\mathbf{1.46 \times 10^{-4}}$	2.06×10^{-4}	$\mathbf{9.66 \times 10^{-5}}$
	Std	2.39×10^{-4}	$\mathbf{1.30 \times 10^{-4}}$	2.03×10^{-4}	$\mathbf{1.84 \times 10^{-4}}$
	Real	2.32×10^{-4}	$\mathbf{8.69 \times 10^{-5}}$	2.37×10^{-4}	$\mathbf{1.84 \times 10^{-4}}$

5.1 Limitations and Perspectives

The comparison is restricted to two architectures selected for structural contrast (Sect. 4.3), and broader architectural validation remains needed [9]. The operational table distribution derives from a single insurance corpus which, despite 859 distinct configurations, may not represent industries. Finally, the threshold $\tau = 0.5$ may be suboptimal in deployment scenarios with asymmetric false positive/false negative costs [5,12].

Future work should extend this analysis to architectures with intermediate frequency conditioning, test chrominance-table and chrominance-DCT conditioning alongside luminance cues, construct benchmarks from pristine PNG documents with controlled recompression and manually edited documents [4,12], and explore domain adaptation techniques that do not require access to operational documents at training time.

Table 4. Pixel-level IoU score on tampered evaluation sets. **Bold**: best result per row within each architecture family.

			FFDN-*Std*	FFDN-*Real*	Mesorch-*Std*	Mesorch-*Real*
DocTamper [15]	Test	*Orig.*	0.882	**0.923**	0.692	**0.762**
		Std	0.579	**0.597**	**0.617**	0.525
		Real	0.653	**0.809**	0.644	**0.647**
	FCD	*Orig.*	0.895	**0.913**	**0.505**	0.469
		Std	0.484	**0.489**	**0.505**	0.432
		Real	0.574	**0.776**	**0.488**	0.480
	SCD	*Orig.*	0.806	**0.837**	0.498	**0.618**
		Std	0.439	**0.466**	**0.430**	0.402
		Real	0.524	**0.699**	0.450	**0.505**
RTM [11]	–	*Orig.*	**0.039**	0.032	0.046	**0.061**
		Std	**0.026**	0.024	**0.037**	0.033
		Real	**0.023**	0.018	0.040	**0.046**
T-SROIE [18]	–	*Orig.*	0.766	**0.827**	**0.803**	0.799
		Std	0.663	**0.720**	**0.755**	0.677
		Real	0.643	**0.753**	**0.761**	0.724
Find-it [1]	–	*Orig.*	0.257	**0.370**	0.275	**0.321**
		Std	0.147	**0.192**	**0.196**	0.167
		Real	0.165	**0.269**	0.223	**0.223**
Find-it Again [13]	Test	*Orig.*	0.096	**0.123**	0.051	**0.158**
		Std	0.036	**0.045**	0.044	**0.057**
		Real	**0.053**	0.048	0.057	**0.092**
	Val	*Orig.*	**0.153**	0.152	0.044	**0.191**
		Std	0.051	**0.057**	**0.049**	0.045
		Real	**0.083**	0.053	0.030	**0.136**

6 Conclusion

This paper investigated the impact of JPEG quantization table distribution shifts on document manipulation localization models deployed in operational insurance workflows. Through a controlled factorial study comparing FFDN [2] and Mesorch [20], selected for contrasting quantization table awareness (Sect. 4.3), we showed that training under operationally calibrated quantization tables yields substantial localization gains—up to 14.5 F1 points on DocTamper [15]—but only for architectures explicitly ingesting the quantization table as input. On the false positive side, this protocol reduces FPR_{pix} on the MAIF operational image corpus by nearly an order of magnitude (8.69×10^{-5} vs. 5.64×10^{-4}). These findings show that standard quality factor augmentation does not adequately proxy operational compression diversity, and that architectures conditioning on quantization tables provide a meaningful robustness advantage for real-world deployment. Additionally, near-zero performance on RTM [11] confirms that the synthetic-to-real manipulation gap remains an open problem orthogonal to compression, motivating future benchmarks built from pristine documents, controlled recompression, and manually edited forgeries.

Acknowledgments. This research was supported by the CIFRE PhD program funded by the ANRT.

Disclosure of Interests. Kylian Ronfleux–Corail and Guillaume Bernard are employed by MAIF, which provided the operational corpus. Mickaël Coustaty and Nicolas Sidère declare no additional competing interests.

References

1. Artaud, C., Sidère, N., Doucet, A., Ogier, J.M., Poulain d'Andecy, V.Y.: Find it! fraud detection contest report. In: 2018 24th International Conference on Pattern Recognition (ICPR), pp. 13–18. IEEE (2018). https://doi.org/10.1109/ICPR.2018.8545428
2. Chen, Z., et al.: Enhancing tampered text detection through frequency feature fusion and decomposition. In: Computer Vision – ECCV 2024. Lecture Notes in Computer Science, vol. 15091, pp. 200–217. Springer, Cham (2025). https://doi.org/10.1007/978-3-031-73414-4_12
3. Dong, L., Liang, W., Wang, R.: Robust text image tampering localization via forgery traces enhancement and multiscale attention. IEEE Trans. Consum. Electron. **70**(1), 3495–3507 (2024). https://doi.org/10.1109/TCE.2024.3367947
4. Du, B., et al.: ForensicHub: a unified benchmark & codebase for all-domain fake image detection and localization (2025). https://doi.org/10.48550/arXiv.2505.11003
5. Guillaro, F., Cozzolino, D., Sud, A., Dufour, N., Verdoliva, L.: TruFor: leveraging all-round clues for trustworthy image forgery detection and localization. In: 2023 IEEE/CVF Conference on Computer Vision and Pattern Recognition (CVPR), pp. 20606–20615. IEEE (2023). https://doi.org/10.1109/CVPR52729.2023.01974
6. Huang, Z., et al.: ICDAR2019 competition on scanned receipt OCR and information extraction. In: 2019 International Conference on Document Analysis and Recognition (ICDAR), pp. 1516–1520. IEEE (2019). https://doi.org/10.1109/ICDAR.2019.00244
7. Jaume, G., Ekenel, H.K., Thiran, J.P.: FUNSD: a dataset for form understanding in noisy scanned documents (2019). https://doi.org/10.48550/arXiv.1905.13538
8. Krawetz, N.: A picture's worth (2007)
9. Kwon, M.J., Nam, S.H., Yu, I.J., Lee, H.K., Kim, C.: Learning JPEG compression artifacts for image manipulation detection and localization. Int. J. Comput. Vision **130**(8), 1875–1895 (2022). https://doi.org/10.1007/s11263-022-01617-5
10. Liu, X., Liu, Y., Chen, J., Liu, X.: PSCC-net: progressive spatio-channel correlation network for image manipulation detection and localization. IEEE Trans. Circuits Syst. Video Technol. **32**(11), 7505–7517 (2022). https://doi.org/10.1109/TCSVT.2022.3189545

11. Luo, D., et al.: Toward real text manipulation detection: new dataset and new solution. Pattern Recogn. **157**, 110828 (2025). https://doi.org/10.1016/j.patcog.2024.110828
12. Ma, X., et al.: IMDL-BenCo: a comprehensive benchmark and codebase for image manipulation detection & localization. In: Advances in Neural Information Processing Systems (NeurIPS 2024), Track on Datasets and Benchmarks, vol. 37, pp. 134591–134613 (2024)
13. Martínez Tornès, B., et al.: Jeu de données de tickets de caisse pour la détection de fraude documentaire. In: Actes de CORIA-TALN 2023, vol. 4. pp. 140–147. ATALA, Paris (2023)
14. Nayef, N., Luqman, M.M., Prum, S., Eskenazi, S., Chazalon, J., Ogier, J.M.: SmartDoc-QA: a dataset for quality assessment of smartphone captured document images – single and multiple distortions. In: 2015 13th International Conference on Document Analysis and Recognition (ICDAR), pp. 1231–1235. IEEE (2015). https://doi.org/10.1109/ICDAR.2015.7333960
15. Qu, C., et al.: Towards robust tampered text detection in document image: new dataset and new solution. In: 2023 IEEE/CVF Conference on Computer Vision and Pattern Recognition (CVPR), pp. 5937–5946. IEEE, Vancouver (2023). https://doi.org/10.1109/CVPR52729.2023.00575
16. Song, Y., Jiang, W., Chai, X., Gan, Z., Zhou, M., Chen, L.: Cross-attention based two-branch networks for document image forgery localization in the metaverse. ACM Trans. Multimedia Comput. Commun. Appl. **21**(2), 55:1–55:24 (2025). https://doi.org/10.1145/3686158
17. Wan, K.: An overview of double JPEG compression detection and anti-detection. J. Inf. Hiding Priv. Protect. **4**(2), 89–101 (2023). https://doi.org/10.32604/jihpp.2022.039764
18. Wang, Y., Zhang, B., Xie, H., Zhang, Y.: Tampered text detection via RGB and frequency relationship modeling. J. Cybersecur. **8**, 29–40 (2022)
19. Wu, Y., AbdAlmageed, W., Natarajan, P.: ManTra-net: manipulation tracing network for detection and localization of image forgeries with anomalous features. In: 2019 IEEE/CVF Conference on Computer Vision and Pattern Recognition (CVPR), pp. 9535–9544. IEEE, Long Beach (2019). https://doi.org/10.1109/CVPR.2019.00977
20. Zhu, X., et al.: Mesoscopic insights: orchestrating multi-scale & hybrid architecture for image manipulation localization (2024). https://doi.org/10.48550/arXiv.2412.13753

Adversarial Attacks on Online Handwriting Using Salience-Based Temporal Editing

Yataro Tamura(✉), Brian Kenji Iwana, and Jiseok Lee

Kyushu University, Fukuoka, Japan
{yataro.tamura,jiseok.lee}@human.ait.kyushu-u.ac.jp,
iwana@ait.kyushu-u.ac.jp

Abstract. Deep learning models for online handwriting recognition have been shown effective and are increasingly deployed in practical applications. However, their vulnerability to adversarial attacks is still a challenge. Existing adversarial methods are predominantly designed for image-based inputs and typically rely on additive spatial perturbations. When applied to online handwriting, which is inherently represented as a time series of pen trajectories, such perturbations often introduce high-frequency jitter and visibly unnatural stroke artifacts. In this work, we propose a novel adversarial attack framework for online handwriting recognition based on salience-guided temporal editing. Instead of adding noise, the proposed method generates adversarial examples by inserting and deleting points at time steps selected according to temporal salience, preserving the shape and smoothness of the original handwriting. Temporal salience is estimated using gradient-based activation mapping, which guides edits toward time steps that strongly support the original class prediction. We evaluate the proposed approach on the Unipen and CASIA-OLHWDB datasets under both white-box and one-shot black-box attack settings. Experimental results demonstrate that while conventional image-based attacks achieve strong white-box performance, they exhibit poor transferability across models. In contrast, the proposed temporal editing attack achieves stronger one-shot black-box transferability while preserving the visual structure of the handwriting. These results indicate that temporal editing is a relevant threat model for online handwriting recognition, particularly in one-shot black-box transfer settings.

Keywords: Online Handwriting Recognition · Adversarial Attacks · Temporal Editing

1 Introduction

Online handwriting recognition differs from offline handwriting recognition in that the input is a sequence of pen-tip coordinates rather than a raster image.

G. A. Fink et al. (Eds.): ICDAR 2026, LNCS 16972, pp. 127–144, 2027.
https://doi.org/10.1007/978-3-032-36023-6_8

Modern online handwriting recognizers build on sequence models such as Recurrent Neural Networks (RNN) [33], Temporal Convolutional Neural Networks (CNN) [21], and Transformers [38], and have achieved high accuracy on this input modality [8,10,16,25,30]. However, adversarial robustness for online handwriting remains less explored than for image-based recognition models.

As these systems are deployed in real-world applications, concerns about their reliability and security have become more pronounced. In particular, adversarial attacks are a widely known weakness of neural networks [3,39]. Adversarial attacks work by adding small perturbations to the input with the aim of causing misrecognition by the models.

However, most adversarial attacks were originally developed for image-based classifiers with perturbations constrained by pixel-wise L_p norms. For example, various gradient-based attack methods have been proposed, such as Fast Gradient Sign Method (FGSM) [9], Projected Gradient Descent (PGD) [27], and the Carlini-Wagner (CW) attack [2]. These methods are designed to degrade recognition performance while remaining imperceptible in the image domain. In comparison, attacks on time series are much less explored, and most adversarial attacks on time series are adapted from image-based methods [7,18].

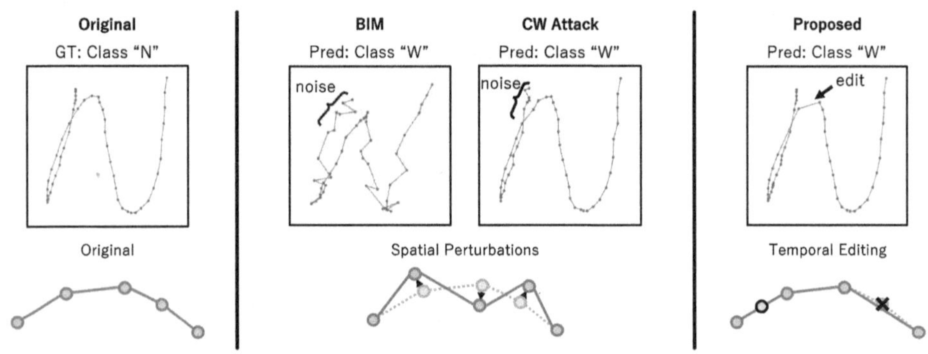

Fig. 1. Comparison between adversarial noise via perturbations and the proposed temporal editing. Adversarial attacks such as BIM and CW use spatial perturbations to generate adversarial examples. The proposed method adds and removes time steps for more realistic characters.

The online handwriting data considered in this study is represented as a sequence of two-dimensional pen-tip coordinates, where each time step t contains a coordinate vector $\mathbf{x}_t \in \mathbb{R}^2$. In the image domain, small perturbations can appear as inconspicuous noise due to the texture present in natural images. However, directly applying image domain adversarial noise to time series data frequently results in abrupt temporal fluctuations and high-frequency artifacts, which create unnatural deformations in online handwriting trajectories, as illustrated in Fig. 1. This indicates that even when perturbations satisfy small L_p

norms, the resulting time series may violate the kinematic smoothness and structural regularities inherent to human handwriting. In other words, the implicit assumption in the image domain that "small noise is difficult for humans to perceive" is not necessarily valid for time series data. This discrepancy highlights a fundamental limitation in current adversarial attacks on time series.

Beyond naturalness issues, most gradient-based adversarial attacks operate under a white-box assumption, requiring full access to model parameters and gradients. In practice, however, deployed handwriting recognition systems are often accessible only through prediction interfaces, making black-box adversarial attacks a more realistic threat model. Commercial handwriting recognition and authentication services typically expose only final recognition outputs, rendering white-box assumptions impractical in many real-world scenarios. Therefore, evaluating robustness under black-box conditions is essential for understanding the true vulnerability of online handwriting recognition systems.

To address these challenges, we propose Adversarial Iterative Temporal Editing (AITE), a novel adversarial attack method that generates adversarial examples through discrete editing operations in the time dimension rather than additive noise in the spatial dimensions. Specifically, AITE iteratively performs element-wise insertion and deletion on the input sequence based on the discriminative importance, or salience, of individual time steps. Salience is estimated using Gradient-weighted Class Activation Maps (Grad-CAM) [35], allowing the attack to target temporally critical regions while preserving the overall geometric structure and kinematic smoothness characteristic of natural handwriting trajectories.

The contributions of this study are summarized as follows:

- We highlight the inadequacy of spatial perturbation-based adversarial noise for online handwriting.
- We propose a novel adversarial attack framework based on salience-guided temporal editing.
- The proposed method is evaluated under both white-box and one-shot black-box attack settings on the Unipen [12] and CASIA-OLHWDB [24] online handwritten character datasets. Experimental results demonstrate that temporal editing achieves stronger visual similarity than conventional spatial perturbation-based noise attacks.
- The black-box attack settings demonstrate that the salience-based editing is effective at attacking unseen models.

Furthermore, for reproducibility, we will release the source code at GitHub[1].

2 Related Work

2.1 Adversarial Attacks

Adversarial attacks were first extensively studied in the context of image classification, where small, carefully crafted perturbations can induce misclassification while remaining visually imperceptible [3]. The vulnerability of deep

[1] https://github.com/yataro0117/AITE.

neural networks to adversarial perturbations was first systematically demonstrated by Szegedy et al. [37], who formulated adversarial example generation as an optimization problem that searches for minimal input perturbations capable of changing a model's prediction. To reduce the computational cost of this approach, Goodfellow et al. [9] proposed FGSM, which approximates the optimization by applying a single gradient step in the direction of the loss. Since then, many methods, such as Basic Iterative Method (BIM) [20], PGD [27], and the CW attack [2], have improved upon gradient-based adversarial attacks through iterative updates and regularized perturbation objectives. A separate line of work targets the black-box transferability of adversarial examples between models. MI-FGSM [5] accumulates a momentum term over the iterative input gradients to stabilize the update direction, while NI-FGSM [23] extends MI-FGSM with Nesterov accelerated gradients by evaluating the gradient at a look-ahead point. TI-FGSM [6] convolves the input gradient with a smoothing kernel before each update so that perturbations become approximately translation-invariant and transfer better across architectures. While these methods form the foundation of adversarial attack research, they were all originally developed for image-based classifiers and rely on additive, pixel-wise perturbations.

2.2 Adversarial Attacks on Time Series

Compared to image recognition, research on adversarial attacks against time series data is an emerging area. Carlini et al. [1] demonstrated that hidden commands can be embedded into automatic speech recognition systems. Fawaz et al. [7] evaluated FGSM and BIM on time series classification datasets and reported the vulnerability of deep learning–based time series classifiers. Karim et al. [18] proposed an Adversarial Transformation Network (ATN) using a distilled model to act as a surrogate to mimic the behavior of classical time series classification models. In addition, there have been black-box models that have been proposed for time series classification [4,31,41]. However, these methods still apply perturbations which may not be appropriate for all time series. Furthermore, many of these methods require feedback from the attacked model [42]. Developing attack methods that take into account structures specific to time series data (e.g., temporal dependency and sequence smoothness) remains an important research challenge [29].

2.3 Adversarial Attacks on Handwriting Recognition

Adversarial attacks on handwriting recognition have mainly been studied in the context of offline handwriting recognition, where adversarial attacks on offline handwriting have been shown to be effective on offline signature verification [11, 13,22] and character recognition in various languages [14,17,36].

In contrast, research on adversarial attacks against online handwriting recognition is very limited. Yamashita et al. [40] showed that spatial perturbation-based attacks work on online handwriting and proposed a defense based on time

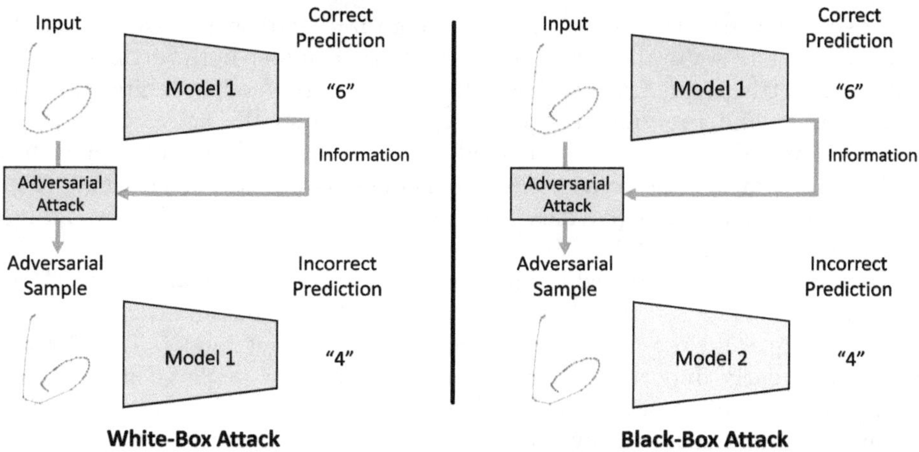

Fig. 2. Example of transferability of adversarial attacks. In the white-box setting (left), the adversarial attack has full knowledge of the target model. In the black-box setting (right), the adversarial sample is generated against one model and is used to attack an unseen target model.

series transformations. Lopresti and Raim [26] showed that a generative attack on online handwriting is possible.

3 Adversarial Attacks

3.1 Threat Model

We consider adversarial attacks against online handwriting recognition systems whose inputs are pen-tip trajectories of handwritten characters. In this study, each character is represented as a time series $\mathbf{X}$ with ground-truth label y:

$$\mathbf{X} = (\mathbf{x}_1, \ldots, \mathbf{x}_t, \ldots, \mathbf{x}_T),$$

where T is the sequence length and $\mathbf{x}_t \in \mathbb{R}^d$ denotes the pen-tip coordinate vector at time step t. In our experiments, we use two-dimensional coordinate sequences, i.e., $d = 2$. A recognition model f_θ with trained parameters θ maps $\mathbf{X}$ to class logits $f_\theta(\mathbf{X}) \in \mathbb{R}^C$. The predicted class label is

$$\hat{y} = \arg\max_c f_\theta(\mathbf{X})_c.$$

The goal of an adversarial attack is to find an adversarial example $\mathbf{X}_{\text{adv}}$ such that

$$\hat{y}_{\text{adv}} = \arg\max_c f_\theta(\mathbf{X}_{\text{adv}})_c, \quad \hat{y}_{\text{adv}} \neq y,$$

while the resulting handwriting remains visually plausible.

Adversarial attacks can be broadly categorized into white-box and black-box attacks, and we evaluate the proposed method under both settings. In the white-box setting, the attacker has full access to the model architecture and parameters θ and the gradients of $f_\theta(\mathbf{X})$. This allows the attacker to design $\mathbf{X}_{\text{adv}}$ so that it directly exploits the model's decision boundary. In the black-box setting, the attacker has no access to the model's parameters or gradients. This setting reflects realistic deployment conditions, as malicious attackers would not have access to the model directly.

Furthermore, as illustrated in Fig. 2, we consider one-shot transferable black-box attacks. In this setting, the adversarial attack method is permitted only a single attempt and receives no feedback from the target model. Transferability is particularly important for real-world scenarios, as deployed handwriting recognition systems typically expose only final predictions, preventing iterative probing or gradient access by an adversary [28]. Our black-box evaluation is therefore transfer-based and one-shot: adversarial samples are generated using a surrogate model and then submitted once to unseen target models. We do not consider query-based black-box attacks, where the adversary adaptively updates the input using repeated feedback from the target model.

3.2 Spatial Perturbation-Based Adversarial Attacks

Spatial perturbation-based adversarial attacks typically generate adversarial examples by adding small-magnitude noise to input representations. As a representative example, FGSM [9] creates an adversarial example $\mathbf{x}_{\text{fgsm}}$ by:

$$\mathbf{X}_{\text{fgsm}} = \mathbf{X} + \epsilon \cdot \text{sign}(\nabla_{\mathbf{X}} L(f_\theta(\mathbf{X}), y)) \,, \tag{1}$$

where $L(f_\theta(\mathbf{X}), y)$ is the loss function between the model prediction for input $\mathbf{X}$ and the label y, $\nabla_{\mathbf{X}} L(f_\theta(\mathbf{X}), y)$ is the gradient of the loss with respect to the input, $\text{sign}(\cdot)$ is the element-wise sign function, and ϵ is a hyperparameter controlling the perturbation magnitude. Iterative methods, such as BIM [20] and PGD [27], repeat this process given constraints.

4 Proposed Method

We propose a temporal editing attack that generates adversarial examples by modifying the time series in the time dimension rather than adding spatial noise. The key idea is to induce misclassification through insertion and deletion of time steps while preserving the overall shape and smoothness of the handwriting. As shown in Fig. 1, this design directly addresses the limitation of image-based adversarial perturbations, which often produce unnatural artifacts when applied to time series, including online handwriting.

4.1 Temporal Editing

In order to modify the time series in the time dimension, we define two basic editing operations, temporal insertion and temporal deletion.

Temporal Insertion. Given input time series $\mathbf{X}$ with $\mathbf{x}_t \in \mathbb{R}^d$, temporal insertion adds a new point between two adjacent points $(\mathbf{x}_i, \mathbf{x}_{i+1})$ with $i \in \{1, \ldots, T-1\}$. Interpolation is used to preserve local continuity and the new sequence becomes:

$$\mathbf{X}_{\text{ins}} = (\mathbf{x}_1, \ldots, \mathbf{x}_i, \frac{\mathbf{x}_i + \mathbf{x}_{i+1}}{2}, \mathbf{x}_{i+1} \ldots, \mathbf{x}_T), \tag{2}$$

increasing the number of time steps by one.

Temporal Deletion. Temporal deletion removes one time step from the sequence. Since removing an endpoint may severely distort the overall shape, we exclude endpoints from deletion candidates. Given a selected time index j, deletion removes the j-th element from $\mathbf{X}$ to form:

$$\mathbf{X}_{\text{del}} = (\mathbf{x}_1, \ldots, \mathbf{x}_{j-1}, \mathbf{x}_{j+1}, \ldots, \mathbf{x}_T). \tag{3}$$

The resulting $\mathbf{X}_{\text{del}}$ has one fewer time step than the original $\mathbf{X}$.

4.2 Adversarial Iterative Temporal Editing (AITE)

To create adversarial examples, an iterative approach is taken. In one step of AITE, we select an insertion position i and a deletion position j, and apply the edit to the current sequence to generate a new sequence. By applying insertion and deletion in equal numbers, the sequence length is always kept at T while locally modifying only the trajectory shape. This editing step is repeated until the predicted label no longer matches the true label y (attack success), or until the maximum number of iterations K is reached.

The following constraints are imposed to ensure stable editing:

- In each iteration, we perform one insertion and one deletion so that the sequence length remains T. This allows the generated adversarial examples to be used for any model without interpolation or similar adjustments.
- The element inserted in the same iteration is not allowed to be deleted. This constraint prevents iterations without changes.
- Endpoints are excluded from deletion to prevent excessive distortion of the trajectory.

4.3 Salience Estimation for Temporal Editing

In order to select the locations of the inserted and deleted elements, we estimate temporal salience using Grad-CAM [35]. Grad-CAM is a post-hoc explainability method that highlights regions of an input that are most influential for a model's prediction by leveraging gradient information from intermediate feature representations. Originally proposed for images, in this work, we adapt Grad-CAM to be used for time series by estimating the element-wise salience instead of pixel-wise salience.

Let $\mathbf{A} \in \mathbb{R}^{M \times T'}$ denote an intermediate temporal feature map, where M is the number of channels and T' is the temporal length of the feature map. We write $A^k \in \mathbb{R}^{T'}$ for the k-th channel of $\mathbf{A}$. Because temporal pooling layers in the CNN can change the temporal resolution, T' may differ from the input length T. Let $z_c = f_\theta(\mathbf{X})_c$ denote the pre-softmax logit for class c. In Grad-CAM, channel weights are defined as:

$$\alpha_k^{(c)} = \frac{1}{T'} \sum_{\tau=1}^{T'} \frac{\partial z_c}{\partial A_\tau^k}. \tag{4}$$

The feature-level temporal contribution is then obtained as:

$$\tilde{s}_\tau^{(c)} = \sum_{k=1}^{M} \alpha_k^{(c)} A_\tau^k, \quad \tau = 1, \dots, T'. \tag{5}$$

Unlike the original Grad-CAM formulation, AITE does not apply the ReLU non-linearity to $\tilde{s}_\tau^{(c)}$, so that both positive and negative contributions to the original class are preserved for temporal edit selection.

Since $\tilde{\mathbf{s}}^{(c)}$ has the temporal resolution of the intermediate feature map, we resize it to the input length before selecting edit positions. Specifically, we linearly interpolate

$$\tilde{\mathbf{s}}^{(c)} = (\tilde{s}_1^{(c)}, \dots, \tilde{s}_{T'}^{(c)}), \tag{6}$$

to length T, yielding the input-level salience sequence $\mathbf{S}^{(c)} = (s_1^{(c)}, \dots, s_T^{(c)})$.

When $T' = T$, this interpolation is the identity operation. The input-level salience sequence $\mathbf{S}^{(c)}$ assigns a score to each original input time step and is used to rank temporal locations for editing.

Temporal Insertion with Grad-CAM. For temporal insertion, we choose the interval between two adjacent input time steps that has the lowest average salience:

$$i^* = \arg \min_{i \in \{1,\dots,T-1\}} \frac{s_i^{(y)} + s_{i+1}^{(y)}}{2}, \tag{7}$$

where y is the original class. We insert the midpoint of the selected interval $(\mathbf{x}_i, \mathbf{x}_{i+1})$ and update the sequence according to Eq. 2.

Temporal Deletion with Grad-CAM. For temporal deletion, after excluding endpoints, the deletion index is selected as:

$$j^* = \arg \max_{t \in \{2,\dots,T-1\}} s_t^{(y)}. \tag{8}$$

5 Experimental Results

5.1 Datasets

To evaluate the proposed method, we used two publicly available online handwriting datasets. The first is Unipen [12], for which we used three subsets: numerical digits (Unipen 1A), uppercase letters (Unipen 1B), and lowercase letters (Unipen 1C). Each sample is represented as a sequence of 2D coordinates, and the sequence length was normalized to $T = 50$ time steps. Unipen 1A has 11,650 training and 1,300 test samples, while Unipen 1B and Unipen 1C have 11,063 training and 1,235 test samples.

The second is CASIA-OLHWDB 1.1 [24], an online handwritten Chinese character dataset with 3,755 classes from 300 writers. For CASIA, we remove stroke-end and character-end sentinel coordinates, concatenate the strokes into a single 2D point sequence, and normalize the coordinates to $[-1, 1]$ per sample and axis. Unlike Unipen, CASIA samples are kept at their original variable lengths. The test split contains 224,559 samples.

5.2 Architecture and Settings

For the experiments, four classifiers are used.

- **1D CNN-3**: A CNN with three convolutional blocks. Each block consists of a 1D convolution, BatchNorm [15], a Rectified Linear Unit (ReLU), and max pooling. The first block uses 64 filters with kernel size 3 and stride 1, while the subsequent two blocks use 128 filters. The network is followed by two fully connected layers: one hidden layer with 512 units and dropout with probability 0.5, and one output layer.
- **1D CNN-4**: This network has the same architecture as 1D CNN-3, except that it contains four convolutional blocks instead of three.
- **BLSTM-2**: To test transferability to a different network structure, we use a Bidirectional LSTM (BLSTM) [34]. Specifically, we use a two-layer BLSTM with 100 units per layer, following the recommendation of [32].
- **Transformer**: To test transferability to an attention-based architecture, we use an encoder-only Transformer [38]. The model consists of a linear input projection, positional encoding, and six Transformer encoder blocks with four attention heads each. The encoder output is averaged over time and passed through a small MLP head to produce the class logits.

For CASIA-OLHWDB, we use analogous classifiers adapted to variable-length inputs and 3,755 classes: a 1D CNN with masked global average pooling in place of 1D CNN-3, together with BLSTM-2 and Transformer counterparts. For CASIA-OLHWDB, we additionally train a second 1D CNN as a black-box target (channels 64→128→128→128, kernel size 3 throughout, 80 epochs, clean test accuracy 94.8%), providing a within-CNN-family transfer target with a different filter configuration from the surrogate (64→128→256→256).

In both white-box and black-box settings, adversarial samples are generated using the gradients or salience of a 1D CNN surrogate model: 1D CNN-3 for

Unipen and the analogous 1D CNN for CASIA-OLHWDB. In the white-box setting, the generated adversarial samples are used to attack the same surrogate model. In the black-box setting, they are transferred to different target models: 1D CNN-4, BLSTM-2, and Transformer for both datasets. We only consider one-shot black-box attacks, where the attacker receives no feedback from the target classifiers.

All models are trained using the Adam optimizer [19] with an initial learning rate of 10^{-3} and 10,000 training iterations.

5.3 Comparison Methods

To evaluate the proposed method, we compare to seven well-established adversarial attacks: four classical attacks (FGSM, BIM, PGD, CW) and three transfer-oriented variants (MI-FGSM, NI-FGSM, 1D-TI-MI-FGSM). The hyperparameters for each attack are as follows:

- **FGSM** [9]: $\epsilon = 0.1$ under the L_∞ constraint.
- **BIM** [20]: $\epsilon = 0.1$, L_∞, step size $\epsilon_{\text{iter}} = 0.001$, and $I = 600$ iterations.
- **PGD** [27]: same as BIM.
- **CW** [2]: L_2 constraint, $I = 600$ iterations, learning rate 5×10^{-3}, 7 binary search steps, initial constant 5×10^{-3}, confidence 0, and early stopping.
- **MI-FGSM** [5]: BIM with an accumulated momentum term over the input gradients to stabilize the update direction. Decay factor $\mu = 1.0$; remaining settings as BIM.
- **NI-FGSM** [23]: MI-FGSM with the gradient evaluated at a Nesterov look-ahead point before each momentum update. Settings as MI-FGSM.
- **1D-TI-MI-FGSM** [6]: MI-FGSM in which the per-step gradient is convolved with a 1D Gaussian kernel (size 7, $\sigma = 3.0$) along the time axis. Remaining settings as MI-FGSM.

Since inputs are normalized to $[-1, 1]$, adversarial examples are also clipped to $[-1, 1]$. For CASIA-OLHWDB, $\epsilon = 0.03$ and $I = 400$ are used to account for the larger number of classes; the CW binary search steps are reduced to 3.

5.4 Results

Table 1 reports recognition accuracy under adversarial attacks for both white-box and one-shot black-box settings across the three Unipen subsets and CASIA-OLHWDB. In the white-box scenario, gradient-based attacks, specifically BIM, PGD, and CW, drastically reduce classification accuracy on the surrogate, with CW achieving the lowest accuracy on every dataset (0.69–5.96% on Unipen and 0.71% on CASIA). MI-FGSM and NI-FGSM behave similarly to BIM in this setting, while 1D-TI-MI-FGSM is comparatively weaker on the surrogate because the smoothing of the input gradient sacrifices in-model optimality in exchange for transferability.

However, strong performance in the white-box setting does not necessarily carry over to the black-box scenario. Although CW is highly effective when

Table 1. Comparative evaluation of adversarial attacks and transferability. Accuracies (%, ↓) on the surrogate (white-box) and on three one-shot black-box targets. Best per column, excluding "No Attack", is in bold; "–" indicates the target is not used.

Dataset	Attack	White-Box	One-Shot Black-Box		
		1D CNN-3	1D CNN-4	BLSTM-2	Transformer
Unipen 1A	No Attack	98.9	98.8	98.8	98.2
	FGSM	73.8	82.8	95.6	92.9
	BIM	10.9	72.9	93.9	91.3
	PGD	12.4	74.2	95.1	92.0
	CW	**0.69**	93.9	98.2	97.1
	MI-FGSM	10.9	70.8	93.2	91.0
	NI-FGSM	9.1	70.5	93.7	91.0
	1D-TI-MI-FGSM	67.5	71.2	80.6	83.3
	AITE (Ours)	25.5	**58.2**	**70.2**	**68.8**
Unipen 1B	No Attack	98.1	97.5	97.7	95.5
	FGSM	68.7	71.5	92.6	88.3
	BIM	11.2	75.4	91.1	85.8
	PGD	13.7	75.8	92.5	86.2
	CW	**4.08**	95.6	97.1	93.4
	MI-FGSM	11.2	71.3	91.1	84.5
	NI-FGSM	9.0	73.4	91.6	85.0
	1D-TI-MI-FGSM	67.7	73.5	78.8	75.9
	AITE (Ours)	15.0	**61.8**	**60.3**	**65.0**
Unipen 1C	No Attack	97.1	97.7	96.5	94.8
	FGSM	64.5	79.4	90.8	85.2
	BIM	20.2	80.1	89.6	85.7
	PGD	30.4	85.8	93.6	89.1
	CW	**5.96**	96.9	96.1	93.8
	MI-FGSM	7.5	74.2	88.9	84.1
	NI-FGSM	6.2	75.6	88.7	85.3
	1D-TI-MI-FGSM	55.6	**68.9**	76.7	73.4
	AITE (Ours)	18.4	70.5	**59.7**	**62.0**
CASIA	No Attack	95.0	94.8	96.5	96.3
	FGSM	65.2	78.5	90.1	90.7
	BIM	53.5	75.3	89.6	90.4
	PGD	53.8	75.2	89.7	90.4
	CW	**0.71**	74.4	90.5	93.0
	MI-FGSM	54.3	75.4	89.6	90.4
	NI-FGSM	53.6	75.2	89.6	90.4
	1D-TI-MI-FGSM	85.3	89.9	94.2	93.5
	AITE (Ours)	29.2	**52.5**	**53.6**	**47.9**

applied directly to the surrogate, it transfers poorly to unseen targets: its impact on the 1D CNN-4, BLSTM-2, and Transformer columns is consistently the smallest among all attacks (e.g., the recognition accuracy on the Transformer target stays above 93% on every dataset). This is because gradient-based attacks rely on perturbations optimized with respect to the surrogate model's gradients, which may not align with the decision boundaries of an unseen target, and the emphasis of CW on minimizing the perturbation magnitude further limits its transferability. MI-FGSM and NI-FGSM, which were designed for improved black-box transferability, do improve over plain BIM/PGD in some cases (e.g., Unipen 1A and 1B on 1D CNN-4), but they still leave the BLSTM-2 and Transformer targets largely intact, with accuracies above 84% on every Unipen subset and CASIA target. 1D-TI-MI-FGSM is the strongest baseline overall and even slightly outperforms AITE on Unipen 1C against 1D CNN-4 (68.9 vs. 70.5). However, it remains substantially weaker than AITE on the BLSTM-2 and Transformer targets across all datasets.

In contrast, the proposed AITE achieves the lowest recognition accuracy in almost every one-shot black-box column, including transfer to the attention-based Transformer target that was never used during attack generation. On CASIA-OLHWDB, AITE reduces the 1D CNN-4, BLSTM-2, and Transformer accuracies to 52.5%, 53.6%, and 47.9%, respectively. Notably, AITE transfers more effectively to the within-CNN-family target (CNN-4) than across architectures, consistent with the intuition that CNN-to-CNN perturbations share more decision-boundary structure. The strongest additive baseline (1D-TI-MI-FGSM) only reduces the three black-box targets to 90.0%, 94.2%, and 93.5%, respectively. Because AITE selects insertion and deletion positions from salience rather than from raw input gradients, the resulting temporal edits target class-discriminative structure that tends to be shared across architectures, including CNN-to-BLSTM and CNN-to-Transformer transfers.

5.5 Visual Similarity

Attack success is usually quantified by the change in the model's prediction. However, for handwriting recognition, misclassification alone does not characterize the practical relevance of an attack: handwritten inputs are interpreted by humans, and visibly distorted adversarial samples are unlikely to represent plausible handwriting. We therefore also assess the visual similarity between original and adversarial samples.

To this end, we compare the visual similarity of the rendered characters. Table 2 reports quantitative results across all attacks. In this table, the online handwritten digits and characters are rendered into 64×64 grayscale images with black strokes and no points. For the visual similarity measures, we use the pixel-space L_2 distance and the Structural Similarity Index Measure (SSIM) between the original samples and the adversarial samples. L_2 measures the strict pixel-wise differences between the images, whereas SSIM emphasizes structural consistency, which better reflects human visual perception.

Table 2. Visual similarity between adversarial and original samples. The best value for each metric and dataset is in bold.

Dataset	Metric	FGSM	BIM	PGD	CW	MI-FGSM	NI-FGSM	1D-TI-MI-FGSM	AITE
Unipen 1A	$L_2 \downarrow$	22.0	20.9	21.0	**13.8**	21.8	21.4	22.8	14.3
	SSIM $\uparrow$	0.550	0.577	0.574	**0.760**	0.554	0.563	0.518	0.746
Unipen 1B	$L_2 \downarrow$	23.5	22.4	22.5	15.0	23.2	22.9	24.3	**11.7**
	SSIM $\uparrow$	0.497	0.525	0.524	0.725	0.502	0.512	0.466	**0.808**
Unipen 1C	$L_2 \downarrow$	22.1	19.2	19.4	13.1	21.7	21.4	22.7	**11.7**
	SSIM $\uparrow$	0.551	0.627	0.624	0.775	0.558	0.567	0.527	**0.809**
CASIA	$L_2 \downarrow$	17.8	17.6	17.6	**11.1**	17.6	17.6	18.3	15.9
	SSIM $\uparrow$	0.585	0.593	0.594	**0.812**	0.592	0.593	0.564	0.618

FGSM, BIM, PGD, MI-FGSM, NI-FGSM, and 1D-TI-MI-FGSM consistently produce large L_2 distortions and low SSIM scores across all datasets. As shown in Fig. 3, this reflects the large amount of jittering in the generated strokes of the characters. In the rendered examples, these attacks introduce visible jitter and local stroke irregularities. The transfer-oriented variants (MI-FGSM, NI-FGSM, and 1D-TI-MI-FGSM) do not qualitatively change this behavior: they share the additive L_∞ formulation of BIM/PGD, so the rendered trajectories exhibit similar jittering and remain easily detectable. In contrast, the perturbations in CW are minimal, but as discussed above, this adversely affects black-box transferability because the minimal perturbations only exploit surrogate-specific decision boundaries.

In contrast, AITE preserves the overall structure of the handwriting while still being effective under black-box transfer. Although online handwriting is represented as a time series, temporal-domain edits such as point insertion or deletion alter the temporal structure of the sequence while leaving the rendered spatial trace largely unchanged, and are typically less perceptible than the spatial jitter introduced by additive perturbations.

5.6 Gradient Versus Salience for Temporal Edit Selection

As a baseline ablation, we replace the Grad-CAM salience with gradient-based importance: $s_t^{grad} = \|\nabla_{\mathbf{x}_t} \mathcal{L}(f_\theta(\mathbf{X}), y)\|_2$, where larger values indicate higher sensitivity of the loss to time step t.

A comparison of the results is shown in Table 3. In the results, the gradient-based selection creates more similar adversarial samples but has a smaller effect on the accuracy. The reason for this is because the edited time steps tend to be adjacent to each other in salient regions. Conversely, time steps selected by gradient magnitudes tend to be distributed more equally over the character, allowing for less disruptive edits. However, both the white-box and black-box attacks with gradient-based selection are not as effective as salience-based selection. This suggests salience is more suitable than gradients for temporal editing.

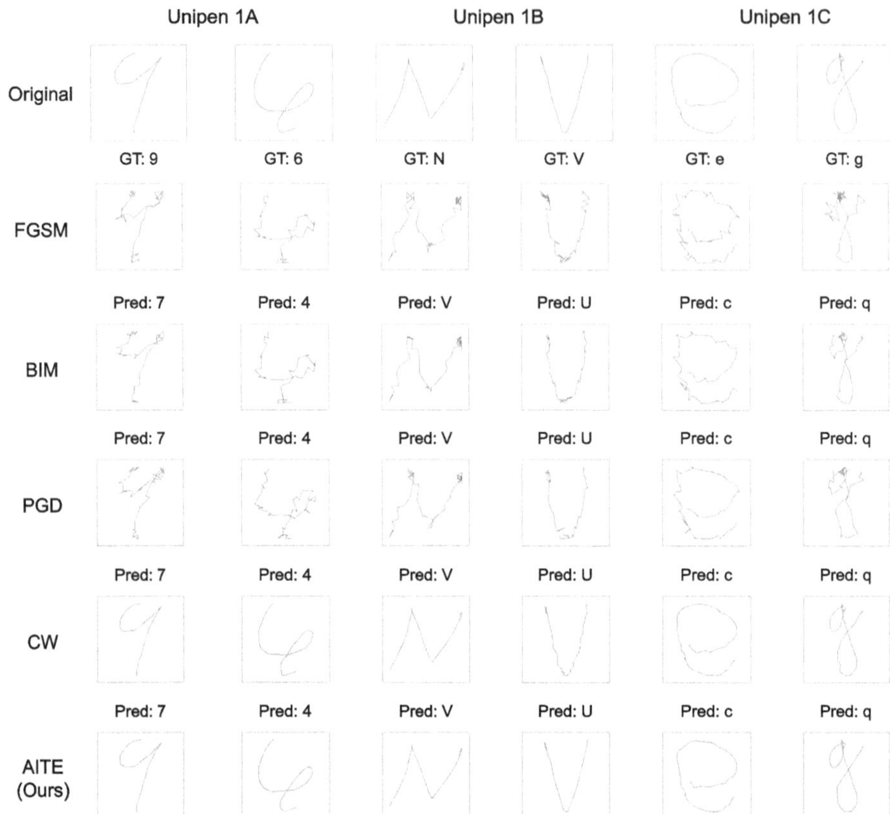

Fig. 3. Qualitative comparison of adversarial handwriting trajectories on Unipen (1A/1B/1C) for representative attacks. Rows show the original sample and adversarial samples generated by PGD, CW, and the proposed AITE. Ground-truth labels (GT) and model predictions (Pred) are shown below each example.

6 Examination of the Temporal Edits

6.1 Single-Iteration Adversarial Samples

An interesting observation from our experiments is that some handwriting samples can be misclassified after only a single iteration of temporal editing. In other words, the class of a character is able to be changed with the removal of a single time step and the addition of a single time step. As demonstrated in Fig. 4, these cases are nearly identical to the original samples. These cases are of interest because they show samples whose correct classification relies on a single time step.

Table 3. Comparative evaluation of AITE selection strategies and transferability. For the white-box experiment, the adversarial samples are generated and tested on 1D CNN-3. For the black-box experiments, the adversarial examples are generated from 1D CNN-3 and tested on 1D CNN-4 and BLSTM-2 models. The best of each dataset is in bold.

Dataset	Selection	Visual Similarity		White-Box	One-Shot Black-Box	
				1D CNN-3	1D CNN-4	BLSTM-2
		L_2 ↓	SSIM ↑	Acc (%) ↓	Acc (%) ↓	Acc (%) ↓
Unipen 1A	Gradient	**9.92**	**0.849**	64.6	76.8	88.8
	Salience	14.3	0.746	**25.5**	**58.2**	**70.2**
Unipen 1B	Gradient	**9.81**	**0.846**	50.1	64.8	72.8
	Salience	11.7	0.808	**15.0**	**61.8**	**60.3**
Unipen 1C	Gradient	**9.46**	**0.855**	53.9	82.1	81.6
	Salience	11.7	0.809	**18.4**	**70.5**	**59.7**

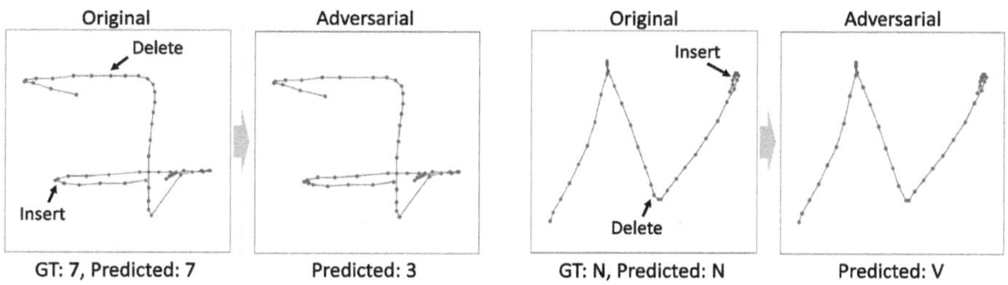

Fig. 4. Example of adversarial samples that only required a single iteration of AITE to change the class. The temporal insertion and deletion are indicated with arrows.

6.2 Class Sensitivity to Temporal Edits

Figure 5 compares the confusion matrices on Unipen 1A under AITE, PGD, and CW. PGD and CW, being gradient-based, produce similar confusion patterns, whereas AITE differs substantially: for instance, "2"→"4" under AITE versus "2"→"7" under the gradient methods, suggesting that "2"s are temporally closer to "4"s but spatially closer to "7"s. "9"s also tend to transition to "1"s or "7"s under AITE, while a few classes behave similarly across methods (e.g., "0"→"8" and "8"→"5").

The matrices also reveal class-wise susceptibility: "1" and "5" are the hardest digits to attack with AITE because their single-stroke shapes leave little room for class change without spatial distortion, while the more complex "6", "8", and "9" are the easiest. This is consistent with the stronger attack effectiveness on the more complex Unipen 1B and 1C subsets in Table 1.

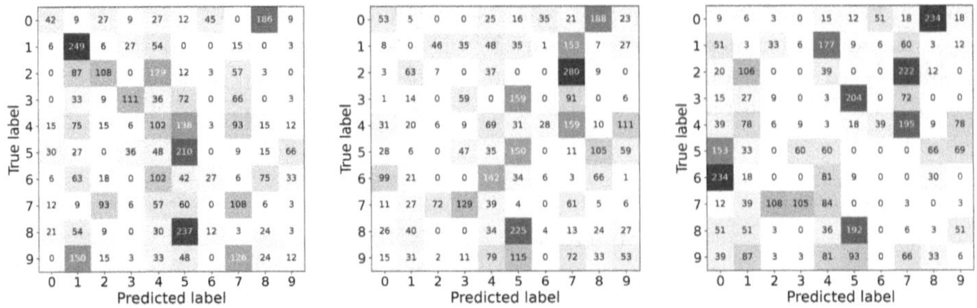

(a) AITE (Ours) (b) PGD (c) CW

Fig. 5. Comparison of the confusion matrices on Unipen 1A under different attacks. Darker cells indicate larger counts.

7 Conclusion

This paper examined adversarial attacks on online handwriting recognition from a temporal perspective. By formulating adversarial example generation as a sequence editing problem, the proposed approach modifies handwriting trajectories through point insertion and deletion guided by temporal salience, rather than through additive spatial perturbations. Experimental evaluations under both white-box and one-shot black-box settings demonstrated that temporal editing produces adversarial examples that preserve the structure and smoothness of natural handwritten characters while remaining effective across different model architectures. Future work will extend this preliminary study to word- and sentence-level online handwriting recognition. Human evaluation should also be conducted to assess the perceived naturalness of adversarial handwriting, and defenses against temporal editing attacks should be investigated.

Acknowledgements. This research was partially funded by JSPS Grant Number 25K22845 and JST CRONOS-JPMJCS24K4.

References

1. Carlini, N., et al.: Hidden voice commands. In: USENIX. USENIX Association (2016)
2. Carlini, N., Wagner, D.: Towards evaluating the robustness of neural networks. In: SP, pp. 39–57. IEEE (2017)
3. Costa, J.C., Roxo, T., Proença, H., Inacio, P.R.M.: How deep learning sees the world: a survey on adversarial attacks & defenses. IEEE Access **12**, 61113–61136 (2024)
4. Ding, D., Zhang, M., Feng, F., Huang, Y., Jiang, E., Yang, M.: Black-box adversarial attack on time series classification. In: AAAI Conference on Artificial Intelligence (AAAI), vol. 37, pp. 7358–7368 (2023)

5. Dong, Y., et al.: Boosting adversarial attacks with momentum. In: IEEE/CVF Conference on Computer Vision and Pattern Recognition (CVPR) (2018)
6. Dong, Y., Pang, T., Su, H., Zhu, J.: Evading defenses to transferable adversarial examples by translation-invariant attacks. In: IEEE/CVF Conference on Computer Vision and Pattern Recognition (CVPR) (2019)
7. Fawaz, H.I., Forestier, G., Weber, J., Idoumghar, L., Muller, P.A.: Adversarial attacks on deep neural networks for time series classification. In: Int.ernational Joint Conference on Neural Networking (IJCNN). IEEE (2019)
8. Ghosh, T., Sen, S., Obaidullah, S.M., Santosh, K., Roy, K., Pal, U.: Advances in online handwritten recognition in the last decades. Comput. Sci. Rev. **46** (2022)
9. Goodfellow, I.J., Shlens, J., Szegedy, C.: Explaining and harnessing adversarial examples. In: International Conference on Learning Representation (ICLR) (2015)
10. Graves, A., Liwicki, M., Fernández, S., Bertolami, R., Bunke, H., Schmidhuber, J.: A novel connectionist system for unconstrained handwriting recognition. IEEE Trans. Pattern Anal. Mach. Intell. (TPAMI) **31**(5), 855–868 (2009)
11. Guo, Z., Li, W., Qian, Y., Arandjelovic, O., Fang, L.: A white-box false positive adversarial attack method on contrastive loss based offline handwritten signature verification models. In: International Conference on Artificial Intelligence and Statistics (AISTATS), pp. 901–909 (2024)
12. Guyon, I.M., Schomaker, L., Plamondon, R., Liberman, M., Janet, S.: Unipen project of on-line data exchange and recognizer benchmarks. In: International Conference on Pattern Recognition (ICPR), vol. 2, pp. 29–33 (1994)
13. Hafemann, L.G., Sabourin, R., Oliveira, L.S.: Characterizing and evaluating adversarial examples for offline handwritten signature verification. IEEE Trans. Inf. Forensics Secur. **14**(8), 2153–2166 (2019)
14. Huynh, T.B., Phan, Q.M., Duong, V.H., Luong, N.H.: Adversarial robustness evaluation of a Vietnamese handwriting OCR system. In: International Conference on Multimedia Analysis and Pattern Recognition (MAPR), pp. 1–6 (2025)
15. Ioffe, S., Szegedy, C.: Batch normalization: accelerating deep network training by reducing internal covariate shift. In: International Conference on Machine Learning, pp. 448–456 (2015)
16. Iwana, B.K., Uchida, S.: Time series classification using local distance-based features in multi-modal fusion networks. Pattern Recognit. **97**, 107024 (2020)
17. Jiang, G., Qian, Z., Wang, Q.F., Wei, Y., Huang, K.: Adversarial attack and defence on handwritten Chinese character recognition. J. Phys: Conf. Ser. **2278**, 012023 (2022)
18. Karim, F., Majumdar, S., Darabi, H.: Adversarial attacks on time series. IEEE Trans. Pattern Anal. Mach. Intell. (TPAMI) **43**(10), 3309–3320 (2021)
19. Kingma, D.P., Ba, J.: Adam: a method for stochastic optimization. In: International Conference on Learning Representation (ICLR) (2015)
20. Kurakin, A., Goodfellow, I., Bengio, S.: Adversarial examples in the physical world. In: Workshop International Conference on Learning Representation (ICLR) (2017)
21. Lang, K.J., Waibel, A.H., Hinton, G.E.: A time-delay neural network architecture for isolated word recognition. Neural Netw. **3**(1), 23–43 (1990)
22. Li, H., Li, H., Zhang, H., Yuan, W.: Black-box attack against handwritten signature verification with region-restricted adversarial perturbations. Pattern Recognit. **111**, 107689 (2021)
23. Lin, J., Song, C., He, K., Wang, L., Hopcroft, J.E.: Nesterov accelerated gradient and scale invariance for adversarial attacks. In: International Conference on Learning Representation (ICLR) (2020)

24. Liu, C.L., Yin, F., Wang, D.H., Wang, Q.F.: CASIA online and offline Chinese handwriting databases. In: International Conference on Document Analysis and Recognition (ICDAR) (2011)
25. Liu, C., et al.: Col-OLHTR: a novel framework for multimodal online handwritten text recognition. In: IEEE International Conference on Acoustics, Speech and Signal Processing, pp. 1–5 (2025)
26. Lopresti, D.P., Raim, J.D.: The effectiveness of generative attacks on an online handwriting biometric. In: International Conference on Audio Video-Based Biometric Person Authentication (AVBPA), pp. 1090–1099 (2005)
27. Madry, A., Makelov, A., Schmidt, L., Tsipras, D., Vladu, A.: Towards deep learning models resistant to adversarial attacks. In: International Conference on Learning Representation (ICLR) (2018)
28. Papernot, N., McDaniel, P., Goodfellow, I.: Transferability in machine learning: from phenomena to black-box attacks using adversarial samples. arXiv preprint arXiv:1605.07277 (2016)
29. Pialla, G., et al.: Smooth perturbations for time series adversarial attacks. In: Pacific-Asia Conference on Knowledge Discovery and Data Mining (PAKDD), pp. 485–496. Springer (2022)
30. Qu, X., Wu, Z.: End-to-end multi-scale attention convolutional recurrent network for online handwritten Chinese text recognition. Expert Syst. Appl. 127626 (2025)
31. Rathore, P., Basak, A., Nistala, S.H., Runkana, V.: Untargeted, targeted and universal adversarial attacks and defenses on time series. In: International Joint Conference on Neural Networks (IJCNN), pp. 1–8 (2020)
32. Reimers, N., Gurevych, I.: Optimal hyperparameters for deep LSTM-networks for sequence labeling tasks. arXiv preprint arXiv:1707.06799 (2017)
33. Rumelhart, D.E., Hinton, G.E., Williams, R.J.: Learning internal representations by error propagation. In: Anderson, J.A., Rosenfeld, E. (eds.) Neurocomputing: Foundations of Research, pp. 673–695. MIT Press (1988)
34. Schuster, M., Paliwal, K.K.: Bidirectional recurrent neural networks. IEEE Trans. Signal Process. **45**(11), 2673–2681 (1997)
35. Selvaraju, R.R., Cogswell, M., Das, A., Vedantam, R., Parikh, D., Batra, D.: Grad-CAM: visual explanations from deep networks via gradient-based localization. In: International Conference on Computer Vision (ICCV), pp. 618–626. IEEE (2017)
36. Shi, L., Wu, H., Ding, X., Xu, H., Pan, S.: A generative adversarial network-based client-level handwriting forgery attack in federated learning scenario. Expert. Syst. **42**(2), e13676 (2025)
37. Szegedy, C., et al.: Intriguing properties of neural networks. arXiv preprint arXiv:1312.6199 (2013)
38. Vaswani, A., et al.: Attention is all you need. In: Advances in Neural Information Processing Systems (NeurIPS), vol. 30 (2017)
39. Xu, H., et al.: Adversarial attacks and defenses in images, graphs and text: a review. Int. J. Autom. Comput. **17**(2), 151–178 (2020)
40. Yamashita, Y., Iwana, B.K.: Test time augmentation as a defense against adversarial attacks on online handwriting. In: International Conference on Document Analysis and Recognition (ICDAR), pp, pp. 156–172 (2024)
41. Yang, W., Yuan, J., Wang, X., Zhao, P.: TSadv: black-box adversarial attack on time series with local perturbations. Eng. Appl. Artif. Intell. **114**, 105218 (2022)
42. Zheng, M., Yan, X., Zhu, Z., Chen, H., Wu, B.: BlackboxBench: a comprehensive benchmark of black-box adversarial attacks. IEEE Trans. Pattern Anal. Mach. Intell. (TPAMI) **47**(9), 7867–7885 (2025)

Document Image Processing

P-HTG: One-Shot Handwritten Text Generation via Prototype-Guided Adaptive Gated Fusion

Heng Wang[1,2], Yiming Wang[1,2], and Hongxi Wei[1,2](✉)

[1] School of Computer Science, Inner Mongolia University, Hohhot, China
cswhx@imu.edu.cn

[2] National and Local Joint Engineering Research Center of Intelligent Information Processing Technology for Mongolian, Hohhot, China

Abstract. Handwritten text generation (HTG) aims to synthesize realistic handwritten text images conditioned on input textual content and a reference style. However, achieving high-quality generation with only a single reference image (one-shot) remains a significant challenge. Existing models often struggle to balance global style consistency (e.g., slant, stroke width) with fine-grained local character details (e.g., ligatures, glyph shapes), leading to unstable or illegible outputs. In this paper, we introduce P-HTG, a novel framework for one-shot handwritten text generation that addresses this dilemma via a Prototype-Guided Adaptive Gated Fusion (AGSF) mechanism. Instead of relying solely on global style vectors, P-HTG introduces a hierarchical dual-stream strategy. We employ a Teacher-Student distillation paradigm where the student network learns to align its generated features with robust style prototypes derived from a many-shot teacher. Crucially, the proposed AGSF module dynamically integrates these stable global prototypes with semantic-aware local features through a learnable gating mechanism. This allows the model to adaptively weigh the reliance on global stability versus local diversity based on the generation context, ensuring both structural accuracy and stylistic coherence. Experimental results on the IAM dataset demonstrate that P-HTG significantly outperforms state-of-the-art methods in terms of visual fidelity and style consistency under challenging one-shot settings.

Keywords: Handwritten Text Generation · One-Shot Learning · Prototypical Representation · Adaptive Gated Fusion

1 Introduction

Handwritten text generation (HTG) is a crucial task in image synthesis, aiming to create realistic handwritten text images that not only match given textual content but also emulate unique writing styles. The ability to generate authentic-looking handwriting from a small number of examples is highly valuable in

G. A. Fink et al. (Eds.): ICDAR 2026, LNCS 16972, pp. 147–164, 2027.
https://doi.org/10.1007/978-3-032-36023-6_9

various applications, including assistive technologies for individuals with motor impairments, data augmentation for training robust Optical Character Recognition (OCR) systems [27], and historical document restoration.

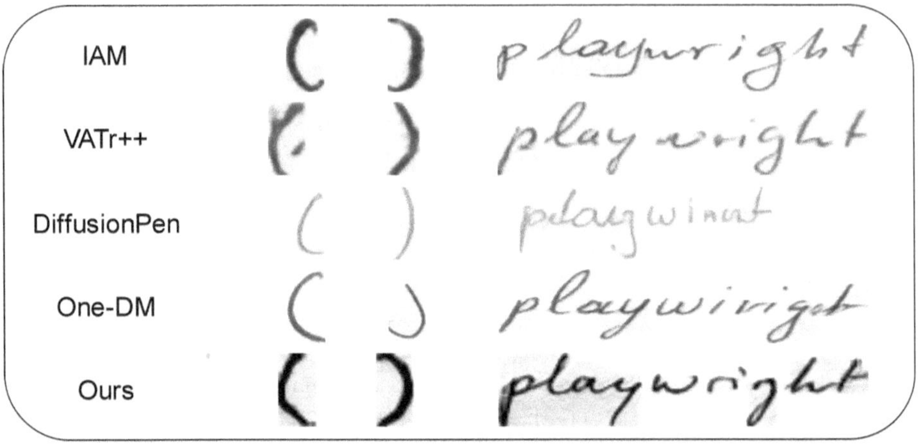

Fig. 1. Qualitative comparison with Transformer-based model (VATr++ [25]) and two diffusion-based models (DiffusionPen [21], One-DM [5]) on single-character and long-word generation. Transformer methods often show unstable local stroke continuity, while diffusion models exhibit structural drift on long sequences due to iterative denoising. Our method maintains smooth strokes and consistent structure across varying lengths.

However, high-fidelity handwritten text generation, especially in the one-shot setting where only a single style reference is available, remains highly challenging. First, the model must capture and reproduce intricate details of an individual's handwriting from only one reference image, such as stroke thickness, slant, character shapes, and ligatures. Second, existing methods often fail to balance global style consistency with fine-grained local character details. Traditional methods typically favor global style vectors, which often leads to the loss of specific glyph shapes, while purely local attention mechanisms can introduce artifacts when the reference image is limited or noisy. Third, the sequential nature of handwriting requires the generator to ensure smooth stroke transitions and coherent line quality; yet, effectively bridging the gap between global stylistic prototypes and local structural details remains an open problem.

To address these challenges, we propose P-HTG, a Prototype-Guided Handwritten Text Generation framework suitable for one-shot scenarios. We adopt a meta-learning inspired episodic training paradigm. To enable the network to extract style prototypes from a single image, we design a dual-network prototype guidance mechanism. Existing one-shot generative models, such as One-DM [5], typically rely on sampling and training from a single reference image to represent the writer's style. While this approach facilitates one-shot synthesis, it inherently suffers from "stylistic bias," where the model overfits to the specific accidental features (e.g., local noise or temporary stroke artifacts) of that single sample rather than capturing the writer's underlying, invariant stylistic characteristics.

In contrast, our approach utilizes a teacher-student paradigm: the student network processes only a single reference image for actual generation, while the teacher network (Prototype Teacher) leverages a query set containing multiple images from the same writer during training to provide a "gold standard" supervisory signal. By distilling knowledge from this many-shot teacher, the student network is guided to suppress sample-specific noise and instead learn a more robust, generalizable style representation that transcends the limitations of a single observation. A key novelty of our framework is the **Adaptive Gated Style-Fusion (AGSF)** module. Unlike standard fusion methods that treat global and local features indiscriminately, AGSF dynamically integrates global style prototypes with semantic-aware local features through a learnable gating mechanism. This allows the model to adaptively weigh the reliance on global stability versus local diversity based on the generation context, ensuring both structural accuracy and stylistic coherence.

Furthermore, we incorporate a Transformer-based decoder architecture to refine the fused features. By leveraging cross-attention mechanisms, the decoder effectively models the long-range dependencies and fine-grained spatial dynamics of handwritten sequences. This design ensures that the generated text maintains consistent stroke transitions and coherent structural layouts, even under the constraints of a single-shot reference. Figure 1 demonstrates the strong generalization and generation capabilities of P-HTG across various scenarios.

Our key contributions are summarized as follows:

- **Prototype-Guided One-Shot Generation:** We propose P-HTG, combining a dual-network prototype guidance mechanism with a meta-learning inspired episodic training paradigm, enabling the model to extract robust and generalizable style prototypes from a single reference image.
- **Adaptive Gated Style-Fusion (AGSF):** We introduce the AGSF module, which dynamically balances global style prototypes and local character details through a learnable gating mechanism, effectively addressing the dilemma between stylistic consistency and structural fidelity in one-shot scenarios.
- **State-of-the-Art Performance:** Through extensive experiments, we demonstrate that P-HTG significantly outperforms state-of-the-art methods on benchmark datasets like IAM under the challenging one-shot setting. The source code for our model and experiments is provided anonymously at https://github.com/wh807088026/PHTG.

2 Related Work

Handwritten Text Generation (HTG) is generally categorized into online and offline methods. While online methods [1,12] leverage dynamic trajectory data, our work P-HTG focuses on the more challenging *offline* setting, which synthesizes text images from static visual samples. The architectural backbone of offline HTG has evolved from early Generative Adversarial Networks (GANs) to recent Transformer, Diffusion, and State-Space paradigms.

Early breakthroughs predominantly relied on GANs [11]. Alonso et al. [2] pioneered the use of GANs for synthesizing fixed-size handwritten words. This was rapidly advanced by ScrabbleGAN [8], which enabled variable-length text generation, and GANwriting [14], which explored disentangling content and style for multi-writer synthesis. Subsequent frameworks like the HiGAN series [9,10] and SLOGAN [18] integrated contextual losses and local patch mechanisms to enhance style consistency. However, these convolutional backbones often struggle to capture the long-range structural dependencies essential for natural handwriting.

Inspired by the success of self-attention, HWT [3] successfully applied Transformers to leverage powerful sequence modeling. Recently, VATr [22] and VATr++ [25] introduced visual archetypes to guide the generation of unseen characters. Meanwhile, autoregressive paradigms like Emuru [24] and Eruku [30] harness the T5 architecture to enable zero-shot styled text generation. However, because these autoregressive models are primarily optimized for typographic and machine-printed fonts, they show limited performance when modeling the high intra-writer variability and intricate continuous stroke connections (ligatures) of handwriting.

Parallel to this, Diffusion-based models [20,31] have pushed the boundaries of visual fidelity. For instance, DiffusionPen [21] improved generation from few style examples, and One-DM [5] achieved competitive one-shot results by sampling from a single reference. Nevertheless, diffusion approaches typically suffer from high computational costs and may overfit to accidental artifacts of the lone reference. Most recently, GL-GAN [29] explores efficient sequence modeling via a Mamba-based encoder to perceive global style features.

Despite these advancements, most existing architectures are inherently designed for many-shot scenarios [2,9,10,14,20,21,26,28,29,31], relying on multiple reference samples to ensure stylistic stability. When constrained to a single-shot reference, these methods often fail to extract a sufficiently representative style descriptor, leading to distorted glyphs or identity loss. Even for specialized one-shot models like One-DM, the reliance on a single observation can lead to overfitting on accidental artifacts. To address this, we propose P-HTG, which replaces the conventional rigid fusion with an Adaptive Gated Style-Fusion (AGSF) module within a Transformer-based decoder. By dynamically adjudicating between stable global prototypes and semantic-aware local details, AGSF effectively bridges the gap between stylistic consistency and structural fidelity under extreme data constraints. Furthermore, we employ a Teacher-Student distillation paradigm, leveraging a many-shot teacher to guide the one-shot student in learning a robust, noise-resilient style representation that transcends the limitations of a single observation.

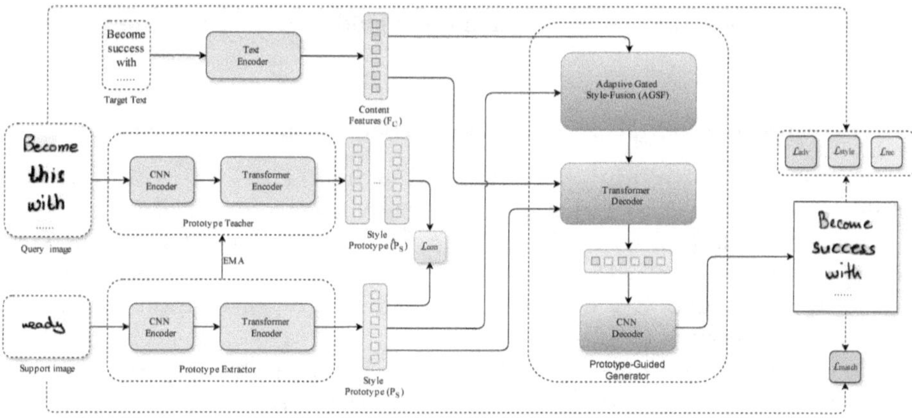

Fig. 2. Overview of the proposed P-HTG framework. The model adopts a Teacher-Student distillation paradigm, where a student network learns to extract robust style prototypes from a single support image under the guidance of a many-shot teacher. The core of the generator is the Adaptive Gated Style-Fusion (AGSF) module, which dynamically balances global stylistic stability and local structural details. The fused features are then processed by a Transformer-based decoder and a CNN-based reconstruction head to synthesize high-fidelity handwritten text.

3 Proposed Method

We propose P-HTG, a Prototype-Guided Handwritten Text Generation framework designed for one-shot scenarios. The framework employs a dual-network prototype guidance mechanism, including a student network (Prototype Extractor) and a teacher network (Prototype Teacher), enabling robust style representation extraction from a single reference image.

As illustrated in Fig. 2, the overall architecture of P-HTG consists of the Prototype Extractor, the Prototype Teacher, a Text Encoder, and the Prototype-Guided Image Generator. During training, the teacher network observes multiple images from the query set of the same writer, providing supervision to the student network and guiding it to learn stable and generalizable style representations through an EMA-based update and a consistency loss $\mathcal{L}_{con}$. After sufficient training, the teacher network is discarded at test time, and only the Prototype Extractor is utilized to extract style prototypes from a single support image for one-shot handwritten text generation.

Our method is grounded in a meta-learning inspired episodic training paradigm. In each training episode, we randomly sample a writer and construct a support set S containing a single reference image for style extraction, while using multiple images from the same writer to form a query set Q for content and style supervision. All input images are resized to a height of 32 pixels while maintaining the aspect ratio. Unlike classic Prototypical Networks which use prototypes for distance-based classification, in our framework, the style prototype acts as a rich conditional sequence P_S that dynamically guides the generator in the image synthesis process.

3.1 Prototype Extractor, Prototype Teacher, and Text Encoder

Prototype Extractor and Prototype Teacher. The Prototype Extractor distills the stylistic essence from a single reference image into a compact, powerful sequence: the style prototype $P_S \in \mathbb{R}^{S \times B \times d}$. Its architecture follows the style encoder in VATr: a ResNet18 backbone followed by a Transformer encoder. The Transformer consists of $L = 3$ layers, each containing 8 self-attention heads and a feed-forward network (MLP).

In addition, we introduce the Prototype Teacher, which has the same architecture as the Prototype Extractor but does not share parameters. During training, the Prototype Teacher is updated via an exponential moving average (EMA) of the Prototype Extractor's weights to provide stable supervision for the student network.

The process begins with a support image of batched dimensions $[B, C, H, W]$. This image is first processed by the ResNet18 backbone to extract low-level spatial features. The resulting feature map is flattened and projected into a sequence of L_s tokens, where L_s depends on the spatial dimensions of the CNN output. Each token has an embedding dimension $d = 512$. The sequence of size $[L_s, B, d]$ is then fed into the Transformer encoder to capture global dependencies among tokens and encode holistic style information. The final style prototype sequence P_S is obtained from the Transformer's output.

Text Encoder. The Text Encoder converts character sequences into a feature representation F_C suitable for the generator. Given a batch of integer-tokenized texts of size $[B, L_{\text{text}}]$, each discrete token is mapped to a continuous embedding of dimension $d = 512$ using a learnable embedding layer. The embeddings are further refined via an MLP. To be compatible with the decoder, the final tensor F_C is permuted to $[L_{\text{text}}, B, d]$.

3.2 Prototype-Guided Image Generator

The generator synthesizes the final image, intricately conditioned by the style prototype P_S and the content features F_C. It adopts a hierarchical serial structure comprising the Adaptive Gated Style-Fusion (AGSF) module, a Transformer-based decoder, and a CNN reconstruction head.

Adaptive Gated Style-Fusion (AGSF) Module. The AGSF module (Fig. 3) is the core innovation of our generator, designed to adaptively fuse global stylistic stability with local character nuances. Given the style sequence P_S and text features F_C, the module extracts two distinct streams:

- **Local Style Retrieval (P_{local}):** We employ a Cross-Attention mechanism where F_C serves as the query and P_S serves as both key and value. This allows the model to retrieve character-specific style details according to the target text.

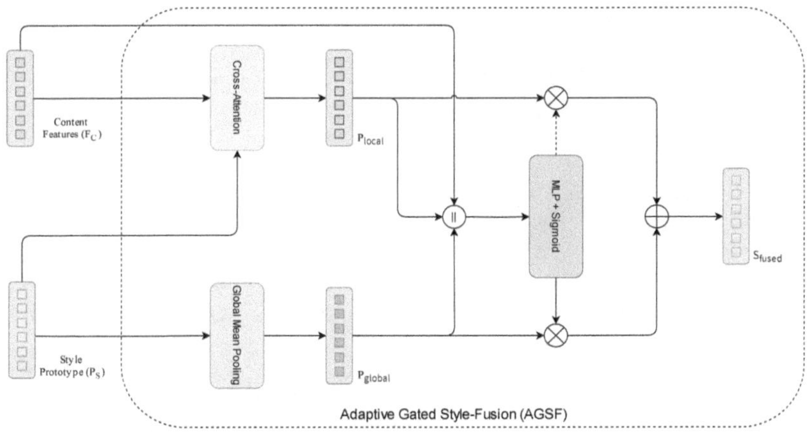

Fig. 3. Detailed architecture of the Adaptive Gated Style-Fusion (AGSF) module. The module utilizes a gating network to dynamically adjudicate the contribution of the global prototype P_{global} and the local style features P_{local}, yielding a fused stylistic representation S_{fused} optimized for one-shot generation.

- **Global Prototype Extraction (P_{global}):** P_{global} is computed via global mean pooling across the stylistic tokens in P_S, representing the writer's invariant stylistic signature.

To achieve dynamic integration, a Gating Network (MLP + Sigmoid) processes the concatenated features of F_C, P_{local}, and P_{global} to produce a learnable gate value $\alpha \in [0, 1]$. The final fused style S_{fused} is defined as:

$$S_{fused} = \alpha \cdot P_{local} + (1 - \alpha) \cdot P_{global} \tag{1}$$

Transformer-Based Sequential Decoder. The fused style S_{fused} is fed into the Transformer Decoder as the initial target sequence (tgt). The decoder consists of 8 layers with 3 self-attention heads and a hidden dimension of 512. Within each layer, S_{fused} undergoes self-attention for sequence consistency and cross-attention with the original style sequence P_S for fine-grained structural refinement. This stage effectively models the long-range dependencies and pen dynamics of handwritten text.

CNN Decoder and Reconstruction. The refined hidden states from the Transformer decoder are reshaped and projected into a 2D latent feature map. The CNN Decoder, following standard designs with residual blocks and three upsampling stages, performs pixel-level reconstruction. A final tanh activation layer generates the output image of size $[B, C, 32, W_{\text{out}}]$, where W_{out} is dynamically determined by the text length. The framework is trained end-to-end using a joint objective including adversarial ($\mathcal{L}_{adv}$), style ($\mathcal{L}_{style}$), reconstruction ($\mathcal{L}_{rec}$), and matching ($\mathcal{L}_{match}$) losses.

3.3 Training Objective

The P-HTG framework is trained end-to-end using a composite objective function to ensure stylistic fidelity, content legibility, and structural consistency. The total loss is defined as a weighted sum:

$$L_{\text{total}} = \lambda_{adv} L_{adv} + \lambda_{style} L_{style} + \lambda_{rec} L_{rec} + \lambda_{match} L_{match} + \lambda_{con} L_{con} \quad (2)$$

where the weights are empirically set to $\lambda_{adv} = 1.0$, $\lambda_{style} = 0.7$, $\lambda_{match} = 1.0$, and $\lambda_{con} = 1.0$. The content loss weight λ_{rec} is dynamically adjusted to 1.0 for in-vocabulary words and 0.5 for out-of-vocabulary words to balance recognition accuracy and generalization.

Adversarial and Matching Losses. To synthesize realistic handwritten images, we employ a Projection Discriminator D trained jointly with the generator. The discriminator extracts a feature embedding D_{embed} through a CNN backbone. The adversarial loss L_{adv} utilizes a hinge loss objective:

$$\begin{aligned} L_{adv} = & \frac{1}{q} \sum_{i=1}^{q} \mathbb{E}_{x_{q,i}} [\max(0, 1 - D_{score}(x_{q,i}))] \\ & + \frac{1}{q} \sum_{i=1}^{q} \mathbb{E}_{(x_s, y_{q,i})} [\max(0, 1 + D_{score}(G(x_s, y_{q,i})))] \end{aligned} \quad (3)$$

where x_s is the support image, $x_{q,i}$ is the i-th real query image, and $y_{q,i}$ is the text label. To further align the generated distribution with the target style, the feature matching loss L_{match} minimizes the distance between the discriminator's output and the style prototype P_S:

$$L_{match} = \frac{1}{q} \sum_{i=1}^{q} \mathbb{E}_{(x_s, y_{q,i})} [1 - \text{sim}(D_{embed}(G(x_s, y_{q,i})), P_S)] \quad (4)$$

where $\text{sim}(\cdot, \cdot)$ denotes cosine similarity.

Style Consistency Loss. A style classifier C_{style} is integrated to enforce writer-specific characteristics. It penalizes the discrepancy between the ground-truth writer ID w and the predicted identity from both real and synthesized images:

$$L_{style} = \mathbb{E}_{x,w}[-\log p(w \mid C_{style}(x))] \quad (5)$$

where x is sampled from either real query images $\{x_q\}$ or generated images $\{G(x_s, y_q)\}$.

Content Recognition Loss. To ensure the generated text is legible, a CRNN-based recognizer R is trained using the Connectionist Temporal Classification (CTC) loss:

$$L_{rec} = \mathbb{E}_{x,y}[\text{CTC}(R(x), y)] \quad (6)$$

This objective ensures that the structural information of characters is correctly preserved during the style-fusion and decoding processes.

Prototype Consistency and Distillation Loss. The core of our one-shot capability lies in the supervision provided by the Prototype Teacher. The prototype consistency loss L_{con} aligns the student's representation with the teacher's "gold standard" prototype $\hat{P}_S$:

$$L_{con} = \|P_S - \hat{P}_S\|_1 \tag{7}$$

where $\hat{P}_S = \frac{1}{q}\sum_{i=1}^{q} f_{T,i}$ is the aggregated style feature from the teacher network observing multiple query images. Furthermore, to guide the AGSF module, we distill the teacher's fused style $S_{fused,T}$ into the student's $S_{fused,S}$ using Mean Squared Error (MSE), ensuring the student learns semantic-aware style retrieval even with a single reference.

Table 1. Image quality assessments on IAM test set (lower is better). Bold = best.

Model	Shot	FID ↓	GS ↓	KID ↓	HWD ↓
GANwriting [14]	few	28.37	5.67×10^{-2}	–	–
ScrabbleGAN [8]	few	20.72	2.56×10^{-2}	–	–
Davis et al. [6]	few	20.65	4.88×10^{-2}	–	–
HiGAN [9]	few	19.77	3.19×10^{-2}	0.95	1.13
HiGAN+ [10]	few	24.09	–	1.26	0.80
HWT [3]	few	19.74	1.01×10^{-2}	0.93	1.13
WordStylist [20]	few	18.58	2.85×10^{-2}	–	–
VATr [22]	few	17.55	2.19×10^{-2}	0.59	0.83
VATr++ [25]	few	16.29	0.23×10^{-2}	0.50	0.74
DiffPen [21]	few	22.54	–	–	–
One-DM [5]	one	15.62	0.20×10^{-2}	0.62	1.36
GL-GAN [29]	few	14.32	0.23×10^{-2}	0.60	0.81
Ours (P-HTG)	one	**12.79**	$\mathbf{0.16\times10^{-2}}$	0.53	**0.74**

Table 2. Comparisons with competitors on the CVL dataset (lower is better). The results demonstrate the cross-dataset robustness of our method.

Model	HWT [3]	VATr [22]	VATr++ [25]	One-DM [5]	**Ours (P-HTG)**
FID ↓	17.09	16.39	15.42	16.28	**14.51**

4 Experiments

4.1 Implementation Details

Datasets and Preprocessing

We evaluate P-HTG on two widely-recognized offline handwriting benchmarks: IAM [19] and CVL [16]. The **IAM** dataset contains 9,862 text lines from 500 writers. Following the official partition, we use 340 writers for training and 160 for testing. The **CVL** dataset is a larger, bilingual (English/German) dataset featuring 310 writers. We adhere to the standard split of 283 writers for training and 27 for testing. For preprocessing, all word images are resized to a fixed height of 32 pixels, with the width adjusted proportionally to maintain the aspect ratio. During training, we construct episodes by sampling one image as the support sample ($k = 1$) and 10 images as the query set ($q = 10$) for each writer.

Training Details

All experiments are conducted on a single NVIDIA RTX 4090 GPU. We utilize the Adam optimizer with an initial learning rate of 2×10^{-5} and an exponential decay schedule. The model is trained for 10,000 iterations with a batch size of 8 episodes. The loss weighting hyperparameters, as described in Sect. 3.3, are set to $\lambda_{adv} = 1.0$, $\lambda_{style} = 0.7$, $\lambda_{match} = 1.0$, and $\lambda_{con} = 1.0$.

4.2 Quantitative Analysis

To demonstrate the effectiveness of P-HTG, we conduct a comprehensive quantitative comparison against state-of-the-art HTG methods. Evaluation metrics include Fréchet Inception Distance (FID) [13], Geometric Score (GS) [15], Kernel Inception Distance (KID) [4], and Handwritten Word Distance (HWD) [23]. Lower values indicate better performance across all metrics.

Table 3. FID scores for seen/unseen words and writers.

Model	IV_S ↓	IV_U ↓	OOV_S ↓	OOV_U ↓
Davis et al. [6]	118.56	128.75	127.11	136.37
GANwriting [14]	120.07	124.30	125.87	130.68
HiGAN+ [10]	117.33	116.95	121.55	121.48
GC-DDPM [7]	99.86	105.73	112.52	118.39
HWT [3]	98.10	104.27	109.45	115.52
WordStylist [20]	109.25	106.90	116.55	113.52
VATr [22]	103.75	101.73	111.64	108.76
VATr++ [25]	100.44	102.92	105.46	107.87
One-DM [5]	89.47	98.36	93.30	102.75
P-HTG (Ours)	**84.38**	**90.90**	**87.01**	**93.73**

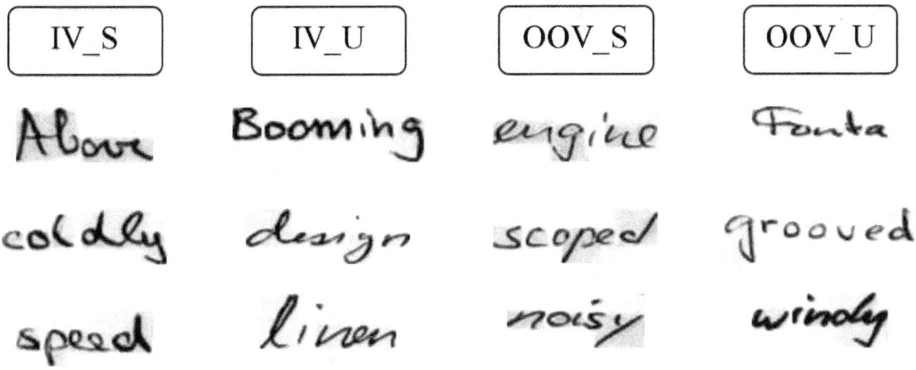

Fig. 4. Presents qualitative results produced by our method across four scenarios: (1) In-Vocabulary words with a previously observed style (IV_S), (2) In-Vocabulary words with an unseen style (IV_U), (3) Out-of-Vocabulary words with a familiar style (OOV_S), and (4) Out-of-Vocabulary words with a novel style (OOV_U).

Overall Quality Comparison. As shown in Table 1, our P-HTG model, operating in a challenging one-shot setting, achieves new state-of-the-art results on the IAM dataset. It consistently outperforms all compared methods across FID, GS, and KID metrics. To demonstrate robustness, we further evaluated our framework on the CVL dataset (Table 2), where P-HTG again surpasses all competitors, including few-shot models like VATr++. This confirms that our AGSF-based architecture possesses superior generation quality and cross-dataset adaptability, even when provided with only a single style reference.

It should be noted that the missing values (denoted by "–") in Table 1 correspond to specific metrics that were not reported in the original papers of certain baseline methods (e.g., ScrabbleGAN [8] and DiffPen [21]). Throughout the evaluation, we strictly utilized official codebases and pre-trained weights to reproduce all baselines. However, we observed significant performance degradation in the reproduction of several recent methods, such as DiffPen, compared to their claimed performance. To ensure absolute *upper-bound fairness*, we opted to report the best results directly from the original papers whenever our reproduced performance was inferior. Similarly, for the CVL dataset, we conduct comparisons only with methods that explicitly reported results on this benchmark to maintain a consistent and fair evaluation standard.

Generalization Ability. We further evaluate the model's generalization capability to unseen writer styles and novel words. The results, presented in Table 3, are categorized into four scenarios: In-Vocabulary words with Seen (IV_S) or Unseen (IV_U) styles, and Out-of-Vocabulary words with Seen (OOV_S) or Unseen (OOV_U) styles. P-HTG demonstrates remarkable performance, achieving the best results across all four challenging settings. More importantly, P-HTG exhibits exceptional stability, with its FID scores showing minimal fluctuation across all scenarios. This suggests that our prototype-guided, AGSF-based archi-

Table 4. Evaluation of OCR Performance on the IAM Test Set. Best results are highlighted in **bold**.

Model	**CER ↓**	**WER ↓**
IAM test set	42.32	62.51
VATr [22]	48.78	51.53
VATr++ [25]	37.99	50.15
DiffusionPen [21]	55.06	72.38
One-DM [5]	27.86	55.07
Ours (P-HTG)	**23.33**	**45.90**

tecture effectively captures the invariant stylistic characteristics of handwriting, allowing it to generalize robustly without overfitting to seen data—a common limitation in competing models. The qualitative results for these four scenarios are illustrated in Fig. 4.

Text Legibility. Beyond visual style, the legibility of the generated text is a critical metric for evaluating the structural integrity of the synthesized images. We assess this using the state-of-the-art TrOCR model [17]. Specifically, we employ the publicly available pre-trained *trocr-base-handwritten* weights, which were fine-tuned on the IAM line-level dataset, providing a robust and challenging benchmark for word-level recognition. As reported in Table 4, P-HTG achieves the lowest Character Error Rate (CER) and Word Error Rate (WER) among all compared methods, significantly outperforming existing one-shot and few-shot baselines. This outstanding result indicates that our AGSF-based model not only produces stylistically coherent images but also maintains precise character structures and connectivity. Similarly, to ensure fair comparison and reproducibility, we only compare with methods that provide state-of-the-art public code and pre-trained weights.

Table 5. Ablation analysis on the core components of P-HTG. We evaluate visual quality, geometric diversity, and stylistic faithfulness using FID, GS, KID, and HWD metrics. Best results are in **bold**.

	AGSF	Prototype	FID ↓	GS ($\times 10^{-2}$) ↓	KID ↓	HWD ↓
(a)			22.73	1.18	0.98	1.11
(b)	✓		14.27	0.21	0.58	0.80
(c)		✓	15.38	0.22	0.64	0.83
(d)	✓	✓	**12.79**	**0.16**	**0.53**	**0.74**

Ablation Studies. To validate the effectiveness of our proposed components, we conduct a multi-dimensional ablation study, with results summarized in

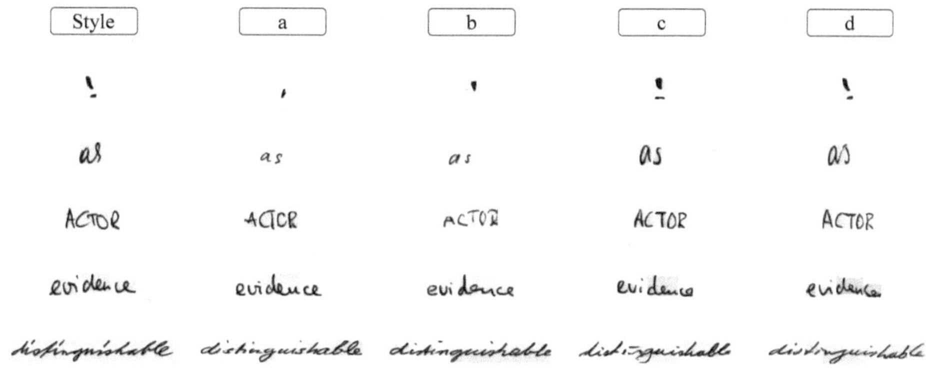

Fig. 5. Visual ablation study. From left to right: results from (a) the baseline model, (b) the model with only the AGSF architecture, (c) the model with only prototype-guided training, and (d) our full P-HTG model. This visualization highlights how each component—the adaptive gated fusion and the prototype guidance—contributes to the final high-quality synthesis, consistent with the quantitative results in Table 5.

Table 5. We analyze the individual contributions of our two core innovations: the prototype-guided episodic training scheme (**Prototype Guidance**) and the AGSF-based generator architecture. The results indicate that integrating either Prototype Guidance into the baseline or adopting the AGSF architecture independently leads to substantial improvements across all metrics (e.g., FID drops from 22.73 to 15.38 and 14.27, respectively). This demonstrates that both components are highly effective on their own in terms of visual quality, geometric diversity (GS), and stylistic faithfulness (HWD). Crucially, the combination of both in our full P-HTG model yields the superior performance, confirming the synergistic effect between our novel training paradigm and the adaptive fusion design.

To provide a more intuitive understanding, we present a visual ablation in Fig. 5 across various word lengths and styles. The baseline struggles to maintain adequate style and content fidelity simultaneously. Introducing the AGSF module alone significantly enhances structural coherence and stroke continuity, particularly for longer words where global style drift is common. Conversely, incorporating only the **Prototype Guidance** enforces a robust stylistic constraint, yielding impressive content accuracy even for rare characters or punctuation. The full P-HTG model, combining these strengths, produces the most faithful and high-quality results, visually confirming the quantitative advantages reported in Table 5.

Computational Cost. Finally, we analyze the computational efficiency of P-HTG in Table 6. Our model achieves state-of-the-art performance while maintaining the highest efficiency among its peers, featuring the lowest parameter count (28.81M) and computational workload (211.27 GFLOPs). While its complexity remains comparable to other Transformer-based architectures, it is an order of magnitude more efficient than recent diffusion-based one-shot mod-

Table 6. Computational cost comparison on the IAM dataset. Best results are highlighted in **bold**.

Model	FLOPs (GFLOPs) ↓	Parameters (M) ↓
HWT [3]	211.82	30.18
VATr++ [25]	211.87	30.32
One-DM [5]	3397.06	149.54
Ours (P-HTG)	**211.27**	**28.81**

els such as One-DM [5]. This highlights a core advantage of our framework: the AGSF-based generator delivers superior one-shot generation quality without the prohibitive computational overhead associated with large-scale diffusion processes, positioning P-HTG as a more practical and scalable solution for real-world handwriting synthesis.

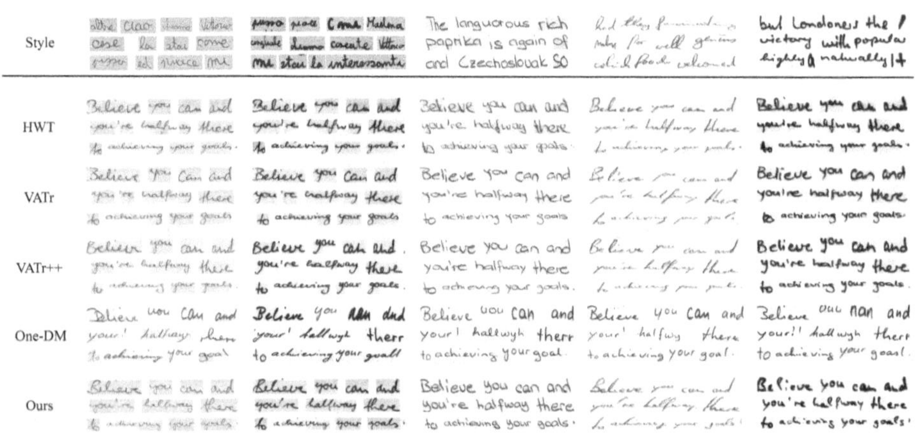

Fig. 6. Visual comparison of handwritten images synthesized by various state-of-the-art methods and our proposed P-HTG. Our method exhibits superior fidelity in both character structure and stylistic texture.

4.3 Qualitative Comparison

Based on the quantitative findings, we selected four representative state-of-the-art methods–HWT [3], VATr [22], VATr++ [25], and One-DM [5]–for visual comparison (Fig. 6). All samples were generated using officially released optimal weights, encompassing two unseen and three seen styles, conditioned on the phrase: "*Believe you can and you're halfway there to achieving your goals.*" Every third line in the figure corresponds to a consistent writing style reference.

While most methods demonstrate reasonable stylistic consistency, maintaining word-level structural integrity remains a challenge for several baselines, particularly the diffusion-based One-DM. In contrast, P-HTG excels in preserving fine-grained stylistic nuances and stroke continuity. For instance, in the third style, the word "*Believe*" maintains its authentic cursive ligatures only in our results, whereas other methods exhibit stroke interruptions or incomplete glyphs. Furthermore, for styles involving colored or textured backgrounds, P-HTG achieves a more robust and faithful reproduction. This underscores the superior capability of our AGSF-based generator in adaptive feature representation and stylistic alignment.

To further assess perceptual quality, we conducted a user preference study involving 200 participants. Each was presented with a ground-truth handwritten sample alongside anonymous synthetic outputs from HWT, VATr++, One-DM, and P-HTG, and asked to identify the most faithful reconstruction. The results indicate a clear preference for our method, with P-HTG being selected 32% of the time, outperforming VATr++ (27%), One-DM (23%), and HWT (18%). This subjective evaluation confirms the practical effectiveness of our framework in producing human-like handwriting.

5 Discussion

The AGSF module further addresses "style washing" and structural drift commonly seen in conventional Transformer- or diffusion-based methods by adaptively balancing global stylistic stability with local character fidelity, as validated by our ablation studies (Sect. 4).

Despite these advancements, several limitations warrant further investigation. First, while P-HTG performs remarkably well in standard one-shot settings, capturing styles from extremely irregular or highly decorative handwriting remains a challenge, as a single reference may not contain sufficient entropy to represent such complex glyph variations. Second, the training of a dual-network prototype system introduces non-negligible memory overhead during the training phase, though it maintains high efficiency during inference. Finally, our current evaluation is limited to Latin-based scripts (English and German). Extending P-HTG to non-Latin scripts, such as Chinese or Arabic, may require additional structural priors to handle complex radical-based glyph compositions.

Future work will explore more efficient prototype distillation strategies, such as lightweight teacher-student architectures, to reduce training costs. We also plan to investigate dynamic adaptation mechanisms to better handle extreme stylistic variations and explore cross-lingual style transfer to enhance the model's versatility across diverse linguistic domains.

6 Conclusion

In this work, we introduced P-HTG, a novel prototype-guided framework designed for high-quality, one-shot handwritten text generation. By distilling robust stylistic prototypes through a teacher-student paradigm and integrating them via our proposed **Adaptive Gated Style-Fusion (AGSF)** module, P-HTG effectively balances global style consistency with fine-grained character-level accuracy. Extensive experiments on standard benchmarks demonstrate that our method significantly outperforms existing state-of-the-art approaches across multiple dimensions, including visual fidelity (FID), geometric diversity (GS), and text legibility (CER/WER), even under the most challenging one-shot constraints.

Furthermore, qualitative evaluations and user studies confirm that P-HTG produces visually compelling and highly readable text that generalizes robustly to unseen writers and novel vocabularies. Notably, our framework achieves this superior performance with a lean parameter count and high computational efficiency, making it an order of magnitude more practical than recent diffusion-based alternatives. Overall, P-HTG sets a new benchmark for handwritten text generation, offering a robust, accurate, and stylistically coherent solution for realistic digital handwriting synthesis.

Acknowledgements. This work was supported by the National Natural Science Foundation of China (62466040) and the Project for Science and Technology of Inner Mongolia Autonomous Region (2025YFHH0079).

References

1. Aksan, E., Pece, F., Hilliges, O.: DeepWriting: making digital ink editable via deep generative modeling. In: Proceedings of the 2018 CHI Conference on Human Factors in Computing Systems, pp. 1–14 (2018)
2. Alonso, E., Moysset, B., Messina, R.: Adversarial generation of handwritten text images conditioned on sequences. In: 2019 International Conference on Document Analysis and Recognition (ICDAR), pp. 481–486. IEEE (2019)
3. Bhunia, A.K., Khan, S., Cholakkal, H., Anwer, R.M., Khan, F.S., Shah, M.: Handwriting transformers. In: Proceedings of the IEEE/CVF International Conference on Computer Vision, pp. 1086–1094 (2021)
4. Bińkowski, M., Sutherland, D.J., Arbel, M., Gretton, A.: Demystifying MMD GANs. In: International Conference on Learning Representations (ICLR) (2018)
5. Dai, G., Zhang, Y., Ke, Q., Guo, Q., Huang, S.: One-DM: one-shot diffusion mimicker for handwritten text generation. In: European Conference on Computer Vision, pp. 410–427. Springer (2024)
6. Davis, B., Tensmeyer, C., Price, B., Wigington, C., Morse, B., Jain, R.: Text and style conditioned GAN for generation of offline handwriting lines. In: Proceedings of the British Machine Vision Conference (BMVC) 2020 (2020)
7. Ding, H., Luan, B., Gui, D., Chen, K., Huo, Q.: Improving handwritten OCR with training samples generated by glyph conditional denoising diffusion probabilistic model. In: International Conference on Document Analysis and Recognition, pp. 20–37. Springer (2023)
8. Fogel, S., Averbuch-Elor, H., Cohen, S., Mazor, S., Litman, R.: ScrabbleGAN: semi-supervised varying length handwritten text generation. In: Proceedings of the IEEE/CVF Conference on Computer Vision and Pattern Recognition, pp. 4324–4333 (2020)

9. Gan, J., Wang, W.: HiGAN: handwriting imitation conditioned on arbitrary-length texts and disentangled styles. In: Proceedings of the AAAI Conference on Artificial Intelligence, vol. 35, pp. 7484–7492 (2021)
10. Gan, J., Wang, W., Leng, J., Gao, X.: HiGAN+: handwriting imitation GAN with disentangled representations. ACM Trans. Graphics (TOG) **42**(1), 1–17 (2022)
11. Goodfellow, I., et al.: Generative adversarial networks. Commun. ACM **63**(11), 139–144 (2020)
12. Graves, A.: Generating sequences with recurrent neural networks. arXiv preprint arXiv:1308.0850 (2013)
13. Heusel, M., Ramsauer, H., Unterthiner, T., Nessler, B., Hochreiter, S.: GANs trained by a two time-scale update rule converge to a local nash equilibrium. In: Advances in Neural Information Processing Systems, vol. 30 (2017)
14. Kang, L., Riba, P., Wang, Y., Rusinol, M., Fornés, A., Villegas, M.: GANWriting: content-conditioned generation of styled handwritten word images. In: Computer Vision–ECCV 2020: 16th European Conference, Glasgow, UK, 23–28 August 2020, Part XXIII 16, pp. 273–289. Springer (2020)
15. Khrulkov, V., Oseledets, I.: Geometry score: a method for comparing generative adversarial networks. In: International Conference on Machine Learning, pp. 2621–2629. PMLR (2018)
16. Kleber, F., Fiel, S., Diem, M., Sablatnig, R.: CVL-database: an off-line database for writer retrieval, writer identification and word spotting. In: 2013 12th International Conference on Document Analysis and Recognition, pp. 560–564. IEEE (2013)
17. Li, M., Lv, T., Chen, J., Cui, L., Lu, Y., Florencio, D., Zhang, C., Li, Z., Wei, F.: TrOCR: transformer-based optical character recognition with pre-trained models. In: Proceedings of the AAAI Conference on Artificial Intelligence, vol. 37, pp. 13094–13102 (2023)
18. Luo, C., Zhu, Y., Jin, L., Li, Z., Peng, D.: SloGAN: handwriting style synthesis for arbitrary-length and out-of-vocabulary text. IEEE Trans. Neural Netw. Learn. Syst. **34**(11), 8503–8515 (2022)
19. Marti, U.V., Bunke, H.: The IAM-database: an English sentence database for offline handwriting recognition. Int. J. Doc. Anal. Recogn. **5**, 39–46 (2002)
20. Nikolaidou, K., et al.: WordStylist: styled verbatim handwritten text generation with latent diffusion models. In: International Conference on Document Analysis and Recognition, pp. 384–401. Springer (2023)
21. Nikolaidou, K., Retsinas, G., Sfikas, G., Liwicki, M.: DiffusionPen: towards controlling the style of handwritten text generation. In: European Conference on Computer Vision, pp. 417–434. Springer (2024)
22. Pippi, V., Cascianelli, S., Cucchiara, R.: Handwritten text generation from visual archetypes. In: Proceedings of the IEEE/CVF Conference on Computer Vision and Pattern Recognition, pp. 22458–22467 (2023)
23. Pippi, V., Quattrini, F., Cascianelli, S., Cucchiara, R.: HWD: a novel evaluation score for styled handwritten text generation. In: Proceedings of the British Machine Vision Conference (BMVC) (2023)
24. Pippi, V., Quattrini, F., Cascianelli, S., Tonioni, A., Cucchiara, R.: Zero-shot styled text image generation, but make it autoregressive. In: Proceedings of the Computer Vision and Pattern Recognition Conference, pp. 7910–7919 (2025)
25. Vanherle, B., Pippi, V., Cascianelli, S., Michiels, N., Van Reeth, F., Cucchiara, R.: VATr++: choose your words wisely for handwritten text generation. IEEE Trans. Pattern Anal. Mach. Intell. **47**(2), 934–948 (2024)

26. Wang, H., Wang, Y., Wei, H.: AFFGANwriting: a handwriting image generation method based on multi-feature fusion. In: International Conference on Document Analysis and Recognition (ICDAR 2023), pp. 302–312. Springer (2023)
27. Wang, Y., Wei, H., Sun, S.: SmartExp: an adaptive data expansion strategy for improving handwritten text recognition. In: ICASSP 2025-2025 IEEE International Conference on Acoustics, Speech and Signal Processing (ICASSP), pp. 1–5. IEEE (2025)
28. Wang, Y., Wei, H., Wang, H.: As-net: Adaptive style-aware network for handwritten text generation. In: ICASSP 2025-2025 IEEE International Conference on Acoustics, Speech and Signal Processing (ICASSP), pp. 1–5. IEEE (2025)
29. Wang, Y., Wei, H., Wang, H., Sun, S., He, C.: GL-GAN: perceiving and integrating global and local styles for handwritten text generation with mamba. In: Proceedings of the 31st International Conference on Computational Linguistics, pp. 2434–2444 (2025)
30. Zaccagnino, C., Quattrini, F., Pippi, V., Cascianelli, S., Tonioni, A., Cucchiara, R.: Autoregressive styled text image generation, but make it reliable. In: Proceedings of the IEEE/CVF Winter Conference on Applications of Computer Vision, pp. 1234–1245 (2026)
31. Zhu, Y., Li, Z., Wang, T., He, M., Yao, C.: Conditional text image generation with diffusion models. In: Proceedings of the IEEE/CVF Conference on Computer Vision and Pattern Recognition, pp. 14235–14245 (2023)

Revisiting Structural Dependency in Autoregressive Multi-task Table Recognition via Order-Independent Cell-Level Representations

Takaya Kawakatsu(✉)

Preferred Networks, Inc., 1-6-1 Otemachi, Chiyoda, Tokyo, Japan
kat.nii.ac.jp@gmail.com
https://researchmap.jp/t.kat

Abstract. Multi-task table recognition jointly addresses table structure prediction, cell localization, and cell content recognition within a unified framework. Existing approaches often rely on autoregressive decoders to generate table structures and reuse their hidden states for cell localization and content recognition. This autoregressive generation process can make cell representations order-dependent, degrading global consistency across cells. This paper proposes a structural refinement module that produces order-independent cell features through non-causal attention. This design enables parallel inference of cell contents while conditioning each cell on global context encoded in the refined features. Experiments on two large datasets demonstrate consistent gains in cell localization and end-to-end recognition, while reducing overall inference time by around threefold.

Keywords: Deep Learning · Table Recognition · Transformer · OCR

1 Introduction

Table recognition aims to convert table images into structured formats, such as HTML, which preserve both geometric layout and textual cell contents. Modern methods [5,10–12] use multi-task models to jointly predict table structure, cell locations, and cell contents. Sharing structural features provides helpful context for cell localization and content recognition.

In practice, many methods [5,13,23,29] rely on autoregressive (AR) decoders to generate structural sequences. However, the causal attention mask allows each cell representation to access only previously generated cells, leaving information from later cells unavailable. The resulting representations reflect generation order rather than intrinsic table structure. The reuse of such order-dependent features in downstream tasks may weaken the global consistency of cell relationships.

G. A. Fink et al. (Eds.): ICDAR 2026, LNCS 16972, pp. 165–180, 2027.
https://doi.org/10.1007/978-3-032-36023-6_10

In this paper, we revisit multi-task table recognition from the perspective of structural representation design. Instead of modifying tokenization strategies or decoder architectures, we separate structural generation from cell-level structural aggregation. We introduce a global structural refinement module which removes generation-order dependency at the cell level. Cell representations extracted from the structural decoder are refined through non-causal self-attention and realigned with visual features, producing order-independent structural features.

As illustrated in Fig. 1, causal masking appears in both structural decoding and cell-content decoding. In these stages, the triangular mask ensures that each token is conditioned only on its past decoding context. The proposed refinement module bridges these two processes by applying non-causal attention to the cell sequence. It transforms the order-dependent cell features into globally aggregated cell representations. These refined representations provide the basis for improved geometric consistency and parallel cell-content inference.

Experimental results on two large-scale table datasets [28,29] show consistent improvements over prior multi-task approaches. The proposed method improves cell localization intersection over union (IoU) by around 3–5 points and increases tree edit distance-based similarity (TEDS), particularly for tables with long cell contents. Furthermore, aggregating structural context before content generation enables parallel cell-level decoding, which achieves a roughly threefold inference speedup over MuTabNet [5] while maintaining comparable recognition accuracy.

These results demonstrate the importance of dependency modeling in shared structural representations and suggest that dependency topology is a promising design axis for structured document recognition.

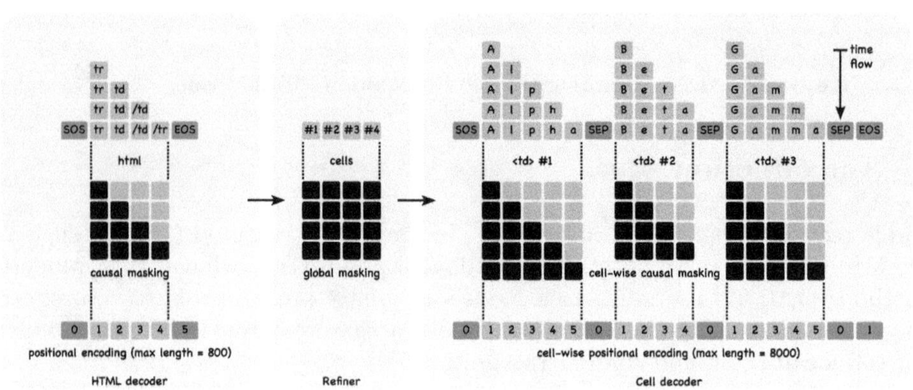

Fig. 1. Key idea of order-independent cell features and parallel cell decoding.

2 Related Work

Table recognition predicts structural elements such as rows, columns, cells, and span attributes from a table image, localizes each cell region, recognizes the text inside each cell, and outputs a table representation. Prior studies can be viewed from three perspectives: geometric detection, image-to-sequence generation, and multi-task modeling.

Geometric methods [4,6,9,15–19,21,25] detect table components such as rows, columns, text blocks, or cells, and infer the table structure from spatial relationships. These methods are effective when table and component boundaries are visually clear, but can be sensitive to ambiguous cell boundaries and complex merged cells.

Image-to-sequence models [1,2,29] generate LaTeX or HTML sequences from table images using encoder–decoder architectures. Recent studies [3,13,23] have extended this formulation with Transformer models. This top-down formulation avoids reliance on hand-designed reconstruction rules, but it linearizes the table structure into a one-dimensional sequence, making structural features dependent on the chosen generation order.

Multi-task frameworks [5,10–12,29] jointly model structure prediction, cell localization, and cell content recognition. These frameworks use shared structural features to promote consistency among logical structure, visual layout, and text recognition. Nevertheless, the shared structural features are typically reused from AR decoder states, leaving the causal dependency among cells unresolved.

Modern vision–language models (VLMs) [7,8,20,26] provide strong reasoning capabilities for document understanding. However, many of them rely on broader training and task settings that differ from standard evaluation protocols for table recognition. Our focus is on end-to-end table recognition without using external OCR or additional training data.

Several studies have improved structural decoders for long-sequence decoding and visual alignment. Local attention [10], token compression [30], and coverage modeling [27] address the cost and instability of long sequences. VAST [3] aligns logical cell representations with image evidence using a visual alignment loss.

In contrast to prior work, this paper examines how the dependency topology of shared structural features affects downstream tasks. Our approach leaves the output markup unchanged and does not rely on external OCR. Instead, it refines structural representations obtained from the structural decoder with non-causal self-attention and cross-attention with image features. This captures spatial and logical relationships among cells more precisely, leading to improved cell content recognition.

3 Methodology

We adopt a multi-task model for joint structure prediction, cell localization, and cell content recognition. Although prior work [5,10] has explored such multi-task models, it remains unclear what type of structural information should be shared across tasks.

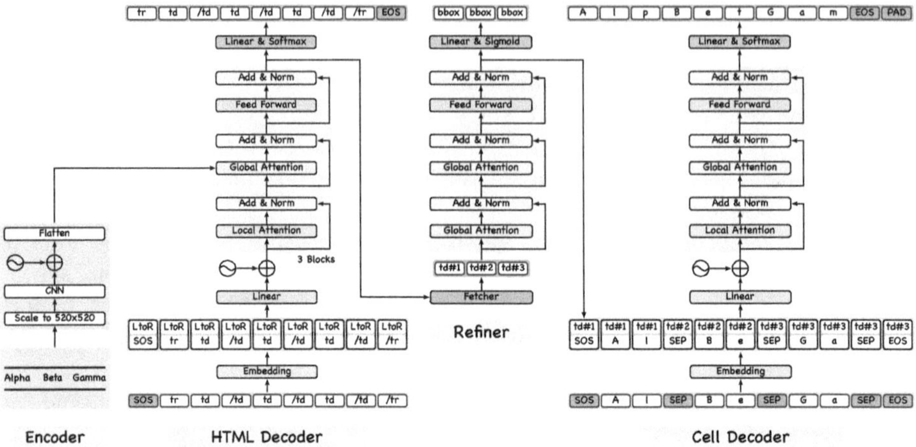

Fig. 2. Overview of the proposed multi-task framework.

Bounding boxes provide geometric supervision for cell localization but encode only position and size. They do not directly encode inter-cell relationships, such as row-wise alignment, column-wise consistency, and global layout regularity. In contrast, the hidden states of an HTML decoder implicitly accumulate structural information during decoding. We therefore focus on extracting and refining such structural representations so that they can be effectively reused for downstream tasks.

However, causal sequence generation prevents earlier cell features from using information from later cells, limiting their structural context. We thus decouple structural generation and structural aggregation. The HTML decoder generates the structural skeleton, while a cell-level refiner reconstructs globally consistent representations through non-causal attention.

3.1 Overall Architecture

Figure 2 provides an overview of the proposed architecture. The model consists of a vision encoder, an HTML decoder, a cell-level refiner, a localization head, and a cell content decoder. The encoder first extracts image features, and the HTML decoder generates HTML tokens sequentially conditioned on them.

Cell-level features associated with non-empty <td> tokens are extracted from the outputs of the HTML decoder. The refiner then updates these features using non-causal self-attention and visual realignment. The refined features are shared with both the localization head and the cell content decoder, enabling structural information to be aggregated independently of token generation.

3.2 Cell-Level Structural Representation

The decoder state corresponding to each non-empty <td> token is extracted and treated as the initial structural representation $\mathbf{h}_c$ of cell c. These representations capture structural information accumulated during HTML decoding but remain conditioned on generation order.

Although AR decoding is required for sequential HTML generation, its causal constraint yields asymmetric structural representations when they are reused for downstream tasks. This asymmetry may limit their suitability for cell localization and cell content recognition, where row and column dependencies are inherently bidirectional.

3.3 Global Structural Refinement

We introduce a dedicated refinement module that operates independently of the HTML decoder to construct order-independent and globally consistent structural features. The refiner is implemented as a cell-level Transformer block consisting of self-attention, cross-attention, and a feed-forward network.

First, all cell representations are globally contextualized through non-causal multi-head self-attention:

$$\hat{\mathbf{h}}_c = \mathrm{MHSA}(\{\mathbf{h}_1, \ldots, \mathbf{h}_C\})_c. \tag{1}$$

In contrast to the HTML decoder, Eq. (1) replaces the causal dependency pattern with fully connected cell-level self-attention, which enables the model to capture table-wide alignment and layout consistency.

Second, the resulting features are realigned with the full encoder feature map through multi-head cross-attention:

$$\bar{\mathbf{h}}_c = \mathrm{MHCA}(\hat{\mathbf{h}}_c, \mathbf{F}), \tag{2}$$

where $\mathbf{F}$ denotes the image features from the encoder. Equation (2) links the globally aggregated structural representations back to the image features, grounding the refined cell features in visual information related to text layout, ruling lines, and cell spacing.

Finally, a position-wise feed-forward network produces the refined structural feature:

$$\tilde{\mathbf{h}}_c = \mathrm{FFN}(\bar{\mathbf{h}}_c). \tag{3}$$

The resulting representation $\tilde{\mathbf{h}}_c$ integrates inter-cell structural dependencies and image-grounded geometric information. These refined representations are shared with both the localization head and the cell content decoder.

This refinement benefits both cell localization and content recognition. When borders are invisible or cell contents span multiple lines, bounding box regression can be ambiguous. The refined cell-level representations provide image-grounded structural context, promoting geometrically coherent localization and improving IoU.

The refiner module is designed to change the dependency structure of shared representations, rather than merely to add model capacity. AR decoding restricts cell-level dependencies to a causal lower-triangular graph, whereas the proposed non-causal attention enables fully connected inter-cell dependencies. We evaluate this topology-focused design choice empirically in Sect. 4.6.

Algorithm 1. Parallel cell decoding

```
Initialize S with SOS and C SEP tokens
Set all cells active
for t = 0, 1, 2, ... do
    Predict next-token probabilities for each cell
    for each active cell c do
        Insert predicted token before cell c's SEP
        if EOS emitted then deactivate cell c
        end if
    end for
    if all cells inactive then break
    end if
end for
```

3.4 Parallel Cell Decoding

MuTabNet [5] decodes cell contents sequentially across cells, allowing preceding cell contents to provide context for subsequent cells. This design improves content recognition through cross-cell information sharing, but makes decoding cost scale linearly with the total number of cell-content tokens. In our approach, structural dependencies are captured during the global refinement stage, reducing the need for sequential cross-cell conditioning during content generation.

We therefore decode cell contents independently and in parallel. At each time step, all active cells predict their next tokens simultaneously. Cells that emit an end-of-sequence token are deactivated, while the others continue decoding. This procedure is summarized in Algorithm 1. No cross-cell attention is performed at this stage, since the refined cell features already encode inter-cell relationships.

Under this design, cell decoding is bounded by the length of the longest cell sequence, rather than by the total number of cell-content tokens. Since cells are decoded in parallel, short cells finish early and no longer contribute to subsequent iterations. The refinement stage adds only a single attention block over cell-level features, with modest overhead compared with sequential concatenated content decoding. This improves scalability while preserving recognition accuracy.

4 Experiments

We conducted experiments on two public table datasets.

4.1 Datasets

We utilized FinTabNet (FTN) [28] and PubTabNet (PTN) [29] for training and evaluation. Both datasets support evaluation of table structure recognition, cell localization, and cell content recognition, and are regarded as de facto standard benchmarks for image-based table recognition. Statistics are shown in Table 1.

Table 1. The statistics of the table image datasets.

	Cells per table			Characters per cell		
Set	Train	Valid	Test	Train	Valid	Test
FTN	45.78	40.38	40.02	16.61	18.15	15.23
PTN	64.86	66.56	72.52	13.37	13.54	12.36
PTN250	144.45	146.42	-	8.59	8.52	-

FinTabNet. FTN is a large-scale dataset of table images from the annual reports of S&P 500 companies. It provides HTML tokens, cell bounding boxes, and cell contents as annotations. The dataset contains approximately 112,000 tables and is split into training, validation, and test sets. We used the *validation* set, which contains 10,656 tables, for evaluation, following prior work [5,10–13,28]. We also prepared four subsets by requiring at least 5, 10, 15, and 20 tokens per cell for the ablation study.

PubTabNet. PTN is a large-scale dataset of scientific tables from the PubMed Central Open Access subset. It provides HTML tokens, cell bounding boxes, and cell contents as annotations. The dataset contains approximately 510,000 images and is split into training and validation sets, but does not include an official test set. The official test set was released by ICDAR for the competition [24].

PTN250 [10] is an unofficial subset of tables containing at least 250 structural tokens. It is comparable in size to FTN and contains many cells per table, making it suitable for the ablation study. Since only 76 samples had more than 15 content tokens per cell, we excluded subsets with higher token thresholds.

4.2 Metrics

Since a typical table has a nested structure of rows, cells, cell contents, and cell attributes, we employed TEDS [29] for evaluation. We parsed the predicted and ground-truth HTML sequences into trees and calculated TEDS as follows:

$$\text{TEDS}(T_1, T_2) = 1 - \frac{\text{TED}(T_1, T_2)}{\max(|T_1|, |T_2|)}, \tag{4}$$

where T_1 and T_2 are the trees, TED is the tree edit distance function that reflects insertions, deletions, and renames, and $|T|$ is the number of nodes in T.

In addition, we report S-TEDS, which evaluates the predicted table structure without cell contents, whereas TEDS evaluates the complete table representation including cell contents.

Following the standard evaluation protocol [29], tables are divided into simple and complex subsets. Simple tables contain no vertically or horizontally merged cells, whereas complex tables contain at least one merged cell.

4.3 Implementation

We built our implementation on MMOCR [14] by extending MuTabNet [5]. We refer to this extended model as MuTabNet M2.

All models were trained for 30 epochs on four GPUs with a global batch size of 8 using the Ranger [22] optimizer. The learning rate started at 10^{-3} and was reduced to 10^{-4} for epochs 26–28 and 10^{-5} for epochs 29–30. Except for the full PTN setting, each experimental condition was trained three times with different random seeds.

Each image was padded with black pixels to preserve its aspect ratio, resized to 520×520 pixels, and normalized in RGB space. Each cell bounding box was represented by its center coordinates and size, with all values normalized to the range $[0, 1]$. No data augmentation was applied for fair comparison with previous work.

Table images were encoded using the same backbone as MuTabNet [5]. Each token was converted into a 512-dimensional embedding. The attention blocks in the two decoders and the refiner were configured with 8 heads. The window size for local attention was set to 300. We used greedy search with maximum lengths of 800 and 8000 for the HTML and cell decoders, respectively.

4.4 Performance

Table 2. Comparison on FinTabNet evaluation set.

	Ablation			S-TEDS	TEDS	
Model	WS	LA	PT	Total	OCR	Total
GTE†	-	-	-	87.14	-	-
	-	-	✓	91.02	-	-
TableFormer†	-	-	-	96.80	-	-
VAST†	-	-	-	98.63	✓	98.21
Zhu et al.†	-	-	-	99.04	✓	96.82
MTL-TabNet†	✓	-	-	98.72	-	95.32
	-	-	-	98.79	-	-
	-	✓	-	98.85	-	95.74
MuTabNet†	-	✓	-	98.87	-	97.69
MuTabNet M2	-	✓	-	98.87	-	97.93
	-	✓	✓	99.07	-	98.08

† denotes results reported in the original papers. WS and LA denote weak supervision and local attention. PT denotes PubTabNet pretraining.
OCR denotes the use of an external OCR engine

Table 3. Comparison on PubTabNet validation set.

Model	Ablation		S-TEDS	TEDS			
	WS	LA	Total	OCR	Simple	Complex	Total
EDD†	-	-	89.90	-	91.20	85.40	88.30
TableFormer†	-	-	96.75	✓	95.40	90.10	93.60
SEM†	-	-	-	✓	94.80	92.50	93.70
LGPMA†	-	-	96.70	✓	-	-	94.60
VCGroup†	-	-	-	✓	-	-	96.26
VAST†	-	-	97.23	✓	-	-	96.31
Zhu et al.†	-	-	98.82	✓	-	-	96.77
MTL-TabNet†	✓	-	-	-	97.89	95.02	96.48
	-	-	-	-	97.92	95.36	96.67
	-	✓	-	-	98.07	95.42	96.77
MuTabNet†	-	✓	97.71	-	98.16	95.53	96.87
MuTabNet M2	-	✓	97.74	-	98.34	95.68	97.04

† denotes results reported in the original papers. WS and LA denote weak supervision and local attention. OCR denotes the use of an external OCR engine

We evaluate the proposed method against representative prior approaches under standard table recognition protocols. Recent VLMs are excluded from the main comparison because they typically require substantially larger training data and broader task settings, as discussed in Sect. 2.

Unless otherwise stated, all models were trained without additional training data. The proposed method does not rely on any external OCR for cell content recognition. Results of the proposed method are averaged over three independent training runs, except for the full PTN setting.

FinTabNet. The proposed model achieves the highest TEDS among end-to-end models as reported in Table 2. VAST [3] reports slightly higher TEDS using an external OCR engine. In contrast, the proposed model jointly performs structure prediction, cell localization, and cell content recognition within a single model.

Compared with previous multi-task approaches [5,10–12], the refiner yields consistent gains in TEDS, suggesting that the improvement mainly comes from better cell content recognition. Although the absolute TEDS gains are moderate due to metric saturation, the improvements are consistent across settings.

PubTabNet. The proposed model achieves competitive or superior performance across both simple and complex tables as shown in Table 3. It also outperforms prior multi-task methods [5,10–12], particularly on complex tables, suggesting improved modeling of inter-cell dependencies.

Table 4. Comparison on ICDAR evaluation set.

	Ablation			TEDS		
Model	WS	LA	OCR	Simple	Complex	Total
XM†	-	-	✓	97.60	94.89	96.27
VCGroup†	-	-	✓	97.90	94.68	96.32
Davar-Lab-OCR†	-	-	✓	97.88	94.78	96.36
MTL-TabNet†	✓	-	-	97.51	94.37	95.97
	-	-	-	97.60	94.68	96.17
	-	✓	-	97.77	94.58	96.21
MuTabNet	-	✓	-	97.94	94.85	96.42
MuTabNet M2	-	✓	-	98.10	95.08	96.62

† denotes results reported in the original papers. WS and LA denote weak supervision and local attention. OCR denotes the use of an external OCR engine

Table 4 further compares our method with top-performing systems from the ICDAR competition [24]. The consistent improvements on both sets demonstrate the robustness of the refinement and parallel decoding strategy. The MuTabNet results are recomputed after removing redundant `html` and `body` elements from the original TEDS evaluation [5].

4.5 Inference Time

Table 5 reports inference time for cumulative task settings, from table structure prediction only to the full setting with bounding box regression and cell content generation. Inference time was measured during AR decoding without KV-cache using four NVIDIA V100 GPUs with a total batch size of 8.

The concatenated AR decoding introduced in MuTabNet [5] scales with the total number of tokens across all cells. Parallel cell decoding instead synchronizes generation across cells, making the number of AR steps depend on the maximum cell length rather than the sum of all cell lengths.

The proposed strategy reduces overall inference time by approximately three times, while maintaining accuracy. The additional cost of the refinement module is small compared with token-level AR decoding.

4.6 Ablation Study

The ablation study is structured along two independent axes. The first axis varies the shared information, where *bbox* uses only cell coordinates and *full* uses dense features from the HTML decoder. The second axis compares no refiner, a causal refiner, and the proposed global refiner.

The causal refiner is introduced as a variant with the same architecture and parameter count as the global refiner, while differing only in the attention mask

Table 5. Inference time on NVIDIA V100 GPUs.

Set	Parallel	Tasks			Time (s)
		HTML	Bbox	Cell	
FTN	-	✓	-	-	0.674
		✓	✓	-	0.679
		✓	✓	✓	3.707
	✓	✓	-	-	0.787
		✓	✓	-	0.792
		✓	✓	✓	1.300
PTN	-	✓	-	-	1.049
		✓	✓	-	1.056
		✓	✓	✓	4.634
	✓	✓	-	-	1.150
		✓	✓	-	1.156
		✓	✓	✓	1.506

configuration. This controlled comparison isolates whether the gains come from increased model capacity or from restructuring the inter-cell dependency graph.

All variants were evaluated on FTN and PTN250 using a unified experimental protocol. FTN contains relatively longer text content per cell, whereas PTN250 contains a larger number of cells per table. We additionally analyze subsets with increasing minimum token length to examine behavior under stronger structural and geometric ambiguity. Table 6 reports TEDS and IoU for these variants and subsets.

Effect of Structural Feature Sharing. Comparing the bbox and full models shows that sharing dense structural features substantially improves localization accuracy. The improvement is substantial on both datasets, with IoU increasing from 55.5 to 81.7 on FTN and from 51.0 to 82.9 on PTN250.

This gain does not come from feeding decoded cell contents back into the cell localization head. Cell localization is performed directly from structural features prior to content decoding. Instead, sharing the same cell-level features with the cell decoder allows the content recognition loss to shape the shared representation during training. This auxiliary supervision encourages the shared representation to encode cell-specific visual cues, thereby improving the feature quality available to the localization head.

The slight TEDS improvement indicates that dense structural feature sharing benefits not only localization but also cell content recognition. The improvement tends to become larger for longer cell contents, where accurate alignment between cell representations and content regions is more important.

Table 6. Ablation study by minimum content tokens per cell.

		TEDS							
		FTN					PTN250		
Model	Refiner	0+	5+	10+	15+	20+	0+	5+	10+
Bbox	-	97.1	97.1	96.5	94.0	91.3	95.6	95.3	94.5
	Causal	97.2	97.2	96.6	94.1	91.5	95.5	95.1	94.7
	Global	97.2	97.2	96.6	94.3	91.6	95.7	95.4	94.7
Full	-	97.9	97.9	97.5	96.0	94.3	95.8	95.5	94.8
	Causal	97.8	97.8	97.5	96.0	94.2	95.7	95.4	94.8
	Global	97.9	97.9	97.7	96.3	94.7	95.9	95.7	95.2
		IoU							
		FTN					PTN250		
Model	Refiner	0+	5+	10+	15+	20+	0+	5+	10+
Bbox	-	55.5	55.4	52.5	46.5	43.2	51.0	51.2	45.6
	Causal	64.4	64.4	62.9	57.8	54.3	51.9	52.0	46.2
	Global	62.6	62.6	60.5	54.3	50.2	51.7	51.8	45.9
Full	-	81.7	81.8	81.4	79.4	77.6	82.9	83.4	83.1
	Causal	84.8	84.9	84.6	82.9	81.5	86.1	86.5	86.4
	Global	85.5	85.6	85.4	83.8	82.4	86.3	86.7	86.7

Effect of Global Feature Refinement. Applying feature refinement improves localization under both shared-information settings. In the bbox setting, adding the global refiner raises IoU from 55.5 to 62.6 on FTN and from 51.0 to 51.7 on PTN250. In the full setting, it improves IoU from 81.7 to 85.5 on FTN and from 82.9 to 86.3 on PTN250.

To assess the role of dependency topology, we compare the causal and global refiners under the full setting. The causal refiner already improves IoU over the no-refiner case, showing the benefit of feature refinement itself. The global refiner further improves IoU over the causal refiner, from 84.8 to 85.5 on FTN and from 86.1 to 86.3 on PTN250. For TEDS, the causal refiner tends to slightly degrade performance, whereas the global refiner consistently recovers or improves it.

In the bbox setting, both refiners improve over the no-refiner baseline, while the causal refiner achieves higher IoU than the global refiner on FTN. Since the cell decoder receives only bounding-box features in this setting, the refined dense features are used mainly for localization. This observation is consistent with the view that global refinement is more effective when the refined structural features are shared with both localization and content recognition, as in the full setting.

Since the causal and global refiners have identical architecture and parameter count, this gap supports our hypothesis that replacing the lower-triangular causal dependency structure with a fully connected inter-cell graph improves structural representation quality. The advantage is more evident on subsets with longer cell contents, where geometric ambiguity and cross-cell dependencies are stronger.

4.7 Case Study

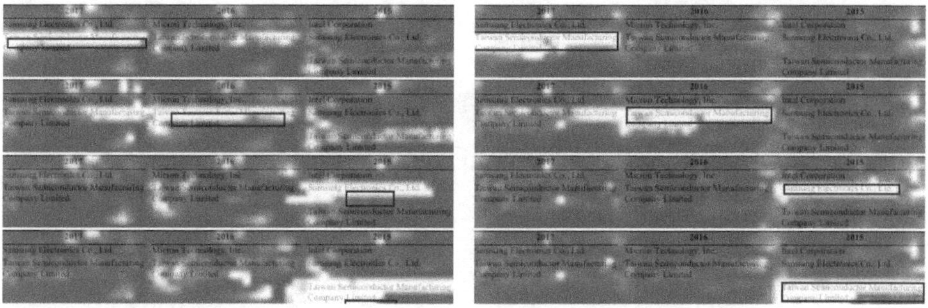

(a) Accepting only bounding boxes. (b) Accepting structural features.

Fig. 3. Cell-decoder attention maps and predicted bounding boxes.

Contribution of Structural Features. Figure 3 visualizes attention maps from the cell decoder and bounding boxes predicted from the refiner. In the bbox-only setting, attention often spreads outside the target regions, leading to inaccurate bounding boxes and truncated multi-line recognition. This suggests that sharing geometric coordinates alone is insufficient for coherent localization and decoding.

With shared and refined dense structural features, attention is more localized within the corresponding cell regions, and the predicted bounding boxes are more accurately aligned with cell boundaries. The results in Sect. 4.6 are consistent with these qualitative observations.

Contribution of Global Refinement. Figure 4 compares cell-level self-attention and entropy profiles in the causal and global refiners. The causal refiner exhibits a lower-triangular dependency pattern due to AR masking, restricting each cell to attend only to previously generated cells. The global refiner enables each cell to interact bidirectionally with all other cells, thereby forming a fully connected inter-cell dependency graph.

The entropy profiles further support this difference. The causal refiner yields a position-dependent entropy profile, as later query positions have access to more keys. In contrast, the global refiner exhibits a uniform entropy profile, indicating position-independent contextual access.

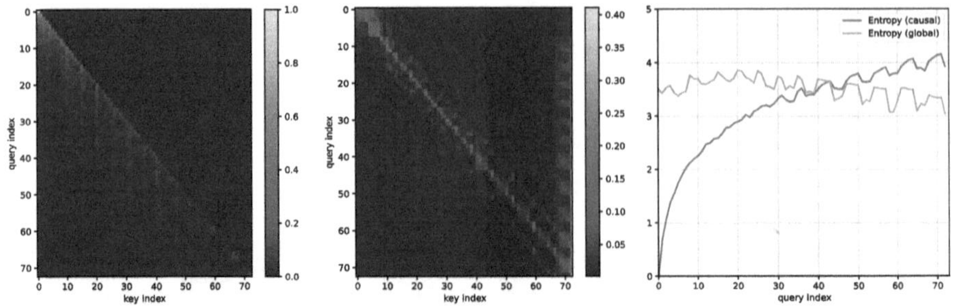

(a) Causal refiner. (b) Global refiner. (c) Entropy.

Fig. 4. Self-attention maps and entropy profiles in the causal and global refiners.

4.8 Limitations

The proposed refiner operates after the HTML decoder has generated the table structure. It improves structural representations for cell localization and content recognition, but it does not directly revise structural errors such as missing cells or incorrect spans. Such errors can degrade the final TEDS score.

5 Conclusion

We investigated how structural representations should be shared across tasks in table recognition. We highlighted generation-order dependency in autoregressive decoder states as a limitation for cell localization and content recognition.

We introduced a structural refinement module that converts order-dependent decoder states into globally coherent cell-level structural representations. Results on large-scale table datasets verify the effectiveness of the proposed method for both localization and end-to-end recognition. The ablation study further suggests that modeling fully connected dependencies, rather than causal dependencies, is key to the observed improvements. The proposed design also enables parallel cell decoding, reducing inference time by approximately threefold without degrading recognition accuracy.

These results underscore the role of dependency topology in shared structural representations and motivate explicit modeling of inter-cell dependencies.

References

1. Deng, Y., Kanervisto, A., Ling, J., Rush, A.M.: Image-to-markup generation with coarse-to-fine attention. In: International Conference on Machine Learning (ICML), vol. 70, pp. 980–989 (2017)
2. Deng, Y., Rosenberg, D., Mann, G.: Challenges in end-to-end neural scientific table recognition. In: International Conference on Document Analysis and Recognition (ICDAR), pp. 894–901 (2019). https://doi.org/10.1109/ICDAR.2019.00148

3. Huang, Y., et al.: Improving table structure recognition with visual-alignment sequential coordinate modeling. In: IEEE/CVF Conference on Computer Vision and Pattern Recognition (CVPR), pp. 11134–11143 (2023). https://doi.org/10.1109/CVPR52729.2023.01071
4. Itonori, K.: Table structure recognition based on textblock arrangement and ruled line position. In: International Conference on Document Analysis and Recognition (ICDAR), pp. 765–768 (1993). https://doi.org/10.1109/ICDAR.1993.395625
5. Kawakatsu, T.: Multi-cell decoder and mutual learning for table structure and character recognition. In: Document Analysis and Recognition - ICDAR 2024, pp. 389–405. Springer Nature Switzerland (2024). https://doi.org/10.1007/978-3-031-70533-5_23
6. Kieninger, T.G.: Table structure recognition based on robust block segmentation. In: Photonics West '98 Electronic Imaging, vol. 3305, pp. 22–32 (1998). https://doi.org/10.1117/12.304642
7. Kim, G., et al.: OCR-free document understanding transformer. In: Computer Vision – ECCV 2022, pp. 498–517 (2022). https://doi.org/10.1007/978-3-031-19815-1_29
8. Lee, K., et al.: Pix2Struct: screenshot parsing as pretraining for visual language understanding. In: 40th International Conference on Machine Learning (2023)
9. Liu, H., et al.: Show, read and reason: table structure recognition with flexible context aggregator. In: 29th ACM International Conference on Multimedia, pp. 1084–1092 (2021). https://doi.org/10.1145/3474085.3481534
10. Ly, N.T., Takasu, A.: An end-to-end local attention based model for table recognition. In: International Conference on Document Analysis and Recognition (ICDAR), pp. 20–36 (2023). https://doi.org/10.1007/978-3-031-41679-8_2
11. Ly, N.T., Takasu, A.: An end-to-end multi-task learning model for image-based table recognition. In: International Joint Conference on Computer Vision, Imaging and Computer Graphics Theory and Applications (VISIGRAPP), pp. 626–634 (2023). https://doi.org/10.5220/0011685000003417
12. Ly, N.T., Takasu, A., Nguyen, P., Takeda, H.: Rethinking image-based table recognition using weakly supervised methods. In: International Conference on Pattern Recognition Applications and Methods (ICPRAM), pp. 872–880 (2023). https://doi.org/10.5220/0011682600003411
13. Nassar, A., Livathinos, N., Lysak, M., Staar, P.: TableFormer: table structure understanding with transformers. In: IEEE/CVF Conference on Computer Vision and Pattern Recognition (CVPR), pp. 4604–4613 (2022). https://doi.org/10.1109/CVPR52688.2022.00457
14. OpenMMLab: MMOCR. https://github.com/open-mmlab/mmocr
15. Prasad, D., Gadpal, A., Kapadni, K., Visave, M., Sultanpure, K.: CascadeTabNet: an approach for end to end table detection and structure recognition from image-based documents. In: IEEE/CVF Conference on Computer Vision and Pattern Recognition Workshops (CVPRW), pp. 2439–2447 (2020). https://doi.org/10.1109/CVPRW50498.2020.00294
16. Qiao, L., et al.: LGPMA: complicated table structure recognition with local and global pyramid mask alignment. In: International Conference on Document Analysis and Recognition (ICDAR), pp. 99–114 (2021). https://doi.org/10.1007/978-3-030-86549-8_7
17. Raja, S., Mondal, A., Jawahar, C.V.: Table structure recognition using top-down and bottom-up cues. In: Computer Vision – ECCV, pp. 70–86 (2020). https://doi.org/10.1007/978-3-030-58604-1_5

18. Raja, S., Mondal, A., Jawahar, C.V.: Visual understanding of complex table structures from document images. In: IEEE/CVF Winter Conference on Applications of Computer Vision, pp. 2543–2552 (2022). https://doi.org/10.1109/WACV51458.2022.00260
19. Schreiber, S., Agne, S., Wolf, I., Dengel, A., Ahmed, S.: DeepDeSRT: deep learning for detection and structure recognition of tables in document images. In: International Conference on Document Analysis and Recognition (ICDAR), pp. 1162–1167 (2017). https://doi.org/10.1109/ICDAR.2017.192
20. Wan, J., et al.: OmniParser: a unified framework for text spotting, key information extraction and table recognition. In: IEEE/CVF Conference on Computer Vision and Pattern Recognition, pp. 15641–15653 (2024). https://doi.org/10.1109/CVPR52733.2024.01481
21. Wang, Y., Phillips, I.T., Haralick, R.M.: Table structure understanding and its performance evaluation. Pattern Recogn. **37**(7), 1479–1497 (2004)
22. Wright, L.: Ranger - a synergistic optimizer (2019). https://github.com/lessw2020/Ranger-Deep-Learning-Optimizer
23. Ye, J., Qi, X., He, Y., Chen, Y., Gu, D., Gao, P., Xiao, R.: PingAn-VCGroup's solution for ICDAR 2021 competition on scientific literature parsing task B: table recognition to HTML (2021). https://doi.org/10.48550/arXiv.2105.01848
24. Yepes, A.J., Zhong, P., Burdick, D.: ICDAR 2021 competition on scientific literature parsing. In: International Conference on Document Analysis and Recognition (ICDAR), pp. 605–617 (2021). https://doi.org/10.1007/978-3-030-86337-1_40
25. Zhang, Z., Zhang, J., Du, J., Wang, F.: Split, embed and merge: an accurate table structure recognizer. Pattern Recogn. **126**, 108565 (2022). https://doi.org/10.1016/j.patcog.2022.108565
26. Zhao, W., et al.: TabPedia: towards comprehensive visual table understanding with concept synergy. In: 38th International Conference on Neural Information Processing Systems (2024)
27. Zhao, W., Gao, L.: CoMER: modeling coverage for transformer-based handwritten mathematical expression recognition. In: Computer Vision – ECCV 2022, pp. 392–408 (2022). https://doi.org/10.1007/978-3-031-19815-1_23
28. Zheng, X., Burdick, D., Popa, L., Zhong, X., Wang, N.X.R.: Global table extractor (GTE): a framework for joint table identification and cell structure recognition using visual context. In: IEEE Winter Conference on Applications of Computer Vision (WACV), pp. 697–706 (2021). https://doi.org/10.1109/WACV48630.2021.00074
29. Zhong, X., ShafieiBavani, E., Yepes, A.J.: Image-based table recognition: data, model, and evaluation. In: Computer Vision – ECCV, pp. 564–580 (2020). https://doi.org/10.1007/978-3-030-58589-1_34
30. Zhu, Z., Zhao, W., Gao, L.: Enhancing transformer-based table structure recognition for long tables. In: Pattern Recognition and Computer Vision, pp. 216–230 (2025). https://doi.org/10.1007/978-981-97-8511-7_16

Improving MLLM Historical Record Extraction with Test-Time Image Augmentation

Taylor Archibald[1,2](✉), Bryan Morse[1], and Tony Martinez[1]

[1] Brigham Young University, Provo, UT, USA
{morse,martinez}@cs.byu.edu
[2] Ancestry.com, Draper, UT, USA
tarchibald@ancestry.com

Abstract. This paper presents a novel ensemble framework that stabilizes MLLM-based text extraction from noisy historical documents. This framework transcribes multiple augmented variants of each image with Gemini 2.0 Flash and fuses these outputs with a custom Needleman–Wunsch-style aligner that yields both a consensus transcription and a confidence score. We contribute a new dataset of 622 Pennsylvania death records, and demonstrate our method improves transcription accuracy by 4% points relative to a single-shot baseline. Results demonstrate that padding and blurring are the most useful for improving accuracy, while grid-warp perturbations are best for separating high- and low-confidence cases. The approach is simple, scalable, and immediately deployable to other document collections and transcription models.

Keywords: Historical Records · Multimodal Large Language Models · Test-Time Augmentation · Confidence Scoring

1 Introduction

While traditional Optical Character Recognition (OCR) and Handwritten Text Recognition (HTR) systems have made significant strides, they often struggle with the variability in historical texts, such as severe degradation, diverse handwriting styles, and non-standard layouts [4]. The advent of Multimodal Large Language Models (MLLMs) presents a new paradigm. Models such as GPT-4V [3,29] and Gemini [11] can perform challenging text extraction tasks in a zero-shot or few-shot manner, significantly reducing the need for extensive, task-specific training data [8,14].

However, despite early promising results, MLLMs are still susceptible to hallucination, struggle with noisy inputs, and provide no explicit or calibrated confidence score. While a model's internal log probabilities are sometimes considered indicators of confidence, these scores can be poorly calibrated or inaccessible through proprietary APIs. Indeed, emerging research suggests that

G. A. Fink et al. (Eds.): ICDAR 2026, LNCS 16972, pp. 181–196, 2027.
https://doi.org/10.1007/978-3-032-36023-6_11

model-generated confidence scores often do not correlate well with accuracy and that models can exhibit overconfidence without a coherent internal sense of their own uncertainty [19]. This unreliability necessitates external validation mechanisms and motivates our work on an ensemble framework that derives confidence from agreement rather than relying on opaque internal model states.

(a) Noise (b) Padding

(c) Grid Warp (d) Blur

Fig. 1. A death record segment under different distortions can be combined to improve predictions and provide a confidence value.

To address this challenge, we introduce a scalable, black-box ensemble framework that uses a single MLLM to generate multiple transcriptions of a document image by applying a set of test-time augmentations (Fig. 1). By design, our method requires only one model, which avoids the complexities of managing distinct models that may have asymmetric costs, performance, or hardware requirements. We adapt the Needleman–Wunsch (NW) algorithm [18] to produce both a single consensus transcription and an agreement-based confidence score. To evaluate our method, we introduce a new dataset of 622 Pennsylvania death records, curated so the evaluated models have not previously been trained on it. On this dataset, our method improves field transcription accuracy by 4% points relative to a standard single-pass baseline. We assess different ensemble selection and augmentation strategies, and identify which transforms are most effective for improving accuracy or increasing confidence reliability.

2 Related Work

Our approach builds upon recent advances in MLLMs, ensemble methods, confidence estimation, and test-time augmentation.

2.1 MLLMs for Document Understanding

MLLMs are increasingly applied to document analysis, combining visual understanding with language-based reasoning [27]. For example, LayoutLM learns joint representations of text and 2D layout information for scanned documents [28] but still requires external OCR processing, while Donut and Dessurt achieve end-to-end document parsing without explicit OCR by directly predicting content from images via a transformer [9,13].

General-purpose MLLMs, including GPT-4V [3,29], Gemini [11], and Claude 3 [2], perform well on OCR and information extraction tasks [8,14], particularly for contemporary printed or neatly handwritten text. They are less reliable on noisy, low-resource, or historical documents, where complex layouts, dense script, degraded scans, or archaic handwriting can lead even state-of-the-art models to misread characters or hallucinate content [6].

2.2 Confidence Score Generation

Estimating the confidence of extracted text is crucial for downstream usability, especially with noisy historical data. While many traditional OCR systems provide confidence scores, proprietary MLLMs usually do not expose reliable, calibrated uncertainty. Methods for estimating uncertainty are often categorized as white-box, requiring access to internal model states such as output-token probabilities, or black-box, relying only on textual output. Because many state-of-the-art models are closed-source, white-box approaches are often infeasible.

Among black-box strategies, prompting a model to verbally state its confidence can be better calibrated than using internal probabilities [21]. However, verbalized confidence can still be poorly calibrated and overconfident [26]. Subsequent work [7] confirmed overconfidence in MLLMs tested on corrupted image data: as image corruption increased, verbalized confidence remained high despite worse performance.

2.3 Ensemble Methods and Consensus Algorithms

Combining outputs from multiple sources is a well-established way to improve accuracy. Traditional methods often rely on multiple OCR/HTR engines: ROVER uses dynamic programming to align outputs and vote for the best hypothesis, often weighting votes by confidence [10]; early OCR ensembles and HMM recognizer combinations showed similar benefits [5,16].

Recent large-model approaches include self-consistency, which samples multiple reasoning paths from one model and takes a majority vote [24], and multi-model combinations [30]. ConfBERT integrates confidence scores from different OCR systems into a language model for error spotting [12], but still requires each OCR system to produce a useful confidence score. Consensus entropy estimates OCR confidence from agreement across distinct MLLMs [32]; however, relying on multiple models introduces operational cost and performance asymmetry, and our NW-consensus method was better suited to the short name fields in our dataset.

2.4 Multi-sample Consensus and Test-Time Augmentation

A key challenge for ensemble methods is generating multiple diverse outputs without running several distinct large models. Test-time augmentation (TTA) applies a single trained model to multiple augmented versions of an input at inference time. TTA is established in computer vision [17] and is being explored for large-model inference [31], including domains such as medical image segmentation where robustness to unseen variations is important [17]. Its use for structured data extraction from historical documents with diverse image transforms on a single MLLM remains less established.

Our framework combines these threads: diverse TTA with one MLLM for structured historical data extraction, producing both improved transcription accuracy and an agreement-based confidence score.

3 Method

3.1 Formal Task Definition

Let $\mathbf{I} \in \mathbb{R}^{H \times W \times C}$ denote a document image drawn from a distribution $\mathcal{P}$, and let $y \in \Sigma^*$ represent its latent ground-truth transcription. Our goal is to construct an estimator h that minimizes the expected Character Error Rate (CER).

Generative Vision Extractor. We utilize an MLLM as a stochastic mapping function $\Gamma : \mathbb{R}^{H \times W \times C} \to \Sigma^*$. Unlike deterministic OCR, Γ is sensitive to minor input variations.

Visual Perturbation Ensemble. We define a space of perturbation operators Θ (spanning affine warps, blurring kernels, and noise injection) that are approximately label-preserving. We construct a Monte Carlo approximation of the local image neighborhood by drawing N parameters $\{\theta_n\}_{n=1}^N$ to generate observations $\mathbf{I}^{(n)} = \mathcal{A}_{\theta_n}(\mathbf{I})$. This yields a set of candidate hypotheses:

$$\mathcal{S} = \{s_n\}_{n=1}^N, \quad \text{where } s_n = \Gamma(\mathbf{I}^{(n)}).$$

Consensus Aggregation. To exploit the ensemble, we define a consensus operator $\mathcal{C}$ that aggregates the hypotheses via character-level majority voting. Assuming the sequences in $\mathcal{S}$ are mapped to a common coordinate system via alignment, the consensus character at position k is:

$$\widehat{y}_k = \text{argmax}_{c \in \Sigma} \sum_{n=1}^{N} \mathbb{I}(s_{n,k} = c)$$

where $s_{n,k}$ denotes the character token at the aligned index k of the n-th hypothesis.

3.2 Theoretical Motivation: Variance Reduction

We motivate this aggregation with a stylized error model. We model the MLLM's error at a specific locus k as a Bernoulli variable $E_n \in \{0, 1\}$ with expected error probability $\varepsilon_k = \mathbb{E}[E_n]$. We adopt a compound-symmetry approximation, treating the pairwise error correlations between different perturbations as a homogeneous scalar ρ.

Since the perturbations $\{\mathbf{I}^{(n)}\}$ are derived from a single source image, $\rho > 0$. The variance of the ensemble mean error $\bar{E}$ is given by:

$$\mathrm{Var}(\bar{E}) = \frac{\varepsilon_k(1 - \varepsilon_k)}{N}\big[1 + (N - 1)\rho\big].$$

This formulation highlights the limitations of naive ensembling. If $\rho \to 1$ (highly correlated hallucinations), the variance remains constant regardless of N. We quantify the ensemble's utility using the effective sample size N_{eff} [15]:

$$N_{\text{eff}} = \frac{N}{1 + (N - 1)\rho}.$$

This model is not intended as a complete account of MLLM errors, which are structured and image-dependent. Its purpose is to clarify the quantity our augmentations should affect: not only individual accuracy, but also correlation among errors. Thus, augmentations with slightly worse individual CER can still improve the consensus if they reduce shared failure modes. This motivates our empirical use of Error Correlation and our preference for augmentation sets that improve validation-set consensus performance.

3.3 Dataset

We utilize a newly contributed dataset comprising 622 scanned images of Pennsylvania death records from the early 20th century. This dataset presents significant challenges typical of historical archives, including varying handwriting styles, image degradation (stains, fading, noise), and inconsistent layouts. These documents have not been part of existing benchmarks, ensuring no risk of training data contamination for the MLLM. The source certificates are public historical records held by the Pennsylvania State Archives, and the official 622-image Pennsylvania death-record dataset and labels are released at github.com/Tahlor/pa-death-records-622. The benchmark release documents the 622-image selection, name-field annotations, and cross-validation fold assignments.

The primary task is structured information extraction, specifically targeting the 3,684 non-blank given-name and surname fields for the decedent, mother, and father. We focus on these fields because they exhibit high variability and require no normalization (unlike dates or places). These fields are crucial for genealogical and historical research, making them a stringent test for extraction accuracy, though the same methods described here can generally be applied to most fields.

3.4 MLLM and Ensemble Generation

Among the MLLMs evaluated in preliminary analysis, Gemini 2.0 Flash (Gemini) performed best overall, so we use it for the experiments reported here. For each source document, we generate 20 distorted images for each distortion category and query the model once for each image, yielding 20 candidate extractions, each returned in a structured JSON format. For temperature experiments, we query the model 20 times on the unaltered image for the given sampling temperature. This approach operates entirely at test time, requires no model fine-tuning, and leverages the MLLM's sensitivity to input variations to explore the solution space.

The practical cost of this design is additional inference: a 10-sample ensemble requires 10 model calls for a document image rather than one. At the Gemini API prices in effect during revision, 2.0 Flash cost roughly an order of magnitude less than 2.5 Pro for text/image input (\$0.10 versus \$1.25 per million tokens for prompts up to 200K), and less still for output (\$0.40 versus \$10.00). See the Gemini API pricing page, accessed May 2026. We therefore frame the method as an accuracy and confidence-reliability tradeoff, most useful when the value of fewer transcription errors or fewer manual reviews exceeds the added inference cost. Because the consensus step only consumes returned text, the framework is model-agnostic in principle; broader evaluation across other MLLMs and larger single-pass baselines remains future work.

3.5 Data Augmentations

To investigate the MLLM's response to various document image transformations, we employ 5 image-distortion strategies and 1 temperature variation strategy. For the image-distortion strategies, we conduct a grid search over 20 configurations and create an ensemble from the top 10. This selection is based on a 5-fold cross-validation performance on a dedicated validation set, with 10 samples generated, one for each chosen configuration.

While preliminary tests with Gemini showed good accuracy on the unaltered, high-resolution (median 3136×2904 pixels) images, we determined that processing at this native resolution was not essential for transcription and mostly incurred higher costs. We swept various resizing factors and found that a 50% downscale was the smallest scale that performed no worse than the original. Consequently, we adopt 50% resizing as a baseline post-processing step for most augmentations, as noted. The strategies we evaluate include:

1. Blur + Resize: This strategy combines Gaussian blurring using various odd-numbered pixel kernels (5–17px) with subsequent input to the MLLM at 100%, 75%, or 50% of the original image resolution (omitting the 17px kernel for 50% scale) to yield 20 distinct configurations.
2. Resize Sweep: This strategy directly explores the effect of image resolution. Images are resized to scale factors ranging from 30% to 130% of their original size, in 5% increments, omitting the original size, to achieve 20 variations.

3. Gaussian Noise: This strategy simulates film grain or sensor noise by adding patch-based Gaussian noise to the full-resolution image, combining patch sizes of {2, 4, 8, 16} pixels with noise standard deviations (σ) of {4, 6, 8, 10, 15}, and then downscaling the image to 50% of its original size.
4. Pixel Shift Padding: We shift image content within a padded canvas using offsets of {8, 16, 32, 64, 128} pixels in the four diagonal directions, then downscale the result to 50% of its original size. We hypothesize that these shifts force the model to re-tile and re-patchify the image, creating "token jitter" that exposes new content at boundaries. This approach leverages the sensitivity of Vision Transformers to patch offsets [22,23] and allows us to probe the model's response to grid alignment changes [20].
5. Grid Warp: This strategy applies a smooth, non-linear distortion to the full-resolution image, simulating physical warping or creasing using mesh intervals of {70, 85, 100, 115, 130} pixels and distortion σ of {1, 2, 3, 4}, based on principles similar to those in [25], then downscaling the image to 50% of its original size.
6. Temperature: This strategy investigates the MLLM's output diversity by querying it with the same baseline image (downscaled to 50% resolution), generating 20 samples each for temperatures of 0.5, 1.0, and 2.0, plus a single sample at temperature 0.0 as a quasi-deterministic reference, with each temperature setting constituting a separate experiment. For all temperature experiments, we set the cumulative probability threshold or top-p to 0.95. We posit that sampling with non-zero temperature can be used to approximate a white-box confidence estimation method, as both explore the model's output distribution.

Notably, we found that even when temperature was set to 0, transcriptions were not completely deterministic, though the variation between runs was small (e.g., the CER differed by less than 0.1 percentage points).

3.6 Shift Analysis

To investigate the periodicity of model errors relative to the input grid, we conducted a fine-grained shift analysis. We shifted the image horizontally and vertically in 2px increments (Fine-grained Shift, range ±32px) and 64px increments (Coarse-grained Shift, range ±384px). We then computed the agreement rate between the shifted and original transcriptions and performed a Fast Fourier Transform (FFT) on the resulting signal to identify dominant periodic components.

3.7 Finding Consensus

For each image, we have N transcriptions obtained from distinct test-time augmentations. Starting with the first string as a reference, we progressively align every new string to the current consensus with a NW [18] dynamic program (*match* = +1, *mismatch* = −1, *gap* = −1). Alignment columns carry a vote tally. After all N strings are processed, the consensus character at column j is the non-gap symbol with the most votes; its confidence is that vote count divided by N. Word-level confidence is the minimum of the character confidences in the word.

This approach extends classic methods like ROVER [10] by incorporating dynamic gap/ambiguity symbols, explicit case handling, and importantly, an intrinsic confidence score that serves as a strong estimator of CER at inference time. Default hyperparameters were robust in cross-validation.

3.8 Evaluation

To assess the effectiveness of our ensemble framework, we evaluate both the accuracy of the transcriptions and the reliability of the confidence scores.

Our primary accuracy metric is Character Error Rate (CER), calculated as the Levenshtein distance between the predicted and ground truth transcriptions, normalized by the length of the ground truth. We also measure Field Accuracy, which is the percentage of transcriptions that are an exact match to the ground truth. For both metrics, we ignore case and exclude punctuation and spaces.

We evaluate performance at both the individual sample level and the consensus level. For a set of augmentations, we report the average CER and Field Accuracy across all individual transcriptions. We then report the same metrics for the single consensus transcription generated by our alignment algorithm.

To understand why certain ensembles are more effective than others, we measure the Error Correlation between pairs of transcriptions generated from different augmentations of the same source image. This is calculated as the Pearson correlation of the binary error vectors based on Field Accuracy (where 1 indicates an exact match failure and 0 indicates success) for each field across the dataset. Lower correlation suggests that augmentations are introducing diverse failure modes, which is desirable for an effective ensemble.

Finally, we assess the calibration of our agreement-based confidence scores. We use several standard metrics:

- Expected Calibration Error (ECE) and Adaptive Calibration Error (ACE), which measure the weighted average difference between confidence and accuracy across a number of bins.
- The Brier Score, which is the mean squared error between the predicted probability and the actual outcome.
- Correlation between confidence scores and the actual accuracy.

For these calibration metrics, we report the values for the raw confidence scores and also after applying isotonic regression, a non-parametric method to recalibrate the scores to better reflect the true likelihood of correctness.

4 Results

Our experiments demonstrate that the ensemble-based approach improves transcription accuracy and provides meaningful confidence scores. We evaluated augmentation strategies and ensemble selection methods to identify which combinations are most effective.

Table 1. Average and Consensus Character Error Rate (CER), Field Accuracy, and Error Correlation by Transform Category

Transform	CER (%)			Field Accuracy (%)		Error Corr.
	Avg.	5 samp.	10 samp.	5 samp.	10 samp.	
Baseline	9.0	9.0	9.0	71.2	71.2	—
Temperature = 0.5	9.3	8.6	8.7	71.3	71.5	0.871
Temperature = 1.0	10.2	9.1	8.8	69.9	70.4	0.780
Temperature = 2.0	11.9	9.5	9.3	69.6	69.5	0.715
Blur and Resize	9.1	8.0	8.0	73.3	73.1	0.737
Gaussian Noise	9.3	8.1	8.0	72.7	73.3	0.745
Grid Warp	12.4	8.7	8.0	71.8	73.3	**0.575**
Resize	**8.8**	8.7	8.7	71.9	71.7	0.973
Pad	9.0	**7.4**	**7.4**	**74.6**	**74.8**	0.726

Baseline results are for 1-sample predictions (no consensus). Transform categories are grouped by experiment type. Lower error correlations indicate better diversity in failure modes

4.1 Performance of Augmentation Strategies

We first evaluated each augmentation category. Five-fold cross-validation selected the best-performing configurations from a validation set to form ensembles of 5 and 10. As shown in Table 1, nearly all augmentation strategies yield a consensus transcription that is more accurate than the single-pass baseline.

The Pad augmentation achieved the lowest consensus CER (7.4%) and highest Field Accuracy (74.8% with 10 samples), improving over the baseline by 1.6 CER points and 3.6 Field Accuracy points. Blur and Resize and Gaussian Noise also perform well. Grid Warp has the highest average individual-sample CER (12.4%) but still produces an effective consensus transcription (8.0% CER), outperforming several stronger individual augmentations.

This counterintuitive result is explained by Grid Warp's low Error Correlation (0.575). Its distortions cause the model to fail in diverse and less-correlated ways, making errors easier to correct during consensus alignment. In contrast, Resize augmentations produce highly correlated errors (0.973), yielding little ensemble gain. Figure 2 shows that most augmentations improve over the single-pass baseline, but the magnitude depends on diversity rather than individual accuracy alone. The "All Experiments" series provides a naive global-accuracy baseline that prioritizes individually strong augmentations without preserving categorical diversity.

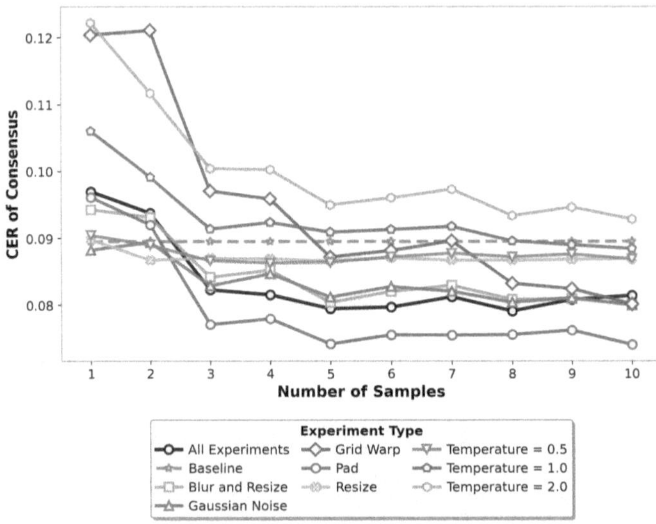

Fig. 2. Consensus CER vs. number of samples by experiment type. The black "All Experiments" line selects the N lowest-validation-CER augmentations from the full pool rather than preserving category diversity.

A small portion of the improvement comes from field recovery. The single-pass baseline failed to extract 7 name fields; depending on the augmentation strategy, the ensembles recovered 4 to 7 of them, reducing overall CER by 0.13 to 0.19 percentage points.

4.2 Ensemble Selection Strategy

The method for selecting augmentations is critical. Choosing augmentations by individual validation CER is suboptimal because their errors are often correlated. Selecting augmentations by their validation-set consensus CER works better: Table 2 and Fig. 3 show that "Best Validation Consensus-CER" reaches 7.2% CER and 75.2% Field Accuracy with 10 samples, a 4-percentage-point accuracy improvement over the 71.2% baseline. This approaches the label-informed Oracle upper bound.

4.3 Confidence and Calibration

A key advantage of our ensemble method is that it generates a confidence score for each transcription. Grid Warp, which creates diverse errors, is particularly effective for calibration. As shown in Table 3, Grid Warp has the highest raw correlation (0.642) between confidence and accuracy, making it a strong choice when confidence reliability is the priority.

The precision-recall tradeoff in Fig. 4 highlights the different strengths of Pad and Grid Warp. We define a positive as a field where the ensemble reached

Table 2. Character Error Rate and Field Accuracy by Ensemble Method, with Oracle as a label-informed upper bound

Method	CER (%)		Field Accuracy (%)	
	5 samp.	10 samp.	5 samp.	10 samp.
Baseline	9.0	9.0	71.2	71.2
Oracle	7.0	6.8	75.6	76.2
Best Validation Consensus-CER	7.5	7.2	74.6	75.2
Best Validation CER	7.9	8.1	73.5	73.3

Oracle greedily selects the augmentation that gives the best label-aware improvement at each step, so it is shown only as an upper-bound reference

Table 3. Reliability Analysis for Specific Transform Categories: Raw vs Isotonic Regression Calibration

Transform	# Exp	# Rec	Correlation		ECE		ACE		Brier	
			Raw	Isotonic	Raw	Isotonic	Raw	Isotonic	Raw	Isotonic
Blur and Resize	20	3684	0.539	0.567	0.2188	0.0218	0.2182	0.0256	0.2138	0.1313
Gaussian Noise	20	3684	0.573	0.598	0.2225	0.0205	0.2225	0.0286	0.2166	0.1287
Grid Warp	20	3684	**0.642**	**0.598**	**0.1677**	0.0476	**0.1677**	0.0328	**0.1746**	**0.1255**
Pad	20	3684	0.521	0.569	0.2043	0.0167	0.2036	0.0319	0.2022	0.1302
Resize	20	3684	0.202	0.205	0.2758	0.0226	0.2755	0.0294	0.2748	0.1880
Temperature = 0.5	12	3684	0.438	0.465	0.2663	**0.0142**	0.2663	**0.0256**	0.2600	0.1639
Temperature = 1.0	12	3684	0.524	0.531	0.2510	0.0255	0.2505	0.0498	0.2437	0.1488
Temperature = 2.0	12	3684	0.589	0.553	0.2402	0.0254	0.2402	0.0531	0.2316	0.1511

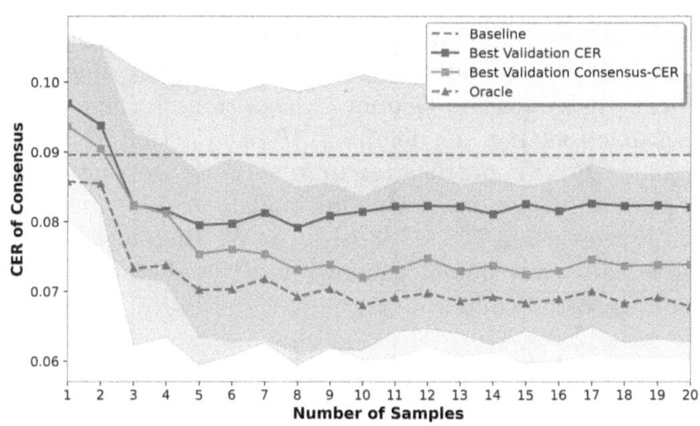

Fig. 3. Comparison of ensemble selection methods. Selecting augmentations by contribution to consensus accuracy (orange) outperforms selecting by individual accuracy (blue) and approaches the Oracle upper bound (green dashed). (Color figure online)

a unanimous prediction. Pad achieves the highest F1 score, while Grid Warp provides superior precision, especially as sample count increases. This is useful for workflows where high-confidence transcriptions are automatically accepted and low-confidence fields are routed to human review.

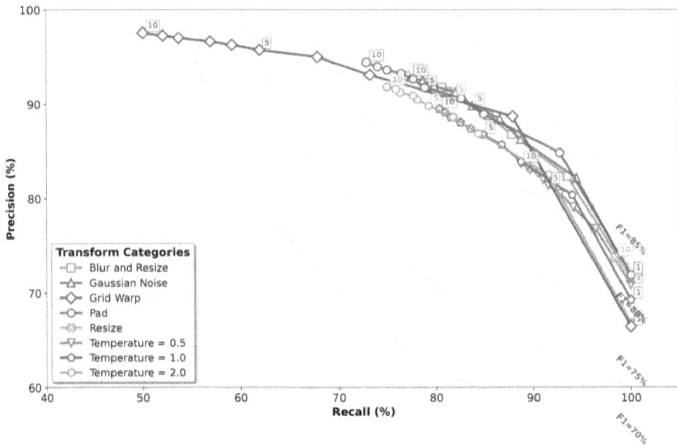

Fig. 4. Precision-recall for unanimous predictions. Pad (blue) gives a strong F1 balance, while Grid Warp (red) gives higher precision, especially with more samples. (Color figure online)

4.4 Shift Analysis and Hierarchical Alignment

We investigated the effectiveness of simple pixel shifts by analyzing agreement rate as a function of shift magnitude. Figure 5 shows that agreement does not decay monotonically; instead, it exhibits periodic signals aligned with Vision Transformer (ViT) patching and Gemini's image-tiling behavior.

In the fine-grained setting, the dominant 16-pixel period in Fig. 5(b) matches the common 16 × 16 ViT patch size. In the coarse setting, agreement also reflects larger structural windows; Gemini models are documented to process high-resolution images using 768 × 768 tiles [1]. We therefore interpret padding gains as marginalizing over patch, window, and tile alignment artifacts rather than simply adding visual context.

4.5 Qualitative Analysis of Errors

The ensemble is not infallible. In an extended analysis using the full pool of 100 augmentation configurations, 870 of 3,684 predicted name fields were unanimous; 7 unanimous fields disagreed with ground truth. These rare failures mainly involved ambiguous characters, spelling autocorrection of unusual names (for

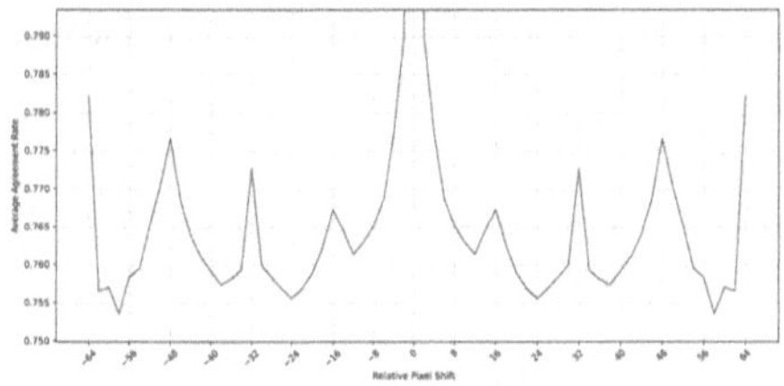

(a) Fine shift agreement (±32px)

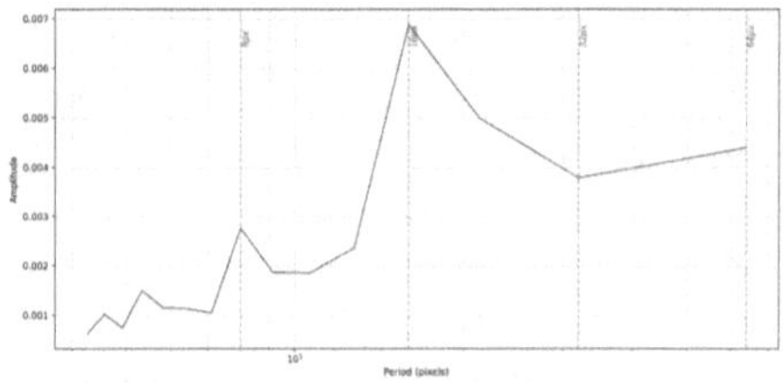

(b) Fine shift FFT period

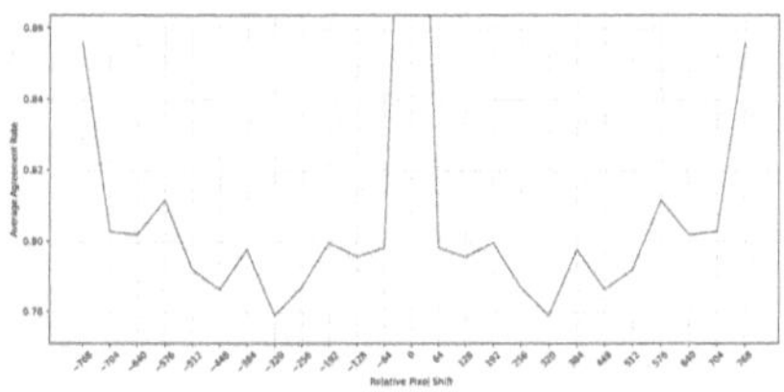

(c) Coarse shift agreement (±384px)

Fig. 5. Effect of pixel shifts on model agreement. Fine-grained shifts recover agreement every 16 pixels, FFT analysis confirms a dominant 16px periodicity, and coarse-grained shifts show broader trends near larger structural windows, including 768px.

example, "Mrytle" to "Myrtle"), and omissions of struck-through or interlineated text. These errors highlight remaining challenges in transcribing complex historical documents and motivate future improvements in MLLM training and adaptation.

5 Conclusion and Future Work

This paper presented a simple yet effective ensemble framework that improves the reliability of MLLM-based text extraction from noisy historical documents. By generating and aligning transcriptions from multiple augmented images, our method boosts accuracy by 4 percentage points and produces a calibrated confidence score without accessing internal model states. Specific augmentations serve distinct purposes: padding is most effective for accuracy, while grid-warp's diverse errors are useful for confidence estimation. The approach is black-box, scalable, and readily applicable to other document collections.

Several avenues for future work remain. The consensus mechanism could be refined beyond simple majority voting to a weighted scheme, perhaps giving more influence to predictions from less severe augmentations or by combining different degradation types into a single, optimized ensemble. Because multi-sample inference increases latency and API cost, future systems should also study adaptive sampling policies that stop early for high-confidence fields and reserve larger ensembles for ambiguous cases. The framework's effectiveness should be validated on a broader range of documents, such as long-form letters or journal collections, and across additional MLLMs and larger single-pass baselines, which would clarify how much of the observed gain is model-specific.

Future work could also investigate the precise mechanics of "token jitter" by logging tile grids under controlled shifts or optimizing padding policies to maximize information retention at patch boundaries.

Acknowledgments. The authors thank the Handwriting Recognition Team at Ancestry.com for providing the dataset for this study.

Disclosure of Interests. Taylor Archibald was affiliated with both Brigham Young University and Ancestry.com while this work was conducted. The authors have no other competing interests to declare that are relevant to the content of this article.

References

1. Image understanding — Gemini API. https://ai.google.dev/gemini-api/docs/image-understanding
2. Introducing the next generation of Claude. https://www.anthropic.com/news/claude-3-family
3. GPT-4V(ision) system card. https://openai.com/index/gpt-4v-system-card/ (2024)
4. AlKendi, W., Gechter, F., Heyberger, L., Guyeux, C.: Advancements and challenges in handwritten text recognition: a comprehensive survey. J. Imaging **10**(1), 18 (2024). https://doi.org/10.3390/jimaging10010018
5. Bertolami, R., Bunke, H.: Hidden Markov model-based ensemble methods for offline handwritten text line recognition. Pattern Recogn. **41**(11), 3452–3460 (2008). https://doi.org/10.1016/j.patcog.2008.04.003
6. Biswas, A., Talukdar, W.: Robustness of structured data extraction from in-plane rotated documents using multi-modal large language models (LLM) (2024)
7. Borszukovszki, M., de Jong, I.P., Valdenegro-Toro, M.: Know what you do not know: verbalized uncertainty estimation robustness on corrupted images in vision-language models (2025). https://doi.org/10.48550/arXiv.2504.03440, arXiv:2504.03440
8. Crosilla, G., Klic, L., Colavizza, G.: Benchmarking large language models for handwritten text recognition (2025). https://doi.org/10.48550/arXiv.2503.15195, arXiv:2503.15195
9. Davis, B., Morse, B., Price, B., Tensmeyer, C., Wigington, C., Morariu, V.: End-to-end document recognition and understanding with Dessurt. In: Computer Vision – ECCV 2022 Workshops: Tel Aviv, Israel, October 23–27, 2022, Proceedings, Part IV, pp. 280–296. Springer-Verlag, Berlin, Heidelberg (2023). https://doi.org/10.1007/978-3-031-25069-9_19
10. Fiscus, J.: A post-processing system to yield reduced word error rates: recognizer Output Voting Error Reduction (ROVER). In: 1997 IEEE Workshop on Automatic Speech Recognition and Understanding Proceedings, pp. 347–354 (1997). https://doi.org/10.1109/ASRU.1997.659110
11. Gemini Team: Gemini: A Family of Highly Capable Multimodal Models (2025). https://doi.org/10.48550/arXiv.2312.11805, arXiv:2312.11805

12. Hemmer, A., Coustaty, M., Bartolo, N., Ogier, J.M.: Confidence-aware document OCR error detection. In: Document Analysis Systems: 16th IAPR International Workshop, DAS 2024, Athens, Greece, August 30–31, 2024, Proceedings, pp. 213–228. Springer-Verlag, Berlin, Heidelberg (2024). https://doi.org/10.1007/978-3-031-70442-0_13
13. Kim, G., et al.: OCR-free document understanding transformer. In: Computer Vision – ECCV 2022: 17th European Conference, Tel Aviv, Israel, October 23–27, 2022, Proceedings, Part XXVIII, pp. 498–517. Springer-Verlag, Berlin, Heidelberg (2022). https://doi.org/10.1007/978-3-031-19815-1_29
14. Kim, S., Baudru, J., Ryckbosch, W., Bersini, H., Ginis, V.: Early evidence of how LLMs outperform traditional systems on OCR/HTR tasks for historical records (2025). https://doi.org/10.48550/arXiv.2501.11623, arXiv:2501.11623
15. Kish, L.: Survey Sampling. Survey Sampling, Wiley, Oxford, England (1965)
16. Lopresti, D., Zhou, J.: Using consensus sequence voting to correct OCR errors. Comput. Vis. Image Underst. **67**(1), 39–47 (1997). https://doi.org/10.1006/cviu.1996.0502
17. Ma, X., Tao, Y., Zhang, Y., Ji, Z., Zhang, Y., Chen, Q.: Test-time generative augmentation for medical image segmentation (2024). https://doi.org/10.48550/arXiv.2406.17608, arXiv:2406.17608
18. Needleman, S.B., Wunsch, C.D.: A general method applicable to the search for similarities in the amino acid sequence of two proteins. J. Mol. Biol. **48**(3), 443–453 (1970). https://doi.org/10.1016/0022-2836(70)90057-4
19. Pawitan, Y., Holmes, C.: Confidence in the reasoning of large language models. Harvard Data Sci. Rev. **7**(1) (2025). https://doi.org/10.1162/99608f92.b033a087
20. Qiao, L., et al.: UniViTAR: unified vision transformer with native resolution (2025). https://doi.org/10.48550/arXiv.2504.01792, arXiv:2504.01792
21. Tian, K., et al.: Just ask for calibration: strategies for eliciting calibrated confidence scores from language models fine-tuned with human feedback. In: Bouamor, H., Pino, J., Bali, K. (eds.) Proceedings of the 2023 Conference on Empirical Methods in Natural Language Processing, pp. 5433–5442. Association for Computational Linguistics, Singapore (2023). https://doi.org/10.18653/v1/2023.emnlp-main.330
22. Touvron, H., Cord, M., El-Nouby, A., Verbeek, J., Jégou, H.: Three things everyone should know about vision transformers. In: Avidan, S., Brostow, G., Cissé, M., Farinella, G.M., Hassner, T. (eds.) Computer Vision – ECCV 2022, pp. 497–515. Springer Nature Switzerland, Cham (2022). https://doi.org/10.1007/978-3-031-20053-3_29
23. Wang, F., et al.: Scaling Laws in Patchification: an image is worth 50,176 tokens and more (2025). https://doi.org/10.48550/arXiv.2502.03738, arXiv:2502.03738
24. Wang, X., et al.: Self-consistency improves chain of thought reasoning in language models. In: The Eleventh International Conference on Learning Representations (2022)
25. Wigington, C., Stewart, S., Davis, B., Barrett, B., Price, B., Cohen, S.: Data augmentation for recognition of handwritten words and lines using a CNN-LSTM network. In: Proceedings of the International Conference on Document Analysis and Recognition, ICDAR, vol. 1, pp. 639–645. IEEE Computer Society (2018). https://doi.org/10.1109/ICDAR.2017.110
26. Xiong, M., et al.: Can LLMs express their uncertainty? An empirical evaluation of confidence elicitation in LLMs. In: The Twelfth International Conference on Learning Representations (2023)

27. Xu, D., et al.: Large language models for generative information extraction: a survey. Front. Comp. Sci. **18**(6), 186357 (2024). https://doi.org/10.1007/s11704-024-40555-y
28. Xu, Y., Li, M., Cui, L., Huang, S., Wei, F., Zhou, M.: LayoutLM: pre-training of text and layout for document image understanding. In: Proceedings of the 26th ACM SIGKDD International Conference on Knowledge Discovery & Data Mining, pp. 1192–1200 (2020). https://doi.org/10.1145/3394486.3403172
29. Yang, Z., et al.: The Dawn of LMMs: preliminary explorations with GPT-4V(ision) (2023). https://doi.org/10.48550/arXiv.2309.17421, arXiv:2309.17421
30. Young, R.J., Matthews, A.M., Poston, B.: Benchmarking multiple large language models for automated clinical trial data extraction in aging research. Algorithms **18**(5), 296 (2025). https://doi.org/10.3390/a18050296
31. Zhang, Q., et al.: A survey on test-time scaling in large language models: what, how, where, and how well? (2025). https://doi.org/10.48550/arXiv.2503.24235, arXiv:2503.24235
32. Zhang, Y., et al.: Consensus Entropy: harnessing multi-VLM agreement for self-verifying and self-improving OCR (2025). https://doi.org/10.48550/arXiv.2504.11101, arXiv:2504.11101

Handwriting Trajectory Recovery with Diffusion Models

Hiroki Nagamatsu(✉), Shoji Toyota, and Seiichi Uchida

Kyushu University, Fukuoka, Japan
hiroki.nagamatsu@human.ait.kyushu-u.ac.jp

Abstract. Recovering online pen trajectories from offline handwriting images, often referred to as handwriting trajectory recovery (stroke recovery), is an offline-to-online conversion task with applications in stroke-level editing and forensic analysis. We propose, to the best of our knowledge, the first diffusion-model-based framework for this task. Our method formulates trajectory recovery as image-conditioned generation and uses a denoising diffusion model to sample pen trajectories consistent with the observed ink trace. Through extensive quantitative evaluations on CASIA-OLHWDB (1.0–1.1), we verify that the proposed approach enables accurate recovery even for complex multi-stroke characters, substantially improving both temporal similarity (DTW/LDTW) and shape fidelity (AIoU) over representative prior methods such as PEN-Net and Cross-VAE. We further show that the model captures general stroke-order tendencies and generalizes to classes unseen during training, exemplified by cross-script transfer: a model trained on Chinese characters can recover reasonable stroke orders for Latin letters to some extent.

Keywords: handwriting trajectory recovery · offline-to-online conversion · diffusion models · conditional generation

1 Introduction

Handwriting trajectory recovery (also referred to as *stroke recovery*) is an *offline-to-online conversion* task: given a completed offline handwriting image, the goal is to reconstruct a time-ordered pen trajectory that is consistent with the observed ink trace. Recovering the trajectory turns a static raster image into a dynamic "digital ink" representation, enabling stroke-level operations (e.g., retouching or deleting individual strokes) and resolution-independent rendering by resampling the continuous path before rasterization. Temporal cues are also valuable for downstream analysis—for example, directionality and pen dynamics can improve comparisons in signature verification and related tasks [12].

Figure 1 illustrates the modality gap between an offline character image and its underlying online trajectory. While rasterizing a trajectory into an image (often called *inking*) is straightforward, the inverse mapping is fundamentally ill-posed: a static image records only the final ink trace, not the writing process that

G. A. Fink et al. (Eds.): ICDAR 2026, LNCS 16972, pp. 197–214, 2027.
https://doi.org/10.1007/978-3-032-36023-6_12

produced it. For simple glyphs (e.g., letters or digits with few strokes and limited pen-up events), heuristic rules can sometimes recover a plausible order [4,6], but ambiguity grows rapidly for complex multi-stroke characters where crossings, tight loops, and stroke concatenations are frequent.

Fig. 1. Example of an offline character image and its corresponding online handwriting trajectory.

The goal of this paper is to experimentally demonstrate that the handwriting trajectory recovery task for complex multi-stroke characters can be solved with high accuracy using *diffusion models* [5,16]. Although diffusion models are widely used for image synthesis, the same denoising framework can also generate 1D time-series data by using a 1D denoiser. For instance, a model trained on many samples of the 14-stroke character “” can stochastically generate diverse trajectories of “” by changing the initial noise.

In trajectory recovery, however, the objective is not to generate diverse trajectories arbitrarily, but to generate the particular trajectory that corresponds to the observed ink trace in a handwriting image. To this end, we propose an image-conditioned diffusion model for time-series generation, where the input handwriting image guides the denoising process. Moreover, to obtain more accurate estimates, we introduce multi-scale conditioning, which gradually injects image evidence from coarse global shape down to fine local details. With this model, even very complex multi-stroke characters can be recovered with high precision.

Several learning-based models for handwriting trajectory recovery have been proposed, including PEN-Net [2] and Cross-VAE [17]. However, they often struggle with the complexity of multi-stroke characters, and their recovered trajectories may fail to preserve the original glyph shape. In particular, the predicted pen path can drift away from the observed ink trace, producing pen trajectories that pass through regions where no stroke exists in the input image. In contrast, by leveraging diffusion models for the trajectory recovery task—to the best of our knowledge, for the first time—we aim to address these limitations.

Experiments demonstrate that the proposed model achieves accurate recovery even for complex characters such as those in Fig. 1. Interestingly, we also observe that recovery accuracy improves as the number of training character classes increases (e.g., from 20 classes to 3,000 classes). This suggests that handwriting trajectories have general trends across character classes. One plausible source of this regularity is the reuse of radicals across characters. Although the characters "," "," and "" are different classes, they share the radical "," and their handwriting trajectories are largely consistent; therefore, more classes and more training data can directly improve recovery accuracy. Beyond such radical-level trends, our experiments indicate that the model also learns more general writing tendencies (e.g., left-to-right and top-to-bottom tendencies). Concretely, a model trained on Chinese characters can recover stroke orders for Latin letters to some extent, demonstrating cross-script generalization.

Our main contributions are as follows.

- We address handwriting trajectory recovery with diffusion models for the first time. Specifically, we propose an image-conditioned diffusion model for time-series generation, and show that multi-resolution conditioning further improves recovery accuracy.
- Experiments on CASIA-OLHWDB show clear improvements over PEN-Net and Cross-VAE in terms of both temporal alignment (DTW/LDTW) and shape fidelity (AIoU).
- We experimentally demonstrate that the learned model generalizes to not only unseen Chinese character classes but also unseen Latin alphabet characters. We also show that it produces plausible recovery results even for "incomplete" characters where parts of strokes are intentionally removed.

2 Related Work

As reviewed in [12], classical handwriting trajectory recovery methods rely on heuristic rules to resolve ambiguous structures such as stroke intersections, overlapping strokes (double-strokes), and loops. In general, these methods (i) detect keypoints such as intersections and endpoints, (ii) decompose the ink trace into sub-strokes based on these keypoints, and (iii) reconnect the sub-strokes according to hand-crafted rules. For instance, a "+"-shaped intersection may be split into four sub-strokes and then reconnected into a horizontal segment "–" and a vertical segment "|," rather than two corner-like fragments (e.g., "⌞" and "⌝"). Such rule-based methods have mainly been applied to single-stroke characters (or single-stroke cursive words) [3,4,6,13], and have rarely been used for more complex multi-stroke characters, because multi-stroke characters introduce many more intersections and ambiguities, making rule-based reconnection impractical.

Deep neural networks have increasingly been used for handwriting trajectory recovery, with different design choices on how strongly the output is constrained by intermediate structures. One line of work explicitly decomposes the input into sub-strokes and then predicts their temporal evolution or ordering, like the above classical methods. Pen Tip Motion Prediction (PTMP) [18] tackles drawing-order recovery through pen-tip motion prediction: a CNN takes partially drawn handwriting images as input and outputs a distribution over the next pen position, which is then used in an iterative framework to generate the pen movement sequence. Setsort [10] further constrains the solution space at the sub-stroke level by first inferring a skeleton with a CNN and cutting it into sub-strokes; a Sub-stroke Encoding Transformer (SET) embeds each sub-stroke point sequence, and a Sub-strokes Ordering Transformer (SORT) predicts the discrete ordering together with the pen state. Such structure-constrained pipelines can produce geometrically precise reconstructions when the extracted skeleton/sub-strokes are reliable, but they inherit the limitations of classical methods on densely written multi-stroke characters.

Another line of work learns a more direct mapping from an offline image to an online coordinate sequence using encoder–decoder architectures, reducing reliance on explicit stroke extraction. DED-Net [1] proposes an end-to-end deep encoder–decoder network that encodes the offline image into features and decodes them into a trajectory point sequence. Kanji-Net [11] adapts encoder–decoder recovery to multi-stroke characters by incorporating an attention mechanism to focus on local context during sequential generation and by using a mixture density output (GMM) to improve robustness to handwriting distortions and writer variation. PEN-Net [2] specifically targets complex glyphs and long trajectories with a parsing-and-tracing encoder–decoder: it introduces a double-stream parsing encoder to capture complementary horizontal/vertical context, and a global tracing decoder that conditions each decoding step on glyph parsing features to mitigate trajectory-point drifting; it also proposes AIoU and LDTW to jointly evaluate glyph fidelity and temporal alignment for complex characters. From a different perspective, Cross-VAE [17] performs bidirectional modality conversion by coupling an image VAE and a trajectory VAE through a shared latent space with an explicit space-sharing loss, enabling stroke recovery by encoding the offline image and decoding it into an online trajectory. Despite these advances, recovering accurate stroke order while maintaining strict geometric consistency with the input ink trace remains challenging for highly complex multi-stroke characters, and several methods still exhibit noticeable drift or inconsistencies in difficult regions.

In contrast to the above pipelines, we introduce diffusion models to handwriting trajectory recovery for the first time, to the best of our knowledge. Our approach performs image-conditioned denoising directly on a 1D trajectory sequence: starting from Gaussian noise, a 1D denoiser iteratively updates the *entire* trajectory while being guided by the input handwriting image through multi-resolution conditioning [5,16]. This design mitigates two recurring limitations of prior methods. First, it avoids fragile intermediate representations such as explicit skeletonization and sub-stroke segmentation by letting the model refer to image evidence throughout the iterative refinement process. Second, unlike single-pass or strictly step-by-step decoding, iterative denoising enables global corrections to the trajectory over multiple refinement steps, which helps suppress geometric drift from the observed ink trace, especially for long and complex multi-stroke characters. As demonstrated in our experiments, the resulting model consistently surpasses existing approaches in both temporal alignment and glyph fidelity on complex characters.

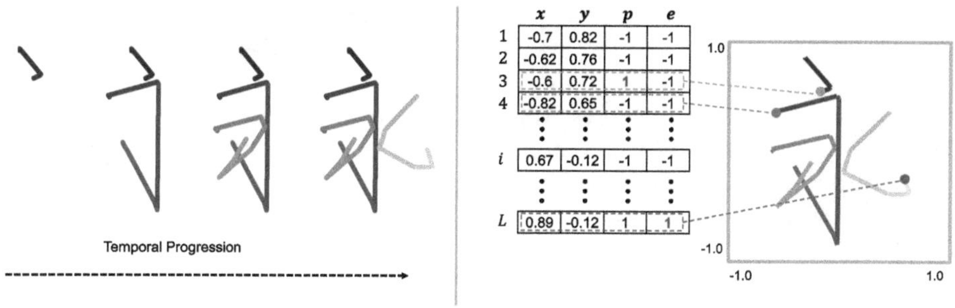

Fig. 2. Visualization of an online trajectory sequence. The color changes from dark to bright along the temporal order.

3 Handwriting Trajectory Recovery with Diffusion Models

3.1 Problem Formulation

We aim to sample plausible online handwriting trajectories conditioned on an offline image. Let $I \in \mathbb{R}^{H \times W}$ be an offline handwriting image with height $H \in \mathbb{N}_{>0}$ and width $W \in \mathbb{N}_{>0}$, and let $\mathbf{X}$ be an online trajectory sequence:

$$\mathbf{X} = \{(x_i, y_i, p_i, e_i)\}_{i=1}^{L}. \tag{1}$$

Figure 2 illustrates $\mathbf{X}$. Here (x_i, y_i) denotes the pen coordinate at sequence index i. $p_i \in \{-1, 1\}$ denotes a binary variable indicating whether the pen is up (1) or down (−1). $e_i \in \{-1, 1\}$ is an end-of-sequence (EOS) flag. Once $e_i = 1$ at the terminal point, we pad the remaining positions up to the maximum length L by repeating the terminal coordinate.

Stroke recovery from an offline image is an ill-posed inverse problem, because multiple plausible writing trajectories can produce visually similar (or even identical) rasterized character images. Therefore, instead of learning a deterministic image-to-sequence mapping, we formulate the task as learning a conditional distribution over trajectories, $p(\mathbf{X} \mid I)$.

For stable learning, we normalize the coordinates (x_i, y_i) to $[-1, 1]$. Although p_i and e_i are inherently discrete variables, we treat them as continuous values in the diffusion model and binarize them using a threshold of 0 at the final output stage. Let $\mathcal{D} = \{(I^{(n)}, \mathbf{X}^{(n)})\}_{n=1}^{N}$ be a set of paired training samples. Our goal is to learn the conditional distribution $p_\theta(\mathbf{X} \mid I)$ and recover trajectories by sampling from it.

3.2 Conditional Diffusion Model for Trajectories

A diffusion model [5,16] is a powerful approach for generating samples from a target distribution. While diffusion models are typically used to generate images or videos, they can also be applied to the generation of handwriting trajectories.

In the diffusion model framework, we define a forward diffusion process $\mathbf{X}_0 \to \mathbf{X}_1 \to \cdots \to \mathbf{X}_T$ over handwriting trajectories by gradually adding Gaussian noise to a clean handwriting trajectory $\mathbf{X}_0 = \mathbf{X}$ over T diffusion steps. Here, $t \in \{1, \ldots, T\}$ denotes the diffusion step, and $\mathbf{X}_t$ represents the noisy handwriting trajectory at step t. Given a variance schedule $\{\beta_t\}_{t=1}^{T}$ and $\alpha_t = 1 - \beta_t$, the noised handwriting trajectory at step t is expressed as

$$\mathbf{X}_t = \sqrt{\bar{\alpha}_t}\,\mathbf{X}_0 + \sqrt{1-\bar{\alpha}_t}\,\boldsymbol{\epsilon}, \quad \boldsymbol{\epsilon} \sim \mathcal{N}(\mathbf{0}, \mathbf{I}), \tag{2}$$

where $\bar{\alpha}_t = \prod_{s=1}^{t} \alpha_s$.

By considering the reverse process $\mathbf{X}_T \to \mathbf{X}_{T-1} \to \cdots \to \mathbf{X}_0$ corresponding to (2), we can generate a handwriting trajectory from the target distribution of $\mathbf{X}_0$. To model the reverse process, it is sufficient to estimate the noise term $\boldsymbol{\epsilon}(\mathbf{X}_t, t)$, which can be learned using a denoising network $\boldsymbol{\epsilon}_\theta(\mathbf{X}_t, t)$ [5].

While the diffusion models described above enable the sampling of random handwriting trajectories, our objective is to generate a trajectory that is aligned with a given offline image I, rather than an arbitrary random sample. To this end, we incorporate I as a conditioning variable into the denoising network. Formally, we define the conditioned denoising network $\boldsymbol{\epsilon}_\theta$ as

$$\hat{\boldsymbol{\epsilon}} = \boldsymbol{\epsilon}_\theta(\mathbf{X}_t, t, f_\phi(I)), \tag{3}$$

where f_ϕ denotes an image encoder applied to I. Using this conditioned denoising network, a handwriting trajectory is generated by iteratively denoising a sample drawn from a Gaussian distribution, with the generation process conditioned on I.

To train the denoising network, we adopt the standard noise-prediction objective:

$$\mathcal{L}(\theta, \phi) = \mathbb{E}_{t,I,\mathbf{X}_0,\boldsymbol{\epsilon}}\left[\|\boldsymbol{\epsilon} - \boldsymbol{\epsilon}_\theta(\mathbf{X}_t, t, f_\phi(I))\|_2^2\right]. \tag{4}$$

The expectation $\mathbb{E}_{t,I,\mathbf{X}_0,\boldsymbol{\epsilon}}$ can be empirically estimated using the training dataset $\mathcal{D} = \{(I^{(n)}, \mathbf{X}^{(n)})\}_{n=1}^{N}$ consisting of offline images and their corresponding trajectories, where the time index t is sampled uniformly from $\{1, \ldots, T\}$ for each training example.

3.3 Multi-scale Image Conditioning

Accurate stroke recovery requires preserving the spatial structure of the input image, because the offline character image provides strong geometric constraints on plausible pen trajectories. Therefore, instead of compressing the image into a single global latent vector, we condition the denoiser on multi-resolution image feature maps that keep spatially localized features. This is particularly important

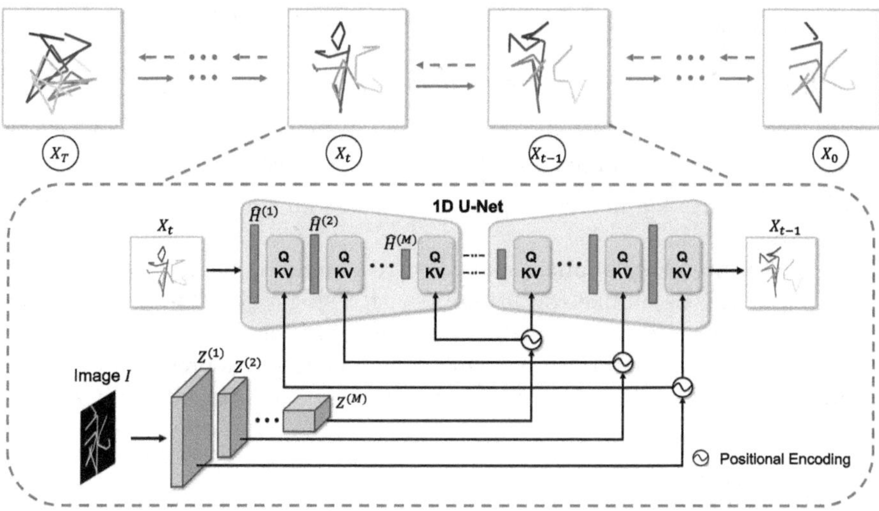

Fig. 3. Overview of the proposed image conditioned diffusion model for handwriting trajectory recovery.

in ambiguous regions (e.g., crossings, closely adjacent strokes, and loops), where local geometric evidence must be aligned with trajectory tokens.

An overview of the full architecture of the proposed model is shown in Fig. 3. Let the image encoder f_ϕ produce M feature maps at different resolutions:

$$\mathcal{Z} = f_\phi(I) = \{Z^{(m)}\}_{m=1}^{M}, \qquad Z^{(m)} \in \mathbb{R}^{H_m \times W_m \times C_m}, \tag{5}$$

where m indexes the feature scale, and H_m, W_m, and C_m denote the height, width, and channel dimensions of the m-th feature map, respectively. For each scale m, we project the feature map to the denoiser hidden dimension d, add a 2D positional encoding, and flatten it into spatial tokens:

$$\tilde{Z}^{(m)} = \text{Flatten}\Big(g^{(m)}(Z^{(m)}) + E_{\text{2D}}^{(m)}\Big) \in \mathbb{R}^{N_m \times d}, \qquad N_m = H_m W_m, \tag{6}$$

where $g^{(m)}(\cdot)$ denotes a scale-specific projection layer used to align the channel dimension to d, and $E_{\text{2D}}^{(m)}$ denotes the 2D positional encoding for the m-th feature map (with the same spatial resolution as $Z^{(m)}$). Flatten$(\cdot)$ reshapes the 2D feature grid into a sequence of N_m spatial tokens.

Our denoiser is a 1D U-Net [15] operating along the trajectory sequence axis, with 1D convolutional blocks for local continuity and self-attention layers for long-range dependencies. To inject image evidence, we insert cross-attention blocks at selected U-Net stages. At each U-Net resolution, we inject the corresponding image-feature scale via cross-attention. Let $\hat{H}^{(m)} \in \mathbb{R}^{L_m \times d}$ denote the trajectory features at the U-Net stage paired with image scale m, where L_m is the trajectory token length at that stage. We then apply cross-attention with trajectory tokens as queries and image tokens as keys and values:

$$H'^{(m)} = \hat{H}^{(m)} + \mathrm{MHCA}\Big(\hat{H}^{(m)} W_Q^{(m)},\ \tilde{Z}^{(m)} W_K^{(m)},\ \tilde{Z}^{(m)} W_V^{(m)}\Big), \tag{7}$$

where MHCA denotes standard multi-head cross-attention, and $W_Q^{(m)}$, $W_K^{(m)}$, and $W_V^{(m)}$ are learnable linear projection matrices for queries, keys, and values, respectively.

This design preserves explicit spatial correspondence between image regions and trajectory tokens, unlike conditioning with a single global vector. Moreover, multi-scale feature injection provides complementary cues: coarse-scale features capture the global character layout, while fine-scale features preserve local stroke geometry. The effect of this design is examined in the ablation study in Sect. 4.5.

4 Experiments

4.1 Experimental Setup

Chinese Handwriting Dataset. We evaluate the performance of our method on large-scale Chinese online handwriting datasets, CASIA-OLHWDB 1.0 and 1.1 [9], which contain samples from 420 and 300 writers, respectively, totaling 720 writers. In this study, we focus on the 3,755 character classes in the GB2312-80 Level-1 set. We create writer-disjoint splits with 576 writers for training and 144 writers for testing. All reported results are computed on samples written by the 144 test writers. To evaluate generalization to unseen character classes, we split the 3,755 classes into 3,000 seen classes and 755 unseen classes. We train the model using only samples from the seen classes. We then report performance on the 3,000 seen classes and the 755 unseen classes.

For fair comparison with prior work, we follow the same offline-image rendering process as PEN-Net [2]. Offline images are rendered from online trajectories. For each trajectory, we rasterize the ink trace into a 64×64 image with a black background and white strokes.

Implementation Details. We use T=1000 diffusion steps with a cosine noise schedule [14]. During testing, we use the DDPM sampler [5] with 1,000 steps. We optimize the model using Adam [7] with a batch size of 512, a learning rate of 8×10^{-5}, and 500,000 training steps. The experiments are conducted on a single A100 GPU.

Evaluation Metrics. We evaluate recovery quality from two complementary viewpoints: temporal correctness and spatial glyph fidelity. For temporal correctness, we use Dynamic Time Warping (DTW) and Length-independent DTW (LDTW) [2] as trajectory distances. DTW aligns two sequences by an optimal warping path and accumulates point-wise Euclidean distances; lower values indicate better temporal similarity. Because DTW increases with sequence length, we also use LDTW. LDTW normalizes the accumulated DTW cost by the warping-path length, enabling fair comparison across characters with different stroke counts. For DTW/LDTW computation, we rescale coordinates to the image coordinate space $[0, 64]$, so LDTW can be interpreted as an average pixel-level deviation from the ground-truth trajectory. In our evaluation, the ground-truth stroke order and direction are defined by the temporal sequence recorded in the online CASIA data, rather than by a canonical dictionary order.

For spatial glyph fidelity, we use Adaptive IoU (AIoU) [2], which measures the overlap between (i) the ink region obtained by rasterizing the recovered trajectory and (ii) the ground-truth ink region. Unlike standard IoU, AIoU adaptively adjusts the rasterization stroke width, making the metric less sensitive to minor stroke-thickness differences and better reflecting pure glyph-shape fidelity [2].

Table 1. Trajectory recovery performance on seen classes (trained on 3,000 classes). Lower is better for DTW/LDTW; higher is better for AIoU.

Method	DTW↓	LDTW↓	AIoU↑
Cross-VAE [17]	713.8	7.029	0.346
PEN-Net [2]	<u>212.3</u>	<u>2.728</u>	<u>0.607</u>
Ours	**111.1**	**1.574**	**0.745**

Baselines. We compare our method with two representative deep learning baselines for handwriting trajectory recovery. We use PEN-Net [2] as a primary baseline because it is one of the most directly comparable methods for complex character trajectory recovery under similar datasets and evaluation metrics. We additionally include Cross-VAE [17] as a complementary baseline; this is because Cross-VAE uses a non-autoregressive decoder, similarly to our approach. Since their official implementations are not publicly available, we re-implement both baselines following the original papers and tune hyperparameters to reproduce the reported behavior.

4.2 Main Results on Seen Classes

We first report the main benchmark results on the seen classes test set. Table 1 shows that our method achieves the best performance across all metrics. In handwriting trajectory recovery, the offline image specifies the target ink layout; altering the glyph shape is invalid because it no longer corresponds to the given input and effectively becomes trajectory synthesis for a different shape. Therefore, evaluation should prioritize fidelity to the given rasterized shape while assessing ordering consistency. DTW/LDTW are affected by both temporal alignment and geometric deviations, so they do not isolate stroke order on their own. In particular, geometric drift can increase DTW/LDTW even when

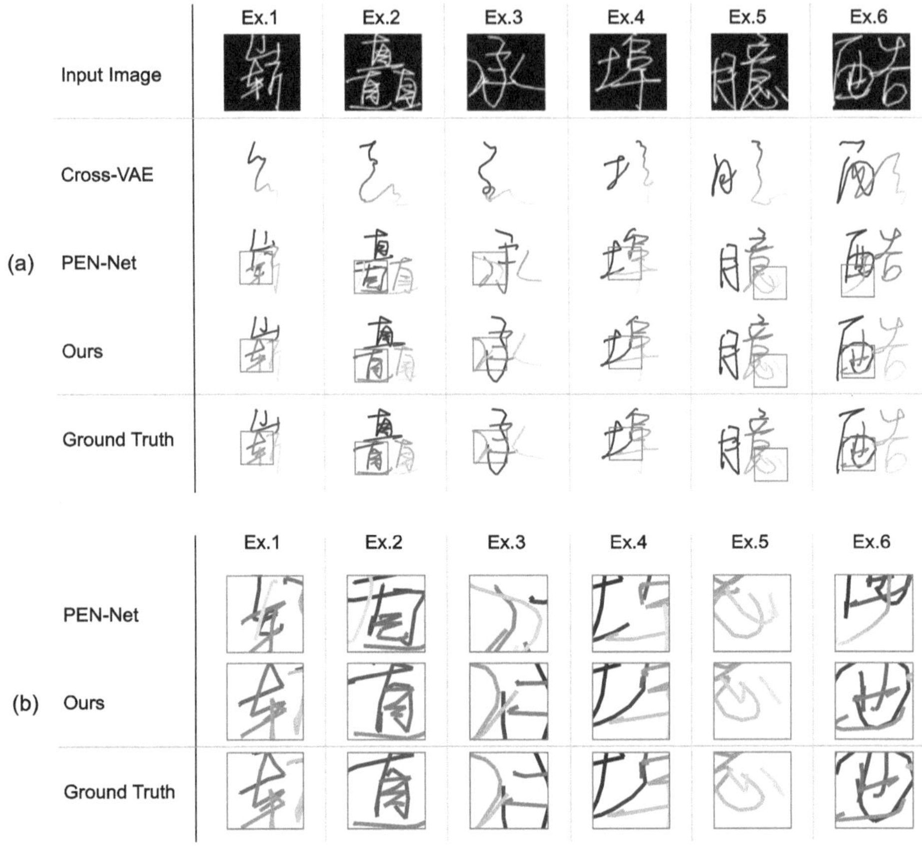

Fig. 4. Qualitative comparison between our method and baselines in the 3,000-character-class setting. (a) Overall trajectory comparisons for representative examples. (b) Zoomed-in views of the green rectangular regions in (a), highlighting local differences in geometric fidelity. (Color figure online)

the stroke order is correct, whereas improved geometry can reduce them without fully resolving ordering errors.

Qualitative comparisons further show that our method preserves fine-grained stroke geometry while maintaining a plausible stroke order. Figure 4 compares recovered trajectories from our method and the baselines on representative test samples. Our method reproduces fine-grained details such as terminal hooks and curvature changes at turning points, and the color-coded temporal progression aligns well with the ground truth, suggesting accurate stroke-order recovery. In contrast, PEN-Net occasionally produces spurious strokes or local mistakes that reduce glyph-shape consistency, and Cross-VAE often fails to reconstruct stroke-level structure for complex characters.

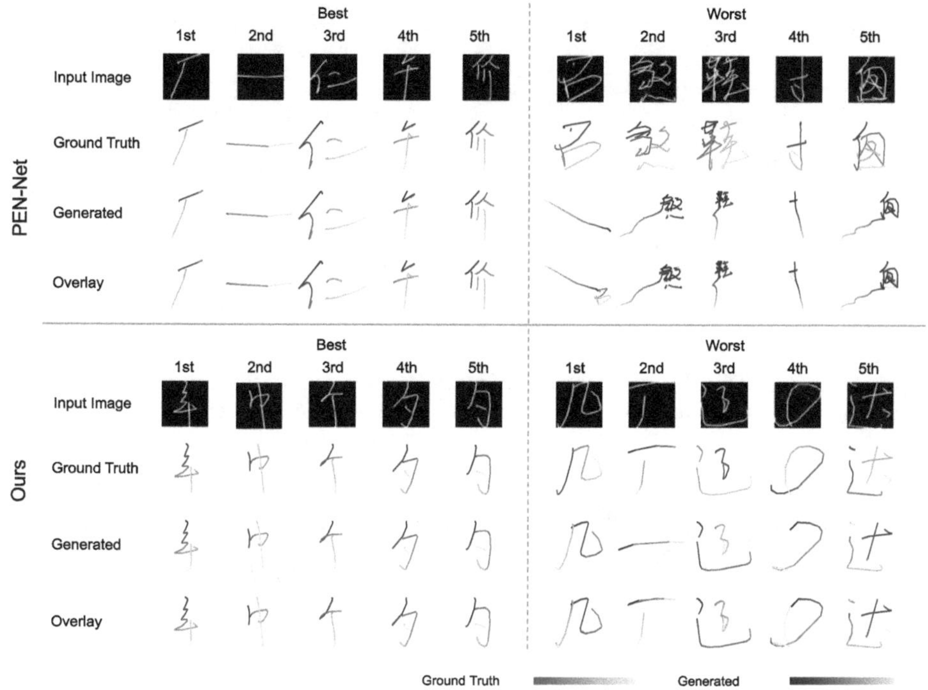

Fig. 5. Five best and five worst qualitative examples ranked by LDTW for our method and PEN-Net (trained on 3,000 classes).

4.3 Error Analysis: Ours vs. PEN-Net

A major difference between our diffusion-model-based method and conventional approaches is that the latter are typically autoregressive: they recursively predict the next pen stroke based solely on the previous ones. In contrast, diffusion-model-based approaches generate pen strokes while incorporating information from both past and future time steps. As a result, in autoregressive methods, once an error occurs in the trajectory, it accumulates progressively over future time steps. By contrast, diffusion models are expected to alleviate such error accumulation, leading to superior performance compared to conventional methods.

To investigate this conjecture, we visualize the five best and five worst test cases ranked by LDTW for both our method and the autoregressive baseline, PEN-Net, as shown in Fig. 5. Both methods perform well on relatively simple characters in the five best examples. However, in the five worst cases, PEN-Net frequently exhibits unnaturally elongated strokes and large local deviations from the ink trace. Because PEN-Net predicts coordinates sequentially conditioned on previous predictions, local mistakes can propagate and amplify over time, eventually causing severe trajectory breakdown.

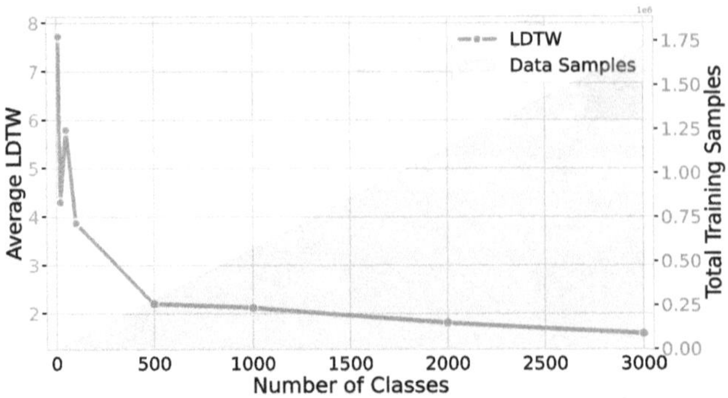

Fig. 6. Test-set LDTW as the number of training character classes increases. The left y-axis shows LDTW, and the right y-axis shows the number of training samples (in 10^6).

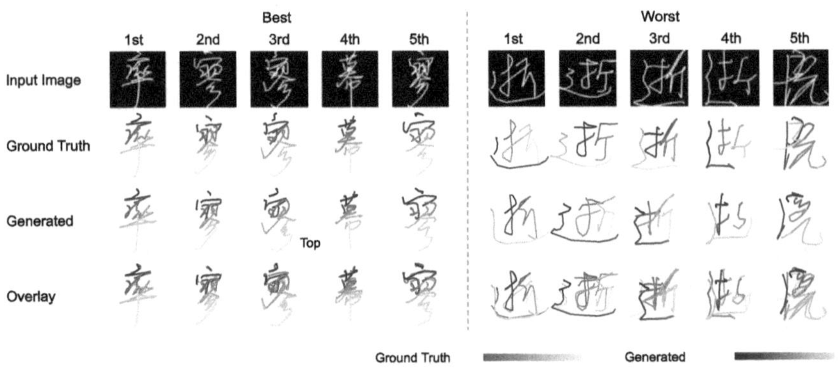

Fig. 7. Qualitative results when training on 20 character classes. We show the five best and five worst ranked by LDTW.

4.4 Effect of the Number of Training Classes

We examine how the number of training character classes affects trajectory recovery performance. We change the number of classes used for training while keeping the model and training settings the same. Using more classes also increases the number of training samples. Figure 6 shows test-set LDTW as we increase the number of training classes, together with the corresponding number of training samples. While the number of training samples increases almost linearly, LDTW does not improve at a constant rate. It drops quickly up to around 500 classes and then decreases more slowly, suggesting smaller gains after the training set becomes sufficiently diverse.

One reason may be that Chinese characters are built from shared parts: many characters reuse the same radicals and local stroke patterns. With small training classes, these common parts appear over and over, so the model can learn them quickly. As we add more classes, many new characters are largely new combinations of parts the model has already seen, so each additional class provides less new information and the improvement slows down.

This trend is also visible in Fig. 7, which shows qualitative results for the smallest training setting (20 classes). The model handles simple characters well, but errors become more frequent for complex ones. In the five worst examples, the recovered trajectories often deviate from the input image, leading to noticeable shape distortions and incorrect stroke order. Overall, these examples suggest that limited class diversity mainly hurts shape fidelity and stroke-order estimation, both of which improve as more classes are included in training.

4.5 Image-Conditioning Injection Strategies

Table 2. Ablation of image-conditioning injection types. Feature map 1/2 and 1/8 inject spatial feature maps extracted at resolutions of 1/2 and 1/8 of the input image, respectively. Global vector uses the final image representation after a global pooling layer, which discards spatial layout information. Multi-scale feature maps inject feature maps from multiple resolutions.

Injection type	DTW↓	LDTW↓	AIoU↑
Global vector	191.3	2.629	0.605
Feature map 1/2	233.8	3.249	0.524
Feature map 1/8	<u>119.7</u>	<u>1.625</u>	<u>0.736</u>
Multi-scale feature maps	**111.1**	**1.574**	**0.745**

Multi-scale spatial conditioning described in Sect. 3.3 is critical for preserving glyph geometry while recovering plausible stroke order. We ablate the image-conditioning design by varying only the form of image features provided to the denoiser: (i) a global pooled image embedding with no spatial layout (Global vector in Table 2), (ii) a single spatial feature map at a fixed encoder scale (Feature map 1/2 or 1/8), and (iii) the proposed multi-resolution feature maps (Multi-scale feature maps). Here, "1/2" and "1/8" denote the spatial resolution of the feature map relative to the input image (i.e., $\frac{H}{2} \times \frac{W}{2}$ and $\frac{H}{8} \times \frac{W}{8}$), corresponding to different downsampling factors in the image encoder.

Table 2 compares image-conditioning injection strategies on the seen classes test set, while keeping the remaining training and sampling settings identical to those used in Table 1. Global vector conditioning performs poorly, indicating that compressing the image into a single global representation discards geometric evidence necessary for accurate trajectory recovery. Replacing the global vector with a spatial feature map substantially improves performance, confirming the importance of preserving 2D image structure in the conditioning pathway. Multi-scale feature-map conditioning achieves the best overall trade-off and the highest glyph fidelity, supporting our design choice of injecting image evidence at multiple resolutions.

Injecting overly fine features alone can hurt because local evidence without global layout increases ambiguity at crossings and junctions. This trend is reflected in the degraded performance of the single 1/2-resolution feature map condition, which is worse than both the 1/8-resolution and multi-scale variants. A plausible explanation is that conditioning only on fine-resolution features makes the denoiser more sensitive to local appearance variations while providing insufficient global structural context for resolving long-range stroke dependencies and stroke-order ambiguity. In contrast, multi-scale conditioning balances coarse and fine cues: coarser features provide robust global layout constraints, whereas finer features refine local geometry, leading to better shape preservation and more stable trajectory recovery.

Table 3. Trajectory recovery performance on unseen character classes (755 classes excluded from training). Lower is better for DTW/LDTW; higher is better for AIoU.

Method	DTW↓	LDTW↓	AIoU↑
Cross-VAE [17]	742.6	7.463	0.337
PEN-Net [2]	272.0	3.178	0.594
Ours	**117.7**	**1.653**	**0.743**

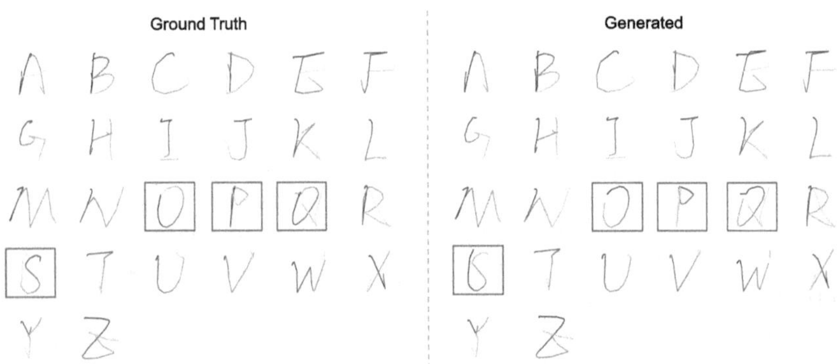

Fig. 8. Qualitative results under cross-script domain shift: English characters are fed to a model trained on Chinese characters. Red boxes highlight examples where the generated trajectories differ from the recorded reference temporal order or direction. (Color figure online)

4.6 Generalization to Unseen Character Classes

Our method generalizes to unseen character classes and maintains strong recovery performance on held-out classes. We evaluate on 755 character classes that are never observed during training. Table 3 reports the quantitative results, showing that our method still outperforms both baselines by a wide margin on the unseen-class test set.

The performance drop from the seen classes setting in Table 1 is relatively small, suggesting that the model learns transferable geometric and temporal primitives rather than memorizing class-specific trajectories. This result is consistent with the compositional structure of Chinese characters, where many classes share reusable substructures and stroke patterns.

We evaluate cross-script generalization by applying a model trained exclusively on Chinese characters to Latin letters in CASIA-OLHWDB. As shown in Fig. 8, we observe stroke-order estimation errors for a few characters (highlighted in red), which are attributable to differences in writing conventions across scripts. Because counter-clockwise stroke directions are uncommon in Chinese handwriting, the recovered trajectories for "O" and "Q" are biased toward clockwise directions. Nevertheless, the predicted stroke order is correct for most letters, and the recovered trajectories remain consistent with the input glyph shapes without noticeable geometric deviation.

4.7 Robustness Analyses

We study robustness under two practical perturbations: stroke-width variation and missing strokes. These analyses test whether the recovered trajectory remains geometrically consistent with the observable ink under degraded or out-of-domain inputs.

Table 4. Robustness to stroke width variation (DTW/LDTW lower is better; AIoU higher is better). The top row shows example images of "" rendered at each stroke width.

	Width=1			Width=2			Width=3		
	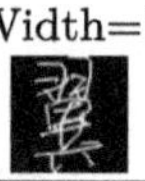								
Method	DTW↓	LDTW↓	AIoU↑	DTW↓	LDTW↓	AIoU↑	DTW↓	LDTW↓	AIoU↑
Cross-VAE [17]	713.8	7.029	0.346	742.6	7.463	0.337	802.2	7.862	0.323
PEN-Net [2]	212.3	2.728	0.607	265.0	3.149	0.593	300.8	3.403	0.586
Ours	**111.1**	**1.574**	**0.745**	**131.1**	**1.805**	**0.741**	**156.9**	**2.121**	**0.729**

Stroke-Width Variation. We evaluate robustness to stroke-width variations because they affect local structures around junctions and crossings. Following the protocol of [2], we train with stroke widths in $\{1, 2, 3\}$ pixels and evaluate separately on each condition. Table 4 shows that all methods degrade as stroke width increases, but our method remains the most accurate across all widths and exhibits only a mild drop in AIoU, indicating stable glyph preservation.

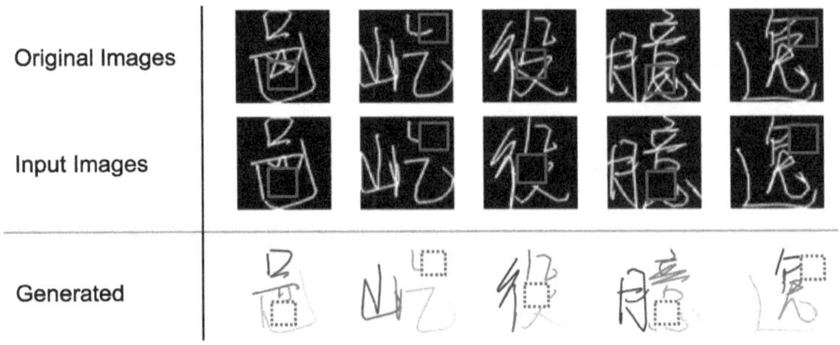

Fig. 9. Recovery results of our method for character images with missing strokes (random rectangular masks). Red boxes indicate the rectangular masked regions. (Color figure online)

Missing Strokes. To test robustness to local missing information, we train on clean images and apply random rectangular masks to the input image at inference time. Figure 9 shows that the model generally avoids generating spurious strokes and keeps the recovered trajectory aligned with the visible ink. In masked regions, the model tends to be conservative: rather than aggressively filling in occluded parts, it often preserves the observed structure, and the trajectory may remain fragmented when large or critical ink regions are removed.

5 Conclusion and Future Work

In this paper, we address the problem of handwriting trajectory recovery using diffusion models. By formulating trajectory recovery as an image-conditioned time-series generation task, we leverage diffusion models to reconstruct temporally ordered pen trajectories that are consistent with a given offline handwriting image. We further introduce a multi-scale conditioning technique to effectively encode image information at different resolutions. This design significantly reduces trajectory drift and improves the performance of handwriting trajectory recovery. Experimental results on CASIA-OLHWDB demonstrate clear improvements over the baselines PEN-Net and Cross-VAE in terms of both temporal alignment (DTW/LDTW) and shape fidelity (AIoU), particularly for complex multi-stroke characters. Furthermore, we confirm that increasing the number of training character classes leads to a significant improvement in recovery performance. These findings suggest that handwriting trajectories exhibit shared structural and dynamic regularities across character classes, and that generative models can learn underlying writing principles rather than merely memorizing character-specific stroke orders.

Several directions remain for future work. First, sampling speed could be improved by leveraging flow matching [8], fewer-step samplers, or network optimization. Second, although our experiments provide preliminary evidence of cross-script transfer, extending the method to a wider range of writing systems remains an important direction. Finally, our current rendering protocol assumes uniform stroke widths, whereas real offline handwriting images may contain local stroke-width variations caused by pen pressure, writing instruments, or scanning conditions. Adapting the model to such realistic image conditions is therefore another important future direction.

Acknowledgments. This work was supported by JSPS KAKENHI Grant Numbers JP25H01149 and JP24K20750, JST CRONOS-JPMJCS24K4, and ACT-X-JPMJAX25C.

References

1. Bhunia, A.K., et al.: Handwriting trajectory recovery using end-to-end deep encoder-decoder network. In: Proceedings of the 24th International Conference on Pattern Recognition (ICPR), pp. 3639–3644 (2018)
2. Chen, Z., et al.: Complex handwriting trajectory recovery: evaluation metrics and algorithm. In: Proceedings of the Asian Conference on Computer Vision (ACCV), pp. 58–74 (2022)
3. Díaz, M., Crispo, G., Parziale, A., Marcelli, A., Ferrer, M.A.: Writing order recovery in complex and long static handwriting. Int. J. Interact. Multimedia Artif. Intell. **7**(4), 171–184 (2022)
4. Doermann, D., Rosenfeld, A.: Recovery of temporal information from static images of handwriting. Int. J. Comput. Vision **15**, 143–164 (1995)
5. Ho, J., Jain, A., Abbeel, P.: Denoising diffusion probabilistic models. In: Advances in Neural Information Processing Systems (NeurIPS), vol. 33, pp. 6840–6851 (2020)
6. Kato, Y., Yasuhara, M.: Recovery of drawing order from single-stroke handwriting images. IEEE Trans. Pattern Anal. Mach. Intell. **22**(9), 938–949 (2000)
7. Kingma, D.P., Ba, J.: Adam: a method for stochastic optimization. arXiv preprint arXiv:1412.6980 (2014)
8. Lipman, Y., Chen, R.T.Q., Ben-Hamu, H., Nickel, M., Le, M.: Flow matching for generative modeling. In: The Eleventh International Conference on Learning Representations (ICLR) (2023)
9. Liu, C., Yin, F., Wang, D., Wang, Q.: CASIA online and offline Chinese handwriting databases. In: Proceedings of the International Conference on Document Analysis and Recognition (ICDAR), pp. 37–41 (2011)
10. Moussa, E.M., Lelore, T., Mouchère, H.: SET, SORT! a novel sub-stroke level transformer for offline handwriting to online conversion. In: Proceedings of the International Conference on Document Analysis and Recognition (ICDAR), pp. 81–97 (2023)
11. Nguyen, H.T., Nakamura, T., Nguyen, C.T., Nakagawa, M.: Online trajectory recovery from offline handwritten Japanese Kanji characters of multiple strokes. In: Proceedings of the 25th International Conference on Pattern Recognition (ICPR), pp. 8320–8327 (2021)
12. Nguyen, V., Blumenstein, M.: Techniques for static handwriting trajectory recovery: a survey. In: Proceedings of the 9th IAPR International Workshop on Document Analysis Systems (DAS), pp. 463–470 (2010)
13. Qiao, Y., Yasuhara, M.: Recover writing trajectory from multiple stroked image using bidirectional dynamic search. In: Proceedings of the 18th International Conference on Pattern Recognition (ICPR), vol. 2, pp. 970–973 (2006)

14. Ramesh, A., et al.: Zero-shot text-to-image generation. In: Proceedings of the International Conference on Machine Learning (ICML), pp. 8821–8831 (2021)
15. Ronneberger, O., Fischer, P., Brox, T.: U-net: convolutional networks for biomedical image segmentation. In: Proceedings of the International Conference on Medical Image Computing and Computer-Assisted Intervention (MICCAI), pp. 234–241 (2015)
16. Sohl-Dickstein, J., Weiss, E.A., Maheswaranathan, N., Ganguli, S.: Deep unsupervised learning using nonequilibrium thermodynamics. In: Proceedings of the International Conference on Machine Learning (ICML), pp. 2256–2265 (2015)
17. Sumi, T., Iwana, B.K., Hayashi, H., Uchida, S.: Modality conversion of handwritten patterns by cross variational autoencoders. In: Proceedings of the International Conference on Document Analysis and Recognition (ICDAR), pp. 407–412 (2019)
18. Zhao, B., Yang, M., Tao, J.: Pen tip motion prediction for handwriting drawing order recovery using deep neural network. In: Proceedings of the 24th International Conference on Pattern Recognition (ICPR), pp. 704–709 (2018)

Benchmarks and Datasets

JaWildText: A Benchmark for Vision-Language Models on Japanese Scene Text Understanding

Koki Maeda[1,2](✉) and Naoaki Okazaki[1,2]

[1] Institute of Science Tokyo, Tokyo, Japan
okazaki@comp.isct.ac.jp
[2] Research and Development Center for Large Language Models, National Institute of Informatics, Tokyo, Japan
koki.maeda@nlp.comp.isct.ac.jp

Abstract. Japanese scene text poses challenges that multilingual benchmarks often fail to capture, including mixed scripts, frequent vertical writing, and a character inventory far larger than the Latin alphabet. Although Japanese is included in several multilingual benchmarks, these resources do not adequately capture the language-specific complexities. Meanwhile, existing Japanese visual text datasets have primarily focused on scanned documents, leaving in-the-wild scene text underexplored. To fill this gap, we introduce **JaWildText**, a diagnostic benchmark for evaluating vision-language models (VLMs) on Japanese scene text understanding. JaWildText contains 3,241 instances from 2,961 images newly captured in Japan, with 1.12 million annotated characters spanning 3,643 unique character types. It comprises three complementary tasks that vary in visual organization, output format, and writing style: (i) Dense Scene Text Visual Question Answering (STVQA), which requires reasoning over multiple pieces of visual text evidence; (ii) Receipt Key Information Extraction (KIE), which tests layout-aware structured extraction from mobile-captured receipts; and (iii) Handwriting OCR, which evaluates page-level transcription across various media and writing directions. We evaluate 14 open-weight VLMs and find that the best model achieves an average score of 0.64 across the three tasks. Error analyses show recognition remains the dominant bottleneck, especially for kanji. JaWildText enables fine-grained, script-aware diagnosis of Japanese scene text capabilities, and we released its evaluation code.

Keywords: Japanese scene text · vision-language models · benchmark · handwriting recognition · key information extraction

1 Introduction

Text is ubiquitous in everyday environments: on street posters, handwritten notes, receipts, and storefronts. For decades, text-centric vision systems relied on

G. A. Fink et al. (Eds.): ICDAR 2026, LNCS 16972, pp. 217–234, 2027.
https://doi.org/10.1007/978-3-032-36023-6_13

a modular workflow that first applied OCR to convert pixels into characters and then fed the text to separate modules for downstream tasks [20]. With the rise of vision-language models (VLMs) such as GPT-4V [33], this workflow is shifting toward an end-to-end approach: VLMs generate outputs of a downstream task directly from natural images. This new workflow is increasingly adopted as a practical alternative to dedicated OCR pipelines.

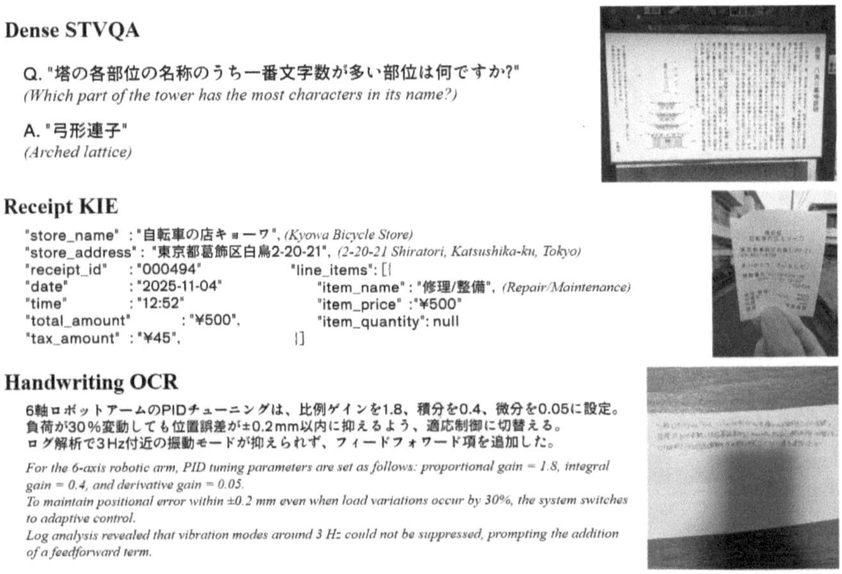

Fig. 1. Overview of the JaWildText benchmark: (i) Dense STVQA, (ii) Receipt KIE, and (iii) Handwriting OCR. We added English translations for readability.

This shift, however, complicates evaluation. When a VLM is evaluated only by downstream task accuracy, it is often unclear whether an error stems from a failure of character recognition or from incorrect reasoning over correctly recognized text. Disentangling these two failure modes is critical because precise reading is a prerequisite for any higher-level text understanding. Benchmarks that assess reading in natural images and separate recognition failures from reasoning failures are therefore essential, yet few existing resources offer this diagnostic capability, particularly for non-English scripts.

Although multilingual benchmarks [29,30,39] have extended scene text evaluation beyond English [4,22,38], they prioritize language breadth over language-specific diagnostics. This matters particularly for Japanese, whose scene text frequently mixes kanji, hiragana, katakana, and Latin alphanumerics. Because the resulting failure patterns differ from those of other scripts, it is difficult to diagnose failures without targeted evaluation. Existing Japanese scene text resources provide recognition data at the character or word level [12,15,28]. In contrast, Japanese VLM benchmarks target scanned documents or knowledge-centric multimodal tasks [31,32] without explicitly measuring the underlying recognition ability. As a result, no existing benchmark can tell whether VLM errors on Japanese scene text arise from recognition or from reasoning.

To fill this gap, we introduce JaWildText, a fine-grained benchmark for evaluating VLMs on Japanese scene text understanding. The benchmark is designed to disentangle reading from reasoning. As shown in Fig. 1, JaWildText consists of three tasks that form a compact yet comprehensive configuration. The tasks are chosen to vary three factors that commonly confound end-to-end evaluation: visual organization (from cluttered to structured), output format (from free-form to verbatim), and writing style (printed or handwritten). This design exposes failure modes that remain conflated when models are scored only by downstream task accuracy. Dense Scene Text Visual Question Answering (Dense STVQA) tests multi-region reading and cross-reference reasoning in cluttered signboards and posters. Receipt Key Information Extraction (Receipt KIE) evaluates layout-aware structured extraction from in-the-wild imagery. Handwriting OCR (page-level transcription) assesses long-context transcription of handwritten text, providing a recognition-dominant setting that complements the reasoning-heavy tasks above. JaWildText contains 3,241 evaluation instances from 2,961 newly collected images in Japan, with 1.12 million annotated characters spanning 3,643 unique characters.

We benchmark 14 open-weight VLMs on JaWildText. The experiments show that the best model achieves an average score of 0.64 across the three tasks. Our error analysis identified distinct bottlenecks: models that read text accurately may still fail at reasoning, and recognition difficulty varies drastically by script type. In summary, these results demonstrate that JaWildText provides fine-grained diagnostic evidence that is invisible in aggregated accuracy alone.

The contributions of this paper are as follows:

1. We introduce JaWildText, to our knowledge, the first benchmark dedicated to evaluating VLMs on Japanese scene text understanding across three complementary tasks grounded in real-world images.
2. We benchmark 14 open-weight VLMs, establish reproducible baselines, and quantify substantial performance gaps across architectures.
3. We provide an error analysis that disentangles recognition from reasoning failures, showing that their relative severity varies markedly across model families.

2 Related Work

2.1 Benchmarking Text Understanding in Natural Images

For English, evaluation resources have matured along three complementary tracks. *Scene text* benchmarks first targeted detection and recognition in natural images [40,45], and then advanced to text-centric VQA [4,18,22,23,38], which requires models to read and reason over recognized text. *Receipt and document understanding* benchmarks such as SROIE [13], CORD [34], and FUNSD [17] evaluate structured key information extraction (KIE), testing whether models can map visually organized fields to predefined categories. *Handwriting recognition* benchmarks, anchored by the IAM Handwriting Database [21], assess verbatim transcription of diverse writing styles; recent work shows that VLM performance degrades substantially on non-English handwriting [7]. Together, these tracks span a range of visual organization, output format, and writing style, forming a comprehensive evaluation ecosystem for English.

Several benchmarks extend this ecosystem to other languages, progressively broadening language coverage and task complexity. The ICDAR MLT challenges [29,30] introduce multilingual scene text detection and recognition across up to ten languages. XFUND [43] extends form understanding to seven languages. Targeting VLMs directly, MTVQA [39] shifted the focus to multilingual text-centric VQA with native annotations. OCRBench [9,19] and CC-OCR [44] broaden the scope to multilingual OCR for VLMs. While these efforts increase language coverage, they treat each language as one among many and provide limited diagnostic depth for language-specific challenges.

Among CJK languages, dedicated benchmarks have emerged for Chinese scene text recognition [5] and Korean text-centric VQA [14]. However, Japanese remains without a comprehensive evaluation despite its unique challenges, notably concurrent use of multiple scripts within a single text, complex layouts, and thousands of distinct characters.

2.2 Japanese Text Understanding

Existing Japanese-specific resources address isolated facets of text understanding. For scene text recognition, existing resources target scene text spotting in omnidirectional video [16], isolated character classification [12], vertical text recognition [36], and comics [2], restricted to a specific visual setting or textual granularity. For receipt understanding, existing resources support training or fine-tuning on mobile-captured receipts and post-OCR correction [10,27], but none serve as a benchmark for assessing general-purpose VLM capabilities. For handwriting, existing datasets provide isolated characters or online stroke data [24–26,28], or target classical cursive [6]; none covers page-level offline recognition of modern handwriting. On the reasoning side, JDocQA [31] addresses question answering over scanned documents, and JMMMU [32] benchmarks multimodal understanding centered on cultural and academic knowledge rather than text recognition ability.

Mapping these resources onto the three evaluation dimensions of visual organization, output format, and writing style reveals that none connect recognition with reasoning for Japanese scene text. JaWildText fills this gap with three complementary tasks that systematically vary these dimensions, enabling fine-grained diagnosis of where and why current VLMs fail on Japanese text in real-world images.

3 Dataset: JaWildText

JaWildText is designed to expose where a model fails, whether in character recognition, layout understanding, or reasoning, rather than reporting only aggregated task accuracy. To this end, it comprises three complementary tasks, each annotated to disentangle recognition errors from reasoning and formatting errors. Because such fine-grained diagnosis requires image diversity and annotation quality that are unavailable in web-scraped corpora, we collected original images and annotations tailored for this work.

3.1 Dense STVQA

Dense STVQA evaluates whether a model can read and reason over dense Japanese scene text, using visually complex real-world images such as signboards, bulletin boards, posters, and product packages.

Image Collection. To test recognition under realistic conditions, we asked workers from a data collection agency in Japan to photograph text-rich scenes with cameras and smartphones, resulting in 745 images. We instructed workers to cover indoor and outdoor locations under both daytime and nighttime lighting and avoid multiple shots of the same subject, ensuring diversity in layout, font style, and visual context. We retained natural artifacts, such as background clutter, partial occlusion, and reflections, to test recognition robustness.

Annotation. Annotations are structured into two layers to separate recognition from reasoning. In the first layer, annotators marked text regions with quadrilateral bounding boxes. They transcribed each region, which is defined as a line-level or column-level sequence of visually recognizable characters. Each image contains 45.1 annotated text regions on average. In the second layer, native Japanese speakers authored open-ended question-answer pairs over these transcribed regions. Annotators were encouraged to write questions that require reasoning across multiple text regions rather than extracting a single string from a single area. We exclude yes/no and multiple-choice formats to minimize chance-level correctness. Crucially, each question is linked to *evidence regions*: the minimal set of text regions necessary and sufficient to derive the answer. This linkage enables automatic diagnosis; if a model fails a question but correctly recognizes the evidence regions, the error is attributable to reasoning rather than recognition. In total, we created 1,025 question-answer pairs.

3.2 Receipt KIE

Receipt KIE evaluates structured field extraction from real-world photographs of Japanese receipts. This setting introduces challenges largely absent from scene text: rigid columnar layouts, mixed use of full-width and half-width characters, and domain-specific abbreviations produced by thermal printers.

Image Collection. Diagnostic value depends on testing under realistic capture conditions; hence, we collected photographs of consumer receipts from everyday transactions rather than flatbed scans. We retained natural artifacts such as creases, folds, and hand-held tilt to ensure that models are evaluated against the geometric and photometric distortions encountered in practical use. To maximize visual diversity, we disallowed multiple receipts from the same store while permitting receipts from different branches of the same chain, yielding 1,151 unique receipt images.

Annotation. We annotate receipts at the individual field level so that evaluation can pinpoint which field types a model struggles with, rather than producing only a single per-receipt score. Our key schema builds on the four header fields of SROIE [13] (`store_name`, `date`, `store_address`, `total_amount`) and extends it in two directions tailored to Japanese receipts: three additional header fields (`receipt_id`, `time`, `tax_amount`) that are usually printed but absent from SROIE, and line-item-level tuples of `item_name`, `item_price`, and `item_quantity`, which test a model's ability to maintain structured alignment across repeated rows. For each field, annotators recorded the text string and a quadrilateral bounding box; fields absent from a receipt are explicitly marked as null, allowing evaluation to distinguish extraction errors from correct recognition of absence. In total, we annotated 56,095 text regions across 1,151 receipts.

3.3 Handwriting OCR

Handwriting OCR evaluates page-level recognition of multi-sentence handwritten Japanese text. Unlike isolated-word or single-line recognition, page-level evaluation requires models to handle layout interpretation, line segmentation, and script mixing simultaneously, reflecting realistic reading scenarios.

Image Collection. To obtain diverse yet controlled handwriting samples, we designed a collection pipeline that separates content generation from handwriting production. First, we defined over 100 genre-keyword pairs spanning everyday topics such as work planning, travel, and cooking. We generated up to 20 prompt texts per pair using a large language model.[1] Each prompt was constrained to approximately 100 characters, long enough to span multiple lines across mixed scripts yet short enough to fit naturally on a single page.

Then, we distributed the prompts to 51 native Japanese writers, who transcribed them onto designated media and photographed the results using their own devices. To systematically introduce visual variation, we specified the writing medium and direction (horizontal or vertical) for each instance. The media include lined paper, plain paper, whiteboards, and tablets. Writers chose line-break positions freely. Each instance is accompanied by metadata recording the

[1] Specifically, we used `openai/gpt-oss-120b` to generate the prompt texts. The generated texts serve only as writing prompts.

writer ID, writing medium, instrument, direction, and ink color, enabling fine-grained analysis of how these factors affect recognition performance.

Annotation. For each image, the target output is a transcription of all visible handwritten text. Annotators transcribed the text line by line and drew a quadrilateral bounding box around each region. A key design principle is that the ground truth should reflect what is visually present in the image rather than what the writer intended. Accordingly, we preserved writer-introduced errors such as misspellings or omitted characters. For characters that a writer started but left incomplete due to writing errors, we assigned a dedicated symbol (□) to mark them explicitly. This annotation policy ensures that model evaluation measures visual recognition fidelity rather than error correction ability. We annotated 1,065 handwriting instances with 6,002 lines, totaling 111,977 characters.

Table 1. Summary statistics of JaWildText. **#Instances** denotes the evaluation unit: an instance is a question–answer pair in Dense STVQA, and a single image in Receipt KIE and Handwriting OCR. **#Regions** denotes the number of annotated quadrilateral text regions.

Subset	#Instances	#Images	#Regions	#Chars	#Unique Chars
Dense STVQA	1,025	745	33,608	571,979	3,313
Receipt KIE	1,151	1,151	56,095	433,558	2,001
Handwriting OCR	1,065	1,065	6,002	111,977	1,830
Total	**3,241**	**2,961**	**95,705**	**1,117,514**	**3,643**

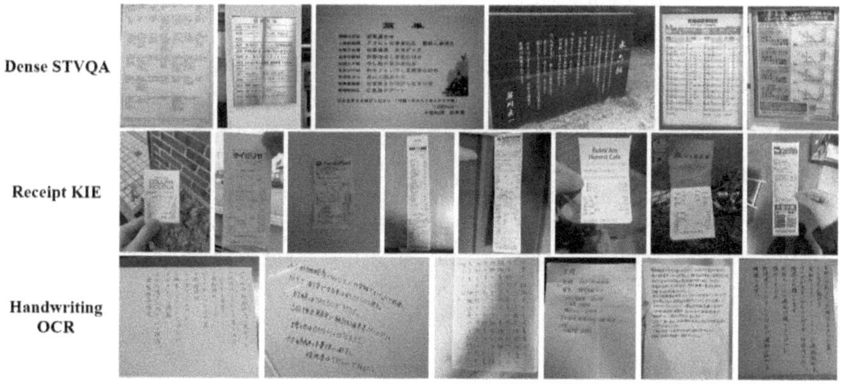

Fig. 2. Representative images from each task of JaWildText. Dense STVQA covers signboards, posters, and product packages under varying conditions. Receipt KIE includes receipts with creases, folds, and diverse perspectives. Handwriting OCR spans multiple writing media and directions.

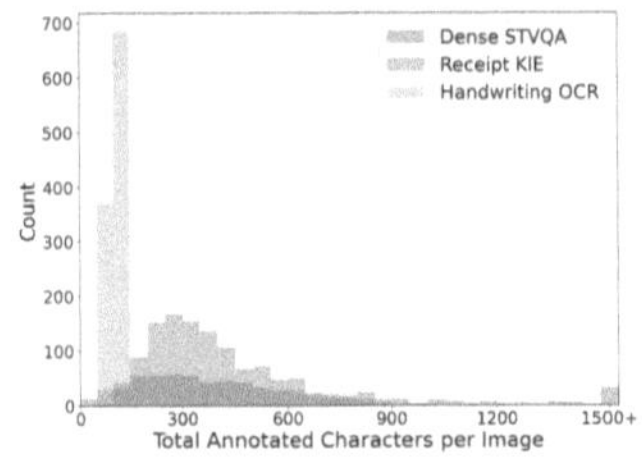

(a) Distribution of total character length per image.

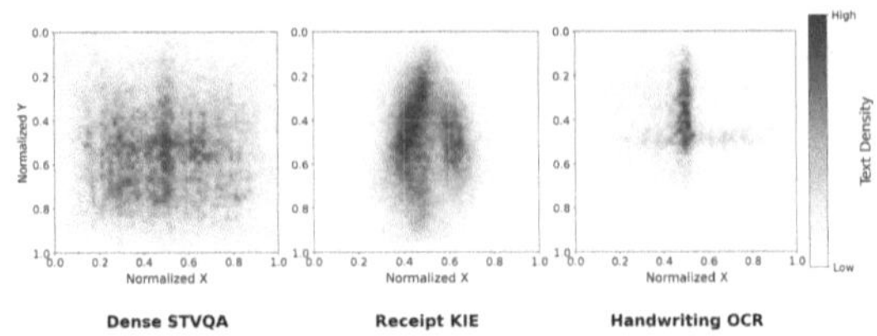

(b) Spatial distribution of text-region centers.

Fig. 3. Image-level text properties in JaWildText.

3.4 Quality Control

Image collection and annotation were conducted by a professional data curation agency with compensated annotators. To calibrate annotation guidelines before full-scale production, the authors and the agency jointly reviewed the first 10% of deliverables and refined the guidelines based on observed inconsistencies. In the main phase, each instance was labeled by one annotator and independently verified by a second; disagreements were resolved through discussion. We excluded images containing non-public personally identifiable information, such as faces, vehicle license plates, or credit card numbers, while retaining publicly displayed information (e.g., store phone numbers on receipts) needed for the benchmark tasks.

We release JaWildText under the Apache 2.0 license, including images, annotations, task definitions, prompts, and evaluation code. The dataset is available on Hugging Face[2], and the evaluation code is available on GitHub[3].

3.5 Dataset Statistics

Table 1 summarizes key statistics of JaWildText. The dataset comprises 3,241 instances drawn from 2,961 unique images, with 95,705 annotated text regions totaling 1,117,514 characters across 3,643 unique characters. By design, each task comprises approximately 1,000 instances, allowing for broadly comparable score precision across tasks. Figure 2 shows representative images from each task, and Fig. 3 visualizes text density and layout properties discussed below.

Text Density and Spatial Layout. Dense STVQA and Receipt KIE are text-dense, often containing several hundred characters per image, whereas Handwriting OCR is intentionally controlled to approximately 100 characters per image (Fig. 3a). Dense STVQA exhibits a long tail beyond 2,000 characters per image, reflecting the high information density of signboards and bulletin boards;

[2] https://huggingface.co/datasets/llm-jp/jawildtext.

[3] https://github.com/llm-jp/jawildtext.

this density forces models to locate and integrate information across many text regions, making the task sensitive to both recognition errors and cross-region reasoning failures. In contrast, the controlled length of Handwriting OCR isolates recognition ability from reasoning, providing a setting in which errors can be attributed almost entirely to character-level reading. The spatial distribution of text-region centers (Fig. 3b) mirrors these design choices: Dense STVQA regions spread broadly across the frame, Receipt KIE regions form a narrow vertical band consistent with elongated receipt layouts, and Handwriting OCR clusters near the page center along both axes.

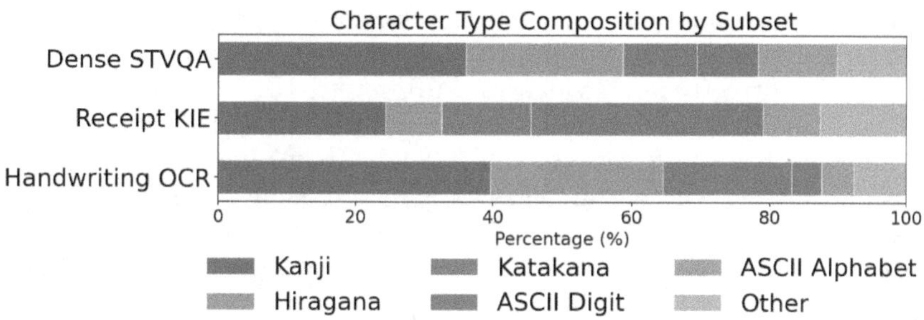

Fig. 4. Stacked character-type composition by task. Japanese scripts dominate Dense STVQA and Handwriting OCR, while Receipt KIE allocates a large fraction to ASCII digits.

Table 2. Question types.

Question type	# (%)
Compositional retrieval	406 (39.6%)
Counting	374 (36.5%)
Calculation	125 (12.2%)
Spatial	84 (8.2%)
Other	36 (3.5%)

Table 3. Receipt fill rates.

Field	Fill Rate
`store_name`	100.0%
`date`	100.0%
`total_amount`	99.8%
`time`	98.3%
`tax_amount`	95.2%
`receipt_id`	92.0%
`store_address`	48.9%

Table 4. Handwriting metadata.

Medium	*# (%)*
Lined paper	470 (43.9%)
Unlined paper	344 (32.3%)
Tablet	130 (12.2%)
Whiteboard	121 (11.4%)
Direction	*# (%)*
Horizontal	4,222 (70.3%)
Vertical	1,780 (29.7%)

Character-Type Composition. Character-type distributions differ markedly across tasks (Fig. 4), which foreshadow the character-type analysis in Sect. 4.3. Dense STVQA and Handwriting OCR are dominated by kanji (36.0% and 39.7%) and hiragana (22.6% and 24.8%), consistent with natural Japanese prose and placing heavy demands on mixed-script recognition. Receipt KIE contains a much larger share of ASCII digits (33.5%), reflecting prices, quantities, and dates; accurate digit reading is therefore a decisive factor for extraction performance in this task. Overall, JaWildText contains 2,866 unique kanji characters. Of the 2,136 Jōyō kanji (the Japanese government's daily-use list), the dataset covers 1,985 (92.9%), and it additionally includes 881 kanji beyond the Jōyō set, meaning that models must generalize beyond standard literacy inventories to handle real-world text.

Task-specific Properties. Each task contributes a distinctive evaluation signal. In Dense STVQA, questions cover a balanced mix of reasoning types: compositional retrieval, counting, calculation, and spatial reasoning (Table 2). Questions are linked to minimal evidence regions, making it possible to separate recognition errors (failing to read individual regions) from reasoning errors (failing to combine correctly read evidence). In Receipt KIE, field fill rates vary substantially (Table 3): `store_name` and `date` are present in all receipts, whereas `store_address` has the lowest fill rate (48.9%), reflecting real-world omission patterns and testing whether models can handle missing fields without hallucinating content. In Handwriting OCR, 51 writers contributed data with controlled numbers of instances per writer, spanning multiple writing media and directions (Table 4), ensuring that recognition performance is evaluated across a range of handwriting variability rather than being biased toward a single style.

4 Experiments

4.1 Experimental Setup

Inference. We evaluate 14 open-weight VLMs from five model families. We include four recent high-performing families: Qwen3-VL [3], InternVL3.5 [41], Gemma3 [11], and Phi-4-Multimodal [1]. For families offering multiple sizes, we select variants with 1B to 38B parameters (see Table 5 for the complete list). We additionally include Sarashina2.2-Vision [37], a model built on a Japanese-centric LLM backbone and trained with Japanese document and OCR data. This selection lets us examine how model scale and language-specific training influence performance on JaWildText. We set the temperature to 0 and the maximum token length to 2,048. Each instance uses a single image at its original resolution; resizing or tiling is applied according to the default preprocessing.

Evaluation. To reliably extract the final answer from model output, we enforce machine-parseable output formats using a fixed prompt template for each task. Models must enclose the final answer in `\boxed{...}` for Dense STVQA, return a single JSON object following the predefined schema for Receipt KIE, and output plain text transcriptions for Handwriting OCR. Any output that cannot be parsed receives a score of 0. Before scoring, we apply Unicode NFKC normalization to both predictions and references to absorb superficial character-form differences. In the Dense STVQA task, answers are open-ended and may vary due to differences in units or paraphrasing. Thus, we adopt judge-based accuracy: an

LLM verifier compares each prediction against the reference and returns a binary correctness label. Combined with the evidence region annotations, this binary signal enables error analysis to attribute each failure to recognition or reasoning. In the Receipt KIE task, outputs are structured, and field boundaries are well-defined. We report overall F1 and field-level accuracy for major header fields. This indicates which field types are most challenging to extract. For Handwriting OCR, we compute character-level similarity as $\max(0,\ 1 - \text{CER})$, where CER is defined as the Levenshtein distance between prediction and reference, divided by reference length. Character-level scoring is suitable for Japanese because it lacks explicit word boundaries. We further report script-type breakdown (e.g., kanji vs. hiragana) in CER. We compute the overall score as the unweighted average of the three task scores.

Extended Leaderboard. The main experiments focus on 14 open-weight VLMs to support detailed error analysis under a fixed evaluation setting. We additionally release an extended leaderboard covering 52 model variants, including recent open-weight and proprietary models. For Dense STVQA, the leaderboard uses the released open-weight `gpt-oss-20b` verifier, which we validated against human judgments and the main verifier; Handwriting OCR and Receipt KIE use the same deterministic metrics as the main experiments.

4.2 Main Results

Table 5 summarizes the performance of all evaluated models on JaWildText. The best model, Qwen3-VL-8B, achieves an overall score of only 0.64, indicating that Japanese scene text understanding remains a substantial challenge for current open-weight VLMs. Dense STVQA exhibits the widest performance spread across models (0.008–0.62), suggesting that this task effectively differentiates models with varying levels of Japanese scene text capability. In Handwriting

Table 5. Results on the JaWildText benchmark.

Model	Params	Avg.	STVQA	KIE	OCR
Qwen3-VL-8B	8B	**0.64**	**0.62**	**0.53**	**0.79**
Qwen3-VL-4B	4B	0.60	0.52	0.50	0.77
Qwen3-VL-2B	2B	0.52	0.31	0.48	0.76
InternVL3.5-38B	38B	0.55	0.44	0.50	0.71
InternVL3.5-14B	14B	0.49	0.30	0.47	0.71
InternVL3.5-8B	8B	0.53	0.39	0.48	0.72
InternVL3.5-4B	4B	0.48	0.31	0.44	0.70
InternVL3.5-2B	2B	0.44	0.23	0.42	0.66
InternVL3.5-1B	1B	0.37	0.11	0.37	0.61
Sarashina2.2-Vision-3B	3B	0.50	0.44	0.40	0.68
Gemma3-27B-IT	27B	0.43	0.37	0.39	0.53
Gemma3-12B-IT	12B	0.40	0.32	0.38	0.51
Gemma3-4B-IT	4B	0.19	0.12	0.23	0.20
Phi-4-Multimodal	14B	0.18	0.008	0.23	0.29

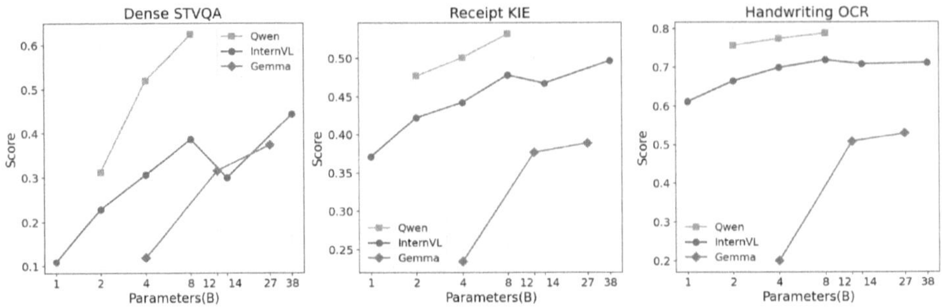

Fig. 5. Performance scaling trends with model size across benchmark tasks. Larger models improve overall performance, but gains differ by task family.

OCR, 10 of 14 models score at least 0.60, indicating that most models already possess a baseline handwriting recognition ability, though a ceiling around 0.80 persists even for the strongest models. Performance differences across model families are significant: Qwen3-VL consistently outperforms InternVL3.5 at similar parameter scales, surpassing InternVL3.5 by 0.11 Overall (0.64 vs. 0.53) at 8B parameters. Gemma3 trails both families despite having up to $\sim 3.4\times$ more parameters than the best-performing model. At the lower end, Phi-4-Multimodal attains near-zero accuracy on Dense STVQA (0.008), struggling to follow the required output format. Notably, raw parameter count does not fully explain these gaps. Sarashina2.2-Vision-3B achieves 0.44 accuracy on Dense STVQA, matching InternVL3.5-38B despite fewer parameters. However, this advantage does not generalize to Receipt KIE or Handwriting OCR, where Sarashina2.2-Vision-3B is comparable to InternVL3.5-2B. This contrast suggests that the benefit of Japanese-centric training data may be task-dependent.

To examine whether each task captures scaling behavior effectively, we plot performance trends within model families as parameter count (Fig. 5).

Within each family, all three tasks show improvement with scale, but their trajectories differ. Dense STVQA and Receipt KIE continue to improve across the parameter range we evaluate, with no clear saturation point. Handwriting OCR, by contrast, plateaus beyond a certain scale within each model family: InternVL3.5 saturates around 0.70 from 4B onward, and Qwen3-VL shows a marginal gain from 2B to 8B (0.76→0.79). Across families, however, scale is not decisive: Qwen3-VL-8B surpasses InternVL3.5-38B by 0.09 Overall, indicating that architecture and training data composition can matter more than parameter count alone.

The header-field accuracies in Table 6 exhibit sharp variation across field types in Receipt KIE. Format-constrained fields such as `time` are relatively well handled, with several top models achieving accuracy above 0.90. In contrast, `store_name` and `store_address` remain difficult, with best accuracies of only 0.16 and 0.55, respectively; these fields often require aggregating non-contiguous text spans across the receipt layout rather than copying a single contiguous line.

Table 6. Receipt KIE field-level accuracy on JaWildText.

Model	Params	Store Name	Address	Receipt ID	Date	Time	Total	Tax
Qwen3-VL-8B	8B	0.14	0.52	0.13	**0.81**	0.91	0.81	0.77
Qwen3-VL-4B	4B	**0.16**	**0.55**	0.11	0.28	0.92	**0.84**	0.80
Qwen3-VL-2B	2B	0.11	0.43	0.08	0.52	0.88	0.82	**0.82**
InternVL3.5-38B	38B	0.15	0.34	0.18	0.52	**0.92**	0.83	0.82
InternVL3.5-14B	14B	0.15	0.35	0.13	0.26	0.89	0.83	0.79
InternVL3.5-8B	8B	0.11	0.31	0.13	0.50	0.89	0.79	0.74
InternVL3.5-4B	4B	0.14	0.29	**0.21**	0.26	0.90	0.77	0.73
InternVL3.5-2B	2B	0.13	0.30	0.11	0.24	0.89	0.77	0.75
InternVL3.5-1B	1B	0.09	0.23	0.10	0.33	0.80	0.69	0.70
Sarashina2.2-Vision-3B	3B	0.11	0.28	0.18	0.28	0.75	0.72	0.67
Gemma3-27B-IT	27B	0.11	0.20	0.08	0.52	0.71	0.72	0.64
Gemma3-12B-IT	12B	0.06	0.21	0.11	0.39	0.67	0.70	0.60
Gemma3-4B-IT	4B	0.03	0.10	0.05	0.20	0.59	0.59	0.35
Phi-4-Multimodal	14B	0.01	0.004	0.07	0.13	0.51	0.51	0.48

This gap indicates that the primary bottleneck in Receipt KIE is not character recognition alone but spatial reasoning over document layout.

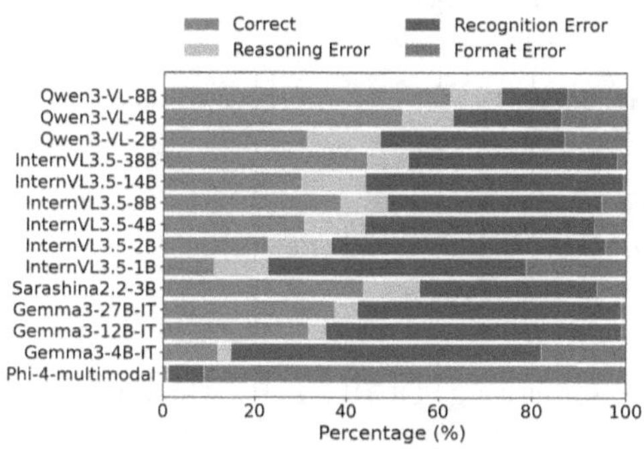

Fig. 6. Error taxonomy on Dense STVQA. Each bar decomposes instances into Correct, Reasoning Error, Recognition Error, and Format Error.

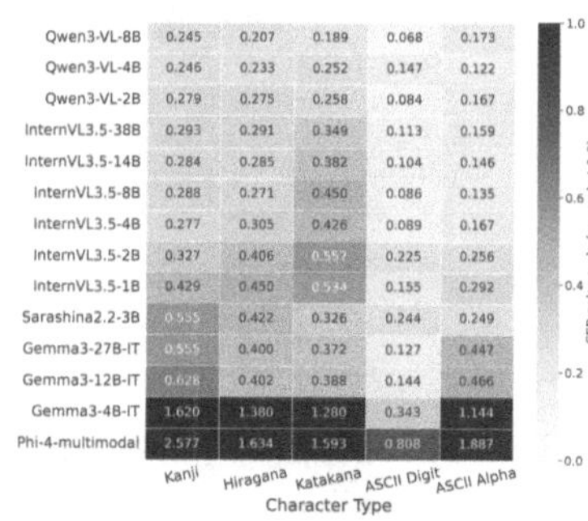

Fig. 7. Character-type CER on Handwriting OCR. Color scale is capped at 1.0; values exceeding 1.0 indicate hallucination-dominant outputs.

4.3 Analysis

Error Taxonomy for Dense STVQA. To disentangle recognition failures from reasoning failures on Dense STVQA, we define an error taxonomy with four categories. For each Dense STVQA image, we separately prompt the model to transcribe all visible text, independently of the QA task. We then compare the resulting transcript against the ground-truth transcriptions of each question's annotated evidence regions: an evidence region is considered "read" if its whole ground-truth string appears as an exact substring in the transcript. Based on this comparison, each instance is assigned one of four outcomes. **Recognition Error**: the answer is not recoverable from the recognized text alone because at least one required evidence region is missing from the transcript. **Reasoning Error**: all evidence regions are present, but the final answer is incorrect. **Format Error**: the output cannot be parsed under the prescribed `\boxed{}` format. **Correct**: the parsed answer matches the reference.

Figure 6 illustrates that Recognition Error is the largest category for most models, indicating that Japanese scene text recognition remains the primary bottleneck. Qwen3-VL-8B, which achieves the highest Correct rate (62.3%), still exhibits a 14.2% Recognition Error rate, showing that even the best-performing model has not fully overcome this bottleneck. Sarashina2.2-Vision-3B shows a relatively low Recognition Error rate (38.0%) compared to other models, such as InternVL3.5-8B (45.9%) and Gemma3-27B-IT (56.7%), suggesting that Japanese-centric training may partially improve scene text recognition capability. InternVL3.5 and Gemma3 remain heavily dominated by Recognition Errors (44.7–58.9% and 56.7–67.3%, respectively). Phi-4-Multimodal is instead dominated by Format Errors, failing to follow the prescribed output format. This taxonomy makes visible the failure stage that aggregate accuracy alone cannot reveal, enabling targeted diagnosis of each model's bottleneck.

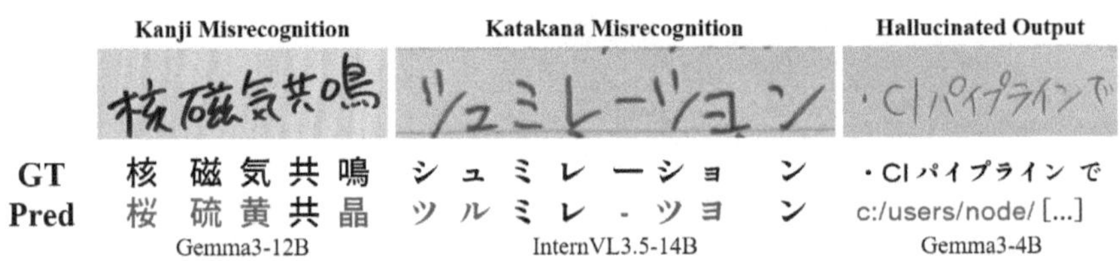

Fig. 8. Representative failure cases on Handwriting OCR. (Left) Gemma3-12B misrecognizes kanji characters, substituting visually dissimilar characters. (Center) InternVL3.5-14B confuses visually similar katakana characters. (Right) Gemma3-4B produces hallucinated output entirely unrelated to the input image. Red bold text indicates erroneous characters in the model predictions. (Color figure online)

Script-Category Analysis on Handwriting OCR. To examine how error rates differ across script categories, we decompose CER by script category (kanji, hiragana, katakana, ASCII digits, and ASCII letters). For each instance, we obtain a character-level alignment between prediction and reference via minimum edit distance backtracing, then compute CER separately for each category.

Figure 7 presents per-category CER for each model, showing that CER varies substantially across script categories. ASCII digits achieve the lowest CER across models, consistent with their small and visually distinct character set. Kanji exhibits the highest error rate, which we attribute to the large character inventory: models must disambiguate among thousands of classes, many with limited per-class training exposure. Since kanji accounts for 39.7% of all reference characters (Fig. 4), its high CER is the dominant contributor to overall scores.

On the other hand, InternVL3.5 models exhibit elevated katakana CER, whereas Gemma3 shows high CER on both kanji and ASCII letters. Figure 8 illustrates these contrasts: InternVL3.5-14B confuses visually similar katakana pairs, while Gemma3-12B misrecognizes kanji with other unrelated kanji characters. These differences likely reflect variation in Japanese script coverage during pre-training. Among the weakest models, Gemma3-4B-IT and Phi-4-Multimodal produce CER values exceeding 1.0, indicating that their edit distances exceed the reference length. As with the Dense STVQA error taxonomy, stratifying evaluation by linguistically meaningful categories reveals distinct failure profiles that aggregate scoring would obscure.

Writing Direction Analysis on Handwriting OCR. To directly test the effect of vertical Japanese writing, we split Handwriting OCR by text direction using the collected metadata. The subset contains 777 horizontal and 286 vertical images. Across 14 open-weight models, mean CER-similarity drops from 0.76 on horizontal images to 0.24 on vertical images. In 43% of vertical predictions, the lines are mostly recognized but emitted in the wrong top-to-bottom order; reordering predicted lines raises vertical CER-similarity to 0.51. Thus, vertical Japanese primarily exposes a layout and reading-order bottleneck rather than only a character-shape recognition bottleneck.

Table 7. Handwriting OCR results for OCR-specialized models.

Model	Params	Handwriting OCR (1-CER)
olmOCR-2-7B	7B	**0.74**
PaddleOCR-VL	0.9B	0.61
DeepSeek-OCR	3B	0.53

Comparison with OCR-Specialized Models. To situate VLM performance on the recognition-dominant Handwriting OCR task, we compare against three OCR-specialized models: DeepSeek-OCR [42], PaddleOCR-VL [8], and olmOCR-2-7B [35], evaluated under the same conditions (Sect. 4.1). As Table 7 shows, the best OCR-specialized model (olmOCR-2-7B, 0.74) falls below the best general-purpose VLM (Qwen3-VL-8B, 0.79) at a comparable parameter scale. OCR-specialized models are often positioned as strong baselines for document OCR and document parsing. Still, they perform similarly or worse than general-purpose VLMs when recognizing handwritten text in real-world environments, where diverse writing media, writing instruments, and imaging conditions differ substantially from scanned documents. This result underscores that robust recognition of handwritten scene text remains an open challenge that cannot be addressed by OCR-specific training alone.

5 Conclusion

We introduced JaWildText, a diagnostic benchmark for evaluating VLMs on Japanese scene text understanding across three complementary tasks: Dense STVQA, Receipt KIE, and Handwriting OCR. Benchmarking 14 open-weight VLMs shows that the best model achieves only 0.64 on our unified score, confirming that robust Japanese text understanding in the wild remains far from solved. Stratifying errors by type and script category reveals that recognition remains the dominant bottleneck, with kanji posing a particular challenge. Closing the remaining gap will require targeted interventions at each stage, informed by the kind of fine-grained diagnosis that JaWildText provides.

Acknowledgements. This work was supported by the "R&D Hub Aimed at Ensuring Transparency and Reliability of Generative AI Models" project of the Ministry of Education, Culture, Sports, Science and Technology (MEXT), Japan, and by JST BOOST, Japan Grant Number JPMJBS2430.

References

1. Abouelenin, A., et al.: Phi-4-Mini technical report: compact yet powerful multimodal language models via mixture-of-LoRAs (2025). arXiv:2503.01743
2. Baek, J., et al.: MangaVQA and MangaLMM: a benchmark and specialized model for multimodal manga understanding (2025). arXiv:2505.20298
3. Bai, S., et al.: Qwen3-VL technical report (2025). arXiv:2511.21631
4. Biten, A.F., et al.: Scene text visual question answering. In: IEEE/CVF International Conference on Computer Vision, pp. 4291–4301 (2019). https://doi.org/10.1109/ICCV.2019.00439
5. Chen, J., et al.: Benchmarking Chinese text recognition: datasets, baselines, and an empirical study (2022). arXiv:2112.15093
6. Clanuwat, T., et al.: Deep learning for classical Japanese literature (2018). arXiv:1812.01718
7. Crosilla, G., et al.: Benchmarking large language models for handwritten text recognition. J. Document. **81**(7), 334–360 (2025). https://doi.org/10.1108/JD-01-2025-0020
8. Cui, C., et al.: PaddleOCR-VL: boosting multilingual document parsing via a 0.9B ultra-compact vision-language model (2025). arXiv:2510.14528
9. Fu, L., et al.: OCRBench V2: an improved benchmark for evaluating large multimodal models on visual text localization and reasoning (2025). arXiv:2501.00321
10. Fujitake, M.: JaPOC: Japanese post-OCR correction benchmark using vouchers (2024). arXiv:2409.19948
11. Gemma Team: Gemma 3 technical report (2025). arXiv:2503.19786
12. Goto, H.: JPSC1400 – Japanese scene character dataset. Dataset (Rev.20201218) (2020). https://www.imglab.org/db/
13. Huang, Z., et al.: ICDAR2019 competition on scanned receipt OCR and information extraction (2021). arXiv:2103.10213
14. Hwang, T., et al.: KRETA: a benchmark for Korean reading and reasoning in text-rich VQA attuned to diverse visual contexts. In: Proceedings of the Conference on Empirical Methods in Natural Language Processing, pp. 33421–33432 (2025). https://doi.org/10.18653/v1/2025.emnlp-main.1696

15. Ishida, T., et al.: ICDAR 2019 robust reading challenge on omnidirectional video. In: International Conference on Document Analysis and Recognition, pp. 1488–1493 (2019). https://doi.org/10.1109/ICDAR.2019.00240
16. Iwamura, M., et al.: Downtown Osaka scene text dataset. In: European Conference on Computer Vision Workshops (ECCVW). LNCS, vol. 9913, pp. 440–455 (2016). https://doi.org/10.1007/978-3-319-46604-0_32
17. Jaume, G., et al.: FUNSD: a dataset for form understanding in noisy scanned documents. In: International Conference on Document Analysis and Recognition Workshops (ICDARW). pp. 1–6 (2019). https://doi.org/10.1109/ICDARW.2019.10029
18. Landeghem, J.V., et al.: Document understanding dataset and evaluation (DUDE). In: IEEE/CVF International Conference on Computer Vision, pp. 19528–19540 (2023). https://doi.org/10.1109/ICCV51070.2023.01789
19. Liu, Y., et al.: OCRBench: on the hidden mystery of OCR in large multimodal models. Sci. China Inf. Sci. **67**(12), 220102 (2024). https://doi.org/10.1007/s11432-024-4235-6
20. Long, S., He, X., Yao, C.: Scene text detection and recognition: the deep learning era. Int. J. Comput. Vision **129**(1), 161–184 (2020). https://doi.org/10.1007/s11263-020-01369-0
21. Marti, U.V., Bunke, H.: The IAM-database: An English sentence database for offline handwriting recognition. Int. J. Doc. Anal. Recogn. **5**, 39–46 (2002). https://doi.org/10.1007/s100320200071
22. Mathew, M., et al.: DocVQA: a dataset for VQA on document images. In: IEEE/CVF Winter Conference on Applications of Computer Vision, pp. 2199–2208 (2021). https://doi.org/10.1109/WACV48630.2021.00225
23. Mathew, M., et al.: InfographicVQA. In: IEEE/CVF Winter Conference on Applications of Computer Vision, pp. 1697–1706 (2022). https://doi.org/10.1109/WACV51458.2022.00264
24. Matsumoto, K., et al.: Collection and analysis of on-line handwritten Japanese character patterns. In: International Conference on Document Analysis and Recognition, pp. 496–500 (2001). https://doi.org/10.1109/ICDAR.2001.953839
25. Matsushita, T., et al.: A database of on-line handwritten mixed objects named Kondate. In: International Conference on Frontiers in Handwriting Recognition, pp. 369–374 (2014). https://doi.org/10.1109/ICFHR.2014.68
26. Nakagawa, M., et al.: Collection of on-line handwritten Japanese character pattern databases and their analyses. Int. J. Doc. Anal. Recogn. **7**(1), 69–81 (2004). https://doi.org/10.1007/s10032-004-0125-4
27. Nathan, S., et al.: Japanese-mobile-receipt-OCR-1.3K: a comprehensive dataset analysis and fine-tuned vision-language model for structured receipt data extraction. TechRxiv (preprint) (2025). https://doi.org/10.36227/techrxiv.175616889.90325672/v1
28. National Institute of Advanced Industrial Science and Technology (AIST): ETL Character Database. Online database (2014). collected 1973–1984. Accessed 18 Dec 2025
29. Nayef, N., et al.: ICDAR2017 robust reading challenge on multi-lingual scene text detection and script identification – RRC-MLT. In: International Conference on Document Analysis and Recognition, pp. 1454–1459 (2017). https://doi.org/10.1109/ICDAR.2017.237

30. Nayef, N., et al.: ICDAR2019 robust reading challenge on multi-lingual scene text detection and recognition – RRC-MLT-2019. In: International Conference on Document Analysis and Recognition, pp. 1582–1587 (2019). https://doi.org/10.1109/ICDAR.2019.00254
31. Onami, E., et al.: JDocQA: Japanese document question answering dataset for generative language models. In: Proceedings of the Joint International Conference on Computational Linguistics, Language Resources and Evaluation, pp. 9503–9514 (2024)
32. Onohara, S., et al.: JMMMU: a Japanese massive multi-discipline multimodal understanding benchmark for culture-aware evaluation. In: Proceedings of NAACL-HLT 2025, pp. 932–950 (2025). https://doi.org/10.18653/v1/2025.naacl-long.43
33. OpenAI: GPT-4 Technical Report. arXiv:2303.08774 (2023)
34. Park, S., et al.: CORD: a consolidated receipt dataset for post-OCR parsing. In: NeurIPS 2019 Workshop on Document Intelligence (2019)
35. Poznanski, J., et al.: OLMOCR 2: unit test rewards for document OCR. arXiv:2510.19817 (2025)
36. Sasagawa, K., et al.: Evaluating multimodal large language models on vertically written Japanese text (2025). arXiv:2511.15059
37. SBIntuitions: Sarashina2.2-Vision. https://huggingface.co/sbintuitions/sarashina2.2-vision-3b (2025)
38. Singh, A., et al.: Towards VQA models that can read. In: IEEE/CVF Conference on Computer Vision and Pattern Recognition, pp. 8317–8326 (2019). https://doi.org/10.1109/CVPR.2019.00851
39. Tang, J., et al.: MTVQA: benchmarking multilingual text-centric visual question answering (2024). arXiv:2405.11985
40. Veit, A., et al.: COCO-text: dataset and benchmark for text detection and recognition in natural images (2016). arXiv:1601.07140
41. Wang, W., et al.: InternVL3.5: advancing open-source multimodal models in versatility, reasoning, and efficiency (2025). arXiv:2508.18265
42. Wei, H., et al.: DeepSeek-OCR: contexts optical compression (2025). arXiv:2510.18234
43. Xu, Y., et al.: XFUND: a benchmark dataset for multilingual visually rich form understanding. In: Findings of the Association for Computational Linguistics: ACL 2022, pp. 3214–3224 (2022). https://doi.org/10.18653/v1/2022.findings-acl.253
44. Yang, Z., et al.: CC-OCR: a comprehensive and challenging OCR benchmark for evaluating large multimodal models in literacy (2024). arXiv:2412.02210
45. Yao, C., et al.: Incidental scene text understanding: recent progresses on ICDAR 2015 robust reading competition challenge **4** (2015). arXiv:1511.09207

ADV-FORMS: A Dataset of Form-Based Historical Documents with Benchmarks for Layout Analysis, HTR and OCR

Bernhard Ortbauer[1,2(✉)], Tobias Doppler[1], Pauline Schmidt[1,4], Lukas Schilcher[3], Wolfgang Göderle[1,5], Malte Rehbein[1], Alexander Werth[4], and Roman Kern[2]

[1] Chair of Computational Humanities, University of Passau, Passau, Germany
[2] Institute of Machine Learning and Neural Computation, Graz University of Technology, Graz, Austria
bernhard.ortbauer@uni-passau.de
[3] Institute of Visual Computing, Graz University of Technology, Graz, Austria
[4] Chair of German Language, University of Passau, Passau, Germany
[5] Department of Digital Humanities, University of Graz, Graz, Austria

Abstract. Automated information extraction from form-based historical records presents unique challenges, including highly variable page layouts, obsolete typefaces, and diverse handwriting styles. We introduce ADV-FORMS, a particularly complex, openly available dataset of form-based historical documents with benchmarks for layout analysis, handwritten text recognition (HTR), and optical character recognition (OCR). ADV-FORMS comprises 1,200 questionnaires in multiple layout templates, collected in the early 1930s, which are filled in with handwritten, typewritten and stamped content, as well as non-textual elements like underscores, drawings, and selection marks. Handwritten text occurs as German Kurrent and Latin script. ADV-FORMS contains manual annotations using bounding boxes capturing 17 recurrent, visually distinct layout categories as well as manual transcriptions of handwritings and all other typewritten or stamped texts. Baseline evaluations using fine-tuned YOLOv11 for layout detection and out-of-the-box TrOCR for HTR/OCR provide valuable insights, with layout detection achieving good precision/recall (>0.9) for most categories. HTR/OCR performance showed high word error rates, which can be addressed with post-filtering, as evidenced by high BERT scores (>0.8). In summary, our dataset provides benchmarks for further model development with each improvement directly bringing benefit to the analysis of historians and research in the Digital Humanities.

Keywords: OCR · HTR · Layout Analysis · Information Extraction · Document Intelligence · Historical Dataset · ADV

G. A. Fink et al. (Eds.): ICDAR 2026, LNCS 16972, pp. 235–251, 2027.
https://doi.org/10.1007/978-3-032-36023-6_14

1 Introduction

As Goody [14,15] has shown, tables and forms are not merely practical tools but cognitive technologies that organize knowledge through spatial arrangement on the page. The role of such paper-based technologies in the constitution of modern bureaucracy and statehood has since been explored from legal-media [44], political [19], and organizational perspectives [9,46]. While preprinted templates became very common in the second half of the 19th century, we observe that handwritten or drawn templates and similar structured schemata are fairly frequent already more than a century earlier. Standardized survey instruments of this kind were central to the production of administrative knowledge about populations [2,13]. Thus, form-based records represent a staple in the category of important primary historical sources. However, they remain underutilised because current technologies often reach their limits when it comes to the automated mass evaluation of historical documents.

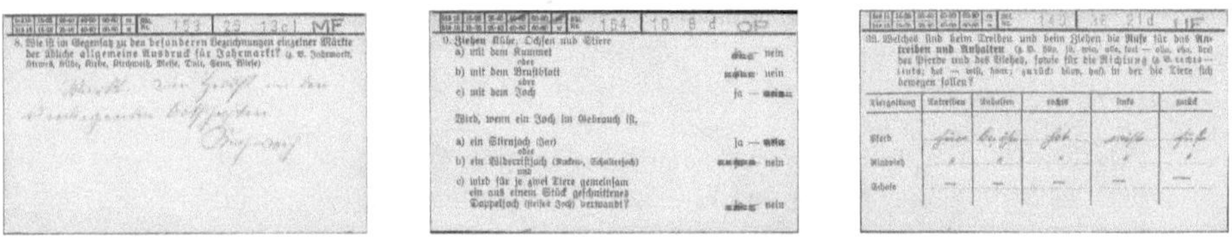

(a) Open questions with free text answers

(b) Closed questions with yes-no selection

(c) Structured answers using tabular forms

Fig. 1. Three question layouts from the ADV survey exemplify the variety of answering modes included in the ADV-FORMS dataset (originals of the images courtesy by the Institut für Volkskunde, Munich).

As Kafka [19] has argued, the gap between the standardizing intent of forms and the actual practices of those who fill them is a constitutive feature of bureaucratic paperwork. Challenges that result from this include: (1) text overflowing the attributed area, (2) corrections applied in retrospect, (3) side- or footnotes outside the designated areas, (4) additional markings with often unclear intention, (5) individual interpretation of how pre-defined values are to be chosen (e.g. underlined versus crossed-out). Examples of these challenges are illustrated in Figs. 5 and 6. Additionally, as with all historical printed and handwritten texts, now obsolete fonts and scripts are often used, and the language might be out of fashion. These factors make form-based historical documents particularly challenging for automated processing. Hence, datasets with gold standard layout annotations and transcriptions are a valuable source for training and evaluation of processing pipelines.

We report on the creation of ADV-FORMS, a novel dataset that captures this complexity of form-based historical records (see examples in Fig. 1). ADV-FORMS contains complex layout annotations as bounding boxes as well as manual transcriptions of handwritten and printed text. ADV stands for *Atlas der deutschen Volkskunde*, which freely translates to "Atlas of German Folklore", a large-scale survey across German speaking regions in Europe in the early 1930s [35]. We show in a pilot study that ADV has significance for humanities research as well as other fields. The 1,200 questionnaires of the ADV-FORMS dataset were created in the context of an ongoing digitization project, which will make large quantities of the original ADV survey material accessible online. In an effort to contribute a benchmark for method development for information extraction from this data and on complex form-based records in general, we provide competitive baselines for the recognition of layout components, HTR, and OCR.

2 Related Work

Advances in automated information extraction have spurred several studies relating to information extraction from complex form-based historical documents such as parish records [22,41], weather records of ship logs [26,33] and census tables [6]. Information extraction from complex historical forms can be approached either as a modular pipeline, separating layout recognition and transcription, or as a more integrated document-understanding task using models such as LayoutLM-style architectures [45] or OCR-free encoder-decoder approaches such as Donut [11]. Regardless of the choice, progress in both settings relies heavily on datasets that provide concurrent layout annotations and ground-truth transcriptions.

Of the 65 historical document image datasets listed by a 2022 survey [27], only a fraction includes ground truth usable for machine learning. Typically, the ground truth provides either transcribed text or complex layout annotations, and only rarely both. Datasets that eye primarily at supporting handwritten text recognition (HTR) like GERMANA [30], the Bentham dataset [37], Digital Peter [29], Hugin-Munin [25] or several ICFHR and ICDAR competition datasets [38–40,42] contain line-wise transcriptions of continuous text. The LION dataset [17] stands out as it transcribes text written in stenography. The simple layouts of these documents do not require layout annotations besides line breaks. HTR datasets with a more complex layout feature side notes, headings, and shorter text blocks with the main information. The ESPOSALLES [32] and the BH2M [7] datasets both contain annotated images from handwritten Spanish marriage lists with side notes as structuring elements from the 15th to the 20th Century and with BH2M offering layout annotations using bounding boxes. Similarly, the OHG dataset [31] contains transcriptions and information on six different layout regions for 596 pages of notarial deeds from the city of Girona written by a single hand in the late 18th century.

Freely accessible historic document image datasets with pre-printed templates most commonly concern tabular layouts. The GloSAT dataset [48] comprises 500 tabulated historical documents from various archival sources with meteorological information. This dataset focuses on layout recognition, as it comprises annotations of cell boundaries, but no transcriptions of the content.

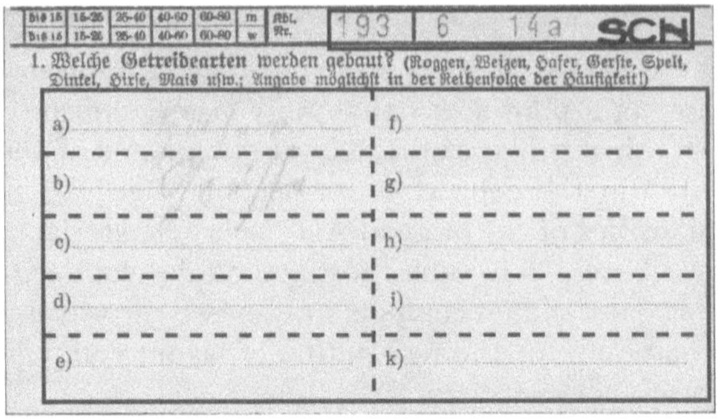

Fig. 2. Example scan of question 1 of the ADV survey. Respondents are asked to list most frequently grown grain species in descending order (green square). The location ID (red square) can be used to localize the municipality on a map. (Color figure online)

Similarly, the dataset prepared for the *ICDAR2019 Competition on Table Detection and Recognition (cTDaR)* contains historical tables from various sources for layout analysis only. The READ ABP WI dataset [8] contains 1,766 pages of complex tabular parish records with instances of non-transcribed handwritten entries marked by bounding boxes.

To our knowledge only three major historic document image datasets with pre-printed templates combine complex layout annotations with expert transcriptions. The POPP dataset [5] contains 209 pages of pre-printed tables from the 1926 Paris census, annotated with row-wise bounding boxes and transcriptions. The ship logs of the HisClima database [33] are built on seven similar table templates as well as pages with continuous handwritten English text from the 19th Century; bounding-box annotations and transcriptions are at line level for continuous text and cell level for tables. The PARES dataset [1] comprises 535 pages of pre-printed tables built on two similar templates, which were filled by manual transcription in modern script during the 1960s and 1970s based on historical records. Bounding boxes are used for row-wise annotation. The corresponding transcriptions additionally encode column boundaries and other layout specific information. POPP, HisClima, and PARES are closely related to ADV-FORMS, as they contain transcribed form-based administrative documents with layout annotations using bounding boxes. However, none of the datasets compares directly with ADV-FORMS in terms of variety of templates, modes of question answering, multitude of writing styles, general complexity, and comprehensiveness and systematisation of the included annotations.

3 Materials' Background: "Atlas Der Deutschen Volkskunde" (ADV)

The Atlas der deutschen Volkskunde (ADV), established by the *Deutsche Forschungsgemeinschaft* (DFG) in 1928, aimed to document everyday culture

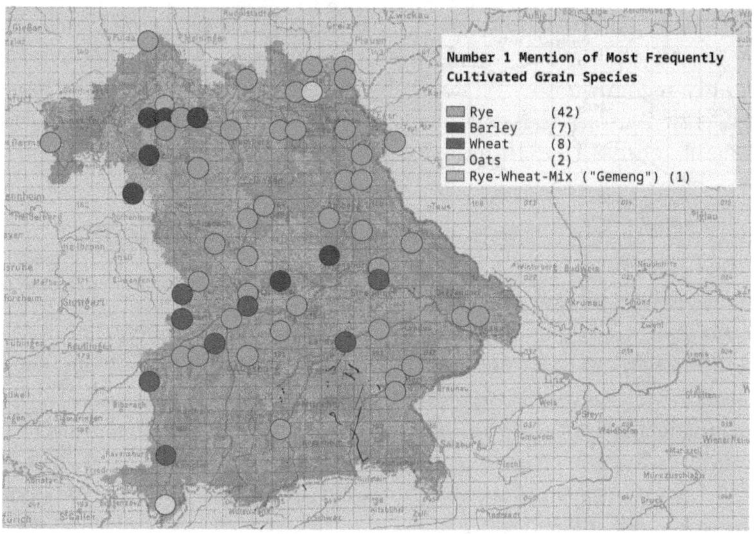

Fig. 3. Evaluation and visualisation of 60 scans of question 1 from the ADV-FORMS dataset. The background is an overlay of a satellite image of Bavaria over the ADV basemap (1937–1939) [16]. The placement of the coloured circles corresponds to the location ID printed on each scan.

across the German-speaking world. Between 1930 and 1935, five questionnaires comprising 243 questions (approx. 750 subquestions) were distributed to informants in about 20,000 municipalities [34]. The questionnaires addressed a broad spectrum of cultural practices and beliefs [36], such as: "Which colours are considered mourning colours?" or "Which basic cradle shape is the most common?".

To enable spatial analysis, the ADV assigned each municipality a unique location ID based on its position within a predefined map grid. The location ID is presented in a very similar way across all scans, whereas the mode of question answering and layout must be considered individually for each survey question. To illustrate the challenges of automated processing, we conducted a small scale study based on the 60 images of question 1 randomly sampled for the ADV-FORMS dataset. As shown in Fig. 2, the question requests the grain species grown in the municipality, ordered by frequency. In our sample study, we identify the prevalent grain species on each of the 60 images and map the information via the first two fields (e.g., "193" and "6") of the header's three-part numerical location ID to the grid of the base map (Fig. 3). The question in Fig. 2 thus produces the green circle in the lower left.

Although the sample is too small for substantive conclusions, the resulting map already reveals plausible regional patterns in grain cultivation. This demonstrates the analytical potential of large-scale automated extraction from ADV materials. An automated analysis of the Bavarian ADV data would yield nearly 2,000 data points, thus allowing for a significantly more detailed analysis. The entire ADV data could extend the analysis to large parts of Central Europe. The ADV constitutes a large-scale historical dataset whose research potential is increasingly recognised, as reflected by current digitisation initiatives such as DigErAdV [12]. Automated extraction would enable analyses at a scale previously infeasible across the Humanities and related disciplines.

4 Dataset Preparation and Annotation Scheme

The ADV-FORMS dataset was prepared in the context of the ADV project at the Univerity of Passau, which pursues digitization and publication of ADV questionnaires from Bavaria, Upper Austria and parts of Hesse. It encompasses 60 random samples of each of 20 questionnaire-templates (out of 243) of the ADV survey, resulting in a total of 1,200 annotated and transcribed images. Each of the 20 templates features a unique combination of pre-printed text, answer fields, and spatial layout, ranging from open-ended questions with textual answers to structured tables and closed questions featuring yes-no selection fields (see examples in Fig. 1). While some questionnaires show structural commonalities (e.g. two questions may both use a tabular layout), no two in the dataset share the exact same pre-printed template.

The ADV-FORMS dataset was created to facilitate automated evaluations of the ADV survey questions. As the pilot study has shown, we need to extract information regarding the location ID and anticipate various modes of question answering. Based on these premisses, we chose to systematically annotate all textual entries, drawings and markings, which are not grid lines of the pre-printed templates using rectangular bounding boxes. Although polygon annotations would have yielded more precise results, their creation would have required considerably greater time and human effort. Consequently, we chose bounding box annotations as a compromise between annotation accuracy and practical resource constraints. The focus in defining the annotation categories was to make the scheme universally applicable to all possible layouts and question types. In combination, the layout annotations and the expert transcriptions should enable various information extraction approaches and evaluation settings.

The 17 layout categories of the layout annotation scheme can be subdivided into (1) layout categories, which are used to annotate any pre-printed text or symbols, (2) layout categories that mark information relating to the stamped location information within the header area of each question, (3) layout categories that are used to distinguish between handwritten and typewritten answer texts and highlight multi-line text instances applied by the respondents and (4) layout categories that are used to annotate frequently occurring patterns of non-textual markings on behalf of the respondents. The full annotation guidelines are publicly available alongside the ADV-FORMS dataset on Zenodo [28]. Due to space constraints, we do not provide a detailed description of them in this paper.

Pre-print. We used rectangular bounding boxes labeled `pre-print` to line-wise annotate any pre-print text that is part of the questionnaire template (Fig. 5). Additionally, the `pre-print-drawing` category is used for the pre-printed pictograms of Questionnaire 13.

Location Information. We manually transcribed the contents of each of the three categories and additionally annotated a container category (`location_ID_full`) comprising the entire location ID to accentuate their coherence. In the same vein we annotated and transcribed the stamped capital letter acronym of the larger administrative region as `location_region_ID`.

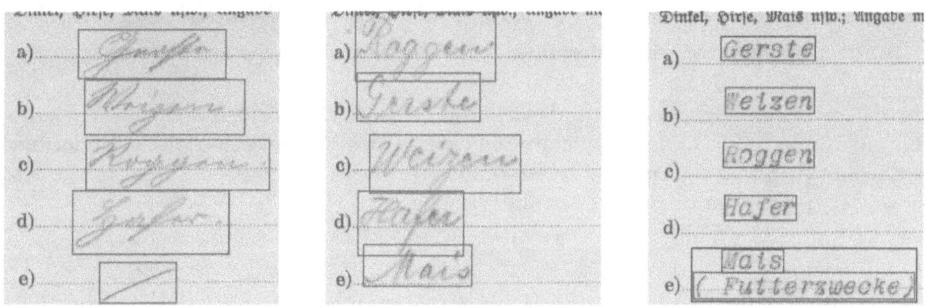

(a) German Kurrent (b) Latin Script (c) Typewriter

Fig. 4. Three examples of different writing styles from the ADV survey: all variants of handwriting, whether they come as (a) German Kurrent in the Sütterlin variant, or (b) Latin handwriting are annotated as **`handwritten_answer`**, while (c) typewritten text answers are labelled as **`typewritten_answer`**.

Text Answers. With textual answers the gap between the standardizing intent of the pre-printed template and the practices of the respondents manifests manifold. Some respondents kept to the bare minimum of necessary text, others, however, filled every available space on the page (compare Fig. 6). Written answers within the ADV-FORMS data occur in a multitude of different hand- and typewriting styles (compare Fig. 4). Transcriptions of the handwritten answers are made more difficult by faded and weakly contrasted entries (as evidenced in Fig. 2) as well as a huge variation in handwriting styles between the several thousand participants of the survey. The ADV survey was conducted at a time when different handwriting scripts were used in parallel. Most frequently handwritten answers occur as German Kurrent of the Sütterlin variant, but there are examples of modern style handwriting using Latin script and various other scripts included in the dataset. See Fig. 4 for illustration. The same does not apply for the typewritten answers. Here the style of the glyphs shows a high consistency. As far as we can tell the same set of typewriting glyphs is used throughout the samples in the dataset. The annotation scheme contains two layout categories, to separate between handwritten answers (`answer_handwritten`) and those printed by typewriter (`answer_typewritten`). If single answer instances span multiple lines, we additionally annotate a multiline container (see Fig. 6). The intent here is to help models learn the patchy nature of the respondents filling-in practices.

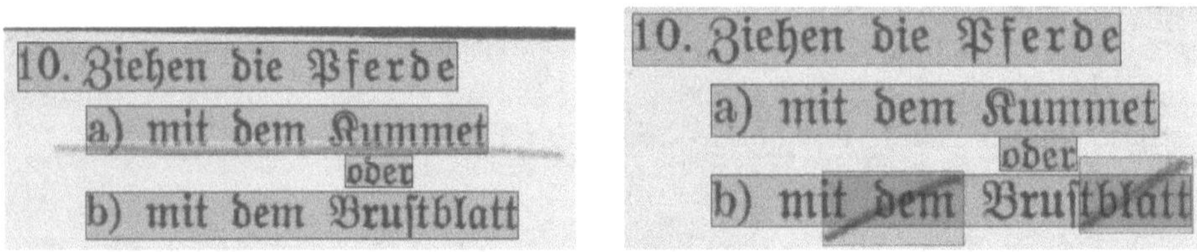

(a) positive selection via underlining (b) negative selection via crossing out

Fig. 5. Two examples of non-standard question answering via non-textual markings. Here, (a) underlining of text typically indicates a positive selection of the underlined text, while (b) crossing out of text amounts to a negative selection. The image also shows the application of `pre-print` annotations.

Non-textual Marks. Same as with text answers, the respondents acted highly inconsistently when using non-textual marks. Although the applied marks exhibit a high degree of variation, we identified seven categories that are clearly discernible by visual characteristics as well as the underlying intentions of the respondents. The label `mark_underline` is assigned, whenever respondents underlined text of the pre-printed template. Mostly this action equals the positive selection of the underlined word. A crossing out of words operates in the opposite direction as a negative selection (`mark_cross_out`). In both instances the bounding boxes closely encompass the respective drawn line (Fig. 5). Bounding boxes with the label `mark_omission` encompass lines across white space. The respondents wanted to signify that they would not or could not give a textual answer. An example can be seen in Fig. 6a. Sometimes the omission line is accompanied by two dots, thus emulating a percentage sign. Two short vertical lines—ditto marks—are often used to indicate that the preceding answer or line of text should be repeated at the location of the `mark_quote`. Curly brackets are often used in combination with a table template. Their intent is to assign an answer to multiple table rows. The respective bounding boxes, labeled as `mark_curly_bracket`, only encompass the curly bracket itself. Bounding Boxes with the label `mark_drawing` encompass any figural drawings or sketches that were applied by a respondent. In some instances respondents used little cross signs, for example to select one out of several pre-printed images. Since these marks did not fit well with any of the other layout categories, the label `mark_selection` was introduced.

Opting for a human-in-the-loop approach we initially annotated five images per question and trained a YOLO [18] model. We then used this model to predict layout annotations for the remaining 1,100 images. These annotations were then manually corrected. The transcriptions were predicted using an out of the box TrOCR model [3] and again manually corrected. The annotation was carried out by multiple annotators: six team members (Digital Humanities/Linguistics backgrounds) contributed to the process, three focusing on layout annotation and YOLO training, three on transcription. Discrepancies were minimised through ongoing discussions and a final double-keying pass in which two authors reviewed and harmonised all annotations, each of which had been processed by at least two annotators.

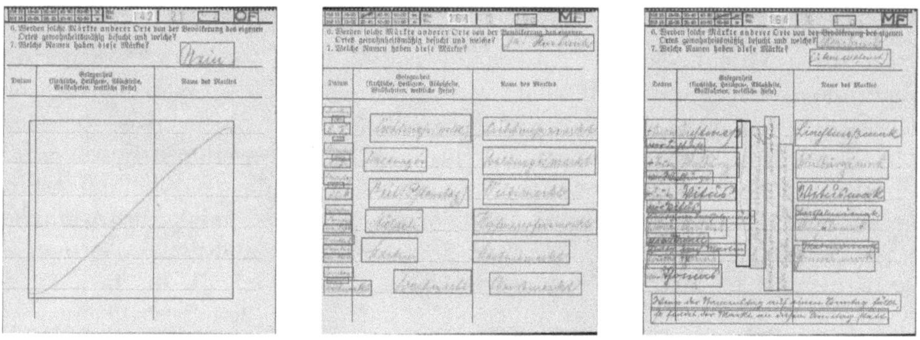

(a) empty table　(b) good adherence to preprinted template　(c) overflowing or rotated text

Fig. 6. Three images exemplifying the range of answers to be expected on a question with pre-printed table. The `pre-print` category is set transparent.

5 Baselines

Recent advances in machine learning have significantly improved the accuracy and efficiency of many computer vision tasks, including layout recognition and HTR/OCR. In this section we present baselines for recognition of layout elements using a fine-tuned YOLOv11 [18] model, and a TrOCR model for HTR/OCR performance. We intentionally keep layout detection and text recognition as separate baseline tasks. This modular setup reflects the structure of the dataset, which includes both bounding-box annotations and transcriptions, and enables future work to evaluate layout analysis, HTR/OCR or end-to-end extraction independently.

5.1 Layout Detection Using YOLOv11

For an initial analysis of the ADV dataset, we decided to train a single general YOLOv11m model applicable to all questions and all categories. YOLO is already used for document layout detection for historical data [20] and layout-aware OCR [4], thus it is a natural choice as a baseline model. To be compatible with YOLO we transferred bounding box labels and coordinates into a compliant dataset according to COCO specifications [24]. Stratification for training, validation and test splits was performed with the goal of proportionate representation of instances for every layout category. We devote 25 training and 5 validation images per question. This results in a total 500 training and 100 validation images overall. The remaining 600 images are defined as test data. Our proposed split is openly accessible via the repository of the dataset [28]. The training procedure was terminated once no further improvement in the validation metrics was observed over 20 consecutive epochs. We report the performance of the baseline model via evaluation on the test split.

Evaluation Measures. Mean Average Precision (mAP) is a standard performance measure for object detection tasks [24]. mAP50 (Mean Average Precision at a IoU threshold of 0.5) is a common default for many object detection evaluations. For our use case we regard a threshold of 50% overlap as insufficient. Especially for hand- and typewritten responses a higher degree of bounding-box overlap is required, so as not to miss parts of a text line (see `pre-print` examples in Fig. 7). With subsequent HTR processing in mind, we additionally evaluated the more nuanced mAP50-95, which averages mAPs from IoU 0.5 to 0.95, in increments of 0.05 and provides additional insight for full area recognition.

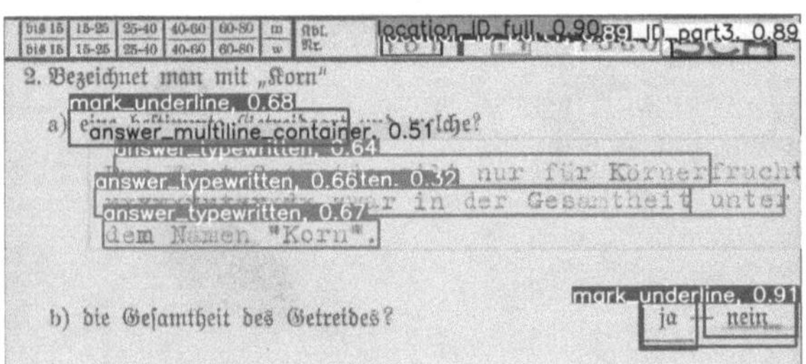

Fig. 7. Typical problems of predictions produced by the trained YOLO model: `answer_typewritten` boxes are too short to cover longer text lines.

Results. Table 1 reports the per-category detection results on the test set. From the results, it is evident that the layout categories concerning the `location-ID`, which appear on every sheet in almost the same place, exhibit high detection accuracy, achieving almost perfect mAP50 values, while even the lowest mAP50–95 value remains above 0.87. Scores for both `pre-print` categories are reported, but not interpreted, since they are likely heavily affected by overfitting as they reappear throughout the dataset. A high recognition rate is particularly important for the `location_ID` and answer categories, as it forms the foundation for subsequent HTR/OCR analyses. mAP50 values exceeding 0.94 for both `answer_handwritten` and `answer_typewritten` categories evidence high recognition rates, but a significant drop below 0.74 for mAP50-95 highlights problems to locate the entire ground truth area (compare Fig. 7). Generally, we can see that layout categories with a support greater than 200 instances all achieve mAP50 scores of 0.899 and above, with `answer_multiline_container` being the only exception. For all categories but the ones of the location-ID we see a more significant drop between mAP50 and mAP50-95, with most categories scoring below 0.7. Low recall values of `mark_selection` (0.6) and `mark_drawing` (0.5) show recognition problems for these least frequent categories.

Table 1. Performance of the baseline model for layout detection for each category on the test set of 600 hand annotated images. Bold highlights the best results. **Images**—num. of test set images the category occurs on; **Instances**—num. of bounding boxes labelled accordingly across test set;

Category	Images	Instances	Precision	Recall	mAP50	mAP50-90	CER	WER	BERTScore
location_ID_part1	599	601	0.999	0.998	**0.995**	0.93	0.026	0.077	-
location_ID_part2	599	601	0.988	0.991	0.994	0.871	0.101	0.163	-
location_ID_part3	599	601	0.995	0.992	**0.995**	0.92	0.456	0.746	-
location_ID_full	599	601	0.991	**1**	**0.995**	**0.944**	0.197	0.492	-
location_region_ID	600	603	0.997	0.998	**0.995**	0.912	0.919	0.982	-
answer_handwritten	354	1,817	0.926	0.92	0.945	0.825	0.37	0.68	0.794
answer_typewritten	73	298	0.917	0.959	0.972	0.733	0.283	0.677	0.819
answer_multiline_con.	159	284	0.641	0.739	0.693	0.547	-	-	-
mark_quote	60	196	0.874	0.776	0.839	0.522	-	-	-
mark_omission	221	472	0.909	0.864	0.92	0.723	-	-	-
mark_cross_out	359	956	0.858	0.848	0.899	0.658	-	-	-
mark_underline	95	187	0.87	0.787	0.87	0.515	-	-	-
mark_selection	21	30	0.867	0.6	0.776	0.604	-	-	-
mark_curly_bracket	27	31	0.916	0.806	0.891	0.62	-	-	-
mark_drawing	10	12	**1**	0.5	0.75	0.686	-	-	-
pre-print	599	13,621	*0.993*	*0.983*	*0.988*	*0.898*	0.262	0.52	0.82
pre-print-drawing	30	120	*0.967*	*1*	*0.97*	*0.961*	-	-	-

5.2 HTR and OCR Using TrOCR

For text recognition, we use the publicly available `dh-unibe/trocr-kurrent` model [3,23]: an encoder-decoder Transformer (TrOCR) fine-tuned on large-scale 19th-century German Kurrent material, evaluated here in a zero-shot transfer setting without any additional training on ADV-FORMS. Transformer-based recognizers of this family have been shown to match or surpass CNN/RNN-CTC architectures on many historical HTR benchmarks [10], making `dh-unibe/trocr-kurrent` a practically relevant baseline for ADV-FORMS.

Evaluation Measures. We evaluate recognition quality using character and word error rates (CER, WER) [43], normalized by collapsing multiple spaces and trimming whitespace before scoring. To complement these edit-distance metrics, we report BERTScore (F1) [47] - where relevant - computed with `bert-base-multilingual-cased` on unnormalized text, preserving casing, punctuation, and diacritics for the embedding model. BERTScore can remain high even when CER and WER are substantial, as contextual embeddings are robust to spelling modernization, near-synonym substitution, and minor reordering. Although the `bert-base-multilingual-cased` model was pre-trained primarily on contemporary corpora, its application to early 20th-century German is justified by the relatively close linguistic proximity of our material to modern Standard German. In addition, Labusch et al. [21] show that Transformer models pre-trained on contemporary text can be effectively adapted to NER on 19th- and early 20th-century German newspapers, and argue that subword tokenisation helps them handle orthographic variation and historical vocabulary.

Results. Table 1 summarizes the performance of `dh-unibe/trocr-kurrent` on ADV-FORMS, aggregated over all transcribed test-set instances. Performance is better on typewritten answers (CER 0.283, WER 0.677, BERTScore 0.819) and on the pre-print text (CER 0.262, WER 0.52, BERTscore 0.82) than on handwritten ones (CER 0.37, WER 0.68, BERTScore 0.794). Considering the variety of handwriting styles and the significant fading of parts of the handwritten texts, the increase of CER and WER is less severe than expected. The comparatively high BERTScore values, despite notable CER and WER, suggest that TrOCR often produces semantically plausible strings that deviate from the diplomatic transcription, for example by modernizing spelling ("thut" → "tut", "daß" → "dass"), slightly changing word boundaries, or substituting near-synonyms ("Ross" → "Pferd"). The `location_region_ID` field yields the highest category-level error rates (CER 0.919, WER 0.982): the ground truth consists of short region codes (e.g. *OF*, *UF*) that the model frequently augments with spurious punctuation tokens (e.g. *OF* → *OF* .), doubling the word count and driving WER near 1.0. Questionnaire 32 performs worst (CER 0.703, WER 0.966, BERTScore 0.779), when comparing different templates, as its answers consist predominantly of onomatopoeic livestock-driving calls (e.g. *hü*, *brr*) absent from the model's training distribution. See discussion below.

6 Discussion

Since precision and recall are above 0.8 for most categories, we can confirm good recognition performance across many different layout categories. As the evaluation results show, the seven categories of non-textual markers can be reproduced using the trained YOLO model. As shown in Fig. 7, it is important that the predicted bounding boxes cover the entire length of a text. Therefore, we consider a high mAP50-95 as a more relevant score compared to mAP50 for any layout category that contains textual transcriptions. The same is true for curly brackets, crossing-outs and underlines, since their informative value also depends on the exact recognition of the relevant area. The comparatively poor performance of the model with regard to `answer_multiline_container` corresponded to expectations, since this layout category is not strictly defined by visual characteristics but rather by semantic coherence of the text.

With regard to HTR/OCR performance, the results evidence that the trans-/former-based `dh-unibe/trocr-kurrent` model does not evaluate well against diplomatic transcriptions. Therefore, the baselines for WER and CER leave significant room for improvement. BERTscores of 0.81 and higher across all textual answers suggest that the model's semantic understanding of the text is significantly better than a low WER might imply. Looking at the results on the per-question level, questions 31 and 32 perform worst across all considered metrics. The reason can be found in the used language, which was often idiomatic and onomatopoeic (i.e. phonetically imitating a natural sound). In Question 31 respondents were asked to write down their calls to lure or scare away different animals. Here, the answers range from *"der Name des Pferdes"* (the name of the horse) to description of sounds like *"Bi-bi-bi!"* and *"Ksch!"*. This further evidences that the transformer model tends to output meaningful words but is less suited to transliterate isolated non-standard expressions.

7 Conclusions

The *Atlas der deutschen Volkskunde* (ADV) survey is an enormous data treasure for Humanities research, and, as we could show in a small-scale pilot study, it may hold valuable historical information for other fields as well. With ADV-FORMS, we presented a manually annotated and transcribed dataset, comprising historical form-based administrative records, compiled from 1,200 individual answer sheets of the ADV survey. This dataset poses unique challenges due to its original intent of being used as a widely distributed questionnaire.

We find that layout analysis with a selected state-of-the-art method is generally able to locate instances of the layout categories defined in the annotation scheme, as it yields satisfactory detection results evidenced by mAP50 scores. However, we identify improvement potential when it comes to the exact coverage of elongated bounding boxes, which is especially relevant for the intended use as preprocessing for HTR and OCR tasks.

Regarding these tasks, we find that an out-of-the-box TrOCR model as a baseline yields high character and word error rates. We find an explanation in the variety of individual handwriting styles, which range from German Kurrent to modern Latin script. More surprisingly, the respective BERTScore provides insights, which indicate that the TrOCR often fails to transcribe the exact text, but still manages to extract the underlying semantics.

In summary, the dataset is designed to motivate research on more sophisticated models and the development of dedicated pre-processing procedures to allow historic scholars and the general public to access the historic information contained in the presented dataset and the vast ADV data once it is published.

We argue that the multitude of pre-printed templates, different handwriting styles, occurrences of alternative modes of question answering and the good overlay of the scans make the dataset also a great candidate to provide a benchmark for general purpose extraction models from complex historical records.

Acknowledgments. We would like to thank Jessica Schneider, Sergei Shevelev and Kilian Baumgartner for their invaluable assistance during data curation and annotation.

Disclosure of Interests. The authors have no competing interests to declare that are relevant to the content of this article.

Funding. This research was funded in whole or in part by the Austrian Science Fund (FWF) 10.55776/PAT1763723 and 10.55776/FG3100, as well as bavarikon (0265) and the Volkswagen Foundation.

Data Availability. The ADV-FORMS dataset is available under Creative Commons Attribution Non Commercial No Derivatives 4.0 International license on Zenodo [28].

References

1. Andrés, J., Wall, C., Tarride, S., Coustaty, M., Toselli, A.H., Vidal, E.: The PARES database: information extraction over historical parish records. Int. J. Doc. Anal. Recognit. (IJDAR) (2025). https://doi.org/10.1007/s10032-025-00531-z
2. Appadurai, A.: Number in the colonial imagination. In: Breckenridge, C.A., van der Veer, P. (eds.) Orientalism and the Postcolonial Predicament: South Asian Perspectives, pp. 314–339. University of Pennsylvania Press, Philadelphia (1993)
3. trocr-kurrent (revision 033b849) (2023). https://doi.org/10.57967/hf/0442. https://huggingface.co/dh-unibe/trocr-kurrent
4. Beyene, F.S., Dancy, C.L.: Layout-aware OCR for black digital archives with unsupervised evaluation (2025). https://arxiv.org/abs/2509.13236
5. Constum, T., et al.: POPP Datasets: Datasets for handwriting recognition from French population census (2022). https://doi.org/10.5281/ZENODO.6581158. https://zenodo.org/doi/10.5281/zenodo.6581158
6. Constum, T., et al.: Recognition and information extraction in historical handwritten tables: toward understanding early 20th century paris census. In: Uchida, S., Barney, E., Eglin, V. (eds.) Document Analysis Systems, vol. 13237, pp. 143–157. Springer, Cham (2022). https://doi.org/10.1007/978-3-031-06555-2_10
7. Fernandez-Mota, D., Almazan, J., Cirera, N., Fornes, A., Llados, J.: BH2M: The Barcelona Historical, Handwritten Marriages Database (2014). https://doi.org/10.1109/ICPR.2014.53. http://ieeexplore.ieee.org/document/6976764/
8. Fiel, S., Kleber, F., Lang, E.M., Fronhöfer, W.: READ ABP WI Dataset - Writer Identification over decades (2018). https://doi.org/10.5281/ZENODO.1421600. https://zenodo.org/record/1421600
9. Gardey, D.: Écrire, calculer, classer: comment une révolution de papier a transformé les sociétés contemporaines. La Découverte, Paris (2008)
10. Garrido-Munoz, C., Rios-Vila, A., Calvo-Zaragoza, J.: Handwritten Text Recognition: A Survey (2025). https://doi.org/10.48550/arXiv.2502.08417. http://arxiv.org/abs/2502.08417. arXiv:2502.08417

11. Geewook, K., et al.: OCR-free document understanding transformer (2022). https://arxiv.org/abs/2111.15664
12. GEPRIS: Digitalisierung und erschließung des archivs atlas der deutschen volkskunde (digeradv) (2025). https://gepris.dfg.de/gepris/projekt/558245029
13. Göderle, W.: Zensus und Ethnizität: Zur Herstellung von Wissen über soziale Wirklichkeiten im Habsburgerreich zwischen 1848 und 1910. Wallstein, Göttingen (2016)
14. Goody, J.: The Domestication of the Savage Mind. Cambridge University Press, Cambridge (1977)
15. Goody, J.: The Logic of Writing and the Organization of Society. Cambridge University Press, Cambridge (1986)
16. Harmjanz, H., Röhr, E.: Atlas der deutschen volkskunde: Grundkarte (1937-1939)
17. Heil, R., Nauwerck, M.: Handwritten stenography recognition and the LION dataset. Int. J. Doc. Anal. Recognit. (IJDAR) **28**(1), 3–18 (2025). https://doi.org/10.1007/s10032-024-00479-6
18. Jocher, G., Qiu, J.: Ultralytics yolo11 (2024). https://github.com/ultralytics/ultralytics
19. Kafka, B.: The Demon of Writing: Powers and Failures of Paperwork. Zone Books, New York (2012)
20. Kiš, M., Hradiš, M., Dvořáková, M., Jiroušek, V., Kersch, F.: Annopage dataset: dataset of non-textual elements in documents with fine-grained categorization (2025). https://arxiv.org/abs/2503.22526
21. Labusch, K., Kulturbesitz, P., Neudecker, C., Zellhöfer, D.: Bert for named entity recognition in contemporary and historical German. In: Proceedings of the 15th Conference on Natural Language Processing, pp. 9–11 (2019)
22. Lang, E., Puigcerver, J., Toselli, A.H., Vidal, E.: Probabilistic indexing and search for information extraction on handwritten German parish records. In: Proceedings of the International Conference on Frontiers in Handwriting Recognition (ICFHR), pp. 44–49. IEEE, Niagara Falls, NY, USA (2018). https://doi.org/10.1109/ICFHR-2018.2018.00017. https://ieeexplore.ieee.org/document/8563224/
23. Li, M., et al.: TrOCR: transformer-based optical character recognition with pre-trained models. In: Proceedings of the Thirty-Seventh AAAI Conference on Artificial Intelligence and Thirty-Fifth Conference on Innovative Applications of Artificial Intelligence and Thirteenth Symposium on Educational Advances in Artificial Intelligence. AAAI'23/IAAI'23/EAAI'23. AAAI Press (2023). https://doi.org/10.1609/aaai.v37i11.26538
24. Lin, T.Y., et al.: Microsoft coco: common objects in context (2015). https://arxiv.org/abs/1405.0312
25. Maarand, M., Beyer, Y., Kåsen, A., Fosseide, K.T., Kermorvant, C.: A comprehensive comparison of open-source libraries for handwritten text recognition in norwegian. In: Uchida, S., Barney, E., Eglin, V. (eds.) Document Analysis Systems, vol. 13237, pp. 399–413. Springer, Cham (2022). https://doi.org/10.1007/978-3-031-06555-2_27
26. Muheki, D., et al.: MeteoSaver v1.0: a machine-learning based software for the transcription of historical weather data (2025). https://doi.org/10.5194/egusphere-2024-3779. https://egusphere.copernicus.org/preprints/2025/egusphere-2024-3779/
27. Nikolaidou, K., Seuret, M., Mokayed, H., Liwicki, M.: A survey of historical document image datasets. Int. J. Doc. Anal. Recognit. (IJDAR) **25**(4), 305–338 (2022). https://doi.org/10.1007/s10032-022-00405-8
28. Ortbauer, B., et al.: ADV-FORMS (2026). https://doi.org/10.5281/zenodo.20682016

29. Potanin, M., et al.: Digital peter: new dataset, competition and handwriting recognition methods. In: The 6th International Workshop on Historical Document Imaging and Processing, pp. 43–48. ACM, Lausanne Switzerland (2021). https://doi.org/10.1145/3476887.3476892. https://dl.acm.org/doi/10.1145/3476887.3476892
30. Pérez, D., Tarazón, L., Serrano, N., Castro, F., Terrades, O.R., Juan, A.: The Germana database. In: 2009 10th International Conference on Document Analysis and Recognition, pp. 301–305. IEEE, Barcelona, Spain (2009). https://doi.org/10.1109/ICDAR.2009.10. http://ieeexplore.ieee.org/document/5277691/
31. Quirós, L., et al.: Oficio De Hipotecas De Girona. A Dataset of Spanish Notarial Deeds (18th Century) for Handwritten Text Recognition and Layout Analysis of Historical Documents (2018). https://doi.org/10.5281/ZENODO.1322666. https://zenodo.org/record/1322666
32. Romero, V., et al.: The ESPOSALLES database: an ancient marriage license corpus for off-line handwriting recognition. Pattern Recognit. **46**(6), 1658–1669 (2013). https://doi.org/10.1016/j.patcog.2012.11.024. https://linkinghub.elsevier.com/retrieve/pii/S0031320312005080
33. Romero, V., Sanchez, J.A.: The HisClima database: historical weather logs for automatic transcription and information extraction. In: Proceedings of the International Conference on Pattern Recognition (ICPR), pp. 10141–10148. IEEE, Milan, Italy (2021). https://doi.org/10.1109/ICPR48806.2021.9412210. https://ieeexplore.ieee.org/document/9412210/
34. Schlenger, H.: Methodische und technische Grundlagen des Atlas der deutschen Volkskunde. No. 27 in Deutsche Forschung. Aus der Arbeit der Notgemeinschaft der deutschen Wissenschaft, Notgemeinschaft der deutschen Wissenschaft (1934)
35. Schmoll, F.: Die Vermessung der Kultur: Der "Atlas der deutschen Volkskunde" und die Deutsche Forschungsgemeinschaft 1928–1980. No. 5 in Studien zur Geschichte der Deutschen Forschungsgemeinschaft, Franz Steiner, Stuttgart (2009)
36. Simon, M.: Volksmedizin im frühen 20. Jahrhundert. Zum Quellenwert des Atlas der deutschen Volkskunde. No. 28 in Studien zur Volkskultur,, Schwalm, Mainz (2003)
37. Sánchez, J.A.: Bentham Dataset R0 (2016). https://doi.org/10.5281/ZENODO.44519. https://zenodo.org/record/44519
38. Sánchez, J.A., Romero, V., Toselli, A.H., Vidal, E.: ICFHR2014 competition on handwritten text recognition on transcriptorium datasets (HTRtS). In: 2014 14th International Conference on Frontiers in Handwriting Recognition, pp. 785–790. IEEE, Greece (2014). https://doi.org/10.1109/ICFHR.2014.137. http://ieeexplore.ieee.org/document/6981116/
39. Sánchez, J.A., Romero, V., Toselli, A.H., Villegas, M., Vidal, E.: Dataset for ICDAR2017 Competition on Handwritten Text Recognition on the Read Dataset (ICDAR2017 HTR) (2017). https://doi.org/10.5281/ZENODO.835489. https://zenodo.org/record/835489
40. Sánchez, J.A., Toselli, A.H., Romero, V., Vidal, E.: ICDAR 2015 competition HTRtS: handwritten text recognition on the transcriptorium dataset. In: 2015 13th International Conference on Document Analysis and Recognition (ICDAR), pp. 1166–1170. IEEE, Tunis, Tunisia (2015). https://doi.org/10.1109/ICDAR.2015.7333944. http://ieeexplore.ieee.org/document/7333944/
41. Tarride, S., Maarand, M., Boillet, M., et al.: Large-scale genealogical information extraction from handwritten quebec parish records. Int. J. Doc. Anal. Recognit. (IJDAR) **26**, 255–272 (2023). https://doi.org/10.1007/s10032-023-00427-w

42. Toselli, A.H., Romero, V., Villegas, M., Vidal, E., Sánchez, J.A.: HTR Dataset ICFHR 2016 (2018). https://doi.org/10.5281/ZENODO.1164045. https://zenodo.org/record/1164045
43. Vaessen, N.: jiwer: evaluate your speech-to-text system with similarity measures such as WER and CER (2026). https://github.com/jitsi/jiwer. Accessed Feb 2026
44. Vismann, C.: Akten: Medientechnik und Recht. Fischer Taschenbuch, Frankfurt a. M. (2000)
45. Xu, Y., Li, M., Cui, L., Huang, S., Wei, F., Zhou, M.: Layoutlm: pre-training of text and layout for document image understanding. In: Proceedings of the 26th ACM SIGKDD International Conference on Knowledge Discovery & Data Mining, KDD 2020, pp. 1192–1200. Association for Computing Machinery, New York, NY, USA (2020). https://doi.org/10.1145/3394486.3403172
46. Yates, J.: Control through Communication: The Rise of System in American Management. Johns Hopkins University Press, Baltimore (1989)
47. Zhang, T., Kishore, V., Wu, F., Weinberger, K.Q., Artzi, Y.: Bertscore: evaluating text generation with BERT (2020). https://arxiv.org/abs/1904.09675
48. Ziomek, J., Middleton, S.E.: GloSAT Historical Measurement Table Dataset: Enhanced Table Structure Recognition Annotation for Downstream Historical Data Rescue, pp. 49–54. ACM, Lausanne Switzerland (2021). https://doi.org/10.1145/3476887.3476890. https://dl.acm.org/doi/10.1145/3476887.3476890

Structure-Aware Text Recognition for Ancient Greek Critical Editions

Nicolas Angleraud(✉), Antonia Karamolegkou, Benoît Sagot, and Thibault Clérice

Inria, Paris, France
{nicolas.angleraud,antonia.karamolegkou,benoit.sagot, thibault.clerice}@inria.fr

Abstract. Recent advances in visual language models (VLMs) have transformed end-to-end document understanding. However, their ability to interpret the complex layout semantics of historical scholarly texts remains limited. This paper investigates structure-aware text recognition for Ancient Greek critical editions, which have dense reference hierarchies and extensive marginal annotations. We introduce two novel resources: (i) a large-scale synthetic corpus of 185,000 page images generated from TEI/XML sources with controlled typographic and layout variation, and (ii) a curated benchmark of real scanned editions spanning more than a century of editorial and typographic practices. Using these datasets, we evaluate three state-of-the-art VLMs under both zero-shot and fine-tuning regimes. Our experiments reveal substantial limitations in current VLM architectures when confronted with highly structured historical documents. In zero-shot settings, most models underperform compared to established off-the-shelf traditional software. Nevertheless, the Qwen3VL-8B model achieves state-of-the-art performance, reaching a median Character Error Rate of 1.0% on real scans. These results highlight both the current shortcomings and the future potential of VLMs for structure-aware recognition of complex scholarly documents.

1 Introduction

Traditional OCR pipelines decompose the task into sequential stages (layout analysis, segmentation, recognition), which are brittle under heterogeneous or complex layouts and noisy scans, and early errors propagate to downstream processing and retrieval. Recent vision–language models (VLMs)—Qwen2.5-VL [1], DeepSeek-OCR-2 [33], and LightOnOCR [31]—address this limitation by performing end-to-end recognition, jointly modeling visual input and text generation. While these models achieve strong results on modern benchmarks [16,19], existing evaluations primarily measure plain-text accuracy and do not isolate failures in document structure. This limitation becomes critical for historical scholarly documents, where layout encodes essential semantic information. In particular, Ancient Greek critical editions combine hierarchical organization, marginalia, and dense reference systems that are central to scholarly use.

G. A. Fink et al. (Eds.): ICDAR 2026, LNCS 16972, pp. 252–268, 2027.
https://doi.org/10.1007/978-3-032-36023-6_15

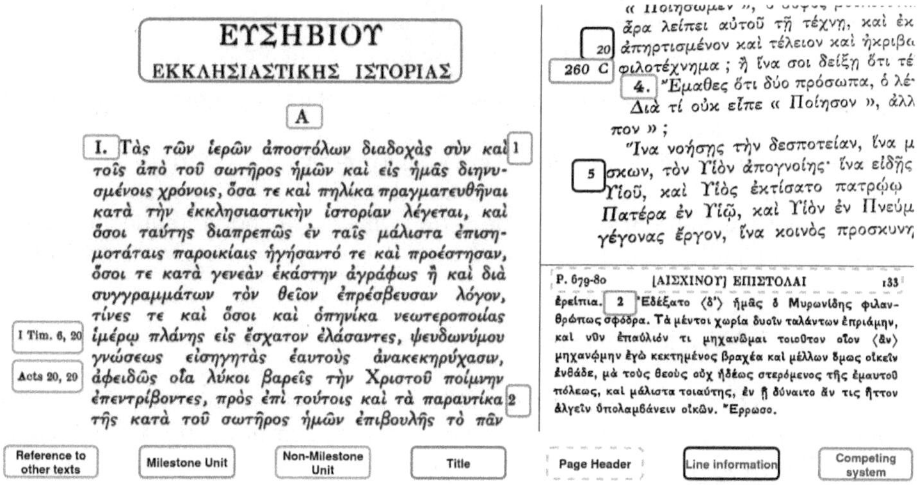

Fig. 1. Example of the richness of the layout in three different pages.

Despite prior OCR work on Ancient Greek [14,29], large-scale computational access remains constrained by two bottlenecks. First, open, machine-actionable corpora remain scarce. Much of the textual heritage is accessible only through licensed resources such as the *Thesaurus Linguae Graecae* (TLG), which contains over 12,000 works but restricts redistribution, bulk analysis, and derivative use [23]. Open corpora derived from printed critical editions remain smaller and uneven in coverage: major alternatives such as the *Perseus Digital Library* [12] and the *Thousand Years of Ancient Greek* [20] focus mainly on the first millennium of Greek, while Late Antique and Medieval Greek are only sparsely represented, despite targeted extensions [30]. Second, Ancient Greek critical editions pose intrinsic OCR challenges because of their dense and explicit document structure (Fig. 1). Beyond polytonic script recognition, they combine section hierarchies, milestone numbering, marginal notes, and critical apparatus entries recording variant readings in the textual tradition. These elements are not peripheral: they form an organizational layer used for citation and navigation and interact directly with OCR performance.

Digitization, therefore, requires not only transcription but also structure recovery, separation of layout artefacts from textual organization, and reconstruction of content fragmented across pages. In practice, this conversion of OCR output into structured TEI/XML representations is often more time-consuming than raw transcription and remains largely manual due to the lack of openly shared ground truth [32]. In this work, we study structure-aware OCR for Ancient Greek critical editions using a lightweight markup scheme for reference and organizational elements. We construct image–transcription benchmarks spanning more than 5,000 works and supporting both training and evaluation. Although our focus is Ancient Greek, the challenges we target—dense layout, intertwined structure, and multi-layer annotations—are representative of a broader class of complex document understanding problems. Our Code and data sources are available at: https://gitlab.inria.fr/almanach/cllg/icdar-2026

- an open large-scale synthetic corpus with controlled typographic and layout variation (185,000 images);
- a reproducible pipeline for generating structure-aware OCR datasets;
- an open curated benchmark of real printed editions (450 images);
- a benchmark analysis of state-of-the-art VLMs in zero-shot and fine-tuned settings, with resulting models below 2% CER(https://huggingface.co/CLLG).

2 Related Work

The reference corpus for Ancient Greek literary texts remains the *Thesaurus Linguae Graecae* (TLG) [23], which offers unmatched coverage from the Archaic period to Byzantium but is distributed under a restrictive license that prohibits redistribution, bulk processing, and derivative datasets. Despite its central role in Greek philology, the TLG therefore cannot support open OCR training, evaluation, or reproducible experimentation. Open alternatives have emerged over the past decades, most notably the Greek collections of the Perseus Digital Library, which established the open dissemination of TEI-encoded Ancient Greek texts. These resources were later consolidated and extended by the *First 1K Years of Greek* (F1KG) project [20], which provides openly licensed literary texts with linguistic annotation and stable citability, and by the *Patristic Text Archive* (PTA) [30], which addresses the under-representation of Late Antique texts in other major open corpora.

Beyond continuous literary texts, other Ancient Greek corpora differ substantially in structure and purpose. Fragmentary corpora focusing on lost works transmitted through quotations do not reflect the layout or continuity of book-length texts [2]. Epigraphical [22] and papyrological [6] corpora provide large quantities of Greek text, but their fragmentary supports, non-linear layouts, and strong editorial intervention make them methodologically distinct from printed critical editions. Overall, although several open Ancient Greek corpora exist [24], their coverage remains partial compared to the TLG, and none provide OCR-oriented ground truth aligned with the structural complexity of critical editions.

Research on automatic processing of Ancient Greek has likewise treated OCR, layout analysis, and post-correction mostly as separate problems. Early OCR work used adaptable frameworks such as Gamera [27] to recognize polytonic Greek in historical prints, while user-facing environments such as Lace [26] emphasized interactive post-correction. More recent work has addressed layout analysis for text-heavy historical commentaries [21], Ancient Greek text recognition [28], and LLM-based post-correction [3,25]. These studies highlight challenges specific to Ancient Greek, including rich morphology, a long diachronic span from approximately 900 BCE to 1500 CE, and strong geographical and dialectal variation. Taken together, prior work either separates layout from text or relies on post-hoc correction, leaving the joint problem of structurally aware OCR for Ancient Greek critical editions largely unexplored.

3 Layout Semantics in Editions of Ancient Texts

What Is Structure in a Critical Edition. Classical texts in Latin and Greek have been for centuries organized into hierarchical units comparable to modern chapters and books [15], and in some cases since their original publication [4, p. 13]. Over time, these divisions have been standardized through a process of canonicalization. As a result, systems of reference have become central to scholarly practice, enabling stable citation and comparison across editions. Their use has persisted and extends to digital academic platforms, making their recognition and encoding in digital versions of critical editions essential.

Some structural systems correspond closely to contemporary notions of textual organization: chapters and sections often align with paragraph boundaries and coherent thematic units. Others, however, are rooted in historical printing practices and reflect the material constraints of early modern editions (such as the Stephanus pagination, a citation system based on an edition of Plato's work in 1578). Despite their arbitrariness from a modern perspective, these systems continue to function as authoritative reference frameworks. We distinguish between two types of structural levels: milestone and non-milestone units. A milestone structural level is defined by its capacity to appear within a paragraph rather than only at its beginning, enabling fine-grained internal referencing. Because different systems respond to different scholarly needs, hybrid structures combining multiple non-milestone levels with a final milestone level are common.

Competing Structures and Marginal Information. The long-standing tradition of canonical referencing has produced a wide diversity of layout practices for encoding structural information. These often coexist with alternative organizational schemes based on content rather than inherited pagination. For example, although the *Patrologia Graeca* citation system remains the scholarly standard, the philological quality of its editions is often considered insufficient, leading modern editors to propose new hierarchical structures based on chapters, sections, or thematic units. Nevertheless, due to the authority of traditional references, these systems are typically preserved in parallel, often in the margins. Such dual referencing extends to texts transmitted through a single manuscript or authoritative witness, where references to original folios are also provided (e.g., *fol. 123r*, *123v*). Margins also contain additional types of information. In Christian texts, they frequently include references to biblical passages or other theological authorities. More generally, even when no canonical reference is present, line numbering is often provided (e.g., at intervals of 5, 10, 15, or more rarely 4, 8, 12). This layering of marginal, heading, and in-text annotations produces a visually dense and semantically complex layout, in which only a subset of elements has canonical use. For non-specialists, interpreting these intertwined systems can require extensive contextual reading to reconstruct relationships between structural units, marginal references, and editorial interventions. From a document analysis perspective, this richness is both a source of information and a major source of complexity. From a corpus-building perspective, identifying structuring reference identifiers among this noise is the most time-consuming step, as it directly impacts resource encoding [10].

Table 1. Comparison of structural characteristics across synthetic and real data. The total count of pages, the median number of words per page, and the proportion of pages containing at least one instance of the respective structural element (`<ref>`, `<note>`, or `markdown header` for titles and subtitles.

Corpus	Pages	Words_{med}	%Ref	%Note	%Head.
Synthetic	185k	242	78.6%	6.1%	55.2%
Real	460	215	83.0%	32.3%	30.7%

4 Constructing Structure-Aware Resources

We construct two complementary resources: (i) a large dataset of synthetically rendered page images derived from existing critical editions, designed to introduce controlled typographic and layout variation, and (ii) a small real-data benchmark of scanned editions for measuring generalization under authentic conditions. Both resources share the same annotation target: structured Markdown with block-level headings and inline tags (Appendix 8).

4.1 Rendering Structured Editions

To enable controlled, large-scale training and evaluation for Ancient Greek OCR under complex scholarly layouts, we generate *synthetic page images* by rendering openly licensed TEI/XML texts into typographically diverse document pages. This preserves the original content while introducing realistic variation in layout and typographical conventions found in printed critical editions. We collect prose texts from TEI/XML corpora of Ancient Greek [7,11,13], including one Latin corpus [8].[1] Where the same work appears in multiple corpora with different encodings, we retain all versions to increase diversity. The pipeline is summarized in Fig. 2.

For each TEI document, we extract a citation-oriented structural description from the TEI tree. This `citeStructure` [5] specifies the hierarchical units used for citation and navigation (e.g., book, chapter, section) and their nesting. We then use a lightweight pseudo-XML annotation scheme to mark the text and its structure. According to this scheme, the hierarchical units are marked in between `<ref>...</ref>` tags, the editorial notes are marked in `<note>...</note>` tags, and all titles are marked with a # following markdown practices. As this paper

[1] The Latin corpus is included to maintain performance for Latin-script material, which frequently appears in Greek editions (e.g., references, abbreviations, editorial conventions).

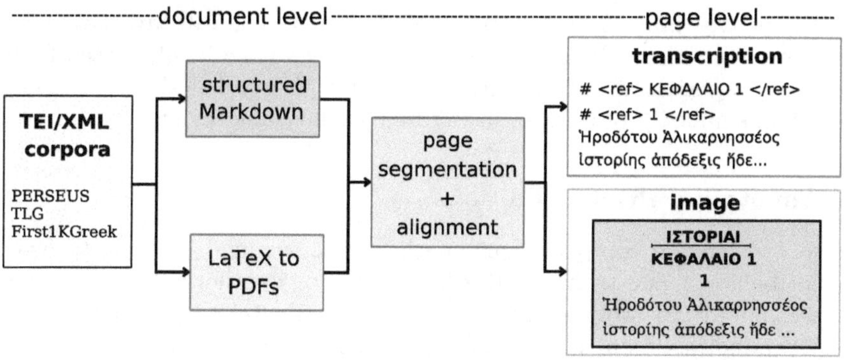

Fig. 2. Overview of the synthetic data generation pipeline. TEI/XML corpora are converted into structured Markdown annotations and rendered into PDF page images via LaTeX, producing aligned text–image pairs.

aims to facilitate the rapid re-encoding of multi-page documents, we also address the issue of paragraphs that are broken across pages. To support their reconstruction, we add a `<tab/>` tag to paragraphs that are indented in the printed layout. This marker enables reliable post-processing and merging of cross-page fragments, ensuring that the synthetic dataset supports both OCR transcription and document-level, structure-aware evaluation.

To introduce controlled visual variation, we surveyed critical editions and, guided by editorial and typographical conventions, constructed a curated set of LaTeX layout configurations and background templates. These modulate page-level and typographic properties, including background color [9], page size, column count, font families, numeral styles, and the positioning of titles and structural markers, as well as higher-level parameters such as paper format and content representation, while preserving the underlying text and annotations. For each document, multiple configurations are randomly sampled and applied during rendering, allowing the same source to be realized under diverse typographic conditions without altering its semantic content. The goal is not to reproduce specific documents, but to cover a representative range of plausible configurations grounded in real practices. Prior to rendering, TEI/XML sources are lightly normalized and then transformed via XSLT into a LaTeX source file with layout placeholders and a parallel pseudo-XML target used for supervision and evaluation. This pipeline ensures consistency between visual output and structural annotation while enabling systematic variation in document appearance.

We generate two PDF versions: one in black text and one with color-coded components. The color-coded component allows for identifying information, such as the footer, which is subsequently masked to prevent the introduction of artificial supervision signals.[2]Page text is extracted from the color-coded PDF via

[2] Footnotes, specifically variants from other manuscripts, are under a different copyright system in multiple countries, including France and Germany.

color filtering to retain only the main-text channel. The full target text is segmented into page-sized chunks using tail matching with a length-based fallback. A page pair is retained only if extracted text and segmented target achieve a Levenshtein similarity of 0.99, ensuring high-fidelity alignment.

4.2 Annotating Printed Editions

We also construct a compact benchmark of real scanned editions to evaluate zero-shot behavior and transfer after fine-tuning on synthetic data. The dataset comprises 30 author–work pairs spanning approximately 1,800 years of Greek literary production, from Hippocrates (5th c. BCE) to Nicolas Cabasilas (14th c. CE). To assess both in-domain and out-of-distribution generalization, half of the authors overlap with the synthetic corpus pool, while the rest are disjoint and increase coverage of post-classical (AD) material. Editions are drawn from diverse editorial traditions: approximately half belong to the *Sources Chrétiennes* series, with additional representation from Oxford/Clarendon, Les Belles Lettres, the *Patrologia Graeca*, the *Griechische Christliche Schriftsteller*, *Zeitschrift für Papyrologie und Epigraphik*, *Revue des études byzantines*, Princeton University Press, and 19th-century scholarly editions (Weidmann, Baillière, Laupp). Publication dates range from 1844 to 2017, covering nearly two centuries of typographic variation.

For each work, we sample 15 pages uniformly at random and manually annotate them following Appendix 8. The annotation scheme mirrors the synthetic data and encodes heading patterns and inline reference tags (`<ref>...</ref>`, `<note>...</note>`). Table 1 summarizes key statistics: 83% of pages contain reference markers and 32.3% marginal notes (vs. 6.1% in synthetic data), confirming the high structural density of critical editions, while titles/subtitles occur on 30.7% of pages. We further classify works into three structural systems based on reference encoding: section systems (21 works), where markers align with block boundaries; milestone systems (7 works), where markers appear within running text; and mixed systems (2 works), combining both. Annotation quality was ensured using a four-eye principle, with each page independently reviewed by two authors; disagreements on `<ref>` and `<note>` were resolved by consulting the full edition.

5 Experimental Design

5.1 Data

We work with two complementary data sources: synthetically rendered page images and real scanned pages. Synthetic images are rendered at high resolution and downsampled to a maximum side length of 1024 pixels; real scans are resized to the same maximum side at inference time. For the synthetic corpus, splits are defined at the level of canonical document groups (i.e., TEI source documents), such that all styled renderings of the same source are assigned to the same split to ensure strict separation of textual content. We allocate 90% of document groups to training and 10% to an evaluation pool. The synthetic corpus comprises approximately 185K pages for training, 518 for validation, and 139 for test. The real-data corpus consists of 300 training pages, 60 validation pages, and 90 test pages spanning 30 author-work pairs. Both test splits were constructed to reflect the range of document-level statistics in the evaluation pool, including variation in page length, reference density, marginal note frequency, and heading density.

5.2 Models and Training Configuration

We evaluate models spanning general-purpose vision-language models (VLMs) and OCR-specialized architectures selected for their performance in leaderboards: Qwen3-VL [1] (2B and 8B), DeepSeek-OCR-2 [33] (3B), and LightOnOCR-2 [31] (1B). Each model is evaluated under four training regimes: (i) `zero-shot`; (ii) `synthetic`, fine-tuned on the synthetic train set; (iii) `real`, fine-tuned directly on the real-data train set; and (iv) `synth→real`, fine-tuned first on synthetic data and subsequently continued on real data, initialized from the best checkpoint.

For fine-tuning, we use parameter-efficient LoRA adapters optimized with AdamW. Hyperparameters were selected via grid search over learning rate, LoRA rank, and gradient accumulation steps, with checkpoint selection based on validation CER. The final configurations are reported in Table 2. All models use greedy decoding with a maximum of 4096 generated tokens, but since we use different model families, there is a difference in how task instructions are provided. LightOnOCR-2 is image-only, without textual prompts, while DeepSeek-OCR-2 uses its built-in *Free OCR* instruction. Qwen3-VL, as a general-purpose VLM, requires an explicit task prompt, which we provide in the form of a structured instruction describing the expected pseudo-Markdown output format. The full prompt is given in Appendix 3.

For baseline OCR performance, we report results under two different settings based on Convolutional Recurrent Neural Networks (CRNN). First, we use Tesseract [17], which requires no additional training and serves as an off-the-shelf baseline. We then use the Kraken software [18], first trained on the synthetic dataset and subsequently fine-tuned using the development set only (60 pages). For evaluation, we report results using automatic segmentation with Kraken base segmentation models, reflecting a fully automated pipeline. All Tesseract and Kraken baselines are evaluated exclusively on textual recognition accuracy and do not take document structure into account.

5.3 Evaluation Metrics

We evaluate transcription quality at two levels: plain-text accuracy and document structure recognition. For the former, we report macro-averaged and median word error rates (WER) and character error rates (CER) computed on raw character sequences after stripping structural markup (markdown and pseudo-XML), applying NFKC normalization, and collapsing whitespace. These

Table 2. LoRA fine-tuning hyperparameters. Synth→Real warm-starts from the best synthetic checkpoint.

Model	Phase	LR	Rank	α	Eff. batch	Epochs
DeepSeek-OCR-2	Synthetic	2e-4	16	16	32	1
	Real	2e-4	32	32	16	20
	Synth→Real	2e-4	32	32	16	20
LightOnOCR-2	Synthetic	2e-4	16	32	32	1
	Real	6e-5	16	32	16	20
	Synth→Real	6e-5	16	32	16	20
Qwen3-VL	Synthetic	2e-4	16	16	64	1
	Real	2e-4	32	32	8	3
	Synth→Real	2e-4	32	32	8	3

metrics are normalized by reference length, allowing values above 100% if predictions exceed the reference. For structure recognition, we evaluate reference markers (`<ref>`) and marginal notes (`<note>`) using longest common subsequence (LCS) F1 in reading order, and header count (`#`) using count-based F1. For paragraph indentation markers (`<tab/>`), we report two accuracy scores: recall (accuracy on paragraphs that should be indented) and specificity (accuracy on paragraphs that should not be indented), measuring whether the model correctly identifies the presence or absence of indentation at the opening paragraph if it could be a continuation from a previous page (i.e., there is no header line before the paragraph). To analyze error patterns, we decompose edit operations at both character and word level into substitutions, insertions, and deletions (reported per 100 reference units). Insertions approximate hallucinations, while deletions capture omitted reference content. Character substitutions are further classified when base Greek letters match, but combining marks differ. We distinguish breathing confusions between `U+0313` and `U+0314`, accent confusions (`U+0300, U+0301, U+0342`), and iota subscript confusions (`U+0345`).

6 Results

Table 3 reports plain-text OCR performance across training regimes and test domains. We observe four main patterns: (i) baseline models have competitive results OCR-wise, (ii) fine-tuning substantially improves transcription accuracy over zero-shot inference, (iii) synthetic supervision transfers effectively to real scanned material, and (iv) sequential synthetic-to-real training yields the strongest overall performance.

Impact of Fine-Tuning. Fine-tuning consistently improves performance over zero-shot across all models. For instance, Qwen3-VL-8B reduces median CER on real data from 5.2% (zero-shot) to 2.1% with real-only training and to

1.0% under the **synth→real** regime. Similar gains hold for Qwen3-VL-2B and LightOnOCR-1B, confirming the importance of domain-adapted supervision. Synthetic-only training already transfers strongly to real scans: Qwen3-VL-2B drops from 10.0% to 2.1% CER without exposure to real images, and for Qwen3-VL-8B, synthetic-only (1.7%) even surpasses real-only (2.1%). Sequential **synth→real** yields the best results for Qwen models, suggesting complementary effects of broad synthetic coverage and real-domain refinement. This pattern is not universal, however: LightOnOCR-1B performs best with real-only training (1.7%), while **synth→real** slightly degrades performance.

Table 3. Pure OCR results with markup stripped before scoring. Scores are percentages (%). Train set indicates fine-tuning data (— for zero-shot).

Model	Train set	Test set	CER_{med}	CER_{mean}	WER_{med}	WER_{mean}
Tesseract	—	real	7.2	9.9	24.3	28.9
Kraken (End to end)	real	real	5.3	6.8	16.9	9.3
DeepSeek-OCR-2	—	synth	3.8	4.6	18.6	20.5
	—	real	10.2	13.8	31.6	35.0
	synth	synth	1.6	98.3	7.9	111.4
	synth	real	5.5	52.7	15.5	71.1
	real	real	4.7	10.4	16.7	21.7
	synth→real	real	4.7	26.7	17.0	40.7
LightOnOCR-1B	—	synth	4.8	10.6	14.5	19.8
	—	real	5.9	10.3	19.2	23.7
	synth	synth	4.6	12.2	11.9	19.4
	synth	real	5.4	14.6	18.6	25.4
	real	real	1.7	2.1	9.4	9.8
	synth→real	real	4.2	8.5	15.1	20.3
Qwen3-VL-2B	—	synth	12.4	23.2	43.1	55.8
	—	real	10.0	13.1	35.4	40.5
	synth	synth	0.9	1.3	4.2	5.4
	synth	real	2.1	2.7	10.3	11.3
	real	real	4.0	4.7	18.1	18.2
	synth→real	real	1.5	2.2	8.2	8.8
Qwen3-VL-8B	—	synth	6.2	11.8	22.8	27.3
	—	real	5.2	8.7	21.6	24.7
	synth	synth	0.4	0.7	2.0	3.0
	synth	real	1.7	1.9	8.4	9.2
	real	real	2.1	2.3	10.8	10.9
	synth→real	real	**1.0**	**1.2**	**5.6**	**6.6**

Model Stability and Error Dispersion. Several settings exhibit substantial gaps between median and mean error rates, particularly for DeepSeek-OCR-2. For instance, under synthetic training evaluated on synthetic data, the model attains a low median CER (1.6%) but an extremely high mean CER (98.3%), indicating catastrophic page-level failures. This divergence suggests that while many pages are transcribed accurately, a subset produces very long or unstable outputs that inflate mean error rates. In contrast, LightOnOCR-1B shows much smaller median-mean discrepancies, indicating more stable behavior. Among all models, Qwen3-VL-8B achieves the strongest performance, reaching a median CER of 1.0% and median WER of 5.6% under the `synth→real` regime on real scanned pages. These results demonstrate that large vision-language models, when combined with structure-aware synthetic supervision and real-domain adaptation, can achieve near-perfect plain-text OCR on complex critical editions.

CRNNs Remain Competitive. An examination of the results, particularly under zero-shot conditions on the real dataset, shows that Tesseract outperforms two out of four models in terms of median CER and three out of four in terms of mean CER.[3] In both cases, Qwen3-VL-8B is the only model that consistently maintains superior performance. Under similar conditions using only real training examples, Kraken achieves better mean CER than DeepSeek, although it is otherwise outperformed within relatively modest margins, especially when considering the respective computational footprints of these models, except for LightOnOCR and Qwen3 models in a `synth→real` fine-tuning scenarios.

Structural Element Recognition. Table 4 reports structure recognition performance on the real test set. Reference markers achieve the highest F1 scores across all models and training regimes, with Qwen3-VL-8B (synth→real) reaching 79.5. Marginal note detection is the most training-sensitive element: synth-only fine-tuning yields near-zero Note F1 for Qwen models (0.0–6.3), consistent with their underrepresentation in the synthetic corpus, while real-domain exposure recovers performance substantially (up to 63.5 for Qwen3-VL-8B synth→real). Header generation remains weak across all models, with only the two Qwen models achieving a count-based F1 score above 75.0 (best: 80.1 for Qwen3-VL-8B synth→real), likely due to the high variability in heading styles across editions. Tab indentation separates instruction-tuned from OCR-specialized models: Qwen models show calibrated specificity/recall trade-offs, while DeepSeek and LightOnOCR almost never predict indentation, yielding trivially perfect specificity at zero recall. Overall, Qwen3-VL-8B under the synth→real regime achieves the strongest structure recognition performance across all element types (Table 4).

[3] It should also be noted that the WER and CER scores of both CRNN-based systems are inherently affected by the overprediction of headers and line numbers. These elements are ignored in ground truth transcriptions for the end-to-end pipelines and therefore introduce systematic noise into the evaluation metrics.

Table 4. Structure recognition (%) on the `real` test set. Headers: count-based F1; references and notes: longest common subsequence F1. $\text{Tab}_{1\text{st}}$: specificity (Spec) and recall (Rec) for the first paragraph with indentation.

Model	Train	Hdr F1	Ref F1	Note F1	$\text{Tab}_{1\text{st}}$ Spec	$\text{Tab}_{1\text{st}}$ Rec
DeepSeek-OCR-2	synth	18.9	47.6	0.0	100.0	0.0
	real	27.1	46.0	13.9	100.0	0.0
	synth→real	39.6	56.7	27.2	100.0	0.0
LightOnOCR-1B	synth	39.0	43.9	7.4	100.0	0.0
	real	27.1	72.8	47.3	92.3	38.5
	synth→real	21.9	51.5	19.9	100.0	0.0
Qwen3-VL-2B	—	0.0	3.3	0.0	42.0	53.3
	synth	41.1	59.2	0.0	78.2	33.3
	real	55.9	65.4	11.8	81.1	78.6
	synth→real	75.8	76.2	39.1	92.8	80.0
Qwen3-VL-8B	—	15.3	15.7	0.0	62.3	40.0
	synth	70.3	63.2	6.3	88.4	53.3
	real	62.0	71.9	14.5	88.4	66.7
	synth→real	80.1	79.5	63.5	88.4	93.3

Error Pattern Analysis. Zero-shot models exhibit high substitution rates at both character (6.10–10.12 per 100 chars) and word level (35–52 per 100 words), indicating substantial lexical instability (Table 5). Fine-tuning consistently reduces substitutions: DeepSeek-OCR-$2_{\texttt{real}}$ achieves the lowest cSub (3.22) and wSub (18.47), roughly halving its zero-shot error. Real fine-tuning also nearly eliminates insertions for Qwen and LightOnOCR models (down to 0.22–0.71), suggesting improved alignment with page content, although DeepSeek-OCR-$2_{\texttt{real}}$ remains insertion-prone. Sequential `synth→real` training further reduces insertion and deletion rates for Qwen3-VL-8B (cIns = 0.22, cDel = 0.35), while also yielding the lowest diacritic confusion rates overall. Manual inspection shows that errors in the best-performing models are primarily orthographic and normalization-related. Typical issues include hyphenation inconsistencies, omitted macrons or breathings, folio insertion artifacts (e.g., embedded page numbers), and occasional Latin look-alike substitutions. Overall, textual grounding remains stable, with errors concentrated at the character level. In contrast, the worst-performing models display elevated substitution and insertion rates driven largely by formatting intrusions and decoding artifacts. These include spurious HTML tags, stray LaTeX fragments, unintended Latin characters, and hallucinated Ancient Greek text. Such artifacts introduce non-source tokens and fragment otherwise correct spans, and often result in length overgeneration, suggesting generative drift during decoding rather than simple orthographic noise.

Table 5. Edit rates and diacritic breakdown on real data. *cSub/cIns/cDel*: per 100 chars; *wSub/wIns/wDel*: per 100 words. *Br./Ac./Is.*: breathing, accent, iota subscript confusions (% of cSub).

Model	Character edits			Word edits			Diacritics		
	cSub	*cIns*	*cDel*	*wSub*	*wIns*	*wDel*	*Br*	*Ac*	*Is*
DeepSeek-OCR-2	8.94	3.15	1.10	44.21	3.85	1.99	4.8	18.2	1.5
LightOnOCR-1B	6.10	5.73	1.11	35.22	6.87	2.94	7.2	6.6	1.5
Qwen3-VL-2B	10.12	3.22	1.86	52.22	4.69	3.73	11.3	24.4	1.6
Qwen3-VL-8B	7.02	3.58	2.12	39.61	2.83	2.94	7.3	15.7	1.3
DeepSeek-OCR-2$_{\text{real}}$	**3.22**	4.57	2.10	**18.47**	5.18	3.17	9.9	17.1	2.4
LightOnOCR-1B$_{\text{real}}$	5.61	0.44	0.60	33.09	1.06	1.49	5.2	3.3	1.2
Qwen3-VL-2B$_{\text{real}}$	6.81	0.71	0.84	36.95	1.58	1.98	6.3	12.8	1.0
Qwen3-VL-8B$_{\text{real}}$	5.81	0.33	0.46	33.42	0.86	**1.20**	4.0	7.2	1.0
DeepSeek-OCR-2$^{\text{synth}\rightarrow\text{real}}$	5.71	11.64	6.59	32.26	15.59	7.72	3.9	7.2	1.1
LightOnOCR-1B$^{\text{synth}\rightarrow\text{real}}$	7.72	1.08	1.79	35.06	2.35	5.61	2.1	3.0	**0.3**
Qwen3-VL-8B$^{\text{synth}\rightarrow\text{real}}$	5.28	0.23	**0.33**	31.34	0.71	2.12	1.7	3.3	0.4
Qwen3-VL-8B$^{\text{synth}\rightarrow\text{real}}$	5.05	**0.22**	0.35	30.39	**0.68**	1.43	**1.1**	**2.1**	**0.3**

7 Discussion

Script Mixing, Orthographic Instability and Symbolic Ambiguity. Across models, we observe systematic confusion between visually similar characters across scripts, including Greek and Latin capitals (e.g., Greek Alpha vs. Latin A). In zero-shot conditions, biases from scientific pretraining corpora occasionally bleed into the output, with LaTeX formatting used to transcribe Greek script. Such substitutions are particularly problematic in critical editions, where case and script distinctions may encode structural or semantic information. More broadly, these errors reflect a deeper symbolic ambiguity: the same glyph may function as a lexical unit, a numeral, or part of a reference marker depending on context. In Greek critical editions, alphabetic numerals reuse letterforms as numeric symbols, and similar ambiguities arise in marginal line numbers and hierarchical markers. When contextual language priors dominate visual evidence, models fail to reliably disambiguate these multifunctional symbols, and errors often propagate to structure recognition. This suggests that structure-aware OCR for historical material requires more explicit modeling of document-level symbolic systems rather than relying solely on token-level language modeling.

Instability and Catastrophic Failures. The divergence between median and mean error rates, especially for DeepSeek-OCR-2, reveals that some models produce catastrophic page-level failures despite strong median performance. These cases are often characterized by long, fluent outputs that deviate from the visual input, suggesting over-reliance on internal language priors. While large VLMs can achieve near-perfect performance on many pages, their behavior remains brittle in structurally dense or symbol-heavy contexts.

Implications for Document Understanding. Beyond Ancient Greek critical editions, our findings carry broader implications for document understanding research. High plain-text accuracy does not guarantee structural competence: models with near-perfect character error rates may still fail to recover hierarchical markers, marginal annotations, or reference systems. Classical OCR baselines remain competitive in transcription while avoiding the catastrophic generative failures observed in some vision–language models, highlighting that accuracy and structural reliability are distinct dimensions. Moreover, the divergence between median and mean error rates shows that generative vision–language models can exhibit unstable behavior, producing fluent but visually ungrounded outputs. Overall, our results suggest that document understanding in historical material requires modeling not only glyph recognition but also the hierarchical and symbolic systems encoded in layout. Structure-aware OCR is therefore a problem of interpretation, not merely transcription.

8 Conclusion

In this paper, we introduced a new document type for joint text and layout recognition, focusing on critical editions of polytonic Greek texts, together with a reusable open-source pipeline and two complementary datasets (synthetic and manually annotated scans). Our results show that most models struggle out-of-the-box on this domain, and that while some vision–language models can adapt to both script and structure, they remain prone to hallucinations that we were not able to fully mitigate. A key takeaway of our study is that classical OCR pipelines remain competitive in terms of efficiency and stability for text recognition. While vision–language models provide clear advantages in adaptability and structure awareness, particularly when trained sequentially on synthetic and real data, these gains come at a substantially higher computational cost. VLM inference requires GPU execution (e.g., 0.09–0.16 kWh per run on A100-class hardware), whereas classical OCR systems operate on CPU with significantly lower energy and runtime requirements. Overall, our findings suggest that vision–language models are powerful but not universally superior solutions for structure-aware OCR. Rather than replacing classical pipelines, they offer complementary strengths. Future work should therefore explore hybrid approaches that combine the structural sensitivity of VLMs with the efficiency and robustness of traditional OCR systems.

Acknowledgments. The project "Corpus Liberatum Linguae Graecae" was supported by the French National Research Agency (ANR) under the France 2030 grant reference number "ANR-24-RRII-0002" operated by the Inria Quadrant Program and is endorsed by the BPI Scribe project.

A Appendix

Annotation Guidelines

Scope. Annotations capture *only* the main textual content, the structural elements and marginal notes. Running headers, running footers, page numbers, and other layout artifacts are explicitly excluded.

Structure. We annotate three types of Markdown-style headings:
- `# TITLE` — section title without an explicit reference marker.
- `# <ref>REF</ref>` — section title consisting solely of a reference marker.
- `# <ref>REF</ref> TITLE` — section title with both a reference and a title.

Inline annotation elements. Inline annotations are restricted to the following two elements:
- `<ref>...</ref>` — reference or milestone markers embedded in running text.
- `<note>...</note>` — marginal notes, glosses, or editorial comments.
- We add `<tab/>` to paragraphs that start on a page with indetation.

Prompt used for Qwen3-VL.

You are an OCR system for Ancient Greek and Latin printed scholarly texts. Transcribe the page exactly in pseudo-Markdown.

Rules: Copy text exactly (retain all characters, accents, ligatures, punctuation, spacing). Do not normalize, correct, or translate. Follow visual reading order. Each paragraph must be output as one line. Add `<tab/>` at the start of paragraphs that clearly begin on this page; do not add it if the paragraph continues from a previous page. Join words split by line-break hyphens. Wrap section markers as `<ref>X</ref>`. Major headers use: `# <ref>X</ref>` or `# <ref>X</ref> TITLE`. Standalone titles use `# TITLE`. Inline and marginal section markers are encoded inline. Margin notes are encoded as `<note>TEXT</note>`. Ignore running heads, footers, page numbers, line numbers, printer marks, and footnote markers. Output only the transcription.

Fig. 3. Instruction prompt provided to Qwen3-VL for structure-aware OCR transcription.

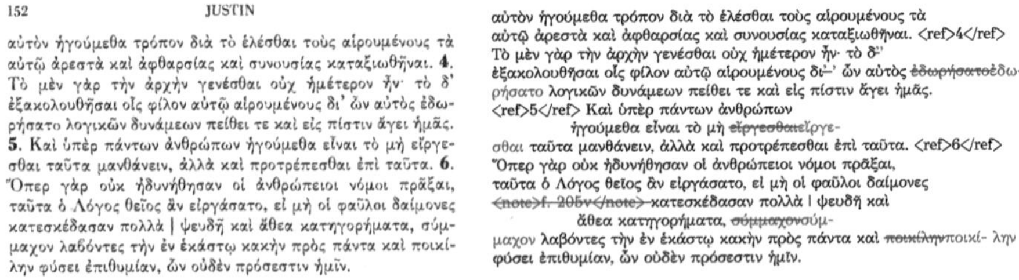

152 JUSTIN

αὐτὸν ἡγούμεθα τρόπον διὰ τὸ ἑλέσθαι τοὺς αἱρουμένους τὰ αὐτῷ ἀρεστὰ καὶ ἀφθαρσίας καὶ συνουσίας καταξιωθῆναι. **4.** Τὸ μὲν γὰρ τὴν ἀρχὴν γενέσθαι οὐχ ἡμέτερον ἦν· τὸ δ' ἐξακολουθῆσαι οἷς φίλον αὐτῷ αἱρουμένους δι' ὧν αὐτὸς ἐδωρήσατο λογικῶν δυνάμεων πείθει τε καὶ εἰς πίστιν ἄγει ἡμᾶς. **5.** Καὶ ὑπὲρ πάντων ἀνθρώπων ἡγούμεθα εἶναι τὸ μὴ εἴργεσθαι ταῦτα μανθάνειν, ἀλλὰ καὶ προτρέπεσθαι ἐπὶ ταῦτα. **6.** Ὅπερ γὰρ οὐκ ἠδυνήθησαν οἱ ἀνθρώπειοι νόμοι πρᾶξαι, ταῦτα ὁ Λόγος θεῖος ἂν εἰργάσατο, εἰ μὴ οἱ φαῦλοι δαίμονες
f. 205v κατεσκέδασαν πολλὰ | ψευδῆ καὶ ἄθεα κατηγορήματα, σύμμαχον λαβόντες τὴν ἐν ἑκάστῳ κακὴν πρὸς πάντα καὶ ποικίλην φύσει ἐπιθυμίαν, ὧν οὐδὲν πρόσεστιν ἡμῖν.

αὐτὸν ἡγούμεθα τρόπον διὰ τὸ ἑλέσθαι τοὺς αἱρουμένους τὰ
αὐτῷ ἀρεστὰ καὶ ἀφθαρσίας καὶ συνουσίας καταξιωθῆναι. <ref>4</ref>
Τὸ μὲν γὰρ τὴν ἀρχὴν γενέσθαι οὐχ ἡμέτερον ἦν· τὸ δ'
ἐξακολουθῆσαι οἷς φίλον αὐτῷ αἱρουμένους δι' ὧν αὐτὸς ~~ἐδωρήσατο~~ἐδω-
ρήσατο λογικῶν δυνάμεων πείθει τε καὶ εἰς πίστιν ἄγει ἡμᾶς.
<ref>5</ref> Καὶ ὑπὲρ πάντων ἀνθρώπων
ἡγούμεθα εἶναι τὸ μὴ ~~εἴργεσθαι~~εἴργε-
σθαι ταῦτα μανθάνειν, ἀλλὰ καὶ προτρέπεσθαι ἐπὶ ταῦτα. <ref>6</ref>
Ὅπερ γὰρ οὐκ ἠδυνήθησαν οἱ ἀνθρώπειοι νόμοι πρᾶξαι,
ταῦτα ὁ Λόγος θεῖος ἂν εἰργάσατο, εἰ μὴ οἱ φαῦλοι δαίμονες
~~<note>f. 205v</note>~~ κατεσκέδασαν πολλὰ | ψευδῆ καὶ
ἄθεα κατηγορήματα, ~~σύμμαχον~~σύμ-
μαχον λαβόντες τὴν ἐν ἑκάστῳ κακὴν πρὸς πάντα καὶ ~~ποικίλην~~ποικί- λην
φύσει ἐπιθυμίαν, ὧν οὐδὲν πρόσεστιν ἡμῖν.

Fig. 4. Example of output of Qwen3.5-VL.8b on the test set.

References

1. Bai, S., Cai, Y., Chen, R., Chen, K., Chen, X., et al.: Qwen3-vl technical report (2025). https://arxiv.org/abs/2511.21631
2. Berti, M.: Historical fragmentary texts in the digital age. In: Digital Classical Philology. Ancient Greek and Latin in the Digital Revolution, pp. 257–276 (2019)
3. Boros, E., Ehrmann, M., Romanello, M., Najem-Meyer, S., Kaplan, F.: Post-correction of historical text transcripts with large language models: an exploratory study. In: Bizzoni, Y., Degaetano-Ortlieb, S., Kazantseva, A., Szpakowicz, S. (eds.) Proceedings of the 8th Joint SIGHUM Workshop on Computational Linguistics for Cultural Heritage, Social Sciences, Humanities and Literature (LaTeCH-CLfL 2024), pp. 133–159. Association for Computational Linguistics, St. Julians, Malta (2024). https://aclanthology.org/2024.latechclfl-1.14/
4. Canfora, L.: Conservazione e perdita dei classici (2016)
5. Cayless, H., Clérice, T., Robie, J.: Introducing citation structures. In: Proceedings of Balisage: The Markup Conference 2021, vol. 26. Washington, United States (2021). https://doi.org/10.4242/BalisageVol26.Cayless01
6. Cayless, H., Roueché, C., Elliott, T., Bodard, G.: EPIGRAPHY IN 2017, pp. 203–222. Gorgias Press, Piscataway, NJ, USA (2010). https://doi.org/10.31826/9781463219222-012
7. Cerrato, L., et al.: Perseusdl/canonical-greeklit: 0.0.22368351567 (2026). https://doi.org/10.5281/zenodo.18763313
8. Cerrato, L., et al.: TDBuck: Perseusdl/canonical-latinlit: 0.0.21268363513 (2026). https://doi.org/10.5281/zenodo.18343976
9. Chagué, A., Scheithauer, H.: Gallicalbum (2023). https://github.com/HugoSchtr/Gallicalbum/
10. Clérice, T.: Les outils CapiTainS, l'édition numérique et l'exploitation des textes. Médiévales. Langues Textes Histoire **73**, 115–131 (2017)
11. Clérice, T., Angleraud, N., Karamolegkou, A., Sagot, B.: Corpus liberatum linguae graecae (2026). https://gitlab.inria.fr/almanach/cllg/freed-corpus
12. Crane, G., Mylonas, E.: The perseus project: an interactive curriculum on classical Greek civilization. Educ. Technol. **28**(11), 25–32 (1988)
13. Crane, G.R., et al.: First1kgreek (2022). https://doi.org/10.5281/zenodo.7015940
14. Evaggelatos, A., Palaiologos, K., Gatos, B., Kaddas, P., Christopoulou, A., Katsouros, V.: Old Greek OCR result correction using LLMs. In: Proceedings of the 2025 ACM Symposium on Document Engineering, DocEng 2025. Association for Computing Machinery, New York, NY, USA (2025). https://doi.org/10.1145/3704268.3748676
15. Fotheringham, L.S.: The numbers in the margins and the structure of cicero's pro murena. Greece Rome **54**(1), 40–60 (2007)
16. Kargaran, A.H., Nikeghbal, N., Diesner, J., Yvon, F., Schütze, H.: Glotocr bench: OCR models still struggle beyond a handful of unicode scripts (2026). https://arxiv.org/abs/2604.12978
17. Kay, A.: Tesseract: an open-source optical character recognition engine. Linux J. (159), 2 (2007)
18. Kiessling, B.: Version 5 of the Kraken ATR Engine for the Humanities. In: International Conference on Document Analysis and Recognition, pp. 443–458. Springer (2025)
19. Li, Y., et al.: Omnibench: towards the future of universal omni-language models (2025). https://arxiv.org/abs/2409.15272

20. Muellner, L.: The free first thousand years of Greek. Digital classical philology, pp. 7–18 (2019)
21. Najem-Meyer, S., Romanello, M.: Page layout analysis of text-heavy historical documents: a comparison of textual and visual approaches. In: Proceedings of the Conference on Computational Humanities Research 2022. CEUR-WS, Antwerp (2022)
22. Orlandi, S., Giberti, L.M.C., Santucci, R.: Eagle: Europeana network of ancient Greek and Latin epigraphy. Making the ancient inscriptions accessible. Lexicon Philosophicum: International Journal for the History of Texts and Ideas (2) (2014)
23. Pantelia, M.: Thesaurus Linguae Graecae (2024). https://stephanus.tlg.uci.edu
24. Papadopoulos, C.: Greekocr: a dataset for historical Greek OCR. In: Proceedings of the Language Resources and Evaluation Conference (LREC), pp. 4567–4573 (2022)
25. Pavlopoulos, J., et al.: Challenging error correction in recognised byzantine Greek. In: Pavlopoulos, J., et al. (eds.) Proceedings of the 1st Workshop on Machine Learning for Ancient Languages (ML4AL 2024), pp. 1–12. Association for Computational Linguistics, Hybrid in Bangkok, Thailand and online (2024). https://doi.org/10.18653/v1/2024.ml4al-1.1
26. Robertson, B.: Optical character recognition for classical philology. Digital Classical Philology, pp. 117–136 (2019)
27. Robertson, B., Boschetti, F.: Large-scale optical character recognition of ancient Greek. Mouseion **14**(3), 341–359 (2017). https://doi.org/10.3138/mous.14.3-3
28. Romanello, M., Najem-Meyer, S., Robertson, B.: Optical character recognition of 19th century classical commentaries: the current state of affairs. In: Proceedings of the 6th International Workshop on Historical Document Imaging and Processing, HIP 2021, pp. 1–6. Association for Computing Machinery, New York, NY, USA (2021). https://doi.org/10.1145/3476887.3476911
29. Sommerschield, T., et al.: Machine learning for ancient languages: a survey. Comput. Linguist. **49**(3), 703–747 (2023). https://doi.org/10.1162/coli_a_00481
30. von Stockhausen, A.: The patristic text archive (PTA): how to build a platform for critical digital editions of ancient Christian texts. In: xIX International Conference on Patristic Studies, Oxford (2024)
31. Taghadouini, S., Cavaillès, A., Aubertin, B.: Lightonocr: a 1B end-to-end multilingual vision-language model for state-of-the-art OCR (2026). https://arxiv.org/abs/2601.14251
32. TEI Consortium: Guidelines for Electronic Text Encoding and Interchange, p5 edn. (2024). https://tei-c.org/P5/, last modified 2024
33. Wei, H., Sun, Y., Li, Y.: Deepseek-ocr 2: visual causal flow (2026). https://arxiv.org/abs/2601.20552

A Text Recognition Dataset from Sahidic Coptic Ancient Manuscripts

Fabio Quattrini, Carmine Zaccagnino, Costanza Bianchi, Silvia Cascianelli(✉), and Rita Cucchiara

University of Modena and Reggio Emilia, Modena, Italy
{fabio.quattrini,carmine.zaccagnino,costanza.bianchi, silvia.cascianelli,rita.cucchiara}@unimore.it

Abstract. In this work, we target Handwritten Text Recognition (HTR) in low-resource scenarios, which arise from underrepresented languages, rare scripts, and degraded visual conditions typical of historical documents. We introduce SCAM (Sahidic Coptic Ancient Manuscripts), a new line-level dataset built from digitized ancient manuscripts written in the extinct Sahidic Coptic dialect. The dataset reflects a realistic and challenging setting, as it combines heterogeneous acquisition conditions across libraries with typical manuscript degradations such as ink fading, bleed-through, and material deterioration. In addition to visual complexity, SCAM poses significant linguistic challenges due to the scarcity of resources for Sahidic Coptic, its uncommon alphabet, and dialect-specific diacritics. To support research in low-resource HTR, we benchmark several state-of-the-art approaches based on different paradigms, highlighting their limitations and strengths in this setting. Our results underline the gap between current HTR performance on well-resourced modern scripts and historically grounded, low-resource scenarios, thus providing a reference point for future developments.

Keywords: Historical Documents · Low-resource Scripts · HTR Datasets

1 Introduction

Handwritten Text Recognition (HTR) has made substantial progress in recent years, particularly due to the success of deep learning and end-to-end trainable architectures [2]. These systems, typically based on CNN-BiLSTM-CTC pipelines or Transformer encoders, achieve good recognition performance when trained on large, well-annotated datasets in modern languages and scripts. However, their performance significantly degrades when applied to low-resource scenarios, where data is limited, noisy, or involves rare scripts and underrepresented languages. Such low-resource settings are common when dealing with historical manuscripts, poorly preserved documents, or extinct language scripts. In these

F. Quattrini and C. Zaccagnino—Equal contribution.

G. A. Fink et al. (Eds.): ICDAR 2026, LNCS 16972, pp. 269–286, 2027.
https://doi.org/10.1007/978-3-032-36023-6_16

contexts, the performance achievable with standard HTR pipelines often drop. To address this limitation, recent work has explored alternative strategies, such as synthetic data generation [8,34], and more recently, script-agnostic frameworks [4]. Nonetheless, some training is still required to achieve satisfactory performance, which realistically enables further usage of the transcribed documents by Humanities scholars.

Fig. 1. Sample images from the manuscript used to obtain the SCAM dataset. The top-four rows are from the SCAM-A subset, the bottom ones are from the SCAM-B.

In light of these considerations, we introduce a new line-level dataset featuring manuscripts in Sahidic Coptic, an underrepresented language of considerable interest to Coptologists and scholars of biblical studies, since it is among the languages into which the Bible was translated at an early stage. Our dataset is obtained from manuscript images preserved in different libraries and digitized with highly differing settings. This characteristic reflects a common fate of Coptic manuscripts from the White Monastery, many of which were dismantled in the 18th and 19th centuries and sold as individual leaves rather than as complete codices. From the point of view of the resulting images, the different acquisition settings and quality adds on the nuisances that are typical of ancient manuscripts (*e.g.*, ink fading, bleed through, scratches, yellowed paper). Moreover, the language of the manuscripts considered is another major challenge of the proposed dataset. In fact, Sahidic Coptic is no longer a spoken language, and the texts in the dataset are written in the Coptic script and employ a supralinear system characteristic of this language.

To promote research in low-resource HTR, especially in the challenging scenario of manuscripts from different libraries, written in extinct languages, we devise our dataset, dubbed Sahidic Coptic Ancient Manuscript (SCAM)[1]. The dataset features text lines extracted from pages preserved in two different archives and digitized under different conditions. Therefore, we organize the dataset into two subsets, dubbed SCAM-A and SCAM-B. Some example images used to build our dataset are reported in Fig. 1. In this paper, we analyze the visual and linguistic characteristics of our dataset, and benchmark different-paradigm State-of-the-Art HTR approaches, both to highlight the challenges of these data and the most promising solutions to face them, and to provide a leaderboard for future works.

2 Related Work

HTR can be approached at various levels of granularity, including characters [18, 19, 31] (especially useful for ideogrammatic languages or rare scripts, such as ciphers), words [8, 29, 53, 55], lines [11, 12, 17, 41, 45, 47, 60], paragraphs [10, 11, 62], or full pages [19, 41]. Among these, line-level HTR is the most widely adopted, typically assuming pre-segmented text lines [12, 17, 45, 47, 60]. These systems can also be integrated with layout analysis and segmentation to achieve paragraph- or page-level recognition [11, 41, 65]. In this work, we focus on pre-segmented line-level HTR, given its popularity and easy integration into page-level pipelines.

Early HTR systems were based on Hidden Markov Models (HMMs) to model sequences of explicitly-isolated character images and n-gram language models to predict textual outputs [57, 58]. Such an explicit character isolation approach was then surpassed by end-to-end approaches, starting from the deep learning-based solution proposed by Graves *et al.* [29], which consists of a multidimensional LSTM network that generates 2D feature maps directly from handwritten images, which are then reduced to 1D sequences and decoded using the Connectionist Temporal Classification (CTC) algorithm. This approach dominated the field [10, 13, 42] until computationally simpler and faster alternatives were proposed [47, 53], using CNNs for feature extraction and 1D-LSTMs for sequence modeling. These backbones combining CNNs and recurrent models remain popular due to their strong performance and efficiency [12, 17, 20, 45, 60]. Nonetheless, a more recent direction models HTR as a sequence-to-sequence problem, where the image is represented as a sequence of patches and decoded into text using encoder-decoder architectures [1, 39, 56, 66]. Models implementing this paradigm can be trained by using cross-entropy loss, CTC loss, or both. Transformers [59] have also been adopted for HTR to replace recurrent neural networks entirely, often leveraging large-scale pretraining on synthetic or real text data [33, 35, 61].

[1] More information and download link at https://scam.silviacascianelli.com.

Finally, it is worth mentioning that the emergence of Large Multimodal Models, such as GPT-4V [64], has opened new avenues for HTR by enabling models to process and understand both visual and textual information. Preliminary explorations suggest that these models can perform HTR tasks with minimal task-specific supervision, although they currently lag behind specialized systems in accuracy and robustness, especially in low-resource settings.

HTR Datasets. Designing effective HTR systems requires large and diverse datasets that reflect the visual complexity of handwriting and the linguistic richness of the task. Widely used modern-language benchmarks include the IAM [38] and RIMES [3] datasets, which contain English and French lines, respectively. These datasets are written by multiple authors under controlled conditions. IAM writers transcribe sentences from the Lancaster-Oslo/Bergen corpus [32], while RIMES contributors follow a script to simulate customer service communication.

For historical HTR, the ICFHR14 [50], ICFHR16 [51], and ICFHR18 [54] datasets are among the most notable. ICFHR14 includes English lines from the Bentham Papers [16], mostly authored by Jeremy Bentham and collaborators. ICFHR16 (also called READ) focuses on ancient German manuscripts from the Ratsprotokolle collection, written over three centuries. ICFHR18 includes documents in German and Italian, from diverse time periods and collections, with test sets defined at the document level. Other historical datasets include Rodrigo [52] and Germana [44], which contain Spanish text written by single authors, and the LAM dataset [15], which has been introduced to support research on ancient Italian single-author manuscripts with temporal variability in writing.

Other small historical datasets exist, which are commonly adopted for exploring low-resource HTR scenarios, which is also the focus of this work. Examples of these datasets are George Washington [26], Parzival [25], Saint Gall [27], Esposalles [49], Leopardi [14], and [28], which contain historical documents written by a few or single authors, often from narrow time spans, in a relatively well-represented language (English, Latin, and Italian). Particularly challenging examples of low-resource scenarios are explored via datasets of ciphered manuscripts, such as Borg and Copiale [5]. These contain text in Latin and German, respectively, rendered in uncommon characters (either existing or abstract symbols). To further explore the low-resource HTR task, we propose a dataset of manuscript written in an underrepresented, extinct language using uncommon symbols. Specifically, we target ancient Sahidic Coptic, and devise a line-level dataset, dubbed SCAM, divided into two sections preserved in different places.

For Coptic material, previous work has explored OCR for printed Coptic fonts through neural or multi-source training [37,40]. Existing digital resources, including CoptOT [22] and Coptic Scriptorium [21], further provide important access points to Coptic textual material. While these platforms focus heavily on biblical text coverage or normalized linguistic corpora, they lack line-level layout annotations and diplomatic transcriptions. Consequently, SCAM fills this gap by introducing a dedicated, manuscript-specific HTR benchmark featuring cross-collection visual variations and intact line-level ground truth.

3 Proposed Dataset

Some recent works have targeted HTR in low-resource scenarios, also proposing datasets to better explore the task. By low-resource, one can refer to the visual appearance of the manuscript (given *e.g.*, by its preservation state, or writer-specific handwriting, or both) or to the language of the manuscript, which is usually the case for historical documents in ancient languages. In this work, we devise a dataset for low-resource HTR that features all the challenges of this scenario. Specifically, our dataset, which we call SCAM, contains lines from 27 leaves, corresponding to 53 pages of an ancient manuscript written by a single author in Sahidic Coptic, an extinct language. Note that by "leaf" we mean a physical parchment sheet, which can accommodate one or two pages, written on both faces. The text is written in *scriptio continua*, meaning that there are no spaces delimiting words. The parchment manuscript, now known as Coptic Literary Manuscript (CLM) 359, is dated to 1002-1003 C.E., as indicated by the colophon, and comes from the White Monastery in Upper Egypt. This manuscript is of considerable importance, as it preserves substantial Coptic translations of Greek canonical and conciliar texts of the fourth and fifth centuries. It is not only a key witness to the development of Coptic canon law, but also an important source for reconstructing earlier stages of the Greek textual tradition. Note that SCAM differs from existing Coptic resources in both source material and benchmark design. [22] mainly targets biblical witnesses and, although several records provide images and transcriptions, these materials are not systematically packaged as line-level HTR samples with polygon annotations and diplomatic line transcriptions. [21] provides rich, normalized, and linguistically annotated Coptic corpora, but its editorial conventions often introduce word segmentation, normalization, punctuation, abbreviation expansion, or removal of supralinear marks, depending on the view and layer. By contrast, SCAM is built from one non-biblical Sahidic manuscript (a commentary on Church councils) whose surviving leaves are dispersed across institutions, preserving the manuscript's diplomatic line content and acquisition variability as the object of study. This makes SCAM complementary to previous Coptic resources: it is not intended to maximize textual coverage, but to isolate a realistic low-resource HTR setting with controlled scribe identity, cross-collection visual shift, and line-level ground truth.

To the best of our knowledge, only 49 leaves of the original manuscript have survived, which are now spread across countries, preserved in various libraries and cultural institutions[2]. As a result, each small set of leaves has been digitized under different conditions. This can significantly hinder the performance of an HTR model trained on samples from a certain collection when applied to samples in a different collection. To study this behavior, we collect samples from leaves preserved in two different libraries and organize the SCAM dataset into two subsets, dubbed SCAM-A and SCAM-B. The 15 leaves in the SCAM-A subset are digitized with professional scanners, under controlled lighting conditions,

[2] https://atlas.paths-erc.eu/manuscripts/359.

and with careful placement of the page in the scanner. On the other hand, the 12 leaves in the SCAM-B collection have been photographed with a relatively good-quality camera, without explicit control over the lighting conditions or the placement of the page. In the following, we describe the process we followed for obtaining the samples from these pages, and analyze the visual and linguistic characteristics of the resulting dataset.

3.1 Data Preparation

First, we isolate the line-level samples from all the page images contained in the collected leaves. Specifically, we resort to the popular open-source computer vision annotation tool CVAT[3] to trace polygons around each line. Note that we prefer using polygons instead of bounding boxes for two main reasons. The first is that some lines are not straight, either because the writer has laid them down without following a perfectly horizontal line or because of digitization artifacts. The second reason is more critical and specific of the manuscript at hand: since the manuscript is in scriptio continua, some of the lines start with bigger, ornate letters to indicate the beginning of a new paragraph or phrase. As a result, straight bounding boxes would lead to glyphs from other line samples be part of the line image of the sample with such slant or beginning letters (see Fig. 1).

Once the lines have been isolated, an expert Coptologist provided their diplomatic transcriptions for each line, which are associated with the image by using CVAT. Note that, in this process, the Coptologist was able to exploit a special keyboard for Sahidic Coptic. The Sahidic Coptic characters are encoded in Unicode, but some of the graphemes are encoded using sequences of multiple code points, called **grapheme clusters**. This makes it possible to introduce variations of a character, such as accents. For model training and prediction, we ask the model to predict grapheme clusters and not individual code points, so that each glyph always corresponds to a single character.

3.2 Dataset Characteristics

Statistics. Our dataset comprises 3240 lines, corresponding to 53 pages. It includes a total alphabet of 99 graphemes, with an average of 11.39 ± 2.04 characters per line. The data are partitioned into training, validation, and test splits as follows: the training set contains 2000 lines from 33 pages (89 unique graphemes), the validation set includes 610 lines from 10 pages (59 unique graphemes), and the test set consists of 630 lines from 10 pages (68 unique graphemes).

As stated above, the images in our dataset are split into two subsets, SCAM-A and SCAM-B, depending on the archive in which the leaves are preserved. The characteristics of these subsets are the following:

- **SCAM-A** contains 1445 lines, corresponding to 24 pages from 12 leaves. It features 87 graphemes, with an average line length of 11.52 ± 2.30 characters.

[3] https://www.cvat.ai/.

Table 1. Visual and linguistic differences between the two subsets (with RGB and grayscale images), expressed in terms of quantitative scores.

	FID	KID	HWD	JSD
SCAM-A $\leftrightarrow$ SCAM-B (RGB)	113.62	0.13	1.06	0.49
SCAM-A $\leftrightarrow$ SCAM-B (grayscale)	65.06	0.07	1.04	

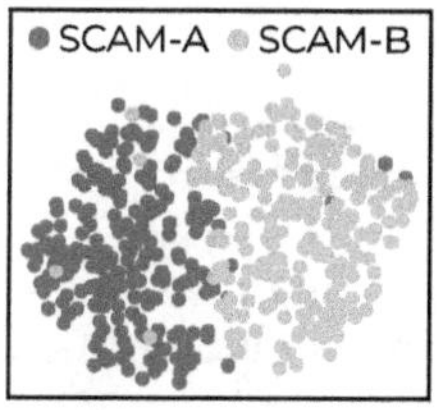

Fig. 2. t-SNE projection of VGG features extracted from RGB images in our dataset. 250 random points per subset plotted.

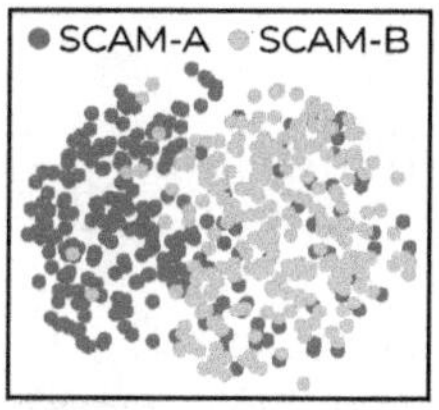

Fig. 3. t-SNE projection of VGG features extracted from grayscale images in our dataset. 250 random points per subset plotted.

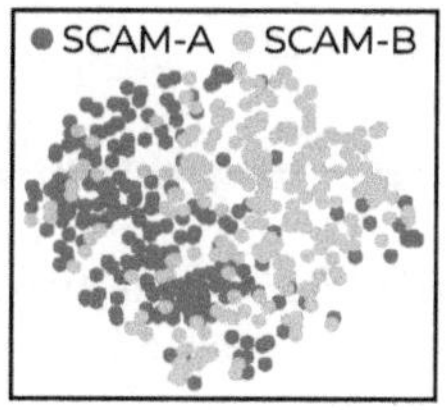

Fig. 4. t-SNE projection of HWD features extracted from RGB images in our dataset. 250 random points per subset plotted.

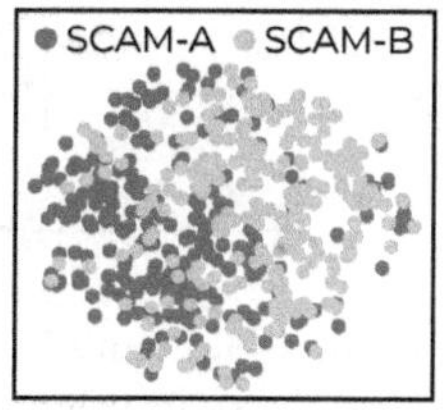

Fig. 5. t-SNE projection of HWD features extracted from grayscale images in our dataset. 250 random points per subset plotted.

The data are partitioned into training, validation, and test sets as follows: the training split includes 943 lines from 16 pages (76 unique graphemes), the validation split comprises 242 lines from 4 pages (48 unique graphemes), and the test split consists of 260 lines from 4 pages (61 unique graphemes).

- **SCAM-B** comprises 1795 lines, corresponding to 29 pages extracted from 15 leaves. It includes 67 distinct graphemes, with an average of 11.29 ± 1.81 characters per line. It is divided into training, validation, and test splits as follows: the training set contains 1057 lines from 17 pages (63 unique graphemes), the validation set includes 368 lines from 6 pages (51 unique graphemes), and the test set consists of 370 lines from 6 pages (50 unique graphemes).

In Table 1 we quantify the visual and linguistic differences between the subsets.

Visual Characteristics. First, we consider the visual characteristics, and quantify them in terms of Fréchet Inception Distance (**FID**) [30] and the Kernel Inception Distance (**KID**) [9]. Note that these scores, commonly adopted in Image Generation evaluation, measure the dissimilarity between the distributions of visual features extracted from two sets of images, and therefore, we deem them suitable to compare the samples in the SCAM-A and SCAM-B subsets in terms of overall visual appearance. Moreover, we include in the analysis the Handwriting Distance (**HWD**) [46], a recently-proposed score which isolates just the differences in writing style between pairs of images. The first row in the table contains the values of such scores computed on the original, RGB images (such as those in Fig. 1).

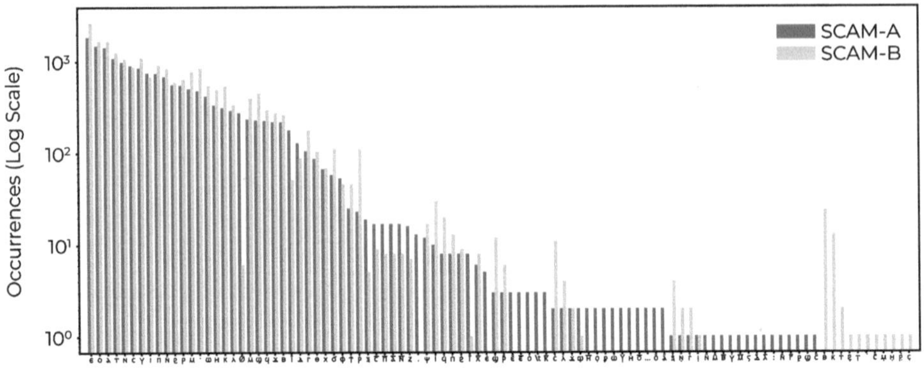

Fig. 6. Graphemes distribution in the two subsets of SCAM, in logarithmic scale. For comparison, graphemes are in descending order of frequency as computed in SCAM-A.

It emerges that the perceptual features are quite distinct and that the writing style in the two subsets is related but not identical. In particular, as shown by [46], typical values for HWD in same-author sets is between 0 and 1.5. This reflects the fact that the manuscript from which SCAM was derived was written by the same person. We deepen our analysis by applying a grayscale conversion to the images of both subsets, as this is a simple transformation that practitioners can easily adopt with the aim of reducing the visual gap between the subsets. Alternatively, document restoration approaches such as binarization [43,48] could be applied, but once sufficiently explored for our data. In fact, blindly applying such approaches can result in text content alteration in the lines. As it can be observed from the values in the second row of Table 1, the difference in visual features is reduced but not entirely eliminated, underlining that the SCAM dataset poses visual challenges that are not trivial to overcome.

These aspects are even clearer when looking at t-SNE projections of visual features extracted from both subsets (Figs. 2 to 5). In particular, we project features extracted by a VGG trained on natural images and by the VGG backbone of the HWD score, trained to distinguish handwriting styles. We argue that the first feature set captures the overall appearance (*e.g.*, color and texture) of the images, while the second set captures the handwriting style characteristics (*e.g.*, thickness, slant, and kerning). Note that both feature sets are extracted from the original RGB images in the subsets and from their grayscale version. From Fig. 2, we can see that the VGG features on RGB images are clearly divided into two separate clusters, which get closer when computed on grayscale images, as shown in Fig. 3. The HWD features, as shown for RGB and grayscale images in Fig. 4 and in Fig. 5 respectively, are less separated. This suggests that, although the handwriting in the two subsets is the same (the images come from a single-author manuscript), the differences between the images in the two subsets in terms of preservation state and digitization settings have also impacted the characteristics of the handwriting (*e.g.*, by altering the thickness).

Linguistic Characteristics. In Table 1 we also report the Jensen-Shannon divergence (**JSD**) computed on n-grams (with $n = 1, ..., 5$) extracted from the textual transcriptions of the SCAM-A and SCAM-B subsets. This score is considered to obtain a measure of the similarity of the text content of the two subsets. The value of the score suggests that, even though the dataset contains text taken from the same manuscript treating the same topic, there are non-negligible differences in the text content.

We deepen this analysis by plotting the distribution of the graphemes on the two subsets (in Fig. 6). It can be observed that, as happens in languages, the most common characters are equally frequent in both subsets. However, both subsets contain unique graphemes that are not present in the other, with SCAM-A having more of such cases (31 graphemes that are not present in SCAM-B, versus 11 graphemes in SCAM-B that are not in SCAM-A).

4 Experimental Analysis

This section presents an experimental analysis of the performance of several state-of-the-art models. The evaluation is performed by training each of them on **SCAM-A**, **SCAM-B** and the full **SCAM** dataset. Then, each model variant is tested on both the SCAM-A and SCAM-B subsets separately and jointly.

Considered Scores. To assess performance, we report the Character Error Rate (**CER**) and the Sequence Error Rate (**SER**). Following standard conventions in Handwritten Text Recognition (HTR), these scores are derived from the Edit Distance between the predicted text and the corresponding ground truth. Specifically, we compute the minimum number of substitutions, deletions, and insertions required to transform the predicted sequence into the ground truth. To determine the CER and SER for the entire test set, we sum the distances of all samples and normalize by the total number of units in the ground truth. For the CER, the distance is computed at the character level, while for the SER, it is computed on the content of the entire line sample, since our dataset is in scriptio continua.

Considered Approaches. We consider a range of model architectures and sizes in order to showcase the challenging nature of this dataset, and highlight which models perform best on it. In particular, we test the following architectures:

- We include the model proposed in [47], dubbed **CRNN**, which combines a convolutional backbone with BLSTM layers and is trained end-to-end to optimize the CTC loss. This model, which represents one of the most popular paradigms for learning-based HTR, serves as our primary recurrent baseline for sequence modeling without explicit alignment.
- We also consider the Sequence-to-Sequence recurrent attention paradigm by including the approach proposed by **Michael et al.** [39]. This approach combines a CNN-BiLSTM encoder with an attention-based LSTM decoder and auxiliary CTC loss, serving as a strong RNN-based attention baseline.

- Additionally, we consider models that rely on the attention operation instead of recurrent components, namely the Vertical Attention Network (**VAN**) proposed in [23] and the Convolutional Self-Attention Network (**C-SAN**) from [24]. These models replace RNNs with gated or self-attention mechanisms to enable parallelization and more efficient training, assessing whether attention alone can model handwriting dependencies.
- Another strategy to replace the recurrent component of the HTR pipeline entails exploiting a Transformer encoder-decoder, as **LT** [6], **VLT** [7], and the one proposed by **Kang et al.** [33], or a Transformer encoder, as done in **HTR-VT** [36]. In both cases, the visual features are extracted by a convolutional backbone, and the Transformer-based decoder improves parallelization and captures long-range dependencies in handwriting.
- Finally, we consider **TrOCR** [35], which represents the fully Transformer-based, end-to-end paradigm. Specifically, the image features are extracted with a Transformer encoder, and the transcription is obtained with a Transformer decoder. Both components have been pre-trained on synthetic data.

Implementation Details. For the experimental evaluation, we finetune TrOCR using the official HuggingFace implementation [63]. For each model size, we initialize the weights from the respective *handwritten* released checkpoint[4]. For the other architectures, we base our code on the HTR models implementations available online[5] and train them from scratch. To this end, we modify the tokenizer, removing the ASCII transliteration step[6], due to its incompatibility with non-Latin scripts and, therefore, our graphemes (Sect. 3.1). Each model is trained (or fine-tuned, in the case of the TrOCR variants) with an early stopping strategy based on the CER computed on the validation set of the dataset used for training, with patience set to 100 epochs.

4.1 Results

In the following, we report an experimental analysis, both quantitative (Tables 3 to 2) and qualitative (Fig. 7) of the considered approaches, and discuss the insights drawn from such analysis.

Architecture Type Matters more than Scale for Rare Scripts. Across all tables, we can observe that VAN achieves the best CER and SER on both subsets (some examples of its best and worst transcriptions over the different setups are reported in Fig. 7). We argue that the gated vertical-attention mechanism is particularly well-suited to the regular layout of the Coptic script glyphs: by attending to vertical strips of the image, VAN naturally aligns with the character-level structure of the lines without relying on explicit language priors. In contrast, TrOCR variants, despite their large-scale synthetic pre-training on Latin-script handwriting, do not achieve competitive results. These models rely

[4] `hf.co/microsoft/trocr-`*`size`*`-handwritten, where size` ∈ {small, base, large}.
[5] https://github.com/carlos10garrido/HTR-OOD/.
[6] https://pypi.org/project/Unidecode/.

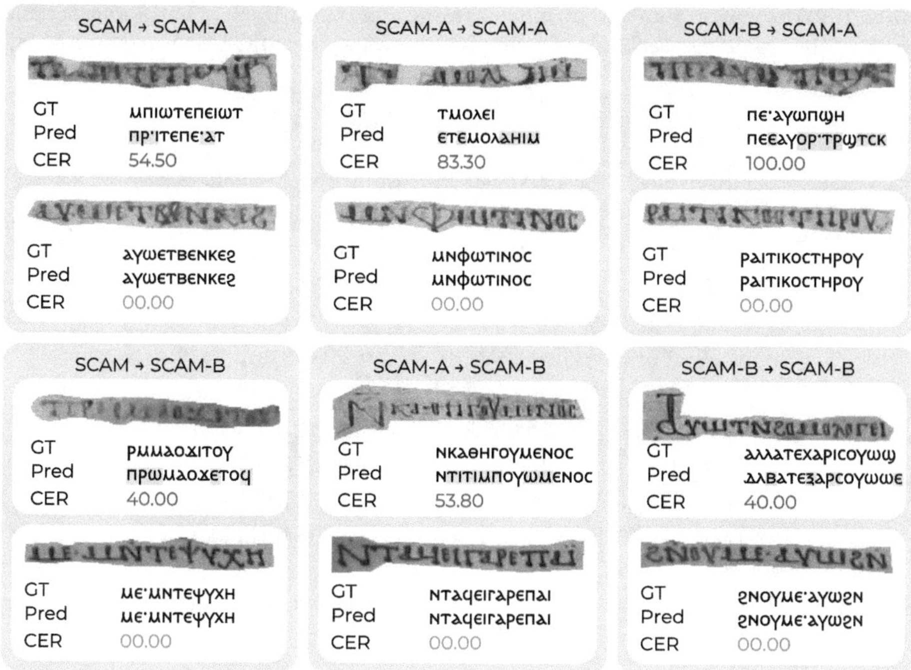

Fig. 7. Qualitative results on the considered setups (Train dataset → Test dataset) of the overall best-performing benchmarked model (VAN). We also report the CER on each considered sample and highlight the errors in red. (Color figure online)

on an autoregressive language prior at decoding time, while CTC-based models condition each predicted frame based on visual features independently. This pattern suggests that Transformer decoders pre-trained on high-resource scripts cannot efficiently adapt their learned language priors to an unseen script under data-scarce fine-tuning, and that model scale is not a substitute for architectural alignment with the task. Additionally, the scriptio continua writing provides a particularly challenging setting, which prevents the model from easily learning word priors. We also note that SER values are consistently high across all models and settings, even when the CER is relatively low. This is a direct consequence of the scriptio continua: with an average line length of 11.39 ± 2.04 characters, even a model achieving $\sim$7.5 CER is expected to produce at least one error in roughly half of all lines, counted as a full-line mismatch in the SER.

Data Breadth Matters more than Model Scale. In Table 2, we report results when training on the full SCAM dataset and testing on each subset. This represents the most favorable scenario, with models trained on annotated samples from multiple archives. Training on SCAM brings substantial improvements with respect to training on just one subset for most models. We argue that these results confirm that the in-domain failures observed in Table 3 are primarily due to data scarcity rather than to a fundamental incompatibility with the Sahidic

Table 2. Performance obtained by models trained on the entire SCAM training set and tested on the whole test set (SCAM) and on each test subset (SCAM-A and SCAM-B).

	SCAM		SCAM-A		SCAM-B	
	CER	SER	CER	SER	CER	SER
CRNN [47]	9.47	49.79	9.00	51.54	9.85	48.38
C-SAN [24]	88.06	100.00	87.52	100	88.50	100.0
VAN [23]	5.25	31.99	5.96	36.15	4.68	28.65
HTR-VIT [36]	10.27	59.62	10.98	66.54	9.70	54.05
Kang et al. [33]	11.43	54.18	11.44	57.69	11.42	51.35
Micheal et al. [39]	33.35	91.16	36.19	95.39	31.07	87.76
LT [6]	22.59	78.74	21.86	78.85	23.18	78.65
VLT [7]	15.03	62.81	14.89	67.31	15.15	59.19
TrOCR-S [35]	13.12	65.99	17.04	86.87	9.96	49.20
TrOCR-B [35]	12.42	74.39	14.42	77.78	10.81	71.66
TrOCR-L [35]	11.33	72.63	12.82	74.50	10.14	71.12

Table 3. Performance obtained by models trained and tested on the same subset. "A" indicates the SCAM-A subset, "B" the SCAM-B.

	A→A		B→B	
	CER	SER	CER	SER
CRNN [47]	13.91	65.39	14.63	64.32
C-SAN [24]	87.52	100.0	9.87	44.05
VAN [23]	7.50	40.77	7.58	41.08
HTR-VIT [36]	15.24	74.23	14.48	71.08
Kang et al. [33]	14.16	65.39	18.38	67.84
Micheal et al. [39]	79.15	98.46	82.00	100
LT [6]	44.28	98.46	46.99	97.30
VLT [7]	21.48	80.00	32.94	88.92
TrOCR-S [35]	30.70	98.23	12.11	56.15
TrOCR-B [35]	18.57	81.57	12.36	75.40
TrOCR-L [35]	14.99	78.54	10.59	68.45

Coptic script. Overall, these results show that using in-domain training data is more important than model capacity, as lightweight VAN and CRNN outperform the pre-trained TrOCR-L.

Domain Generalization Across Subsets. In Table 4, we evaluate the performance of models trained on one subset and tested on the other. This setting directly mirrors a real-world scenario, where a scholar might annotate leaves from one library, train a model, and then apply it to newly-digitized leaves from

Table 4. Performance obtained by models trained on a subset and tested on a different subset. "A" indicates the SCAM-A subset, "B" the SCAM-B.

	A→B		B→A	
	CER	SER	CER	SER
CRNN [47]	26.63	76.76	31.28	93.85
C-SAN [24]	88.50	100	23.54	88.46
VAN [23]	14.44	55.41	19.94	79.61
HTR-VIT [36]	29.76	89.46	25.31	94.23
Kang et al. [33]	11.44	57.69	32.92	93.08
Micheal et al. [39]	81.48	100	80.51	99.62
LT [6]	64.03	99.73	62.03	99.62
VLT [7]	37.50	88.92	47.28	98.85
TrOCR-S [35]	45.46	98.13	30.37	98.49
TrOCR-B [35]	24.18	92.25	18.36	90.15
TrOCR-L [35]	14.15	79.95	18.36	90.15

Table 5. Performance obtained by models trained on a subset and tested on a different subset, both in grayscale. "A" indicates the SCAM-A subset, "B" the SCAM-B.

	A→B		B→A	
	CER	SER	CER	SER
CRNN [47]	26.22	78.11	28.63	89.23
C-SAN [24]	87.50	100.00	85.36	100.00
VAN [23]	12.69	52.93	17.96	77.31
HTR-VIT [36]	27.65	85.68	25.52	90.77
Kang et al. [33]	26.91	81.08	35.91	96.54
Micheal et al. [39]	80.81	100.00	81.35	100.00
LT [6]	63.29	98.92	60.01	99.62
VLT [7]	34.13	88.65	51.74	99.62
TrOCR-S [35]	22.73	92.42	25.38	84.23
TrOCR-B [35]	17.73	82.58	18.98	89.31
TrOCR-L [35]	14.84	76.52	15.77	83.70

a different archive. Note that a scholar does not always have physical access to the archives and, therefore, control over digitization conditions. New leaves might have different acquisition conditions and contain graphemes absent from the training data (*e.g.*, SCAM-A and SCAM-B have only partially overlapping grapheme vocabularies of 87 and 67 symbols, respectively, as shown in Fig. 6). This is an interesting setting that can be explored with the SCAM dataset. We can observe that training on SCAM-B and testing on SCAM-A is consistently harder than doing the opposite for almost every model. We attribute this to the lower grapheme count of SCAM-B. Training on the SCAM-A, which has broader glyph coverage and cleaner images, provides a richer and more generalizable foundation. This asymmetry shows that high-quality library scans with broad grapheme coverage lead to data that can be used to train models that generalize better to field photographs than vice versa. This result offers a guiding principle for practitioners planning transcription pipelines, and defines a challenge for scholars who want to tackle domain generalization in scenarios similar to the one in our dataset

Image Preprocessing Helps Reduce Domain Gap. In Table 5 we evaluate the effect of a simple image preprocessing step, namely conversion to grayscale, on the inter-subset performance. From the results, we can see that this step consistently improves performance across most models. As a notable example, we observe that, in this setting, TrOCR-L reaches comparable performance *w.r.t.*VAN in CER values, with slightly worse SER values, revealing that bridging the color gap is not sufficient to achieve reliable line-level transcription. This is consistent with the dataset statistics (Table 1): converting to grayscale nearly halves the FID, but the HWD score barely changes, indicating a residual visual gap. Furthermore, in a scriptio continua text, a single character error in a long boundary-free line leads to a full-line SER mismatch.

Hybrid Decoders Underperform CTC-Only Approaches. A consistent finding across all three experimental settings (intra-subset - Table 3, inter-subset - Table 4, full-dataset - Table 2) is the unsatisfactory performance of the three *hybrid* architectures (Michael et al., LT, and VLT), which combine a CTC branch with an additional attention or cross-entropy decoder, compared to approaches relying solely on CTC decoding. From the results, we can infer that the joint CTC+decoder training objective does not converge well on the proposed SCAM dataset. Pure CTC-based models (CRNN, VAN) do not rely on an explicit autoregressive prior at decoding time and are agnostic to word delimiters, making them the most reliable paradigm in this low-resource setting.

5 Conclusion

In this paper, we have introduced SCAM, a new line-level dataset in the underrepresented Sahidic Coptic language, derived from the 11th-century CLM 359 manuscript and manually annotated by domain experts. By gathering sources from different libraries and digitized under heterogeneous acquisition conditions, we have modeled a realistic and challenging low-resource scenario featuring visual variability, degradation phenomena, and linguistic complexity.

Our experimental evaluation of several state-of-the-art HTR architectures across different sub-splits reveals several findings of broader relevance for Document AI. First, pure CTC-based architectures proved to be the most robust choice for scriptio continua manuscripts in resource-scarce languages: by not relying on an explicit autoregressive decoder, they are less prone to error cascades and scale better with limited data than hybrid models, which combine

CTC with attention or cross-entropy decoders. Second, the domain gap between SCAM's two subsets results in models trained on SCAM-B consistently generalizing worse to SCAM-A than vice versa. Third, grayscale preprocessing helps narrow the cross-archive gap when multi-domain data is unavailable but is not a substitute for training on diverse, multi-archive data. When such data is available, models like VAN provide the best accuracy across all settings. When only a single archive is available, simpler architectures like CRNN with grayscale preprocessing can combine robustness and adaptability to cross-domain deployment.

More broadly, these findings underscore the value of SCAM's dual-archive structure, exposing the domain asymmetry, the grapheme coverage gap, or the architectural sensitivity of certain models to specific visual distributions. We release SCAM as a benchmark to support future research on low-resource, rare-script, and cross-domain handwritten text recognition, and we hope it will encourage the development of HTR methods better suited to the challenges posed by real historical manuscript collections.

Acknowledgments. This work was supported by the PNRR project Italian Strengthening of ESFRI RI Resilience (ITSERR) funded by the European Union – NextGenerationEU, the "AI for Digital Humanities" project funded by "Fondazione di Modena", the GCP Credit Award, the Google Cloud Research Credits program with the award GCP19980904, and the FARD2025 project (CUP E93C25000370005). We acknowledge EuroHPC Joint Undertaking and ISCRA for awarding us access to LUMI at CSC, Finland, LEONARDO at CINECA, Italy, and MareNostrum5 at BSC, Spain.

Disclosure of Interests. The authors have no competing interests to declare that are relevant to the content of this article.

References

1. Aberdam, A., et al.: Sequence-to-sequence contrastive learning for text recognition. In: CVPR (2021)
2. AlKendi, W., Gechter, F., Heyberger, L., Guyeux, C.: Advancements and challenges in handwritten text recognition: a comprehensive survey. J. Imaging (2024)
3. Augustin, E., Carré, M., Grosicki, E., Brodin, J.M., Geoffrois, E., Prêteux, F.: RIMES evaluation campaign for handwritten mail processing. In: IWFHR, pp. 231–235 (2006)
4. Baena, R., Kalleli, S., Aubry, M.: General detection-based text line recognition. In: NeurIPS (2024)
5. Baró, A., Chen, J., Fornés, A., Megyesi, B.: Towards a generic unsupervised method for transcription of encoded manuscripts. In: DATeCH (2019)
6. Barrere, K., Soullard, Y., Lemaitre, A., Coüasnon, B.: A light transformer-based architecture for handwritten text recognition. In: DAS (2022)
7. Barrere, K., Soullard, Y., Lemaitre, A., Coüasnon, B.: Training transformer architectures on few annotated data: an application to historical handwritten text recognition. In: IJDAR (2024)
8. Bhunia, A.K., Das, A., Bhunia, A.K., Kishore, P.S.R., Roy, P.P.: Handwriting recognition in low-resource scripts using adversarial learning. In: CVPR (2019)

9. Bińkowski, M., Sutherland, D.J., Arbel, M., Gretton, A.: Demystifying MMD GANs. In: ICLR (2018)
10. Bluche, T.: Joint line segmentation and transcription for end-to-end handwritten paragraph recognition. In: NeurIPS (2016)
11. Bluche, T., Louradour, J., Messina, R.: Scan, attend and read: end-to-end handwritten paragraph recognition with MDLSTM attention. In: ICDAR (2017)
12. Bluche, T., Messina, R.: Gated convolutional recurrent neural networks for multilingual handwriting recognition. In: ICDAR (2017)
13. de Buy Wenniger, G.M., Schomaker, L., Way, A.: No padding please: efficient neural handwriting recognition. In: ICDAR (2019)
14. Cascianelli, S., Cornia, M., Baraldi, L., Piazzi, M.L., Schiuma, R., Cucchiara, R.: Learning to read L'Infinito: handwritten text recognition with synthetic training data. In: ICPR (2021)
15. Cascianelli, S., et al.: The LAM dataset: a novel benchmark for line-level handwritten text recognition. In: ICPR (2022)
16. Causer, T., Wallace, V.: Building a volunteer community: results and findings from transcribe bentham. Digit. Humanit. Q **6**(2) (2012)
17. Chowdhury, A., Vig, L.: An efficient end-to-end neural model for handwritten text recognition. In: BMVC (2018)
18. Cilia, N.D., De Stefano, C., Fontanella, F., di Freca, A.S.: A ranking-based feature selection approach for handwritten character recognition. Pattern Recognit. Lett. **121**, 77–86 (2019)
19. Clanuwat, T., Lamb, A., Kitamoto, A.: KuroNet: pre-modern Japanese Kuzushiji character recognition with deep learning. In: ICDAR (2019)
20. Cojocaru, I., Cascianelli, S., Baraldi, L., Corsini, M., Cucchiara, R.: Watch your strokes: improving handwritten text recognition with deformable convolutions. In: ICPR (2020)
21. Coptic Scriptorium: Coptic Scriptorium: Digital research in coptic language and literature. https://copticscriptorium.org
22. CoptOT: CoptOT: Digital edition of the coptic old testament. https://coptot.manuscriptroom.com
23. Coquenet, D., Chatelain, C., Paquet, T.: End-to-end handwritten paragraph text recognition using a vertical attention network. IEEE Trans. PAMI (2023)
24. d'Arce, R., Norton, T., Hannuna, S., Cristianini, N.: Self-attention networks for non-recurrent handwritten text recognition. In: ICFHR (2022)
25. Fischer, A., Frinken, V., Fornés, A., Bunke, H.: Transcription alignment of Latin manuscripts using hidden Markov models. In: HIP (2011)
26. Fischer, A., Keller, A., Frinken, V., Bunke, H.: Lexicon-free handwritten word spotting using character HMMs. Pattern Recognit. Lett. **33**(7), 934–942 (2012)
27. Fischer, A., et al.: Automatic transcription of handwritten medieval documents. In: VSMM (2009)
28. Levenson, M.G.: Towards a general open dataset and models for late medieval castilian text recognition (HTR/OCR). J. Data Min. Digit. Humanit. (2023)
29. Graves, A., Schmidhuber, J.: Offline handwriting recognition with multidimensional recurrent neural networks. In: NeurIPS (2008)
30. Heusel, M., Ramsauer, H., Unterthiner, T., Nessler, B., Hochreiter, S.: GANs trained by a two time-scale update rule converge to a local nash equilibrium. In: NeurIPS (2017)
31. Jaderberg, M., Simonyan, K., Zisserman, A., Kavukcuoglu, K.: Spatial transformer networks. In: NeurIPS (2015)

32. Johansson, S., Leech, G.N., Goodluck, H.: Manual of information to accompany the Lancaster-Oslo/Bergen Corpus of British English, for use with digital computer. Department of English, University of Oslo (1978)
33. Kang, L., Riba, P., Rusiñol, M., Fornés, A., Villegas, M.: Pay attention to what you read: non-recurrent handwritten text-line recognition. Pattern Recognit. (2022)
34. Kang, L., Riba, P., Wang, Y., Rusiñol, M., Fornés, A., Villegas, M.: GANwriting: content-conditioned generation of styled handwritten word images. In: ECCV (2020)
35. Li, M., et al.: TrOCR: transformer-based optical character recognition with pre-trained models. In: AAAI (2023)
36. Li, Y., Chen, D., Tang, T., Shen, X.: HTR-VT: handwritten text recognition with vision transformer. Pattern Recognit. (2025)
37. Lincke, E.S., Bulert, K., Büchler, M.: Optical character recognition for coptic fonts: a multi-source approach for scholarly editions. In: DATeCH (2019)
38. Marti, U.V., Bunke, H.: The IAM-database: an English sentence database for offline handwriting recognition. IJDAR **5**(1), 39–46 (2002)
39. Michael, J., Labahn, R., Grüning, T., Zöllner, J.: Evaluating sequence-to-sequence models for handwritten text recognition. In: ICDAR (2019)
40. Miyagawa, S., Bulert, K., Büchler, M., Behlmer, H.: Optical character recognition of typeset coptic text with neural networks. Digit. Scholarsh. Humanit **34**(Supplement 1), i135–i141 (2019)
41. Moysset, B., Kermorvant, C., Wolf, C.: Full-page text recognition: learning where to start and when to stop. In: ICDAR (2017)
42. Moysset, B., Messina, R.: Are 2D-LSTM really dead for offline text recognition? Int. J. Doc. Anal. Recognit. (IJDAR) **22**(3), 193–208 (2019). https://doi.org/10.1007/s10032-019-00325-0
43. Otsu, N.: A threshold selection method from gray-level histograms. IEEE Trans. Syst. Man Cybern. (1979)
44. Pérez, D., Tarazón, L., Serrano, N., Castro, F., Ramos Terrades, O., Juan, A.: The GERMANA database. In: LREC (2010)
45. Pham, V., Bluche, T., Kermorvant, C., Louradour, J.: Dropout improves recurrent neural networks for handwriting recognition. In: ICFHR (2014)
46. Pippi, V., Quattrini, F., Cascianelli, S., Cucchiara, R.: HWD: a novel evaluation score for styled handwritten text generation. In: BMVC (2023)
47. Puigcerver, J.: Are multidimensional recurrent layers really necessary for handwritten text recognition? In: ICDAR (2017)
48. Quattrini, F., Pippi, V., Cascianelli, S., Cucchiara, R.: Binarizing documents by leveraging both space and frequency. In: ICDAR (2024)
49. Romero, V., et al.: The ESPOSALLES database: an ancient marriage license corpus for off-line handwriting recognition. Pattern Recognit. **46**(6), 1658–1669 (2013)
50. Sánchez, J.A., Romero, V., Toselli, A.H., Vidal, E.: ICFHR2014 competition on handwritten text recognition on transcriptorium datasets (HTRtS). In: ICFHR (2014)
51. Sanchez, J.A., Romero, V., Toselli, A.H., Vidal, E.: ICFHR2016 competition on handwritten text recognition on the READ dataset. In: ICFHR (2016)
52. Serrano, N., Castro, F., Juan, A.: The RODRIGO database. In: LREC (2010)
53. Shi, B., Bai, X., Yao, C.: An end-to-end trainable neural network for image-based sequence recognition and its application to scene text recognition. IEEE Trans. PAMI **39**(11), 2298–2304 (2016)
54. Strauß, T., Leifert, G., Labahn, R., Hodel, T., Mühlberger, G.: ICFHR2018 competition on automated text recognition on a READ dataset. In: ICFHR (2018)

55. Such, F.P., Peri, D., Brockler, F., Paul, H., Ptucha, R.: Fully convolutional networks for handwriting recognition. In: ICFHR (2018)
56. Sueiras, J., Ruiz, V., Sanchez, A., Velez, J.F.: Offline continuous handwriting recognition using sequence to sequence neural networks. Neurocomputing **289**, 119–128 (2018)
57. Toselli, A.H., et al.: Integrated handwriting recognition and interpretation using finite-state models. Int. J. Pattern Recognit. Artif. Intell. **18**(04), 519–539 (2004)
58. Toselli, A.H., Vidal, E.: Handwritten text recognition results on the Bentham collection with improved classical n-gram-HMM methods. In: HIP (2015)
59. Vaswani, A., et al.: Attention is all you need. In: NeurIPS (2017)
60. Voigtlaender, P., Doetsch, P., Ney, H.: Handwriting recognition with large multidimensional long short-term memory recurrent neural networks. In: ICFHR (2016)
61. Wick, C., Zöllner, J., Grüning, T.: Rescoring Sequence-to-Sequence Models for Text Line Recognition with CTC-Prefixes. arXiv preprint arXiv:2110.05909 (2021)
62. Wigington, C., Tensmeyer, C., Davis, B., Barrett, W., Price, B., Cohen, S.: End-to-end full-page handwriting recognition. In: ECCV (2018)
63. Wolf, T., et al.: Transformers: state-of-the-art natural language processing. In: EMNLP (2020)
64. Yang, Z., et al.: The Dawn of LMMs: Preliminary Explorations with GPT-4V(ision). arXiv preprint arXiv:2309.17421 (2023)
65. Yousef, M., Bishop, T.E.: OrigamiNet: weakly-supervised, segmentation-free, one-step, full page text recognition by learning to unfold. In: CVPR (2020)
66. Zhang, Y., Nie, S., Liu, W., Xu, X., Zhang, D., Shen, H.T.: Sequence-to-sequence domain adaptation network for robust text image recognition. In: CVPR (2019)

Graphics Representation

Optical Music Recognition for Real-World Manuscripts with Synthetic Data

Jiří Mayer[1], Martina Dvořáková[2], Vojtěch Dvořák[1], Markéta Herzánová Vlková[2], Filip Jebavý[2], Pavel Pecina[1], Samuel Šomorjai[2], Petr Žabička[2], and Jan Hajič jr.[1](✉)

[1] Faculty of Mathematics and Physics, Charles University, Prague, Czech Republic
{mayer,hajicj}@ufal.mff.cuni.cz
[2] Moravian Library, Brno, Czech Republic

Abstract. Optical Music Recognition (OMR) has seen major progress in model design, with end-to-end methods now capable of recognising notation at all levels of complexity. However, the impact of this progress has been limited by the visual domains of available training datasets, which are largely born-digital. Existing large collections of sheet music in libraries and other heritage institutions contain predominantly manuscripts, whose visual domains are highly diverse and different, so existing OMR systems fail when applied in the real world. These institutions are often resource-constrained, so large in-domain datasets cannot be expected. We provide a first baseline on real-world manuscripts with complex piano notation in the resource-constrained scenario. Using fine-grained Music Notation Graph (MuNG) annotations and the Smashcima synthesis tool, we then show that while some direct transcriptions of in-domain data remain essential, domain adaptation using synthetic musical manuscript images brings significant improvement. Furthermore, the symbols used do not need to be in-domain, so the expensive fine-grained annotation can be avoided. We thus bring OMR closer to one of its stated goals: preserving and promoting musical cultural heritage.

Keywords: Optical Music Recognition · Synthetic Data · Manuscript Recognition · Digital Libraries

1 Introduction

Optical Music Recognition (OMR), the field that aims to automatically read music notation, has been a particularly difficult sub-field of document recognition [6]. The difficulties related to the non-sequential and non-local nature of music notation have recently been largely overcome with attention mechanisms [35,39, 43,44]. However, as these approaches require a substantial amount of supervised training data, a more insidious challenge remains: applying these models to real collections in resource-constrained setting.

Music notation documents are highly diverse (see Fig. 1). As tools for digital production of new music notation are becoming increasingly more and more

G. A. Fink et al. (Eds.): ICDAR 2026, LNCS 16972, pp. 289–305, 2027.
https://doi.org/10.1007/978-3-032-36023-6_17

Fig. 1. Examples of Common Western Music Notation in a library collection.

integrated in musical workflows, the most likely large-scale users of OMR are increasingly music libraries and memory institutions, and perhaps music education institutions, rather than individual end users. The role of OMR in cultural heritage preservation, access and study thus requires systems to be able to process the collections of such institutions. The collections of music notation documents in these institutions hold many manuscripts: the RISM database of musical sources in memory institutions records 1,346,175 manuscripts vs. 261,162 prints.[1] To provide, e.g., retrieval services [12,13,19], OMR needs to work across a variety of visual domains, including handwritten music. Domain adaptation has recently been identified in surveys as a key challenge for OMR [7,10].

However, OMR currently has no dataset of real-world musical manuscripts publicly available. And because producing in-domain ground truth for training OMR on musical manuscripts is an expensive endeavour [50], it is unlikely that building sufficiently large in-domain datasets will be within the means of user institutions any time soon.

An obvious next-best solution in the absence of "real" data is synthetic data. A system for OMR data synthesis needs to output two components necessary for supervised learning: (1) the musical content that should be available in notation and (2) the image that corresponds to the music notation encoding this musical content. The first component presents few issues. Public-domain datasets of machine-readable representations of diverse music that can serve as targets for end-to-end learning, in symbolic formats such as MusicXML [17,27], ABC [52], or `**kern`,[2] are available. However, the problem is on the side of actually producing score images. Many tools for rendering music notation from its symbolic encoding exist: regular notation editors such as MuseScore[3] (which has a batch

[1] https://rism.online/ as of February 6, 2026.
[2] https://kern.ccarh.org/.
[3] https://musescore.com/.

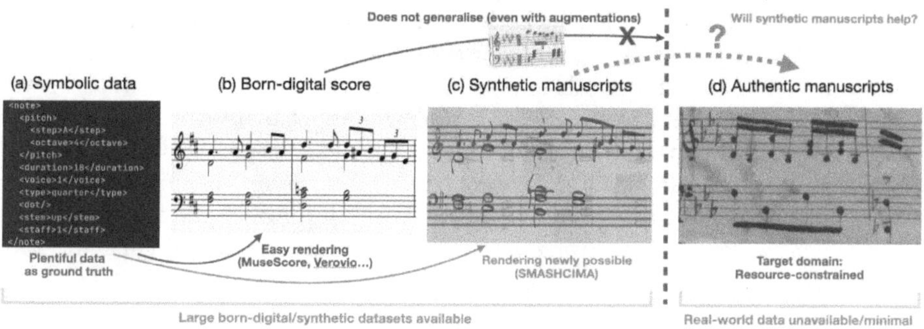

Fig. 2. We experiment with newly available synthetic manuscript images of sheet music to explore whether they help with domain adaptation to authentic (real) manuscripts.

processing mode), Sibelius[4] or Dorico,[5] the TeX-based LilyPond system, or the Verovio renderer for the MEI format. However, all of these (naturally) only render print-like notation that furthermore does not even entirely generalize to scans of music prints, even with data augmentation [35]. A certain "glass ceiling" has been observed [1].

None of this notation software is capable of rendering images of musical manuscripts. Attempts have been made in style transfer from print to manuscript [22,45,48], for a workflow where the manuscript image is obtained as a purely visual transformation of a rendered print image while not "damaging" the musical semantics, but these have not led to a workable solution so far.

The remaining pathway is therefore to implement a music notation rendering engine that could render correct manuscript images directly from symbolic representations. Previous experiments with limited synthetic manuscript data have shown some promise [33,34]. However, these systems have been limited to very simple music and still unrealistic images, and this "promise" has never been confronted with real-world scores in realistic settings.

Recently, the Smashcima system has been introduced [32] that is capable of rendering full-page manuscript images from (nearly) arbitrary MusicXML inputs. It extracts glyph libraries from the Music Notation Graph (MuNG) representation, introduced in the MUSCIMA++ dataset [21]. In combination with the MuNG Studio tool for efficiently annotating sheet music in this representation [31], the infrastructure is now in place to leverage manuscript synthesis for domain adaptation to musical manuscripts, even for diverse collections of resource-constrained memory institutions. However, experimental work to show whether this pathway is in fact useful is still missing.

As indicated in Fig. 2, this is what we contribute. Simulating the scenario of large available symbolic datasets and very limited visually diverse in-domain data (primarily manuscripts) from the musical collections of a resource-constrained

[4] https://www.avid.com/sibelius.
[5] https://www.steinberg.net/dorico/.

library,[6] we find that realistic synthetic manuscript data lead to useful improvement in domain adaptation on real-world diverse manuscripts.

After reviewing related work in Sect. 2, we describe the data synthesis setup in Sect. 3, the datasets used in Sect. 4, and the experimental setup in Sect. 5. Section 6 describes experiment results and Sect. 7 discusses the implications of the experimental results for OMR and cultural heritage preservation, limitations of the current setup, sketch out future work, and highlights the interaction between generic machine learning methods and adequate domain-specific software infrastructure that makes this possible.

2 Related Work

An essential context for our work are the successes of attention-based sequence-to-sequence learning methods in dealing with the complexities of Common Western Music Notation (CWMN) [35,43], even at the full page level [44]. (Vision large language models have so far not been able to contribute to OMR beyond replicating state-of-the-art results on a born-digital synthetic dataset [5,47].) OMR models are thus no longer limited by what is in the music—rather, they are limited by what the target music looks like.

Despite manuscripts making up the majority of CWMN collections, however, none of these experiments were performed with manuscript data. So far, beyond the first attempts on CWMN manuscripts based on the object detection pipeline [18,36], there has been just one attempt with domain adaptation and BLSTMs [2], and these works were limited to the MUSCIMA++ dataset, which is binarized and contains notation written in the 21st century [15,21]. Therefore, how the advance in OMR models maps to real-world conditions in music libraries is unknown. We believe this foregrounds domain adaptation as a leading current challenge for OMR.

This is fundamentally connected to data availability: there is no publicly available dataset of CWMN manuscript images available for training OMR systems (the monophonic 20th-century FMT dataset that has been used is not available [11]). Hence, no study on the impact of CWMN synthetic data on real-world OMR has been performed.

However, the potential of data synthesis for a resource-poor field such as OMR, especially on handwritten notation, has been recognised for some time. First attempts at synthesizing handwritten music images come from Baró et al. [2] with their approach of measure shuffling the MUSCIMA++ dataset. The logical extension of this idea to individual symbols was performed by Mayer et al. with the Mashcima system [33]. Many of its limitations (monophonic music, non-standard encoding, no customisability) were recently overcome with the successor Smashcima system [32].

[6] In this work, the term "domain" refers to the visual styles within one target sheet music collection, e.g., the holdings of a single library. The musical content across most memory institutions' collections is comparatively consistent: predominantly tonal music from the 18th–20th centuries.

The avenue of deep-learning synthesis approaches has been explored by many different people. The team around Elona Shatri has done experiments with generative adversarial networks [22,45,48]. Autoencoders were used to synthesize individual symbols [23]. Recently, researchers from the MALer lab have built a system that can produce synthetic images of music notation through an RQVAE decoder [24]. Still, all have problems preserving the musical content in the image.

Manuscript data for earlier European notations (medieval and renaissance-period) is available; experiments across multiple datasets of mensural notation, of which two were handwritten, indicated that synthetic data may indeed help domain adaptation [30], and that unsupervised adaptation has potential as well [28]. Outside of OMR, the widespread use of synthetic data for manuscripts has been reviewed in 2024 [46], noting that image generation models were increasingly being used to generate synthetic data for low-resource settings. Content and style conditioned synthesis was achieved in handwriting [25], but it still remains elusive in the music domain.

3 OMR Data Synthesis Process

As with any data for supervised learning, a synthesis engine for OMR works in the opposite direction of an OMR system: it produces a sheet music image correctly reflecting a given ground truth encoding of music notation, such as MusicXML. The synthesiser essentially engraves the given music.

Due to the nature of music notation as a writing system [6], this is a very complex task. Music notation editors such as MuseScore, Sibelius, or Verovio do this; however, their outputs are restricted to their own fonts, which naturally prioritise a "clean", consistent look with just one shape per glyph type, certainly not simulating manuscripts with their inherent variability.

Fortunately, the recent Smashcima tool [32] acts like a music notation rendering engine with built-in randomness that utilizes a large and diverse symbol set. This finally opens an opportunity to perform visual domain adaptation without being restricted to oversimplified notation styles (such as in [33] or [2]).

Smashcima requires three inputs to operate. Most crucially, it needs symbol sets from the target domain. These are supplied in the MuNG format, established in the MUSCIMA++ dataset [21]. An example MuNG annotation is shown in Fig. 3. Annotating music documents in the MuNG format is cca 10x as expensive as transcribing them in MusicXML [50]. The second input Smashcima needs are several samples of expected backgrounds, which can easily be taken from in-domain documents. The third input are MusicXML files encoding the sheet music to be rendered. These are plentiful, e.g., the ground truth of datasets with born-digital typeset images can be used. Smashcima can also apply a number of standard document image augmentation techniques as postprocessing steps (noise, ink simulation, bleedthrough, deformations, scribbles, shadows and other camera effects, etc.). Example outputs of synthesising the same MusicXML in different styles with Smashcima are shown in Fig. 4.

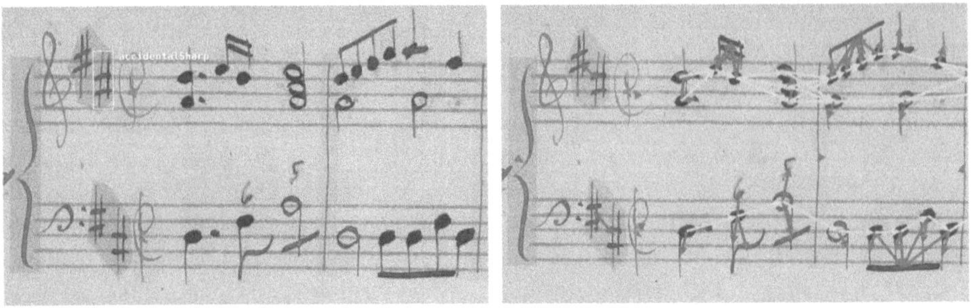

(a) Authentic-MuNG (b) Authentic-MuNG with edges shown

Fig. 3. Example annotation in the MuNG format. On the left, just the symbols are shown; note the accuracy of the symbol masks (e.g., the G-clef or the 8th flag in the first measure of the bottom staff). On the right, the annotation is shown with edges.

3.1 Perception Study

In a small user study, we have found that images rendered by Smashcima are barely recognisable for humans from authentic scores. We ran a survey with 10 authentic and 10 synthetic images, asking for each image whether it was authentic, or synthesised. The survey was shuffled randomly for each participant to avoid priming effects. A total of 30 participants marked each presented image as synthetic or authentic, leading to 600 binary responses in total, out of which 300 were on synthetic examples. While participants still could recognise synthetic images at higher than chance levels (posterior CDF for H_0: $P(synthguessed \mid synthtrue) \leq 0.5$ was < 0.001 under the Beta-Bernoulli model with prior $\alpha, \beta = 1$), their overall accuracy was merely 0.61 (effect size 0.11), and participants who were able to read music notation (23 out of the 30) were in fact worse, with overall accuracy 0.57 (effect size 0.07, but still above chance level, with posterior CDF for H_0 at 0.013).

4 Datasets

The composition of datasets used in this experiments reflects the "resource-constrained library" scenario outlined above: large out-of-domain datasets with born-digital images, and very small in-domain ("authentic") datasets. The one

Fig. 4. Different renderings of the same MusicXML file using Smashcima.

exception is in-domain data in MuNG format for synthesis, which would require a significant resource investment: in the spirit of providing practical recommendations, we attempt to find out whether this extra investment (rather than just transcribing to MusicXML) leads to significantly better results than using non-authentic (out-of-domain) MuNG data.

To ensure the complexity of real music notation is accounted for, and to facilitate comparability with the most recent end-to-end OMR experiments, the setup focuses on piano notation [6]. We also operate on single-system[7] data, as most of end-to-end OMR has (with a recent exception of the SMT++ [44]).

In order to measure how synthetic data help in domain adaptation, we need three types of datasets:

- Large out-of-domain datasets, representing the plentiful born-digital typeset data,
- Small in-domain (authentic) datasets, representing outputs of the resource-constrained annotation: cheaper end-to-end ground truth (MusicXML) for fine-tuning, and possibly expensive data (MuNG) for data synthesis.
- Large synthetic handwritten datasets, that are created by combining the out-of-domain ground truth with in-domain symbol data.

4.1 Out-of-Domain Datasets

We use two existing datasets of the first type: GrandStaff [41] and OLiMPiC [35]. Both datasets define default train/test splits that we follow.

The **GrandStaff** dataset [41] is a subset of piano music from the KernScores database[8] that contains scores of 474 compositions by 6 composers. All compositions are transposed into 3 additional keys, segmented into chunks of 3–6 measures, and each of these chunks was rendered by Verovio [38] as a JPEG image. Each chunk serves as one data point, for a total dataset size of 53,882 samples. The **Camera-GrandStaff** dataset is exactly the same as GrandStaff, except that each rendered JPEG image is further distorted [8]. In order to make GrandStaff ground truth compatible with other datasets, we use its version encoded in LMX (Linearised MusicXML [35]).

The **OLiMPiC** dataset [35] is derived from OpenScore Lieder [16,17], a corpus of music encodings. The dataset consists of piano accompaniments for 1,295 different French and German 19th-century songs by 107 distinct composers, also cut into systems for a total of 17,945 data points consisting of a rendered image of a system (using MuseScore 3.6.2).[9] As with GrandStaff, we encode the MusicXML ground truth in LMX.

[7] A system in music notation means one line of music of all instruments or voices that play simultaneously. In solo piano music, a system consists of music on two staffs: one for the right hand, one for the left.

[8] https://kern.ccarh.org/.

[9] This is in fact a more musically diverse dataset than GrandStaff: without transpositions, GrandStaff has 13,470 distinct lines of music.

Fig. 5. Examples of the variety of out-of-domain datasets.

Fig. 6. Examples of images from the Authentic dataset.

The **MUSICMA++** dataset [21] serves as an alternative source of handwritten symbols—out-of-domain—for creating synthetic data with Smashcima. It consists of 91,255 annotated notation symbols. In this work, it is used to evaluate the importance of using in-domain symbols as inputs for the data synthesis process, compared to generic handwritten symbols. Samples from these out-of-domain datasets are shown in Fig. 5.

4.2 In-Domain Datasets

The **Authentic**[10] dataset (Fig. 6) is at the heart of the experiment setup: it represents the small in-domain dataset for end-to-end recognition used for fine-tuning experiments and, most importantly, evaluating the quality of domain adaptation. It comes from the collections of the Moravian library, as a representative sample selected by its music librarians. It consists of 159 piano systems that have been transcribed in MuseScore, exported to MusicXML and encoded in LMX. A 59:50:50 train-validation-test split is defined by assigning the systems randomly. The total effort to transcribe this dataset was 25 h.

The **Authentic-MuNG** dataset serves as the source of input symbols for creating synthetic data. For simplicity, we exploit the same 59 samples from the training (or fine-tuning) portion of the Authentic dataset, but annotated in the

[10] The data is derived from the MusiCorpus v1.0 UFAL.OmniOMR segment [51].

Table 1. List of datasets split into sections according to the description above. Datasets with bold names are those that had to be created by the library in our scenario. The "Function" column describes how the dataset is used in our experiments: "synthesis" means that the dataset is used as input to Smashcima and its ground truth are symbols annotated as MuNG rather than LMX-encoded music transcriptions. "BD" in Production stands for born-digital.

Name	Domain	Production	Function	Size	Unit
GrandStaff	out	typeset(BD)	training	53,882	systems
Camera-GrandStaff	out	typeset(BD)	training	53,882	systems
OLiMPiC	out	typeset(BD)	training	17,945	systems
Authentic	in	manuscript	eval+fine-tuning	159	systems
Authentic-MuNG	in	manuscript	synthesis	39,376	symbols
MUSCIMA++	out	manuscript	synthesis	91,255	symbols
Synth-GS-Auth	in	manuscript(BD)	training	44,192	systems
Synth-OL-Auth	in	manuscript(BD)	training	16,273	systems
Synth-GS-MPP	out	manuscript(BD)	training	44,191	systems
Synth-OL-MPP	out	manuscript(BD)	training	16,273	systems

MuNG format. The total number of annotated symbols across all splits is 39,376 but only 15,011 are in the training split, which is used for synthesis. The total effort to annotate this dataset was approximately 260 h.

4.3 Synthetic Datasets

Finally, we describe the handwritten datasets synthesised using the Smashcima tool. As described in Sect. 3, it requires three inputs:

- MusicXML – musical content; taken from the GrandStaff and OLiMPiC datasets.
- Music notation glyphs – capture the visual domain of a document; taken from MUSCIMA++ and the Authentic-MuNG datasets.
- Background patches – we use 13 patches from the MZKBlank set [14].

The **Synth-GS-Auth** dataset is created by running the Smashcima system with MusicXML inputs from the GrandStaff dataset and using in-domain symbol glyphs from the **Authentic-MuNG** dataset. The **Synth-GS-MPP** dataset is the same, but uses symbols from MUSCIMA++ instead. Similarly **Synth-OL-Auth** and **Synth-OL-MPP** are constructed in the same manner from MusicXML data from the OLiMPiC dataset. The two datasets that use MUSCIMA++ symbols are not considered in-domain since they do not use Authentic symbols: we use them to estimate the benefit of having in-domain symbols in the synthetic data. When running Smashcima, we lose 9 % of OLiMPiC samples and 18 % of GrandStaff samples due to incompatibilities and crashes (e.g., unsupported rare symbol classes; some of these come up more often in GrandStaff because of its transpositions). The full overview of datasets used in this paper is given in Table 1.

5 Experiments

Experiments consist of three phases. **First,** we take a randomly-initialized model and train it on a base dataset (Camera-GrandStaff, OLiMPiC, or one of the synthetic datasets). This training takes the longest amount of time (cca 3 d on a A40 GPU). **Second,** we "re-train" this base model, continuing the training on synthetic data (only for non-synthetic base datasets). This phase is meant to explore adaptation of existing pre-trained OMR models. This re-training phase is shorter, but still takes around 2 to 8 h. **Third,** we fine-tune on the Authentic dataset. This is the fastest phase (cca 10 min) because the Authentic dataset is small and prone to overfitting.

In the second and third phase we do not use any data from the previous phase (that is, we do not re-train with replay). This way the model risks catastrophic forgetting of the previously learned data. As our aim is to adapt to the target domain, not to build a general-purpose model, this does not matter. In the last experiment, however, we perform re-training *with* replay[11] to see what effect it has on the final result.

The initial training phase uses learning rate of $1e^{-3}$ with cosine decay over 150 or 500 epochs on GrandStaff or OLiMPiC respectively (GrandStaff is about 3x the size of OLiMPiC). The two re-training and fine-tuning phases use constant learning rate of $1e^{-4}$.

The re-training phase stores a model snapshot after each epoch and then we find the best epoch. This turns out to be very short without replay (2–10 epochs) due to the forgetting-learning tension; with replay, the re-training can proceed for 30 epochs with no issues.

The final fine-tuning phase always takes 20 epochs, which we estimated by observing the symbol error rate (SER) stopping decreasing on the validation split of the Authentic dataset. This is stable regardless of which model is being fine-tuned.

5.1 Model

We use the Zeus model, which holds (together with a fine-tuned PaliGemma 2) the reported state-of-the-art SER on the GrandStaff dataset [35]. The only relevant difference between Zeus and the Sheet Music Transformer (SMT) [43] is the use of soft attention with recurrent layers rather than Transformer self-attention and positional encodings, and Zeus led to a nearly 50 % relative error reduction on the GrandStaff dataset [35], we use this model. However, in principle, one could also use SMT; the experimental conditions are not different models but different dataset choices. The exact settings with which the model is run are the same as in [35].

[11] Mixing the synthetic images with the born-digital images.

5.2 Evaluation

We perform all evaluation on the test split of the Authentic dataset, as it represents the unseen manuscripts from the library we want to adapt to. We use the Symbol Error Rate (SER), a metric reported in most end-to-end OMR works [1,9,24,35,40,43,44]. Evaluation of OMR is a notoriously hard problem [4,20,49] but SER has been shown to correlate well with suggested improvements [29,49]. Previous in-domain results on GrandStaff and OLiMPiC reported at a SER around 2.0 and 11.0 respectively [24,35,43,47].

The test split of the Authentic dataset has 50 systems of music. This is of course a small sample with high variance, however, we are interested in qualitative leaps, not in competing for the final few percentage points, and we believe the variance in this case is acceptable; in any case, no larger real-world dataset of musical manuscripts is currently available. Cross-validating the best-performing model across all 6 permutations of the fine-tuning, validation, and test portion of the Authentic dataset, we find that the standard deviation across the Authentic dataset is 4.2, so a difference of some 8.4 SER on this dataset is 95% likely meaningful (and a difference of 4.2 SER is 68% likely meaningful).

6 Results

We perform a total of 11 experiments grouped into four scenarios as indicated in Table 2.[12] Experiments 1 and 2 act as baselines where we only fine-tune existing OMR models on our Authentic data: one trained on the Camera-GrandStaff dataset, the other on OLiMPiC. The far better base model is the Zeus OLiMPiC model from 2024: from a SER of 88.0, when naively applied to our Authentic test set, it improves to 41.3 after fine-tuning (while the base model on Camera-GrandStaff fine-tunes to 85.2). The OLiMPiC base model result is stable within 1 SER when re-run with different seeds, thus forming the baseline for all future models that use synthetic data. If none of them drop significantly below 41.3 SER, synthetic data brings no advantage compared to just fine-tuning.

Experiments 3 to 6 try building a base model on synthetic data only. We see that applying these models naively results in SER of 56.9–76.3, somewhat better than the 88.0 from Experiment 2; however, fine-tuning only improves this to 51.4–58.5 SER. Though these models seem better suited for handwritten music, the Zeus OLiMPiC model proves more adaptable. However, from pre- and post- fine-tuning evaluation, it is clear that Authentic symbols outmatch MUSCIMA++ symbols. This makes a case for using in-domain glyphs when building the synthetic dataset.

Experiments 7 to 10 take the pre-trained Zeus OLiMPiC model and then train it further on our four synthetic datasets. This re-training happens without replay of OLiMPiC data, which causes a tension between forgetting of typeset music and learning of handwritten music. The optimal re-training epoch is between 2 and 6, based on the lowest SER (after fine-tuning) on the Authentic validation

[12] The code for all experiments is available at https://github.com/Jirka-Mayer/icdar26.

Table 2. Main experimental results: For each experiment a new model is created, trained on the first dataset, then re-trained (without replay) on the second one, and finally fine-tuned (no replay) on the Authentic dataset (training split). Symbol error rate (SER) is reported after each step on the Authentic dataset (test split). Number of training epochs is shown in parentheses. Fine-tuning is done for 20 epochs. (SER >100 can happen when the model generates many extra tokens.)

Ex	Trained on	SER	Re-trained on	SER	SER ft
1	Camera-GrandStaff (150e)	122.9	–	–	85.2
2	OLiMPiC (500e)	88.0	–	–	**41.3**
3	Synth-GS-MPP (150e)	70.9	–	–	56.3
4	Synth-GS-Auth (150e)	56.9	–	–	52.5
5	Synth-OL-MPP (500e)	76.3	–	–	58.5
6	Synth-OL-Auth (500e)	62.8	–	–	51.4
7	OLiMPiC (500e)	88.0	Synth-GS-MPP (2e)	63.4	34.7
8	OLiMPiC (500e)	88.0	Synth-GS-Auth (2e)	51.8	34.6
9	OLiMPiC (500e)	88.0	Synth-OL-MPP (6e)	63.1	34.8
10	OLiMPiC (500e)	88.0	Synth-OL-Auth (2e)	49.9	34.4
11	OLiMPiC (500e)	88.0	Synth-OL-Auth w/ replay (30e)	**38.1**	**33.2**

split. Before fine-tuning, the re-trained models have SER cca 7.0 lower compared to Experiments 3 to 6, and authentic glyphs are still preferred. However, *after* fine-tuning, the preference disappears: models improve to around 34.6 SER in all experiments. This is the first setup that managed to beat the 41.3 SER fine-tuning baseline (Ex 2).

To avoid the tension of Experiments 7 to 10, we designed Experiment 11 to include data replay. The re-training dataset is the OLiMPiC and Synth-OL-Auth datasets combined into one (leading to twice as many samples per epoch). This lets us not worry about early stopping (before forgetting important OLiMPiC traits) and we can now train for 30 epochs (equivalent to 60 epochs in Experiments 7 to 10). With this setup, we just about outperform the Ex 2 baseline without even fine-tuning, at 38.1 SER. If we now also fine-tune, we improve to 33.2 SER, which is technically the best result (though by little).

7 Discussion and Conclusions

For the first time, we have run OMR experiments with real-world manuscripts of piano music, the most complex class of Common Western Music Notation. We have thus established a baseline for this most difficult, but most impactful, setting for OMR: low-resource adaptation to diverse historical manuscripts.

A SER of 33.2 is of course much higher than the previously reported results on simple datasets such as GrandStaff or OLiMPiC with SER approx. 2.0 and

11.0 respectively [35,43,47]. However, these results were achieved for clean born-digital images, with limited augmentations. While these datasets are good benchmarks for model comparison, these results are not representative of expected performance in application settings at all. Already on the scanned version of OLiMPiC evaluation data, which are all high-quality scans of high-quality prints, SER deteriorates to 17.7 [35]. In other words—the OMR *models* are fine, but the task of making OMR *systems* valuable is just beginning.

To our main question: Are high-quality synthetic data in turn valuable for training OMR systems in difficult low-resource settings?

On the one hand, the answer is yes. The presence of synthetic handwritten data did lead to significantly better results than when this type of data was absent. On the other hand, the results are certainly not the qualitative leap one may have imagined on the basis of impressions of the data synthesis results (Fig. 4), and of the near-inability of people (incl. those capable of reading music) to distinguish Smashcima-generated from real images.

However, the experiments revealed a silver lining. After fine-tuning, the advantage of using in-domain MuNG data for Smashcima over using the MUSCIMA++ symbols disappears. Therefore, it seems that to get the available mileage out of Smashcima, one actually does not need to take on the expense of annotating extra in-domain MuNG data. One needs authentic fine-tuning data anyway, and the final system quality will likely correlate to the amount and diversity of this data. But adding a synthetic dataset into the training process does provide a significant boost, especially when fine-tuning data is scarce; at least, this implies that less fine-tuning data will be needed to achieve the same performance when Smashcima is used. Therefore, a recommendation for OMR system development would be not to treat the synthetic data as part of domain adaptation, but to mix them directly into the base model training.

As evidenced by the impact of fine-tuning with even a very small authentic dataset, Smashcima is still not able to entirely simulate the domain of real manuscript music notation. The limitations mostly concern how symbols connect: we noticed, e.g., that stems only attach to noteheads on the rightmost point, while many historical writing styles attach on the top of the notehead. However, the software will likely never be able to simulate the broad diversity of musical manuscript in its rendering engine. We therefore see most future potential in combining its manuscript images with unsupervised domain adaptation techniques [26], either diffusion models [3,37], or moment matching methods exploiting batch normalisation, which have seen some success on OMR [42]. Also, the large impact of fine-tuning even with small data may be due to domain shift in the underlying music itself, so further improvement may be found by synthesising the MusicXML ground truth itself to match the target collection.

Despite these limitations, and with the limited evaluation data in mind, we still conclude that high-quality synthetic manuscript images can lead to a significant improvement towards real-world applications of OMR for cultural heritage preservation and access.

Acknowledgments. This study was funded by the Ministry of Culture of the Czech Republic (project OmniOMR of the NAKI III programme, no. DH23P03OVV008).

Disclosure of Interests. The authors have no competing interests to declare that are relevant to the content of this article.

References

1. Alfaro-Contreras, M., Iñesta, J.M., Calvo-Zaragoza, J.: Optical music recognition for homophonic scores with neural networks and synthetic music generation. Int. J. Multimed. Inform. Retrieval **12**(1), 12 (2023). https://doi.org/10.1007/s13735-023-00278-5
2. Baró, A., Riba, P., Calvo-Zaragoza, J., Fornés, A.: From optical music recognition to handwritten music recognition: A baseline. Pattern Recogn. Lett. **123**, 1–8 (2019). https://doi.org/10.1016/j.patrec.2019.02.029
3. Benigmim, Y., Roy, S., Essid, S., Kalogeiton, V., Lathuilière, S.: One-shot unsupervised domain adaptation with personalized diffusion models. In: Proceedings of the IEEE/CVF Conference on Computer Vision and Pattern Recognition, pp. 698–708 (2023)
4. Byrd, D., Simonsen, J.G.: Towards a standard testbed for optical music recognition: definitions, metrics, and page images. J. New Music Res. **44**(3), 169–195 (2015). https://doi.org/10.1080/09298215.2015.1045424
5. Calvo-Zaragoza, J., Fuentes-Martınez, E., Luna-Barahona, N., Rıos-Vila, A.: Can multimodal large language models read music score images? In: 6th International Workshop on Reading Music Systems, pp. 4–6 (2024)
6. Calvo-Zaragoza, J., Hajič jr., J., Pacha, A.: understanding optical music recognition. ACM Comput. Surv. **53**(4), 77 (2020). https://doi.org/10.1145/3397499
7. Calvo-Zaragoza, J., Martinez-Sevilla, J.C., Penarrubia, C., Rios-Vila, A.: Optical music recognition: recent advances, current challenges, and future directions. In: Coustaty, M., Fornés, A. (eds.) Document Analysis and Recognition - ICDAR 2023 Workshops, pp. 94–104. Springer Nature Switzerland, Cham (2023)
8. Calvo-Zaragoza, J., Rizo, D.: Camera-primus: neural end-to-end optical music recognition on realistic monophonic scores. In: 19th International Society for Music Information Retrieval Conference (ISMIR). pp. 248–255. Paris, France (2018). http://ismir2018.ircam.fr/doc/pdfs/33_Paper.pdf
9. Calvo-Zaragoza, J., Rizo, D.: End-to-end neural optical music recognition of monophonic scores. Appl. Sci. **8**(4), 606 (2018). https://doi.org/10.3390/app8040606
10. Castellanos, F.J., Gallego, A.J., Fujinaga, I.: Deep learning for optical music recognition: a review. TechRxiv (Feb 2025). https://doi.org/10.36227/techrxiv.174077177.78767136/v1
11. Castellanos, F.J., Garrido-Munoz, C., Ríos-Vila, A., Calvo-Zaragoza, J.: Region-based layout analysis of music score images. Expert Syst. Appl. **209**, 118211 (Dec 2022). https://doi.org/10.1016/j.eswa.2022.118211
12. Crawford, T., Lewis, D., Porter, A.: Exploring early vocal music and its lute arrangements: Using f-tempo as a musicological tool. In: Proceedings of the 10th International Conference on Digital Libraries for Musicology, pp. 77–81 (2023)
13. Diet, J.: Optical music recognition in der Bayerischen Staatsbibliothek. Bibliothek Forschung und Praxis **42**(2), 319–323 (2018). https://doi.org/10.1515/bfp-2018-0030

14. Dvořák, V., Hajič jr., J., Mayer, J.: Staff layout analysis using the YOLO platform. In: 6th International Workshop on Reading Music Systems (WoRMS), pp. 18–22. Online (2024). https://doi.org/10.48550/arXiv.2411.15741
15. Fornés, A., Dutta, A., Gordo, A., Lladós, J.: CVC-MUSCIMA: a ground truth of handwritten music score images for writer identification and staff removal. Int. J. Document Anal. Recogn. (IJDAR) **15**, 243–251 (2011). https://doi.org/10.1007/s10032-011-0168-2
16. Gotham, M., Jonas, P., Bower, B., Bosworth, W., Rootham, D., VanHandel, L.: Scores of scores: an openscore project to encode and share sheet music. In: 5th International Conference on Digital Libraries for Musicology (DLfM), pp. 87–95. Paris, France (2018). https://doi.org/10.1145/3273024.3273026
17. Gotham, M.R.H., Jonas, P.: The openscore lieder corpus. In: Music Encoding Conference, pp. 131–136., Alicante, Spain (2022). https://doi.org/10.17613/1my2-dm23
18. Hajič jr., J., Dorfer, M., Widmer, G., Pecina, P.: Towards full-pipeline handwritten OMR with musical symbol detection by u-nets. In: 19th International Society for Music Information Retrieval Conference (ISMIR), pp. 225–232. Paris, France (2018). http://ismir2018.ircam.fr/doc/pdfs/175_Paper.pdf
19. Hajič jr, J., Kolárová, M., Pacha, A., Calvo-Zaragoza, J.: How current optical music recognition systems are becoming useful for digital libraries. In: Proceedings of the 5th International Conference on Digital Libraries for Musicology, pp. 57–61 (2018)
20. Hajič jr., J., Novotný, J., Pecina, P., Pokorný, J.: Further steps towards a standard testbed for optical music recognition. In: 17th International Society for Music Information Retrieval Conference (ISMIR), pp. 157–163. New York, USA (2016). https://wp.nyu.edu/ismir2016/event/proceedings/
21. Hajič, jr., J., Pecina, P.: The MUSCIMA++ dataset for handwritten optical music recognition. In: 14th IAPR International Conference on Document Analysis and Recognition (ICDAR), pp. 39–46. Kyoto, Japan (2017). https://doi.org/10.1109/ICDAR.2017.16
22. Hander, P., Shatri, E., Timms, B., Fazekas, G.: Towards artificially generated handwritten sheet music. In: 5th International Workshop on Reading Music Systems (WoRMS), pp. 25–30. Milan, Italy (2023). https://doi.org/10.48550/arXiv.2311.04091
23. Havelka, J., Mayer, J., Pecina, P.: Symbol generation via autoencoders for handwritten music synthesis. In: 5th International Workshop on Reading Music Systems (WoRMS), pp. 20–24. Milan, Italy (2023). https://doi.org/10.48550/arXiv.2311.04091
24. Jung, J., et al.: U-must: A unified framework for cross-modal translation of score images, symbolic music, and performance audio. IEEE Transactions on Audio, Speech and Language Processing, pp. 1–16 (2025). https://doi.org/10.1109/TASLPRO.2025.3648794
25. Kang, L., Riba, P., Wang, Y., Rusiñol, M., Fornés, A., Villegas, M.: Ganwriting: Content-conditioned generation of styled handwritten word images. In: 16th European Conference on Computer Vision (ECCV), pp. 273–289. Glasgow, UK (2020). https://doi.org/10.1007/978-3-030-58592-1_17
26. Kang, L., Rusinol, M., Fornés, A., Riba, P., Villegas, M.: Unsupervised writer adaptation for synthetic-to-real handwritten word recognition. In: Proceedings of the IEEE/CVF Winter Conference on Applications of Computer Vision, pp. 3502–3511 (2020)

27. Long, P., Novack, Z., Berg-Kirkpatrick, T., McAuley, J.: PDMX: A large-scale public domain MusicXML dataset for symbolic music processing. In: ICASSP 2025 - 2025 IEEE International Conference on Acoustics, Speech and Signal Processing (ICASSP), pp. 1–5. IEEE (Apr 2025). https://doi.org/10.1109/icassp49660.2025.10890217
28. Luna-Barahona, N.N., Rosello, A., Alfaro-Contreras, M., Rizo, D., Calvo-Zaragoza, J.: Unsupervised synthetic-to-real adaptation for optical music recognition. In: Proceedings of the 25th International Society for Music Information Retrieval Conference, ISMIR 2024, San Francisco, California, USA and Online, November 10-14, 2024, pp. 462–469 (2024). https://doi.org/10.5281/ZENODO.14877375
29. Martinez-Sevilla, J.C., et al.: Sheet music benchmark: Standardized optical music recognition evaluation. In: 25th International Society for Music Information Retrieval (ISMIR) (2025)
30. Matrinez-Sevilla, J., Roselló, A., Rizo, D., Calvo-Zaragoza, J.: On the performance of optical music recognition in the absence of specific training data. In: 24th International Society for Music Information Retrieval (ISMIR), pp. 319–326. Milan, Italy (2023). https://doi.org/10.5281/ZENODO.10265289
31. Mayer, J., Jebavý, F., Vlková, M., Dvořáková, M., Pecina, P., Hajič jr., J.: MuNG studio: annotation tool for music notation graph. In: de Luca, E. (ed.) Proceedings of the 12th International Conference on Digital Libraries for Musicology, pp. 114–118. Association for Computing Machinery, Association for Computing Machinery, New York, NY, United States (2025)
32. Mayer, J., Pecina, P., Hajič jr., J.: Smashcima: Full-page handwritten music document synthesizer. In: de Luca, E. (ed.) Proceedings of the 12th International Conference on Digital Libraries for Musicology, pp. 119–123. Association for Computing Machinery, Association for Computing Machinery, New York, NY, United States (2025)
33. Mayer, J., Pecina, P.: Synthesizing training data for handwritten music recognition. In: 16th International Conference on Document Analysis and Recognition (ICDAR), pp. 626–641. Lausanne, Switzerland (2021). https://doi.org/10.1007/978-3-030-86334-0_41
34. Mayer, J., Pecina, P.: Obstacles with synthesizing training data for OMR. In: 4th International Workshop on Reading Music Systems (WoRMS), pp. 15–19. Online (2022). https://doi.org/10.48550/arXiv.2211.13285
35. Mayer, J., Straka, M., Hajič jr., J., Pecina, P.: Practical end-to-end optical music recognition for pianoform music. In: 18th International Conference on Document Analysis and Recognition (ICDAR), pp. 55–73. Athens, Greece (2024). https://doi.org/10.1007/978-3-031-70552-6_4
36. Pacha, A., Calvo-Zaragoza, J., Hajič jr., J.: Learning notation graph construction for full-pipeline optical music recognition. In: 20th International Society for Music Information Retrieval (ISMIR), pp. 75–82. Delft, Netherlands (2019). https://doi.org/10.5281/zenodo.3527744
37. Peng, D., Ke, Q., Ambikapathi, A., Yazici, Y., Lei, Y., Liu, J.: Unsupervised domain adaptation via domain-adaptive diffusion. IEEE Trans. Image Process. **33**, 4245–4260 (2024). https://doi.org/10.1109/tip.2024.3424985
38. Pugin, L., Zitellini, R., Roland, P.: Verovio: A library for engraving MEI music notation into SVG. In: 15th International Society for Music Information Retrieval Conference (ISMIR), pp. 107–112. Taipei, Taiwan (2014), https://archives.ismir.net/ismir2014/paper/000221.pdf

39. Rios-Vila, A., Fuentes-Martinez, E., Castellanos, F.J.: An implicit layout-aware transformer for full-page end-to-end optical music recognition. Int. J. Multimed. Inform. Retrieval **14**(4) (Oct 2025). https://doi.org/10.1007/s13735-025-00385-5
40. Ríos-Vila, A., Iñesta, J.M., Calvo-Zaragoza, J.: End-to-end full-page optical music recognition of monophonic documents via score unfolding. In: 4th International Workshop on Reading Music Systems (WoRMS), pp. 20–24. Online (2022), https://sites.google.com/view/worms2022/proceedings
41. Ríos-Vila, A., Rizo, D., Iñesta, J.M., Calvo-Zaragoza, J.: End-to-end optical music recognition for pianoform sheet music. International Journal on Document Analysis and Recognition (IJDAR) **26**(3), 347–362 (2023). https://doi.org/10.1007/s10032-023-00432-z
42. Roselló, A., Fuentes-Martínez, E., Alfaro-Contreras, M., Rizo, D., Calvo-Zaragoza, J.: Source-Free Domain Adaptation for Optical Music Recognition, pp. 3–19. Springer Nature Switzerland (2024). https://doi.org/10.1007/978-3-031-70552-6_1
43. Ríos-Vila, A., Calvo-Zaragoza, J., Paquet, T.: Sheet music transformer: End-to-end optical music recognition beyond monophonic transcription (2024). https://arxiv.org/abs/2402.07596
44. Ríos-Vila, A., Calvo-Zaragoza, J., Rizo, D., Paquet, T.: End-to-end full-page optical music recognition for pianoform sheet music (2024). https://doi.org/10.48550/ARXIV.2405.12105
45. Shatri, E., Palavala, K.R., Fazekas, G.: Synthesising handwritten music with gans: a comprehensive evaluation of cyclewgan, progan, and DCGAN. In: IEEE International Conference on Big Data, BigData 2024, Washington, DC, USA, December 15-18, 2024, pp. 3208–3217. IEEE (2024). https://doi.org/10.1109/BIGDATA62323.2024.10825834
46. de Sousa Neto, A.F., Bezerra, B.L.D., de Moura, G.C.D., Toselli, A.H.: Data augmentation for offline handwritten text recognition: A systematic literature review. SN Comput. Sci. **5**(2) (Feb 2024). https://doi.org/10.1007/s42979-023-02583-6
47. Steiner, A., et al.: PaliGemma 2: A family of versatile VLMs for transfer (2024). https://arxiv.org/abs/2412.03555
48. Tirupati, N., Shatri, E., Fazekas, G.: Crafting handwritten notations: Towards sheet music generation. In: 6th International Workshop on Reading Music Systems (WoRMS). pp. 50–56. Online (2024). https://doi.org/10.48550/arXiv.2411.15741
49. Torras, P., Biswas, S., Fornés, A.: A unified representation framework for the evaluation of optical music recognition systems. Int. J. Document Anal. Recogn. (IJDAR) **27**, 379–393 (2024). https://doi.org/10.1007/s10032-024-00485-8
50. Torras, P., et al.: Two journeys: Insights on the annotation of large-scale optical music recognition datasets. In: Proceedings of the 17th International Symposium on Computer Music Multidisciplinary Research. University College London, The Laboratory PRISM, Marseille, France (2025)
51. Torras, P., et al.: A dataset for the recognition of historical and handwritten music scores in western notation (2026). https://arxiv.org/abs/2605.18436
52. Wang, Y., et al.: NotaGen: advancing musicality in symbolic music generation with large language model training paradigms. In: Proceedings of the Thirty-Fourth International Joint Conference on Artificial Intelligence, pp. 10207–10215 (2025)

EAGLE: Explicit Anchoring and Graph Reasoning with Diagram Structure Priors for Multimodal Geometry Problem Solving

Jie Zhang[1,2,3], Xiangren Wang[1], Ziming Li[1,2,3], Minzhi Zhang[1,2,3], Xingxiang Zhou[1,2,3], Zhi Chen[1,2,3], Guanglai Gao[1,2,3], and Xiangdong Su[1,2,3](✉)

[1] College of Computer Science, Inner Mongolia University, Hohhot, China
32409173@mail.imu.edu.cn, cssxd@imu.edu.cn

[2] National & Local Joint Engineering Research Center of Intelligent Information Processing Technology for Mongolian, Hohhot, China

[3] Inner Mongolia Key Laboratory of Multilingual Artificial Intelligence Technology, Hohhot, China

Abstract. Multimodal geometry problem solving requires translating a sparse diagram and a clause-structured statement into a verifiable symbolic solution. Current neural solvers often rely on implicit cross-modal fusion, which can attend to visually salient but irrelevant regions and produce unstable diagram–text grounding, leading to brittle reasoning under strict geometric constraints. We view geometry solving as a staged process—perceiving diagram primitives, explicitly anchoring each textual clause to the diagram, and performing relational deduction—and present EAGLE, a unified perceive–align–reason framework for this setting. EAGLE learns a lightweight Euclid Field prior to highlight geometry-bearing regions (e.g., points, segments, arcs, and symbols), constructs an explicit clause–structure anchoring matrix that makes grounding observable, and performs graph-based constraint propagation over the induced clause–structure graph before decoding a solution program. The generated solution program is executable, enabling direct verification of the final numeric answer or the selected option. By decoupling grounding from reasoning, EAGLE provides a diagnosable intermediate representation and improves robustness without test-time oracle structure. Experiments on Geometry3K and PGPS9K show consistent gains in Top-1 reliability over strong program-generation baselines under realistic settings without test-time oracle diagram structure, together with structure-aware grounding visualizations and proxy diagnostics.

Keywords: Geometry problem solving · Multi-modal fusion · Graph reasoning · Diagram structure priors

1 Introduction

Plane geometry problem solving asks a machine to answer a question given a diagram and a text statement, ideally by producing a faithful chain of infer-

G. A. Fink et al. (Eds.): ICDAR 2026, LNCS 16972, pp. 306–322, 2027.
https://doi.org/10.1007/978-3-032-36023-6_18

ences rather than a guess. Beyond classroom tutoring, this capability matters whenever diagrams and symbolic annotations are the primary carrier of meaning, where language alone is insufficient [12]. What makes geometry particularly challenging is not "more pixels" or "more words", but the mismatch between sparsity and rigor: the decisive evidence occupies only a few primitives and tiny labels (points, segments, arcs, symbols), yet the required reasoning is compositional and unforgiving—one wrong grounding can invalidate the entire solution [11,37]. Moreover, the text is rarely a single query; it is typically a set of role-specific statements (conditions, intermediate facts, and the goal) that must be tied to different parts of the diagram [20,36]. As a result, a robust solver must simultaneously answer two questions that humans resolve almost instinctively: where to look in the diagram and what each statement refers to, before any global deduction can be trusted [24].

Recent progress has largely followed an end-to-end paradigm: encode the diagram and the text, fuse them with cross-modal attention, and directly generate an answer or a solution program [28,38]. These models are attractive for their simplicity and strong supervision signal. However, their fusion often remains implicit: the model is asked to discover structure and grounding solely from final supervision while treating the diagram as a generic image with uniformly processed patches [10]. In practice, this leaves two geometry-critical signals under-exploited. First, the model is not explicitly told which regions are geometry-bearing versus background; attention can drift to locally salient but irrelevant textures or numbers [37]. Second, the linkage between individual textual statements and specific diagram sub-structures is buried inside black-box fusion weights, making the learned alignment unstable and difficult to diagnose [26,33]. These limitations are not merely interpretability concerns: when the solver generates solution programs, incorrect or drifting grounding tends to produce non-executable solution program candidates or systematically mis-ranked beams, directly harming reliability.

We take inspiration from how humans solve geometry: they rarely "fuse everything at once". Instead, a student typically follows a staged routine—scan the diagram to pick out meaningful primitives and labels, point each statement to the relevant part of the figure, and then combine the anchored constraints to infer the unknown [29]. This cognitive routine motivates our central design principle: if geometry requires "look–link–reason", then a model should expose these steps as explicit computations rather than hiding them inside a single fusion block. Guided by this idea, we propose EAGLE (Explicit Anchoring and Graph reasoning with a Learned structural prior for Euclidean geometry), a framework where each component superficially mirrors a human step, while essentially addresses a missing piece in prior end-to-end solvers (Fig. 1).

Concretely, EAGLE first makes the "where to look" signal explicit via a learned Euclid Field, a four-channel geometry-structure prior rather than a mathematical field, that highlights geometry-bearing regions (points/lines/circles-or-arcs/symbols) [39]. This prior is predicted from the diagram at inference time, so EAGLE does not rely on oracle intermediate representations at test

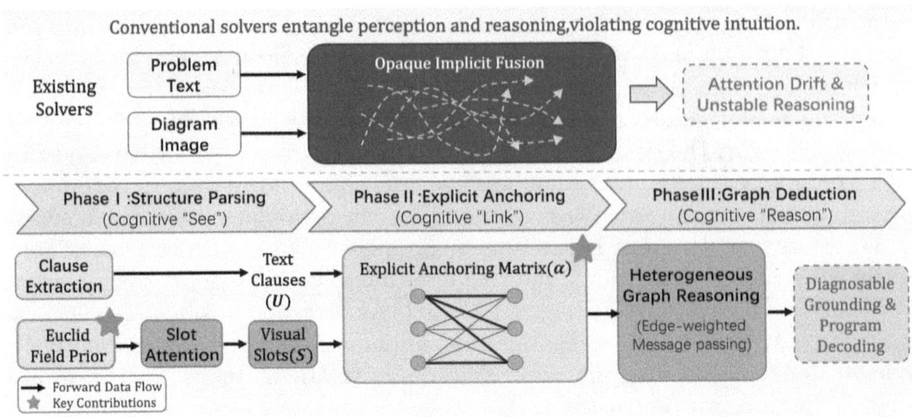

Fig. 1. Conceptual comparison of geometry solvers. Top: Conventional implicit fusion models often suffer from attention drift. Bottom: Our EAGLE framework explicitly decouples reasoning through diagram-guided structure parsing, anchoring, and graph-based deduction.

time, while providing a dedicated spatial cue that regularizes visual representation learning and stabilizes downstream cross-modal interactions by encouraging the solver to build its evidence on primitives and labels rather than accidental saliency. Next, EAGLE operationalizes "what to link" with SOAR (Structure-Observing Alignment and Reasoning), which decomposes cross-modal interaction into explicit stages: it summarizes the diagram into a small set of geometry-aware structure units (slots) and computes an explicit clause–slot anchoring that indicates which diagram structure units are most relevant to each textual clause. By turning what is typically an opaque attention pattern into a diagnosable intermediate interface, this explicit anchoring makes downstream reasoning less brittle and failure cases easier to trace [19]. Finally, EAGLE performs heterogeneous graph reasoning over the induced clause–structure graph, enabling relation-centric inference before program generation. In doing so, it decouples anchoring from reasoning and yields a clearer information flow from diagram/text evidence to solution program candidates, improving stability and interpretability simultaneously.

In summary, our main contributions are:

- We propose EAGLE, a geometry solver guided by a learned Euclid Field that explicitly captures "where to look" in sparse diagrams by emphasizing primitives and symbolic regions, improving the structure-awareness of visual evidence used for solving.
- We introduce SOAR, a human-inspired staged cross-modal reasoning pipeline that explicitly performs structure parsing, statement-to-structure anchoring, and graph-based deduction, turning implicit fusion into diagnosable intermediate computations.

- We validate EAGLE on public geometry problem benchmarks, showing strong performance together with qualitative diagnostic evidence that the model's attention and grounding align better with geometry-relevant structure.

2 Related Work

2.1 Geometry Problem Solving

Geometry problem solving (GPS) generally spans two closely related directions: geometric theorem proving and geometric numerical calculation [16]. In this work, we focus on geometric numerical calculation, which has gained renewed attention with program-supervised benchmarks [7].

Early GPS systems were largely symbolic: they parse text and diagrams into explicit constraints and apply rule-driven search [24]. While logically grounded and interpretable, their performance is highly sensitive to constraint extraction errors, and scaling them requires substantial manual engineering [22].

Recent work increasingly formulates GPS as multimodal program generation with neural encoder-decoder models [38]. Despite strong empirical progress, existing neural solvers face two critical bottlenecks [6]. First, they often implicitly treat a geometry diagram as an ordinary image, processing all patches uniformly [9]. This ignores the extreme sparsity of geometric diagrams, leaving the model vulnerable to background clutter [39]. Second, they entangle perception, alignment, and reasoning within opaque implicit fusion [21]. This often causes attention drift toward visually salient yet irrelevant regions, making grounding errors hard to trace. Recent work such as SCANS studies content-aware attention and adaptive fusion for efficient GPS, further connecting geometry solving with document analysis and recognition [8]. To bridge this gap, our proposed EAGLE introduces diagram structure priors (the Euclid Field) to explicitly guide the encoder toward geometry-bearing regions (points, lines, symbols). Furthermore, it replaces black-box fusion with SOAR, a staged pipeline that decouples the solving process to support stable and diagnosable reasoning [27].

2.2 Multimodal Reasoning and Grounding

Multimodal reasoning integrates information across modalities to answer questions or generate structured outputs. A common paradigm is implicit fusion [10]. In this paradigm, cross-attention directly mixes modalities into a shared latent representation. While this design underlies many powerful vision-language models, GPS still requires more precise, structure-aware grounding cues for rigorous deduction [2].

An alternative paradigm is explicit reasoning, where models construct intermediate structured representations (e.g., graphs or object-centric variables) for step-wise inference [15]. Graph-based message passing is particularly effective for relational modeling [23]. However, in diagram-centric reasoning tasks, the key difficulty lies in defining reliable cross-modal edges between language conditions and visual evidence. Existing graph-based solvers often leave these links implicit, leading to brittle multi-step reasoning and limited traceability [27]. Motivated by this, our work completely decouples anchoring from reasoning [31]. Instead of relying on implicit cross-attention, EAGLE performs unitization (structure-slot parsing) to form geometry-aware visual slots, learns an explicit clause-slot anchoring matrix to serve as diagnosable cross-modal edges, and finally conducts edge-weighted graph reasoning. This explicitly structured grounding makes the cross-modal links observable and easier to diagnose [18].

3 Method

3.1 Problem Formulation and Overview

Plane-geometry problem solving takes as input a diagram image and a problem text, and outputs a solution program that yields the final numeric answer or the correct choice. Following the look–link–reason routine discussed in the Introduction, EAGLE exposes the solving process as explicit intermediate computations rather than a single opaque fusion block. To make the correspondence to our motivation explicit, we summarize EAGLE as a read–look–link–reason–decode workflow: Read (clause structuring), Look (Euclid Field prior learning and injection), Link (Stage1–2 structure-slot parsing and explicit clause anchoring), Reason (Stage3 heterogeneous graph deduction), and Decode (solution-program generation).

Inputs and Outputs. Given a problem instance, we take as input a diagram image $I \in \mathbb{R}^{3 \times H_I \times W_I}$ and a tokenized problem text sequence $W = \{w_1, \dots, w_L\}$. Our goal is to generate a symbolic solution program (expression) token sequence $\hat{Y} = \{\hat{y}_1, \dots, \hat{y}_T\}$ that can be executed to obtain the final numeric answer (for completion) or the correct option (for choice). Here L and T denote the lengths of the text and solution-program sequences, respectively.

Overview. Built on top of the PGPSNet-style [38] program generation backbone, we propose EAGLE, which introduces two key designs: (i) Euclid Field, a learned diagram-structure prior that answers "where to look", and (ii) SOAR, a three-stage reasoning pipeline that explicitly performs (Stage1) structure-slot parsing, (Stage2) clause grounding, and (Stage3) heterogeneous graph reasoning to answer "what to link" and "how to reason". The full pipeline flows from (I, W) to $\hat{Y}$, as detailed below. Figure 2 provides an overview of the full architecture and information flow.

3.2 Text Encoding and Clause Structuring

Story Connection ("read" to enable "link"). Humans typically parse the statement into role-specific pieces before pointing them to the diagram. We mimic this by structuring the text into clause-level units that serve as the anchors for later grounding.

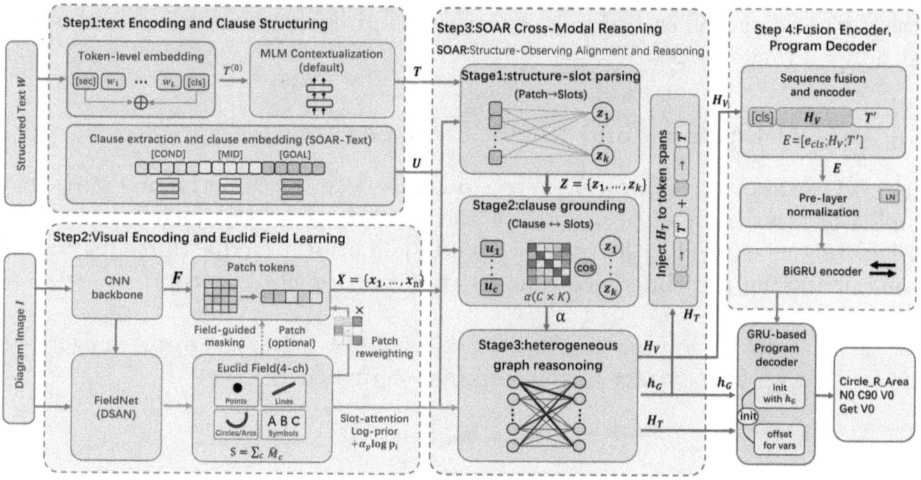

Fig. 2. Overall architecture of EAGLE's explicit read–look–link–reason workflow. The process consists of text encoding (Read), visual encoding via Euclid Field (Look), SOAR cross-modal reasoning (Link & Reason), and sequence fusion for program generation (Decode).

Token-Level Embedding. Each text position i is associated with its token identity w_i and two discrete tags: a section tag s_i (e.g., `[STRU]`/`[COND]`/`[PROB]`) and a class tag c_i (e.g., `[GEN]`/`[POINT]`/`[NUM]`/`[ARG]`/`[ANGID]`). We form a tag-aware token embedding by summing the corresponding embeddings:

$$\mathbf{t}_i^{(0)} = \mathbf{e}(w_i) + \mathbf{e}^{\text{sec}}(s_i) + \mathbf{e}^{\text{cls}}(c_i), \tag{1}$$

where $\mathbf{t}_i^{(0)} \in \mathbb{R}^d$ is the embedding of the i-th text token, d is the embedding dimension, and $\mathbf{e}(\cdot), \mathbf{e}^{\text{sec}}(\cdot), \mathbf{e}^{\text{cls}}(\cdot)$ are learnable embedding lookups.

MLM Contextualization (default). By default, we contextualize the token sequence using an MLM-pretrained Transformer encoder $f_{\text{MLM}}(\cdot)$ [25]:

$$\mathbf{T} = \{\mathbf{t}_1, \dots, \mathbf{t}_L\} = f_{\text{MLM}}\left(\mathbf{T}^{(0)}\right), \qquad \mathbf{T}^{(0)} = \{\mathbf{t}_1^{(0)}, \dots, \mathbf{t}_L^{(0)}\}. \tag{2}$$

The contextualized token features $\mathbf{t}_i \in \mathbb{R}^d$ are used for clause pooling and subsequent multimodal fusion. (When f_{MLM} is disabled in ablation, we directly set $\mathbf{t}_i = \mathbf{t}_i^{(0)}$.)

Clause Extraction and Clause Embedding (SOAR-Text). We split the token sequence W into C clauses using punctuation and keyword-based rules, producing clause spans $\{(b_c, e_c)\}_{c=1}^{C}$ with clause types (condition/middle/goal) [4]. The clause representation is obtained by average pooling over its span:

$$\mathbf{u}_c = \frac{1}{e_c - b_c} \sum_{i=b_c}^{e_c - 1} \mathbf{t}_i, \tag{3}$$

where $\mathbf{u}_c \in \mathbb{R}^d$ is the embedding of clause c, and $[b_c, e_c)$ is its left-closed, right-open index range in W.

3.3 Visual Encoding and Euclid Field Learning

Story Connection ("look"). When humans scan a geometry diagram, they implicitly prioritize sparse primitives and symbols. EAGLE makes this "where to look" signal explicit by learning a Euclid Field prior from diagram supervision, while predicting it at inference time (no test-time oracle structure).

Backbone Feature and Patch Tokens. Given the input diagram image $I \in \mathbb{R}^{3\times H_I \times W_I}$, we first extract a convolutional feature map:

$$\mathbf{F} = f_{\text{vis}}(I), \tag{4}$$

where $f_{\text{vis}}(\cdot)$ is a CNN backbone and $\mathbf{F} \in \mathbb{R}^{C_v \times H_p \times W_p}$ is the output feature map. We flatten $\mathbf{F}$ into $N = H_p W_p$ patch tokens, add a learnable 2D positional encoding, and project to the model dimension d, resulting in a patch sequence $\mathbf{X} = \{\mathbf{x}_1, \ldots, \mathbf{x}_N\}$ with $\mathbf{x}_i \in \mathbb{R}^d$.

Four-Channel Geometry Structure Supervision. For each annotated diagram, we rasterize a four-channel structure map $\mathbf{M} \in [0,1]^{4\times H_I \times W_I}$, where channels correspond to: (1) point blobs, (2) line strokes, (3) circle/arc outlines, and (4) symbol/text bounding boxes. After applying the same geometric augmentation as the diagram, we downsample $\mathbf{M}$ to the patch-grid resolution, yielding $\mathbf{M}^{\downarrow} \in [0,1]^{4\times H_p \times W_p}$ [37]. These maps supervise FieldNet during training and are not used as oracle inputs at inference time.

Euclid Field Prediction and Distribution Supervision. We attach a lightweight distribution-aware spatial attention network (DSAN), dubbed FieldNet, on top of $\mathbf{F}$ to predict a four-channel Euclid Field map:

$$\hat{\mathbf{M}} = \sigma\left(g_{\text{field}}(\mathbf{F})\right), \tag{5}$$

where $g_{\text{field}}(\cdot)$ denotes the FieldNet module, $\sigma(\cdot)$ is the sigmoid function, and $\hat{\mathbf{M}} \in [0,1]^{4\times H_p \times W_p}$ is the predicted Euclid Field distribution. We supervise it using a SmoothL1 distribution loss:

$$\mathcal{L}_{\text{dist}} = \text{SmoothL1}\left(\hat{\mathbf{M}}, \mathbf{M}^{\downarrow}\right). \tag{6}$$

Injecting the Euclid Field. Let $\mathbf{R} \in \mathbb{R}^{H_p \times W_p}$ be the summed field intensity $\mathbf{R} = \sum_{c=1}^{4} \hat{\mathbf{M}}_c$, and let $\mathbf{r} = \text{vec}(\mathbf{R}) \in \mathbb{R}^N$ be its vectorized form. We convert it into a patch prior distribution $\{p_i\}_{i=1}^N$ by clamping and normalization:

$$p_i = \frac{\max(r_i, \epsilon)}{\sum_{j=1}^N \max(r_j, \epsilon)}, \qquad \sum_{i=1}^N p_i = 1, \tag{7}$$

where ϵ is a small constant for numerical stability. We exploit $\{p_i\}$ in the following ways: (i) Patch reweighting (enhancing structure-related patches), (ii) Slot-attention log-prior (biasing structure-slot parsing, Eq. 8), and (iii) Field-guided patch masking (optional robustness regularization after warm-up). In our default training, we apply random patch masking on patch tokens and enable field-guided masking only when explicitly specified [32].

3.4 SOAR Cross-Modal Reasoning

Story Connection ("link" then "reason"). After "looking" at the diagram, humans point each clause to relevant parts and combine anchored constraints. SOAR operationalizes this as explicit (Stage2) grounding and (Stage3) graph reasoning on top of geometry-aware structure units (Stage1).

Stage1: structure-slot parsing (Patch → Slots). We parse patch tokens into K structure slots using a Slot-Attention style aggregator [15], producing slot embeddings $\mathbf{Z} = \{\mathbf{z}_1, \ldots, \mathbf{z}_K\}$ and slot-to-patch attention $\mathbf{A} \in \mathbb{R}^{K\times N}$. The attention weights are computed as:

$$a_{k,i} = \text{Softmax}_i\left(\frac{\mathbf{q}_k^\top \mathbf{k}_i}{\sqrt{d}} + \alpha_p \log p_i\right), \tag{8}$$

where $a_{k,i}$ is the attention of slot k over patch i, $\mathbf{q}_k = W_q\,\mathbf{z}_k$ and $\mathbf{k}_i = W_k\,\mathbf{x}_i$ are the query/key projections, α_p controls the strength of the Euclid Field log-prior, and p_i is the normalized patch prior derived from $\hat{\mathbf{M}}$ (as defined in Eq. 7). Slot embeddings $\mathbf{z}_k$ are iteratively updated via weighted aggregation and a GRU-FFN update (standard Slot Attention update).

Stage2: clause grounding (Clause ↔ Slots). We compute a clause–slot similarity matrix using cosine similarity:

$$S_{c,k} = \cos(W_u\mathbf{u}_c,\ W_z\mathbf{z}_k), \tag{9}$$

where $S_{c,k} \in [-1, 1]$ measures the compatibility between clause c and slot k, and W_u, W_z are learnable linear projections (followed by ℓ_2 normalization). We then obtain soft groundings from clause to slot:

$$\alpha_{c,k} = \text{Softmax}_k(S_{c,:})_k, \tag{10}$$

where $\alpha_{c,k}$ is the assignment probability. This explicit clause–slot matrix $\boldsymbol{\alpha}$ is the "link" interface used in Stage3.

Since explicit clause–structure grounding labels are not available, $\mathcal{L}_{\text{match}}$ is not direct grounding-correctness supervision but a self-bootstrapped regularizer for the clause–slot interface. We take $k_c^* = \arg\max_k S_{c,k}$ as the current pseudo-positive slot for clause c and minimize a cross-entropy matching loss over valid clauses:

$$\mathcal{L}_{\text{match}} = -\sum_{c\in\mathcal{V}} \log\left(\text{Softmax}\left(\frac{S_{c,:}}{\tau}\right)_{k_c^*}\right), \tag{11}$$

where $\mathcal{V}$ denotes the set of valid (non-padded) clauses and τ is a temperature hyperparameter. This term is used jointly with the program loss, Euclid prior, and balance regularizer to reduce arbitrary assignment collapse.

To prevent assignment collapse (e.g., many clauses selecting the same slot), we regularize the average slot usage. Let $\bar{\alpha}_k = \frac{1}{|\mathcal{V}|}\sum_{c\in\mathcal{V}} \alpha_{c,k}$, then:

$$\mathcal{L}_{\text{bal}} = \sum_{k=1}^{K} \bar{\alpha}_k \log(\bar{\alpha}_k K), \tag{12}$$

where $\mathcal{L}_{\text{bal}} = \text{KL}(\bar{\boldsymbol{\alpha}} \,\|\, \text{Uniform}(K))$ written in closed form.

Optional Alignment Regularization. To further stabilize "look" and "link", we optionally align the mean slot-attention distribution over patches with the Euclid-field prior:

$$\bar{a}_i = \frac{1}{K}\sum_{k=1}^{K} a_{k,i}, \qquad \mathcal{L}_{\text{align}} = \text{KL}(\bar{\mathbf{a}} \,\|\, \mathbf{p}), \tag{13}$$

where $\mathbf{p} = (p_1, \ldots, p_N)$ is the Euclid-derived patch prior and $\bar{\mathbf{a}} = (\bar{a}_1, \ldots, \bar{a}_N)$.

Stage3: Heterogeneous Graph Reasoning. We perform multi-layer reasoning on a bipartite graph with clause nodes and slot nodes [5]. Crucially, node features have explicit origins: we initialize clause-node features from clause embeddings and slot-node features from structure slots, i.e., $\mathbf{h}_c^{T,(0)} = \mathbf{u}_c$ and $\mathbf{h}_k^{V,(0)} = \mathbf{z}_k$. Each layer contains (i) intra-type self-attention and (ii) cross-modal message passing guided by $\boldsymbol{\alpha}$. The cross-modal messages are:

$$\mathbf{m}_c^T = \sum_{k=1}^{K} \tilde{\alpha}_{c,k}\, \mathbf{h}_k^V, \qquad \mathbf{m}_k^V = \sum_{c\in\mathcal{V}} \tilde{\beta}_{k,c}\, \mathbf{h}_c^T, \tag{14}$$

where $\mathbf{h}_c^T$ and $\mathbf{h}_k^V$ are the current clause/slot node features, $\tilde{\alpha}$ is the row-normalized grounding weights (equivalent to $\boldsymbol{\alpha}$ over valid clauses), and $\tilde{\beta}$ is the column-normalized reverse weights (each slot sums to 1 over valid clauses).

After stacking multiple layers, we obtain refined clause features $\mathbf{H}_T = \{\mathbf{h}_c^T\}$, refined slot features $\mathbf{H}_V = \{\mathbf{h}_k^V\}$, and a global graph feature $\mathbf{h}_G$ computed by pooling over valid clause nodes and all slot nodes [13]. The refined slot features replace $\mathbf{Z}$ for downstream fusion. To make the "reason" outcome available to token-level decoding, we inject each refined clause feature back to the tokens in its span:

$$\mathbf{t}_i' = \mathbf{t}_i + W_{\text{inj}}\mathbf{h}_c^T, \qquad \forall i \in [b_c, e_c), \tag{15}$$

where W_{inj} is a learnable projection. We denote the resulting enhanced token sequence as $\mathbf{T}' = \{\mathbf{t}_i'\}_{i=1}^{L}$ [23].

3.5 Fusion Encoder, Solution Program Decoder, and Objective

Sequence Fusion and Encoder. We build the final encoder input by concatenating a learnable [CLS] token, the refined structure slots, and the enhanced text token sequence:

$$\mathbf{E} = [\mathbf{e}_{\text{cls}}; \mathbf{H}_V; \mathbf{T}']. \tag{16}$$

We apply a pre-layer normalization and feed $\mathbf{E}$ to a sequence encoder (bidirectional GRU in our default setting) to obtain contextualized encoder outputs.

Solution-Program Decoding and Output. A GRU-based solution-program decoder generates the expression tokens auto-regressively. To keep variable references consistent, we shift variable positions by the offset introduced by [CLS] and slot tokens. When available, the global graph feature $\mathbf{h}_G$ is linearly projected to initialize the decoder hidden state; otherwise, we use the encoder hidden state. The decoder outputs the predicted solution-program sequence $\hat{Y}$.

Overall Objective. The total training loss is a weighted sum of the program generation loss and auxiliary losses from Euclid Field learning and Stage2 interface regularization:

$$\mathcal{L} = \mathcal{L}_{\text{prog}} + \lambda_{\text{dist}}\mathcal{L}_{\text{dist}} + \lambda_{\text{match}}\mathcal{L}_{\text{match}} + \lambda_{\text{bal}}\mathcal{L}_{\text{bal}} + \lambda_{\text{align}}\mathcal{L}_{\text{align}}, \tag{17}$$

where $\mathcal{L}_{\text{prog}}$ is the masked cross-entropy loss for solution-program tokens (padding steps ignored), $\mathcal{L}_{\text{dist}}$ is the FieldNet distribution supervision (Eq. 6), $\mathcal{L}_{\text{match}}$ and $\mathcal{L}_{\text{bal}}$ are defined in Eq. 11–12, and $\mathcal{L}_{\text{align}}$ is defined in Eq. 13. The scalars $\lambda_{\cdot}$ are hyperparameters.

4 Experiments

4.1 Experimental Setup

Datasets and Protocol. We evaluate EAGLE on two public benchmarks for plane-geometry problem solving: Geometry3K [16] and PGPS9K [38]. We follow the official splits and evaluation scripts adopted by PGPSNet [38] for fair comparison. Geometry3K contains 8,432 training samples and 589 test samples, and PGPS9K contains 8,021 training samples and 1,000 test samples. Both datasets provide ground-truth solution programs for supervised learning. When training the Euclid-Field module, we render four-channel structure maps from the provided diagram annotations; at inference time, the Euclid Field is predicted from the raw diagram image by FieldNet and no ground-truth structure is used. Thus, "non-oracle" refers to no oracle diagram structure at inference time. Importantly, these diagram annotations are publicly released as part of Geometry3K and PGPS9K and are equally available to all compared methods; we further report an ablation without $\mathcal{L}_{\text{dist}}$ (Table 3) to isolate the effect of this supervision.

Evaluation Metrics. We report three standard metrics. Completion evaluates whether the first *executable* solution program in the beam returns the correct final answer. Choice reports accuracy on multiple-choice questions; when no valid program can be executed, we follow the common protocol and use a random guess. Top-3 checks whether any correct solution program appears among the top three decoded candidates.

Implementation. Unless otherwise stated, each diagram is resized to 256×256. We use a ResNet10 visual backbone (stride 32), yielding an 8×8 patch grid ($N = 64$), and apply learnable 2D positional embeddings to patch tokens. SOAR uses Slot Attention with $K = 8$ slots and 3 iterative updates for Stage1, and a 2-layer heterogeneous graph reasoning module with 4 attention heads for Stage3. FieldNet predicts 4-channel Euclid Field maps (points/lines/circles-or-arcs/symbol regions), which guide patch reweighting and slot parsing (Eq. 8). During training, we use patch-level random masking with ratio 0.10; we additionally evaluate a field-guided masking variant (enabled after a warm-up stage) in the ablation study.

Training and Decoding. We optimize all models using AdamW (weight decay 0.01). The learning rate is initialized to 10^{-3} with a 40-epoch warm-up, and then decayed by a factor of 0.5 at epochs 160/280/360/440/500. We train for 540 epochs with batch size 90. The maximum input length is 400 and the maximum output length is 40. We decode with beam size 10. All experiments are conducted using PyTorch DDP on 2× NVIDIA Tesla V100 GPUs (16GB).

Baselines. We compare with representative symbolic/hybrid solvers, neural program solvers, and general-purpose multimodal LLMs following the standard protocol [35]. Our primary baseline is PGPSNet [38], a strong program-generation framework that encodes the diagram and the problem text and decodes a symbolic program. Compared with PGPSNet, EAGLE explicitly (i) emphasizes geometry-relevant regions via the learned Euclid Field, and (ii) deploys SOAR to decompose cross-modal reasoning into three stages (structure-slot parsing, clause grounding, and graph reasoning).

4.2 Main Results

Specialized Solvers. Table 1 compares EAGLE with symbolic, neural, and general multimodal baselines. EAGLE improves over PGPSNet and PGPSNet-v2-S on both benchmarks. On Geometry3K, it reaches 68.4/82.2/82.2, with a clear Choice gain over PGPSNet-v2-S (82.2 vs. 79.8). On PGPS9K, it reaches 65.5/74.2/83.1, consistently outperforming PGPSNet (62.7/70.4/79.5) and PGPSNet-v2-S (63.4/71.7/81.2). InterGPS (All GT) has slightly higher Geometry3K Completion (69.0 vs. 68.4), but it uses oracle intermediate representations, whereas EAGLE uses no test-time oracle and obtains much higher Choice accuracy (82.2 vs. 75.9). Overall, the largest gains appear on Choice/Top-3, indicating better executable-program ranking under beam search. This is

Table 1. Results on Geometry3K and PGPS9K (%). Dashes indicate that the metric is not applicable or not reported.

Method	Geometry3K			PGPS9K		
	Comp.	Choice	Top-3	Comp.	Choice	Top-3
interGPS (predict) [16]	44.6	56.9	–	–	–	–
InterGPS(DiagramGT) [16]	64.2	71.7	–	59.8	68.0	–
InterGPS (All GT) [16]	**69.0**	75.9	–	–	–	–
GeoDRL [20]	–	68.4	–	–	–	–
GPT-4V [1]	38.6	42.3	–	31.8	40.3	–
Qwen-VL [3]	22.1	26.7	–	20.1	23.2	–
LLaVA-v1.5 [14]	7.6	11.2	–	6.3	9.1	–
mPLUG-Owl2 [30]	12.1	17.4	–	10.1	13.1	–
NGS [7]	35.3	58.8	62.0	34.1	46.1	60.9
Geoformer [6]	36.8	59.3	62.5	35.6	47.3	62.3
PGPSNet [38]	65.0	77.9	80.7	62.7	70.4	79.5
PGPSNet-v2-S [36]	68.2	79.8	81.6	63.4	71.7	81.2
DenseGPS [17]	66.7	77.4	80.2	60.7	70.1	79.9
GeoX [28]	58.6	72.5	69.4	52.7	63.3	65.4
EAGLE	68.4	**82.2**	**82.2**	**65.5**	**74.2**	**83.1**

important because GPS evaluation rewards not only plausible answers but also valid symbolic programs ranked above competing beams. The consistent improvement on PGPS9K further suggests that the explicit look–link–reason design is not limited to a single benchmark split.

Metric Trade-off. EAGLE sharpens structure-aware evidence selection, which improves top-ranked reliability but may slightly reduce beam diversity; thus oracle-assisted or relaxed variants can be marginally better on isolated metrics. This trade-off is consistent with the role of the Euclid Field and SOAR: they bias decoding toward structurally supported hypotheses instead of allowing many loosely grounded candidates. As a result, Completion and Choice benefit more directly, while diversity-oriented Top-3 may not always increase monotonically.

General-Purpose Multimodal LLMs. General-purpose multimodal LLMs perform much worse because GPS requires small primitive/symbol perception and executable multi-step constraint reasoning, which generic multimodal pretraining does not directly provide. Tiny labels, arcs, intersections, and symbolic relations are easily missed without task-specific structure priors, while final answers must be verified through executable programs.

Table 2. Component ablations on Geometry3K and PGPS9K (%). S1/S2/S3 denote Stage1/2/3 in SOAR; Field denotes the Euclid Field. Best results in each column are in bold.

Setting	Geometry3K			PGPS9K		
	Comp.	Choice	Top-3	Comp.	Choice	Top-3
w/o S1	66.21	78.78	80.81	**65.80**	73.60	82.10
w/o S2	65.87	79.63	82.34	64.80	72.90	82.50
w/o S3	65.53	79.80	81.83	63.40	73.30	82.10
w/o Field	67.40	81.49	**83.53**	65.10	73.50	82.50
Full (EAGLE)	**68.42**	**82.17**	82.17	65.50	**74.20**	**83.10**

Table 3. Loss and masking ablations on Geometry3K (%). Best results in each column are in bold.

Setting	Completion	Choice	Top-3
w/o $\mathcal{L}_{\text{dist}}$	67.91	81.83	**83.70**
w/o $\mathcal{L}_{\text{bal}}$	66.04	79.46	82.00
w/o $\mathcal{L}_{\text{match}}$	67.06	80.14	82.00
w/o $\mathcal{L}_{\text{align}}$	67.23	80.31	82.86
w/ field-guided masking	65.53	79.29	81.66
Full (EAGLE)	**68.42**	**82.17**	82.17

4.3 Ablation Results

Table 2 shows that removing S1/S2/S3 generally reduces Completion and Choice, confirming the value of structure-slot parsing, clause grounding, and graph reasoning. Stage1 provides compact structure-aware visual units, Stage2 builds the explicit clause–slot interface, and Stage3 performs relation-centric propagation over the induced graph. S3 causes the largest Geometry3K Completion drop, while S1 causes the largest Choice drop, suggesting that graph propagation improves top-ranked correctness and slot unitization stabilizes candidate ranking. Removing the Euclid Field lowers Completion/Choice on both datasets, although Top-3 may increase on Geometry3K, reflecting a diversity–sharpness trade-off under beam search.

Table 3 shows that $\mathcal{L}_{\text{bal}}$ and $\mathcal{L}_{\text{match}}$ help stabilize the clause–slot interface: the former prevents degenerate slot usage, while the latter provides useful self-bootstrapped matching regularization. Field-guided masking performs worse than random masking, likely because sparse geometry evidence can be removed when high-field primitives or symbols are masked.

4.4 Proxy Grounding Diagnostics

Since manual clause–slot labels are unavailable, we report proxy diagnostics rather than direct grounding accuracy. Field Soft-F1 and Soft-IoU evaluate Euclid-Field prediction quality, and Clause Grounding Lift measures clause-conditioned grounding concentration over a baseline.

Table 4. Proxy grounding diagnostics on Geometry3K and PGPS9K.

Dataset	Field Soft-F1	Field Soft-IoU	Clause Grounding Lift
Geometry3K	0.6441	0.5298	2.3155
PGPS9K	0.6431	0.5297	2.2196

Table 4 provides additional evidence for the grounding mechanism. The similar Field Soft-F1 and Soft-IoU scores on both datasets indicate that FieldNet learns a consistent structure prior, and the Grounding Lift values above 2 suggest that clause-conditioned grounding is more concentrated than the baseline even without manual clause–slot labels. These proxy results do not replace direct human grounding annotations, but they support the claim that EAGLE produces a more observable and structured intermediate interface.

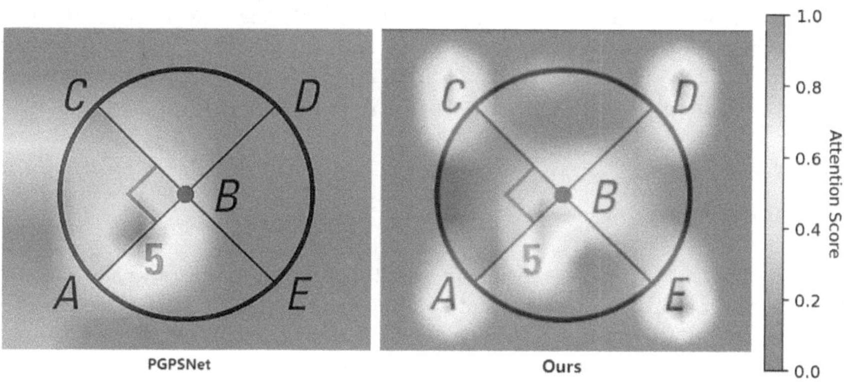

Fig. 3. Attention/region heatmaps for the same example comparing PGPSNet and EAGLE. Left: PGPSNet. Right: EAGLE. Colors indicate normalized attention scores in [0, 1].

4.5 Qualitative Results

Figure 3 shows that PGPSNet concentrates on a locally salient region, whereas EAGLE highlights task-relevant numbers together with critical symbols and arcs/segments. This indicates that EAGLE attends not only to the numeric cue but also to labeled points and circle structure, supporting the role of the Euclid Field and SOAR in promoting diagnosable grounding [34].

5 Conclusion

We propose EAGLE, a multimodal geometry solver that makes the look–link–reason process explicit. EAGLE learns a four-channel Euclid Field prior to highlight geometry-bearing regions, grounds clauses through an explicit clause–slot anchoring matrix, and performs heterogeneous graph reasoning before decoding executable solution programs. Experiments on Geometry3K and PGPS9K show consistent gains over strong program-generation baselines, while proxy diagnostics and qualitative results indicate more structure-aware and diagnosable grounding. Limitations include training-time reliance on public diagram annotations, rule-based clause segmentation, and a fixed slot budget. Future work will explore weaker-supervision structure priors, learned clause segmentation, adaptive slot allocation, and verification-aware decoding.

Acknowledgements. This work was funded by the National Natural Science Foundation of China (Grant No. 62366036), the Outstanding Youth Fund Project of Inner Mongolia Autonomous Region (Grant No. 2025JQ010), the Program for Young Talents of Science and Technology in Universities of Inner Mongolia Autonomous Region (Grant No. NJYT24033), the Major Science and Technology Projects of Inner Mongolia Autonomous Region (Grant No. 2025ZDSF0029), the Key R&D and Achievement Transformation Program of Inner Mongolia Autonomous Region (Grant Nos. 2025YFDZ0011, 2025YFDZ0026, 2025YFSH0021, 2025YFHH0073), and the Hohhot Science and Technology Project (Grant No. 2023-Zhan-Zhong-1).

References

1. Achiam, J., et al.: Gpt-4 technical report (2023). arXiv preprint arXiv:2303.08774
2. Anderson, P., et al.: Bottom-up and top-down attention for image captioning and visual question answering. In: Proceedings of the IEEE Conference on Computer Vision and Pattern Recognition, pp. 6077–6086 (2018)
3. Bai, J., Bai, S., Yang, S., Wang, S., Tan, S., Wang, P., Lin, J., Zhou, C., Zhou, J.: Qwen-VL: A versatile vision-language model for understanding, localization. Text Reading, and Beyond **2**(1), 1 (2023)
4. Cao, J., Xiao, J.: An augmented benchmark dataset for geometric question answering through dual parallel text encoding. In: Proceedings of the 29th International Conference on Computational Linguistics, pp. 1511–1520 (2022)
5. Chen, D., O'Bray, L., Borgwardt, K.: Structure-aware transformer for graph representation learning. In: International Conference on Machine Learning, pp. 3469–3489. PMLR (2022)
6. Chen, J., et al.: Unigeo: Unifying geometry logical reasoning via reformulating mathematical expression. In: Proceedings of the 2022 Conference on Empirical Methods in Natural Language Processing, pp. 3313–3323 (2022)

7. Chen, J., et al.: Geoqa: a geometric question answering benchmark towards multimodal numerical reasoning. In: Findings of the Association for Computational Linguistics: ACL-IJCNLP 2021, pp. 513–523 (2021)
8. Chen, Z., et al.: Scans: An efficient geometric problem solver with content-aware attention and adaptive fusion. In: International Conference on Document Analysis and Recognition, pp. 327–343. Springer (2025)
9. Da, C., Luo, C., Zheng, Q., Yao, C.: Vision grid transformer for document layout analysis. In: Proceedings of the IEEE/CVF International Conference on Computer Vision, pp. 19462–19472. (2023)
10. Kim, W., Son, B., Kim, I.: Vilt: Vision-and-language transformer without convolution or region supervision. In: International Conference on Machine Learning, pp. 5583–5594. PMLR (2021)
11. Lee, K., Joshi, M., Turc, I.R., Hu, H., Liu, F., Eisenschlos, J.M., Khandelwal, U., Shaw, P., Chang, M.W., Toutanova, K.: Pix2struct: Screenshot parsing as pretraining for visual language understanding. In: International Conference on Machine Learning, pp. 18893–18912. PMLR (2023)
12. Liu, F., et al.: Deplot: one-shot visual language reasoning by plot-to-table translation. In: Findings of the Association for Computational Linguistics: ACL 2023, pp. 10381–10399. (2023)
13. Liu, H., Li, X., Liu, B., Jiang, D., Liu, Y., Ren, B.: Neural collaborative graph machines for table structure recognition. In: Proceedings of the IEEE/CVF Conference on Computer Vision and Pattern Recognition, pp. 4533–4542 (2022)
14. Liu, H., Li, C., Wu, Q., Lee, Y.J.: Visual instruction tuning. Adv. Neural. Inf. Process. Syst. **36**, 34892–34916 (2023)
15. Locatello, F., et al.: Object-centric learning with slot attention. Adv. Neural. Inf. Process. Syst. **33**, 11525–11538 (2020)
16. Lu, P., et al.: Inter-gps: Interpretable geometry problem solving with formal language and symbolic reasoning. In: Proceedings of the 59th Annual Meeting of the Association for Computational Linguistics and the 11th International Joint Conference on Natural Language Processing (Volume 1: Long Papers), pp. 6774–6786 (2021)
17. Ma, B., Jian, P., Pan, C., Wang, Y., Ma, W.: A geometric neural solving method based on a diagram text information fusion analysis. Sci. Rep. **14**(1), 31906 (2024)
18. Mao, J., Gan, C., Kohli, P., Tenenbaum, J.B., Wu, J.: The neuro-symbolic concept learner: Interpreting scenes, words, and sentences from natural supervision (2019). arXiv preprint arXiv:1904.12584
19. Ning, M., Wang, Q.F., Huang, K., Huang, X.: A symbolic characters aware model for solving geometry problems. In: Proceedings of the 31st ACM International Conference on Multimedia, pp. 7767–7775 (2023)
20. Peng, S., Fu, D., Liang, Y., Gao, L., Tang, Z.: Geodrl: a self-learning framework for geometry problem solving using reinforcement learning in deductive reasoning. In: Findings of the Association for Computational Linguistics: ACL 2023, pp. 13468–13480 (2023)
21. Perez, E., Strub, F., De Vries, H., Dumoulin, V., Courville, A.: Film: Visual reasoning with a general conditioning layer. In: Proceedings of the AAAI Conference on Artificial Intelligence, vol. 32 (2018)
22. Sachan, M., Dubey, K., Xing, E.: From textbooks to knowledge: a case study in harvesting axiomatic knowledge from textbooks to solve geometry problems. In: Proceedings of the 2017 Conference on Empirical Methods in Natural Language Processing, pp. 773–784 (2017)

23. Santoro, A., et al.: A simple neural network module for relational reasoning. In: Advances in Neural Information Processing Systems, vol. 30 (2017)
24. Seo, M., Hajishirzi, H., Farhadi, A., Etzioni, O., Malcolm, C.: Solving geometry problems: Combining text and diagram interpretation. In: Proceedings of the 2015 Conference on Empirical Methods in Natural Language Processing, pp. 1466–1476 (2015)
25. Su, J., Ahmed, M., Lu, Y., Pan, S., Bo, W., Liu, Y.: Roformer: enhanced transformer with rotary position embedding. Neurocomputing **568**, 127063 (2024)
26. Tang, Z., et al.: Unifying vision, text, and layout for universal document processing. In: Proceedings of the IEEE/CVF Conference on Computer Vision and Pattern Recognition, pp. 19254–19264 (2023)
27. Wu, W., et al.: E-GPS: Explainable geometry problem solving via top-down solver and bottom-up generator. In: Proceedings of the IEEE/CVF Conference on Computer Vision and Pattern Recognition, pp. 13828–13837 (2024)
28. Xia, R., et al.: Geox: Geometric problem solving through unified formalized vision-language pre-training, (2024). arXiv preprint arXiv:2412.11863
29. Xiao, T., et al.: Learning to solve geometry problems via simulating human dual-reasoning process, (2024). arXiv preprint arXiv:2405.06232
30. Ye, Q., et al.: mPLUG-Owl2: revolutionizing multi-modal large language model with modality collaboration. In: Proceedings of the IEEE/CVF Conference on Computer Vision and Pattern Recognition, pp. 13040–13051 (2024)
31. Yi, K., Wu, J., Gan, C., Torralba, A., Kohli, P., Tenenbaum, J.: Neural-symbolic vqa: Disentangling reasoning from vision and language understanding. Adv. Neural. Inf. Process. Syst. **31** (2018)
32. Yin, B.W., Cao, J.L., Cheng, M.M., Hou, Q.: Dformerv2: Geometry self-attention for RGBD semantic segmentation. In: Proceedings of the IEEE/CVF Conference on Computer Vision and Pattern Recognition, pp. 19345–19355 (2025)
33. Yu, Z., Yu, J., Cui, Y., Tao, D., Tian, Q.: Deep modular co-attention networks for visual question answering. In: Proceedings of the IEEE/CVF Conference on Computer Vision and Pattern Recognition, pp. 6281–6290 (2019)
34. Zhang, H., Su, X., Zhou, X., Gao, G.: Ssan: A symbol spatial-aware network for handwritten mathematical expression recognition. In: Proceedings of the AAAI Conference on Artificial Intelligence, vol. 39, pp. 22398–22406 (2025)
35. Zhang, J., Moshfeghi, Y.: Gold: Geometry problem solver with natural language description. In: Findings of the Association for Computational Linguistics: NAACL 2024, pp. 263–278 (2024)
36. Zhang, M.L., Li, Z.Z., Yin, F., Lin, L., Liu, C.L.: Fuse, reason and verify: Geometry problem solving with parsed clauses from diagram. arXiv preprint arXiv:2407.07327 (2024)
37. Zhang, M.L., Yin, F., Hao, Y.H., Liu, C.L.: Plane geometry diagram parsing, (2022). arXiv preprint arXiv:2205.09363
38. Zhang, M.L., Yin, F., Liu, C.L.: A multi-modal neural geometric solver with textual clauses parsed from diagram (2023). arXiv preprint arXiv:2302.11097
39. Zhao, J., Zhang, T., Sun, J., Tian, M., Huang, H.: Pi-GPS: enhancing geometry problem solving by unleashing the power of diagrammatic information. In: Proceedings of the IEEE/CVF International Conference on Computer Vision, pp. 1526–1536 (2025)

Stroke-Level Connectivity Verification: Grounding Vision-Language Models Against Topology Hallucination in Diagram Understanding

Abdullah Ibne Hanif Arean[1,2], Niamul Hassan Samin[1], Md Arifur Rahman[1,3], Renu Akter Suity[4], Juena Ahmed Noshin[5], and Md Ashikur Rahman[1(✉)]

[1] The KOW Company, Dhaka, Bangladesh
{arean,samin,arifur.rahman,ashikur}@thekowcompany.com
[2] University of Dhaka, Dhaka, Bangladesh
[3] Technische Universität Dortmund, Dortmund, Germany
[4] Bangladesh University of Business & Technology, Dhaka, Bangladesh
[5] American International University-Bangladesh, Dhaka, Bangladesh
juena@aiub.edu

Abstract. When a vision-language model declares that two components in a technical diagram are connected, what constitutes proof? Current models infer connectivity from semantic plausibility and spatial proximity-cues that yield fluent but structurally unfaithful predictions, a failure mode we term *connectivity hallucination.* We introduce **Stroke-Level Connectivity Verification (SLCV)**, a formal verification framework that treats each predicted edge as a proof obligation: an edge is accepted if and only if a continuous path through the diagram's stroke skeleton connects the corresponding terminals without traversing any intervening symbol interior. Under topology-preserving skeletonization and accurate symbol localization, this conservative criterion provides a verifiable connectivity criterion: every accepted edge is grounded in the physical stroke topology of the original document. We embed SLCV within **PaRCO** (Parallel Reconstruction and Constraint-Oriented reasoning), a three-stage architecture that decouples structural reconstruction, semantic interpretation, and constraint reconciliation into independently auditable layers. On *LogicBench-1K*, a new benchmark of 1,000 annotated digital logic circuit diagrams with topology-level ground truth, PaRCO achieves a connectivity F1 of 0.98 under full supervision. Applied as a training-free verification layer to off-the-shelf VLMs, SLCV consistently improves connectivity F1 by 8–10 percentage points across all evaluated models (Claude 3.5 Sonnet 0.81 → 0.89; GPT-4o 0.78 → 0.88; Gemini 1.5 Pro 0.76 → 0.86) without any fine-tuning; a gain that replicates on CGHD handwritten circuits (+7 pp). The core components of SLCV (skeleton extraction, constrained path search, and provenance tracking) are domain-general, making pixel-grounded verification a reusable validation primitive for multimodal document understanding.

G. A. Fink et al. (Eds.): ICDAR 2026, LNCS 16972, pp. 323–340, 2027.
https://doi.org/10.1007/978-3-032-36023-6_19

Keywords: Diagram topology verification · Connectivity hallucination · Hybrid vision-language systems · Document structure analysis · Formal verification for Document AI

1 Introduction

Modern vision-language models (VLMs) produce fluent descriptions of technical diagrams [18,22], yet they routinely misrepresent the most safety-critical property of such diagrams: *connectivity.* A VLM may assert that two logic gates are wired together because they are semantically compatible or spatially proximate, even when no physical wire exists. In circuit applications, a single spurious edge corrupts the netlist and invalidates every downstream analysis-from functional simulation to design-rule checking. This paper addresses one question: *Can we verify VLM-suggested connections by demanding pixel-level evidence, thereby eliminating hallucinated edges while preserving the model's semantic understanding?*

Connectivity Hallucination. We define *connectivity hallucination* as the prediction of an edge between two diagram components that is *not supported by any continuous stroke path* in the underlying image. Unlike semantic or factual hallucinations studied in natural-language generation, connectivity hallucination is a *verifiable* error: a correct connection must correspond to an unbroken sequence of foreground pixels linking the terminals of both components. This verifiability is what makes structural grounding both possible and practical.

Key Insight. If connectivity is a structural fact encoded in strokes, then every predicted edge is a proposition that can be *proved or refuted* by examining the diagram's pixel topology. Stroke-Level Connectivity Verification (SLCV) operationalises this principle: it extracts a topology-preserving skeleton from the binary stroke map and accepts an edge only when a breadth-first search (BFS) finds a connected path between the corresponding terminal regions-without crossing any intervening symbol interior. This conservative criterion yields 99.2 % precision with near-zero false positives, at the cost of modest recall loss from broken strokes. As a concrete illustration (Fig. 3, top row): two spatially adjacent gates lead the VLM to predict a connection between them. SLCV extracts the stroke skeleton, runs BFS from one terminal region toward the other, finds no continuous path through the pixel topology, and discards the predicted edge, replacing a proximity-driven hallucination with a verified structural fact, without any retraining.

Contributions. Our work offers four contributions:

- **A formal verification framework for diagram connectivity.** SLCV reduces structural claims to path-existence proofs in the stroke topology, providing a conservative correctness guarantee in which every accepted edge is anchored to pixel-level evidence. The criterion is binary, auditable, and model-agnostic.

- **A separation-of-concerns architecture for diagram understanding.** PaRCO decouples perception, structural verification, and semantic interpretation into independently auditable stages. It supports both fully supervised deployment (F1 = 0.98 on LogicBench-1K) and zero-shot enhancement of foundation models (+8–10 pp F1 on LogicBench-1K, +7 pp on CGHD), giving practitioners a choice between specialisation cost and accuracy.
- **A topology-focused diagram benchmark.** LogicBench-1K provides 1,000 digital logic circuit diagrams with ground-truth graph annotations and controlled degradation variants (digital-native and scanned-legacy splits), for systematic evaluation of topological fidelity.
- **Empirical evidence for verification as a reusable primitive.** SLCV delivers consistent +8–10 pp gains on LogicBench-1K and +7 pp on CGHD across all evaluated architectures (Claude 3.5 Sonnet, GPT-4o, Gemini 1.5 Pro) without fine-tuning, and structural verification functions as a complementary validation layer for document AI.

2 Related Work

Graphic Recognition and Structural Analysis. Early graphic recognition systems identified basic shapes and analysed simple spatial relationships such as adjacency and alignment [1,2]. Modern pipelines employ deep-learning-based symbol detection followed by structural construction using geometric constraints [3,4]. This separation of detection from structural reasoning has proven effective: detection handles visual variability, while geometric reasoning enforces constraints such as junction alignment and terminal contact.

Circuit Diagram Parsing. Handwritten circuit diagram analysis is anchored to the CGHD benchmark [5,6], which provides images and ground-truth graph labels. Irregular strokes, similar-looking symbols, and ambiguous wire crossings make both learned and rule-based systems error-prone. More recent datasets (Digitize-HCD and JUHCCR-v1 [7,8]) increase diversity, but the core challenge persists: even when symbol recognition is strong, connectivity prediction remains unreliable.

End-to-end systems typically follow a three-stage decomposition: (i) component detection with Faster R-CNN or YOLO [15,16]; (ii) wire extraction via U-Net segmentation or morphological thinning; and (iii) connectivity inference through junction detection and endpoint association. Bayer et al. [10,11] demonstrate a modular approach on CGHD that combines instance segmentation with rule-based graph building, achieving strong results through hybrid integration. Recent deep-learning-based approaches, Netlistify [12] and SINA [13], attempt end-to-end netlist generation but still struggle with connectivity precision on noisy inputs. Across all these systems, purely geometric methods fail on broken strokes, while purely learned methods hallucinate plausible-but-incorrect edges [9,11].

Vision-Language Models for Document AI. Large multimodal models have advanced document understanding by integrating text, layout, and image modalities. LayoutLMv3 [21] fuses these three signals; Donut [22] eliminates OCR by processing images directly; Pix2Struct [23] learns from rendered screenshots. Benchmarks such as DocVQA [17], AI2D [19], IconQA [20], and MMMU [18] show that models excel at semantic reasoning but remain imprecise at pixel-level structural verification-the exact gap our work targets.

Verification-Based Approaches. Despite advances in both circuit parsing and multimodal AI, no prior work has proposed a principled verification layer that provides verifiable connectivity criteria for the structural predictions of VLMs on diagram connectivity. Existing hybrid systems combine heuristics with learned features but lack a formal criterion that separates verified from unverified edges. SLCV fills this gap by casting connectivity as a proof obligation grounded in the stroke topology. Crucially, SLCV is *complementary* to prior rule-based, graph-based, and structure-constrained parsers [10–12]: it can wrap any predictor as a post-hoc filter rather than replacing existing parsing pipelines.

3 Methodology

PaRCO follows a three-stage design (Fig. 1): *Stage A* reconstructs a verified structural graph from pixel evidence; *Stage B* builds semantic meaning on top of that graph; and *Stage C* reconciles the two under hard constraints. Digital logic circuits serve as the evaluation domain because their fixed symbol vocabulary and deterministic connection rules provide an objective ground truth.

3.1 Stroke-Level Connectivity Verification (SLCV)

Core Principle. SLCV treats connectivity as a *proof obligation*: a connection between components u and v is accepted only when a continuous path γ within the diagram's stroke skeleton links the terminals of u and v while avoiding the interior of every other component. If no such path can be traced in the actual pixels, no connection is admitted.

Definitions. Let $\mathcal{S} = (N_s, E_s)$ be the stroke skeleton graph extracted from image I via topology-preserving thinning [14], where N_s are skeleton pixels and E_s are 8-connectivity adjacencies. Let $V = \{v_1, \ldots, v_n\}$ denote the detected symbols with terminal regions $\mathcal{T}(v_i) = \{t_i^1, \ldots, t_i^{k_i}\}$. Let $\mathcal{M}_k \subseteq N_s$ be the interior mask of symbol v_k.

Connectivity Verification Criterion. An edge (v_i, v_j) is *verified*, written $(v_i, v_j) \in E_{\text{verified}}$, if:

$$\exists \gamma \subseteq \mathcal{S},\ \exists t_i \in \mathcal{T}(v_i),\ \exists t_j \in \mathcal{T}(v_j) :\ \text{path}(\gamma, t_i, t_j) \ \wedge\ \gamma \cap \Big(\bigcup_{k \neq i,j} \mathcal{M}_k \Big) = \emptyset, \quad (1)$$

where $\text{path}(\gamma, t_i, t_j)$ denotes a connected walk in $\mathcal{S}$ from terminal region t_i to t_j.

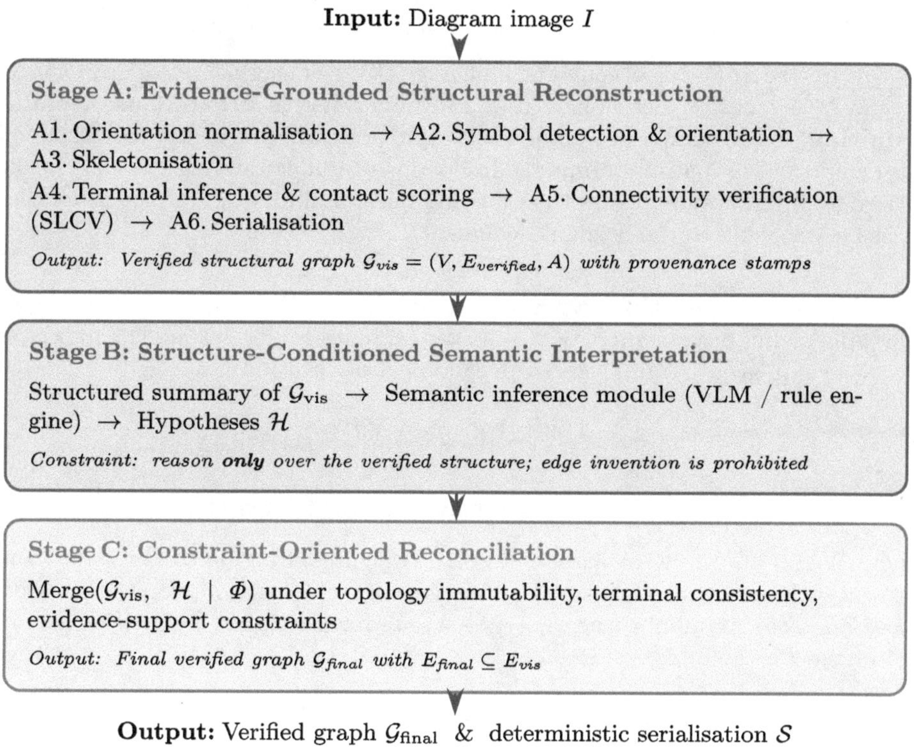

Fig. 1. The PaRCO framework. Three independently auditable stages: **Stage A** produces a conservative, high-precision structural graph via pixel-level verification (SLCV); **Stage B** interprets semantics conditioned on the verified structure; **Stage C** reconciles both under hard constraints, guaranteeing $E_{\text{final}} \subseteq E_{\text{vis}}$.

Correctness Property. The criterion is *conservative*:

$$\forall (u, v) \in E_{\text{verified}} \implies \exists \text{ physical stroke connecting } u \text{ and } v \text{ in } I.$$

Under topology-preserving skeletonisation [14], connectivity in the skeleton implies connectivity in the original image. SLCV consequently achieves 99.2 % precision (near-zero false edges), accepting a controlled recall loss from broken strokes (35 % of remaining errors) to avoid structurally invalid connections.

Semantic Immutability Constraint. Let $\mathcal{G}_{\text{vis}} = (V, E_{\text{ver}}, A)$ be the verified structure and $\mathcal{H}$ the semantic hypotheses. The final output satisfies:

$$E_{\text{final}} \subseteq E_{\text{verified}} \quad \text{(no hallucinated edges).} \tag{2}$$

This constraint prevents any probabilistic module-including VLMs-from introducing topologically unsupported connections.

3.2 Stage A: Evidence-Grounded Structural Reconstruction

The reconstruction stage converts image I into a conservative, high-precision graph. A key design decision is that **connectivity is treated as a claim requiring evidence**: edges enter the graph only when a contiguous skeleton path connects compatible terminals. In document understanding systems, incorrect edges are more damaging than missing ones, and all downstream semantics must be traceable to the original document.

A1. Global Orientation Normalisation. Scanned or embedded diagrams may appear under arbitrary rotations. We evaluate four right-angle candidates $\{I_r\}_{r\in\{0°,90°,180°,270°\}}$ and select the canonical orientation:

$$S(r) = \lambda_1 \sum_{g\in G_r} \mathrm{conf}(g) + \lambda_2 \,\mathrm{WA}(I_r) + \lambda_3 \,\mathrm{CP}(I_r), \tag{3}$$

$$r^* = \arg\max_r S(r), \tag{4}$$

where $\mathrm{WA}(I_r)$ is the wire-alignment score (weighted fraction of horizontal and vertical skeleton segments) and $\mathrm{CP}(I_r)$ is the corner-projection score (grid coherence of detected symbol corners); $\lambda_1, \lambda_2, \lambda_3$ are fixed weights. This yields canonical image I^*.

A2. Symbol Detection. An orientation-aware detector outputs (b_i, t_i, θ_i) per instance on I^*, where orientation is not merely a label but a *constraint variable* that determines terminal geometry and restricts valid edge directions. After confidence filtering and NMS: $G = \{(b_i, t_i, \theta_i)\}_{i=1}^n$.

A3. Stroke Substrate Extraction and Skeletonisation. A binary wiring mask W is obtained via foreground extraction. Topology-preserving thinning reduces W to a near-single-pixel skeleton that retains all connectivity, forming the substrate on which graph search verifies connections without crossing symbol boundaries.

A4. Terminal Inference and Contact Scoring. Rather than rigid geometric rules alone, we compute a contact score by measuring skeleton intersections within narrow sampling bands near each candidate terminal plane. This yields confident contact sets C_i comprising both unambiguous and marginal connections; the scores serve as provenance attributes for auditing and constraint reconciliation.

A5. Evidence-Verified Connectivity. For each symbol pair (v_i, v_j), connectivity is verified via constrained BFS on $\mathcal{S}$ (Fig. 2). The search starts from every skeleton pixel inside terminal region $t_i \in \mathcal{T}(v_i)$, expands through 8-connected neighbours, and succeeds if any pixel in $t_j \in \mathcal{T}(v_j)$ is reached-provided the path never enters the interior mask of any other symbol.

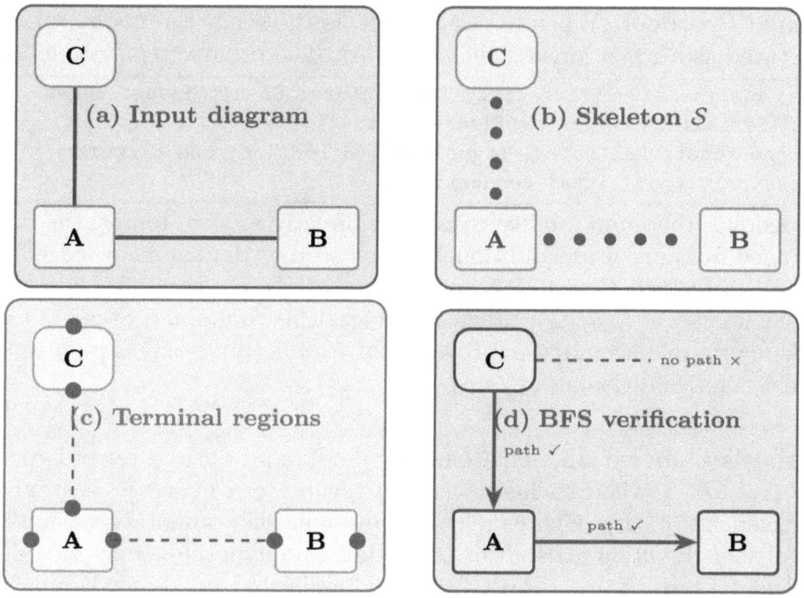

Fig. 2. SLCV verification illustrated. (a) Input with three components (A and B connected by wire; C connected to A). **(b)** Skeleton $\mathcal{S}$ via Zhang-Suen thinning reveals pixel-level paths. **(c)** Terminal regions (red dots) identified for each component. **(d)** BFS on $\mathcal{S}$: paths A→B and C→A are verified (✓); a hypothetical edge C→B is rejected (×) because no skeleton path exists.

A6. Provenance Stamps and Stable Serialisation. Every node and edge is tagged with provenance metadata (detection confidence, contact score, path length, skeleton coverage). The graph $\mathcal{G}_{\text{vis}}$ is serialised into a deterministic, netlist-style canonical form by ordering nodes according to topological depth and sorting within each layer by normalised position. This serialisation is not specific to any LLM; it serves as a *document representation model* that downstream modules-whether neural or rule-based-can consume.

3.3 Stage B: Structure-Conditioned Semantic Interpretation

Stage B builds semantic meaning on top of $\mathcal{G}_{\text{vis}}$, treating the verified structure as *immutable ground truth.* By supplying the semantic module with a structured summary rather than raw pixels, PaRCO prevents unconstrained completions that could re-introduce hallucinated edges.

Structured Input Representation. The verified graph is serialised into a compact summary $\mathcal{S}_{\text{in}}$: component inventory (id, type, bbox, orientation, confidence); verified adjacency list with provenance scores; layout metadata; and unresolved marginal contacts ($0.3 < \text{score} < 0.7$) flagged for disambiguation.

Inference Protocol. When a VLM serves as the semantic module, the query is structured as a *constrained completion task*. The prompt template follows:

```
System: Analyse a verified circuit structure; do not invent edges.
Input: [Serialised S_in + verified edge list]
Determine functional roles, signal-flow direction, and circuit
patterns only from listed connections.
```

By embedding the immutability constraint directly in the prompt, the model is discouraged from inventing additional edges based on domain knowledge. Empirically, unconstrained Claude 3.5 Sonnet proposes on average 3.7 unsupported edges per image ($\notin E_{\text{verified}}$); Stage B reduces this to 0.4 and Stage C to near-zero. Semantic-role accuracy improves from 79.3 % (pixel-only input) to 94.1 % with structure conditioning (+14.8 pp).

Hypothesis Generation. The semantic module produces a set of hypotheses $\mathcal{H} = \{h_1, \ldots, h_m\}$, where each $h_k = (\textit{claim}, \ \textit{evidence}, \ \textit{confidence})$ is grounded in specific elements of $\mathcal{G}_{\text{vis}}$. Claims may include signal-flow annotations, sub-circuit labels, or logic-function attributions. Crucially, no hypothesis may introduce an edge absent from E_{verified}. PaRCO is not predicated on heavy language-only reasoning; the structural graph itself is the primary output-something that can be verified and directly consumed by document analysis systems.

3.4 Stage C: Constraint-Oriented Reconciliation

The final stage merges $\mathcal{G}_{\text{vis}}$ and $\mathcal{H}$ under a set of hard constraints Φ:

- **Topology immutability:** the semantic module may not add or remove edges; $E_{\text{final}} \subseteq E_{\text{vis}}$.
- **Terminal consistency:** inferred edge directions must agree with terminal roles established in Stage A.
- **Evidence support:** every semantic claim must map to measurable properties of $\mathcal{G}_{\text{vis}}$-motif matches, fan-in/fan-out patterns, or provenance scores.

Unsupported hypotheses are discarded or marked as uncertain. The final output is:

$$\mathcal{G}_{\text{final}} = \text{Merge}\big(\mathcal{G}_{\text{vis}}, \ \mathcal{H} \mid \Phi\big), \qquad E_{\text{final}} \subseteq E_{\text{vis}}. \tag{5}$$

Precision-Recall Trade-off. Precision remains extremely high and improves with skeleton quality (99.6 % on clean data, 99.2 % on noisy data). Recall is more sensitive to stroke discontinuities (96.8 % on noisy data), with ~35 % of failures arising from broken strokes:

$$\text{Precision}(E_{\text{verified}}) \to 1 \quad \text{as skeleton fidelity} \to 1, \tag{6}$$

$$\text{Recall}(E_{\text{verified}}) \le 1 - p_{\text{break}}, \tag{7}$$

where p_{break} is the probability that noise or artefacts break a stroke. This trade-off is deliberate: in safety-critical settings, a missing connection is less harmful than a fabricated one.

4 Experiments and Results

Experimental Design. We evaluate on LogicBench-1K (digital logic circuits) and CGHD (handwritten circuits). *Important caveat on comparison fairness:* PaRCO is a supervised, domain-specific system, whereas the VLM baselines are used zero-shot. This is an asymmetric comparison by design-it reflects the real-world deployment question: *when is it worth building a specialised system versus prompting a general-purpose model?* We include a fine-tuned VLM baseline (LLaVA-1.5 with LoRA) to partially address this asymmetry.

We benchmark across two paradigms:

1. **Generalist probabilistic baselines:** zero-shot VLMs (Tiers I-III).
2. **Specialised deterministic verification:** PaRCO with domain-specific detection and SLCV.

4.1 Experimental Setup

Dataset: LogicBench-1K. All evaluations use *LogicBench-1K*, a curated corpus of 1,000 digital logic circuit schematics explicitly constructed to assess *topological fidelity.* Source schematics from educational materials were annotated by two experts (inter-annotator $\kappa = 0.91$ for gates, 0.88 for connectivity). The dataset is partitioned into:

- **Digital-native ($\mathcal{D}_{\textbf{clean}}$):** vector-derived raster images serving as a controlled reference.
- **Scanned-legacy ($\mathcal{D}_{\textbf{noisy}}$):** historical schematics with scanning artefacts, affine skew ($\pm 15°$), and salt-and-pepper noise.

Benchmark Tasks. LogicBench-1K defines three tasks: (i) *Symbol Detection* (localisation and classification), (ii) *Graph Recovery* (directed connectivity reconstruction), and (iii) *Functional Equivalence* (logical correctness verification). This paper reports Symbol Detection and Graph Recovery; functional equivalence is reserved for future work.

Evaluation Metrics.

Gate Accuracy. Symbol-level classification accuracy over correctly localised components (IoU ≥ 0.5).

Netlist Edit Distance (NED). Minimum-cost graph edit distance between predicted and ground-truth graphs:

$$\text{NED}(\mathcal{G}_{\text{pred}}, \mathcal{G}_{\text{gt}}) = \min_{\gamma \in \Gamma} \sum_{op \in \gamma} c(op), \tag{8}$$

with edit costs $c(\text{node\_ins/del}) = 1$, $c(\text{edge\_ins/del}) = 1$, $c(\text{node\_subst}) = 0.5$. Lower NED indicates better topological fidelity.

Connectivity F1. Edge-level precision, recall, and F1 over directed edges:

$$\mathrm{F1} = \frac{2 \cdot |E_p \cap E_g|}{|E_p| + |E_g|}. \tag{9}$$

Edge matching requires both endpoints to correspond to the same symbol instances via bipartite bounding-box matching, with correct direction.

Baselines. *Tier I* (proprietary): Gemini 1.5 Pro, GPT-4o, Claude 3.5 Sonnet. *Tier II* (open): LLaMA 3.2 Vision (11 B), Qwen3-VL (8 B). *Tier III* (lightweight): Moondream2 (1.6 B), LLaVA 7B. *Fine-tuned*: LLaVA-1.5 (7 B) adapted on 700 LogicBench-1K samples via LoRA. *PaRCO (Std)* uses the full pipeline with standard CPU inference; *PaRCO (Opt)* replaces the primary detector with a lightweight variant (25 M parameters) for maximum throughput at a small accuracy cost.

4.2 Quantitative Results

Table 1 presents the comprehensive results. Figure 3 shows representative examples of VLM hallucination versus SLCV-verified topology.

Table 1. Performance on LogicBench-1K. **NED** = Netlist Edit Distance (lower is better). All VLM experiments on NVIDIA A100 40 GB; PaRCO on Intel Xeon Gold 6248R CPU. PaRCO is trained on LogicBench-1K, whereas most VLM baselines are evaluated zero-shot; results should be interpreted accordingly.

Model	Type	Params	Avg Time (s)	Suc. Rate ↑	Gate Acc. ↑	NED ↓	Conn. F1 ↑
Zero-Shot VLM Baselines							
Moondream2	Open	1.6 B	**0.79**	66.0%	0.0%	18.4	0.05
LLaVA 7B	Open	7 B	2.77	50.0%	64.1%	8.7	0.22
Qwen3-VL	Open	8 B	89.06	40.0%	82.3%	5.1	0.45
LLaMA 3.2 Vision	Open	11 B	11.08	100.0%	75.0%	3.2	0.58
Gemini 1.5 Pro	Closed	–	4.20	98.5%	95.8%	0.8	0.76
GPT-4o	Closed	–	8.50	100.0%	94.2%	1.2	0.78
Claude 3.5 Sonnet	Closed	–	12.10	100.0%	96.5%	0.9	0.81
Claude 3.5 Sonnet (text)‡	Closed	–	12.10	100.0%	96.1%	–	0.84
Fine-Tuned VLM Baseline (Supervised)							
LLaVA-1.5 (FT)	Open	7 B	3.20	98.0%	92.3%	1.5	0.87
Hybrid: VLM + SLCV Verification (Zero-Shot VLM)							
Gemini 1.5 Pro + SLCV	Hybrid	–	4.40	98.5%	95.8%	–	0.86
GPT-4o + SLCV	Hybrid	–	8.80	100.0%	94.2%	–	0.88
Claude 3.5 Sonnet + SLCV	Hybrid	–	12.50	100.0%	96.5%	0.9	**0.89**
Specialised Pipeline (Supervised)							
PaRCO (Std)	Hybrid	–	2.00	100.0%	**99.1%**	**0.1**	**0.98**
PaRCO (Opt)	Hybrid	–	**0.14**	100.0%	98.5%	0.2	0.97

‡Text-only: Claude receives serialised $\mathcal{G}_{\text{vis}}$ (no image); SLCV still adds +5 pp over this text-grounded baseline.

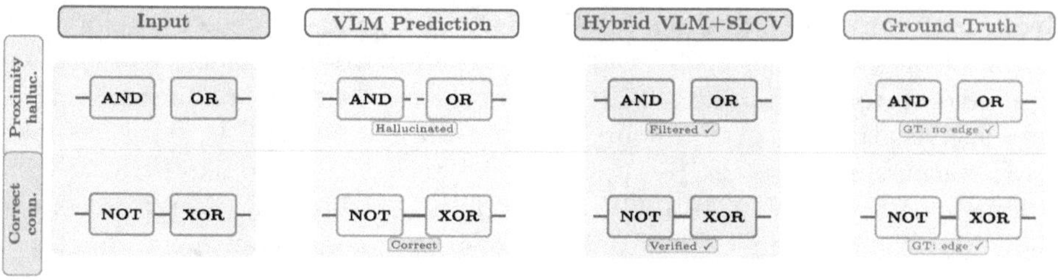

Fig. 3. VLM hallucination vs. SLCV verification (with ground truth). *Top row:* two disconnected gates that are spatially proximate. The VLM hallucinates an edge (red dashed); the hybrid system filters it; ground truth confirms no edge exists. *Bottom row:* two gates with an actual wire. The VLM correctly identifies the connection; SLCV verifies it via skeleton path search; ground truth confirms the edge. (Color figure online)

Analysis of Structural Integrity.

Performance Stratification. PaRCO achieves **Connectivity F1 = 0.98**, an improvement of **+11 pp** over fine-tuned LLaVA (0.87) and **+17 pp** over zero-shot Claude 3.5 Sonnet (0.81). Fine-tuning substantially improves VLM performance (+6 pp over zero-shot), but specialised verification architectures add further gains.

The Hybrid Result. The most practically significant finding is the **zero-shot hybrid** result. SLCV acts as a *post-hoc structural filter*: it accepts every VLM-predicted edge that passes skeleton-path verification and discards the rest, using the same endpoint, prompt, and test set as the standalone VLM; only the SLCV layer differs. The improvement is consistent across all evaluated models: Claude 3.5 Sonnet F1 0.81 → 0.89 (+8 pp), GPT-4o 0.78 → 0.88 (+10 pp), Gemini 1.5 Pro 0.76 → 0.86 (+10 pp), none requiring retraining.

Interpretation. PaRCO's advantage stems from task-specific inductive biases, hard topological constraints, and deterministic verification. Zero-shot VLMs predict connections between spatially close but unconnected components; fine-tuning reduces but does not eliminate these errors.

Failure Modes of Open-Source Baselines.

- **Semantic blindness:** Moondream2 (0.79 s) lacks the vocabulary for logic gates (0 % gate accuracy).
- **Stochastic instability:** Qwen3-VL suffered a 60 % timeout rate; LLaVA 7B frequently returned empty responses.
- **Counting errors:** LLaMA 3.2 Vision generated fluent descriptions but consistently miscounted components (75 % accuracy); language fluency does not imply visual-counting precision.

Cross-Benchmark Validation: CGHD. To validate generalisability, we evaluated on CGHD [5] (200 test samples, handwritten circuits). Adaptation required detector fine-tuning (40 GPU-hours transfer learning) and terminal template specification (30 person-hours); **zero changes to SLCV**. Table 2 shows PaRCO achieving F1 = 0.94 (−4 pp vs. LogicBench-1K due to handwriting variability). The hybrid VLM + SLCV gain replicates: F1 = 0.86 (+7 pp over standalone Claude 3.5 Sonnet), confirming cross-benchmark complementarity.

Table 2. Cross-benchmark validation on CGHD (200 test samples). SLCV verification generalises to handwritten circuits with minimal adaptation.

Model	Gate Acc. ↑	NED ↓	Conn. F1 ↑	Adaptation
Claude 3.5 Sonnet (zero-shot)	91.2%	1.4	0.79	0
Claude 3.5 Sonnet + SLCV	**93.5%**	**0.7**	**0.86**	0 (zero-shot VLM)
PaRCO (adapted)	**97.8%**	**0.3**	**0.94**	40 GPU-h + 30 p-h

Clean vs. Noisy Subsets. Table 3 reports LogicBench-1K performance stratified by image quality. PaRCO maintains high precision on both subsets but shows reduced recall on noisy images due to broken strokes. VLMs exhibit similar precision/recall patterns across conditions, suggesting their errors are dominated by inference failures rather than perception failures.

Table 3. Precision and recall on clean ($\mathcal{D}_{\text{clean}}$) vs. noisy ($\mathcal{D}_{\text{noisy}}$) subsets. PaRCO trades recall for precision on degraded inputs.

Model	Clean		Noisy	
	Prec. ↑	Rec. ↑	Prec. ↑	Rec. ↑
Claude 3.5 Sonnet	0.816	0.805	0.809	0.798
LLaVA-1.5 (FT)†	0.932	0.895	0.941	0.985
PaRCO (Std)	**0.996**	**0.989**	**0.992**	**0.968**

†Training data: 70 % noisy-augmented samples (affine skew, noise).

4.3 Latency and Efficiency

Table 4 details computational efficiency. All timings reflect single-sample inference on the hardware specified in Table 1.

PaRCO (Optimised) processes images in 0.14 s (636× faster than Qwen3-VL). There is an explicit trade-off: PaRCO requires up-front supervision (∼120 person-hours + 8 GPU-hours), whereas VLMs can be prompted immediately. For large-scale archival digitisation, PaRCO's 600–1,000× cost reduction suggests specialisation is cost-effective at scale.

Table 4. Inference efficiency. Rel. Latency = time relative to PaRCO (Optimised); higher means slower.

Model	Latency (s) ↓	Rel. Latency ↓	Throughput (img/h) ↑
PaRCO (Opt)	**0.14**	1.0×	25 700+ img/h
Moondream2	0.79	5.6×	4 500
PaRCO (Std)	2.00	14×	1 800
Gemini 1.5 Pro	4.20	30×	850
GPT-4o	8.50	61×	420
LLaMA 3.2 Vision	11.08	79×	320
Claude 3.5 Sonnet	12.10	86×	290
Qwen3-VL	89.06	636×	40

4.4 Ablation Studies

Table 5 reports systematic ablations on a held-out validation subset (100 samples).

Table 5. Ablation study. Each row removes or replaces one PaRCO module. Baseline = full PaRCO (Standard).

Configuration	Gate Acc. ↑	NED ↓	Conn. F1 ↑	Time (s)
Full PaRCO	99.1%	0.1	0.98	2.00
– Orientation (A1)	96.3%	0.5	0.92	1.85
YOLO-v5 detector (A2)	97.8%	0.2	0.95	1.60
Raw binary, no skel. (A3)	99.0%	1.8	0.84	1.95
Geometric terminals (A4)	98.9%	0.2	0.94	1.92
Proximity heuristic (A5)	98.7%	2.3	0.79	1.70
Sub-parameter sensitivity within SLCV				
Hilditch thinning (A3-alt)	99.0%	0.1	0.97	2.10
4-connected BFS (A5-alt)	99.1%	0.2	0.97	1.95
BFS reach threshold ×0.5 (A5-alt)	99.1%	0.1	0.97	1.90
Hybrid: VLM + SLCV	96.5%	0.9	0.89	12.5

Key Findings. **(A5) Path verification** is the single largest contributor: replacing skeleton-based BFS with proximity heuristics causes F1 to drop from 0.98 to 0.79 (−19 pp), validating the necessity of pixel-level verification. **(A3) Skeletonisation** is the second most critical component: operating on raw binary images without thinning reduces F1 to 0.84 (−14 pp). **(A1) Orientation normalisation** contributes −6 pp when removed. **(A4) Contact scoring** provides a -4 pp gain over geometric-only terminals. **(A2) Detector choice** has the smallest individual impact (-1.3 % accuracy with YOLO-v5), but offers a 20 % speed-up. Sub-parameter choices within SLCV shift F1 by 0.3–1.1 pp, confirming that path-existence criterion dominates over skeleton-pipeline details.

4.5 Failure Mode Analysis

Manual analysis of 20 failure cases (F1 < 0.90) identified four patterns (Fig. 4):

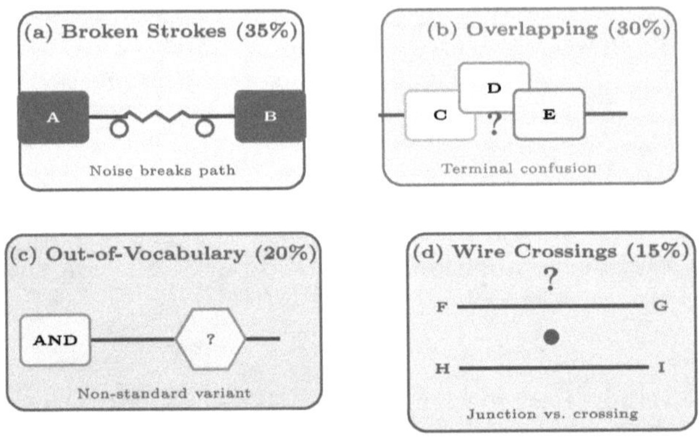

Fig. 4. Failure modes. (a) Broken strokes: noise disconnects the skeleton, causing false negatives. **(b)** Overlapping symbols: dense layouts confuse terminal inference. **(c)** Out-of-vocabulary: non-standard symbol variants cause detector errors. **(d)** Wire crossings: ambiguous junction-vs-crossing interpretation.

VLM failures differ qualitatively: predominantly *false positives* (spurious plausible edges) versus PaRCO's *false negatives* (missing faint connections).

4.6 Limitations

L1: Domain Specificity. PaRCO requires training data for each new domain (~500-1,000 annotated samples, terminal geometry definitions, and connectivity rules; ~120 person-hours). VLMs, by contrast, work zero-shot.

L2: Benchmark Scope. Evaluation covers digital logic circuits (LogicBench-1K) and handwritten circuits (CGHD). Performance on analog circuits, flowcharts, P&ID diagrams, or severely warped multi-layered schematics remains unexplored. The core SLCV components are designed for transferability: adapting to a new domain requires ~40 GPU-hours for detector retraining and ~30 person-hours for terminal specification, while the verification algorithms remain unchanged.

L3: Precision-Recall Asymmetry and Scope. The current SLCV formulation targets solid wire connections; dashed lines, arrowhead-only links, and label-only connections lie outside the present scope. Wire crossings are distinguished from junctions using local degree and orientation continuity, yielding a ∼7 pp improvement on CGHD, though ambiguous cases contribute 15 % of failure cases. Broken strokes account for ∼35 % of recall failures; complete topology recovery may require combining SLCV with learning-based stroke-gap filling.

L4: Detector Dependency. ∼20 % of connectivity failures trace back to detection errors, particularly for non-standard symbol variants. Improving robustness to out-of-vocabulary symbols remains open.

5 Discussion

Verification as a Paradigm for Document AI. The central thesis of this work is that multimodal document understanding benefits from an explicit *verification layer* that separates structural facts from probabilistic beliefs. SLCV fills a role analogous to type checking in compiled languages: the VLM acts as the programmer, producing plausible but potentially erroneous output; SLCV acts as the compiler's type system, rejecting outputs that violate structural invariants. This analogy is not merely metaphorical: under topology-preserving skeletonization and accurate symbol localization, SLCV provides a verifiable connectivity criterion (Eq. 1) in which every accepted edge is backed by a physical stroke path, and the constraint in Eq. 2 ensures that no downstream module can circumvent this check.

When to Specialise vs. Generalise. Our results delineate a practical decision boundary. For archival digitisation demanding high-precision topology extraction (legacy circuit migration, patent analysis, design verification), PaRCO's +11 pp F1 gain over fine-tuned VLMs justifies the engineering investment. For rapid prototyping or low-stakes applications, the zero-shot hybrid (VLM + SLCV) offers a compelling balance: +8–10 pp with no supervision cost.

Towards Domain-General Verification. Although evaluated on circuit diagrams, the three core SLCV primitives (skeleton extraction, constrained path search, and provenance tracking) are not circuit-specific. Any stroke-based diagram whose connectivity is encoded in foreground pixels (flowcharts, P&ID diagrams, architectural plans, UML activity diagrams) could benefit from the same verification principle. Extending SLCV to these domains requires only a domain-appropriate detector and terminal specification; the verification logic itself is reusable without modification. Future work includes learned gap-filling before skeleton extraction and tighter SLCV–VLM decoding-loop coupling.

6 Conclusion

Foundation models can produce plausible but structurally incorrect diagram topologies by predicting connections with no physical basis in the image. We introduced Stroke-Level Connectivity Verification (SLCV), which grounds every connectivity claim in pixel-level evidence, and PaRCO, a three-stage architecture that separates perception, verification, and interpretation into independently auditable layers.

We draw three main findings. First, the supervised PaRCO pipeline achieves high-fidelity topology recovery (F1 = 0.98 on LogicBench-1K), setting a strong performance reference for verification-based methods on this domain. Second, applying SLCV as a training-free verification layer to off-the-shelf VLMs improves connectivity F1 by 8–10 pp on LogicBench-1K and +7 pp on CGHD without any fine-tuning. Structural verification and probabilistic reasoning are complementary, not competing, strategies. Third, the core primitives of SLCV (skeleton extraction, constrained path search, and provenance tracking) are domain-general: adapting to a new diagram type requires retraining only the detector and specifying terminal geometry, while the verification logic remains unchanged.

Together, these results position pixel-grounded verification as a reusable validation primitive for multimodal document AI: a structural soundness check that adds reliability to probabilistic reasoning without constraining adaptability.

Acknowledgments. SLCV and LogicBench-1K (images, JSON ground truth, 700/100/200 split) are at https://github.com/tkcl-research/LogicBench1k. Inter-annotator agreement on a 100-image subset: Cohen's $\kappa = 0.91$ (symbols), $\kappa = 0.88$ (connectivity).

Disclosure of Interests. The authors have no competing interests to declare that are relevant to the content of this article.

References

1. Gross, M.D.: Recognizing and interpreting diagrams in design. In: Proceedings of the AVI '94, pp. 88–94. ACM (1994)
2. Alvarado, C., Davis, R.: SketchREAD: A multi-domain sketch recognition engine. In: Proc. UIST, pp. 23–32. ACM (2004). https://doi.org/10.1145/1029632.1029637
3. Moreno-Garc'ıa, C.F., Elyan, E., Jayne, C.: New trends on digitisation of complex engineering drawings. Neural Comput. Applic. **31**(6), 1695–1712 (2019). https://doi.org/10.1007/s00521-018-3583-1
4. Mani, S., Haddad, M., Constantini, D., Douhard, W., Li, Q., Poirier, L.: Automatic digitization of engineering diagrams using deep learning and graph search. In: Proc. IEEE/CVF Conference on Computer Vision and Pattern Recognition Workshops (CVPRW), pp. 673–679. IEEE (2020)
5. Thoma, F., Bayer, J., Li, Y., Dengel, A.: A public ground-truth dataset for handwritten circuit diagram images. In: Barney Smith, E.H., Pal, U. (eds.) ICDAR 2021. LNCS, vol. 12916, pp. 20–27. Springer, Cham (2021). https://doi.org/10.1007/978-3-030-86198-8_2

6. Thoma, F., Bayer, J., Li, Y.: CGHD1152. Zenodo (2021). https://doi.org/10.5281/zenodo.6385814
7. Ahmed, N., et al.: Digitize-HCD: A dataset for digitization of handwritten circuit diagrams. Data Brief **59**, 111315 (2025). https://doi.org/10.1016/j.dib.2025.111315
8. Roy, A., et al.: JUHCCR-v1: A database for hand-drawn electrical and electronics circuit component recognition. Sci. Reports **15**, 38617 (2025). https://doi.org/10.1038/s41598-025-22404-5
9. Rachala, R.R., Panicker, M.R.: Hand-drawn electrical circuit recognition using object detection and node recognition. SN Comput. Sci. **3**, 244 (2022). https://doi.org/10.1007/s42979-022-01159-0
10. Bayer, J., Roy, A.K., Dengel, A.: Instance segmentation based graph extraction for handwritten circuit diagram images. In: Proceedings of the International Conference on Pattern Recognition Applications and Methods (ICPRAM), pp. 926–931. SCITEPRESS (2023)
11. Bayer, J., van Waveren, L., Dengel, A.: Modular graph extraction for handwritten circuit diagram images. arXiv preprint arXiv:2402.11093 (2024)
12. Huang, C.-Y., et al.: Netlistify: transforming circuit schematics into netlists with deep learning. In: Proceedings of the ACM/IEEE Symposium on Machine Learning for CAD (MLCAD), pp. 1–8. IEEE (2025). https://doi.org/10.1109/MLCAD65511.2025.11189145
13. Aldowaish, S., Karumanchi, Y., Chiang, K.-C., Noorzad, S., Fayazi, M.: SINA: A circuit schematic image-to-netlist generator using artificial intelligence (2026). arXiv preprint arXiv:2601.22114
14. Zhang, T.Y., Suen, C.Y.: A fast parallel algorithm for thinning digital patterns. Commun. ACM **27**(3), 236–239 (1984). https://doi.org/10.1145/357994.358023
15. Ren, S., He, K., Girshick, R., Sun, J.: Faster R-CNN: towards real-time object detection with region proposal networks. IEEE Trans. Pattern Anal. Mach. Intell. **39**(6), 1137–1149 (2017). https://doi.org/10.1109/TPAMI.2016.2577031
16. Redmon, J., Divvala, S., Girshick, R., Farhadi, A.: You only look once: Unified, real-time object detection. In: Proceedings of the IEEE Conference on Computer Vision and Pattern Recognition (CVPR), pp. 779–788. (2016)
17. Mathew, M., Karatzas, D., Jawahar, C.V.: DocVQA: A dataset for VQA on document images. In: Proceeding of the IEEE/CVF Winter Conference on Applications of Computer Vision (WACV), pp. 2200–2209 (2021)
18. Yue, X.: MMMU: A massive multi-discipline multimodal understanding and reasoning benchmark for expert AGI. In: Proceedings IEEE/CVF Conference on Computer Vision and Pattern Recognition (CVPR), pp. 9556–9567 (2024)
19. Kembhavi, A., Salvato, M., Kolve, E., Seo, M., Hajishirzi, H., Farhadi, A.: A diagram is worth a dozen images. In: Leibe, B., Matas, J., Sebe, N., Welling, M. (eds.) ECCV 2016. LNCS, vol. 9908, pp. 235–251. Springer, Cham (2016). https://doi.org/10.1007/978-3-319-46493-0_15
20. Lu, P., et al.: IconQA: A new benchmark for abstract diagram understanding and visual language reasoning. In: Advances in Neural Information Processing Systems, Datasets and Benchmarks Track (2021)
21. Huang, Y., Lv, T., Cui, L., Lu, Y., Wei, F.: LayoutLMv3: pre-training for document AI with unified text and image masking. In: Proceedings of the ACM International Conference on Multimedia (ACM MM), pp. 4083–4091. ACM (2022). https://doi.org/10.1145/3503161.3548112

22. Kim, G., et al.: OCR-free document understanding transformer. In: Computer Vision – ECCV 2022, Lecture Notes in Computer Science, vol. 13688, pp. 498–517. Springer (2022). https://doi.org/10.1007/978-3-031-19815-1_29
23. Lee, K., et al.: Pix2Struct: Screenshot parsing as pretraining for visual language understanding. In: Proc. International Conference on Machine Learning (ICML), vol. 202, pp. 18893–18912. PMLR (2023) Proceedings of Machine Learning Research

CPAgent: A Tool-Augmented Agentic Framework for Chart Parsing

Chen-Yu Xie[1,2], Xiao-Hui Li[2], Boran Wang[4,5], Fei Yin[2,3], and Cheng-Lin Liu[1,2,3](✉)

[1] School of Information Science and Technology, ShanghaiTech University, Shanghai 201203, People's Republic of China
xiechy2023@shanghaitech.edu.cn, liucl@nlpr.ia.ac.cn
[2] State Key Laboratory of Multimodal Artificial Intelligence Systems, Institute of Automation of Chinese Academy of Sciences, Beijing 100190, People's Republic of China
{xiaohui.li,fyin}@nlpr.ia.ac.cn
[3] School of Artificial Intelligence, University of Chinese Academy of Sciences, Beijing 100049, People's Republic of China
[4] College of Artificial Intelligence, Nankai University, Tianjin 300350, China
wangbr@mail.nankai.edu.cn
[5] Zhongguancun Academy, Beijing, China

Abstract. Chart data extraction (Chart2Table) aims to transform chart images into structured tables, facilitating downstream analysis and understanding. Existing multimodal models often suffer from reading noise and hallucination on charts with sparse textual cues or missing explicit numeric annotations, while their intermediate reasoning steps remain largely opaque. We present CPAgent (Chart Parsing Agent), a tool-augmented agentic framework built on Google's Agent Development Kit (ADK) that reformulates chart parsing as an observable, traceable, and self-correctable task-driven pipeline. CPAgent provides a modular tool library comprising 16 visual and reasoning tools for horizontal bar, vertical bar, line, and pie charts, covering OCR, keypoint detection, axis calibration, and graphical elements reading. Within a single-run session, the agent dynamically orchestrates tool invocations and produces a structured table, a verifiable annotated chart, and complete Logs/Trace for reproducibility. Experiments demonstrate that CPAgent applies effectively to both general-purpose MLLMs and chart-specialized models. Taking `CPAgent(Qwen3-VL-8B)` as an example, it achieves an average RMS_{v2} improvement of +6.97% across Chart2Table benchmarks (ChartQA, DeChart, ChartBench) and an average accuracy gain of +3.42% across Chart QA benchmarks (ChartQA-*Total*, ChartBench, PlotQA), demonstrating the benefits of structured tables for chart question answering, particularly in settings with dense numerical reasoning. Source code are available at https://github.com/starbilibili/CPAgent.

Keywords: Chart Data Extraction · Agentic Framework · Chart QA

G. A. Fink et al. (Eds.): ICDAR 2026, LNCS 16972, pp. 341–358, 2027.
https://doi.org/10.1007/978-3-032-36023-6_20

1 Introduction

Charts are an indispensable medium for conveying data-driven insights in scientific research, business analytics, and decision support [1]. They appear in diverse forms such as bar charts, line charts, pie charts, and others, each encoding quantities and trends through visual primitives such as heights, positions, angles, and proportions. Automatically interpreting such charts has given rise to several core tasks in chart understanding: chart structure recognition and data extraction (Chart2Table), which converts chart images into structured tables; chart question answering (chart QA), which reasons over chart content to answer natural-language questions; and related tasks such as chart summarization and trend analysis. These tasks are of both academic interest, as they require joint visual and symbolic reasoning, and practical value, as they enable automated report parsing, dashboard interpretation, and numeric evidence retrieval from figures in documents.

Multimodal large language models (MLLMs) have advanced chart understanding [2–5], yet they rely heavily on explicit annotations and degrade on charts with sparse textual cues [6]. Chart2Table addresses this by producing verifiable structured tables, but prior methods—from rule-based and keypoint pipelines to Transformer and image-to-sequence models [7–12]—often remain text-dependent or lack traceable intermediate steps. Tool-augmented agents [13–16] improve observability, yet chart-oriented systems such as ChartAgent target open-ended QA via multi-round orchestration, where geometric evidence may be re-serialized across rounds, causing context drift and inconsistent alignments, without enforcing calibration-before-reading for full-table extraction. High-precision Chart2Table therefore still benefits from an explicit, calibrated, and traceable numeric pipeline.

Based on these observations, we focus on Chart2Table and propose **CPAgent (Chart Parsing Agent)**, a tool-augmented agentic framework that reconstructs chart parsing as an observable, traceable, and self-correctable tool-call chain. In contrast to multi-round, QA-oriented orchestration, CPAgent prioritizes direct data extraction. It enforces a strict calibration-before-reading workflow, stores intermediate evidence in structured state, and feeds tool outputs into the reasoning module. Specifically, CPAgent first constructs an explicit pixel-to-value mapping from axis ticks, then performs type-aware element-level reading to extract datapoints for horizontal/vertical bar, line, and pie charts. We implement a modular library of **16** visual and reasoning tools (OCR, chart-type recognition, axis calibration, and element-level reading) and produce a structured table together with a verifiable annotated chart and complete Logs/Trace for verification. As a training-free system, CPAgent requires no extra training or fine-tuning for its backbone LLM or orchestration logic, and tool-side models remain modular and easily replaceable. When calibration evidence proves insufficient, the agent automatically switches to end-to-end generation so that an output is always produced.

We build CPAgent on Google's **Agent Development Kit (ADK)** using an "LLM agent + tool registration" design. ADK treats a single chart-parsing request as a single-run session where the agent makes consecutive decisions, invokes multiple tools, and accumulates intermediate evidence within a unified context and state, avoiding repeated context stitching across external LLM calls. This single-run execution improves robustness and interpretability on weak-text charts by ensuring that later steps can directly access prior calibration and reading evidence.

To validate the effectiveness of our method, We conduct extensive experiments on Chart2Table benchmarks (ChartQA, DeChart, ChartBench [6,17,18]) and downstream chart QA benchmarks (ChartQA, ChartBench, PlotQA [6,17,19]). The results show that CPAgent consistently improves both chart-specialized models and general-purpose MLLMs when generating tables, and that using generated tables as computable evidence improves chart QA accuracy, especially for numerically intensive weak-text cases.

Our main contributions are as follows:

- **An agentic framework built on Google ADK**: we propose CPAgent, an agentic framework with adaptive tool-usage mechanisms, decomposing chart structure recognition into an observable, traceable, and self-correctable tool-call chain.
- **A modular tool library for mainstream chart types**: we build a library of 16 tools for bar/line/pie charts, supporting calibration and type-aware element-level reading for Chart2Table.
- **Transferability and downstream effectiveness**: we demonstrate that CPAgent applies effectively to both chart-specialized models and general-purpose MLLMs, and improves the accuracies of Chart2Table and chart QA significantly.

2 Related Work

2.1 Chart2Table Conversion

Chart structure recognition (Chart2Table) converts chart images into structured tables by recovering axes, ticks, legends, and graphical primitives, enabling downstream numerical reasoning and QA. Early rule-based pipelines [7,20] are brittle to layout diversity; keypoint+OCR methods [8] generalize better but propagate errors; Transformer-based models [9,21] learn more unified image-to-structure mappings. More recently, encoder–decoder image-to-sequence models generate tables directly from images [10–12,22]. Despite the end-to-end convenience, they still rely on explicit textual annotations and provide opaque, unverifiable intermediate steps, motivating tool-augmented, traceable decomposition.

2.2 MLLMs for Chart Structure Recognition

Multimodal large language models (MLLMs) pretrained on large-scale image–text data are widely used for chart tasks [2–5]. Several studies further adapt

MLLMs to chart-centric parsing through instruction tuning, data synthesis, and knowledge distillation [23–25]. However, when charts lack textual annotations, MLLMs often fail at geometric measurement and scale mapping, causing large performance drops [6].

2.3 Agentic Framework and Google ADK

Agentic methods decompose the targeted task into verifiable tool-invocation steps with dynamic action selection and refinement. General frameworks such as ReAct [13], Toolformer [14], and AutoGen [26] enable tool use, yet they typically lack structured output constraints, numeric mapping validation, and artifact-centric organization required for chart parsing. Chart-oriented agents such as VproChart [27] and ChartAgent [15,16] explore tool chaining and support Chart2Table. The method ChartAgent [15] employs multi-round orchestration with cost–gain tool selection, where each round is a new LLM invocation that consumes prior tool outputs and decides the next action. This may yield context drift when precise geometric evidence (e.g., tick–value pairs and pixel coordinates) is re-serialized and carried across rounds, leading to truncation or inconsistent re-interpretation. We build CPAgent on Google ADK, which provides agent–tool–state abstractions and logs tool outputs as traceable session events. It supports both LM-driven agents and workflow agents for adaptive tool use. CPAgent leverages ADK's single-run session to keep structured state and visual elements as programmatic objects, following a calibration-before-reading workflow. ChartAgent can also support open-ended chart QA with higher-level semantic tools that assemble question-specific evidence via Chart2Table.

3 CPAgent

This section presents the method design and implementation of CPAgent. Our key contribution is an integration of mature tool modules with Google ADK's agentic framework, which reformulates Chart2Table from an end-to-end black-box generator into an observable, traceable, and self-correctable parsing chain. Concretely, the LLM performs structured analysis, tool orchestration, and structured table synthesis; vision tools provide verifiable geometric measurements and produce intermediate visual elements; ADK manages unified context accumulation, state, and end-to-end Logs/Trace within a single-run session, ensuring that each value in the final table can be traced back to pixel-level visual cues—tick locations, bar endpoints, line intersections, and sector angles/proportions. For reference, we index the vision tools as T1–T16; Table 1 summarizes the tool library used by CPAgent.

3.1 Overview

CPAgent comprises an LLM analyzer, a modular tool library, and a dynamic orchestration controller. Figure 1 illustrates the end-to-end task-driven pipeline,

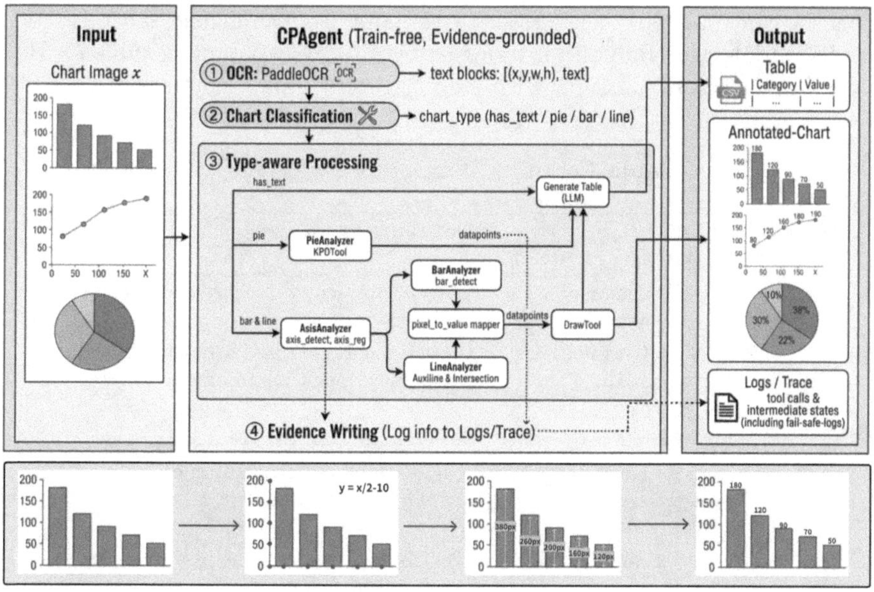

Fig. 1. Overview of CPAgent. In a single-run session, the system performs OCR, chart-type recognition, type-aware tool execution, and elements write-back, producing a structured table, a verifiable annotated chart, and complete Logs/Trace.

and Listing 1.1 provides the single-run control flow with five type-aware branches (`has_text`/`pie`/`horizontal_bar`/`vertical_bar`/`line`). In one ADK `run`, CPAgent (i) performs OCR and chart-type recognition, (ii) triggers calibration and element-level reading tools to produce datapoints, (iii) renders datapoints onto the chart to produce a verifiable annotated image, and (iv) generates the final table from the datapoints and annotated image—falling back to end-to-end generation from the original image only when critical evidence is missing, while logging complete Logs/Trace for reproducibility.

CPAgent is training-free, meaning that no additional training or fine-tuning of the backbone LLM or orchestration logic is required. Tool-side models are modular and replaceable; those requiring training are `bar_detect` (T8) and the keypoint detector used by `axis_detect` (T4) and `pie_detect` (T12), while `ocr_detect` (T1) uses PaddleOCR off-the-shelf. See Sect. 3.2 for model architectures, training data, and implementation details. Beyond the synthesized table, CPAgent also outputs annotated chart and Logs/Trace for verification, error attribution, and reproducible replay. Representative qualitative examples (a–c) and typical failure cases (d) are shown in Fig. 3.

3.2 Tool Library

CPAgent treats tools as replaceable perception modules and follows three design principles: modular composability, task-driven outputs, and backup behavior (fallback when tools fail or intermediate results are insufficient, with Trace logging). The tools are grouped into four categories by capability (Table 1). Below we describe each tool's design, with emphasis on trainable detectors.

Table 1. List of CPAgent tools (T1–T16).

Category	ID	Tool (interface/module)	Description
General	T1	OCR Engine (`ocr_detect`)	Detect and recognize chart text for parsing and calibration.
	T2	Text Classifier (`text_cls`)	Classify OCR text into semantic roles (title/legend/ticks/data labels, etc.).
	T3	Chart Classifier (`chart_cls`)	Classify chart type: `has_text`, `pie`, `horizontal_bar`, `vertical_bar`, `line`.
AxisAnalyzer	T4	Axis Keypoint Detector (`axis_detect`)	Hourglass-based localization of axis tick keypoints for tick pairing.
	T5	Axis Role (`axis_role`)	Identify axis types: value axis vs. category axis.
	T6	Axis Locator (`axis_locator`)	Select x/y-axis label text blocks from OCR via bbox clustering.
	T7	Pixel-to-Value Mapper (`pixel2value_mapper`)	Fit linear mapping from value-axis tick coordinates and labels (RANSAC, OLS fallback).
BarAnalyzer	T8	Bar Segmentation (`bar_detect`)	YOLOv11-based bar detection; color-based box extension for incomplete detections.
	T9	Bar Value Calculator (`bar_value_calc`)	Extract bar start/end from boxes, infer sign from axis, map to datapoints via T7.
LineAnalyzer	T10	Auxiliary Line Drawer (`auxiline_detect`)	Draw vertical guide lines at category ticks for intersection localization.
	T11	Line Intersection Locator (`line_intersection_loc`)	Localize intersection pixels, map to datapoints; pixel-based handling for dashed lines.
PieAnalyzer	T12	Pie Keypoint Detector (`pie_detect`)	Hourglass-based sector-boundary keypoints (instance distinct from T4).
	T13	Circle Geometry Fitter (`circle_fit`)	Least-squares fit of circle center and radius for proportion estimation.
	T14	Sector Proportion Calculator (`sector_prop_calc`)	Compute sector proportions and align them with category text to output datapoints.
General	T15	Datapoints-to-Chart Renderer (`draw_tool`)	Render datapoints back onto the chart to produce a verifiable annotated chart.
	T16	Table Generator (`table_gen`)	Generate the final table from datapoints and optional image context.

Text and Structure (T1–T3). `ocr_detect` (T1) uses PaddleOCR [28] off-the-shelf for chart text detection and recognition. `text_cls` (T2) assigns OCR text to semantic roles (title, legend, ticks, data labels); `chart_cls` (T3) classifies chart type into `has_text`, `pie`, `horizontal_bar`, `vertical_bar`, or `line`. Both T2 and T3 are implemented by the backbone LLM—no additional classifiers are required.

Geometric Calibration (T4–T7). We refer to the axis encoding numeric quantities (e.g., 0, 10, 20) as *value axis* and the axis encoding categorical labels (e.g., Jan, Feb) as *category axis*. `axis_detect` (T4) localizes axis tick keypoints via the Hourglass detector (see below). `axis_role` (T5) identifies value vs. category axes by heuristics: an axis whose labels are all numeric is preferred as the value axis; when both qualify, chart-type defaults apply (e.g., y-axis for vertical bar/line, x-axis for horizontal bar). `axis_locator` (T6) selects x- and y-axis label text blocks from OCR results by clustering bounding boxes. `pixel2value_mapper` (T7) fits a linear slope/intercept mapping from value-axis tick keypoint coordinates paired with their numeric labels, using RANSAC (residual threshold 10 px, `min_samples`=2) to suppress outliers; it falls back to ordinary linear regression on failure.

Datapoint Extraction (T8–T14). `bar_detect` (T8) uses YOLOv11 [29] trained on PlotQA and supplementary ChartQA bar annotations to improve style transfer. A color-based box-extension post-process expands incomplete detections for bars with extreme aspect ratios. `bar_value_calc` (T9) extracts start/end pixel coordinates from each detected box, infers sign from axis position, and maps to datapoints via T7. `auxiline_detect` (T10) draws vertical guide lines at category ticks. `line_intersection_loc` (T11) localizes intersection pixels and maps to datapoints, with pixel-based handling for dashed lines. `pie_detect` (T12) uses an Hourglass-based detector instantiated with parameters and trained separately from T4 (see below). `circle_fit` (T13) fits center and radius via least-squares. `sector_prop_calc` (T14) computes sector proportions and aligns them with category text.

Hourglass-Based Keypoint Detector. T4 and T12 use separate Hourglass instances with distinct parameters and training. The shared architecture is shown in Fig. 2. Built upon Stacked Hourglass [30] with *nstack*=4, our implementation adds: (1) a Group Relation Head that outputs $2E$ channels of relation embeddings (query/key, E=64) for keypoint grouping; (2) a deeper regression branch (multiple Conv+BN+ReLU layers) replacing the default 1×1 regression to improve localization on low-resolution feature maps. For pie-sector keypoints, we annotate 330 additional samples on DeChart; for axis keypoints, we annotate 510 additional bar/line samples. We apply data augmentation to expand the effective training set. For the heatmap output, focal loss is used to mitigate the severe positive–negative sample imbalance (keypoints vs. abundant background pixels), while coordinate regression uses standard L2 loss. We will release annotation guidelines and model parameters for reproducibility.

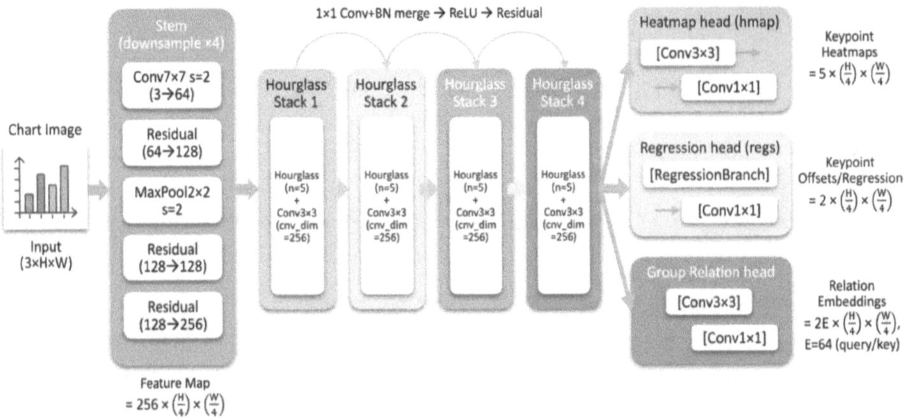

Fig. 2. Architecture of Hourglass-based keypoint detector (Stacked Hourglass, *nstack*=4). The network uses three output heads: keypoint heatmaps (`hmap`), coordinate regression (`regs`), and relation embeddings (Group Relation Head, E=64).

Structured Outputs (T15–T16). `draw_tool` (T15) renders datapoints onto the chart image to produce the annotated chart. `table_gen` (T16) is implemented by the backbone LLM; it receives datapoints and optional image context and produces the final table.

3.3 Tool Orchestration

Orchestration is fully agent-driven. The agent receives the tool list (Table 1)—each tool's interface, input schema, and purpose—plus a single instruction encoding invocation triggers. Within one ADK session, the LLM dynamically decides which tools to call based on the current structured state and accumulated evidence; there is no hard-coded pipeline. This design enables adaptive failure handling. For example, when `circle_fit` (T13) fails on a pie chart (e.g., due to outlier keypoints), the agent detects the failure, filters keypoints, and reinvokes the fit; when too few inliers remain for tick fitting, it may relax the filtering and retry. Such dynamic scheduling and recovery are a central advantage of ADK.

```
Input: chart image (path)
Output: table, annotated_image, trace/logs

Init: structured state S + elements store E + Trace
1) OCR: ocr_detect(T1) -> ocr_results
2) chart_cls(T3) -> chart_type
3) If has_text: (optional) text_cls(T2) -> table_gen(T16)
4) If pie: pie_detect(T12) -> circle_fit(T13) -> sector_prop_calc(T14) ->
     datapoints
5) If horizontal_bar, vertical_bar, or line: axis_detect(T4) + axis_role(
    T5) + axis_locator(T6)
            -> pixel2value_mapper(T7)
            -> if horizontal_bar/vertical_bar: bar_detect(T8) +
                bar_value_calc(T9);
            -> if line: auxiline_detect(T10) + line_intersection_loc(T11
                ) -> datapoints
            -> datapoints -> draw_tool(T15) -> annotated_image ->
                table_gen(T16)
6) Backup: if critical evidence missing, record Trace and fall back to
    end-to-end table_gen(T16) with image
```

Listing 1.1. CPAgent single-run workflow with ADK.

Listing 1.1 outlines the standard flow. A session maintains three shared states: (i) a structured state (chart type, text roles, axis roles, tick candidates, legend info, datapoints), (ii) an element store (keypoints, segmentation, pixel-to-value mapping, intersections, overlay paths), and (iii) a Trace (tool inputs, output summaries, failure causes). OCR (T1) and chart-type classification (T3) run first; chart type and optionally text roles (T2) persist in the structured state. The orchestration branches by chart type: `has_text` uses optional T2 then T16; `pie` runs T12→T13→T14→datapoints→T15→T16; bar and line charts run T4–T7 (axis calibration) before T8–T9 (bar) or T10–T11 (line).

For bar and line charts, calibration precedes reading. T4–T7 establish axis roles, tick positions, and a linear pixel-to-value mapping—all retained for downstream tools. T9 maps the bar baseline and top endpoints, takes their difference, and converts to values; T10 draws vertical auxiliary lines at category ticks, T11 localizes intersections with the polyline and maps pixels to values. This decomposes error into "geometric measurement error + mapping error," reducing drift on weakly annotated charts. Datapoints from T8–T14 are written to the structured state and passed to T15 (annotated chart render) and T16 (table generation). When evidence is insufficient (e.g., fewer than two valid ticks, an empty set of datapoints), the agent logs the cause in the Trace and falls back to `table_gen` (T16) with only the chart image. Lazy loading (PaddleOCR/Hourglass/YOLOv11) and robust parsing of LLM outputs prevent single-point failures from breaking the pipeline.

Table 2. Characteristics of benchmark datasets. **Annotations**: available (√) or absent (×). **Source**: Real vs. Synthetic-dominant. **Text**: Rich (abundant labels) vs. Weak (sparse labels). **Test**: total and per-type (bar/line/pie) sample counts.

Dataset	Annotations			Source		Text Cues	Test (T/B/L/P)	Main Use
	QA	Table	Fine-gr. Text	Real	Synth.			
ChartQA [17]	√	√	×	√	√	Rich	1509 (1220/211/78)	Chart2Table + QA
DeChart [18]	×	√	√	√	√	Weak	1940 (1474/328/136)	Chart2Table
ChartBench [6]	√	√	×	×	√	Weak	700 (450/200/50)	Chart2Table + QA
PlotQA [19]	√	√	×	×	√	Weak	2816 (1876/940/—)	QA

4 Experimental Protocol and Details

4.1 Datasets

We evaluate on four benchmarks—ChartQA, DeChart, ChartBench, and PlotQA, which cover diverse supervision (QA vs. table-only), sources (real vs. synthetic-dominant), and difficulty regimes (text-rich vs. weak-text that requires geometric reading and numerical reasoning). As summarized in Table 2, ChartQA provides QA pairs and table annotations with richer text cues, DeChart offers Chart2Table annotations plus fine-grained text-role labels for diagnosing text–structure alignment, and ChartBench/PlotQA contain many weak-text samples that stress text calibration and numeric robustness. ChartBench spans more diverse chart types (e.g., radar and area charts); we retain only bar/line/pie samples to align the task with our tool library's scope. PlotQA contains only bar and line charts (no pie); due to the large original test set, we sample 2,816 instances for evaluation, taking the first 10% of each chart type. Unless otherwise specified, all results are reported on the test split of each dataset.

4.2 Evaluation Metrics

For Chart2Table, we report two RMS scores (higher is better): RMS_{v1}, the table-similarity metric used in DePlot [11], and RMS_{v2}, an improved variant introduced with the DeChart [18] benchmark. In simple terms, RMS_{v1} is stricter and can be sensitive to header/format variations (e.g., units, separators, header order) even when values are correct, while RMS_{v2} uses more robust normalization and flexible alignment, making it more tolerant to such semantically equivalent formatting differences. Reporting both provides complementary views under strict formatting match versus structural equivalence.

For downstream chart QA, we follow the official ChartQA evaluation protocol [17] and report accuracy (Accuracy, %) on the *Human* and *Augmented* splits, as well as *Total*, defined as the average of *Human* and *Augmented*. For consistent comparison across datasets (ChartQA/ChartBench/PlotQA), we use a unified answer-matching criterion, with exact match for textual answers and a 5% relative error tolerance for numeric answers.

4.3 Integration Protocols

To demonstrate transferability, we adopt two integration protocols:
(1)
Agent Backbone Replacement. For general-purpose MLLMs (e.g., `Qwen3-VL-8B` and `Doubao-Seed-1.6`), we use them directly as the CPAgent backbone, completing analysis, tool execution, and table generation within a single session (e.g., `CPAgent(Qwen3-VL-8B)`). The structured table produced by CPAgent is **directly concatenated to the prompt** as computable evidence for downstream chart QA.
(2) Intermediate Elements Injection. For chart specialized models (e.g., `UniChart` and `DePlot`) that are hard to run as an agent backbone, we use CPAgent as a pre-processor by parsing charts with `CPAgent(Qwen3-VL-8B)` to produce an annotated chart (with readings written back onto the image), and replace the original input with this enhanced image only for no-text/weak-text samples before feeding it to `UniChart`/`DePlot`. We inject no additional prompt text, but only the annotated image, to make implicit readings explicit via verifiable elements.

4.4 Implementation Details

We build CPAgent on Google ADK and use the tool-side modules described in Sect. 3.2 (PaddleOCR [28], YOLOv11 [29], Hourglass). PaddleOCR is used off-the-shelf. Both YOLOv11 and the Hourglass keypoint detector can be trained on a single 24GB GPU. YOLOv11 (`bar_detect`) is trained on 19,887 bar-instance samples for 200 epochs (∼15 h total). The Hourglass detector (`axis_detect`, `pie_detect`) uses batch size 8 and 15,000 steps for both axis keypoints (510 samples) and pie-sector keypoints (330 samples), taking ∼4 h per instance. For backbone replacement, we use the official checkpoints of `Qwen3-VL-8B` [5], `Doubao-Seed-1.6`, and `InternVL3.0-14B` [4] with default inference settings. For Chart QA, the structured table is concatenated to the question as plain text; we use greedy decoding. For intermediate elements injection, the annotated chart replaces the original image only for no-text/weak-text samples (ChartBench, PlotQA, and ChartQA-*Human*); ChartQA-*Augmented* uses the original image. Each chart is processed in a single ADK run. We will release code, pretrained weights, and configuration files for reproduction.

5 Results and Analysis

This section presents the experimental results of CPAgent on chart structure understanding (Chart2Table) and downstream chart question answering (ChartQA), followed by detailed analysis and discussion.

5.1 Performance Comparison

Table 3 reports Chart2Table results on ChartQA, DeChart, and ChartBench, while Table 4 presents downstream chart QA results. On Chart2Table, CPAgent consistently improves a variety of backbone models (e.g., `UniChart`, `DePlot`, `Qwen3-VL-8B`, and `Doubao-Seed-1.6`), with larger gains on weakly labeled or unlabeled charts—consistent with our tool-augmented parsing chain that calibrates before reading. For downstream QA, we directly concatenate the structured tables produced by CPAgent to the prompt as computable evidence during answering. The improvements are more pronounced on settings that require precise reading and numerical reasoning (e.g., ChartBench and PlotQA), suggesting that high-quality Chart2Table effectively complements backbone models' weaknesses in geometric measurement and scale mapping under weak-text cues.

Table 3. Performance comparison on Chart2Table task across ChartQA, DeChart, and ChartBench benchmarks. Results are reported using RMS_{v1} and RMS_{v2}. CPAgent is integrated via backbone replacement (`Qwen3-VL-8B`, `Doubao-Seed-1.6`, `InternVL3.0-14B`) and intermediate elements injection (`UniChart`, `DePlot`).

Model	RMS_{v1}			RMS_{v2}		
	ChartQA	DeChart	ChartBench	ChartQA	DeChart	ChartBench
`UniChart` [12]	70.76	58.35	20.61	74.27	63.52	32.33
`UniChart(CPAgent)`	80.26↑	65.06↑	22.82↑	83.49↑	71.22↑	32.81↑
`DePlot` [11]	88.74	70.86	39.05	89.65	77.32	45.74
`DePlot(CPAgent)`	89.24↑	78.07↑	36.86↓	90.03↑	80.14↑	43.74↓
`TinyChart` [22]	92.54	79.07	50.52	92.84	81.39	57.13
`PaddleOCR-VL` [31]	87.85	78.05	69.32	89.70	81.31	72.19
`Gemini3` [2]	82.85	76.56	**78.84**	85.67	79.65	**79.94**
`ChartAgent(Qwen3-235b)` [15]	91.53	–	–	–	–	–
`Doubao-Seed-1.6`	80.31	73.60	72.53	84.43	77.66	74.30
`CPAgent(Doubao-Seed-1.6)`	82.31↑	75.91↑	72.49↓	86.44↑	80.68↑	74.39↑
`Qwen3-VL-8B` [5]	68.44	65.84	60.16	79.65	74.77	66.38
`CPAgent(Qwen3-VL-8B)`	87.41↑	78.45↑	64.06↑	89.40↑	81.92↑	70.38↑
`InternVL3.0-14B` [4]	93.97	83.83	71.91	94.71	86.13	73.81
`CPAgent(InternVL3.0-14B)`	**94.11**↑	**84.14**↑	72.72↑	**94.93**↑	**86.56**↑	74.40↑

5.2 Transferability and Effectiveness of CPAgent

Table 3 shows that CPAgent yields clear gains for both chart specialized models (`UniChart`, `DePlot`) and general-purpose models (`Qwen3-VL-8B`, `Doubao-Seed-1.6`, `InternVL3.0-14B`), demonstrating strong transferability of the proposed agentic framework and tool library. From an engineering perspective, CPAgent can either run as an agent via backbone replacement or

Table 4. Accuracy comparison on downstream chart question answering across multiple benchmarks – ChartQA, ChartBench, and PlotQA.

Model	ChartQA			ChartBench	PlotQA
	Human	Augmented	Total		
UniChart	38.72	85.76	62.24	33.14	25.40
MATCHA [10]	36.96	86.72	61.84	5.27	19.80
ChartGemma [32]	33.20	90.32	61.76	44.23	25.62
ChartAst [33]	40.17	89.23	64.70	38.21	20.46
TinyChart	58.16	94.48	76.32	61.03	71.50
InternVL3.0-14B	78.64	**95.44**	**87.04**	89.44	91.20
CPAgent(InternVL3.0-14B)	**79.28**↑	93.88↓	86.58↓	91.77↑	91.31↑
Doubao-Seed-1.6	66.08	83.56	74.82	90.68	70.49
CPAgent(Doubao-Seed-1.6)	69.68↑	83.28↓	76.48↑	**94.48**↑	81.08↑
Qwen3-VL-8B	73.92	93.76	83.84	89.89	86.23
CPAgent(Qwen3-VL-8B)	76.88↑	94.00↑	85.44↑	92.76↑	**92.01**↑

serve as a pre-processor that outputs enhanced images with written-back readings, improving specialized models' utilization of explicit annotations without modifying model parameters. End-to-end parsing with CPAgent(Qwen3-VL-8B) takes approximately 70 s per chart; because each chart is processed as an independent session, concurrent batch processing can reduce the average per-chart response time to the second level. Notably, with CPAgent, Qwen3-VL-8B yields Chart2Table performance comparable to specialized models (e.g., RMS_{v2}=89.40% on ChartQA, close to DePlot's 89.65%). This trend is even more visible for downstream QA: on weak-text, numerically intensive settings such as ChartBench/PlotQA, CPAgent brings larger gains for general backbones (Table 4), while the improvement on ChartQA-*Augmented* is marginal, aligning with the fact that weak-text scenarios rely more on the geometric chain of "calibration → reading → structured output." For Chart2Table, CPAgent improves results consistently across datasets, although the gain also depends on annotation formats and chart-style differences.

When considering stronger baselines, InternVL3.0-14B is among the best open-source models for chart understanding [4]. When integrated with CPAgent, it further attains RMS_{v2} scores of 94.93%, 86.56%, and 74.40% on ChartQA, DeChart, and ChartBench, respectively. However, performance gains saturate in style-diverse settings. On ChartBench, the RMS_{v2} scores of InternVL3.0-14B (73.81%), CPAgent(Doubao-Seed-1.6) (74.39%), and CPAgent(InternVL3.0-14B) (74.40%) are comparable, showing limited incremental gains. The effectiveness of CPAgent is constrained by two factors: (1) noise from annotation-format inconsistencies and header misalignment, where semantically equivalent but misaligned headers can still degrade strict table-matching scores such as RMS_{v1}; and (2) tool failures under chart-style domain shifts, such as 3D or perspective bar charts in ChartBench (see Fig. 3(d)). These observations suggest that tool-based agent approaches often require additional training data or redesigned or even replaced tools to achieve robust generalization to novel chart styles. On the positive side, agent-based methods support convenient tool replacement and plug-and-play, where interface alignment can be completed automatically without the extensive code modifications typically required in pipeline-based systems.

5.3 Qualitative Examples and a Typical Failure Case

We visualize intermediate evidence and write-back annotations produced by the observable parsing chain. Figure 3 shows representative examples (a–c) and a typical failure on 3D-styled bars (d), consistent with the limits discussed above.

Although the overall trend is positive, partial fluctuations can appear on style-diverse synthetic charts such as ChartBench. For example, in Table 3, `DePlot(CPAgent)` shows a small drop in RMS_{v1} on ChartBench compared with DePlot (39.05%→36.86%). CPAgent validates parsing results and falls back to end-to-end table generation when evidence is insufficient; however, this fallback does not apply under the intermediate-elements-injection integration used for DePlot and UniChart, where the backbone receives only the write-back annotated image and the original chart is not passed as input. When tool outputs are biased, the erroneous annotations are thus baked into the annotated image, and there is no path to fall back to the unannotated image. Errors propagate without the benefit of fallback, explaining the observed drop.

5.4 The Role of Structured Tables in Improving ChartQA

The main challenge of chart QA often lies in precise reading, scale mapping, and numerical computation. Under weakly annotated or unannotated settings, answering directly from pixels leads to visual errors and hallucination. The structured tables produced by CPAgent serve as a computable intermediate representation that converts continuous visual signals—positions, lengths, angles—into discrete, verifiable numeric units while explicitly preserving category–value alignment. We concatenate these tables to the QA prompt for deterministic lookup and computation, and write back key readings onto the image as an annotated chart, giving evidence in both structured form (computation) and visual form (verification). Table 4 shows that this mechanism is particularly beneficial on numerically intensive and weak-text benchmarks. For `Qwen3-VL-8B`, CPAgent improves ChartQA-*Total* by +1.60% (Human +2.96%, Augmented +0.24%), while the gains are larger on ChartBench and PlotQA (+2.87% and +5.78%). A similar trend holds for `Doubao-Seed-1.6` (ChartQA-*Total* +1.66%; ChartBench +3.80%, PlotQA +10.59%), aligning with our hypothesis that when questions require cross-axis reading, cross-category comparison, or computational reasoning, structured evidence reduces uncertainty and improves answer stability. In contrast, `CPAgent(InternVL3.0-14B)` yields limited gains (ChartQA Total 86.58% vs. 87.04%; ChartBench +2.33%, PlotQA +0.11%), because InternVL

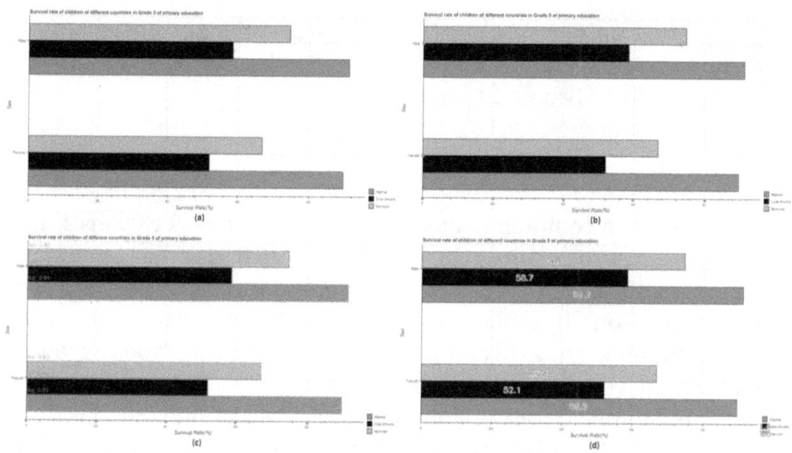

(a) Horizontal bar chart.

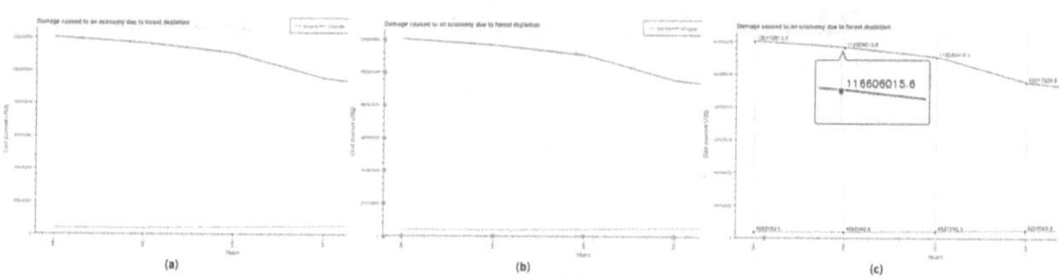

(b) Line chart.

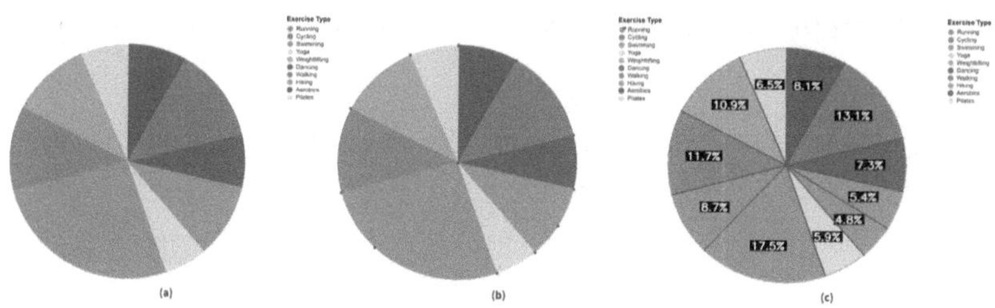

(c) Pie chart.

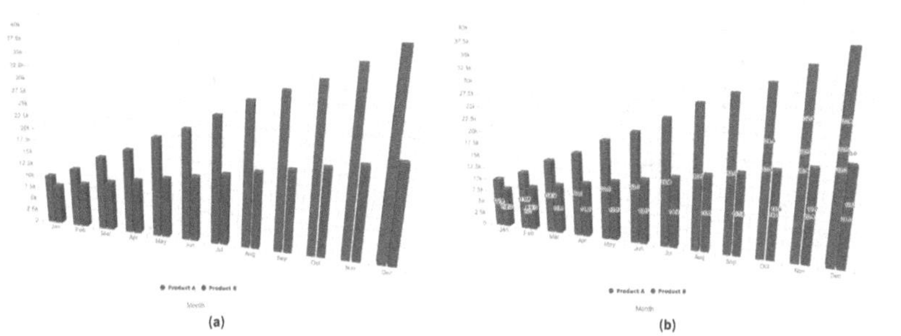

(d) Typical failure on 3D-styled bars: perspective and shadows cause unstable boundaries and biased readings.

Fig. 3. Qualitative examples with CPAgent write-back annotations (a–c) and a typical failure case (d).

is already fine-tuned on ChartQA with format-specific expectations. Our table injection without further fine-tuning thus provides less benefit than for general-purpose backbones like Qwen and Doubao.

6 Conclusion

We propose CPAgent, a tool-augmented agentic framework for Chart2Table that integrates a modular, type-aware tool library (T1–T16; Table 1) spanning OCR, detection, calibration, bar/line/pie reading, and table synthesis into a Google ADK single-run session. It replaces black-box generation with an observable, elements-driven pipeline (Logs/Trace): the LLM orchestrates tools, grounds each table value in pixel-level cues, and falls back when evidence is insufficient. Experiments on multiple public datasets show that CPAgent applies effectively to both general-purpose MLLMs and chart specialized models via backbone replacement and intermediate elements injection, substantially improving Chart2Table and downstream chart QA (Tables 3 and 4). Particularly, `CPAgent(Qwen3-VL-8B)` improves average RMS_{v2} by +6.97% on three Chart2Table benchmarks and boosts QA accuracy by +3.42% on three QA benchmarks. Despite these gains, performance remains constrained by annotation misalignment noise and tool failures triggered by significant chart style shifts. To address these limitations, future work will focus on: (1) expanding the tool library to additional chart types (3D, perspective, radar, and area charts); (2) lightweight fine-tuning or few-shot adaptation of tool-side modules for improved robustness; and (3) self-correction mechanisms for intermediate-element injection to mitigate pre-processor error propagation.

Acknowledgement. This work is supported by the National Natural Science Foundation of China (NSFC) Grant U23B2029.

References

1. Wang, H., et al.: Scientific discovery in the age of artificial intelligence. Nature **620**(7972), 47–60 (2023)
2. Team, G., et al.: Gemini: a family of highly capable multimodal models. arXiv preprint arXiv:2312.11805 (2023)
3. Liu, H., Li, C., Wu, Q., Lee, Y.J.: Visual instruction tuning. Adv. Neural. Inf. Process. Syst. **36**, 34892–34916 (2023)
4. Zhu, J., Wang, W., Chen, Z., Liu, Z., Ye, S.: InternVL3: exploring advanced training and test-time recipes for open-source multimodal models. arXiv preprint arXiv:2504.10479 (2025)
5. Bai, S., et al.: Qwen3-VL technical report. arXiv preprint arXiv:2511.21631 (2025)
6. Xu, Z., Du, S., Qi, Y., Xu, C., Yuan, C., Guo, J.: ChartBench: a benchmark for complex visual reasoning in charts. arXiv preprint arXiv:2312.15915 (2023)
7. Yan, P., Ahmed, S., Doermann, D.: Context-aware chart element detection. In: International Conference on Document Analysis and Recognition. LNCS, vol. 14187, pp. 218–233. Springer (2023). https://doi.org/10.1007/978-3-031-41676-7_13

8. Luo, J., Li, Z., Wang, J., Lin, C.Y.: ChartOCR: data extraction from charts images via a deep hybrid framework. In: Proceedings of the IEEE/CVF Winter Conference on Applications of Computer Vision, pp. 1917–1925 (2021)
9. Cheng, Z.Q., Dai, Q., Hauptmann, A.G.: ChartReader: a unified framework for chart derendering and comprehension without heuristic rules. In: Proceedings of the IEEE/CVF International Conference on Computer Vision, pp. 22202–22213 (2023)
10. Liu, F., et al.: Matcha: enhancing visual language pretraining with math reasoning and chart derendering. In: Proceedings of the 61st Annual Meeting of the Association for Computational Linguistics (Volume 1: Long Papers), pp. 12756–12770. Association for Computational Linguistics, Toronto, Canada (2023)
11. Liu, F., et al.: DEPLOT: one-shot visual language reasoning by plot-to-table translation. In: Findings of the Association for Computational Linguistics: ACL 2023, pp. 10381–10399. Association for Computational Linguistics, Toronto, Canada (2023)
12. Masry, A., Kavehzadeh, P., Do, X.L., Hoque, E., Joty, S.: Unichart: a universal vision-language pretrained model for chart comprehension and reasoning. In: Proceedings of the 2023 Conference on Empirical Methods in Natural Language Processing, pp. 14662–14684. Association for Computational Linguistics, Singapore (2023)
13. Yao, S., et al.: REACT: synergizing reasoning and acting in language models. In: International Conference on Learning Representations ICLR (2023)
14. Schick, T., et al.: ToolFormer: language models can teach themselves to use tools. Adv. Neural. Inf. Process. Syst. **36**, 68539–68551 (2023)
15. Wang, B., et al.: ChartAgent: a chart understanding framework with tool integrated reasoning. arXiv preprint arXiv:2512.14040 (2025)
16. Kaur, R., Srishankar, N., Zeng, Z., Ganesh, S., Veloso, M.: ChartAgent: a multimodal agent for visually grounded reasoning in complex chart question answering. arXiv preprint arXiv:2510.04514 (2026)
17. Masry, A., Long, D.X., Tan, J.Q., Joty, S., Hoque, E.: ChartQA: a benchmark for question answering about charts with visual and logical reasoning. In: Findings of the Association for Computational Linguistics: ACL 2022, pp. 2263–2279. Association for Computational Linguistics, Dublin, Ireland (2022)
18. Xie, C.Y., Li, X.H., Yin, F., Liu, C.L.: DeChart: a benchmark and text-enhanced chart-to-table conversion method with multimodal LLMs. In: 20th International Conference on Document Analysis and Recognition (ICDAR 2026). LNCS. Springer (2026)
19. Methani, N., Ganguly, P., Khapra, M.M., Kumar, P.: PLOTQA: reasoning over scientific plots. In: Proceedings of the IEEE/CVF Winter Conference on Applications of Computer Vision, pp. 1527–1536 (2020)
20. Rane, C., Subramanya, S.M., Endluri, D.S., Wu, J., Giles, C.L.: ChartReader: automatic parsing of bar-plots. In: 2021 IEEE 22nd International Conference on Information Reuse and Integration for Data Science (IRI), pp. 318–325. IEEE (2021)
21. Xue, W., Chen, D., Yu, B., Chen, Y., Zhou, S., Peng, W.: ChartDETR: a multi-shape detection network for visual chart recognition. arXiv preprint arXiv:2308.07743 (2023)
22. Zhang, L., et al.: TinyChart: efficient chart understanding with program-of-thoughts learning and visual token merging. In: Proceedings of the 2024 Conference on Empirical Methods in Natural Language Processing, pp. 1882–1898. Association for Computational Linguistics, Miami, Florida, USA (2024)

23. Han, Y., et al.: ChartlLAMA: a multimodal LLM for chart understanding and generation. arXiv preprint arXiv:2311.16483 (2023)
24. Kim, W., Park, S., In, Y., Han, S., Park, C.: Simplot: enhancing chart question answering by distilling essentials. In: Findings of the Association for Computational Linguistics: NAACL 2025, pp. 573–593. Association for Computational Linguistics, Albuquerque, New Mexico (2025)
25. Masry, A., Shahmohammadi, M., Parvez, M.R., Hoque, E., Joty, S.: ChartinStruct: Instruction tuning for chart comprehension and reasoning. In: Findings of the Association for Computational Linguistics: ACL 2024, pp. 10387–10409. Association for Computational Linguistics, Bangkok, Thailand (2024)
26. Wu, Q., et al.: AutoGen: enabling next-gen LLM applications via multi-agent conversations. In: First Conference on Language Modeling (2024)
27. Huang, M., Zhang, L., Lai, H., Wu, W., Zhang, X., Liu, J.: VPROCHART: answering chart question through visual perception alignment agent and programmatic solution reasoning. In: Proceedings of the AAAI Conference on Artificial Intelligence, vol. 39, pp. 3689–3696 (2025)
28. Cui, C., et al.: PADDLEOCR 3.0 technical report. arXiv preprint arXiv:2507.05595 (2025)
29. Khanam, R., Hussain, M.: YOLOv11: an overview of the key architectural enhancements. arXiv preprint arXiv:2410.17725 (2024)
30. Newell, A., Yang, K., Deng, J.: Stacked hourglass networks for human pose estimation. In: Leibe, B., Matas, J., Sebe, N., Welling, M. (eds.) European Conference on Computer Vision, pp. 483–499. Springer (2016). https://doi.org/10.1007/978-3-319-46484-8_29
31. Cui, C., et al.: PADDLEOCR-VL: boosting multilingual document parsing via a 0.9b ultra-compact vision-language model. arXiv preprint arXiv:2510.14528 (2025)
32. Masry, A., Thakkar, M., Bajaj, A., Kartha, A., Hoque, E., Joty, S.: Chartgemma: visual instruction-tuning for chart reasoning in the wild. In: Proceedings of the 31st International Conference on Computational Linguistics: Industry Track, pp. 625–643 (2025)
33. Meng, F., et al.: ChartAssistant: a universal chart multimodal language model via chart-to-table pre-training and multitask instruction tuning. In: Findings of the Association for Computational Linguistics: ACL 2024, pp. 7775–7803. Association for Computational Linguistics, Bangkok, Thailand (2024)

Handwriting Recognition

InkTree: A Unified Representation of Structured Online Ink

Jakob Seitz, Tobias Lengfeld, and Radu Timofte(✉)

Computer Vision Lab, CAIDAS and IFI, University of Würzburg, Würzburg, Germany
{jakob.seitz,tobias.lengfeld,radu.timofte}@uni-wuerzburg.de

Abstract. Research on online handwriting recognition, especially handwritten mathematics with structural annotation, relies on a heterogeneous set of data formats. W3C InkML reliably stores digital ink, but structure and relations are often split across additional layers and connected via ID-based cross-references to trace groups and external markup. This multi-layer setup increases parsing and tooling complexity, hinders reproducible experimentation, and makes fine-grained relations between strokes, symbols, and higher-level structures harder to access, an issue that becomes more pronounced beyond single-line expressions (e.g., multi-line formulas, tabular layouts, mixed text and mathematics, and diagrammatic relations).

We introduce **InkTree**, a lightweight ML-oriented representation that unifies ink, structure, and spatial relations in a single self-describing sample. Each instance is encoded as a self-contained hierarchical object with strokes attached directly to structural nodes and relations represented explicitly and extensibly. This eliminates costly cross-reference resolution and enables consistent derivation of structural views, including relational and graph-based views, from the same source. We also provide a reference implementation that converts common online-handwriting source formats to InkTree, alongside lightweight Python tools for visualization and dataset access. Across 15 benchmark configurations spanning seven dataset families, InkTree consistently reduces file size and accelerates loading compared with each dataset's original source format.

1 Introduction

Online handwriting expression recognition operates on pen trajectories rather than rasterized images. Each sample is represented as ordered traces (typically x, y, often with pen-state and optional timing), preserving stroke order, pen-lift boundaries, and writing dynamics that are largely lost in offline renderings. These signals are directly useful for segmentation and relation modeling, where

---

Code and data: Reference implementation, dataset loaders, benchmark scripts, and results are available at https://github.com/JakobSeitz/InkTree.

This work was partially supported by an EXIST project and by the Alexander von Humboldt Foundation.

G. A. Fink et al. (Eds.): ICDAR 2026, LNCS 16972, pp. 361–377, 2027.
https://doi.org/10.1007/978-3-032-36023-6_21

spatial and temporal locality provide complementary cues [27,29]. In contrast, image-only pipelines must infer stroke structure indirectly from pixels and are more sensitive to rendering choices (resolution, line thickness, antialiasing) before structural reasoning begins.

Among online tasks, handwritten mathematical expression recognition (OHMER) is a particularly strong stress test for representation design: expressions are intrinsically two-dimensional and require explicit spatial semantics such as superscript/subscript, fractions, roots, under/over attachments, and matrix-like layout regions [27,31]. Unlike plain sequence transcription, correct interpretation depends on modeling both symbols and their relations.

This structure-first perspective is not specific to mathematics. The same representational need appears in broader online document settings with mixed object types and layout-dependent semantics (text, formulas, diagrams), and it naturally extends to other structured notation domains such as chemical expressions or technical markup [12,19]. In these settings, tree/graph views are often the natural substrate for training and evaluation.

In parallel, model design is shifting from narrow task-specific systems toward relation-aware and transferable approaches. For OHMER, structure-centric formulations (e.g., SLT/SLG/SRT-style views and GNN-based variants) are increasingly explored because they align with the underlying 2D composition. Representative examples include label-graph and structure-aware evaluation settings [27,33]; recent SRT and edge-aware GNN results further support this direction [28,30]. More broadly, handwriting research increasingly investigates pretrained and cross-task transfer settings: transformer-based pretrained recognizers are common in offline HTR, and recent work begins to map online ink into representations compatible with generalist vision-language backbones [4,7,16]. Cross-dataset robustness, out-of-distribution behavior, and domain generalization are also becoming explicit targets [6,20].

Despite these trends, online handwriting workflows still suffer from format fragmentation. Even within InkML, annotation practice is not uniform: CROHME-style files often center `traceGroup` plus MathML-linked annotations, whereas IAMonDo-style files use nested `traceView` hierarchies with typed annotation blocks and document-level object structure [12,22,29]. Beyond InkML, dataset families use incompatible native serializations (e.g., JSON folder hierarchies, SQL exports, archive-based text segments, NDJSON vectors, or binary record formats), each requiring separate parser and adapter logic [9,10,17]. In *atomic-ink datasets*, each sample is typically a single symbol, character, or sketch without explicit inter-symbol relations. This is practical for single-task setups, but it hinders cross-dataset experimentation and transfer pipelines that require consistent sequence, graph, and layout views.

This paper introduces **InkTree**, a lightweight representation for structured online ink. InkTree keeps the essential information needed by modern handwriting pipelines, but removes format friction by making each sample self-contained and semantically explicit. The schema is intentionally open-ended: each node carries a short `type` string and semantic child keys, and unknown types decode gracefully via a generic fallback, which makes it straightforward to extend the format to new structural granularities such as words, sentences, or graph-level annotations without breaking existing decoders. Our contributions are:

1. A compact, extensible hierarchical schema that unifies traces, structure, and relations in one object, with named semantic child keys and a type-string design that supports future node types without format versioning.
2. A deterministic conversion pipeline from common online-handwriting source formats to InkTree (including InkML and non-InkML formats), with matching decoder support and dataset-specific loaders.
3. An empirical evaluation across 15 benchmark configurations covering CROHME, CROHME+ [24], MathWriting+, IAMonDB, Unipen, DeepWriting, and Detexify, comparing InkTree against each dataset's original source format and demonstrating consistent reductions in storage cost and loading time.

2 Related Work

InkML is the dominant standard for online digital ink exchange and the foundation of many recognition benchmarks [29]. Its design deliberately separates raw stroke data (`trace`, `traceFormat`) from semantic grouping and annotation (`traceGroup`, `context`, `annotation`, `annotationXML`), allowing the same ink to carry multiple interpretations without modifying the underlying traces. This flexibility is well suited for archival interchange, but for ML loading it means that structure must be reassembled on every read by resolving cross-references between trace IDs, traceGroup hierarchies, and external markup. UNIPEN, an earlier handwriting exchange format, anticipated the same tension and resolved it differently: it was designed as a simple ASCII flat file with human-readable keywords, deliberately keeping the format easy to parse and extend [1,13]. Format conversion studies confirm that bridging these representations typically requires substantial engineering adapter code [11].

For handwritten mathematics, CROHME established widely used datasets in which InkML traces are paired with a MathML annotation layer that encodes the symbol-layout structure via `annotationXML` elements linked to `traceGroups` by XML IDs [3,18,21–23]. This multi-layer design follows naturally from InkML's general-purpose architecture—the format intentionally allows structural interpretations to be overlaid without altering the raw ink—but it means that loading a single sample requires parsing both layers and resolving their cross-references. A further consequence of grounding structure in MathML is that the annotation vocabulary is inherently scoped to mathematical notation: MathML defines elements for fractions, roots, scripts, and operators, but provides no vocabulary for document-level objects (paragraphs, headings, diagrams), chemical structure, or arbitrary relational graphs. Datasets outside the mathematical domain therefore cannot reuse this annotation layer and must introduce separate structural conventions, contributing to the format fragmentation observed across the broader online handwriting landscape. Structural quality metrics and label-graph formulations have further emphasised that symbol-only accuracy is insufficient, reinforcing the need for explicit relational views [32,33]. MathWriting preserves this ecosystem while extending to larger scale [8], and structure-enriched variants augmenting trace-group and relation annotations support structural recognition pipelines more directly [24,25].

Beyond formula-level tasks, datasets diverge further in format. IAMonDo uses InkML's `traceView` element to encode hierarchical document layouts—DOCUMENT → TEXTBLOCK → TEXTLINE → WORD, alongside DRAWING, DIAGRAM, LIST, and editorial MARKING layers—demonstrating that InkML can represent full document structure [12]. The result is expressive but deeply nested, with delta-encoded four-channel traces (X, Y, T, F) and rich author metadata embedded throughout, making per-sample extraction non-trivial. IAMonDB and DeepWriting instead use ad-hoc JSON folder hierarchies where strokes, character ranges, word-level labels, and per-point timestamps are packed into large shared JSON files alongside user IDs, image paths, and curation flags [2,5]. Detexify stores symbol samples as a PostgreSQL dump with JSON-encoded stroke arrays, requiring a full table scan to extract any subset [14,15]. Unipen is distributed as a `.tgz` archive of ASCII segment and include files [13]. Each of these choices reflects different optimisation goals—archival completeness, annotation richness, or database convenience—but none is designed for efficient per-sample random access in training pipelines.

Quick, Draw! and CASIA-OLHWDB represent *atomic-ink* settings (single-instance samples, no inter-symbol relations) and map naturally to an InkTree subset containing only strokes and a label field [9,10,17]. Kondate further extends the document setting to mixed text/formula/diagram pages [19], a granularity that InkTree's open type system supports without a schema change.

In OHMER, recent literature increasingly evaluates structure- and relation-aware formulations rather than pure sequence decoding [27,31]. Symbol Layout Trees (SLT) encode symbols with explicit spatial relations in a layout-preserving tree view [31], while Stroke Label Graphs (SLG) represent stroke/symbol assignments and relation labels as graph structure [33]. In online settings, Symbol Relation Trees (SRT) derive symbol relations from stroke streams, and GNN variants extend this with edge-aware attention for joint node/edge prediction [27,28,30]. InkTree is designed as an engineering bridge between these structural modelling demands and the heterogeneous existing data ecosystem.

3 InkTree Format

InkTree represents one sample as:

```
{ "version": "1.0", "label": "<latex>", "node": <root> }
```

where `node` is a recursively defined tree with typed relation nodes. Multiple samples are stored one per line in a gzip-compressed JSONL file (`.inktree.jsonl.gz`), enabling sequential streaming without loading entire datasets into memory.

3.1 Design

The format follows four practical design decisions. First, node types are short canonical identifiers (e.g., `sym`, `row`, `frac`, `sup`). Second, structural roles are explicit semantic keys instead of positional child conventions (`numer`/`denom`, `base`/`sub`/`sup`). Third, strokes are embedded directly in symbol and structural nodes, including fraction-bar and radical strokes, which avoids external ID resolution. Fourth, the `type` string is an open identifier: any new node type (e.g., `word`, `sentence`, `graph`) is valid JSON and decodes gracefully via a generic fallback in older decoders, allowing schema extension without a format version bump.

3.2 Node Semantics

The schema supports common math relations (`frac`, `sub`, `sup`, `subsup`, `sqrt`, `root`, `under`, `underover`), generic containers (`row`, `any`, `noisy`), and a multi-line grouping node (`line`) that bundles rows into multi-line expressions such as step-by-step calculations. This set mirrors relation types already produced by existing CROHME-style parsers in our pipeline, enabling lossless structural conversion. The generic `children` list used by `row`/`any` nodes is the natural building block for extended granularity: for example, a word node can be represented with fields `type=word`, `text=...`, and `children=[...]` (symbol nodes), while sentences can be represented as rows of word nodes within the same file format.

3.3 Illustrative Example

The following compact sample and Fig. 1 refer to the same expression ($\frac{21}{5}$): numerator as a row of two symbols, denominator as one symbol, plus an explicit fraction-bar trace group.

```
{
  "label": "\frac{2 1}{5}",
  "node": {
    "type": "frac",
    "numer": {
      "type": "row",
      "children": [
        {"type": "sym", "label": "2", "strokes": ["x":[...],"y":[...]]},
        {"type": "sym", "label": "1", "strokes": ["x":[...],"y":[...]]}
      ]
    },
    "denom": {"type": "sym", "label": "5", "strokes":
          ["x":[...],"y":[...],"x":[...],"y":[...]]},
    "bar": ["x":[...],"y":[...]]
  }
}
```

3.4 Rationale

InkTree is not a replacement for any of the archival source formats described above; it is a shared ML-oriented substrate that normalises their training-relevant content into one consistent form. Each sample is locally complete and

(a) Rendered InkTree sample from CROHME 2023 test (one color per trace group).

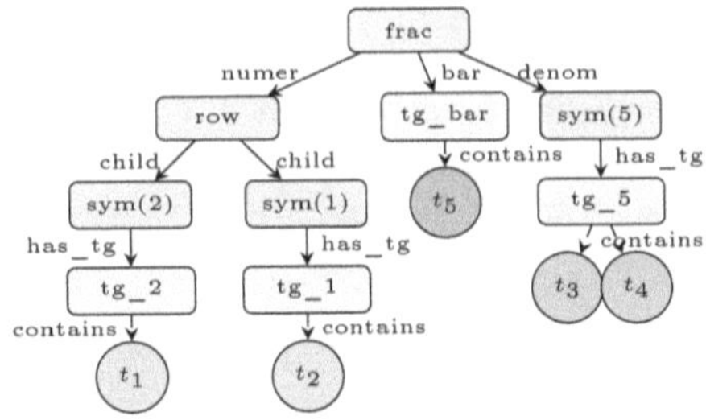

(b) Graph view matching the same $\frac{21}{5}$ sample: explicit node → trace-group → trace edges.

Fig. 1. InkTree example: real stroke rendering and explicit connectivity structure.

self-describing: relation roles are explicit in named keys rather than inferred from positional conventions or cross-referenced IDs, and the open type system covers the full structural range—from single-stroke symbol classification to formula trees and document-level hierarchies—without per-dataset conventions or schema changes. Dataset-level context (coordinate units, license, writer demographics, device specifications) is documented once in companion metadata rather than embedded in every sample.

Provenance metadata—writer IDs, user IDs, source-system identifiers, image file paths—is excluded per sample because it does not contribute to the ink signal; it belongs in dataset-level companion documentation. *Parallel encodings* such as MathML (CROHME, MathWriting+) are not stored verbatim: they are re-generated on demand from the decoded node tree via the built-in `get_math_ml()` method (implemented for all relation node types), and the LaTeX string is likewise re-derivable via `node.latex()`. *Curation scaffolding*—segmentation validity flags and OCR correctness scores (DeepWriting, IAMonDB)—belongs to dataset preparation rather than training. For *temporal data*, the format natively supports a `t` field per stroke point; timestamps are retained for all datasets that provide them (MathWriting+, Detexify, DeepWriting, IAMonDB). Pen pressure (IAMonDB) is absent from most other datasets and not widely used in current pipelines. Word-level labels from DeepWriting and IAMonDB can be retained as nested `word` nodes with label fields; the format supports this without a schema change. Writer IDs, acquisition dates, DPI, and character counts belong to dataset-level companion documentation rather than per-sample records, and can be stored in a sidecar metadata file alongside the InkTree split.

4 Conversion Tools

We provide a reference Python implementation with three components:

1. **Encoder**: converts relation graphs to InkTree JSON objects by recursively mapping node classes to schema types and semantic child keys, including trace-group strokes for structural nodes (`FracNode`, `SqrtNode`, `RootNode`).
2. **Decoder**: reconstructs relation-node graphs from InkTree, enabling round-trip use in existing tooling.
3. **I/O layer**: stores and loads gzip-compressed JSONL files (`.inktree.jsonl.gz`), one sample per line.

The conversion pipeline processes CROHME and MathWriting+ splits directly from InkML input and writes a single compressed InkTree file per split. For other datasets, dedicated loaders parse each original format directly: Deep-Writing and IAMonDB JSON folders are loaded by a shared folder-walking parser that assembles per-character SymbolNodes from range-indexed point lists and groups them into RowNodes; Detexify samples are extracted from an SQL dump; and Unipen characters are assembled from the `.tgz` segment and include files. The pipeline supports graceful handling of undefined relations (`AnyRelationNode`) and preserves formula-level labels when available.

This design keeps integration cost low: researchers can continue using their parser front-end and only swap intermediate storage and dataset loading steps. We also provide scripts for benchmark generation and descriptive dataset statistics to support reproducible reporting. The code and benchmark scripts are publicly available at https://github.com/JakobSeitz/InkTree. Beyond format conversion, the implementation exposes practical utilities for downstream research: plotting and visual inspection tools for traces and relation graphs, programmatic extraction of trace groups and symbol/relation subgraphs for task-specific setups (e.g., classification or segmentation), and node-level formula export interfaces (LaTeX and MathML) for all supported relation node types. Figure 2 illustrates converted samples across content types and source formats, rendered with these tools: one color per trace group (segmentation) and arrows for the encoded relations.

5 Experiments

We evaluate format-level efficiency across 15 benchmark configurations drawn from seven dataset families. For InkML-based datasets (CROHME, CROHME+, MathWriting+), we measure the cost of full XML parsing and MathML relation-graph construction against InkTree loading. For other datasets, we compare against their native source format: JSON folder hierarchies for DeepWriting and IAMonDB, a PostgreSQL dump (`.sql`) for Detexify, and the Unipen `.tgz` CD-ROM archive. All paired comparisons use the same set of samples.

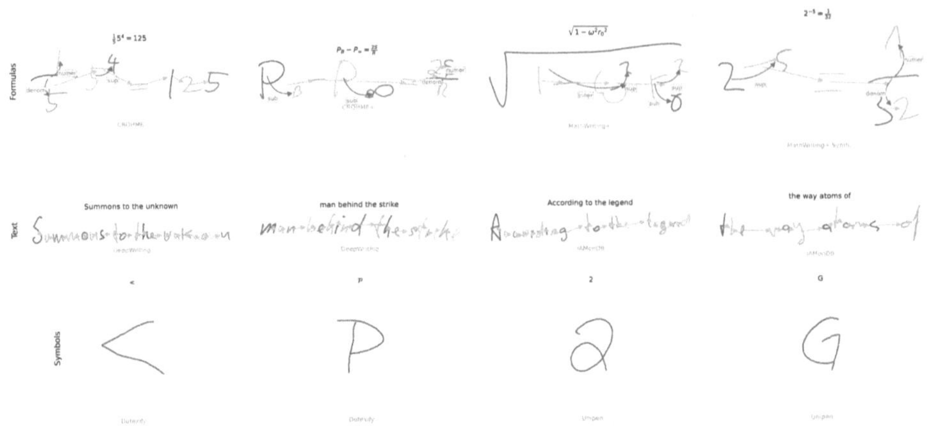

Fig. 2. Converted InkTree samples across content types and heterogeneous sources: math formulas (CROHME, CROHME+, MathWriting+; InkML), text lines (DeepWriting, IAMonDB; JSON), and isolated symbols (Detexify, SQL dump; Unipen, `.tgz` archive). Colors distinguish trace groups (segmentation); labeled dark arrows show semantic relations (`numer`, `denom`, `sup`, `inner`, ...), light gray arrows show reading order within rows. (Color figure online)

5.1 Experimental Setup

Benchmarks were run on a single machine (Intel Core i7-14700KF, 32 GB RAM, NVMe SSD) using Python 3.10 with our reference implementation. Timing uses `time.perf_counter()` and is reported as wall-clock time including object construction. InkTree files are gzip-compressed JSONL (`.inktree.jsonl.gz`). For InkML datasets, source size is the uncompressed XML on disk. For native formats, source size is reported on the same sample universe as the compared InkTree split (i.e., proportional normalization when the loader yields a strict subset of raw entries), while retaining JSON metadata bytes where present. Large splits (MathWriting+ Train: 143 106 files; MathWriting+ Synthetic: 85 962 files; CROHME+ Synthetic: 124 325 files) were evaluated on a random sample of N = 2,000; DeepWriting (N = 5,159), IAMonDB (N = 11,242), Detexify (N = 210,454) and Unipen (N = 79,452) were benchmarked over their full converted splits. For context, these benchmarks use the converted InkTree splits stored in our benchmark folder; they may represent a strict subset of the full dataset releases when conversion is restricted to samples with available structural annotations and successful parsing.

To assess measurement reproducibility, we repeated seven representative configurations five times each (three times for Detexify, given its 1 GB source file). The coefficient of variation (CV) of the speedup ratio was below 14% across all configurations (median 7.0%), confirming that the relative advantage of InkTree is stable under repeated measurements. Because InkTree and source-format loads share the same hardware and I/O path in each comparison, systematic hardware-specific factors such as CPU clock, SSD throughput, and memory bandwidth affect both sides proportionally and largely cancel in the ratio. The reported speedups are therefore expected to transfer across hardware configurations with similar I/O characteristics; absolute load times will scale with hardware but the relative ordering is robust.

5.2 Compared Representations

1. **InkML**: raw XML (full parse, trace extraction, and MathML relation-graph construction via ID cross-references). Used for CROHME and MathWriting+.
2. **Native format**: `.json` folders (DeepWriting, IAMonDB), `.sql` (Detexify), `.tgz` (Unipen). Each requires format-specific parsing code.
3. `.inktree.jsonl.gz`: proposed compact hierarchical format (this paper).

5.3 Results: InkML-Based Datasets

Table 1 reports results for CROHME, CROHME+, and MathWriting+ splits. InkTree consistently reduces file size to 0.20–0.32× of the raw InkML footprint (i.e., 68–80% reduction) and loads 1.2–9.5× faster across all configurations. The MW+ Symbols split achieves the largest speedup (9.5×) and size reduction (0.20×) because individual-symbol files carry disproportionate XML header overhead. The CROHME Real Train split (N = 12,010) illustrates the practical gain: InkTree reduces disk access from 113 MB to 28 MB and load time from 5.0 s to 2.6 s per full scan, a 1.9× improvement that compounds over training epochs. The CROHME+ Synthetic split [24] (N = 2,000 sampled from 124 325 files) achieves consistent 3.4× speedup and 23% size ratio, in line with the other CROHME splits.

5.4 Results: Other Source Formats

Table 2 reports results for the four non-InkML datasets. DeepWriting and IAMonDB are distributed as JSON folder hierarchies where each `.json` file contains multiple word samples together with per-sample metadata (image paths, user IDs, validation flags, etc.) that InkTree does not retain. InkTree stores stroke sequences, per-point timestamps, and character labels, but discards the metadata not needed for training, yielding large size reductions: DeepWriting shrinks from 588 MB (matched-sample equivalent of the JSON source) to 13.2 MB (2.2%) and IAMonDB from 2 064 MB to 45.7 MB (2.2%), with 2.2× and 2.2× load speedups respectively. Detexify's SQL dump must be scanned in full to extract any number of samples; InkTree eliminates this overhead entirely, yielding a 10.9× speedup over the full N = 210,454 split; the InkTree file is 181.6 MB (17.2%), which includes per-point timestamps for all 210 454 samples. For Unipen (N = 79,452, full converted split), InkTree reduces storage from 57.1 MB (matched-sample equivalent of the archive) to 9.3 MB (16.2%) and loads 10.3× faster. This is achieved by a single-pass streaming loader that caches segment file contents (~4 MB) and include file text (~490 MB) during one sequential forward scan of the gzip archive, avoiding the backward-seek penalty of random access.

Table 1. InkML vs. InkTree: file size and loading performance. Size % = InkTree / InkML; ms/s = ms per sample; ↑ = higher is better; ↓ = lower is better.

Dataset	N	InkML		InkTree		Size %↓	Speed × ↑
		MB	ms/s	MB	ms/s		
CROHME 2023 Test	2 300	28.95	0.491	8.06	0.232	27.8%	2.1×
CROHME 2019 Test	1 199	9.27	0.505	2.21	0.155	23.9%	3.2×
CROHME 2016 Val	1 147	9.28	0.448	2.32	0.169	25.0%	2.7×
CROHME 2023 Val	555	7.23	0.398	2.01	0.339	27.8%	1.2×
CROHME Real Train	12 010	113.04	0.419	28.16	0.219	24.9%	1.9×
CROHME+ Synth.*	2 000	9.46	0.214	2.19	0.063	23.2%	3.4×
MW+ Test	5 739	55.52	0.583	17.61	0.242	31.7%	2.4×
MW+ Val	9 336	85.52	0.410	26.21	0.185	30.7%	2.2×
MW+ Symbols	6 276	11.82	0.190	2.39	0.020	20.2%	9.5×
MW+ Train*	2 000	20.26	0.372	6.09	0.169	30.1%	2.2×
MW+ Synthetic*	2 000	29.19	0.570	7.88	0.280	27.0%	2.0×

Table 2. Original source format vs. InkTree for non-InkML datasets. Size % = InkTree / source; ms/s = ms per sample; † = full split. For DeepWriting and Unipen, source MB is normalized to the matched sample universe. JSON source sizes (DeepWriting, IAMonDB) include metadata beyond stroke data not stored in InkTree.

Dataset	Fmt	N	Source		InkTree		Size %↓	Speed ×↑
			MB	ms/s	MB	ms/s		
DeepWriting†	`.json`	5 159	588.3	0.562	13.2	0.257	2.2%	2.2×
IAMonDB†	`.json`	11 242	2063.6	0.731	45.7	0.339	2.2%	2.2×
Detexify†	`.sql`	210 454	1058.2	0.547	181.6	0.050	17.2%	10.9×
Unipen†	`.tgz`	79 452	57.1	0.147	9.3	0.014	16.2%	10.3×

5.5 Summary Figures

Figures 3 and 4 visualize size ratio and loading speedup for all benchmark configurations. InkTree is smaller than the source format in all 15 cases. DeepWriting and IAMonDB achieve the smallest size ratios (2.2% each), because their JSON source files contain extensive metadata beyond stroke data that InkTree does not retain (InkTree stores stroke data and per-point timestamps but not image paths, user IDs, or curation flags). Across all 15 configurations, Detexify shows the largest speedup (10.9×), closely followed by Unipen (10.3×) and MW+ Symbols (9.5×); InkML-based datasets (including CROHME+) see consistent 1.2–9.5× speedups.

5.6 Dataset Structure Snapshot

For the CROHME 2023 test split ($N = 2{,}300$), the converted InkTree file contains 21 684 symbol nodes, 29 850 strokes, and 101 unique symbols. Mean formula statistics are 12.65 nodes, 9.43 symbol nodes, 12.98 strokes, and depth 2.17. SymbolNode dominates node frequency (74.5%), followed by RowNode (11.7%), SupNode (5.0%), SubNode (3.3%), and FracNode (3.0%). These values confirm that the benchmark reflects realistic structural variation and not only trivial expressions.

6 Discussion

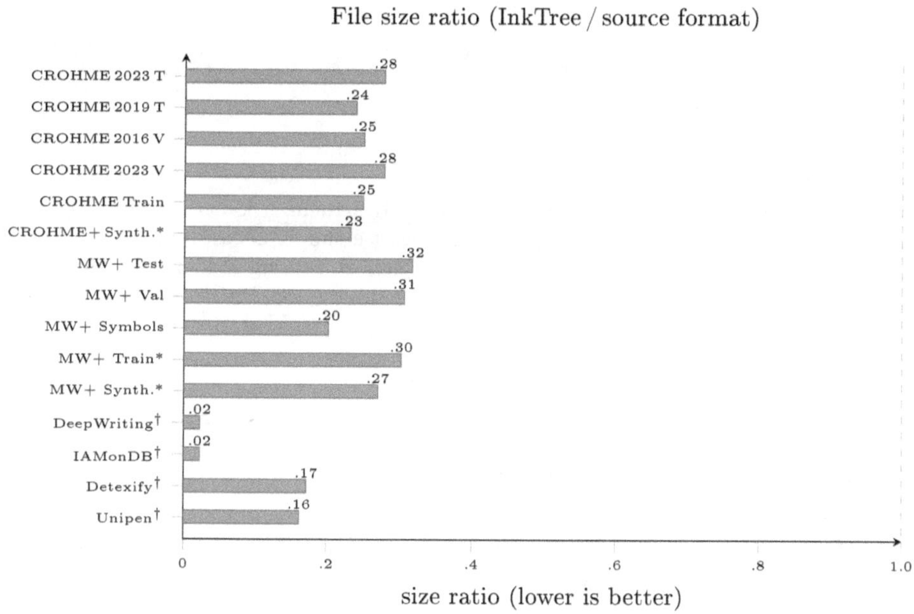

Fig. 3. File-size ratio (InkTree / source) for all 15 benchmark configurations (lower is better). Bottom four rows use original proprietary source formats (with matched-sample normalization where needed); top eleven use InkML. DeepWriting and IAMonDB achieve very small ratios because their JSON source files contain additional metadata beyond stroke data. Entries marked * use a random sample; † = full split.

The results support the main engineering claim: when traces and structure are co-located with explicit semantic roles, downstream loading and manipulation become cheaper. The gain is modest per sample but accumulates in iterative training workflows, especially when datasets are repeatedly scanned for batching, augmentation, or graph extraction. For the CROHME Real Train split alone

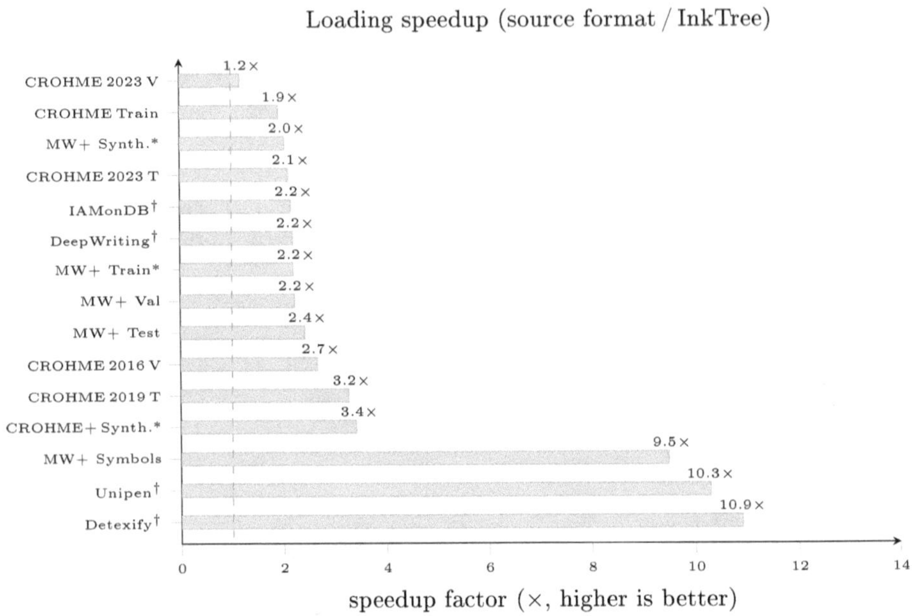

Fig. 4. Loading speedup (source / InkTree) for all 15 benchmark configurations (higher is better). Detexify (†) yields the largest gain because the SQL dump must be scanned in full; Unipen (†) achieves 10.3× via a single-pass streaming loader that eliminates backward seeks; all InkML-based datasets (including CROHME+) show consistent 1.2–9.5× speedups.

(N = 12,010), switching from InkML to InkTree saves ∼85 MB of disk access and reduces load time from 5.0 s to 2.6 s per full scan, a 1.9× improvement that compounds across training epochs.

A second observation is methodological. By making relations explicit at the serialization level, InkTree lowers the amount of format-specific adapter code needed between datasets and model families. This is relevant for hybrid pipelines that combine sequence and graph views, and for future tasks beyond formula transcription. The four non-InkML datasets (DeepWriting, IAMonDB, Detexify, Unipen) demonstrate this generality: the same InkTree schema accommodates isolated symbol samples (Detexify, Unipen) and cursive-text sequences with character segmentation (DeepWriting, IAMonDB) without per-dataset format conventions.

This also aligns with relation-aware OHMER pipelines: explicit relations remove cross-layer ID resolution, and consistent node/edge graph views can be derived directly for GNN training and evaluation [26,27,30].

To isolate where InkML loading time is spent, we profiled the InkML pipeline phase-wise across all InkML-based subdatasets used in this work: (i) parser/frontend work (InkML initialization, trace extraction/grouping, and preprocessing) and (ii) relation-graph construction (`get_relation_graph` + `fix`). Figure 5 reports a compact grouped view (MW Symbols, MW Real, MW Synthetic, CROHME Train/Val/Test, and CROHME+). Across all groups, parser/frontend work remains dominant ($\sim$ 88% to 93%), while relation-graph construction is consistently smaller ($\sim$ 6% to 11%). This supports the claim that the main practical bottleneck in InkML-centric pipelines is format parsing and layer resolution rather than relation decoding itself.

The per-sample comparison also reveals that the speedup depends on source-format complexity. MW+ Symbols (0.190 ms/sample InkML, 0.020 ms/sample InkTree, 9.5×) are significantly cheaper than full expressions (CROHME Train: 0.419 ms/0.219 ms, 1.9×), reflecting the overhead of MathML relation-graph construction in InkML. Detexify (10.9×) and Unipen (10.3×) show the largest gains because their original formats are not designed for per-sample random access. DeepWriting and IAMonDB achieve small size ratios (2.2% each) because their JSON source files bundle substantial metadata—image paths, user identifiers, validation flags—that InkTree does not retain; the speedup is more modest (2.2× each) as JSON parsing is faster than SQL or `.tgz` scanning.

At the same time, benchmark values should be interpreted as implementation-dependent rather than absolute. Our focus is reproducible relative behavior within one controlled pipeline. A binary container format (e.g., `.npz` or `.h5`) can store the same arrays faster, but would still require an external shared convention for tree structure and relation roles, reintroducing the format fragmentation InkTree removes. InkTree's value is a self-describing, human-inspectable, interoperable encoding of structured ink—not minimal byte count.

7 Limitations

Our evaluation covers one implementation stack and does not yet provide cross-language parsers, extensive ablation on compression strategies, or direct training-time impact on end-task accuracy. We do not yet report peak memory or end-to-end training throughput, which are important in large-scale learning pipelines. Regarding *random-access latency* (single-sample fetch): our default storage uses gzip-compressed JSONL, which is a streaming format without native seek support. Therefore, accessing sample i requires scanning from the beginning ($O(i)$). In a micro-benchmark on CROHME 2023 test ($N = 2{,}300$), this yields $\approx$47 ms mean latency per random fetch in our Python implementation, compared with $\approx$1 ms when opening a single InkML file (numbers depend on hardware and parsing stack). In common training pipelines this is typically not limiting because data loading is predominantly sequential; for interactive/random-access workloads, simple mitigations include caching the split in memory after the first pass or using seekable gzip variants (e.g., indexed gzip/BGZF) without changing the per-sample representation. Large splits (MathWriting+ Train, Synthetic) are evaluated on random samples of $N = 2{,}000$; full-split results may differ slightly but per-sample metrics are expected to generalize. Detexify, Unipen, DeepWriting, and IAMonDB are evaluated over their full splits. The size advantage for DeepWriting and IAMonDB over their JSON sources partly reflects metadata not stored in InkTree; size comparisons against a stroke-only JSON baseline would be smaller. While the schema includes generic node categories (`any`, `noisy`), edge cases in heterogeneous document domains may require additional normalization conventions.

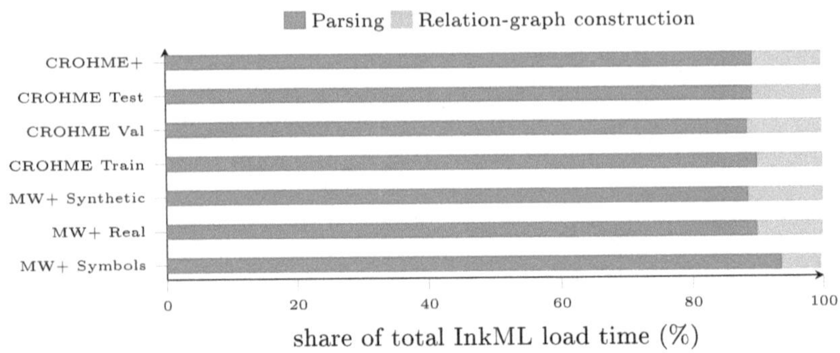

Fig. 5. Phase-wise decomposition of InkML loading time (weighted by sample count within each group). Parsing dominates in all groups; explicit relation-graph construction remains a smaller share.

Additional per-point channels available in some sources are not yet propagated: IAMonDB pen pressure and word-level OCR labels could be incorporated as a loader extension without schema changes. InkTree is intended as an operational representation for model development; the original source formats remain authoritative for full archival fidelity. A direct next step is to evaluate InkTree as a shared substrate for cross-dataset pretraining and multi-task learning, including adaptation of generalist backbones (e.g., vision-language models) via tokenized ink representations [4,6].

8 Conclusion

InkTree is a compact, relation-oriented representation for structured online handwriting that unifies traces, structure, and semantic roles in a single self-contained object. Across 15 benchmark configurations spanning seven dataset families, InkTree reduces file size by up to 98% and accelerates loading by 1.2×–10.9× compared with each dataset's original source format.

The proposed schema and tools target practical research throughput: less parsing glue, fewer fragile cross-references, and easier derivation of relational views for modern models. Companion tooling (see repository link above) includes conversion, plotting, graph/symbol extraction utilities, and LaTeX/MathML export interfaces, with relation-consistency checks and extended node coverage as the next release step. Future work will extend evaluation to additional languages, study the direct impact on end-to-end training and recognition performance, and investigate schema extensions for multi-line and diagrammatic content.

References

1. Agrawal, M., Bali, K., Madhvanath, S., Vuurpijl, L.: Upx: A new xml representation for annotated datasets of online handwriting data (2004). https://scispace.com/pdf/upx-a-new-xml-representation-for-annotated-datasets-of-1ptjbakl9m.pdf, pDF available online; Accessed 14 June 2026
2. Aksan, E., Pece, F., Hilliges, O.: DeepWriting: Making digital ink editable via deep generative modeling. In: Proceedings of the ACM CHI Conference on Human Factors in Computing Systems (2018). https://doi.org/10.1145/3173574.3173779
3. CROHME 2023 Organizers: Icdar 2023 crohme datasets (2023). https://zenodo.org/records/8428035, zenodo record Accessed 27 Feb 2026
4. Fadeeva, A., et al.: Representing online handwriting for recognition in large vision-language models (2024)
5. FKI / HEIA-FR: Iam on-line handwriting database (iam-ondb) — download page and xml format description. https://fki.tic.heia-fr.ch/databases/download-the-iam-on-line-handwriting-database (2026). Accessed 27 Feb 2026
6. Garrido-Munoz, C., Calvo-Zaragoza, J.: On the generalization of handwritten text recognition models. In: IEEE/CVF Conference on Computer Vision and Pattern Recognition, CVPR 2025, Nashville, TN, USA, June 11-15, 2025, pp. 15275–15286. Computer Vision Foundation / IEEE (2025). https://doi.org/10.1109/CVPR52734.2025.01423
7. Garrido-Munoz, C., Ríos-Vila, A., Calvo-Zaragoza, J.: Handwritten text recognition: A survey. CoRR **abs/2502.08417** (2025). https://doi.org/10.48550/ARXIV.2502.08417
8. Gervais, P., Fadeeva, E., et al.: Mathwriting: A dataset for handwritten mathematical expression recognition (2024). https://arxiv.org/abs/2404.10690, arXiv preprint; Accessed 14 June 2026
9. Google Creative Lab: The quick, draw! dataset (repository and download instructions) (2017). https://github.com/googlecreativelab/quickdraw-dataset, Accessed 27 Feb 2026
10. Ha, D., Eck, D.: A neural representation of sketch drawings (2017). https://doi.org/10.48550/arXiv.1704.03477
11. Hu, R., Watt, S.M.: From mit sketchml to inkml or there and back again. In: Document Analysis Systems (DAS) (2012). https://cs.uwaterloo.ca/~smwatt/pub/reprints/2012-das-sketchml-inkml.pdf, Accessed 14 June 2026
12. Indermühle, E., Liwicki, M., Bunke, H.: Iamondo-database: An online handwritten document database with non-uniform contents. In: Proceedings of the International Workshop on Document Analysis Systems (DAS) (2010). https://doi.org/10.1145/1815330.1815343
13. International Unipen Foundation: Unipen data set of on-line (vectorial) handwriting (1998). https://zenodo.org/records/1195803, zenodo record (published 2018); Accessed 27 Feb 2026
14. Kirsch, D.: Detexify: Latex symbol classification using neural networks. In: International Workshop on Document Analysis Systems (DAS) (2011). see also https://detexify.kirelabs.org
15. Kirsch, D., contributors: detexify-data: Instructions and license for detexify's sample data (2014). https://github.com/kirel/detexify-data, Accessed 27 Feb 2026
16. Li, M., et al.: Trocr: Transformer-based optical character recognition with pre-trained models. CoRR **abs/2109.10282** (2021)

17. Liu, C.L., Yin, F., Wang, D.H., Wang, Q.F.: Casia online and offline chinese handwriting databases. In: Proceedings of the International Conference on Document Analysis and Recognition (ICDAR) (2011). https://doi.org/10.1109/ICDAR.2011.17
18. Mahdavi, M., Zanibbi, R., Mouchère, H., Viard-Gaudin, C., Garain, U.: Icdar 2019 crohme + tfd: Competition on recognition of handwritten mathematical expressions and typeset formula detection. In: Proceedings of ICDAR (2019). https://www.cs.rit.edu/~rlaz/files/CROHME%2BTFD%E2%80%932019.pdf, Accessed 14 June 2026
19. Matsushita, T., Nakagawa, M.: A database of on-line handwritten mixed objects named Kondate. In: Proceedings of the International Conference on Frontiers in Handwriting Recognition (ICFHR) (2014). https://web.tuat.ac.jp/~nakagawa/pub/2014/pdf/Matsushita_ICFHR2014.pdf, Accessed 14 June 2026
20. Mehralian, P., BabaAli, B., Mohammadi, A.G.: Self-supervised representation learning for online handwriting text classification (2023)
21. Mouchère, H., Viard-Gaudin, C., Zanibbi, R., Garain, U.: Crohme 2012: Competition on recognition of online handwritten mathematical expressions. In: Proceedings of the International Conference on Frontiers in Handwriting Recognition (ICFHR) (2012). https://www.iapr-tc11.org/dataset/CROHME/CROHME_ICFHR_2012.pdf, Accessed 14 June 2026
22. Mouchère, H., Viard-Gaudin, C., Zanibbi, R., Garain, U., Kim, D.H., Kim, J.H.: Icdar 2013 crohme: Third international competition on recognition of online handwritten mathematical expressions. In: Proceedings of the International Conference on Document Analysis and Recognition (ICDAR) (2013). https://doi.org/10.1109/ICDAR.2013.288
23. Mouchère, H., Zanibbi, R., Garain, U., Viard-Gaudin, C.: Advancing the state-of-the-art for handwritten math recognition: The crohme competitions, 2011–2014 (2014). https://www.cs.rit.edu/~rlaz/files/CROHME_revision.pdf, overview manuscript. PDF available from RIT; Accessed 14 June 2026
24. Seitz, J.: Crohme+ (2025). https://zenodo.org/records/14968570, https://doi.org/10.5281/zenodo.14968570, zenodo record; Published Mar 2025 (v1). Derived from CROHME 2023 data
25. Seitz, J.: Mathwriting+ (2025). https://zenodo.org/records/15091575, https://doi.org/10.5281/zenodo.15091575, zenodo record; Published Jun 2025 (v2). Structural annotations over MathWriting
26. Seitz, J., Lengfeld, T., Timofte, R.: The return of structural handwritten mathematical expression recognition (2025). https://doi.org/10.48550/arXiv.2508.19773
27. Truong, T.N., Nguyen, C., Zanibbi, R., Mouchère, H., Nakagawa, M.: A survey on handwritten mathematical expression recognition. Pattern Recogn. (2024). https://doi.org/10.1016/j.patcog.2024.110531
28. Truong, T.N., Nguyen, H.T., Nguyen, C.T., Nakagawa, M.: Learning symbol relation tree for online mathematical expression recognition (2021). https://doi.org/10.48550/arXiv.2105.06084
29. W3C: Ink markup language (inkml) (2011). https://www.w3.org/TR/InkML/, w3C Recommendation, 20 Sep 2011
30. Xie, Y., Zanibbi, R., Mouchère, H.: Local and global graph modeling with edge-weighted graph attention network for handwritten mathematical expression recognition (2024)
31. Zanibbi, R., Blostein, D.: Recognition and retrieval of mathematical expressions. Int. J. Document Anal. Recogn. (IJDAR) **15**(4), 331–357 (2012). https://doi.org/10.1007/s10032-011-0174-4

32. Zanibbi, R., Mouchère, H., Viard-Gaudin, C.: Stroke-based performance metrics for handwritten mathematical expression recognition (2011). https://www.cs.rit.edu/~rlaz/files/MathMetricICDAR2011.pdf, iCDAR 2011. PDF available from RIT; Accessed 14 June 2026
33. Zanibbi, R., Mouchère, H., Viard-Gaudin, C.: Evaluating structural pattern recognition for handwritten math via primitive label graphs. In: Document Recognition and Retrieval (DRR) (2013). https://www.cs.rit.edu/~rlaz/files/ZMV_DRR2013.pdf, Accessed 14 June 2026

Democratizing the Medieval English Legal Tradition

Michael Zhang[1(✉)], Elise Wang[2], Charlotte Whatley[3], Seth Strickland[4], and Dylan Bannon[5]

[1] University of Chicago, Chicago, IL 60637, USA
mzzhang2014@gmail.com
[2] California State University, Fullerton, CA 92831, USA
[3] University of Wisconsin-Madison, Madison, WI 53706, USA
[4] Carnegie Mellon University, Pittsburgh, PA 15213, USA
[5] Brooklyn, USA
https://astro.uchicago.edu/ mz/

Abstract. The record of the beginning of the most widespread legal system in the world is contained in millions of pages of handwritten text. Most of the records of the first centuries of the Anglo-American legal system are hand-written in a highly abbreviated form of medieval Latin which only a few dozen scholars in the world are trained to read. In this interdisciplinary project, we construct a dataset of 4029 lines of text across 193 medieval criminal and civil cases. We then use the dataset to train an open-source end-to-end pipeline for transcribing these manuscripts. We first train standard neural network architectures for line segmentation and handwriting recognition (R-Blla and CNN+LSTM with CTC decoding, respectively) and show that they can already achieve 79% word accuracy, despite the relatively small training set and the challenge of expanding abbreviations. We then demonstrate that simple post-processing significantly boosts accuracy: adding an n-gram language model to the CTC decoder improves word accuracy to 82%, while asking Gemini Pro 3 to correct mistakes boosts accuracy to 88%. Finally, we compare the CNN+LSTM architecture with TrOCR, a transformer-based OCR architecture, demonstrating that TrOCR shows comparable word accuracy but worse character accuracy due to its over-willingness to guess, making it harder for humans to infer the correct reading. We incorporated our pipeline into a web portal (glyphmachina.com), opening up the English legal tradition to legal scholars, medievalists, and students.

Keywords: Handwritten Text Recognition · Latin

1 Introduction

The English common law emerged as a unified legal system over the course of the twelfth and thirteenth centuries. Common law functions through the accumulation of cases; in the most basic sense, each case provides guidance for how

G. A. Fink et al. (Eds.): ICDAR 2026, LNCS 16972, pp. 378–394, 2027.
https://doi.org/10.1007/978-3-032-36023-6_22

similar, future cases ought to be decided. As a result, records are of paramount importance to the system, and were meticulously preserved, both at the king's court and in local archives. Over centuries, this body of law formed the basis for the common law system that persists to this day and has left a lasting legacy on every continent in the world. It is the basis of the legal system not just in the United Kingdom, but also in many of its former colonies, including the United States, India, and Tanzania. It is estimated that 35% of the world's population live under either a common law system or a mixed system with common law as one of the main components.[1]

Despite the immense impact of this tradition, very few historians are able to directly study the medieval cases on which it is based. In recent years, this has not been due entirely to lack of access. In fact, most of the extant legal record from before 1500 (preserved on parchment in a handwritten script) has been digitized and made available online at the Anglo-American Legal Tradition (AALT) website (http://aalt.law.uh.edu/). To read these sources, however, one must be trained in medieval legal Latin, in the idiosyncratic system of abbreviation that these records use, and in deciphering the handwriting of scribes who were writing more for speed than long-term legibility. Furthermore, mere access is not enough to use the archives – scholars need an instinct for what is important, decades of experience in the currently available records, and support or budget to comb through these millions of important pages. Even scholars who have such access rarely use the online archives alone, since they operate better as a digital reference database than a browse-able archive. Thus, though eleven million pages of the legal manuscript records at the National Archives in Kew, London are available online, only a few dozen scholars in the world can make use of them. Around twenty of these rolls are transcribed, translated, and published, but this is a tiny minority of the extant corpus. This means that these records are largely inaccessible to students not yet proficient in paleography, legal scholars who specialize in other periods, and even medievalists who do not regularly work with these archives.

These manuscripts pose a number of unique challenges. Some are written in Anglo-Norman, the language of prestige in post-Conquest England, but the majority are written in Latin. Due to the scribes' sometimes limited Latin proficiency, the syntax and spelling are not always correct. The script is highly abbreviated, so that around half of the words have one or multiple letters omitted. As is typical of medieval scribal productions, some abbreviations are predictable but most are not standardized. Sometimes abbreviations are indicated with a bar or a loop; sometimes they are implicit. Even when words are not abbreviated, letter forms often merge and are often difficult or impossible to distinguish. For example, "i", "n", and "m" are usually indicated with 1, 2, and 3 short vertical lines ("minims") respectively, making it impossible to correctly read a succession of minims without combining the linguistic and syntactic contexts. To top off the challenges, even within a script family, different scribes have different handwriting styles, abbreviation choices, and letter forms, requiring adaptive

[1] https://www.juriglobe.ca/eng/syst-demo/tableau-dcivil-claw.php.

recontextualization even for expert users. These difficulties mean that even a trained paleographer can require around five minutes to decipher a single line of text, making archival searches painstaking and time consuming.

In this work, we use artificial intelligence to enable an end-to-end pipeline to transcribes medieval English legal manuscripts quickly and accurately. We also release the dataset we used to train the AI in PageXML format. To our knowledge, it is the only publicly available dataset of medieval English legal text amenable to training neural networks.

2 Related Works

Handwriting text recognition (HTR) has long been a field of active research. We highlight some milestones here, but refer the reader to [7] for a more detailed survey. The first commercial products incorporating handwriting recognition were introduced in the 1980s. In the 1990s, the harder problem of recognizing handwritten text from an image was usually tackled using a combination of character-level segmentation and handcrafted features (e.g. [11]). The advent of powerful programmable GPUs ushered in the deep learning era, enabling the training of neural networks to transcribe entire lines of text with impressive accuracy without any handcrafted features. In 2015, [16] first proposed the deep learning architecture that would become standard: a convolutional neural network feeding into a bidirectional long short-term memory (LSTM) recurrent neural network, all trained with CTC loss. In 2017, the attention-based transformer architecture [17] ushered in a new paradigm and revolutionized deep learning. The first work to utilize the transformer architecture for HTR was [9], who obtained state-of-the-art error rates by using a ResNet CNN as a visual feature extractor and a series of multi-head attention layers in both the encoder and decoder. [5] obtained even better performance with a CNN backbone coupled with a self-attention encoder and a CTC decoder. In 2021, Microsoft released TrOCR [12], which dispenses with CNNs entirely and uses the transformer architecture for both image processing and text generation. In 2023, [6] introduced DTrOCR, which achieves substantially lower error rates than all previous models by removing the encoder in TrOCR and using only the decoder. The input image is broken up into 16×16 patches, which are converted into vectors with a patch embedding before being fed to a decoder initialized with GPT-2 weights. The model is pre-trained with two billion synthetic line images in 5427 handwritten fonts, and then fine-tuned on real data. The pre-trained DTrOCR model is, however, not public.

In parallel to these algorithmic improvements, researchers have developed HTR platforms such as Transkribus [8], kraken,[2] and PyLaia [15]. These platforms allow users unfamiliar with ML-specific tools such as PyTorch or TensorFlow to either train a neural network on their own data, or use a pretrained network from the literature. Transkribus has seen especially wide adoption across many disciplines [13].

[2] https://kraken.re/main/index.html.

A few works have focused specifically on handwriting recognition with medieval manuscripts. [2] trained a CNN + LSTM architecture with CTC loss on 120,000 lines of French and Latin, obtaining a final word error rate of 16% for the Textualis script and 28% for the Cursiva script. [3] released CATMuS, a dataset of 160,000 text lines in 10 different languages with Latin scripts, spanning the 8th to 16th centuries. They also released kraken and PyLaia models trained on their dataset. [1] constructed an open-source corpus of medieval and early modern manuscripts (TRIDIS) spanning the 12th to 16th centuries and trained two HTR models, TrOCR and MiniCPM-Llama3-V. They achieve a WER of 21–25% and a CER of 9–11% with TrOCR.

Recently, [4] evaluated state-of-the-art off-the-shelf models on a new dataset, including models trained on the CATMuS and TRIDIS datasets. They find that these models are still far from providing a generalizable transcription solution, but provide a good foundation for fine-tuning to task-specific datasets.

3 Dataset

For our dataset, we transcribed 4029 lines (3602 for training, 427 for testing) from cases drawn from the rolls digitized on the AALT archive. Each case was handwritten in abbreviated Latin in a script called *cursiva anglicana* or chancery cursive on a "membrane" of parchment. These membranes were then sewn together at the top (like a legal pad) into "roll," which were organized by the court they came from, the type of case, and the term in which they occurred. In general, we sought cases that would represent the archive as it is most useful to a researcher. This meant that we focused on the period from Edward I to Henry VI (1272-1461), which uses a consistent script and relatively consistent forms. These records are far more numerous and far less organized than those of the twelfth and early thirteenth centuries, and as a result only a small percentage of them have been transcribed.

Among Latin datasets, our dataset offers an unusual corpus that, despite its size, is a better representation of "everyday" documentary Latin than other existing datasets. Latin was the documentary language of the medieval European world, which meant that almost all legal, financial, trade, and property documents were written in Latin. Most datasets (and collections of Latin texts, like the *Corpus Corporum*) collect texts that have been important to humanities researchers—literature, philosophy, and political treatises—but that are not terribly representative of everyday life.

Among manuscript HTR datasets, ours is again unique, in part due to the formulaic nature of these documents, the standardization of the vocabulary, and the large size of the corpus. Scripts change over time, sometimes rapidly, but in this case, because these records formed the basis of a legal system that constantly referenced its own history, the script stayed relatively stable for at least 200 years. The specialized vocabulary and abbreviations in these records are legion but also relatively stable, and as a result, the 4000 lines we extracted are representative enough of a corpus of at least 220 million (11 million images, each with around 20–35 lines on them).

3.1 Data Preparation

Rather than working at the roll or membrane level, we chose individual cases, because most researchers read the rolls in that way, and because it allowed us to sample the dataset much more widely. On average, we have not included more than three cases from the same roll. In our selection, we chose cases that would give us two types of variety: case type (varying the crime, severity, etc.) and "hand" (the individual handwriting style). In search of a representative sample, we drew from three types of rolls: the King's Bench Rolls (KB), the Common Plea Rolls (CP), and the eyre rolls (JUST1). We also included the 51 example lines that AALT uses to showcase Latin palaeography.[3]

Table 1. Number of cases, lines, and words from each type of roll in the dataset.

Roll Type	Cases	Lines	Words
King's Bench (KB)	54	916	15,495
Common Pleas (CP)	39	887	14,746
Justices Itinerant (JUST1)	91	2175	38,619
Palaeography Assistance	9	51	892

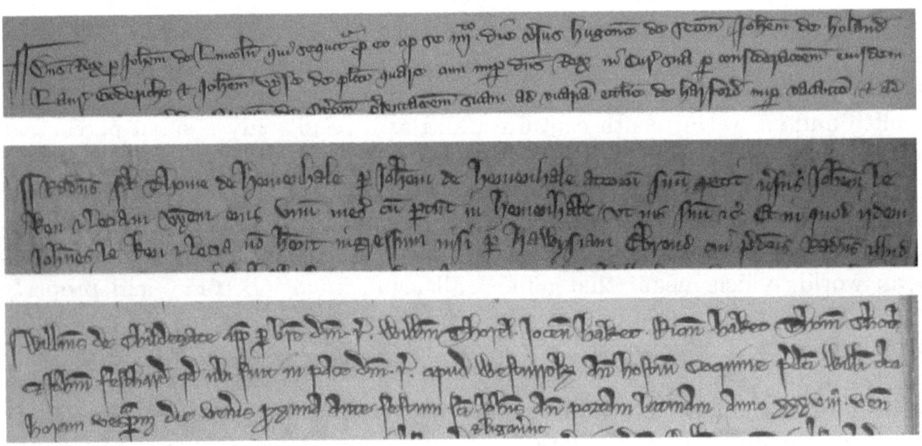

Fig. 1. Samples from our dataset: KB27 349m21, CP40 169m62, JUST1 235m13.

The King's Bench was the most senior criminal court of England. We drew primarily from KB27, known as the *Coram Rege* rolls, which provide a mix of

3 http://aalt.law.uh.edu/Pal1.html, http://aalt.law.uh.edu/Pal2.html, http://aalt.law.uh.edu/Pal3.html.

subject, place, and stage of process. The Common Pleas Rolls were also crown business, but include a wider variety of cases. The scribes of the KB and CP rolls were well-trained and tend to be regular, use standard abbreviations, and follow a familiar structure (top panel of Fig. 1). Because the KB and CP rolls are especially important to the formation of the common law, about half of our data is drawn from these rolls. The JUST1 rolls were the records of the justices in eyre, or the court that the crown sent across the island to try serious cases at each county. These records are more idiosyncratic in almost all ways: in their organization and indexing, in their vocabulary, and in the types of cases they recorded. JUST1 hands tend to be messier, more inconsistent in abbreviation, and sometimes follow English word order because the scribe translated on the fly into Latin (see bottom panel). Most scholars accept Ralph Pugh's estimate that criminal cases in eyre took on average fifteen minutes [14], during which time the scribe would also complete his record. All of these features make them harder to transcribe for the average paleographer, so we included a large number of them in order to train the model on this more difficult type with the aim of making the type accessible (Table 1).

3.2 Constraints in Data Preparation

A major constraint on the dataset has been human labor. EW and CW performed the transcriptions by hand and cross-checked each other's work, but this process is extremely time-consuming. Both estimate that a case of around fifteen lines took at minimum an hour to transcribe correctly, and another fifteen minutes to check for correctness. SS is trained in paleography, but primarily in medieval English and non-legal Latin. A case might take him hours; unfamiliarity with legal abbreviations in these cases would make repeated consultation with EW or CW necessary, thus limiting the efficacy of a transcriber ostensibly trained in a similar skillset. We point this out to indicate just how specialized working knowledge of these scripts are. Predictably, the KB and CP rolls sometimes went faster, and the JUST1 rolls much slower.

Drawing from the AALT archive has been fruitful but has also resulted in several constraints. First, because we are working from another team's digitization efforts which produced images of varying quality, we have had to select cases more for image quality than for interest or hand. This has meant that while (for example) we might have wished to draw more from coroners' rolls because they offer a wide variety of hands, the quality of the images on the AALT made them poor training data. Second, the AALT has only digitized crown cases, which means that cases from manorial and other local courts have been excluded. However, because both Latin and the documentary script used for its records were standard for almost all legal records of the period, whether local or crown, the AALT examples nevertheless end up providing a good representation for the wider documentary practice of the period. We thus believe that this tool has immediate application on manorial and local records yet to be digitized.

3.3 Test Set

We chose a test set of twenty cases comprised of about half "standard" cases: ones where the handwriting and abbreviations were of average difficulty and where the subject was not very unusual. We imagine such cases would make up most of what researchers would be interested in (for an example, see KB27 263m21). The other half was made up of "difficult" and "unusual" cases, ones that covered less common subject matter (like JUST1 633m52, a complex case of debt and book theft) or had especially difficult hands (like JUST1 230m2). Although we imagine researchers will use the tool to handle more cases like the former than cases like the latter, we included the "unusual" half of the test set to test the model's effectiveness on edge cases that are difficult even for human paleographers. The test set that we chose does not overlap with the training set, but there might be a few cases in which the same scribe was responsible for both a test case and a training case, because many of the scribes (in particularly for the King's Bench) wrote many rolls. Among the JUST1 and CP examples this is highly unlikely.

4 The Pipeline: Design and Training

Our pipeline is available as a web portal at https://glyphmachina.com. The user can upload an image, crop it to the relevant case, have a neural network identify the bottoms of all lines of text (the "baselines"), and have another neural network transcribe the text. All of the above takes mere seconds. The user can then send it to Gemini Pro 3 to correct transcription errors, which takes a few minutes. The user can also send the resulting Latin transcription to Gemini for translation into modern English. In this section, we briefly describe the main steps of the pipeline, with a focus on the transcription step.

4.1 Segmentation

For our purposes, segmentation only involves extracting images of individual lines of text from a manuscript image. The vast majority of medieval English legal documents have a simple layout, and our web portal allows users to crop out only the block of text they are interested in. For these reasons, we adopt a simple segmentation algorithm which works well in practice.

We isolate line images by training a baseline segmentation model using the OCR engine kraken.[4] We use the default R-Blla architecture [10], except that all images are resized to a width of 1800 pixels instead of a height of 1800 pixels. The R-Blla architecture consists of 5 convolutional layers connected to two LSTM layers, one for the x dimension and one for the y dimension, followed by a 1×1 convolution bottleneck and another two LSTMs (for the y and x directions). The output consists of 3 heat maps: one indicates the beginnings of baselines, one indicates the baselines themselves, and the third indicates the ends. During

[4] https://github.com/mittagessen/kraken.

inference, kraken finds the baselines from these heat maps and draws polygons around the text lines. We optimized this portion of the pipeline, reducing its runtime from 52 s to 2.6 s on an Intel Core i9 13900k CPU and a RTX 4090 GPU. Most of the reduced runtime came from removing the polygon-drawing algorithm; most of the rest came from offloading CPU-intensive image processing tasks to the GPU using cuCIM.[5]

After identifying the baselines, we extract line images by defining regions less than $0.73H$ pixels above and less than $0.23H$ pixels below the baseline as part of the text line, where H is the median spacing between baselines. We then rectify the line images by shifting the pixels in each column up or down using linear interpolation so that the baseline in the rectified image is horizontal. Finally, we convert the image to grayscale, normalize it by subtracting the pixel value median and dividing by the difference between the 90th and 10th percentiles, and invert the result so that text is bright while the background is near 0.

Fig. 2. A line image generated by our algorithm (top) and by kraken (bottom). We obtain 4% lower word error rate using our own algorithm. Image from: JUST1 734m1, line 19.

Our line image extraction method is far simpler than *kraken*'s sophisticated algorithm, which uses the baseline as a starting point to draw a polygon around the text line, maximizing the text included inside the polygon while minimizing overlap with adjacent lines. kraken's algorithm can handle lines of very different heights, and includes ascenders and descenders to the maximum extent possible while zeroing out strokes from adjacent lines (see Fig. 2). We expected superior transcription accuracy using kraken's line images, but observed the opposite: the word error rate of the CNN + LSTM model (Table 3) was 4% lower using our custom line images. We suspect this is due to a combination of the strong gradient between the zero filling and the line image, variability the vertical location of the baseline, and inconsistency of the unsubtracted background in kraken's line images. We encourage more research in this area, because nearly every other publicly available dataset of medieval Latin handwriting (e.g. TRIDIS, CATMuS) uses line images generated by a very similar algorithm to kraken's.

4.2 Transcription

We implemented two very different models to transcribe the line images. The first is a classic CNN + LSTM, with a CTC decoder coupled with an n-gram language model. The second is the transformer-based TrOCR.

[5] https://github.com/rapidsai/cucim.

CNN + LSTM. This model is based on the architecture in [2]. Line images are resized to a height of 128 (preserving their aspect ratio) and fed into four convolutional blocks. Each block consists of a 2D convolution layer, a ReLU activation function, a 2D batch normalization step, and a max pool with a 2×2 kernel size and 2×2 stride. We replaced the dropout layers in [2] with batch normalization layers because we find that this change greatly improves ease of training while having no perceptible negative impact on the test set character error rate or word error rate.

We feed the output of the four convolutional blocks into three bidirectional LSTM layers with dropout layers sandwiched in between. Each LSTM layer has a latent dimension size of 512 per direction (twice the size of the LSTMs in [2]), meaning that the final output is of size 1024 x N. A fully connected layer converts this to $((N_{\text{char}} + 1) \times N$ logits, indicating the likelihood that a given column is occupied by a given character (the N_{char} rows) or marks the beginning of a new character (the remaining row). The details of the architecture are laid out in Table 2.

Table 2. The fiducial model architecture (CNN + LSTM). All convolutional blocks include ReLU and BatchNorm2d, and all convolution blocks except the last end with a 2×2 max pool with 2×2 stride. The LSTM layers include Dropout ($p = 0.3$). Total parameters: 19M.

Stage	Layer Type	Configuration
1	Conv Block	$1 \rightarrow 32$ ch, k: 4×16, pad: 1×7, Max Pool
2	Conv Block	$32 \rightarrow 32$ ch, k: 4×16, pad: 1×7, Max Pool
3	Conv Block	$32 \rightarrow 64$ ch, k: 3×8, pad: 1×3, Max Pool
4	Conv Block	$64 \rightarrow 64$ ch, k: 3×8, pad: 1×3
5	Bi-LSTM (×3)	Latent size: 512 (per direction), Input: 960
6	Linear	Input: $1024 \rightarrow$ Output: $N_{\text{char}} + 1$

The neural network is implemented in PyTorch and trained on the ~4000 line training set using the AdamW optimizer and Connectionist Temporal Classification (CTC) loss function. We start with a learning rate of 10^{-3} and weight decay of 10^{-2} but reduce the learning rate by a factor of 3 every time the WER stops decreasing for 10 consecutive epochs, down to a minimum learning rate of 10^{-5}. We train for 250 epochs, but the test set CER and WER change negligibly after ~100 epochs.

We use data augmentation during training to improve performance. During each epoch of training, each line image in the training dataset is randomly jittered in brightness, contrast, saturation, and hue by up to 50%. We then apply a random affine transformation: each image is rotated by up to $\pm 0.7°$, translated by up to $\pm 0.01W$ horizontally and $\pm 0.02H$ vertically, and scaled by up to 2%. In the final step, we either do nothing (1/4 probability), sharpen the image by a factor of 2 (1/4 probability), or apply a Gaussian blur with a randomly chosen standard deviation of 1–6 pixels.

For inference, we find that greedy CTC decoding performs well and beam search has a negligible effect on the results. However, coupling the beam search with a language model lowers the word error rate by 3%. We used pyctcdecode[6] to perform the beam search decoding with the help of a KenLM[7] n-gram model. We adopted n=2 and trained the KenLM model solely on the text we used to train the neural network. Training the n-gram model on the massive *Corpus Corporum*, a Latin library containing 215 million words, did not improve results; in fact, it increased the WER by 1%.

TrOCR. TrOCR [12] is an entirely transformer-based model which comes in three sizes: small (62M parameters), base (334M parameters), and large (558M parameters). Training even the small model from scratch is infeasible with our limited training set. [12] pretrained TrOCR with 684M textlines from PDF files on the Internet, followed by 17.9M synthetic textlines in 5427 handwritten fonts, with the text taken from Wikipedia.

Because our training data is far more limited, we finetune the TrOCR model of [1], which in turn was finetuned from the TrOCR_{LARGE} model using 178,000 lines of medieval and early modern handwriting. We use an initial learning rate of 5×10^{-5} which linearly decays to zero and a batch size of 8, and train for 30 epochs. We tried training for 250 epochs and found that both CER and WER start increasing well before the end of training, indicating overfitting. Data augmentation is performed on the training data using the same procedure as for the CNN + LSTM architecture.

4.3 Correction by Gemini

After obtaining transcriptions of all textlines in the image from our neural network, we call the Gemini API and ask Gemini Pro 3 to correct obvious errors. We do not provide Gemini the image to be transcribed, only the preliminary transcription. We provide the following system prompt:

I ran a handwriting recognition neural network on a medieval English legal case, written in Latin. Please try to correct the mistakes. The neural network doesn't typically add or remove entire words, so try to keep a similar word count, if possible. Also, preserve the line breaks at all costs. If you see a very short line, don't delete it or merge it with the previous or next line, because it's likely an interlinear addition. Keep in mind that this is medieval, not classical, Latin. Please don't add punctuation unless necessary, because these documents usually don't have much punctuation. Do not output anything other than the corrected transcription–no explanations, comments, or alternatives.

---

[6] https://github.com/kensho-technologies/pyctcdecode.

[7] https://kheafield.com/code/kenlm/.

5 Results

In this section, we use the character error rate (CER) and word error rate (WER) to evaluate our models. We also include qualitative insights from expert paleographers, which show that CER/WER is not the whole story.

Table 3. Performance comparison of different models

Model	CER	WER
CATMuS	47	81
TRIDIS (CNN+LSTM)	29	60
TRIDIS (TrOCR)	38	64
CNN + LSTM	6.5	20.6
CNN + LSTM + KenLM	6.0	17.8
CNN + LSTM + KenLM + Gemini	**3.8**	**11.9**
TrOCR	6.9	16.3
TrOCR + Gemini	5.9	14.9

5.1 Quantitative Evaluation

No off-the-shelf model is capable of producing a useful transcription of our dataset, but our models perform well and offer significant time savings to paleographers. This is evident in Table 3, which shows the character error rate and word error rate achieved by both our own models (below the horizontal line) and by recently published models trained on medieval handwriting (above the horizontal line). The standard CNN + LSTM architecture with CTC loss results in an acceptably low word error rate of 21%. Adding a simple bigram language model reduces the WER to 18%, while correction by Gemini 3 Pro dramatically improves the transcription, reducing WER to only 12%. The transformer-based TrOCR architecture outperforms CNN + LSTM + KenLM in WER, but not in CER. After correction by Gemini, the CNN + LSTM + KenLM transcriptions significantly outperform the TrOCR transcriptions in both CER and WER. This suggests that the classical architecture's errors, even if slightly more numerous, are easier to recognize and correct than TrOCR's errors–a hypothesis confirmed by human examination of the transcriptions (see next subsection).

Table 4 shows the WER achieved by the fiducial model on each case within the test set, both with and without Gemini correction. Gemini almost always improves the transcription, but curiously, there is one case where it increases the error rate (KB27-344Rm14). As expected, some cases are easier than others for the neural network to transcribe. However, all transcriptions are accurate enough to be usable, and the error rate is not dramatically different across the three types of manuscript (Common Pleas, Justices in Assize, King's Bench).

Table 4. Word error rates achieved by the fiducial model (CNN + LSTM + KenLM) on each individual case. We also compare to the WERs achieved by Gemini alone (see Subsect. 5.3).

Case	Type	No Gemini	With Gemini	Gemini only
CP40-236m280	Standard	15	7	32
CP40-58m14b	Difficult	25	10	Hallucinated lines
CP40-648m81a	Standard	9	4	No output
CP40-649m116a	Difficult	17	9	20
JUST1-230m2	Difficult	33	21	58
JUST1-235m13	Difficult	27	20	45
JUST1-633m52	Difficult	12	8	No output
JUST1-633m5a	Difficult	9	8	27
JUST1-633m60d	Difficult	15	9	Hallucinated lines
JUST1-734m17da	Difficult	28	17	Hallucinated lines
JUST1-734m24d	Difficult	14	10	28
JUST1-734m4	Difficult	16	9	Hallucinated lines
KB27-263m21	Standard	24	21	54
KB27-645m22a	Standard	37	16	Hallucinated lines
KB27-342Rm22	Standard	14	12	Hallucinated lines
KB27-345Rm87a	Standard	13	10	Hallucinated lines
KB27-345Rm87b	Standard	9	7	88
KB27-359Rm8d	Standard	12	9	Hallucinated lines
KB27-344Rm14	Standard	14	19	Hallucinated lines
KB27-352Rm13	Standard	14	12	94

They are also not dramatically different between the cases we selected to be "standard" and those we selected to be "difficult": the former has a WER of 16.1% prior to Gemini correction, compared to 19.6% for the latter. These rates drop to the nearly identical 11.7% and 12.1% after Gemini correction.

5.2 Qualitative Evaluation

Our team includes three medievalists trained in paleography: EW, CW, and SS. To supplement our quantitative evaluation, we present their subjective evaluation of the quality and usefulness of the transcriptions.

Some aspects of usefulness to the user are more difficult to characterize quantitatively, and these demonstrate that though the Gemini-corrected model has a lower error rate, it might also mislead researchers where the model will not. To illustrate this, we chose an example of a case from a roll the model had never seen before, from about fifty years before our dataset's range (see Fig. 3). We chose this example to simulate how a user might stretch the model's capabilities;

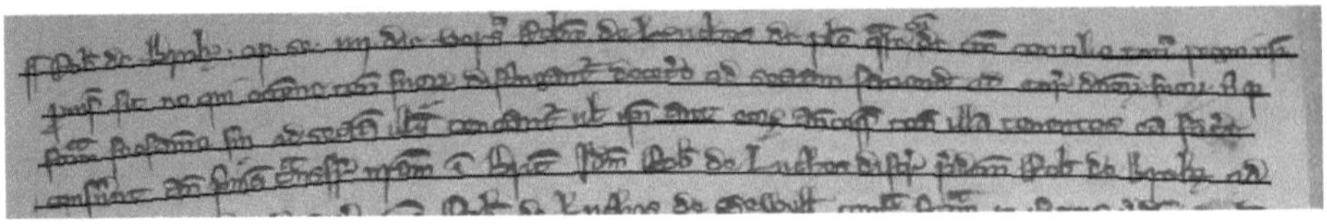

Fig. 3. JUST1 701m6.

the case is outside our time period, the handwriting is not very clear, and it has both idiosyncratic abbreviations and smudges. This difficulty was reflected in the accuracy of the transcription the model provided—85% (in 161 words, the model made 24 errors)—and of the Gemini corrections—94% (10 errors).

For the purposes of the user, there are roughly three types of errors, in ascending order of severity:

1. A word that is misspelled but easily comprehensible (highlighted in green)
2. An incomprehensible word (highlighted in yellow)
3. An incorrect word that is semantically appropriate (underlined and highlighted in red)

The first is easily corrected, and the second cues the user to return to the manuscript. But the third is dangerous to understanding, because it does not signal a possible error and might escape notice.

In the example below, we have taken the first four lines of the test case and highlighted the errors according to their type for both the model's transcription and the Gemini-corrected version (though we have highlighted the entire word, often the mistake is a single letter, as with communi/commune). We then selected a few of the errors to better illustrate the recognition characteristics in a table.

1. Robertus de Brok optulit se iiij die versus Roberto de Luches de placito quare de tonmunti consilio toti regni nostri
2. peris sit ne qui occasione tenementa suorum distringantur decetero ad seatm faciendam ad Curiam dominorum suorum nisi per
3. formam feoffamenti sui ad sectam illam tencantur vel ipsi aut eorum antetesie tenementa illa tenentes ea facere
4. consuerunt ante primam transfretacionem nostram et bricis Idem Robertus de Luches distringet predictum Robertum de Brok ad

1. Robertus de Brok optulit se iiij die versus Robertum de Luches de placito quare cum de communi consilio [] regni nostri
2. provisum sit ne qui occasione tenementorum suorum distringantur decetero ad sectam faciendam ad Curiam dominorum suorum nisi per
3. formam feoffamenti sui ad sectam illam teneantur vel ipsi aut eorum antecessores tenementa illa tenentes eam facere
4. consueverint ante primam transfretacionem nostram in Britanniam idem Robertus de Luches distringit predictum Robertum de Brok ad

Table 5. Selected errors with Ground Truth, model, and Gemini outputs.

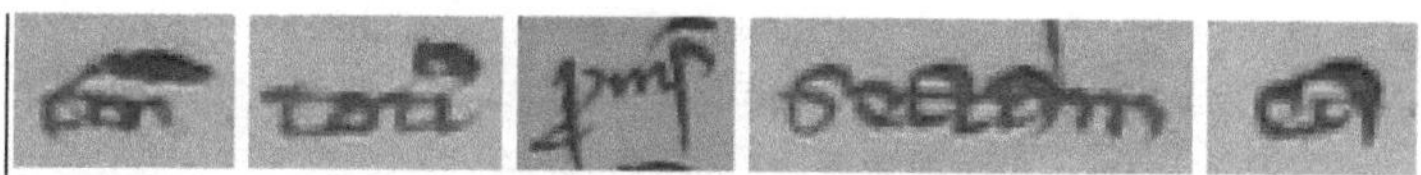

Ground Truth	commune	totius	provisum	sectam	ea
Model	tonmunti	toti	peris	seatm	ea
Gemini	communi	[]	provisum	sectam	eam

Of the ten errors the model made in these lines, nine were of the first two types (fixable by the user), and only one was of the misleading third type. The Gemini-corrected version made only four errors, but three were misleading, including a total elision (totius) and the introduction of an error where the model made none and where the manuscript is especially clear (ea). For this reason, we recommend that users do not use the Gemini-corrected version without reference to the model's version. Interestingly, the transformer-based OCR model that we trained (TrOCR) suffers from similar tendencies: it would often guess a grammatically correct Latin word that is not relevant to the case at hand, thus achieving similar word accuracy as our fiducial model, but worse character accuracy (Table 5).

Tests on legal rolls other than JUST1, CP, and KB suggest that the model produces comprehensible transcriptions, but may occasionally be less accurate. It does very well on JUST2 and JUST3 rolls, as well as other KB and CP rolls that were not included in the dataset. It has a lower accuracy on the Exchequer of Pleas rolls, or E rolls, which recorded debts owed to the crown and their payment, possibly due to differences of vocabulary and the many numbers involved.

Testing has shown that the model transcribes texts written earlier than those in our dataset (like Fig. 4) with a higher degree of accuracy than texts written later than our dataset. This is likely because the hands and many of the legal forms were standardized before the period of our data set, but after our data set, growing literacy rates, the advent of the printing press, and the prevalence of the vernacular (English and French) in official life increased the variation in the hands and introduced more of the vernacular.

Currently, the model is trained on Latin only and will not provide accurate transcriptions of documents in vernacular languages. On occasion, however, English words, often job titles such as "skynner" or "yeoman", will appear in predominantly Latin legal enrollments. In such cases, the model will usually transcribe these words correctly, especially if the word is written in full.

5.3 Can Gemini 3 Pro Transcribe Cases by Itself?

When we began the current project in October 2023, no LLMs that we tested were at all capable of transcribing medieval English legal documents. As of December 2025, GPT 5.1 continues to be unable to transcribe our dataset. It refuses to even make an attempt for longer cases, and when prompted to attempt just the first few lines, it hallucinates a transcription with no relationship to the manuscript. Claude Sonnet 4.5 likewise refuses to attempt a transcription, while Grok 4.1 hallucinates a transcription.

Gemini Pro 3, however, performs remarkably well. We have noticed that for some cases, it produces a remarkably good transcription, even as it fails catastrophically on others. In keeping with its inconsistency across cases, Gemini Pro 3's accuracy also alters across iterations of attempts on the same case. To quantify its performance, we used the Gemini API to request a transcription for each case in the test set, with the following prompt:

Attached is a medieval English legal case, written in Latin. Please transcribe it line by line in vertical order, and don't output anything other than the transcription. Interlinear additions count as independent lines. Transparently expand all abbreviations and don't put the expansion in square brackets. Don't add punctuation that isn't in the manuscript. When you're not sure, take your best guess instead of putting [?] or indicating multiple possible alternatives.

Table 4 shows the resulting WER for each case. For 3 cases, Gemini produces an acceptable transcription (WER $<$ 30%), although it does not outperform our model. For 2 cases, Gemini returns no output. For 9 cases, Gemini returns the wrong number of text lines. We examined the output for these 9 cases and found that the mismatch is not the result of minor mistakes, but of Gemini hallucinating text not found anywhere in the manuscript.

We conclude that frontier LLMs are not yet capable of consistently producing reliable transcriptions of medieval English legal cases. However, Gemini is making rapid progress–Gemini Pro 2.5 was completely unable to produce any reasonable transcriptions, while Gemini Pro 3 sometimes produces transcriptions with a word error rate of only 20%. It is not impossible that the next version of Gemini will excel at this task, at least for some scrolls.

6 Conclusion

We release a dataset containing $\sim$ 4000 lines of medieval legal Latin. This dataset is a sample from the millions of manuscript pages that comprise the English legal tradition, which the legal systems of much of the English-speaking world is based on. The formulaic nature of these documents, the specialized vocabulary and abbreviations, and the broad cross section of everyday life that these legal cases address make our dataset unique in the literature. These distinctive features pose both problems and opportunities for machine learning models.

We find that a CNN + LSTM + CTC model trained on this dataset achieves excellent accuracy, especially when coupled with a bigram model. The model is sufficient to transcribe the majority of the medieval English legal corpus to a high degree of comprehensibility, and most of the errors the model does make do not impact understanding. Gemini can correct many of the remaining errors, but at the cost of sometimes making detectable errors undetectable. We release this model to the public in the form of an easy-to-use web portal, vastly increasing the accessibility of the medieval English legal tradition to students and scholars.

7 Data and Code Availability

All the code and data required to reproduce these results are publicly available on GitHub. The data are also hosted on Hugging Face.

Acknowledgments. CW thanks the Ames Foundation for funding her work on this project.

Disclosure of Interests. The authors have no competing interests to declare that are relevant to the content of this article.

References

1. Aguilar, S.T.: TRIDIS: A comprehensive medieval and early modern corpus for HTR and NER. CoRR **abs/2503.22714** (2025). https://doi.org/10.48550/ARXIV.2503.22714
2. Aguilar, S.T., Jolivet, V.: Handwritten text recognition for documentary medieval manuscripts. J. Data Mining Digital Human. **Historical Documents and automatic text recognition**, 6 (Dec 2023). https://doi.org/10.46298/jdmdh.10484, https://jdmdh.episciences.org/10484
3. Clérice, T.: Catmus medieval: a multilingual large-scale cross-century dataset in Latin script for handwritten text recognition and beyond. In: Barney Smith, E.H., Liwicki, M., Peng, L. (eds.) Document Analysis and Recognition - ICDAR 2024, pp. 174–194. Springer Nature Switzerland, Cham (2024)
4. Coll Ardanuy, M., Berganzo-Besga, I., Sarobe, R., Cuadrada, C.: Evaluating handwritten text recognition in medieval notarial manuscripts: A new dataset and comprehensive analysis. In: Yin, X.C., Karatzas, D., Lopresti, D. (eds.) Document Analysis and Recognition - ICDAR 2025, pp. 340–357. Springer Nature Switzerland, Cham (2026)

5. Diaz, D.H., Qin, S., Ingle, R.R., Fujii, Y., Bissacco, A.: Rethinking text line recognition models. CoRR **abs/2104.07787** (2021). https://arxiv.org/abs/2104.07787
6. Fujitake, M.: Dtrocr: Decoder-only transformer for optical character recognition. In: IEEE/CVF Winter Conference on Applications of Computer Vision, WACV 2024, Waikoloa, HI, USA, January 3-8, 2024, pp. 8010–8020. IEEE (2024). https://doi.org/10.1109/WACV57701.2024.00784
7. Garrido-Munoz, C., Rios-Vila, A., Calvo-Zaragoza, J.: Handwritten Text Recognition: A Survey. arXiv e-prints arXiv:2502.08417 (Feb 2025). https://doi.org/10.48550/arXiv.2502.08417
8. Kahle, P., Colutto, S., Hackl, G., Mühlberger, G.: Transkribus - a service platform for transcription, recognition and retrieval of historical documents. In: 2017 14th IAPR International Conference on Document Analysis and Recognition (ICDAR). vol. 04, pp. 19–24 (2017).https://doi.org/10.1109/ICDAR.2017.307
9. Kang, L., Riba, P., Rusiñol, M., Fornés, A., Villegas, M.: Pay attention to what you read: Non-recurrent handwritten text-line recognition. Pattern Recognit. **129**, 108766 (2022)
10. Kiessling, B.: A modular region and text line layout analysis system. In: 2020 17th International Conference on Frontiers in Handwriting Recognition (ICFHR), pp. 313–318 (2020). https://doi.org/10.1109/ICFHR2020.2020.00064
11. Kim, G., Govindaraju, V.: A lexicon driven approach to handwritten word recognition for real-time applications. IEEE Trans. Pattern Anal. Mach. Intell. **19**(4), 366–379 (1997)
12. Li, M., et al.: Trocr: Transformer-based optical character recognition with pre-trained models. In: Williams, B., Chen, Y., Neville, J. (eds.) Thirty-Seventh AAAI Conference on Artificial Intelligence, AAAI 2023, Thirty-Fifth Conference on Innovative Applications of Artificial Intelligence, IAAI 2023, Thirteenth Symposium on Educational Advances in Artificial Intelligence, EAAI 2023, Washington, DC, USA, February 7-14, 2023. pp. 13094–13102. AAAI Press (2023). https://doi.org/10.1609/AAAI.V37I11.26538
13. Nockels, J., Gooding, P., Ames, S., Terras, M.: Understanding the application of handwritten text recognition technology in heritage contexts: a systematic review of transkribus in published research. Arch. Sci. **22**(3), 367–392 (2022)
14. Pugh, R.B.: The duration of criminal trials in medieval England. Presented at the (1983) Law, Litigants and the Legal Profession
15. Puigcerver, J., Mocholí, C.: Pylaia. https://github.com/jpuigcerver/PyLaia (2018)
16. Shi, B., Bai, X., Yao, C.: An end-to-end trainable neural network for image-based sequence recognition and its application to scene text recognition. IEEE Trans. Pattern Anal. Mach. Intell. **39**(11), 2298–2304 (2017)
17. Vaswani, A., et al.: Attention is all you need. In: Guyon, I., et al., (eds.) Advances in Neural Information Processing Systems 30: Annual Conference on Neural Information Processing Systems 2017, December 4-9, 2017, Long Beach, CA, USA, pp. 5998–6008 (2017). https://proceedings.neurips.cc/paper/2017/hash/3f5ee243547dee91fbd053c1c4a845aa-Abstract.html

Limits of Style-Transferred Synthetic Data for Ad-Hoc Personalization in Handwritten Text Recognition

Said Yasin and Torsten Zesch(✉)

Center of Advanced Technology for Assisted Learning and Predictive Analytics
FernUniversität, Hagen, Germany
said.yasin@studium.fernuni-hagen.de, torsten.zesch@fernuni-hagen.de

Abstract. Personalizing handwritten text recognition can significantly enhance recognition accuracy, but requires a substantial number of handwritten samples, limiting its practical relevance. In this work, we investigate the potential of style-transferred synthetic samples for ad-hoc personalization, i.e. adapting a generic handwriting recognition system to a new writer using only a few samples. We show that one-shot and few-shot generators can produce visually similar handwriting samples, but our personalization experiments show that style-transferred data provides no measurable benefit for the tested system. This finding holds under a fair comparison, using the same number of samples as in real-data personalization, and also when substantially larger amounts of synthetic data are used. It also holds when controlling for non-stylistic influences.

Keywords: Handwriting Recognition · Personalization · Style Transfer

1 Introduction

Offline handwritten text recognition (HTR) is an important tool for supporting tasks like grading handwritten university exams [29] or giving feedback in primary education [20]. The person writing is usually known (e.g. exam written by a specific student; sentences attempted by a specific language learner), which means that the quality of HTR can be considerably improved by personalization [8,16,30], i.e. by adapting a generic HTR system to a specific writer's handwriting style. However, the amount of handwritten samples required for this purpose is a major factor limiting practical usefulness. One possible way to address this limitation is to use the few available real handwriting samples and apply style transfer to generate a large number of synthetic samples for personalizing an HTR system. Figure 1 illustrates how synthetic handwriting samples can be generated by transferring a writer's style onto arbitrarily chosen text.

While recent work has shown remarkable progress in one-shot and few-shot handwriting style transfer [3,25], it remains unclear if such data is useful for ad-

G. A. Fink et al. (Eds.): ICDAR 2026, LNCS 16972, pp. 395–411, 2027.
https://doi.org/10.1007/978-3-032-36023-6_23

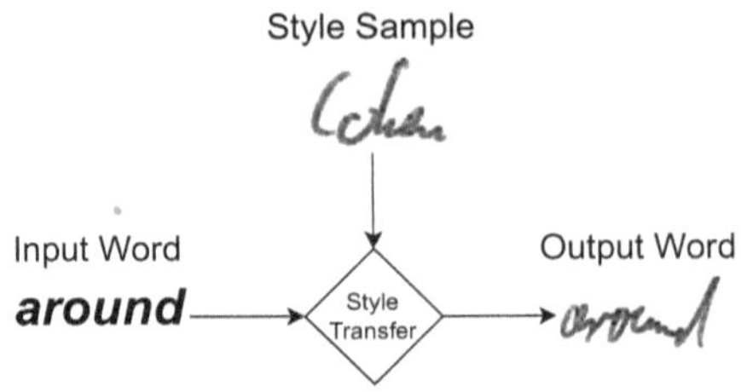

Fig. 1. Style transfer example.

hoc personalization. To assess whether such style-transferred data can be successfully used for personalization, we first conduct an intrinsic evaluation of state-of-the-art one-shot and few-shot handwriting generators with respect to their sensitivity to reference sample selection, and then use the generated synthetic data to evaluate their effectiveness in personalizing HTR systems. Our findings reveal that while style-transferred synthetic data might look similar, it has no measurable personalization effect. We substantiate this conclusion through experiments that rule out other influences like data quantity, specific vocabulary, the complete lack of real data in personalization, and non-stylistic image properties. We make our experimental code publicly available at https://github.com/said702/htr-style-transfer-personalization, with links to all generated datasets and the baseline model checkpoint.

2 Style Transfer

Before evaluating the effect of synthetic data for HTR personalization, we first assess the intrinsic quality of the style transfer itself. We quantify how closely generated samples match each writer's real handwriting and analyze how sensitive generation quality is to the choice (and number) of reference samples.

2.1 One-Shot Vs Few-Shot

Style transfer methods can be classified into one-shot generation [6], relying on a single reference, and few-shot generation [11], using multiple references [3]. Style transfer for unknown styles is achieved by utilizing a learned continuous latent space, where each point corresponds to a known writer, enabling interpolation between styles. Earlier approaches have predominantly been using GAN-based methods [6,12,21] in both one-shot and few-shot settings. Recent few-shot models leveraging encoder-decoder transformers aim to improve character-level style variation through cross-attention [1,27]. Meanwhile, denoising diffusion-based models are becoming increasingly popular for one-shot handwriting generation [3,5,24], as they yield higher image quality, broader distribution coverage, and more stable training [4].

In comparison, one-shot learning is more user-friendly, as few-shot learning is inconvenient and time-consuming due to its reliance on multiple samples [3].

However, one-shot learning is likely more challenging, as it lacks robustness to input variations due to the absence of additional reference samples. This trade-off motivates our experiments, where we first evaluate one-shot and few-shot style transfer quality before assessing their usefulness for personalization.

2.2 Data and Evaluation

We use the **GoBo dataset** [8] to evaluate style transfer quality. The dataset comprises English words from the following sources: random words from the Brown Corpus, pseudowords from the ARC Nonword Database, balanced CEDAR letter samples, and two domain-specific word lists. In sum, it contains 37,000 words written by 40 individuals, with an average of 926 words per writer.

Following prior work [3, 11, 25, 27], we use the **Fréchet Inception Distance** (FID), the standard metric for evaluating the similarity between real and synthetic handwriting [9].

2.3 Experimental Setup

We evaluate two state-of-the-art generators trained on the IAM dataset for one-shot and few-shot style transfer, respectively, both reported to outperform previous approaches in terms of quality and adaptability.

For the **one-shot** setting, we use **One-DM**, which integrates high-frequency component analysis into a conditional diffusion model. It employs two parallel encoders to extract spatial and high-frequency style features and a style–content fusion module to perform style transfer from a single reference sample [3].

For the **few-shot** setting, we use **VATr**, which combines a Transformer-based encoder–decoder architecture with visual archetypes. It leverages a pre-trained convolutional backbone trained on a large synthetic dataset to enable robust style extraction and style transfer from a small set of reference samples [27].

The generation of all synthetic datasets was carried out on Google Colab using an NVIDIA A100, requiring approximately 380 GPU-hours in total.

Quantitative Analysis. To evaluate the dependence on the reference samples in the one-shot and few-shot settings, we generated synthetic datasets with One-DM and VATr, based on the vocabulary of 40 writers from the GoBo dataset and randomly selected reference images for style transfer.[1]

We then evaluate these synthetic datasets against real handwritten data using the FID metric. The results (see Fig. 2) show that One-DM's one-shot performance varies significantly with the chosen reference. A broad FID range indicates high intra-writer variation, making handwriting harder to map consistently in the latent space, whereas a narrow range suggests a more uniform handwriting

[1] Note that in the few-shot setting, each shot configuration is sampled only once; however, using multiple references results in more stable style representations than in the one-shot setting.

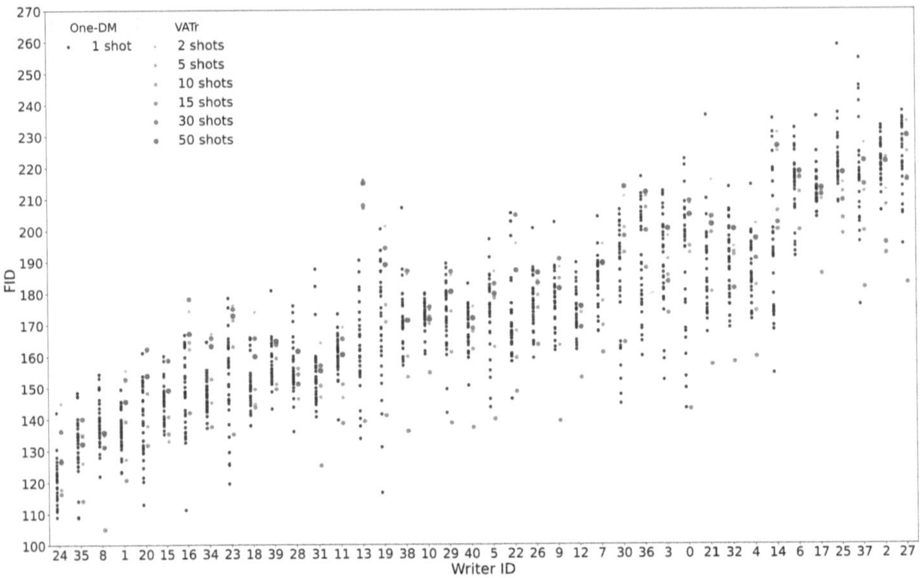

Fig. 2. Style transfer quality (in terms of FID values – lower values better) per writer. The spread of the blue one-shot dots shows the variance due to selection effects (it matters considerably, which sample is used). The red few-shot results show no clear trend for more shots improving results. (Color figure online)

with minimal deviations across references. Notably, few-shot generation outperforms the best one-shot reference for 21 out of 40 writers, predominantly when using 15 reference samples (19 cases), indicating that the use of multiple references to capture a writer's handwriting style can reduce input variation effects. However, additional references beyond this point do not lead to consistent improvements, as for 37 out of 40 writers the lowest few-shot FID is achieved with 15 reference samples, while the remaining cases favor lower shot counts. One possible explanation is that VATr is optimized for this sample size.

Qualitative Analysis. Fig. 3 shows the writers with the highest and lowest FID ranges, highlighting their best and worst one-shot references for style transfer. While lower FID values should generally reflect better style transfer, this is not always the case. This discrepancy likely arises because the FID metric measures distributional similarity between synthetic and real datasets, rather than capturing stylistic differences at the level of individual images. For the writers with the lowest FID spread, the best and worst style-transferred samples for *well* and *voacsed* closely match the original handwriting. Table 1 presents few-shot examples, showing that increasing the number of shots improves similarity to the original handwriting. For instance, in the word *will*, the two final *l* differ in the 2-shot setting but closely match the original in the 15-shot setting.

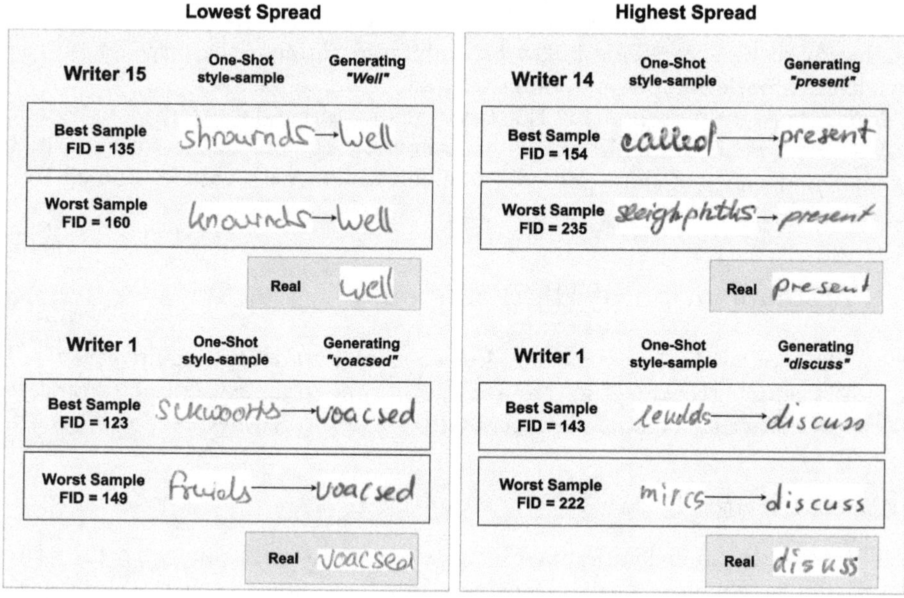

Fig. 3. Examples with highest and lowest FID spread in the one-shot setting.

Table 1. Few-shot examples for writer ID 19

Shots	Real	2	5	10	15
Samples	will	will	will	will	will
	provide	provide	provide	provide	provide
	followed	followed	followed	followed	followed
	as	as	as	as	as
	and	and	and	and	and
FID	0	**321**	**283**	**287**	**290**

2.4 Discussion of Style Transfer Quality

Our quantitative analysis demonstrated that style transfer can achieve a certain degree of similarity with real writer samples while capturing specific handwriting characteristics. However, it also revealed significant variations in effectiveness across writers and, in one-shot learning also within individual writers due to the strong dependence on the chosen reference. Overall, it remains unclear whether the observed similarity is sufficient for effective HTR personalization. We hypothesize that style-transferred synthetic data may contribute to reducing

CERs but is likely less effective than real handwriting due to deviations from the true handwriting style. This might be countered through the larger quantities in which synthetic samples can be generated.

Next, we begin personalizing HTR systems with real data to demonstrate its effectiveness. Then, through various experiments, we personalize with synthetic few-shot and one-shot data and compare the results with those obtained from real data.

3 Personalization of HTR Systems

Personalization of HTR systems, also known as writer adaptation, involves transforming a generic recognizer into a user-specific one to improve recognition rates [16]. This can either be done with real writer data or synthetic data.

3.1 Real Writer Data

Fine-Tuning. If a sufficient number of transcribed samples from a specific writer is available, personalization can be achieved by fine-tuning a generic model. Gold et al. [8] fine-tuned an HTR system with 528 writer-specific words, reducing mean CER from 14.1% to 8.0% across 40 writers. Similarly, Kienzle & Chellapilla [16] showed that character-level personalization with 2,000 writer-specific samples (20 per character) reduced mean CER from 10.2% to 4.4% across 21 writers. The READ 2018 competition on historical handwritten documents showed that fine-tuning with 16 transcribed target-document pages reduced the best-performing system's average CER from 31% to 9% [32]. Using the same benchmark, Soullard et al. [31] adapted both the image-based recognizer and the language model, reducing CER from 26.6% to 5.8% with 16 specific pages. In all cases, the fine-tuning requires rather large quantities of transcribed samples for each writer.

Style Embeddings. Instead of directly fine-tuning the model, the transcribed samples can also be used to learn style embeddings for each writer. Kohút et al. [19] show that for known writers, this approach reduced CER by 9.2%, but for unknown writers, neither selecting the closest existing embedding nor fine-tuning a new one proved effective; instead, standard fine-tuning outperformed both.

Weakly-Supervised. In real-life situations like exams, there are often many handwritten samples that are personalized but not labeled. This makes it possible to use weakly supervised learning methods, e.g. deep transfer mapping with linear transformations [35], or extracting writer-specific stylistic features [26,34]. Kang et al. [13] propose domain adaptation from synthetic to real HTR, aligning the feature distributions of both domains. This reduces average CER from 24.3% to 18.3% on IAM validation data, showing that incorporating untranscribed data improves results a bit, but overall performance remains unsatisfactory.

3.2 Synthetic Data

Data Augmentation. Using augmented data has been shown to be beneficial for improving recognition of historical handwriting in general [7,18]. For personalization, results are rather negative. Jemni et al. [10] experiment with Arabic data, but the augmentation only improves the generic model and hinders personalization. A general limitation of data augmentation methods is that they rely on transforming existing data and cannot generate unseen words [21].

Style Transfer. To the best of our knowledge, style transfer has not been applied to the personalization for individual writers. Existing studies focus on improving generic HTR through synthetic data generated from diverse writer styles rather than a single writer with results showing limited effectiveness when entirely synthetic data is used. For example, Pippi et al. [28] show that transformer-based synthetic training data in the target test style perform poorly, but adding only 1–5% of real data yields substantial CER gains. Similarly, Dai et al. [3] and Pippi et al. [27] applied diffusion- and transformer-based style transfer, where only training on synthetic data resulted in CERs of 11.7% and 11.9% on IAM, compared to 5–6% with real data. Likewise, Muth et al. [23] showed that GAN-generated handwriting underperformed real samples, but fine-tuning a synthetically pre-trained model reduced real data needs by 70–80% with comparable performance. Similarly to Muth et al. [23], Cascianelli et al. [2] also found that halving real data and supplementing with font-based synthetic data achieved similar results on the Leopard dataset. Kang et al. [11] investigate Spanish number recognition with GAN-based style transfer, showing that 160 real samples augmented with synthetic data outperform 298 real samples alone. Other GAN-based studies also found gains with mixed real and synthetic data [21,33,36]. Contrary to these findings, Ding et al. [5] found that diffusion-generated synthetic training data outperformed real samples on the IAM dataset (4.1% vs. 7.3% CER), with a combination of the two further improving to 3.8%.

Overall the studies show a mixed picture for training generic HTR systems with style-transferred data. As has been noted above, style transfer has not been applied to the personalization for individual writers.

4 Personalized HTR Experimental Setup

We outline our style transfer approach in Fig. 4 (bottom) in comparison to personalization with real data (top).

In personalization with real data, a generic HTR model – trained on a large, writer-independent dataset – is directly fine-tuned using a small number of real samples from a specific writer. In contrast, our approach additionally uses the same large, generic dataset to train a handwriting generator alongside the generic HTR model. The HTR model learns general handwriting recognition, while the generator builds a continuous latent space to synthesize new handwriting styles. Using this handwriting generator, we can use a small number of samples from

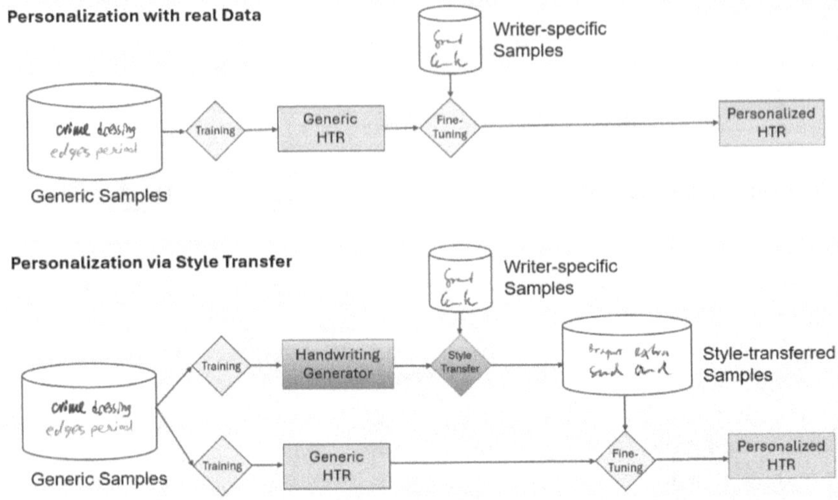

Fig. 4. Personalization with real data vs. personalization via style transfer.

unseen writers to create a large style-transferred synthetic dataset for personalization. For both conditions, we use **AttentionHTR** as our **generic HTR** model, an end-to-end HTR system that leverages ResNet for feature extraction and bidirectional LSTMs for sequence modeling, incorporating a content-based attention mechanism within an encoder–decoder architecture [15]. AttentionHTR outperforms state-of-the-art approaches such as TrOCR in isolated word-level recognition [14]. Also, we prioritize depth over breadth, enabling a more extensive set of controlled experiments. We train AttentionHTR on the IAM dataset [22] and validate it using 5-fold cross-validation. The generic HTR achieves a mean CER (Character Error Rate) of 3.2% ± 0.11%.

This value is reasonably close to the current state of the art to serve as the starting point for our personalization experiments. At the time of writing, Papers with Code[2] lists 18 systems for this setup with CERs between 2.4 and 7.6%.

Table 2. Statistics of personalized handwriting datasets

Dataset	Words	Unique Words	Writers	avg. words per writer
CVL-en	84,514	253	310	273
GoBo	37,000	437	40	926
IAM	115,320	10,841	657	176

[2] https://paperswithcode.com/sota/handwritten-text-recognition-on-iam.

To evaluate personalization, we require handwriting datasets that provide multiple samples per writer in order to adapt a generic HTR model to individual writing styles. Several such datasets exist, such as **IAM** [22], the English subset of **CVL** [17], and **GoBo** [8]. Table 2 summarizes their key properties. Across all three datasets, personalization is potentially beneficial because recognition performance varies substantially across writers. Figure 5 illustrates this writer dependence by evaluating our generic HTR model across writers in the seen dataset IAM and the unseen datasets GoBo and CVL.[3]

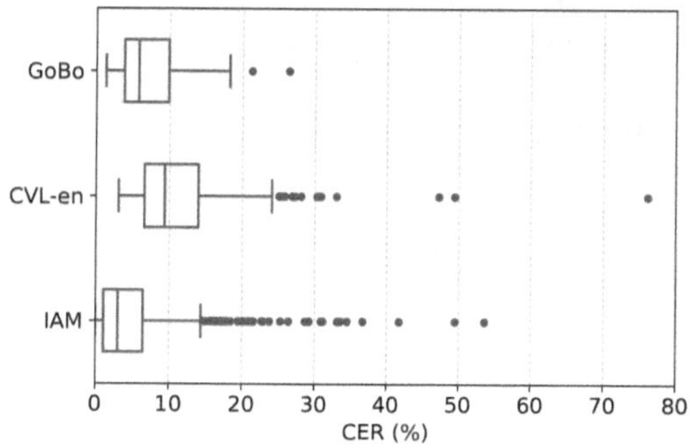

Fig. 5. Distribution of CER for individual writers.

Across all three datasets, the CER distributions consistently exhibit high-error outliers. For example, in IAM, the median CER is about 3%, meaning that half of the writers have a CER of at most this value, while the highest-error writers reach approximately 15–54%. From these datasets, we select GoBo for our personalization experiments, as it provides by far the highest number of samples per writer (Table 2) and has been previously used for personalization experiments [8].

For all personalization experiments, we fine-tune the generic model to each writer's handwriting style for five epochs with a batch size of 10 and a learning rate of 0.001. For this purpose, we generate synthetic data according to the findings in Sect. 2 using 15 reference samples for few-shot and the best reference sample for one-shot (see Fig. 2).[4] All experiments were conducted on Google Colab using NVIDIA L4 and T4 GPUs and required approximately 300 GPU-hours.

[3] Note that the 70% outlier in CVL results from a transcription inconsistency: the author wrote entirely in uppercase, while transcriptions are case-normalized.

[4] Since the best one-shot reference is unknown in practice, this setting provides an upper bound: if personalization fails even with this reference, it is unlikely to succeed under realistic sampling.

5 Personalization with Real Data

As we have shown in the previous section, CER values vary considerably between writers. Before we turn to style transfer in the next section, we now establish how much improvement is possible when fine-tuning with real samples from the GoBo dataset. This is at the same time a replication of the results from Gold et al. [8], but with the use of AttentionHTR instead of an HTR approach based on a Convolutional Recurrent Neural Network (CRNN). For this purpose, we fine-tune the baseline model individually for each of the 40 writers in the GoBo dataset, i.e. starting from the same pretrained model and creating 40 separately adapted models. At each step, 10 user-specific samples are added incrementally, reaching a total of 528 instances (maximum from the GoBo training data), and the performance is evaluated on the remaining 398 test samples.

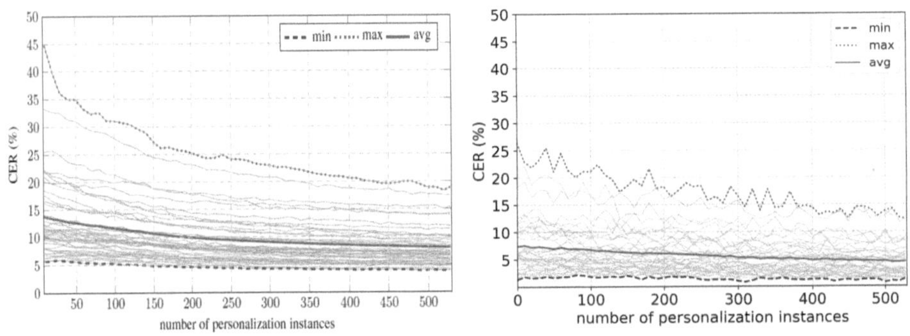

(a) Results from Gold et al. [8] (b) Our replication with AttentionHTR

Fig. 6. Personalization results on the GoBo dataset.

Figure 6 presents the results by Gold et al. [8] (a screenshot from that paper) and ours (copying their diagram style for better comparison). The slope of the personalization curves is very similar, but our results start and arrive at much lower CER values due to the overall performance advantages of AttentionHTR compared to the CRNN used in [8]. Our replication shows that personalization with a few hundred user-specific samples is feasible. In Table 3, we present examples of fully correct predictions after personalization for challenging writers with the highest CER.

To further assess the contribution of different word types to personalization, we repeated the experiment for each of the four word groups in GoBo (see Fig. 7). The results show that all word groups except nonwords contributed to personalization. As we are controlling for word overlap in the test data, this cannot be a vocabulary effect, but probably a result of forcing writers to write unusual character combinations leading to an 'impersonal' writing style.

Table 3. Recognition results for selected challenging samples

Writer	**17**	**10**	**1**	**26**
Input	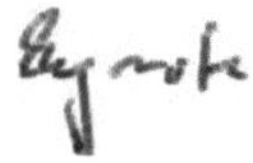	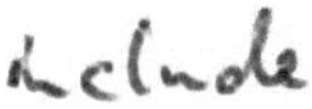		
Generic model	Suymort	malnde	sundons	dissmubue
Personalized model	keynote	include	scholars	disputed

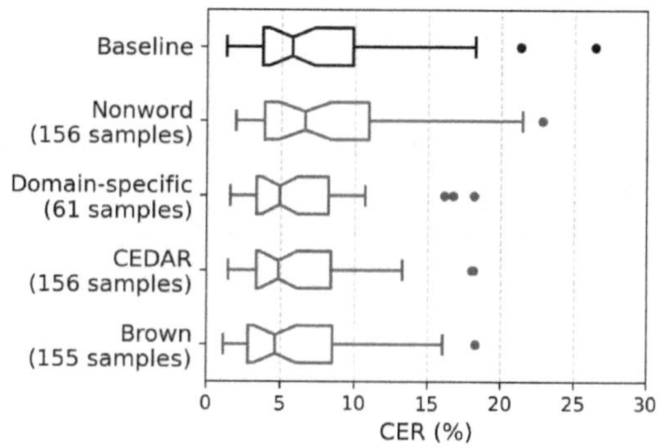

Fig. 7. Personalization by training data.

5.1 Importance of Real Writer Data

There is a theoretical possibility that CER values improve not because we are adding data for a specific writer, but in general due to fine-tuning with more data (as the vocabulary for each writer in the dataset is fixed). To test this, we rerun the personalization experiment using for each writer samples from a different writer for 'personalization'.[5] The (bad) results in Fig. 8 clearly show that the improvements can indeed be attributed to personalization with samples from a specific writer.

6 Personalization via Style Transfer

We now repeat the personalization experiment from the previous section, but instead of real data, we use style transfer to generate one-shot and few-shot synthetic data, as outlined in Sect. 4. To ensure a fair comparison, we generate the words from the GoBo dataset so that the personalization with real data and the personalization with style transfer use exactly the same vocabulary. We find that neither one-shot nor few-shot synthetic data improve recognition, and both yield similar outcomes, leading us to exclude one-shot from further experiments, since it fails even when using the best-performing one-shot reference sample.

[5] For reproducibility, the assignment of target writers to randomly selected training writers is fixed using a random seed in the published code.

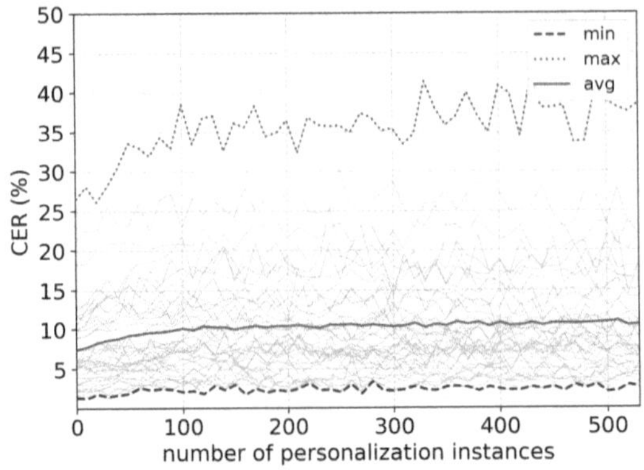

Fig. 8. Failed 'personalization' when using samples from a different writer.

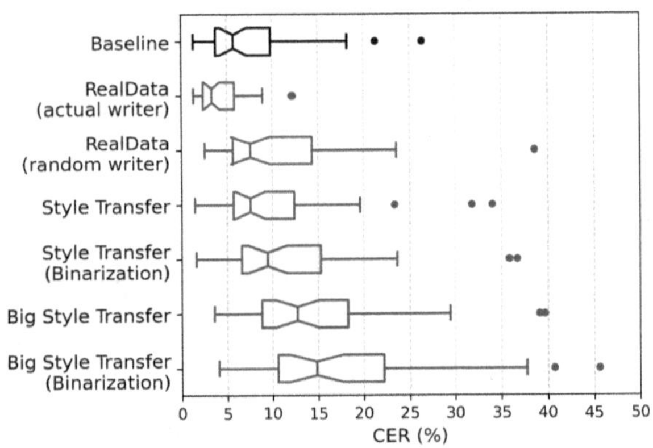

Fig. 9. Overview of personalization results.

Figure 9 shows the results for personalization with few-shot style-transferred data under 'Style Transfer' and for comparison also shows the results for the (unpersonalized) baseline as well as using real data from the actual writer and from a random writer.[6] We find that style transfer performs on the level of the unpersonalized baseline, i.e. it has no measurable effect.

6.1 Mixing Real and Synthetic Data

Synthetic style-transferred data alone may not be sufficient, as previous work has successfully used combinations of real and synthetic data [23,33,36]. We therefore further investigate whether adding synthetic samples to real personalization data improves performance.

[6] The remaining entries in the figure are results from additional experiments described later in this paper.

Table 4. Effect of real/synthetic training ratios on personalization performance

Training ratio (%)		CER (%)		
Real	Synthetic	Min	Mean	Max
100	0	1.3	4.4	12.2
90	10	1.0	4.5	14.7
80	20	1.2	4.7	14.8
50	50	1.1	5.4	16.8
20	80	1.4	6.5	18.8
10	90	1.6	7.9	27.1
0	100	1.4	10.2	34.0

To this end, we evaluate several real/synthetic training ratios, ranging from fully real training data (100/0) to fully synthetic training data (0/100).[7] Table 4 summarizes the personalization results across all writers for each real/synthetic training ratio. The results show that increasing the proportion of synthetic data consistently worsens performance, with the mean CER increasing from 4.4% for 100% real data to 10.2% for 100% synthetic data. This indicates that adding style-transferred synthetic samples does not improve personalization and that the performance gains mainly come from real writer-specific data.

6.2 More Synthetic Data

Although our previous experiments showed that synthetic data does not contribute significantly to performance, we want to rule out the possibility that data quantity is the limiting factor. Therefore, we extended our experiments from 500 to 10,000 few-shot synthetic samples based on a dictionary of frequent English words[8], using a 30% validation split to select the checkpoint with the lowest CER (see Fig. 10). The results show that increasing the amount of synthetic data does not improve personalization and instead leads to higher CERs. This is consistent with the mixing-ratio experiment (see Table 4), where performance degrades as the proportion of synthetic data increases.

6.3 Non-stylistic Influences

Finally, we want to rule out non-stylistic influences (i.e. noise that humans do not pick up when visually inspecting style similarity, but machines might) in the synthetic data as a reason for failed personalization. An example of such

[7] We also evaluated an ordered setup, where the model is first personalized with real samples and then further trained with synthetic samples. Since the results were similar to mixed training, we only report the mixed setup.

[8] https://wortschatz.uni-leipzig.de/en/download/English.

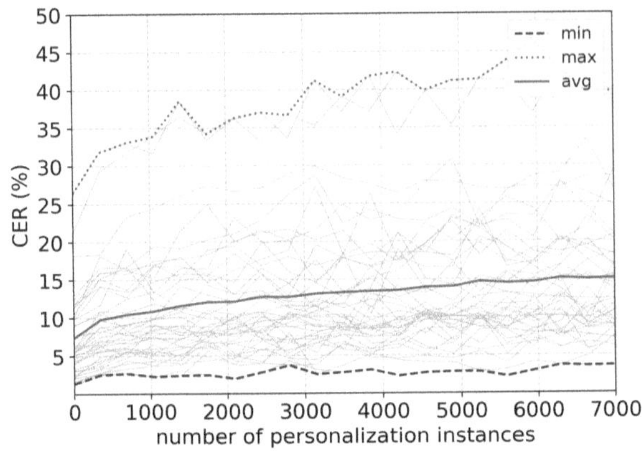

Fig. 10. Setup 'Big Style Transfer' - Personalization using a much bigger dataset of 10,000 style-transferred samples (7000 training / 3000 validation).

a non-stylistic influence can be observed in Table 1 for the word *will* in the 10-shot setting, where gray background artifacts are clearly observable. To this end, we reduce potential generator-induced noise by binarizing the generated images with global Otsu thresholding. This results in images that contain only black letter shapes on a white background. We re-ran the style transfer experiments in the replication setting ('Style Transfer (Binarization)') and with the extended amount of data ('Big Style Transfer (Binarization)') and show results again in Fig. 9. In both settings, the system gets worse (not better) likely due to the loss of grayscale information through binarization. The more important result though is that personalization still fails, so we can rule out any non-stylistic influence. For completeness, we also binarized the real training and test data, which had only a marginal effect and thus cannot explain failed personalization on synthetic data.

7 Conclusion

We started with the hypothesis that personalization of HTR systems (like AttentionHTR in this case) could be achieved through the use of style transfer. However, while synthetic handwriting often appears visually similar to real handwriting, this similarity was not sufficient in our experiments to enable successful personalization. We ruled out potential confounders, including data quantity, specific vocabulary, the absence of real data in personalization, and non-stylistic image properties, as explanations for this outcome.

While it cannot be ruled out that a small set of samples (1–15 in our case) may be insufficient to fully capture the natural variation within a single writer's handwriting, the more likely reason is that style generators need to be improved to produce synthetic handwriting that more closely resembles the authentic style.

Future research should thus focus more on understanding what makes handwriting styles similar and to identify suitable metrics for measuring this similarity. As our results show, existing style-transfer approaches, while scoring high on the similarity metrics, fail to adequately capture the writer-specific characteristics required for personalization.

Limitations. We only evaluate on one language (English), mainly due to the sparsity of suitable data for other languages. However, synthetic data generation was shown to also work for other languages [3], so the overall setup should be applicable as well - possibly with even more room for improvement starting from less well performing baselines. Another limitation is that our FID-based style transfer evaluation is restricted to 40 writers in the GoBo dataset. While this dataset was designed to be diverse [8], handwriting is highly individual, and some variations, especially outlier styles, may not be fully represented. We evaluate only one HTR system (AttentionHTR), for which we show that personalization with real data is effective and outperforms prior approaches on the same data. Therefore, we prioritize depth over breadth to better understand the factors that may limit the effectiveness of style-transferred synthetic data for personalization, rather than reporting additional benchmark results.

Ethical Considerations. Being able to synthesize handwriting from just a few samples may pose significant risks, as it can be exploited for fraudulent activities such as identity theft, forgery of signatures on legal documents, manipulation of handwritten records, and social engineering scams that deceive individuals by mimicking authentic handwriting. Storing writing samples is thus a potential security risk. Our research into ad-hoc personalization is a step towards solving the storage issue, as no writing samples (and derived style-transfer models) have to be stored if ad-hoc personalization from a few, directly obtained and then discarded, samples is possible.

References

1. Bhunia, A.K., Khan, S., Cholakkal, H., Anwer, R.M., Khan, F.S., Shah, M.: Handwriting transformers. In: Proceedings of ICCV, pp. 1086–1094 (2021)
2. Cascianelli, S., Cornia, M., Baraldi, L., Piazzi, M.L., Schiuma, R., Cucchiara, R.: Learning to Read L'Infinito: handwritten text recognition with synthetic training data. In: Computer Analysis of Images and Patterns, pp. 340–350 (2021)
3. Dai, G., Zhang, Y., Ke, Q., Guo, Q., Huang, S.: One-DM: one-shot diffusion mimicker for handwritten text generation. In: Leonardis, A., Ricci, E., Roth, S., Russakovsky, O., Sattler, T., Varol, G. (eds.) Computer Vision - ECCV 2024, pp. 410–427. Springer Nature Switzerland, Cham (2024)
4. Dhariwal, P., Nichol, A.: Diffusion models beat GANs on image synthesis. In: Proceedings of NIPS'21, Curran Associates Inc, Red Hook, NY, USA (2021)
5. Ding, H., Luan, B., Gui, D., Chen, K., Huo, Q.: Improving handwritten OCR with training samples generated by glyph conditional denoising diffusion probabilistic model. In: Document Analysis and Recognition - ICDAR 2023, pp. 20–37 (2023)

6. Gan, J., Wang, W., Leng, J., Gao, X.: HiGAN+: handwriting imitation GAN with disentangled representations. ACM Trans. Graph. **42**(1), 1–17 (2022)
7. Arias, E. G., Pai, V., Schöffel, M., Heumann, C., Aßenmacher, M.: Automatic transcription of handwritten old Occitan language. In: Proceedings of EMNLP, pp. 15416–15439. Association for Computational Linguistics, Singapore (2023)
8. Gold, C., van den Boom, D., Zesch, T.: Personalizing handwriting recognition systems with limited user-specific samples. In: Document Analysis and Recognition – ICDAR 2021, pp. 413–428. Springer International Publishing, Cham (2021)
9. Heusel, M., Ramsauer, H., Unterthiner, T., Nessler, B., Hochreiter, S.: GANs trained by a two time-scale update rule converge to a local Nash equilibrium. In: Proceedings of the 31st International Conference on Neural Information Processing Systems, pp. 6629–6640 (2017)
10. Jemni, S.K., Ammar, S., Kessentini, Y.: Domain and writer adaptation of offline Arabic handwriting recognition using deep neural networks. Neural Comput. Appl. **34**(3), 2055–2071 (2022)
11. Kang, L., Riba, P., Rusiñol, M., Fornés, A., Villegas, M.: Content and style aware generation of text-line images for handwriting recognition. IEEE Trans. Pattern Anal. Mach. Intell. **44**(12), 8846–8860 (2022)
12. Kang, L., Riba, P., Wang, Y., Rusiñol, M., Fornés, A., Villegas, M.: Ganwriting: Content-conditioned generation of styled handwritten word images. In: European Conference on Computer Vision (ECCV). vol. 12368, p. 273–289. Springer (2020)
13. Kang, L., Rusiñol, M., Fornés, A., Riba, P., Villegas, M.: Unsupervised adaptation for synthetic-to-real handwritten word recognition. In: 2020 IEEE Winter Conference on Applications of Computer Vision (WACV, pp. 3491–3500 (2020)
14. Karlsson, V., Åkerman, A.: Real-time detection of spelling mistakes in handwritten notes (2022). Master's Thesis
15. Kass, D., Vats, E.: AttentionHTR: handwritten text recognition based on attention encoder-decoder networks. In: Document Analysis Systems, pp. 507–522. Springer International Publishing, Cham (2022)
16. Kienzle, W., Chellapilla, K.: Personalized handwriting recognition via biased regularization. In: Proceedings of ICML, pp. 457–464. ACM (2006)
17. Kleber, F., Fiel, S., Diem, M., Sablatnig, R.: CVL-DataBase: an off-line database for writer retrieval, writer identification and word spotting. In: 12th International Conference on Document Analysis and Recognition, pp. 560–564 (2013)
18. Koch, P., et al.: A tailored handwritten-text-recognition system for medieval Latin. In: Proceedings of the Ancient Language Processing Workshop, pp. 103–110 (2023)
19. Kohút, J., Hradiš, M., Kišš, M.: Towards writing style adaptation in handwriting recognition. In: Document Analysis and Recognition - ICDAR 2023, pp. 377–394. Springer Nature Switzerland, Cham (2023)
20. Laarmann-Quante, R.: Towards a tool for automatic spelling error analysis and feedback generation for freely written German texts produced by primary school children. In: 7th ISCA Workshop on Speech and Language Technology in Education (SLaTE 2017), pp. 36–41 (2017)
21. Luo, C., Zhu, Y., Jin, L., Li, Z., Peng, D.: SloGAN: handwriting style synthesis for arbitrary-length and out-of-vocabulary text. IEEE Trans. Neural Netw. Learn. Syst. **34**(11), 8503–8515 (2023)
22. Marti, U.V., Bunke, H.: The IAM-database: an English sentence database for offline handwriting recognition. In: International Journal on Document Analysis and Recognition, vol. 5, pp. 39–46 (2002)

23. Muth, M., Peer, M., Kleber, F., Sablatnig, R.: Maximizing data efficiency of HTR models by synthetic text. In: Document Analysis Systems, pp. 295–311. Springer Nature Switzerland, Cham (2024)
24. Nikolaidou, K., et al.: WordStylist: styled verbatim handwritten text generation with latent diffusion models. In: Document Analysis and Recognition - ICDAR 2023, pp. 384–401 (2023)
25. Nikolaidou, K., Retsinas, G., Sfikas, G., Liwicki, M.: DiffusionPen: towards controlling the style of handwritten text generation. In: Computer Vision - ECCV 2024, pp. 417–434. Springer Nature Switzerland, Cham (2025)
26. Nosary, A., Heutte, L., Paquet, T.: Unsupervised writer adaptation applied to handwritten text recognition. Pattern Recogn. **37**(2), 385–388 (2004)
27. Pippi, V., Cascianelli, S., Cucchiara, R.: Handwritten text generation from visual archetypes. In: IEEE/CVF Conference on Computer Vision and Pattern Recognition (CVPR, pp. 22458–22467 (2023)
28. Pippi, V., Cascianelli, S., Kermorvant, C., Cucchiara, R.: How to choose pretrained handwriting recognition models for single writer fine-tuning. In: IEEE International Conference on Document Analysis and Recognition, pp. 330–347 (2023)
29. Rowtula, V., Oota, S.R., C.V., J.: Towards automated evaluation of handwritten assessments. In: Proceedings of ICDAR, pp. 426–433 (2019)
30. Scius-Bertrand, A., Ströbel, P., Volk, M., Hodel, T., Fischer, A.: The Bullinger dataset: a writer adaptation challenge. In: Document Analysis and Recognition - ICDAR 2023, pp. 397–410. Springer Nature Switzerland, Cham (2023)
31. Soullard, Y., Swaileh, W., Tranouez, P., Paquet, T., Chatelain, C.: Improving text recognition using optical and language model writer adaptation. In: 2019 International Conference on Document Analysis and Recognition (ICDAR), pp. 1175–1180 (2019)
32. Strauß, T., Leifert, G., Labahn, R., Hodel, T., Mühlberger, G.:). ICFHR 2018 competition on automated text recognition on a read dataset. In: 16th International Conference on Frontiers in Handwriting Recognition (ICFHR), pp. 477–482 (2018)
33. Wang, Y., Wei, H., Wang, H., Sun, S., He, C.: GL-GAN: perceiving and integrating global and local styles for handwritten text generation with mamba. In: Proceedings of COLING, pp. 2434–2444. Abu Dhabi, UAE (2025)
34. Wang, Z.R., Du, J.: Fast writer adaptation with style extractor network for handwritten text recognition. Neural Netw. **147**, 42–52 (2022)
35. Yang, H.M., Zhang, X.Y., Yin, F., Sun, J., Liu, C.L.: Deep transfer mapping for unsupervised writer adaptation. In: 16th International Conference on Frontiers in Handwriting Recognition (ICFHR), pp. 151–156 (2018)
36. Yeleussinov, A., Amirgaliyev, Y., Cherikbayeva, L.: Improving OCR accuracy for Kazakh handwriting recognition using GAN models. Appl. Sci. **13**(9), 5677 (2023)

Character Template Representation for Confidence Learning in Handwritten Text Recognition

Yangyang Liu[1,2(✉)], Heng Zhang[1,2], Fei Yin[1,2], and Cheng-Lin Liu[1,2]

[1] State Key Laboratory of Multimodal Artificial Intelligence Systems (MAIS), Institute of Automation, Chinese Academy of Sciences, Beijing 100190, China
{liuyangyang2021,heng.zhang}@ia.ac.cn, {fyin,liucl}@nlpr.ia.ac.cn

[2] School of Artificial Intelligence, University of Chinese Academy of Sciences, Beijing 100049, China

Abstract. In human pattern recognition, a template of each class is learned and recognition is performed by matching the templates to observed patterns in representation space. We take this inspiration into deep neural network learning, and propose **C**haracter **T**emplate **R**epresentation (CTR), which presets a template for each character and uses template embedding as the prototype for each character. Through a template matching mechanism, samples are classified into the character class whose embedding is closest. CTR can be applied to character classification and text line recognition, with character classification regarded as a local task of text line recognition. We conducted extensive experiments on Chinese and English handwriting datasets. The empirical evidence shows that our template representation achieves better performance in out-of-distribution detection, misclassification detection, and domain generalization.

Keywords: Handwritten Text Recognition · Confidence Learning · Prototype Classifier · Template Representation Learning

1 Introduction

In the era of deep learning, visual pattern recognition have mostly used convolutional neural networks (CNNs) [12] trained with cross-entropy loss, aimed to discriminate different classes in feature space. However, human cognition of characters usually proceed by template matching, under a a distance metric in a representation space. When learning to read, humans typically begin by familiarizing themselves with standard printed characters, referred to as character templates. Upon encountering a new character sample, the brain utilizes a template-matching mechanism [17] to associate the sample with its closest template, as illustrated in Fig. 1(a). Clear samples are given high confidence, ambiguous samples are given low confidence, and samples of unseen classes are

G. A. Fink et al. (Eds.): ICDAR 2026, LNCS 16972, pp. 412–430, 2027.
https://doi.org/10.1007/978-3-032-36023-6_24

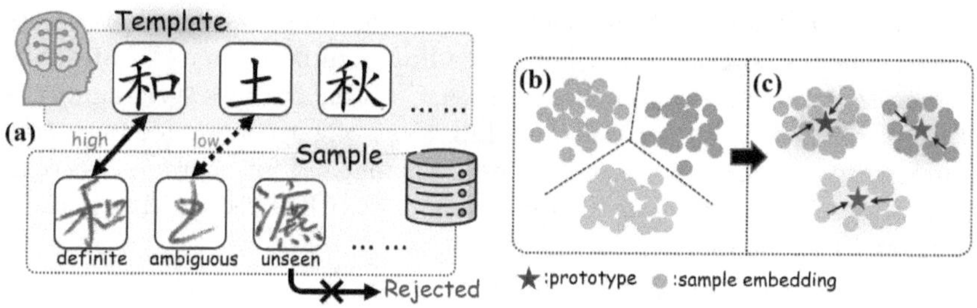

Fig. 1. **(a)** The template matching mechanism of characters. A high or low confidence is assigned to character samples based on the degree of matching with templates and rejects samples of unseen classes. **(b)** The feature space of a conventional softmax-based DNN classifier. **(c)** The distance-based softmax feature space of our CTR. Samples are distributed around the prototypes of each class.

rejected. Templates can be seen as prior knowledge, representing the most confident samples of a certain character class according to human cognition. In contrast, deep neural networks trained with cross-entropy primarily focus on finding decision boundaries between features of training characters, as illustrated in Fig. 1(b). By merely fitting the training set, the model learns to classify but does not understand **"what makes a character itself."** This brings overconfidence: overfitting to the training set results in low capability to exclude misclassified and out-of-distribution samples, as the model still tends to predict them with confidence close to 1. Considering the gap between the distributions of training samples and standard templates, the confidence of model outputs under this paradigm is often unreliable and has poor transferability.

Template-based recognition in the human brain is more confident because the brain learns templates as prior knowledge and stores the abstract features of these templates as prototypes for each class. This mechanism naturally supports rejecting ambiguous or unseen samples.

To leverage this mechanism, we propose *Character Template Representation* (CTR), a unified template-prototype formulation for both *single-character classification* and *text line recognition*. In CTR, each character class is assigned a template representing standard forms (we use printed characters in this paper), and the template embedding serves as the class prototype. Predictions are made by matching an input embedding to these prototypes, yielding a distance-based posterior and reliable confidence: OOD samples can be detected by their large distances to all prototypes. Moreover, because the embedding is learned under classification objectives, CTR also excels in closed-set recognition and misclassification detection.

CTR also applies to text line recognition, where we extract a feature map of the input line and generate a 1D sequence by sliding a window along the horizontal direction (the window spans the full feature-map height with stride 1). Each window corresponds to one time step and is classified by template matching mechanism; the resulting posterior sequence is then decoded (e.g., by

CTC) into the final string. In this view, single-character classification is a special case with sequence length L=1, where decoding is unnecessary. Additionally, by anchoring predictions to standard templates, CTR encourages style-agnostic representations of each character class, leading to stronger confidence performance, as demonstrated in our experiments. Our CTR method for character recognition has been presented in [25], and the extension to text line recognition is the main contribution of this paper.

2 Related Work

2.1 Character Classification

Character classification has been studied since early OCR systems. Before deep learning, it mainly relied on hand-crafted features and conventional classifiers, e.g., hybrid features for handwritten English [1] and SVM-based recognizers [2]. For Chinese characters, classical pipelines evolved from LVQ/MQDF-style models [21] to scalable designs such as hierarchical classification and candidate reduction [37], often combined with classifier ensembles [42]. In the deep learning era, CNN feature extraction with softmax training dominates [12], with improvements from architecture and training strategies [41]. Chinese-specific designs further exploit structural cues such as radicals [39], and recent studies also explore challenging scripts (e.g., oracle bone characters) [49] and the use of printed characters as standard patterns [46].

2.2 Text Line Recognition

Text line recognition transcribes a character sequence from a word/line image and is often implemented as *sliding-window classification* plus sequence decoding. Early deep models such as CRNN combine CNN features with recurrent sequence modeling and CTC to enable end-to-end training without explicit segmentation [11,31]. For handwritten Chinese text, a representative sliding-window framework unifies character segmentation and recognition by extracting candidate character features on the feature map and decoding them into the final string [47]. Later work improves robustness to irregular layouts via rectification, e.g., RARE [32] and ASTER [33]. More recent recognizers strengthen context modeling with language-model-based refinement (ABINet) or transformer-style sequence modeling. For online handwritten Chinese text recognition, prototype-based models have also been explored, where convolutional prototype networks are used to improve the learning of character representations [4,5].

2.3 Confidence Learning

Confidence learning mitigates over-confidence in deep models and is commonly evaluated on misclassification detection (MisD) and out-of-distribution (OOD) detection. Maximum softmax probability (MSP) [14] is a widely used baseline, effective for many MisD cases but often weak for OOD. MisD (ambiguity rejection) has classical foundations such as the optimal reject rule for Bayes classifiers [6], and modern approaches either improve selectivity during training [50] or learn confidence with auxiliary predictors (e.g., ConfidNet using TCP targets) [19]. OOD methods include post-hoc scoring on pretrained networks [20,23] and training-time regularization or dynamics-based techniques [40]. To leverage additional data, outlier exposure (OE) [15] and follow-ups [28] use auxiliary outliers, while synthetic outliers (e.g., VOS) [8] reduce the need for manual collection.

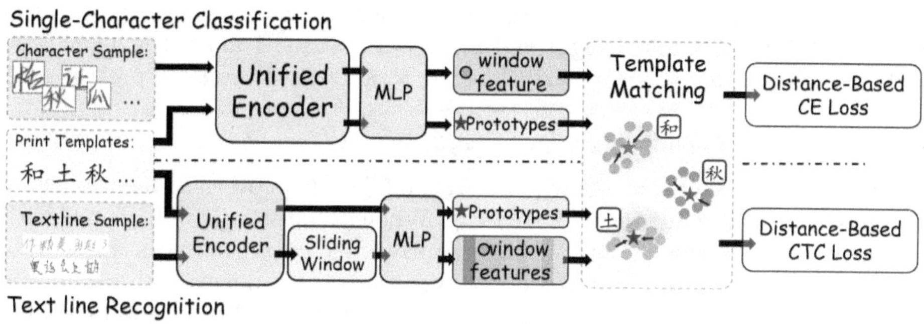

Fig. 2. Overview of our framework. CTR provides a shared template-matching mechanism that can be plugged into both character classification (upper) and text line recognition (lower).

3 Method

Our framework for text line recognition is shown in Fig. 2. We follow a sliding-window classification pipeline, where a text line image is encoded into a feature map and then converted into a 1D sequence by sliding windows. Each window produces a candidate/window feature that is classified into character categories, and the resulting posterior sequence is finally decoded (e.g., by CTC) into the output string. This formulation is widely used in handwritten Chinese text recognition, bridging explicit and implicit segmentation by generating dense candidate character features via sliding windows [47].

In this pipeline, CTR specifies *how the per-window classifier is implemented*: we learn template prototypes and compute a distance-based posterior for each window, which naturally yields confidence scores for rejection and reliability estimation.

3.1 Sliding-Window Text Line Recognition Framework

Overall Pipeline. We adopt a sliding-window based handwritten text line recognition framework [47], which is intermediate between explicit and implicit segmentation: it produces dense candidate character features with coarse locations while avoiding precise character boundary extraction. Given a text line image $\boldsymbol{x}$, we first normalize the input height and apply a CNN-based encoder $\mathcal{F}(\boldsymbol{x};\theta)$ to obtain a 2D feature map.

Window Features From the Feature Map. Following the sliding-window paradigm, we slide windows on the feature map along the width direction to obtain a 1D sequence of candidate (window) features. Intuitively, the feature map can be viewed as being divided into a set of horizontal grid positions, where each position corresponds to a local rectangular region (receptive field) in the original image. The t-th window feature is thus associated with a local region centered at the t-th grid position. In our implementation, the window spans the full feature-map height and moves with stride 1 along the width; the pooled window feature is further projected to a D-dimensional representation by:

$$\boldsymbol{u}_t = \mathrm{MLP}(\mathrm{SlidingWindow}(\mathcal{F}(\boldsymbol{x};\theta))_t) \in \mathbb{R}^D, \qquad t = 1, \ldots, L. \tag{1}$$

This yields a dense sequence of window features $\{\boldsymbol{u}_t\}_{t=1}^{L}$ covering the whole line.

Sliding-Window Alignment Property. When a character is well aligned with the center of a sliding window (i.e., the receptive-field center matches the character center), the corresponding window feature tends to produce a high-confidence character prediction. When the window center and the character center are misaligned, the model tends to output a blank label or a low character score. This behavior enables sequence decoding (e.g., CTC) to aggregate reliable window predictions while suppressing redundant or misaligned windows.

Efficiency and Single-Character Special Case. Because the feature map is computed once and all windows are processed in a fully-convolutional manner, sliding-window recognition allows high parallelization and efficient inference. For single-character classification, the input can be regarded as the special case with a single time step (L=1): we obtain one window feature and directly perform classification without sequence decoding.

3.2 CTR as the Window Classifier

Template Prototypes. We define one template for each character class (see Sec. *Templates Fonts*). Let $\{\boldsymbol{s}_i\}_{i=1}^{C}$ denote the template set for C classes. Templates are embedded by the same encoder, and their embeddings serve as class prototypes:

$$\boldsymbol{m}_i := \boldsymbol{f}(\boldsymbol{s}_i) = \mathcal{F}(\boldsymbol{s}_i;\theta) \in \mathbb{R}^D, \qquad i = 1, \ldots, C. \tag{2}$$

Distance-Based Posterior for Each Window. Given a window feature $\boldsymbol{u}_t$, CTR computes its squared Euclidean distances to all prototypes:

$$d_i(\boldsymbol{u}_t) = \|\boldsymbol{u}_t - \boldsymbol{m}_i\|_2^2, \qquad i = 1, \ldots, C, \tag{3}$$

and converts distances into a softmax-like posterior:

$$p(y \mid \boldsymbol{u}_t) = \frac{\exp\left(-\frac{\gamma}{2} d_y(\boldsymbol{u}_t)\right)}{\sum_{i=1}^{C} \exp\left(-\frac{\gamma}{2} d_i(\boldsymbol{u}_t)\right)}. \tag{4}$$

In all experiments we set $\frac{\gamma}{2} = 1$ for simplicity. The predicted class for each window follows the minimum-distance rule:

$$\hat{y}(\boldsymbol{u}_t) = \arg\max_{y} p(y \mid \boldsymbol{u}_t) = \arg\min_{y} d_y(\boldsymbol{u}_t). \tag{5}$$

Blank Prototype for CTC. For text line recognition with CTC, we additionally introduce a blank prototype $\boldsymbol{m}_{\text{blank}}$ to obtain posteriors over $C+1$ symbols. We obtain $\boldsymbol{m}_{\text{blank}}$ in two ways: (1) **Template-Based:** embed a blank image as the blank prototype; (2) **Learning-Based:** initialize a learnable vector in the feature space and optimize it during training (similar to prototype-based classifiers).

According to the mechanism of human cognition rejection by comparing similarity/distance with a threshold, we also attempt using a learnable threshold to replace the blank prototype.

Confidence Estimation and Rejection. CTR provides confidence at the *window level.* For misclassification/ambiguity rejection, we use the maximum posterior $\max_y p(y \mid \boldsymbol{u}_t)$ as a confidence score. For OOD detection, we use the minimum distance $\min_i d_i(\boldsymbol{u}_t)$: a large minimum distance indicates the window feature is far from all class prototypes. These window-level confidence scores can be aggregated along the decoded sequence to support rejection and reliability estimation for text line recognition, and directly apply to single-character classification when L=1.

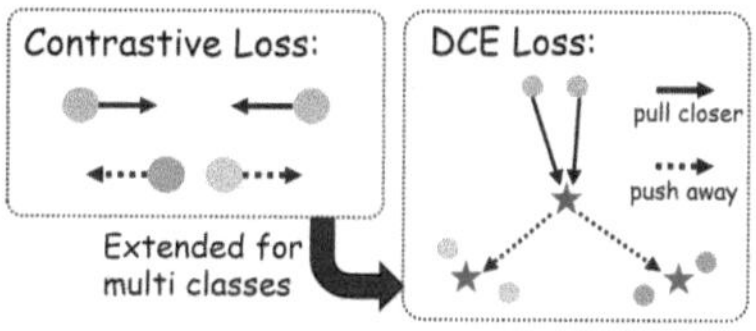

Fig. 3. DCE Loss can be viewed as extended contrastive loss.

3.3 Distance-Based Loss

In this subsection, we describe the losses for learning character template representations.

Distance-Based Cross Entropy Loss (DCE): For single-character classification, we train the model with the standard cross-entropy objective, but compute class posteriors using the distance-based probability in Eq 4. Concretely, given an input sample $\boldsymbol{u}$ with label y, we maximize its posterior $p(y \mid \boldsymbol{u})$ (equivalently, minimize $-\log p(y \mid \boldsymbol{u})$). This encourages the embedding of $\boldsymbol{u}$ to move closer to its corresponding prototype while moving away from other prototypes.

Relation to Contrastive Learning: DCE can be interpreted as a "prototype-aware" extension of contrastive learning. Similar to contrastive objectives, it pulls an input embedding toward its positive prototype (the ground-truth class) and pushes it away from negatives (other class prototypes). Different from pairwise contrastive loss, DCE uses the normalized posterior over *all* classes, so the update strength is automatically modulated by how confusing the competing classes are. As illustrated in Fig. 3, this leads to simultaneous attraction to the correct prototype and repulsion from all others in a single objective, without explicit sampling of negative pairs.

Distance-Based CTC Loss: For text line recognition, we use CTC training. By replacing the standard per-step posterior with the distance-based posterior, the CTC loss becomes:

$$\mathcal{L}_{\text{CTC}}(P, Y) = -\log\left(\sum_{\pi \in \mathcal{B}^{-1}(Y)} \prod_{t=1}^{L} \text{softmax}(-\boldsymbol{d}_t)_{\pi_t}\right), \tag{6}$$

where Y is the ground-truth label sequence, π is a CTC alignment path, $\mathcal{B}(\cdot)$ denotes the standard CTC decoding operation that maps an alignment path to a label sequence by removing repeated labels and blanks, and $\boldsymbol{d}_t \in \mathbb{R}^{C+1}$ is the distance vector at time step t (including the blank prototype). $\text{softmax}(-\boldsymbol{d}_t)_{\pi_t}$ denotes the distance-based posterior probability of symbol π_t at time step t. During training, Eq. (6) marginalizes over all valid alignment paths, so no single alignment path needs to be specified. During inference, we adopt standard CTC best-path decoding by selecting $\hat{\pi}_t = \arg\max_k \text{softmax}(-\boldsymbol{d}_t)_k$ at each time step and obtaining the final recognition result as $\hat{Y} = \mathcal{B}(\hat{\pi})$.

3.4 Template Fonts

In our method, we select character templates in printed fonts. Printed fonts represent the most standard form of each character and thus have the highest confidence within each character class. We use these as templates and their embeddings in the feature space as prototypes. We conducted experiments on both Chinese and English characters. For Chinese fonts, we used **Kai**, **Song**, and **Shu** fonts, and for English fonts, we used **Times New Roman**, **Comic**, and **Freestyle Script**. Visualizations of these fonts are shown in Fig. 4. We conducted comparative experiments on these fonts in the experimental section.

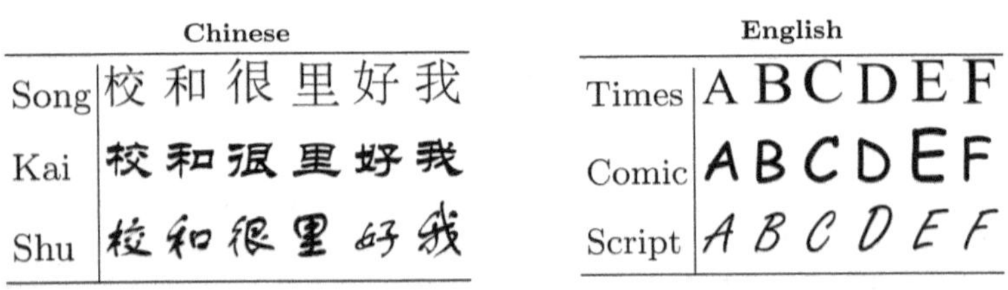

Fig. 4. Visualization of templates. *Times* means Times New Roman; *Script* means Freestyle Script.

3.5 Classification Rules

For misclassification detection (MisD), we follow the maximum softmax probability (MSP) paradigm [14], which uses the highest posterior probability among all classes as prediction confidence. Samples with confidence below a threshold are rejected.

For out-of-distribution detection (OOD), we implement the distance-based rejection (DR) rule [44]. A sample is considered OOD if the minimum distance to all prototypes exceeds a threshold.

The decision rules are:

$$\text{MisD}: \ \hat{y}(\boldsymbol{u}) = \begin{cases} \underset{i}{\operatorname{argmax}}\, p(i|\boldsymbol{u}), & \text{if } \max_i p(i|\boldsymbol{u}) > \delta_{\text{MisD}}, \\ \text{Misclassification}, & \text{otherwise.} \end{cases} \tag{7}$$

$$\text{OOD}: \ \hat{y}(\boldsymbol{u}) = \begin{cases} \underset{i}{\operatorname{argmax}}\, p(i|\boldsymbol{u}), & \text{if } \min_i d_i(\boldsymbol{u}) < \delta_{\text{OOD}}, \\ \text{OOD}, & \text{otherwise.} \end{cases} \tag{8}$$

The performance metrics we use in experiments are threshold-independent, so we do not need to set specific thresholds in the experiments.

4 Experiment

To demonstrate the effectiveness of our CTR approach for confidence learning, we conducted experiments across multiple tasks. For character classification, we performed out-of-distribution detection (OOD), misclassification detection (MisD), and domain generalization experiments. For text line recognition, we carried out character-level rejection experiments. Additionally, we conducted ablation studies to validate the necessity of the template.

4.1 Datasets

EMNIST [7] is a handwritten character dataset containing letters and digits, with 814,255 labeled samples.
CASIA-HWDB [22] is an offline handwritten Chinese database. CASIA-HWDB1.0–1.2 contains 3,895,135 isolated character samples from 1,020 writers, and CASIA-HWDB2.0–2.2 contains 52,230 text lines from 5,091 pages.
CASIA-AHCDB [43] is an ancient Chinese handwritten character dataset, containing over 2.2 million annotated samples from 10,658 classes.
ICDAR-2013 [45] is a handwritten Chinese text recognition benchmark, containing 3,342 text lines from 300 handwritten pages.
Chars74K [3] is a character recognition dataset containing English and Kannada characters. We use its English handwritten character subset, which includes 62 classes: uppercase letters, lowercase letters, and digits.
IAM [27] is a widely used English handwritten text recognition dataset with word- and line-level annotations. We use its text line subset.

4.2 Evaluation Metrics

AUROC [14] measures the ability to distinguish positive and negative samples as the area under the ROC curve and is invariant to the choice of positive/negative labels, making it a standard metric for OOD detection.
AUPR [26] measures the area under the Precision–Recall curve and is more informative than AUROC when class distributions are highly imbalanced.
AURC [10] is selective prediction using the Risk–Coverage curve, where coverage denotes the fraction of accepted samples and risk denotes their error rate.
E-AURC [10] is the excess area under the Risk–Coverage curve over the optimal one, thus removing the influence of recognition accuracy on MisD detection.
FPR95 [9] is the false positive rate when the true positive rate reaches 95%.

4.3 Implementation Details

For character classification, we use ResNet-18 for English characters and a compact Chinese character recognition network [41] for Chinese characters. All character images are resized to 64×64. Models are trained with Adam using a batch size of 32 and an initial learning rate of 1×10^{-1}, which is decayed by 0.1 at 30%, 50%, and 80% of the total training iterations.

For text line recognition, we follow the sliding-window framework in [47]. Real text line data are weakly annotated with transcriptions, while synthetic data provide character-level positions. Text line images are resized to a height of 64 while preserving the aspect ratio. Models are trained on four Titan RTX GPUs with Adam, using a batch size of 8 per GPU and an initial learning rate of 1×10^{-1}, decayed by 0.1 at 50% and 80% of the total training iterations.

4.4 Out-of-distribution Detection

Out-of-distribution detection (OOD) aims to distinguish OOD samples while maintaining accurate classification on in-distribution (InD) samples. For English characters, we use EMNIST [7], where letters are treated as InD classes and digits as OOD classes. For Chinese characters, we use HWDB [22] for training and characters extracted from ICDAR2013 [45] for testing, where the 3,755 first-level Chinese characters [34] are treated as InD classes and the others as OOD.

We evaluate AUROC, AUPR, FPR95, and InD classification accuracy, with InD samples as positives and OOD samples as negatives. The results are reported in Table 1 and Table 2. We compare CTR with MSP [14], post-processing methods including ODIN [20], EBO [23], ReAct [35], and MaxLogit [13], as well as training-time methods including Mixup [38], Label Smoothing [30], Relation [36], and CPN [44]. As shown in Table 1 and Table 2, CTR achieves better OOD detection performance and competitive InD accuracy. Compared with CPN, a prototype classifier without templates, CTR further demonstrates the benefit of using character templates as class prototypes.

Table 1. Performance of OOD detection on EMNIST. "Acc" means the classification accuracy on in-distribution classes. The metrics are all percentages.

Method	InD:Letter OOD:Number			
	AUROC↑	AUPR↑	FPR95↓	Acc↑
MSP [14]	75.07	72.28	78.14	93.42
ODIN [20]	76.61	76.04	77.95	93.42
EBO [23]	75.39	73.22	78.49	93.42
ReAct [35]	73.64	70.17	82.56	93.42
MaxLogit [13]	76.71	75.03	79.16	93.42
Mixup [38]	75.37	73.91	78.66	93.26
LS [30]	75.08	75.49	78.14	93.42
Relation [36]	77.86	76.17	78.04	93.11
CPN [44]	76.31	73.23	80.11	93.45
CTR	**77.78**	**76.51**	**78.10**	**93.49**

Table 2. Performance of OOD detection on HWDB and ICDAR2013. The backbone is specially designed for Chinese character [41].

Method	InD:First-level chars OOD:The rest			
	AUROC↑	AUPR↑	FPR95↓	Acc↑
MSP [14]	90.59	97.60	49.51	92.30
ODIN [20]	91.94	98.36	46.00	92.30
EBO [23]	91.60	98.27	44.89	92.30
ReAct [35]	91.11	98.11	44.20	92.30
MaxLogit [13]	92.62	98.28	45.27	92.30
Mixup [38]	90.90	97.89	48.71	92.59
LS [30]	88.37	97.22	51.65	92.42
Relation [36]	92.51	98.70	45.86	92.57
CPN [44]	91.23	98.71	47.56	92.13
CTR	**92.66**	**98.91**	**42.68**	**92.61**

4.5 Misclassification Detection

In Closed-Set Recognition, ambiguous images are likely to be misclassified and are desired to be detected based on confidence (posterior probability). We conducted experiments on the English (EMNIST) and Chinese (HWDB and ICDAR2013) datasets for Misclassification detection (MisD).

The metrics used are AUROC, FPR95, AURC, E-AURC, Accuracy. The results are shown in Table 3 and Table 4. The correctly classified samples are regarded as positive samples, and the incorrectly classified samples are regarded

Table 3. Performance of MisD on EMNIST. AUROC and FPR95 are percentages. The rest are permillages.

Method	AUROC↑	FPR95↓	AURC↓	E-AURC↓
MSP [14]	81.29	73.78	42.35	31.56
ODIN [20]	81.30	73.20	43.33	32.64
EBO [23]	82.09	72.76	42.72	32.03
ReAct [35]	80.61	78.37	44.71	34.02
MaxLogit [13]	81.92	73.47	42.48	31.79
Mixup [38]	80.93	78.08	43.74	33.17
LS [30]	81.10	72.87	42.05	31.34
Relation [36]	82.07	71.99	41.37	30.79
CPN [44]	81.39	71.47	41.83	31.45
CTR	**82.35**	**71.27**	**40.56**	**30.23**

Table 4. Performance of MisD on HWDB and ICDAR2013.

Method	AUROC↑	FPR95↓	AURC↓	E-AURC↓
MSP [14]	89.96	47.77	15.16	12.35
ODIN [20]	90.03	44.73	14.50	11.69
EBO [23]	88.33	46.69	15.10	12.29
ReAct [35]	84.98	51.46	20.62	17.81
MaxLogit [13]	87.37	49.96	16.33	13.52
Mixup [38]	89.09	47.95	11.51	8.99
LS [30]	89.47	45.88	11.10	8.06
Relation [36]	92.74	43.88	10.11	7.66
CPN [44]	90.55	44.09	11.08	8.55
CTR	**93.19**	**40.76**	**8.72**	**6.40**

as negative samples. Among them, E-AURC eliminates the impact of accuracy on MisD performance.

As the experimental results show, CTR predicts lower confidence for samples likely to be misclassified better than other methods. Regularization-based methods, such as Mixup [38] and Label Smoothing [30], aim to smooth the model's feature space distribution to alleviate overconfidence. However, the experimental results indicate that regularization does not positively affect overconfidence in character datasets. In contrast, the proposed CTR outperforms most methods, including CPN, which has no templates. It is worth noting that compared to the baseline, CTR performs better on Chinese characters than on English characters.

4.6 Rejection in Text Line Recognition

In text line recognition, rejection is performed at the character level by discarding low-confidence predictions. For Chinese datasets, we follow [24]: a prediction is correct only when both its character class and position match the ground truth (Fig. 5). For English datasets without character-level position annotations, we use minimum-edit-distance alignment to determine correct and incorrect predictions.

We use the same metrics as MisD, where AUPR-S treats correctly predicted characters as positives and AUPR-E treats errors as positives [14]. The results on Chinese and English text lines are reported in Table 5 and Table 6, respectively. CTR outperforms existing calibration methods, including the template-free prototype classifier CPN, showing that template priors provide more reliable character-level confidence.

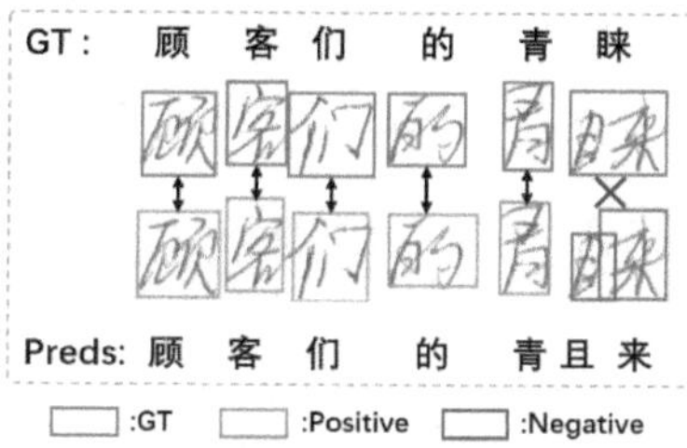

Fig. 5. Positive and negative sample definition.

Prototype Embedding for the Blank Class: Since CTC [11] requires a blank class, we compare two blank prototype strategies: **Template-Based**, which uses a blank image as the blank prototype, and **Learning-Based**, which uses a learnable vector optimized during training. As shown in Table 5 and Table 6, Learning-Based performs better on Chinese text lines, while the two strategies perform similarly on English text lines. This indicates that the blank prototype design can affect CTR's rejection performance, especially when the number of character classes is large.

The results also show that Learning Threshold performs comparably with the strategies of learning blank prototype. Learning Threshold works because it conforms to human cognition and the actual samples of blank class (covering all non-character and background patterns) are divergent.

4.7 Domain Generalization

To evaluate style-independent feature learning, we trained CTR on a source domain and tested it directly on a target domain with the same classes but different styles, without fine-tuning. For Chinese, the source was HWDB, and the targets were CASIA-AHCDB [43] and CTW [48]; for English, the source was

Table 5. Performance of rejection in Chinese text line recognition. In our CTR which uses CTC decoding, the prototype for the blank class is obtained using two methods: Template-Based and Learning-Based. Learning Threshold denotes setting a learnable threshold to classify between acc.

Method	AUROC ↑	AUPR-S ↑	AUPR-E ↑	FPR95 ↓	AURC ↓	E-AURC ↓
MSP [14]	88.76	99.26	38.50	45.97	62.42	60.65
LS [30]	88.85	98.56	41.22	48.23	62.39	60.69
FOCAL [29]	90.36	98.54	42.92	47.89	61.96	60.31
FMFP [50]	89.05	99.28	35.79	48.48	62.33	60.54
CASLS [18]	90.01	99.30	40.57	48.92	62.75	60.99
CPN [44]	90.42	99.31	44.05	43.74	62.01	60.31
Learning Threshold	91.81	99.47	45.71	41.28	60.89	59.06
CTR(Template Blank)	91.69	99.46	45.55	41.83	61.15	59.37
CTR(Learning Blank)	**91.87**	**99.51**	**45.90**	**41.05**	**60.73**	**58.92**

Table 6. Performance of rejection in English text line recognition on IAM [27].

Method	AUROC ↑	AUPR-S ↑	AUPR-E ↑	FPR95 ↓	AURC ↓	E-AURC ↓
MSP [14]	52.35	94.12	7.09	92.90	58.94	57.34
LS [30]	52.42	94.19	7.13	92.85	58.88	57.23
FOCAL [29]	52.18	93.87	6.94	93.58	59.74	58.19
FMFP [50]	52.49	94.23	7.13	92.68	58.82	57.11
CASLS [18]	52.84	94.25	7.16	92.58	58.63	56.99
CPN [44]	52.30	94.25	7.18	93.08	59.20	57.63
Learning Threshold	53.95	94.41	7.55	92.27	59.31	55.91
CTR(Template Blank)	53.91	94.40	7.61	**92.25**	**58.01**	**55.80**
CTR(Learning Blank)	**53.97**	**94.43**	**7.52**	92.32	58.23	56.08

EMNIST and the target was Chars74K. Table 7 shows that CTR outperforms MSP [14] and CPN [44] on these target domains, demonstrating that using standard templates enables CTR to learn style-independent class representations robust to domain shifts.

4.8 Ablation Study

We performed ablation experiments to verify the necessity of using templates (Fig. 6). The baseline is the prototype classifier without templates ("No Temp.", i.e., the prototype learning method of [44]). Two additional controls were included: 1) **Origin**: a random sample from each class in the training set serves as the template; 2) **Shuffle**: the template labels are randomly shuffled, so each class is paired with a template from another class. For example, the class "A"

Table 7. Performance of domain generalization of CTR.

source domain	EMNIST	HWDB
target domain	Chars74K	CASIA-AHCDB
MSP [14]	44.67	91.95
CPN [44]	44.52	93.88
CTR	**45.31**	**94.64**

may use the template of "B" instead of its own template. For both, we report averaged results over multiple runs.

Results show that using templates aligned with human prior knowledge substantially improves OOD and MisD performance. Shuffle templates yield no meaningful gain, indicating that the improvement stems from matching templates, not the training approach itself. Origin templates perform comparably to printed templates, though dataset variability (e.g., similar characters like O" vs o") can slightly reduce performance. Among printed templates, Times New Roman and Comic are better for English, while Song and Kai excel for Chinese, highlighting the benefit of neat, consistent templates for confident classification.

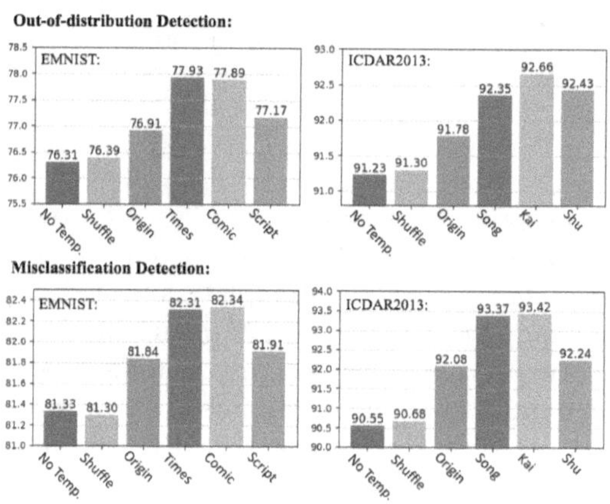

Fig. 6. AUROC of OOD and MisD tasks on English and Chinese datasets with different templates. The left column is on EMNIST dataset, the right column is on ICDAR2013 dataset. CPN (without templates) performs as the baseline for the ablation.

4.9 Visualization

Prototypes are class embeddings in the feature space, visualized by t-SNE [16] in Fig. 7. Samples cluster around their corresponding prototypes, and those farther from the prototypes receive lower confidence. This supports the effectiveness of

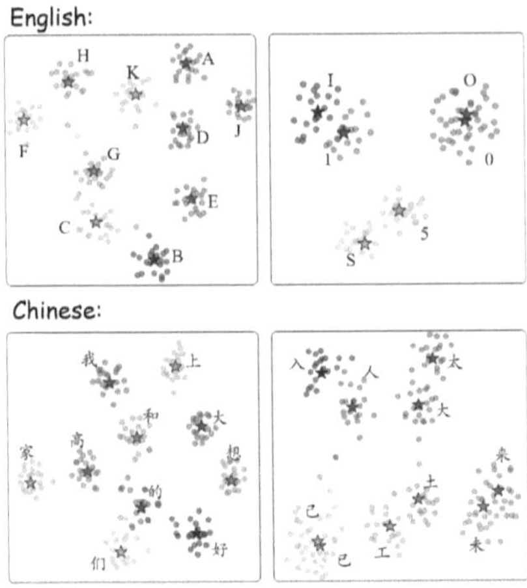

Fig. 7. Visualization of prototypes and sample embeddings. The samples are embedded around the prototypes of their classes. The right column displays the embedding distribution of similar classes.

CTR for confidence learning. The right column of Fig. 7 further shows that CTR maintains discriminative ability for visually similar classes, while Chinese characters exhibit larger style variations than English characters.

5 Conclusion

In this paper, we propose Character Template Representation (CTR), a template-based prototype framework for handwritten text line recognition. CTR assigns each character class a standard template and uses its embedding as the class prototype, enabling classification by matching input samples to template prototypes in the feature space. This design encourages samples to be close to their corresponding templates while assigning lower confidence to ambiguous or out-of-distribution samples. Experiments on English and Chinese character datasets demonstrate that CTR improves confidence estimation in OOD detection, misclassification detection, text-line rejection, and domain generalization. These results show that character templates can serve as useful human-prior prototypes for improving the reliability (particularly, the ability of rejecting ambiguous character and non-character patterns) of text recognition models.

Acknowledgment. This work is supported by the National Natural Science Foundation of China (NSFC) Grant U25A20538.

References

1. Ayyaz, M.N., Javed, I., Mahmood, W.: Handwritten character recognition using multiclass SVM classification with hybrid feature extraction. Pakistan J. Eng. Appl. Sci. **10**, 57–67 (2012)
2. Camastra, F.: A SVM-based cursive character recognizer. Pattern Recogn. **40**(12), 3721–3727 (2007). https://doi.org/10.1016/j.patcog.2007.03.014
3. de Campos, T.E., Babu, B.R., Varma, M.: Character recognition in natural images. In: Proceedings of the Fourth International Conference on Computer Vision Theory and Applications - Volume. 2: VISAPP, (VISIGRAPP 2009), vol. 2, pp. 273–280. INSTICC, SciTePress (2009). https://doi.org/10.5220/0001770102730280
4. Chen, Y., Zhang, H., Liu, C.L.: Improved learning for online handwritten Chinese text recognition with convolutional prototype network. In: Proceedings of the 2023 International Conference on Document Analysis and Recognition (ICDAR 2023). LNCS, vol. 14190, pp. 38–53. Springer (2023). https://doi.org/10.1007/978-3-031-41685-9_3
5. Chen, Y., Zhang, H., Ren, M.S., Liu, C.L.: Recognition of online handwritten Chinese texts in any writing direction via stroke classification based over-segmentation. In: Pattern Recognition, ICPR 2024, Kolkata, India, December 1–5, 2024, Proceedings, Part VII (ICPR 2024). LNCS, vol. 15307, pp. 375–391. Springer (2023). https://doi.org/10.1007/978-3-031-78183-4_24
6. Chow, C.K.: On optimum recognition error and reject tradeoff. IEEE Trans. Inf. Theory **16**(1), 41–46 (1970). https://doi.org/10.1109/TIT.1970.1054406
7. Cohen, G., Afshar, S., Tapson, J., van Schaik, A.: EMNIST: extending MNIST to handwritten letters. In: Proceedings of the 2017 International Joint Conference on Neural Networks (IJCNN 2017), pp. 2921–2926. IEEE (2017). https://doi.org/10.1109/IJCNN.2017.7966217
8. Du, X., Wang, Z., Cai, M., Li, Y.: VOS: learning what you don't know by virtual outlier synthesis. In: International Conference on Learning Representations (2022). OpenReview.net (2022). https://openreview.net/forum?id=TW7d65uYu5M
9. Flach, P., Kull, M.: Precision-recall-gain curves: PR analysis done right. In: Advances in Neural Information Processing Systems, vol. 28, pp. 838–846 (2015)
10. Geifman, Y., Uziel, G., El-Yaniv, R.: Bias-reduced uncertainty estimation for deep neural classifiers. In: International Conference on Learning Representations (2019). OpenReview.net (2019). https://openreview.net/forum?id=SJfb5jCqKm
11. Graves, A., Fern'andez, S., Gomez, F., Schmidhuber, J.: Connectionist temporal classification: labelling unsegmented sequence data with recurrent neural networks. In: Proceedings of the 23rd International Conference on Machine Learning (ICML 2006), vol. 148, pp. 369–376. ACM Press (2006). https://doi.org/10.1145/1143844.1143891
12. He, K., Zhang, X., Ren, S., Sun, J.: Deep residual learning for image recognition. In: Proceedings of the IEEE Conference on Computer Vision and Pattern Recognition (CVPR 2016), pp. 770–778. IEEE (2016). https://doi.org/10.1109/CVPR.2016.90
13. Hendrycks, D., et al.: Scaling out-of-distribution detection for real-world settings. In: Proceedings of the 39th International Conference on Machine Learning (ICML 2022, vol. 162, pp. 8759–8773. PMLR (2022)
14. Hendrycks, D., Gimpel, K.: A baseline for detecting misclassified and out-of-distribution examples in neural networks. In: International Conference on Learning Representations (2017). OpenReview.net (2017). https://openreview.net/forum?id=Hkg4TI9xl

15. Hendrycks, D., Mazeika, M., Dietterich, T.G.: Deep anomaly detection with outlier exposure. In: International Conference on Learning Representations (2019). OpenReview.net (2019). https://openreview.net/forum?id=HyxCxhRcY7
16. Hinton, G.E., Roweis, S.T.: Stochastic neighbor embedding. In: Advances in Neural Information Processing Systems, vol. 15, pp. 857–864. MIT Press (2002)
17. Hirai, Y.: A template matching model for pattern recognition: self-organization of templates and template matching by a disinhibitory neural network. Biol. Cybern. **38**(2), 91–101 (1980). https://doi.org/10.1007/BF00356035
18. Huang, S., Luo, Y., Zhuang, Z., Yu, J.G., He, M., Wang, Y.: Context-aware selective label smoothing for calibrating sequence recognition model. In: Proceedings of the 29th ACM International Conference on Multimedia (MM 2021), pp. 4591–4599. ACM (2021). https://doi.org/10.1145/3474085.3475618
19. Jiang, H., Kim, B., Guan, M.Y., Gupta, M.: To trust or not to trust a classifier. In: Advances in Neural Information Processing Systems, vol. 31, pp. 5546–5557. Curran Associates, Inc (2018)
20. Liang, S., Li, Y., Srikant, R.: Enhancing the reliability of out-of-distribution image detection in neural networks. In: International Conference on Learning Representations (2018). OpenReview.net (2018). https://openreview.net/forum?id=H1VGkIxRZ
21. Liu, C.L., Nakagawa, M.: Evaluation of prototype learning algorithms for nearest-neighbor classifier in application to handwritten character recognition. Pattern Recogn. **34**(3), 601–615 (2001). https://doi.org/10.1016/S0031-3203(00)00018-2
22. Liu, C.L., Yin, F., Wang, D.H., Wang, Q.F.: CASIA online and offline Chinese handwriting databases. In: Proceedings of the 2011 International Conference on Document Analysis and Recognition (ICDAR 2011), pp. 37–41. IEEE (2011). https://doi.org/10.1109/ICDAR.2011.17
23. Liu, W., Wang, X., Owens, J.D., Li, Y.: Energy-based out-of-distribution detection. In: Advances in Neural Information Processing Systems, vol. 33, pp. 21464–21475. Curran Associates, Inc (2020)
24. Liu, Y., Chen, Y., Yin, F., Liu, C.L.: Context-aware confidence estimation for rejection in handwritten Chinese text recognition. In: Proceedings of the 2024 International Conference on Document Analysis and Recognition (ICDAR 2024). LNCS, vol. 14804, pp. 134–151. Springer (2024). https://doi.org/10.1007/978-3-031-70533-5_9
25. Liu, Y., Yin, F., Liu, C.L.: Confident classification via template representation learning. Neurocomputing **682**, 133411 (2026). https://doi.org/10.1016/j.neucom.2026.133411
26. Manning, C.D., Sch"utze, H.: Foundations of Statistical Natural Language Processing, The MIT Press, Cambridge, MA, USA (1999)
27. Marti, U.V., Bunke, H.: The IAM-database: an English sentence database for offline handwriting recognition. Int. J. Doc. Anal. Recogn. **5**(1), 39–46 (2002). https://doi.org/10.1007/s100320200071
28. Ming, Y., Fan, Y., Li, Y.: POEM: out-of-distribution detection with posterior sampling. In: Proceedings of the 39th International Conference on Machine Learning (ICML 2022, vol. 162, pp. 15650–15665 (2022). PMLR
29. Mukhoti, J., Kulharia, V., Sanyal, A., Golodetz, S., Torr, P.H.S., Dokania, P.K.: Calibrating deep neural networks using focal loss. In: Advances in Neural Information Processing Systems, vol. 33, pp. 15288–15299. Curran Associates, Inc (2020)
30. M"uller, R., Kornblith, S., Hinton, G.E.: When does label smoothing help? In: Advances in Neural Information Processing Systems, vol. 32, pp. 4694–4703. Curran Associates, Inc (2019)

31. Shi, B., Bai, X., Yao, C.: An end-to-end trainable neural network for image-based sequence recognition and its application to scene text recognition. IEEE Trans. Pattern Anal. Mach. Intell. **39**(11), 2298–2304 (2017). https://doi.org/10.1109/TPAMI.2016.2646371
32. Shi, B., Wang, X., Lyu, P., Yao, C., Bai, X.: Robust scene text recognition with automatic rectification. In: Proceedings of the IEEE Conference on Computer Vision and Pattern Recognition (CVPR 2016), pp. 4168–4176. IEEE (2016). https://doi.org/10.1109/CVPR.2016.452
33. Shi, B., Yang, M., Wang, X., Lyu, P., Yao, C., Bai, X.: ASTER: an attentional scene text recognizer with flexible rectification. IEEE Trans. Pattern Anal. Mach. Intell. **41**(9), 2035–2048 (2019). https://doi.org/10.1109/TPAMI.2018.2848939
34. Standardization Administration of China: GB 2312-1980: Code of Chinese graphic character set for information interchange – primary set (1980). http://www.staroceans.org/e-book/GB2312-1980.pdf
35. Sun, Y., Guo, C., Li, Y.: ReAct: out-of-distribution detection with rectified activations. In: Advances in Neural Information Processing Systems, vol. 34, pp. 144–157 (2021)
36. Sung, F., Yang, Y., Zhang, L., Xiang, T., Torr, P.H.S., Hospedales, T.M.: Learning to compare: relation network for few-shot learning. In: Proceedings of the IEEE/CVF Conference on Computer Vision and Pattern Recognition (CVPR 2018), pp. 1199–1208. IEEE (2018). https://doi.org/10.1109/CVPR.2018.00131
37. Tang, Y.Y., Tu, L.T., Liu, J., Lee, S.W., Lin, W.W., Shyu, I.S.: Off-line recognition of Chinese handwriting by multifeature and multilevel classification. IEEE Trans. Pattern Anal. Mach. Intell. **20**(5), 556–561 (1998). https://doi.org/10.1109/34.682186
38. Thulasidasan, S., Chennupati, G., Bilmes, J.A., Bhattacharya, T., Michalak, S.: On mixup training: improved calibration and predictive uncertainty for deep neural networks. In: Advances in Neural Information Processing Systems, vol. 32, pp. 13888–13899. Curran Associates, Inc (2019)
39. Wang, T.Q., Yin, F., Liu, C.L.: Radical-based Chinese character recognition via multi-labeled learning of deep residual networks. In: Proceedings of the 2017 International Conference on Document Analysis and Recognition (ICDAR 2017), vol. 1, pp. 579–584. IEEE (2017). https://doi.org/10.1109/ICDAR.2017.100
40. Wei, H., Xie, R., Cheng, H., Feng, L., An, B., Li, Y.: Mitigating neural network overconfidence with logit normalization. In: Proceedings of the 39th International Conference on Machine Learning (ICML 2022, vol. 162, pp. 23631–23644. PMLR (2022)
41. Xiao, X., Jin, L., Yang, Y., Yang, W., Sun, J., Chang, T.: Building fast and compact convolutional neural networks for offline handwritten Chinese character recognition. Pattern Recognition **72**(C), 72–81 (2017). https://doi.org/10.1016/j.patcog.2017.06.032
42. Xu, L., Krzyzak, A., Suen, C.Y.: Methods of combining multiple classifiers and their applications to handwriting recognition. IEEE Trans. Syst. Man Cybern. **22**(3), 418–435 (1992). https://doi.org/10.1109/21.155943
43. Xu, Y., Yin, F., Wang, D.H., Zhang, X.Y., Zhang, Z., Liu, C.L.: CASIA-AHCDB: a large-scale Chinese ancient handwritten characters database. In: Proceedings of the 2019 International Conference on Document Analysis and Recognition (ICDAR 2019), pp. 793–798. IEEE (2019). https://doi.org/10.1109/ICDAR.2019.00132
44. Yang, H.M., Zhang, X.Y., Yin, F., Yang, Q., Liu, C.L.: Convolutional prototype network for open set recognition. IEEE Trans. Pattern Anal. Mach. Intell. **44**(5), 2358–2370 (2022). https://doi.org/10.1109/TPAMI.2020.3045079

45. Yin, F., Wang, Q.F., Zhang, X.Y., Liu, C.L.: ICDAR 2013 Chinese handwriting recognition competition. In: Proceedings of the 2013 12th International Conference on Document Analysis and Recognition (ICDAR 2013), pp. 1464–1470. IEEE (2013). https://doi.org/10.1109/ICDAR.2013.218
46. Yu, H., Wang, X., Li, B., Xue, X.: Chinese text recognition with a pre-trained CLIP-like model through image-IDS aligning. In: Proceedings of the IEEE/CVF International Conference on Computer Vision (ICCV 2023), pp. 11943–11952. IEEE (2023). https://doi.org/10.1109/ICCV51070.2023.01097
47. Yu, M.M., Zhang, H., Yin, F., Liu, C.L.: An approach for handwritten Chinese text recognition unifying character segmentation and recognition. Pattern Recogn. **151**(C), 110373 (2024). https://doi.org/10.1016/j.patcog.2024.110373
48. Yuan, T.L., Zhu, Z., Xu, K., Li, C.J., Mu, T.J., Hu, S.M.: A large Chinese text dataset in the wild. J. Comput. Sci. Technol. **34**(3), 509–521 (2019). https://doi.org/10.1007/s11390-019-1923-y
49. Zhang, Y.K., Zhang, H., Liu, Y.G., Yang, Q., Liu, C.L.: Oracle character recognition by nearest neighbor classification with deep metric learning. In: Proceedings of the 2019 International Conference on Document Analysis and Recognition (ICDAR 2019), pp. 309–314. IEEE (2019). https://doi.org/10.1109/ICDAR.2019.00057
50. Zhu, F., Cheng, Z., Zhang, X.Y., Liu, C.L.: Rethinking confidence calibration for failure prediction. In: Computer Vision – ECCV 2022: 17th European Conference, Tel Aviv, Israel, October 23–27, 2022, Proceedings, Part XXV (ECCV 2022). LNCS, vol. 13685, pp. 518–536. Springer (2022). https://doi.org/10.1007/978-3-031-19806-9_30

Historical Document Analysis

Blind Image Decomposition for Recovering Overlapping Text Layers on Palimpsests

Baharan Pourahmadi[1](✉), Panagiotis Leontaridis[2], Paolo Scattolin[2], and Mads Toudal Frandsen[3]

[1] Department of Culture and Language, University of Southern Denmark, Odense, Denmark
baharan@sdu.dk

[2] Department of Cultures and Civilisations, University of Verona, Verona, Italy
{panagiotis.leontaridis,paolo.scattolin}@univr.it

[3] Department of Physics, Chemistry and Pharmacy, University of Southern Denmark, Odense, Denmark
frandsen@cp3.sdu.dk

Abstract. Recovering the *scriptio inferior* (undertext) from palimpsests remains a significant challenge in digital heritage and historical document analysis due to the severe degradation of the scraped ink and complex interference from the visible *scriptio superior* (overtext). A prime example is the collection at the Biblioteca Capitolare of Verona; despite previous digitization efforts, the undertext in these manuscripts often remains illegible to Optical Character Recognition (OCR) models and, in severe cases, even to expert paleographers. Traditional enhancement techniques frequently fail to disentangle these overlapping signals effectively. This paper presents a novel approach to this problem using Blind Image Decomposition (BID). We modify the standard BID architecture to incorporate asymmetric decoders, allowing for specialized processing of the distinct text layers. To address the erosion of the undertext, we introduce a feature consistency loss to ensure robust latent representations, alongside a connectivity-preserving loss designed to enforce structural continuity in faint, fragmented strokes. Furthermore, to bridge the domain gap between training and inference, we employ a masked pre-training strategy on synthetic mixtures that leverages the spatial priors of the undertext script. Experimental results show superior separation performance compared to the baseline BID method. Finally, qualitative evaluation on the degraded *Codex Veronensis* XL (38) from the Biblioteca Capitolare of Verona confirms the model's efficacy as a tool for paleographic analysis. Code and data available at: https://github.com/bhrnprhmd-has/BID-palimpsest

Keywords: Blind image decomposition · Palimpsest · Overlapping text separation

G. A. Fink et al. (Eds.): ICDAR 2026, LNCS 16972, pp. 433–449, 2027.
https://doi.org/10.1007/978-3-032-36023-6_25

1 Introduction

1.1 Palimpsests at the Capitolare

Derived from the Greek *palin* ("again") and *psao* ("scrape"), a palimpsest is a manuscript born from scarcity [1]: a parchment whose original text (*scriptio inferior*) was erased to allow for the inscription of a new layer (*scriptio superior*). These stratified documents act as accidental time capsules, often preserving historically significant texts that were considered obsolete by medieval scribes. However, recovering these lost layers is not straightforward. The undertext is not just hidden, it is also physically degraded by the scraping attempts of the original scribe, and in some cases, chemically corrupted by modern "restoration" attempts [2].

To investigate this challenge, we focus on the collection of the Biblioteca Capitolare of Verona. Documented as active as early as 517 AD [3], the Capitolare is the oldest continuously operating library in the Western world. It stands as a testament to resilience, having survived the plague of 1630 (which decimated the Chapter and erased the memory of its "hidden" codices), the Napoleonic seizures of 1797, and the Allied bombings of 1945 [3]. Among its survivors is the *Codex Veronensis* XL (38) (Fig. 1), a manuscript that embodies the extreme limits of digital readability.

The Codex XL (38) is a chaotic mixture of eras. The visible overtext preserves St. Gregory the Great's *Moralia in Job*, written in a dense 7th- or 8th-century Luxeuil minuscule. Beneath this lies a far older treasure: a 5th-century witness to Virgil's *Bucolica, Georgica, and Aeneis*, written in Rustic Capitals. The fragment holds exceptional philological value as one of the seven surviving *codices Vergiliani antiquiores*, i.e., early witnesses to the Virgilian corpus dating from the 4th to the 6th centuries.

Fig. 1. A folio from Codex XL (38) photographed under visible light. A portion of the palimpsest is shown at higher magnification to improve the visibility of the overlapping texts. The dense overtext is written in Luxeuil minuscule, while the degraded undertext is in Rustic Capitals. The image is oriented according to the overtext, with the undertext appearing inverted.

The difficulty of recovering the Codex XL (38) is compounded by what is described as an "archaeology of destruction" [4]. Upon the rediscovery of the undertext in 1817 [6], 19th-century philologists resorted to aggressive chemical treatments to make the faded ink visible. Angelo Mai (an important figure in palimpsest studies), recognizing the value of the scholia, treated the margins with oak-gall tincture, an alcohol-based reagent that temporarily enhanced the metallic ink but permanently stained the parchment and corroded the text over time. Later interventions introduced even more damaging substances, such as Giobert's tincture, creating a complex chemical stratification that defies linear separation [2].

With the parchment continuing to deteriorate, digitization is a matter of urgent preservation [5].

1.2 The Challenge in Recovering the Text

The conventional approach to recovering hidden undertext relies on non-invasive imaging techniques, such as Multispectral Imaging (MSI), often coupled with statistical decomposition methods like Principal Component Analysis (PCA) or Independent Component Analysis (ICA) [10]. These methods operate on the premise that the overtext and undertext possess distinct spectral signatures that can be decorrelated. While successful in cases of distinct ink compositions, these techniques often fail when addressing the complex, non-linear degradations found in palimpsests like the Codex Veronensis XL (38). As demonstrated in the preliminary analysis of Codex XL (38), using a Phase One Rainbow MSI system [15] following the CHARISMA guidelines for cultural heritage imaging [8], the undertext could not be effectively isolated in specific Infrared (IR) or Ultraviolet (UV) bandwidths (Fig. 2). This is largely due to the severe degradation of the parchment. Consequently, the undertext remains entangled with the overtext in the visible spectrum.

A parallel body of research addresses the problem of overlapping text as bleed-through removal [11]. These computational models typically employ physically inspired constraints to identify and suppress text seeping through from the verso side of a page. However, their primary objective is the preservation of the foreground (overtext), often treating the background signal as noise to be discarded. This formulation is ill-suited for palimpsest research, where the undertext is the signal of interest. Furthermore, these rigid physical models struggle to capture the high variability of degradation on palimpsests.

Deep generative models have attempted to disentangle text layers by learning data-driven priors. Starynska et al. [12] proposed a framework that treats separation as an inverse problem, using a GAN pre-trained on an external clean manuscript to guide an iterative reconstruction of the undertext. While pioneering, this analysis-by-synthesis approach suffers from high computational latency, as it requires hundreds of optimization iterations for every image patch. Furthermore, it relies heavily on the availability of a perfectly matching external proxy for training, which is rare for unique ancient codices. It also often struggles to capture the structural continuity of continuous scripts. Other attempts to frame separation as an inpainting task have emerged [13,14], though they typically lack rigorous quantitative evaluation on real historical degradation.

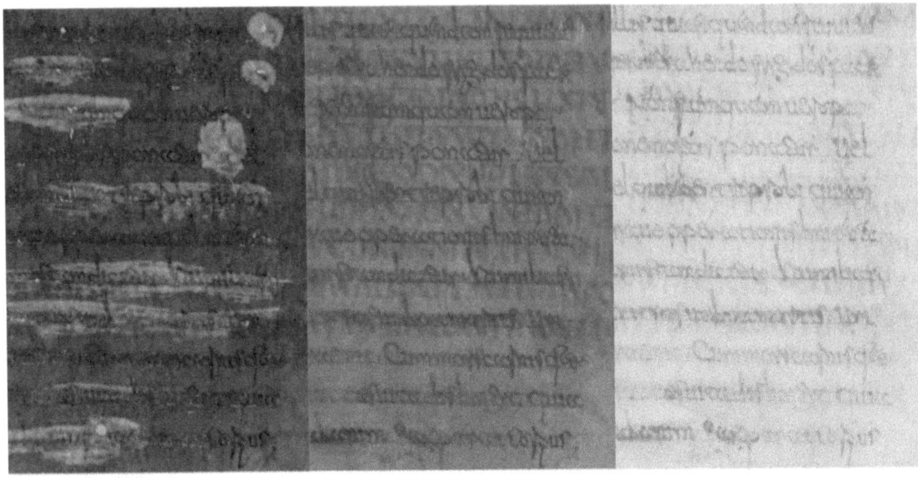

Fig. 2. A portion of Codex XL (38) photographed under Ultraviolet (UV), Visible (VIS), and Infrared (IR) illumination. The UV image reveals traces of a chemical substance previously applied for preservation purposes.

1.3 Contributions

To address the limitations of current imaging and computational techniques in revealing palimpsest undertext, a novel Blind Image Decomposition (BID) framework [16] is proposed, tailored specifically to the challenges of recovering undertext from palimpsests. The main contributions of this work are as follows:

1. Asymmetric BID Architecture: Standard symmetric decoders are replaced with an asymmetric generator tailored to the morphological imbalance of palimpsests. The undertext head is equipped with CBAM and Dilated Residual Blocks, for reconstruction of faint, fragmented undertext strokes.
2. Specialized Loss Functions: A feature consistency loss is implemented to ensure the distinct latent features of the undertext are captured by the network. Additionally, a connectivity loss is applied during fine-tuning to bridge gaps in eroded ink and ensure continuous character strokes.
3. Two-Stage training: A tailored training pipeline is introduced. Masked pre-training is first applied to the undertext decoder to learn the structural priors of the target script, which is subsequently followed by end-to-end network fine-tuning.
4. Historical Validation: The practical efficacy of the model is demonstrated on degraded patches from the Codex Veronensis XL (38), establishing its viability as a non-invasive computational tool for paleographical analysis.

2 Blind Image Decomposition

Blind Image Decomposition (BID) is a computer vision task of decomposing a single superimposed image into its constituent underlying components in a blind setting, meaning both the source components and the mixing mechanism are unknown [16]. Unlike conventional decomposition tasks that assume a fixed linear combination or a known number of layers, BID treats the observed image I as an arbitrary function of latent sources:

$$I = f(x_1, x_2, \ldots, x_n) \tag{1}$$

where f is an unknown mixing function and $\{x_i\}$ are the source components. Recent work, such as the Blind Image Decomposition Network (BIDeN) [16], utilizes Generative Adversarial Networks (GANs) [17] to solve this by learning to predict both the source components involved and their separated forms from a single input.

Overlapping text in historical documents, such as palimpsests, reused bindings, show-through, or offset ink, naturally fits the BID formulation for several reasons. First, the interaction between ink, substrate, aging processes, and imaging conditions results in complex, often non-linear mixing effects that are difficult to model explicitly. BID addresses this by learning the separation implicitly, without prescribing a specific mixing function. Second, overlapping text layers typically contribute unequally to the observed image, with secondary or erased writing often exhibiting very low contrast. BID is explicitly designed to handle such imbalanced mixtures, where some components contribute only weak visual signals. Finally, ground-truth separated layers are rarely available for historical material. The BID framework reduces reliance on pixel-wise paired supervision by combining adversarial, perceptual, and reconstruction-based objectives, making it suitable for palimpsest analysis with limited or indirect supervision.

Therefore, the overlapping historical text can be seen as a special case of blind image decomposition, where each textual stratum constitutes a latent source component and the writing surface acts as an unknown mixing medium. Within this perspective, BID provides a principled and flexible framework for separating superimposed text without imposing unrealistic assumptions on degradation physics.

3 Proposed Method

We approach the recovery of palimpsests text layers as a decomposition task trained on synthetic data. Our goal is to train a deep network that takes a single observed RGB image $I \in R^{H\times W\times 3}$ and predicts the two distinct layers: the overtext (scriptio superior) S_{over} and the undertext (scriptio inferior) S_{under}. The proposed BID framework is constrained to a two-component decomposition ($N = 2$). In this formulation, the background of the image mixture is not treated as a separate, recoverable source. Instead, it is formulated as unstructured noise that must be actively discarded by the network.

3.1 Network Architecture: Asymmetric BID with Attention

We modify the baseline Blind Image Decomposition Network (BIDeN) architecture [16] to address the morphological disparity between the text layers. Our model features a Shared Encoder followed by two specialized Asymmetric Decoders.

Multi-scale Encoder (E). A deep convolutional encoder processes the input image I, extracting a shared high-dimensional feature representation $Z = E(I)$. Similar to the BIDeN [16], the encoder (E) employs a multi-branch architecture. The resulting multi-scale feature maps are subsequently concatenated along the channel dimension to form a single, fused latent representation.

Asymmetric Decoders. To disentangle the layers, we employ two decoders:

Overtext Head (H_{over}). A standard convolutional decoder designed to reconstruct the dominant S_{over}.

Undertext Head (H_{under}). Because the undertext is typically faint and structurally broken, H_2 employs a specialized architecture. It first processes the latent features through a Convolutional Block Attention Module (CBAM) [9] to apply channel and spatial attention, highlighting faint text signals. This is followed by a Context Aggregation stream containing four Residual Block with Dilation [7] modules with exponentially increasing dilation rates ($d = 1, 2, 4, 8$). This drastically expands the receptive field without resolution loss, allowing the network to connect fragmented text strokes before standard upsampling.

Discriminator (D). The discriminator is similar to BIDeN. It features a shared feature-extraction trunk that splits into two functional branches. The separation branch (D_S) classifies inputs as real or generated. Concurrently, the prediction branch (D_P) operates as a multi-label classifier, identifying the exact source components present within the mixed input image I.

The proposed architecture is presented in Fig. 3.

3.2 Loss Functions

Training is done using synthetic mixtures where the ground truth components ($S^{gt}_{over}, S^{gt}_{under}$) are known. A compound objective function is applied directly to the predicted outputs. The original BIDeN framework relies on an adversarial loss ($\mathcal{L}_{GAN}$) [17], a perceptual VGG loss ($\mathcal{L}_{VGG}$) [18], and a global pixel-wise L_1 reconstruction loss ($\mathcal{L}_{rec}$). To ensure a better reconstruction of the degraded undertext scripture, the following two specialized loss functions are also included.

Feature Consistency Loss ($\mathcal{L}_{feat}$). To encourage the multi-scale encoder (E) to actively filter out noise, erosion artifacts, and overtext interference in the

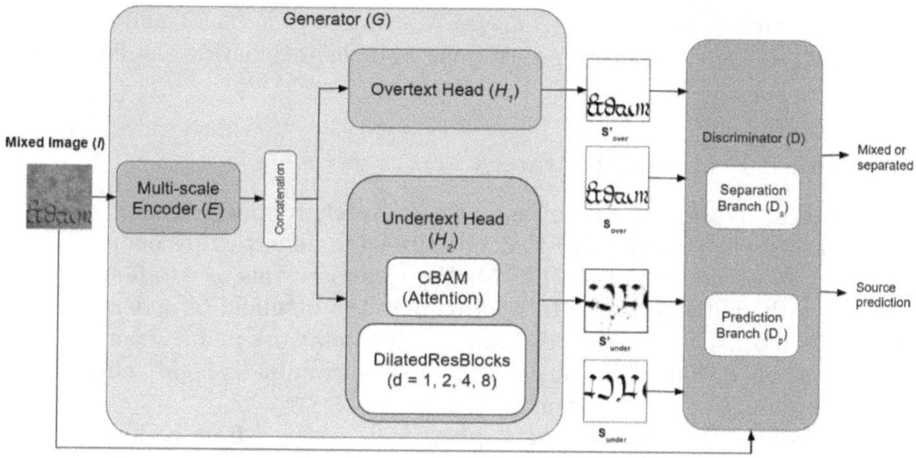

Fig. 3. The architecture of the proposed model.

latent space, a feature consistency regularization term is applied. It explicitly penalizes the L_1 distance between the encoded latent representation of the clean target undertext (S^{gt}_{under}) and the encoded representation of the noisy, mixed palimpsest (z_{mixed}):

$$\mathcal{L}_{feat} = ||E(S^{gt}_{under}) - E(z_{mixed})||_1 \tag{2}$$

Feature consistency has been previously implemented successfully in the context of text image reconstruction [20].

Connectivity-Preserving Loss ($\mathcal{L}_{conn}$). To enforce the structural and topological integrity of the reconstructed handwriting, we incorporate a Connectivity-preserving Loss ($\mathcal{L}_{conn}$) [21,22]. Unlike standard pixel-wise metrics, this objective explicitly targets morphological artifacts by performing connected component analysis on the predicted text masks. The final loss is computed as a weighted interpolation between the base pixel-wise error ($\mathcal{L}_0$) and the topologically penalized error, ensuring the generated undertext maintains legible, continuous character trajectories:

$$\mathcal{L}_{\text{conn}} = (1-\alpha)\,\mathbb{E}\left[\mathcal{L}_0\right] + \alpha\,\mathbb{E}\left[\mathcal{L}_0 \odot W_{\text{topology}}\right] \tag{3}$$

where α dictates the strength of the topological focus, and $W_{topology}$ is the dynamically generated weight map containing the heavy penalties for bridge and gap pixels.

The complete, compound objective function for the generator is formulated as a weighted sum of the individual loss components:

$$\mathcal{L}_G = \lambda_{GAN}\mathcal{L}_{GAN} + \lambda_{rec}\mathcal{L}_{rec} + \lambda_{VGG}\mathcal{L}_{VGG} + \lambda_{feat}\mathcal{L}_{feat} + \lambda_{conn}\mathcal{L}_{conn} \tag{4}$$

During training, $\mathcal{L}_{GAN}$, $\mathcal{L}_{VGG}$, $\mathcal{L}_{rec}$, $\mathcal{L}_{feat}$ are set to 1, 10, 30, and 10 respectively. $\mathcal{L}_{conn}$ is set to 0 during pre-training and set to 1 during the fine-tuning phase.

3.3 Masked Pretraining Strategy

To address the ill-posed nature of separating severely fragmented text, we introduce a masked pretraining stage that initializes the undertext decoder (D_{under}) with a robust structural prior (Fig. 5). We formulate this as a self-supervised task, where the network learns to approximate the manifold of undertext writing style by reconstructing occluded inputs. Previous work in document analysis has demonstrated that self-supervised masked pretraining is highly effective for learning representations of text images [19].

The mask is a binary mask $M \in \{0,1\}^{H \times W}$ sampled from a Bernoulli distribution with parameter $p = 0.5$, spatially structured as random rectangular blocks (Fig. 4). The corrupted input $\tilde{S}$ is obtained via the Hadamard product:

$$\tilde{S} = S_{under} \odot M \tag{5}$$

D_{under} is optimized to predict the original uncorrupted signal S_{under} from the partial view $\tilde{S}$. The objective function calculates loss only on the masked regions, forcing the model to infer missing topology from global context rather than copying visible pixels:

$$\mathcal{L}_{MIM} = ||(S_{under} - D_{under}(\tilde{S})) \odot (\mathbf{1} - M)||_1 \tag{6}$$

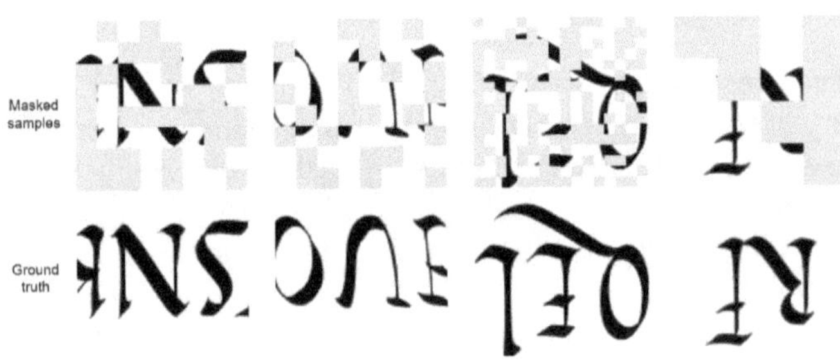

Fig. 4. Examples of masked samples and their ground truth.

4 Experimental Setup and Training Details

4.1 Synthetic Data Generation

To overcome the lack of paired ground-truth data, the network is trained on synthetic palimpsest mixtures generated on-the-fly. The textual content is rendered using a curated selection of medieval Latin fonts. For instance, Luxeuil

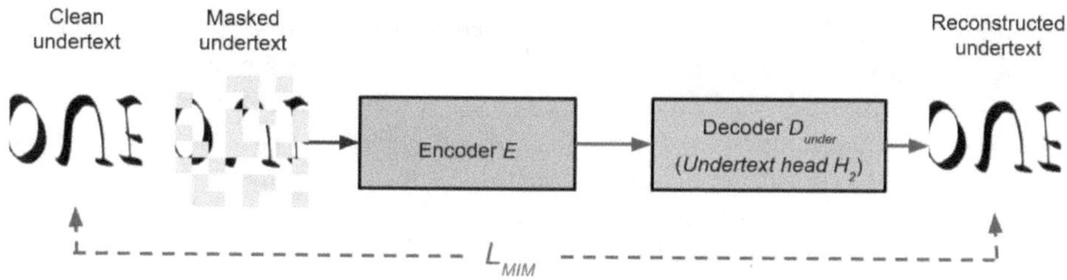

Fig. 5. Masked pretraining on the undertext head.

minuscule [30] is utilized to generate solid, high-contrast overtext, while Rustic Capitals [23] is employed for the undertext. To physically replicate the severely degraded nature of a historical palimpsest, the overtext remains structurally intact, whereas the undertext undergoes a multi-stage degradation process. It is first processed through the BSRGAN [24] degradation pipeline to introduce complex and non-uniform degradation. Subsequently, it is subjected to a randomized flaking mask that drops out between 30% and 50% of its remaining ink pixels. Finally, these text layers are composited onto complex backgrounds, sampled either from text-free manuscript patches or synthetic noise, using luminance-based subtractive alpha-blending to mimic how historical inks absorb into uneven parchment. Examples of synthetic training samples are presented in Fig. 6.

Fig. 6. Synthetic training samples.

4.2 Training Details

The network was implemented in PyTorch [25] and trained utilizing a single NVIDIA A100 GPU. The network weights were initialized using Xavier initialization [26] with a scaling gain of 0.02, and Instance Normalization [27] was applied across both the generator and discriminator networks to mitigate batch dependency. During training, input images were resized to 286 × 286 pixels and randomly cropped to 256 × 256 pixels to provide spatial data augmentation, utilizing a batch size of 4. The model was optimized using the Adam optimizer [28] with an initial learning rate of 1×10^{-4} and momentum parameters $\beta_1 = 0.5$ and $\beta_2 = 0.999$. To stabilize the adversarial training process, a Least Squares GAN (LSGAN) [29] formulation was utilized. The training was executed over a total of 50 epochs, systematically split into 10 epochs of pretraining followed by 40 epochs of full end-to-end training.

5 Evaluation

Since the ground truth does not exist for real palimpsests, the evaluation is divided into two phases. A quantitative analysis was done using synthetic data, followed by a qualitative visual assessment on patches from Codex XL (38). The results were compared to the performance of the standard BIDeN model.

5.1 Quantitative Evaluation on Synthetic Data

Quantitative evaluation was performed on a test set of 1,000 synthetic palimpsest images with known ground truths. Image reconstruction quality was assessed using Peak Signal-to-Noise Ratio (PSNR), Structural Similarity Index Measure (SSIM) [32], and Learned Perceptual Image Patch Similarity (LPIPS) [33]. Furthermore, because the ultimate goal of palimpsest restoration is transcription, Character Error Rate (CER) was calculated using the TrOCR [31] model to explicitly quantify the legibility of the recovered undertext and overtext. Table 1 summarizes the results. Examples of the reconstructed outputs of the synthetic samples are shown in Fig. 7 (for undertext), and in Fig. 8 (for overtext).

Table 1. Quantitative comparison of reconstruction performance on overtext and undertext.

Overtext				
Method	PSNR (dB)↑	SSIM↑	LPIPS↓	CER (%)↓
Standard BID	19.622	0.861	0.157	11.30
Ours	**22.441**	**0.927**	**0.061**	**9.12**
Undertext				
Standard BID	8.714	0.652	0.572	100.00
Ours	**21.221**	**0.924**	**0.114**	**21.07**

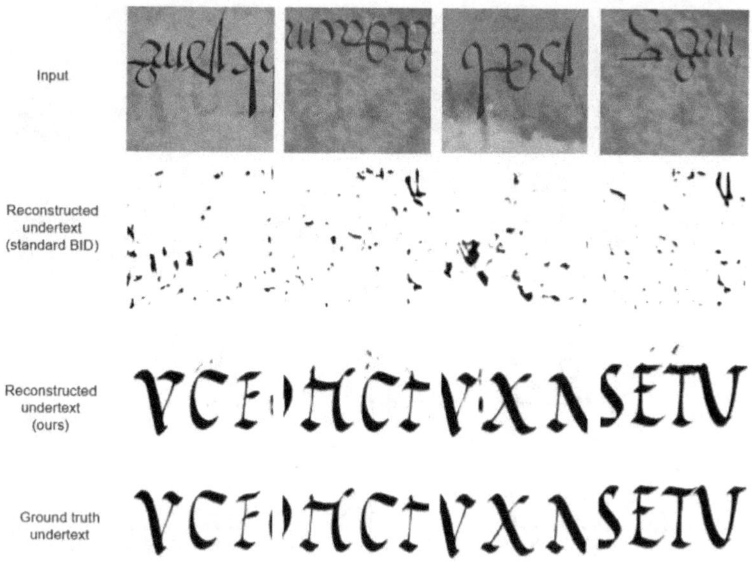

Fig. 7. Reconstructed undertext from synthetic inputs.

5.2 Qualitative Comparison on Codex XL (38)

To validate the model's practical effectiveness, a qualitative comparison was performed on a randomly selected set of patches from Codex XL (38). The outputs are presented in Fig. 9 and Fig. 10. The visual assessment is done against the outputs from the standard BID model (BIDeN). We utilized different crops for the overtext and undertext specifically to maximize the number of samples shown in the paper. Further qualitative performance is presented in Fig. 12.

5.3 Ablation Study

To validate the specific architectural and methodological contributions of our proposed framework, we conducted a systematic ablation study. We isolated the key components designed to handle the severe degradation of the undertext and evaluated their individual impact on both quantitative metrics (using our synthetic test set) and qualitative legibility (on real manuscript patches). Table 2 summarizes the results. The qualitative assessment is presented in Fig. 11.

5.4 Discussion and Limitations

Quantitative and qualitative evaluations demonstrate that the proposed method outperforms the standard BID baseline, with particularly substantial gains in undertext reconstruction (Table 1 and Fig. 9). However, visual inspections reveal that our method occasionally underperforms the baseline in reconstructing the overtext (Fig. 10). This trade-off is primarily caused by the Feature Consistency

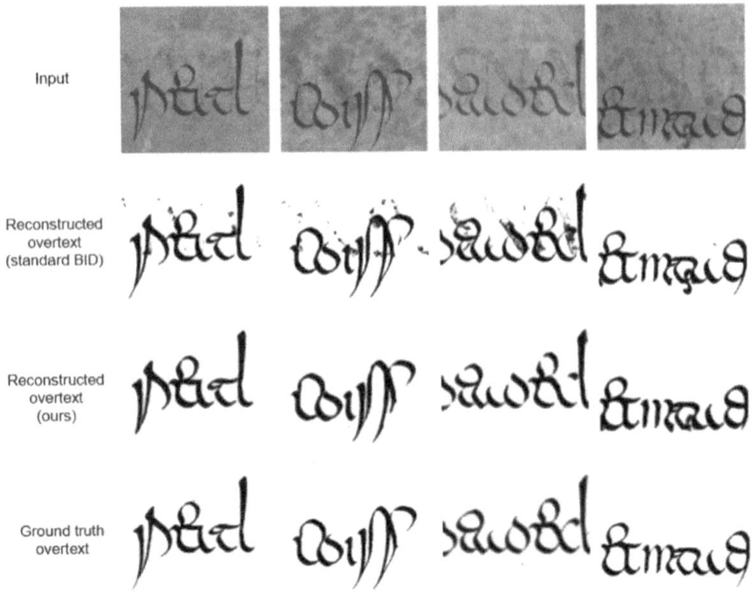

Fig. 8. Reconstructed overtext from synthetic inputs.

Table 2. Ablation study on undertext reconstruction.

Model	Excluded Component	PSNR↑	SSIM↑	LPIPS↓	CER (%)↓
Baseline	All proposed modules	8.714	0.652	0.572	100.00
Model A	Masked pretraining	17.708	0.891	0.200	29.59
Model B	CBAM & dilated ResBlocks	9.394	0.732	0.512	91.05
Model C	$\mathcal{L}_{feat}$ & $\mathcal{L}_{conn}$	18.856	0.906	0.132	22.53
Ours	None (full method)	**21.221**	**0.924**	**0.114**	**21.07**

Loss ($\mathcal{L}_{feat}$), which explicitly forces the shared encoder to prioritize undertext features, slightly diminishing its capacity to represent the top text layer. Furthermore, the ablation study confirms the individual effectiveness of the proposed components (Table 2). The inclusion of CBAM attention modules and Dilated Residual Blocks yielded the most significant performance contribution. Masked pretraining also provided substantial gains in overall stroke continuity. Finally, while the specialized objective functions ($\mathcal{L}_{feat}$ and $\mathcal{L}_{conn}$) exhibited a smaller quantitative effect, they still provide regularizations that improved the functional legibility of the undertext (Fig. 11).

A further limitation of this work is the difficulty of establishing direct comparisons with previous palimpsest reconstruction methods. Reproducing these methods was not feasible due to the lack of publicly available trained models, processed datasets, synthetic data generation procedures, and key imple-

Fig. 9. Reconstructed undertext from patches of Codex XL (38). The image is oriented according to the undertext for better visibility.

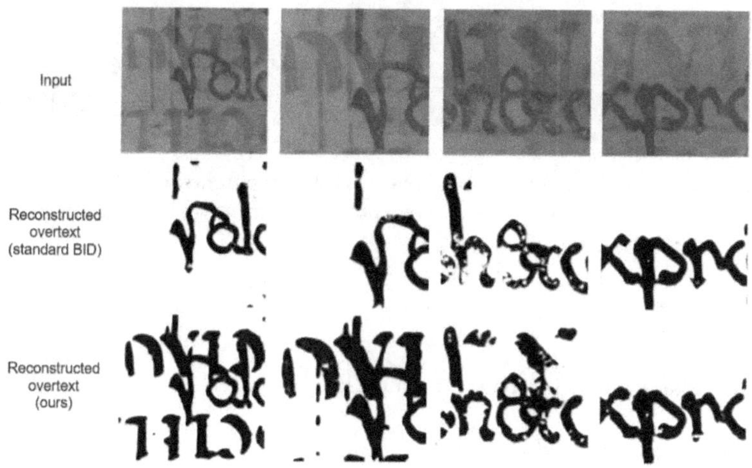

Fig. 10. Reconstructed overtext from patches of Codex XL (38).

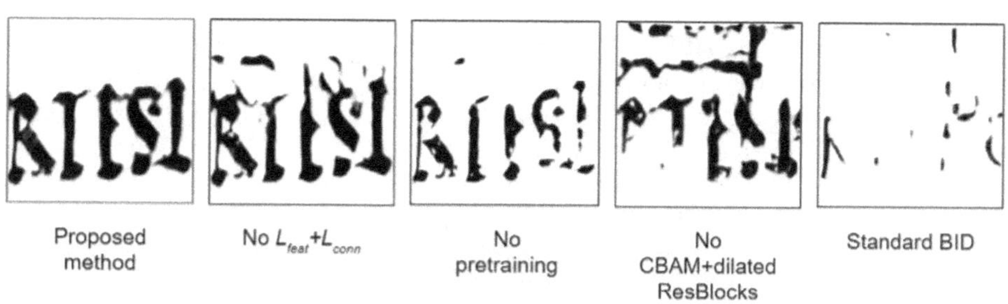

Fig. 11. Effect of each proposed component on reconstruction of undertext from a real palimpsest patch.

mentation details. Furthermore, available benchmarking datasets such as SGP [34] are not directly compatible with our evaluation framework. In SGP, the ground-truth undertext intentionally retains the simulated parchment background, whereas our method is explicitly designed to suppress background interference and recover a clean undertext layer. Consequently, conventional image-quality metrics such as PSNR and SSIM would unfairly penalize our reconstructions when compared against noise-inclusive ground truth. The absence of clean, background-free targets in SGP also complicates the pretraining strategy employed in this work. These limitations highlight the need for standardized datasets, reproducible baselines, and openly available implementations to facilitate more comprehensive comparisons in future research.

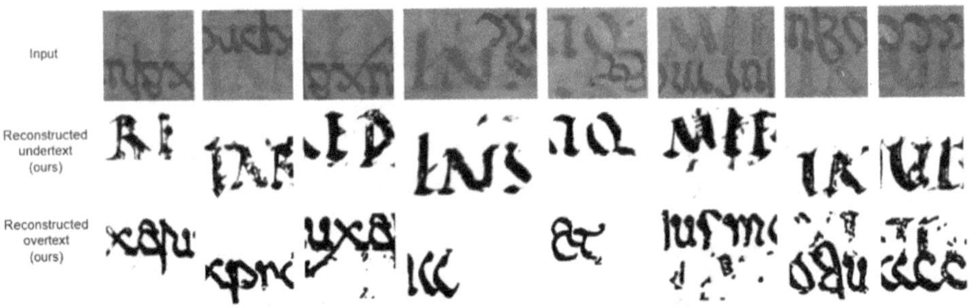

Fig. 12. Reconstructed undertext and overtext from patches of Codex XL (38). For better visibility, the input patches are oriented according to the undertext, with the overtext appearing inverted.

6 Conclusion and Future Work

In this paper, we presented an enhanced Blind Image Decomposition (BID) network specifically tailored for the complex task of palimpsest restoration. Unlike many traditional approaches that rely on Multispectral Imaging (MSI) data, our proposed method separates overlapping text layers using only standard visible-light (VIS) images. To address the severe degradation of undertext, we introduced a context-aware decoder integrated with Convolutional Block Attention Modules (CBAM) and Dilated Residual Blocks, effectively expanding the network's spatial receptive field to bridge massive structural gaps. Furthermore, a masked pretraining phase and specialized loss functions, Connectivity Loss ($\mathcal{L}_{conn}$) and Feature Consistency Loss ($\mathcal{L}_{feat}$), were implemented to explicitly enforce continuous character trajectories. Quantitative evaluation on a synthetic dataset (Table 1), combined with qualitative assessments on manuscript patches from the Codex XL (38) palimpsest (Fig. 9), demonstrated that our proposed method outperforms the standard BID baseline. The most notable achievement was the improvement in the legibility of the degraded undertext. The visual assessments highlighted a trade-off where overtext reconstruction slightly diminished in favor of isolating the undertext (Fig. 10), primarily due to the prioritization enforced by $\mathcal{L}_{feat}$. Given that the main focus was given to recovering the faded undertext, this can be seen as an acceptable compromise.

Deep learning tools like this represent a step forward for cultural heritage preservation. By enhancing the readability of degraded artifacts to the point where automatic transcription becomes viable, this methodology breaks down traditional barriers. It ensures that historians, researchers, and members of the general public without specialized paleographic background knowledge can easily access and study lost historical materials. Furthermore, the proposed method and dataset establish a benchmark for future palimpsest restoration research. Future work will focus on refining feature disentanglement to balance the reconstruction quality of both text layers equally.

Acknowledgments. This work is part of the project *"AntCom: From Antiquity to Community."* The AntCom project has received funding from the European Union's Horizon 2021 research and innovation programme under the Marie Sklodowska-Curie grant agreement No 101073543.

Disclosure of Interests. The authors have no competing interests to declare that are relevant to the content of this article.

References

1. Rossetto, G.: Greek palimpsests at Saint Catherine's monastery. Three euchologia as case studies. Austrian Academy of Sciences Press, Sinai) (2023)
2. Albrecht, F.: Between boon and bane: The use of chemical reagents in palimpsest research in the nineteenth century. Presented at the (2012) ed.) Care and conservation of manuscripts 13
3. Marchi, G.P.: Breve Discorso Storico sulla Biblioteca Capitolare di Verona. In: Marchi, S. (ed.) I manoscritti della Biblioteca Capitolare di Verona. 1st edn., pp. 9–23. Mazziana (1996)
4. Zingg, E.: Written sources on the use of reagents in the palimpsests veronenses XV, XL, and LXII: towards an archaeology of destruction. In: Gippert, J., Maksimczuk, J., Sargsyan, H. (eds.) Palimpsests and Related Phenomena across Languages and Cultures, pp. 21–56. De Gruyter (2024). https://doi.org/10.1515/9783111552170-002
5. Gysembergh, V., Zingg, E.: Les fragments inédits du palimpseste Veronensis XL, 38: La découverte d'un résumé antique des dialogues platoniciens en latin. Les Belles Lettres (2023). https://hal.science/hal-04324293v1
6. Baschera, C.: Gli scolii veronesi a Virgilio: Introduzione, edizione critica e indici. 1st edn. Casa Ed. Mazziana (1999)
7. Yu, F., Koltun, V., Funkhouser, T.: Dilated residual networks. In: Proceedings of the IEEE Conference on Computer Vision and Pattern Recognition, pp. 472–480. IEEE (2017)
8. Dyer, J., Verri, G., Cupitt, J.: Multispectral imaging in reflectance and photo-induced luminescence modes: a user manual (2023)

9. Woo, S., Park, J., Lee, J.Y., Kweon, I.S.: CBAM: convolutional block attention module. In: Proceedings of the European Conference on Computer Vision (ECCV), pp. 3–19. Springer (2018)
10. Tonazzini, A., et al.: Analytical and mathematical methods for revealing hidden details in ancient manuscripts and paintings: a review. J. Adv. Res. **17**, 31–42 (2019)
11. Tonazzini, A., Savino, P., Salerno, E.: A non-stationary density model to separate overlapped texts in degraded documents. SIViP **9**(Suppl 1), 155–164 (2015). https://doi.org/10.1007/s11760-014-0735-3
12. Starynska, A., Messinger, D., Kong, Y.: Revealing a history: palimpsest text separation with generative networks. IJDAR **24**, 181–195 (2021). https://doi.org/10.1007/s10032-021-00379-z
13. Jampour, M., Mohammed, H.A., Gippert, J.: Enhancing the readability of palimpsests using generative image inpainting. In: Proceedings of the 13th International Conference on Pattern Recognition Applications and Methods (ICPRAM), pp. 687–694. SciTePress (2024). https://doi.org/10.5220/0012347100003654
14. Jampour, M.: Revealing palimpsests with latent diffusion models: a generative approach to image inpainting and handwriting reconstruction. In: Proceedings of the Winter Conference on Applications of Computer Vision (WACV) Workshops, pp. 279–286. IEEE (2025)
15. Phase One Multispectral Imaging Camera Systems. https://www.phaseone.com/heritage-solutions/multispectral-imaging/. Accessed 25 Feb 2026
16. Han, J., et al.: Blind image decomposition. In: Proceedings of the European Conference on Computer Vision (ECCV), pp. 218–237. Springer (2022)
17. Goodfellow, I., et al.: Generative adversarial nets. In: Advances in Neural Information Processing Systems, vol. 27 (NeurIPS), Curran Associates, Inc (2014)
18. Johnson, J., Alahi, A., Fei-Fei, L.: Perceptual losses for real-time style transfer and super-resolution. In: Proceedings of the European Conference on Computer Vision (ECCV), pp. 694–711. Springer, Cham (2016)
19. Souibgui, M.A., et al.: Text-DIAE: a self-supervised degradation invariant autoencoders for text recognition and document enhancement. In: Proceedings of the AAAI Conference on Artificial Intelligence (AAAI), AAAI Press (2023)
20. Liu, Y., Wang, Z., Jin, H., Wassell, I.: Synthetically supervised feature learning for scene text recognition. In: Proceedings of the European Conference on Computer Vision (ECCV), pp. 435–451. Springer (2018)
21. Grim, A., Chandrashekar, J., Sümbül, U.: Efficient connectivity-preserving instance segmentation with super voxel-based loss function. In: Proceedings of the AAAI Conference on Artificial Intelligence, vol. 39, pp. 3167–3175. AAAI Press (2025). https://doi.org/10.1609/aaai.v39i3.32326
22. Sterzinger, R., Lin, T., Sablatnig, R.: Few-shot connectivity-aware text line segmentation in historical documents. In: Wallraven, C., He, R., Lovell, B., Chakraborty, P. (eds.) Pattern Recognition and Computer Vision. ACPR 2025. LNCS, vol. 16174, Springer, Singapore (2026). https://doi.org/10.1007/978-981-95-4395-3_8
23. MKapitalisRusticaMedium Font. https://www.ffonts.net/MKapitalisRusticaMedium.font. Accessed 26 Feb 2026
24. Zhang, K., Liang, J., Van Gool, L., Timofte, R.: Designing a practical degradation model for deep blind image super-resolution. In: Proceedings of the IEEE/CVF International Conference on Computer Vision (ICCV), pp. 4791–4800. IEEE (2021)

25. Paszke, A., et al.: PyTorch: an imperative style, high-performance deep learning library. In: Advances in Neural Information Processing Systems, vol. 32 (NeurIPS). Curran Associates, Inc. (2019)
26. Glorot, X., Bengio, Y.: Understanding the difficulty of training deep feedforward neural networks. In: Proceedings of the Thirteenth International Conference on Artificial Intelligence and Statistics (AISTATS), JMLR Workshop and Conference Proceedings, pp. 249–256 (2010)
27. Ulyanov, D., Vedaldi, A., Lempitsky, V.: Instance normalization: the missing ingredient for fast stylization (2016). arXiv:1607.08022 arXiv preprint
28. Kingma, D.P., Ba, J.: Adam: a method for stochastic optimization (2014). arXiv:1412.6980 arXiv preprint
29. Mao, X., Li, Q., Xie, H., Lau, R.Y., Wang, Z., Smolley, S.P.: Least squares generative adversarial networks. In: Proceedings of the IEEE International Conference on Computer Vision (ICCV), pp. 2794–2802. IEEE (2017)
30. Luxeuil WF Font. https://fontsgeek.com/luxeuil-wf-font. Accessed 26 Feb 2026
31. Li, M., et al.: TrOCR: transformer-based optical character recognition with pre-trained models. In: Proceedings of the AAAI Conference on Artificial Intelligence, vol. 37, pp. 13094–13102. AAAI Press (2023)
32. Wang, Z., Bovik, A.C., Sheikh, H.R., Simoncelli, E.P.: Image quality assessment: from error visibility to structural similarity. IEEE Trans. Image Process. **13**(4), 600–612 (2004)
33. Zhang, R., Isola, P., Efros, A.A., Shechtman, E., Wang, O.: The unreasonable effectiveness of deep features as a perceptual metric. In: Proceedings of the IEEE Conference on Computer Vision and Pattern Recognition (CVPR), pp. 586–595. IEEE (2018)
34. Jampour, M., Mohammed, H., Gippert, J.: Synthetic MSI Images of Georgian Palimpsests (SGP Dataset). UHH-FDR Research Data Repository (2023) https://doi.org/10.25592/uhhfdm.13378

Quality Prediction for Large Scale HTR Confidence Is All You Need

Erik Lenas(✉), Viktoria Löfgren, and Olof Karsvall

Swedish National Archives, Stockholm, Sweden
{erik.lenas,viktoria.lofgren,olof.karsvall}@riksarkivet.se

Abstract. Quality prediction of handwritten text recognition (HTR) addresses the challenge of estimating HTR quality in production where no ground-truth data are available. Focusing on pipeline-based HTR (instance segmentation for regions/lines followed by line-level recognition), we formulate quality prediction as a supervised page-level regression task: given a single document image and the runtime outputs of the HTR pipeline, predict full-page quality metrics that reflect both line/region segmentation and transcription performance. We extract and investigate four families of features available in production: (i) handcrafted image features, (ii) learned document-image embeddings from a pretrained document transformer, (iii) text-based features from external resources, including lexicon matching and character n-gram language model coverage/perplexity, and (iv) model-native features derived directly from the pipeline outputs, specifically confidence scores. A key contribution is to move beyond mean confidence and instead model confidence distributions via histogram features computed over tokens, lines, and region/line segmentation outputs. We find that confidence distributions consistently improve quality prediction and, when combined with standard confidence features, largely subsume the information provided by image- and language-based features. This implies that an effective quality prediction system can be built using low-cost model-native features.

Keywords: Quality Prediction · Handwritten Text Recognition

1 Introduction

1.1 Handwritten Text Recognition at Scale

Handwritten text recognition (HTR) has improved substantially with deep neural sequence models and, more recently, transformer-based recognizers that leverage large-scale pretraining [10,12]. As cultural heritage institutions and public agencies digitize archival collections, HTR enables document images to become searchable, allowing researchers and the general public to access the material in a more data-driven way. However, recognition errors propagate to downstream applications such as information retrieval and NLP-based analysis, where degraded text can significantly reduce performance [16].

G. A. Fink et al. (Eds.): ICDAR 2026, LNCS 16972, pp. 450–466, 2027.
https://doi.org/10.1007/978-3-032-36023-6_26

A common approach in large-scale HTR remains pipeline-based recognition: document images are segmented into regions and lines, line images are transcribed by a text recognizer, and the outputs are ordered using heuristics or trained reading-order models [20]. Although end-to-end transformer approaches that reduce reliance on separate segmentation models are emerging [8,15], we focus on pipeline-based HTR for three reasons: (i) it remains widely used in real-world digitization workflows, for instance in large-scale HTR processes at cultural heritage institutions, and HTR projects on the Transkribus platform [11]; (ii) explicit region/line/word segmentation preserves a direct alignment between image evidence and transcribed text, which is beneficial for downstream use-cases; and (iii) the pipeline approach exposes model-native features (e.g., segmentation and recognition confidence statistics) that can be exploited for quality prediction.

1.2 Quality Prediction Without Ground Truth

HTR systems are typically evaluated with ground-truth (GT) data using error metrics such as character error rate (CER) and word error rate (WER). In production, however, GT is to a large extent unavailable, and the variation in layout, handwriting style, scanning conditions, and language use encountered in, for instance, the vast historical collections of a national archive is impossible to fully cover in the annotated training data. As a result, held-out test performance may not reflect actual performance during large-scale inference.

Quality prediction addresses this gap by estimating page-level HTR quality from features available at runtime. The benefits of predicting the quality of HTR in production are numerous:

- Model selection for a material in production for which GT is not available [14]
- To provide more reliable estimation about the transcription quality to end-users, e.g. researchers or the general public.
- Filter out document images with low-quality transcriptions for further annotation, as part of the iterative retraining of the pipeline models, thereby improving the accuracy and generalization of the models.
- To estimate the amount of manual correction needed to meet the quality requirements.

Previous work has investigated the estimation of OCR/HTR quality for large-scale digitization workflows and GT-free evaluation proxies based on confidence scores and language resources [5,14]. Building on this line of work, we formulate quality prediction as supervised page-level regression: given a document image and the runtime outputs of a pipeline HTR system, predict full-page quality metrics that reflect both segmentation and transcription quality.

1.3 Contributions

The contributions of this work are as follows:

1. We propose a standardized approach for training and deploying quality prediction in production as an integrated component of training and re-training a pipeline-based HTR system.
2. We analyze four feature families relevant to quality prediction and available in production—(i) handcrafted image descriptors, (ii) learned document image embeddings from a pretrained document image transformer [9], (iii) text-based features from external resources (lexicons and character n-gram language models), and (iv) model-output features derived directly from the HTR pipeline—and find that model-output features, especially confidence histograms and complementary distributional summaries (e.g., min/max, standard deviation, coefficient of variation, and entropy), account for most of the predictive features and can match the performance of the full feature model while reducing reliance on image- and language-based features. To the best of our knowledge, this is also the first time embeddings from a pretrained document transformer have been used as features for quality prediction, and we think the results show promise and merit future work.
3. We evaluate quality prediction under two deployment-oriented settings: an in-domain scenario where training and test material are drawn from the same distribution, and an out-of-domain scenario using an external test suite with substantially different layouts and writing styles, to quantify robustness under domain shift.

2 Previous Work

2.1 Page-Level and Layout-Aware Evaluation Metrics

Classical OCR/HTR evaluation assumes line-aligned predictions and ground truths, making CER/WER straightforward to compute for line-based text recognition. However, for end-to-end page-level transcriptions, segmentation and reading order errors can invalidate linewise alignment and obscure which failure modes drive the final error. Vidal et al. analyze this issue and advocate page-level evaluation where order-sensitive and order-insensitive measures are reported together; notably, bag-of-words-style word error measures provide an order-agnostic view that better isolates recognition and content-coverage errors in full-page settings [18]. For segmentation and layout analysis, standard detection metrics (precision/recall/F1, IoU thresholds, mAP) remain central, but their interpretation depends on confidence thresholding and the distribution of layout types, motivating complementary proxy targets that are robust to layout variability [17].

2.2 Document Image Quality Assessment (DIQA)

Document image quality assessment (DIQA) is relevant to HTR quality prediction because document degradations such as blur, noise, bleed-through, uneven illumination, warping, and low resolution can affect recognition performance.

However, DIQA and HTR quality prediction address different prediction targets. DIQA primarily estimates the quality or usability of the input document image, often before recognition or independently of a specific HTR pipeline. In contrast, the present work predicts page-level output quality after recognition, using features available at inference time from the document image, the predicted text, the region- and line-segmentation, and the recognition outputs of a specific pipeline.

We therefore treat DIQA as complementary background rather than as a direct solution to the task studied here. Image-quality features may contribute to output-quality prediction, especially under severe degradation or domain shift, but they do not by themselves estimate the page-level transcription and line-segmentation quality targets considered in this work. Our experiments explicitly test this relationship by including handcrafted image features and document-image transformer embeddings alongside text-resource and model-output features.

2.3 GT-Free Quality Estimation and Quality Prediction

There is not a lot of previous work on the challenge of estimating HTR or OCR quality without ground truth, but there are some notable works that have investigated (i) GT-free evaluation proxies for model selection and (ii) supervised quality prediction models trained on pages where GT is available.

On the GT-free side, confidence scores and language-based features are commonly used as proxies for correctness. For historical print OCR, confidence values and lexicon/lexicality measures have been shown to correlate with accuracy and can support model selection, although the strength of the correlation can depend on model and data conditions [13]. More recent work on HTR explicitly compares a range of GT-free metrics, from lexicon-based ratios to n-gram language models and masked language models, and studies their suitability for both quality estimation and model selection when multiple recognizers are available [14].

On the supervised side, Clausner et al. propose an OCR quality prediction system trained to estimate downstream quality targets using a heterogeneous feature set spanning metadata, handcrafted image/layout descriptors, textual features, and confidence scores. Their framework predicts bag-of-words style text quality as well as layout-analysis quality [5]. In contrast to GT-free proxies, supervised quality prediction learns a mapping from runtime features to a task-specific target metric, enabling calibrated, page-level estimates when training labels exist for a representative subset.

3 Data

3.1 HTR-Pipeline Data

Training and Evaluation Split. The line-based text recognition model is trained on a dataset consisting of 43 126 annotated full-page document images spanning the time-period of **1550–1939**. The dataset is predominantly historical Swedish handwriting, but with substantial portions of historical Norwegian handwriting [2] and historical Danish handwriting [1]. Most pages contain running text, e.g. court protocols, while a minority contain more complex layouts such as handwritten lists and tables. The Swedish part of the dataset—the Swedish Lion dataset—was created through a collaboration between multiple institutions and government agencies, where transcriptions have been made by experts and through crowd-sourcing initiatives [7]. The region and line annotations of the Swedish Lion dataset were produced on the Transkribus platform and then manually corrected. The region and line segmentation models are trained only on the Swedish part of the dataset to ensure consistency in the annotation principles.

Out-of-Domain (OOD) Test Suite. To evaluate generalization under domain shift, we use an OOD test suite containing 338 fully annotated pages sampled from material not present in the HTR training data. The test suite contains a variety of layouts (running text, tables, lists) and spans approximately **1500–1900**, with pages sampled to be roughly evenly distributed across the time period. We run the trained HTR pipeline on this test-suite to simulate large-scale inference on previously unseen material, and use the predictions along with the ground truth to build a test set for quality prediction. Table 1 shows a summary of the HTR datasets.

Table 1. Dataset summary for the HTR models

Dataset	Language	Time-period	Pages
Swedish Lion Dataset	Swedish	1550–1900	20,727
Norhand	Norwegian	1800–1920	11,203
Danish Handwriting	Danish	1841–1939	11 196
Swedish Lion Test suite	Swedish	1500–1900	338

3.2 Lexicons and Normalization Resources

To compute lexical coverage features over predicted tokens, we use historical Swedish lexical resources, including Dalin's dictionary, historical editions of *Svenska Akademiens ordlista* (SAOL), and a mapping from a large newspaper corpus to lexicon base forms, all openly available via Språkbanken [3].

4 Method

4.1 HTR-Pipeline

We consider a pipeline-based HTR system consisting of (i) region segmentation, (ii) line segmentation within regions, and (iii) line-level text recognition. Given a document image, a YOLO-based instance segmentation model first segments text regions [19]. For each detected region, a second YOLO-based instance segmentation model segments individual text lines. Each segmented line is cropped from the original page image and transcribed by a line-level TrOCR recognizer [10]. Finally, the transcribed lines are ordered using heuristics based on their spatial coordinates.

4.2 Feature Extraction

Quality prediction is performed at the page level. For each page, we compute a feature vector using only features available at inference time in production. We group the features into four families:

1. **Handcrafted image features** computed directly from the page image.
2. **Document-image embeddings** extracted from a pre-trained document image transformer (DiT) [9].
3. **Text-resource features** derived from external resources applied to the predicted text (lexicons and character n-gram language models).
4. **Model-output features** derived directly from the HTR pipeline outputs (confidence statistics, counts, and geometry).

Handcrafted Image Features. We compute a set of low-level image descriptors designed to capture scan quality and layout difficulty, inspired by previous work on OCR quality prediction [5]. Examples include intensity statistics, foreground–background ratio, blur/noise proxies, connected-component statistics, and skew/rotation estimates.

Document-Image Embeddings. In addition to handcrafted descriptors, we extract dense document-image embeddings using a pretrained DiT model. We run the model on the full page image and use a final-layer representation as features. To reduce dimensionality, we apply Principal Component Analysis (PCA) fitted on the quality-prediction training split only; the learned projection is then applied consistently at inference time.

Text-Resource Features. We compute two families of text-based features from the predicted transcription: lexical coverage and character n-gram language-model statistics.

Lexical coverage. Given historical lexical resources (Sect. 3.2), we compute ratios such as the fraction of predicted words that appear in the lexicons. These features provide signals of out-of-vocabulary language, spelling variation, and recognition noise.
Character n-gram features. To compute language model-based features, we train a character 5-gram language model on reference text from the training split of the Swedish Lion dataset. We explicitly avoid using the evaluation split for

language-model training to prevent information leakage from n-gram features into the quality prediction training set. The trained n-gram model is then applied to the predicted text of each page to derive features such as n-gram coverage and perplexity. Intuitively, low coverage or high perplexity may indicate recognition errors or domain shift.

Model-Output Features. We derive features from region/line segmentation and recognition outputs, including confidence values, counts, and geometric statistics. For region and line segmentation, we use instance confidences and structural properties such as the number of detected regions and lines, region coverage, line heights, line widths, and spacing statistics. For text recognition, we use token-level confidence values and aggregate them to line-level scores (e.g., length-normalized means), along with text-shape indicators such as digit-to-letter ratio and length of predicted words.

Confidence distribution features. Beyond standard aggregates, we represent confidence distributions with histogram features computed over (i) region detection confidences, (ii) line detection confidences, and (iii) token and line recognition confidences. For each confidence source, we discretize confidence values into ten bins and use, as features, the fraction of values falling into each bin. Bin edges are computed on the quality-prediction training split only, using training quantiles (nine cut points), and then kept fixed for validation, testing, and production inference.

4.3 Targets: Page-Level Quality Metrics

We predict full-page quality metrics that reflect both segmentation and transcription performance. All targets are computed by aligning page-level predictions to GT and aggregating over the page.

Bag-of-Words Metrics. We use the reading order agnostic bag-of-words (BoW) metric as the primary HTR full-page quality target [18]. Given the multiset of predicted words and the multiset of GT words on a page, BoW precision is the proportion of predicted words that also occur in the GT, BoW recall is the proportion of GT words that are recovered by the predictions, and BoW F1 is the harmonic mean of precision and recall. These metrics quantify how well the pipeline transcribes the page, regardless of reading order.

Segmentation Metrics (IoU/mAP). To target segmentation quality more directly, we also consider standard overlap-based and precision–recall-based segmentation metrics. We report Intersection Over Union (IoU)-based measures at threshold 0.5 (IoU@0.5) and mean average precision (mAP0.5). We use these standard metrics because we think that they suffice for the main motivations for quality prediction, as stated in Sect. 1.2, although other work has investigated alternative metrics for layout analysis quality [17].

4.4 Quality Prediction Model

We model quality prediction as supervised regression from page-level features to page-level targets using gradient-boosted decision trees, XGBoost [4]. For each target variable, we train an independent regressor on the feature representation described above. Each training sample corresponds to one page.

Training Split and Validation. The quality prediction dataset is built on the evaluation split of the Swedish Lion dataset described in Sect. 3.1, with the split performed at the page level rather than the line level. We focus on the Swedish part of the HTR dataset because it matches our intended domain for quality prediction, and we use only the evaluation split because model-native features, particularly confidence values, are typically inflated on training pages, which would skew the quality prediction signal. [6]. Each page yields a feature vector computed from four sources described in Sect. 4.2. Targets are computed by aligning predictions of the HTR pipeline with the corresponding GT and aggregating to page-level metrics (BoW precision/recall/F1 and segmentation metrics such as IoU and mAP).

We evaluate the quality prediction model in two ways to simulate two common production scenarios:

- **In-domain validation.** The quality prediction model is trained on pages from the evaluation split of the Swedish Lion main dataset and evaluated on a held-out subset from the same distribution. This simulates deployment on material similar to what is represented in the available annotated data.
- **Out-of-domain (OOD) validation.** The trained quality prediction model is evaluated on the external test suite (Sect. 3.1), which contains samples from volumes and layouts not present in the HTR training/evaluation data. This simulates deployment on previously unseen material at scale, although still in the domain of historical Swedish.

Table 2. In-domain validation results (MAE; lower is better)

Model	BoW (F1)	IoU@0.5-lines (F1)	mAP@0.5-lines
Mean baseline	0.1077	0.1594	0.2033
Full model	0.0449	0.0832	0.1144
Full model+FS	**0.0435**	**0.0787**	**0.1055**

Table 3. Out-of-domain test suite results (MAE; lower is better)

Model	BoW (F1)	IoU@0.5-lines (F1)	mAP@0.5-lines
Mean baseline	0.0986	0.1675	0.1752
Full model	0.0439	0.1146	0.1430
Full model+FS	**0.0430**	**0.1128**	**0.1356**

Hyperparameter Optimization. We perform random search over XGBoost hyperparameters and use early stopping based on validation Mean Absolute Error (MAE). The best configuration is selected per target (and per feature set in ablations, see Sect. 5.4). Models are serialized for reproducible inference.

Feature Selection. We also do feature selection based on permutation importance, repeated over multiple sub-samples for stability. We then retrain an additional model on the selected subset.

Evaluation and Logging. We report MAE for each target under both in-domain and OOD validation. For analysis, we store per-page predictions, feature-importance summaries, and feature-selection reports.

5 Results

5.1 Baselines

We compare all models against a baseline that predict the mean target value from the training split for every page. A quality prediction model is only meaningful if it improves upon the mean baseline.

5.2 Main Results Across Targets

Tables 2 and 3 summarize the performance on both the in-domain evaluation split and the out-of-domain test suite for the full model and the model with feature selection (Full+FS), using all feature families. We report results for BoW F1 and for segmentation targets IoU@0.5 (F1) and mAP@0.5 for line segmentation specifically. We focus on line segmentation since it is nested within regions, and therefore also depends on accurate region segmentation.

Findings. Across targets, the full XGBoost model consistently improves over the mean baseline, showing that page-level HTR quality can be predicted from features derived during run-time. Performance drops on the out-of-domain test suite for segmentation-oriented targets (IoU@0.5-lines and mAP@0.5-lines). Predicting segmentation quality is probably harder because the test suite contains more varied and less predictable layouts (e.g., dense tables and lists) that are not well represented in the in-domain training distribution. In contrast, the bag-of-words target is more robust under this shift, supporting its use as a quality prediction target in deployment settings where layout variability is high.

5.3 Error Analysis

Figure 1 and Fig. 2 compare predicted and ground truth values for the Full+FS model on the evaluation split (left panel) and the OOD test suite (right panel), on BoW F1 and IoU@0.5-lines (F1) respectively.

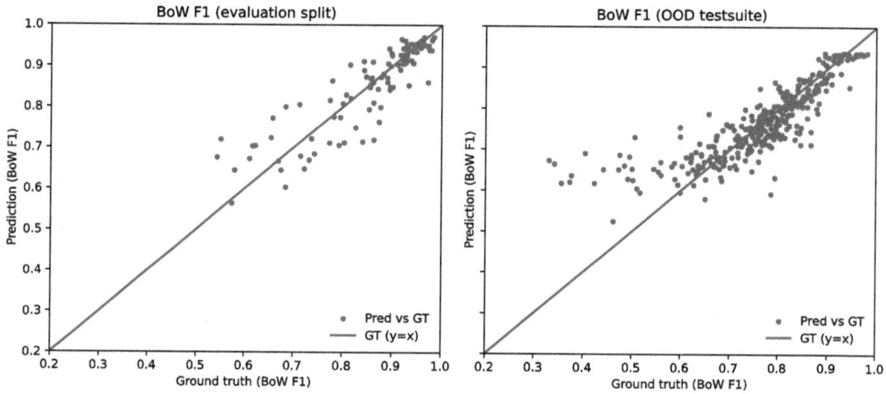

Fig. 1. Predicted vs. ground truth for BoW F1 (Full+FS).

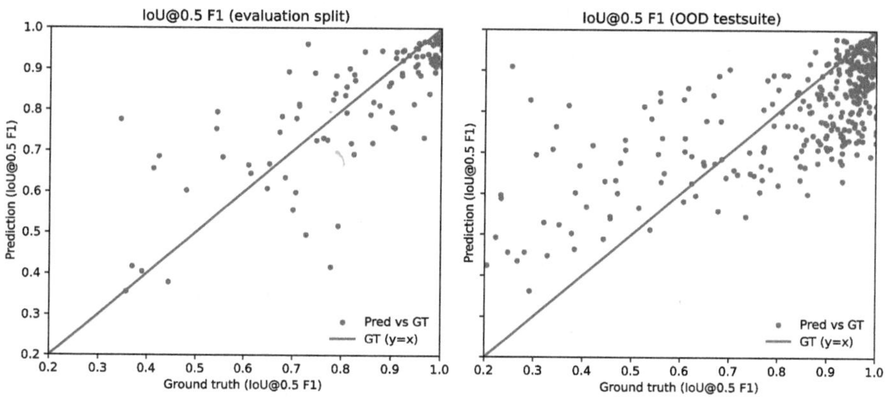

Fig. 2. Predicted vs. ground truth for line IoU@0.5 (F1) (Full+FS). Left panel: evaluation split. Right panel: OOD test suite.

Findings. On the in-domain validation split, predictions for BoW F1 track the ground truth closely, though errors increase as ground-truth values move into the lower range. On the OOD test suite this tendency is more pronounced, especially when BoW F1 drops below roughly 0.6. Such low-quality pages typically correspond to pipeline failure modes that are underrepresented in the HTR training and evaluation material-most notably structured and semi-structured layouts (e.g., handwritten tables and lists)-and are therefore also scarce in the quality-prediction training data. Despite this, the BoW F1 predictions remain highly informative and suitable for the use-cases described in Sect. 1.2.

For the segmentation target IoU@0.5-lines, predictions are noticeably less reliable. Although still somewhat informative, the model exhibits substantially higher scatter than for BoW F1, indicating that segmentation quality is harder to predict robustly under layout variability and domain shift.

5.4 Ablation Study

To quantify the contribution of each feature family, we train models on restricted feature subsets: (i) model-output features (confidence statistics, confidence histograms, counts and geometry), (ii) n-gram features, (iii) lexical matching features, (iv) handcrafted image features, and (v) DiT embeddings. Table 4 reports MAE per target for each feature subset on the OOD test suite.

Table 4. MAE across targets on the OOD test suite for the full model with feature selection (Full+FS) and feature-family-only models.

Model	BoW (F1)	IoU@0.5-lines (F1)	mAP@0.5-lines
Full model+FS	**0.0430**	0.1128	0.1356
Model output	0.0443	**0.1084**	**0.1310**
n-gram	0.1016	0.1379	0.1443
Lexical matching	0.0978	0.1538	0.1704
Image feat.	0.0974	0.1535	0.1504
DiT emb.	0.0879	0.1500	0.1720

Findings. Model-output features provide the strongest predictive signal across targets, and a quality prediction model using only model-output features performs on par with the full feature model, and on segmentation-specific targets even better. This supports the central claim that inexpensive model-native features capture most of the information needed for quality prediction, largely subsuming image- and language-based feature families. We also note that embeddings from an off-the-shelf pretrained DiT provide a stronger signal than handcrafted image features.

5.5 Feature Importance (OOD Test Suite)

We analyze which features the model relies on using permutation importance, a post-hoc measure that quantifies how much predictive performance degrades when the values of a single feature are randomly permuted across pages while all other features are kept unchanged. Features whose permutation causes a larger increase in error are deemed more important. Table 5 shows the ten most important features for two representative targets: BoW F1 (transcription-centric, but including segmentation quality) and IoU@0.5-lines (F1) (segmentation-centric). It should be noted that permutation importance and histogram-based features can be affected by feature correlations; while the qualitative conclusions are stable, exact feature rankings should be interpreted with care. Table 6 sum the permutation importance of features within each feature family, and Table 7 breaks down the model-output feature family further.

Table 5. Top-10 permutation-importance features on the OOD test suite (mean importance), shown side-by-side for BoW F1 and line IoU@0.5-lines (F1). Abbreviations: tr = text recognition; cv = coefficient of variation; conf = confidence; seg = segmentation; hist = histogram.

Rank	BoW F1	IoU@0.5-lines (F1)
1	`token tr conf mean`	`line seg conf mean`
2	`token tr conf hist bin 1`	`line char count mean`
3	`token tr conf hist bin 2`	`line width cv`
4	`line tr confidence mean`	`line spacing cv`
5	`token tr conf hist bin 6`	`line seg conf cv`
6	`token tr conf hist bin 5`	`region seg conf mean`
7	`line tr conf hist bin 1`	`line word count mean`
8	`line tr conf hist bin 3`	`line char count cv`
9	`token tr conf std`	`lines per region max`
10	`line spacing max`	`region seg conf max`

Table 6. Grouped permutation importance by feature family on the OOD test suite. Importance is summed over positive mean permutation importances; shares are relative to the total summed importance *per target.*

Feature family	Share BoW F1 (%)	Share IoU@0.5-lines (%)
Model output	90.1	79.0
DiT embeddings	8.2	17.1
Lexical matching	1.4	0.8
Handcrafted image features	0.2	3.0
n-gram LM	0.1	0.1

Findings. Feature importance cleanly differentiates transcription and segmentation centric targets. For BoW F1, the top-ranked features are dominated by recognition confidence statistics and multiple confidence-histogram bins, indicating that distributional properties of token/line confidence are highly informative for end-to-end textual quality (Table 5). Aggregating by family, BoW F1 is overwhelmingly explained by model-output features, with only a small additional contribution from DiT embeddings and negligible impact from text-resource and handcrafted image features (Table 6, left). In contrast, IoU@0.5-lines depends more strongly on segmentation-derived features: geometry/count descriptors and detection-confidence aggregates dominate both the top-10 list and the model-output breakdown (Table 5 and Table 7, right). Compared to BoW F1, IoU@0.5-lines also assigns a larger share of importance to visually grounded features (DiT embeddings and handcrafted image descriptors), consistent with layout structure being a key driver of segmentation quality under domain shift (Table 6, right).

Table 7. Breakdown of the model-output feature family on the OOD test suite. Shares are relative to the total summed model-output importance *per target*.

Model-output grouping	Share BoW F1 (%)	Share IoU@0.5-lines (%)
Confidence histograms	50.5	7.8
Confidence aggregates	44.2	33.1
Geometry and counts	5.3	59.1

5.6 Summary

Overall, quality prediction of BoW F1 as an encompassing metric for overall transcription quality is effective when applied on in-domain material and remains highly informative under out-of-domain conditions. Across all experiments, model-output features, especially confidence-distribution (histogram) features, provide the most predictive signal and largely subsume image- and language-based feature families. This supports deploying lightweight, model-native quality prediction at scale, and motivates targeted data collection of rare failure modes to improve robustness under domain shift. Segmentation-specific targets are much harder to predict, and produces more unreliable results.

6 Discussion

6.1 Comparative Results

Although our numbers are not directly comparable to previous work due to different data, targets, and evaluation protocols, it is still useful to contextualize our results. On historical *newspaper OCR*, Clausner et al. report an MAE of **0.061** for a BoW F1 target on the full dataset, and note that *layout analysis* quality is harder, with best MAE of **0.113** on the full dataset, consistent with our observation that segmentation/layout targets are harder to predict [5]. Their evaluation scheme best corresponds to our in-domain evaluation scheme where we get a BoW F1 MAE of **0.044** and IoU@50-lines MAE of **0.079**.

6.2 From Quality Estimation to Operational Decisions

The central motivation for quality prediction is operational: large-scale HTR-pipelines must be monitored and improved without exhaustive ground truth. In this setting, a useful quality predictor is not only one that achieves low MAE, but one that supports concrete actions: selecting which pages to send to correction or re-annotation, prioritizing which material, or subsets of a material, would benefit from model retraining, and providing more reliable information about transcription quality to end-users. From this perspective, our results suggest two practical takeaways. First, page-level quality can be reliably estimated from information already produced by the pipeline at runtime, making quality prediction feasible as a lightweight component in production workflows. Second, the choice of target metric determines which failure modes are emphasized. We recommend using BoW F1 as a reliable target that reflects both segmentation and recognition errors.

6.3 Why Confidence Distributions Matter

A recurring pattern across experiments is that confidence distributions provide information beyond confidence means. A page with a reasonable mean token confidence can still contain a small number of very uncertain lines or tokens (e.g., marginal notes, degraded words, or partially segmented lines) that disproportionately affect BoW F1. Histogram features make this heterogeneity explicit: they encode whether uncertainty is concentrated in a low-confidence tail, whether scores are bimodal (e.g., a mixture of easy and hard lines), or whether a page is uniformly uncertain. This is a plausible explanation for why histogram bins appear prominently among the most important features for BoW F1 and why ablations that retain standard and distributional confidence features remain competitive with the full model.

6.4 Why Out-of-Domain Outliers Are Harder

The performance drop on low quality outliers of the OOD test suite shows that failure modes the model has not trained on are hard to predict accurately. The OOD test suite introduces layouts and document conditions that are rare or absent in the quality prediction training distribution: dense tables and lists with unusual alignment. These conditions create feature combinations that are undersampled during training. As a result, the model tends to be conservative with the outliers: very low-quality pages are often over-estimated because the predictor has not seen enough similarly bad examples to learn reliable decision boundaries.

This points to a practical strategy for improving robustness: the limiting factor is not the availability of features, but the scarcity of labeled examples for rare failure modes. Active-learning-style sampling guided by the quality predictor itself is therefore a natural next step: periodically identify pages that the predictor assigns low quality and invest annotation effort in these regions of the input space.

6.5 Do We Really Need Image and Language Features?

A central claim of this work is that model-output features, particularly confidence statistics and confidence histograms, largely subsume the contribution of image- and language-based features for the targets considered. The broader implication is not that image and language features are useless, but that they appear to be secondary when strong model-native features are available. An interesting result though, is that embeddings from a pretrained document image transformer provide a stronger signal than handcrafted image features. We think this is a promising approach to make segmentation-specific targets more reliable.

Future work could investigate fine-tuning a vision language model or a document image transformer for classification of archival images, and use last layer embeddings as features for quality prediction of segmentation oriented targets. As for lexicons there is an added difficulty of working with historical text. It is difficult to find lexical resources that covers all historical inflections and spelling variations in a time-span as long as the one investigated in this work.

6.6 Confidence Is All You Need—as a Deployment Principle

Our findings support a deployment-oriented interpretation of the title claim. If a quality prediction system can achieve most of its performance using features already produced by, or derived from, the HTR pipeline, then quality prediction becomes inexpensive, scalable, and easy to integrate into existing inference workflows. This is particularly important for national-scale digitization projects where the marginal cost of additional per-page processing matters. Confidence distributions extend this principle: they require no additional models, only simple aggregation of already-available scores, yet they capture uncertainty heterogeneity that strongly correlates with page-level quality.

At the same time, "confidence is all you need" should be read as a guideline rather than a universal statement. Under severe domain shift or for tasks where confidence calibration fails, auxiliary image- and language-based features may be essential. The practical conclusion is therefore two-tiered: (i) start with model-output confidence features (including distributional summaries) as a strong default, and (ii) add auxiliary feature families primarily to improve robustness and generalization.

6.7 Limitations

This study has several limitations. First, since model-outputs, specifically confidence values, behave differently between different architectures, training methods and training datasets, the trained quality predictor will be tied to a specific pipeline. We therefore suggest training the quality prediction model as a natural part of training and retraining the pipeline. Section 6.8 outlines general recommendations for this process. Second, we investigate a pipeline approach where both segmentation- and recognition information is available at runtime. While quality prediction of end-to-end methods could benefit from the conclusions of this work, several adjustments would have to be made. Third, the OOD test suite is relatively small compared to the scale of real deployment, which limits fine-grained analysis of rare failure categories. Fourth, an off-the-shelf DiT was employed without further fine-tuning for automatic extraction of image features. Although the results showed promise, better results can possibly be achieved with a model fine-tuned for the explicit purpose of extracting features for downstream quality prediction. Fifth, we have only used openly available lexical resources for historical Swedish. More encompassing lexical resources could yield stronger lexical features. Finally, we explicitly focus on historical Swedish, and do not address multilingual HTR quality prediction.

6.8 Practical Recommendations

Based on the results and analysis, we suggest the following practical recommendations for large-scale HTR deployments:

- Use confidence distributions (histograms) in addition to confidence means to capture uncertainty heterogeneity.
- Train the quality prediction model on the evaluation set of the HTR Pipeline training dataset, with the split being made on full pages.
- Monitor quality prediction under domain shift using an external test suite, and iteratively expand the training set with hard OOD examples.
- Treat image and language features as robustness add-ons rather than first-line features when strong model-native confidences are available.

Disclosure of Interests. The authors have no competing interests.

References

1. Aarhus City Archives: Historical danish handwriting (2025). https://huggingface.co/datasets/aarhus-city-archives/historical-danish-handwriting, handwritten Danish council minutes (1841–1939). License: CC BY 4.0. Dataset card states curated by Aarhus City Archives
2. Beyer, Y., Solberg, P.E.: Norhand V3 / dataset for handwritten text recognition in Norwegian (2023). https://doi.org/10.5281/zenodo.10255840. License: CC BY 4.0
3. Borin, L., Adesam, Y., Holmer, L.: Investigating lexical change with diachronic lexical resources and corpora. In: Dannélls, D., Blensenius, K., Borin, L. (eds.) Sixty years of Swedish computational lexicography /, pp. 233–260. De Gruyter, Berlin (2025)
4. Chen, T., Guestrin, C.: XGBoost: a scalable tree boosting system. In: Proceedings of the 22nd ACM SIGKDD International Conference on Knowledge Discovery and Data Mining, pp. 785–794. Association for Computing Machinery, New York, NY, USA (2016). https://doi.org/10.1145/2939672.2939785
5. Clausner, C., Pletschacher, S., Antonacopoulos, A.: Quality prediction system for large-scale digitisation workflows. In: 2016 12th IAPR Workshop on Document Analysis Systems (DAS), pp. 138–143 (2016). https://doi.org/10.1109/DAS.2016.82
6. Guo, C., Pleiss, G., Sun, Y., Weinberger, K.Q.: On calibration of modern neural networks. In: Proceedings of the 34th International Conference on Machine Learning (ICML) (2017)
7. Karsvall, O.: Building the Swedish Lion. In: A Transkribus Tribute: Celebrating our First Five Years as a Cooperative (2019–2024), pp. 46–48 (2024)
8. Kim, G., et al.: OCR-free document understanding transformer. In: Computer Vision – ECCV 2022. Springer (2022). https://doi.org/10.1007/978-3-031-19815-1_29
9. Li, J., Xu, Y., Lv, T., Cui, L., Zhang, C., Wei, F.: DIT: self-supervised pre-training for document image transformer. In: Proceedings of the 30th ACM International Conference on Multimedia, pp. 8080–8089. Association for Computing Machinery, New York, NY, USA (2022). https://doi.org/10.1145/3503161.3547911

10. Li, M., et al.: TROCR: transformer-based optical character recognition with pre-trained models. Proc. AAAI Conf. Artif. Intelli. **37**(11), 13094–13102 (2023). https://doi.org/10.1609/aaai.v37i11.26538
11. Nockels, J., Gooding, P., Ames, S., Terras, M.: Understanding the application of handwritten text recognition technology in heritage contexts: a systematic review of Transkribus in published research. Arch. Sci. **22** (2022). https://doi.org/10.1007/s10502-022-09397-0
12. Shi, B., Bai, X., Yao, C.: An end-to-end trainable neural network for image-based sequence recognition and its application to scene text recognition (2015). https://arxiv.org/abs/1507.05717
13. Springmann, U., Fink, F., Schulz, K.U.: Automatic quality evaluation and (semi-) automatic improvement of OCR models for historical printings (2016). https://doi.org/10.48550/arXiv.1606.05157, https://arxiv.org/abs/1606.05157, arXiv preprint
14. Ströbel, P.B., Volk, M., Clematide, S., Schwitter, R., Hodel, T., Schoch, D.: Evaluation of HTR models without ground truth material. In: Calzolari, N., (eds.) Proceedings of the Thirteenth Language Resources and Evaluation Conference, pp. 4395–4404. European Language Resources Association, Marseille, France (2022). https://aclanthology.org/2022.lrec-1.467/
15. Taghadouini, S., Cavaillès, A., Aubertin, B.: Lightonocr: A 1B end-to-end multilingual vision-language model for state-of-the-art OCR (2026). https://doi.org/10.48550/arXiv.2601.14251, https://arxiv.org/abs/2601.14251
16. van Strien, D., Beelen, K., Ardanuy, M.C., Hosseini, K., McGillivray, B., Colavizza, G.: Assessing the impact of OCR quality on downstream NLP tasks. In: Proceedings of the 12th International Conference on Agents and Artificial Intelligence - Volume 1: ARTIDIGH, pp. 484–496. INSTICC, SciTePress (2020). https://doi.org/10.5220/0009169004840496
17. Vesalainen, A., Tolonen, M., Ruotsalainen, L.: Document layout error rate (dler) metric to evaluate image segmentation methods. Machine Learning with Applications **18**, 100606 (2024). https://doi.org/10.1016/j.mlwa.2024.100606
18. Vidal, E., Toselli, A.H., Ríos-Vila, A., Calvo-Zaragoza, J.: End-to-end page-level assessment of handwritten text recognition. Pattern Recogn. **142**, 109695 (2023). https://doi.org/10.1016/j.patcog.2023.109695
19. Wang, C.Y., Yeh, I.H., Liao, H.Y.M.: YOLOV9: learning what you want to learn using programmable gradient information (2024). arXiv:2402.13616 arXiv preprint
20. Wang, Z., Xu, Y., Cui, L., Shang, J., Wei, F.: LayoutReader: pre-training of text and layout for reading order detection. In: Proceedings of the 2021 Conference on Empirical Methods in Natural Language Processing (EMNLP), pp. 4735–4744 (2021)

Beyond Labels: Visual Invariance in Self-supervised Learning for Aramaic Bowls

Nour Atamni[1(✉)], Boraq Madi[2], Islam Amar[1], Raid Saabni[2], and Jihad El-Sana[1]

[1] Ben-Gurion University of the Negev, Beer-Sheva, Israel
{atamni,islamaym}@post.bgu.ac.il, jelsana@bgu.ac.il
[2] The Academic College of Tel Aviv-Yaffo, Tel Aviv-Yaffo, Israel
{boraqma,raidsa}@mta.ac.il

Abstract. Aramaic incantation bowls from Late Antiquity present a highly challenging visual domain characterized by severe material degradation, irregular layouts, and ambiguous letter morphology. Under such conditions, supervised learning is limited by the scarcity and instability of reliable paleographic labels. Self-supervised learning (SSL) offers an alternative to learning visual representations without explicit annotations; however, its effectiveness depends critically on the invariance assumptions encoded by data augmentations. In this work, we systematically evaluate seven SSL frameworks on a curated letter-level dataset derived from 32 Aramaic incantation bowls and organized around seven manually annotated letter classes. We formulate augmentation strategies as explicit hypotheses about visual identity: H1 (appearance-based transformations), H2 (structure-aware modifications), and H3 (historically and materially motivated degradations). Using controlled classification probes and quantitative geometric metrics, we analyze how these hypotheses shape embedding topology, cluster compactness, and inter-class separability. Our results show that the augmentation design substantially influences the quality of the representation. In particular, structural and historically grounded degradation hypotheses (H2, H3) consistently outperform appearance-only transformations (H1) and, in several cases, exert an effect comparable to or stronger than the choice of the SSL algorithm itself. These findings highlight the central role of domain-aware invariance design as an inductive bias for representation learning in degraded historical scripts. Project page: https://ssl-aramaic-bowls.github.io/.

Keywords: Self-Supervised Learning · Historical Document Analysis · Aramaic Incantation Bowls · Augmentation Design · Representation Geometry

---

N. Atamni and B. Madi—Both authors contributed equally to this work.

G. A. Fink et al. (Eds.): ICDAR 2026, LNCS 16972, pp. 467–484, 2027.
https://doi.org/10.1007/978-3-032-36023-6_27

1 Introduction

Aramaic incantation bowls from Late Antiquity contain dense handwritten inscriptions arranged in circular and spiral layouts over curved clay surfaces. These artifacts exhibit extreme variability in letter morphology, irregular spacing, non-uniform backgrounds, and substantial material degradation. Unlike standardized manuscripts, they reflect heterogeneous ritual practices and informal writing conventions, making visual interpretation particularly challenging.

From a computational perspective, two factors complicate learning from such material. First, the visual signal is highly unstable due to erosion, partial occlusions, and surface distortions. Second, reliable linguistic ground truth is scarce, and paleographic interpretation remains subject to scholarly debate. Under these conditions, fully supervised learning pipelines risk embedding strong and potentially unjustified labeling assumptions.

Self-supervised learning (SSL) offers a natural alternative by learning representations directly from the data without relying on manual annotation. However, SSL performance critically depends on the notion of visual invariance encoded through data augmentations. In practice, these augmentations are often adopted from natural image benchmarks (e.g., ImageNet) without examining their suitability for historical scripts. Such transformations implicitly define what aspects of visual form are treated as identity-preserving, yet this assumption is rarely analyzed explicitly in domain-specific contexts.

In this work, we treat augmentation design as an explicit hypothesis about visual identity in historical letter forms. We define three augmentation hypotheses: **H1** (appearance-based transformations), **H2** (structure-aware stroke-level modifications), and **H3** (historically and materially motivated degradations simulating erosion, surface damage, and partial observability). Rather than focusing on end-to-end supervised objectives, we evaluate how these hypotheses shape the learned embedding space.

Using controlled probing protocols and quantitative geometric analysis, we examine the impact of H1–H3 on classification accuracy, cluster compactness, and inter-class separation. Our results suggest that the design of invariance assumptions can substantially influence embedding geometry—sometimes exerting a stronger effect than the choice of self-supervised algorithm itself.

Our contributions are threefold:

- We formulate self-supervised representation learning for Aramaic incantation bowl letters under conditions of extreme visual variability and limited supervision.
- We introduce and systematically analyze three augmentation hypotheses (H1-H3) as explicit models of visual invariance, and evaluate their effect using controlled probing and quantitative geometric metrics.
- We release a curated dataset of segmented Aramaic bowl letters with multi-level visual annotations to support future research in historical document analysis.

While our experiments focus on Aramaic incantation bowls, the proposed perspective generalizes to other historical documents characterized by degradation and sparse metadata. By explicitly modeling invariance assumptions, this work provides a principled framework for representation learning in culturally and materially complex visual domains.

2 Related Work

Historical documents present a challenging visual domain due to long-term and non-uniform degradation processes, including material aging, ink erosion, and surface irregularities. These processes introduce structured variability that affects both the appearance and geometry of written forms and cannot be reduced to simple image noise [2,15].

In many collections, reliable metadata are limited and paleographic conventions are not consistently standardized, complicating the definition of stable visual categories [20]. Although numerous methods have been proposed for enhancement and binarization of degraded manuscripts [25,29], such approaches primarily address surface restoration. They do not solve the deeper challenge of defining visual identity under the effects of material degradation, preservation effects, and paleographic variability.

Computational Analysis of Historical Scripts (Supervised Paradigm). A substantial body of prior work has applied supervised machine learning to historical script analysis. Common tasks include script and style classification [1,21,28], writer identification [4,14], handwriting retrieval and verification [16], and supervised OCR for complex historical layouts [12]. These approaches assume the availability of reliable annotations defining script types, writer identities, or character labels.

Early methods relied on handcrafted features capturing stroke geometry, texture, and layout cues, whereas recent work increasingly employs deep neural architectures such as convolutional networks, Siamese models, and transformers to learn discriminative representations directly from image data [4,14,16]. Recent work on damaged historical inscriptions has also explored YOLO and Vision Transformer models for detecting and deciphering medieval Armenian inscriptions, highlighting the potential of more expressive architectures for visually complex epigraphic material [27]. Across these studies, historical samples are modeled as instances of predefined visual categories.

Within this paradigm, visual variability caused by degradation, writing conditions, and stylistic diversity is typically treated as nuisance variation to be normalized or suppressed during training [21,28]. While supervised approaches achieve strong performance on curated and well-annotated datasets, their assumptions become less stable in historical contexts where visual categories are ambiguous or weakly defined. In such settings, defining consistent ground truth for script boundaries or letter identity remains inherently challenging.

Self-supervised Learning for Historical Documents. Self-supervised learning (SSL) has recently emerged as a promising alternative for historical document analysis in settings where labeled data is scarce or unreliable. In this context, SSL is primarily employed to learn visual representations that support downstream supervised tasks, such as writer identification, script classification, manuscript reconstruction, and fragment retrieval [6,7,18,23]. Most approaches adopt contrastive or metric-learning objectives, leveraging augmented views of the same image to enforce representation invariance without explicit annotations.

In existing studies, SSL is typically evaluated indirectly through downstream task performance, such as classification accuracy or retrieval metrics [11,18,23]. The internal structure of the learned embedding space, however, receives comparatively limited analysis. Moreover, augmentation strategies are often inherited from natural image benchmarks, with limited discussion of their suitability for the material and stylistic characteristics of historical handwriting.

While these approaches demonstrate the effectiveness of SSL for document representation learning, open questions remain regarding how different augmentation choices shape representation geometry under material degradation and paleographic variability.

Augmentation as Inductive Bias. In both supervised and self-supervised settings, data augmentation is often treated as a practical design choice. However, in self-supervised learning, augmentations define the transformations under which two instances are considered equivalent, thereby imposing an implicit inductive bias on visual identity. For historical handwriting, where variability stems from material degradation, stylistic diversity, and uncertain category boundaries, this notion of equivalence is not self-evident. Yet augmentation strategies are typically adopted from natural image benchmarks, without systematic examination of their alignment with historical document variability.

To our knowledge, prior work has not explicitly formulated augmentation design as hypotheses about visual invariance in historical letter forms, nor examined how such assumptions influence representation geometry. This gap motivates a systematic investigation of augmentation-induced invariance in historical letter representations.

3 Dataset

3.1 Historical Context and Source

The dataset is derived from a collection of 32 Aramaic incantation bowls dated to Late Antiquity (4th7th centuries CE) [19,24]. These bowls contain handwritten inscriptions arranged in circular and spiral trajectories over concave ceramic surfaces. Unlike flat manuscript pages, the curved geometry introduces spatial distortion, non-uniform illumination, and variations in stroke visibility across the surface.

Preservation quality varies substantially between bowls. Some inscriptions remain relatively clear and well-defined, while others exhibit erosion, fading ink, and surface damage caused by long-term material aging. High-resolution images were obtained from curated museum collections and archival sources, preserving the visual characteristics of authentic historical material.

Figure 1 presents representative examples illustrating the variability in preservation quality and stroke visibility across different bowls.

Fig. 1. Representative samples from the dataset. The examples illustrate variability in preservation quality, ranging from clearly legible inscriptions to heavily degraded surfaces.

3.2 Dataset Overview, Structure, and Annotations

The dataset is organized at the bowl level and comprises 32 bowls containing approximately 8,600 extracted letter-like units in total (about 260 units per bowl on average).

Each bowl is associated with:

- The original high-resolution image.
- Baseline annotations describing text-line trajectories, obtained from a previously published line extraction method [22].
- Letter bounding box detections obtained using a domain-adapted object detection model.
- Segmentation masks refining letter boundaries.
- Cropped letter patches used for representation learning.

All letter instances are extracted through an automatic multi-stage pipeline to ensure consistent processing across bowls. Each extracted patch is stored together with metadata, including bowl identifier, line index, and spatial location within the original image.

For representation learning experiments, letter crops are converted to binary format to emphasize structural information and reduce illumination bias.

Figure 2 illustrates the dataset structure, including line annotations, letter detections, and segmentation outputs overlaid on a representative bowl image.

(a) Original Image (b) Line Annotations (c) Letter Detections (d) Segmentation Masks

Fig. 2. Dataset structure overview. (a) Original bowl image. (b) Text-line annotations. (c) Letter detections. (d) Segmentation masks. These stages illustrate the hierarchical organization from bowl-level imagery to individual letter instances.

3.3 Automatic Multi-stage Pipeline

Letter instances are extracted through an automatic multi-stage pipeline designed to operate consistently across bowls with diverse preservation conditions.

First, baseline annotations are used to localize text-line trajectories within each bowl. These line-level cues restrict the search space and provide structural guidance for subsequent letter detection.

Next, candidate letter regions are detected using a domain-adapted YOLO-based object detection model [26]. The resulting bounding boxes are filtered through simple post-processing steps to remove overlaps and obvious false positives.

To refine spatial boundaries, segmentation masks are generated using the Segment Anything Model (SAM) [17]. These masks improve glyph isolation by separating foreground strokes from surrounding background regions when applicable.

Finally, each refined region is cropped and converted into a standardized binary patch. This representation emphasizes structural information while reducing sensitivity to illumination and surface variability, producing the final letter instances used for self-supervised learning.

Figure 3 provides an overview of the complete processing workflow.

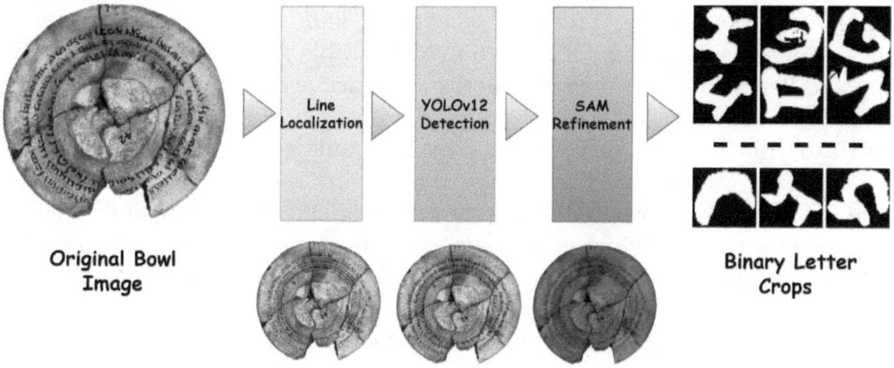

Fig. 3. Overview of the automatic multi-stage pipeline used to extract standardized letter-level patches from bowl images.

4 Methodology

4.1 Self-supervised Learning Models

To evaluate the proposed augmentation hypotheses across diverse representation learning paradigms, we employ a representative set of self-supervised learning (SSL) models grouped into three major families: contrastive, non-contrastive predictive, and redundancy-reduction / clustering-based methods.

This categorization allows us to analyze how different learning objectives interact with augmentation-driven invariance assumptions under a controlled experimental setup.

Contrastive Models. Contrastive methods learn representations by bringing augmented views of the same image (positive pairs) closer in embedding space, while pushing representations of different images (negative samples) apart. This objective is commonly implemented using the InfoNCE loss. Because contrastive learning explicitly defines similarity through augmented pairs, it is particularly sensitive to the choice and strength of data augmentations, which directly determine the invariances learned by the model.
SimCLR [8]: performs instance discrimination using in-batch negative samples and a non-linear projection head. Its performance is strongly influenced by the augmentation design, as positive pairs are entirely defined by stochastic transformations of the same image.
MoCoV2 [9]: extends contrastive learning by introducing a momentum encoder and a dynamic queue of negative samples, enabling a large and consistent set of negatives without requiring large batch sizes. This mechanism stabilizes training while preserving sensitivity to augmentation-defined invariances.

Non-contrastive Models Non-contrastive SSL methods avoid the explicit use of negative samples. Instead, they prevent representation collapse through

architectural asymmetry, predictor networks, and stop-gradient or momentum mechanisms. These models often exhibit stable training behavior and can respond differently to augmentation-induced invariances compared to contrastive approaches.
SimSiam [10]: learns representations by matching predictions from one augmented view to a stop-gradient target derived from another view of the same image. Collapse is prevented through stop-gradient operation and predictor asymmetry, without relying on negative samples or momentum encoders.
BYOL [13]: follows a similar self-prediction paradigm, but introduces a momentum-updated target network. The online network predicts the representation of the target network under different augmentations, allowing stable training without negative samples.

Redundancy Reduction and Clustering-Based Models. This family approaches representation learning by encouraging informative and non-redundant features without relying on explicit negative samples. Instead of contrasting instances, these methods regularize feature statistics or enforce consistency through prototype assignments. As such, they provide a complementary perspective on how augmentation-induced invariances shape the global structure of the embedding space.
Barlow Twins [30]: promotes invariance by aligning representations of augmented views while simultaneously reducing redundancy across feature dimensions. This is achieved by driving the cross-correlation matrix between views toward the identity, encouraging both invariance and decorrelation.
VICReg [3]: explicitly regularizes variance, invariance, and covariance of learned features. By maintaining sufficient feature variance while enforcing view consistency and decorrelation, it provides a stable framework for learning non-collapsed representations without negative samples.
SwAV [5]: adopts an online clustering strategy in which representations are assigned to learnable prototypes. The model enforces consistency between cluster assignments of different augmented views through a swapped prediction objective, implicitly shaping the global organization of the embedding space.

4.2 Augmentation Hypotheses for Historical Letter Representation

In self-supervised learning, data augmentations implicitly define which visual variations are considered irrelevant to identity. They therefore determine what invariances the model is encouraged to learn. For historical handwritten letters-where reliable labels are scarce and visual variability is substantial-the design of augmentations becomes especially critical.

Rather than treating augmentations as purely technical transformations, we formalize them as explicit hypotheses about letter identity. Each hypothesis reflects a different assumption regarding which visual properties can vary while preserving the underlying glyph form.

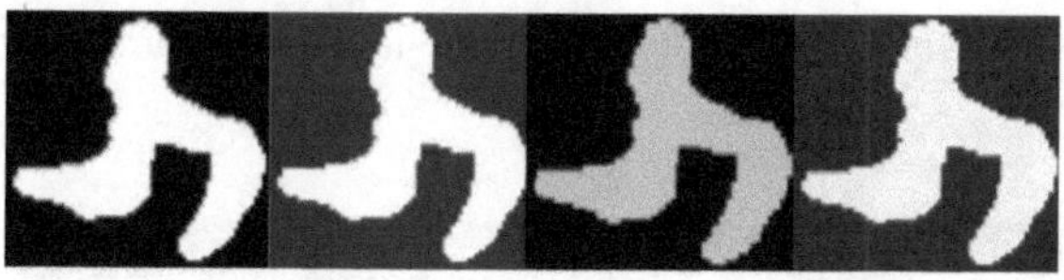

Fig. 4. Examples of the appearance invariance hypothesis (H1). Surface-level visual properties vary, while structural configuration remains stable.

The appearance invariance hypothesis assumes that letter identity remains stable under photometric and surface-level changes. These include variations in illumination, contrast, background texture, scanning artifacts, and mild sensor noise.

Under this assumption, identity is defined at the visual appearance level, provided that the structural configuration of strokes remains unchanged. The model is therefore encouraged to ignore superficial imaging variations while preserving overall letter structure, as illustrated in Fig. 4.

H2: Structural Invariance. The structural invariance hypothesis assumes that letter identity is preserved despite geometric and stroke-level variations. Such variations arise naturally from differences in writing instruments, execution speed, stylistic conventions, and scribal habits.

Under this view, identity shifts from pixel-level similarity to structural topology. Moderate changes in stroke thickness, curvature, or spatial configuration are tolerated, provided that the overall geometric organization of the letter remains coherent, as illustrated in Fig. 5.

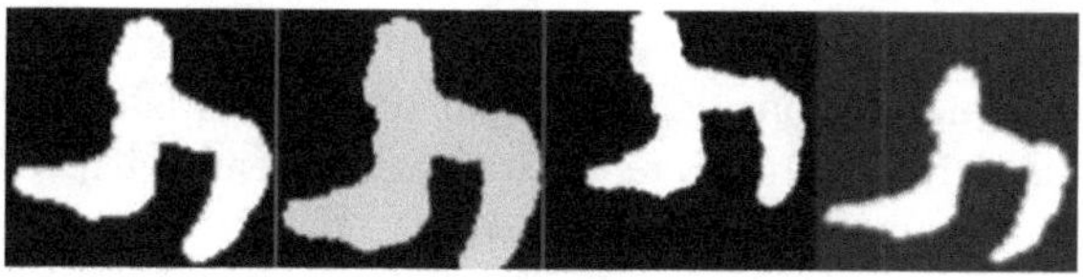

Fig. 5. Examples of the structural invariance hypothesis (H2). Stroke geometry varies while preserving the topological coherence of the letter form.

H3: Historically and Materially Motivated Degradations. The historically and materially motivated degradation hypothesis assumes that letter identity persists under degradation and visual uncertainty characteristic of historical material. These include ink erosion, staining, material decay, partial occlusion, and mechanical damage.

Unlike generic noise injection, these transformations are designed to approximate historically and materially plausible degradation patterns. This hypothesis challenges the model to extract stable semantic structure from incomplete or distorted visual evidence, encouraging higher-level abstraction, as illustrated in Fig. 6.

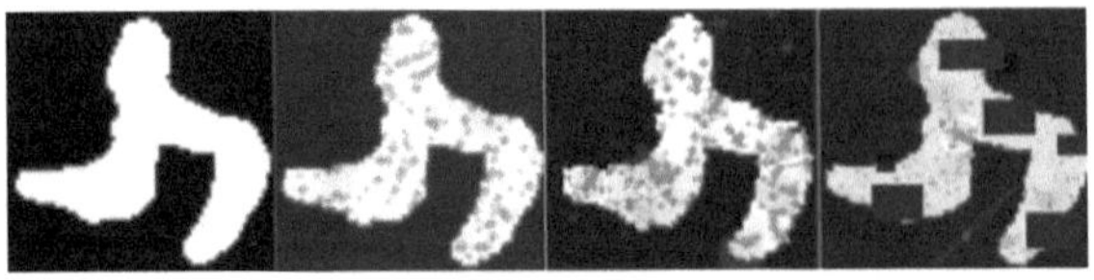

Fig. 6. Examples of the historically and materially motivated degradation hypothesis (H3), simulating degradation commonly observed in historical material.

Collectively, these hypotheses define three distinct notions of letter identity, each imposing a different invariance prior during self-supervised training. In the experimental section, we evaluate how these augmentation-defined assumptions influence representation separability and the geometric organization of the learned embedding space across diverse SSL objectives.

5 Experiments

5.1 Experimental Setup

Dataset Split: The dataset is split at the **bowl level** to prevent stylistic leakage across partitions. A total of 32 bowls are used, with 25 bowls allocated for self-supervised training and 7 bowls held out for evaluation. No validation split is used for model selection.

Approximately 6,500 automatically extracted letter instances from the training bowls are used for self-supervised representation learning. All evaluation letters originate exclusively from bowls unseen during training.

Backbone and Architecture: All methods employ a **ResNet-18** backbone to ensure fair comparison. The backbone output dimension is 512 and serves as the final embedding used for evaluation. A projection head with output dimension 512 is attached during self-supervised training, but is discarded during downstream evaluation.

Training Protocol: All models (SimCLR, SimSiam, BYOL, MoCoV2, SwAV, VICReg, and Barlow Twins) are trained under identical backbone and optimization settings. Only method-specific parameters and augmentation hypotheses differ.

Training is performed for 300 epochs using SGD with a learning rate of 0.05, momentum of 0.9, and weight decay of 10^{-4}. The batch size is 64 and mixed-precision training is enabled. A fixed learning rate schedule is used. All experiments are conducted with a fixed random seed (42) on a single CUDA-enabled GPU.

Method-Specific Parameters: For contrastive methods, the temperature parameter is set to 0.2. MoCoV2 uses a queue size of 16,384 and momentum coefficient of 0.999. SwAV employs 3,000 prototypes, 3 Sinkhorn iterations, and an entropy regularization parameter $\epsilon = 0.05$.

5.2 Evaluation Protocol

We evaluate representation quality using a controlled probe-based protocol designed to measure both class separability and intrinsic embedding geometry.

Probe Construction. A set of manually annotated probes is constructed using seven representative letter classes selected by an expert paleographer, shown in Fig. 7. Approximately 500 instances are collected exclusively from bowls that were not used during self-supervised training, ensuring strict bowl-level separation between training and evaluation data. The dataset is approximately balanced across classes.

Fig. 7. The seven annotated Aramaic letters used in the probe set.

The probe set is divided into 70% training and 30% testing partitions. The training split is used exclusively for classifier fitting and KNN memory construction, while the held-out test split is used only to report performance. No validation split is used for hyperparameter tuning, as all evaluation parameters are fixed prior to experimentation.

Classification-Based Evaluation We assess linear separability of the learned embeddings using two complementary protocols:
K-Nearest Neighbors. KNN classification is performed directly in the embedding space using cosine similarity. The probe training split serves as the memory bank, and predictions are made for probe test samples based on nearest-neighbor voting. This protocol measures local neighborhood consistency without introducing additional trainable parameters.
Linear Probing. A single fully connected layer is trained on frozen embeddings extracted from the probe training split. Only the linear classifier parameters are optimized, while the backbone representation remains fixed. Performance on the held-out probe test split reflects global linear separability under minimal supervision.

Geometric Structure Analysis. Beyond classification accuracy, we quantify the intrinsic structure of the embedding space using clustering-based metrics computed on probe test embeddings.

We report the Silhouette score, Davies–Bouldin index, mean intra-cluster distance, mean inter-cluster distance, and the inter/intra distance ratio. These measures capture cluster compactness, inter-class separation, and overall structural organization of the learned representation.

Higher Silhouette scores and inter/intra ratios indicate improved separation, whereas lower Davies–Bouldin values reflect better cluster compactness.

5.3 Classification Results

Table 1 reports classification accuracy using KNN and linear probing under augmentation hypotheses H1–H3.

Across nearly all methods, structural augmentation (H2) and historically and materially motivated degradation (H3) consistently outperform appearance-only enhancement (H1). This trend holds for both evaluation protocols, indicating improved local consistency and stronger global linear separability when structural invariances are enforced.

SwAV and VICReg achieve the strongest overall performance. SwAV yields the highest KNN accuracy under H2, while VICReg achieves the best linear probing accuracy under H3. This difference reflects their distinct geometric biases: SwAV promotes compact cluster formation, whereas VICReg encourages variance preservation and inter-class decorrelation, supporting improved linear separability under degradation.

Notably, the gap between KNN and linear probing remains small for top-performing configurations, suggesting that the learned embeddings are already well-structured prior to supervised adaptation.

Overall, the results highlight that augmentation design has a substantial impact on representation quality, often exceeding the influence of the specific self-supervised framework.

Table 1. Classification-based evaluation of self-supervised representations under different augmentation hypotheses. For each hypothesis, the two columns report KNN and linear probing accuracy, respectively. Results are reported as accuracy (%). The top-3 results in each column are highlighted in bold.

Model	H1		H2		H3	
SimSiam	78.12	74.89	**84.39**	**84.33**	**84.97**	**83.67**
SimCLR	75.72	76.85	78.74	79.81	80.09	81.75
SwAV	**80.15**	**77.67**	**88.09**	**84.67**	**86.91**	83.12
VICReg	**81.23**	**82.42**	**86.21**	**87.32**	**87.10**	**89.05**
BYOL	77.71	75.40	82.13	83.89	80.44	79.12
MoCoV2	76.21	75.20	79.33	76.91	78.55	78.12
Barlow Twins	**79.06**	**81.21**	82.11	**84.10**	84.10	**85.12**

5.4 Geometric Structure Analysis

Table 2 reports clustering-based metrics computed on probe test embeddings.

Across all models, structural hypotheses (H2) and materially grounded degradation (H3) consistently improve geometric organization compared to appearance-only enhancement (H1). This trend is reflected in higher Silhouette scores, lower Davies–Bouldin indices, and increased inter/intra distance ratios.

Among the evaluated methods, SwAV exhibits the strongest geometric structure under all augmentation hypotheses, achieving the highest Silhouette scores and inter/intra separation ratios, as well as the lowest Davies–Bouldin values. VICReg and Barlow Twins also demonstrate consistent improvements under H2 and H3 compared to H1.

Notably, H2 frequently yields the most compact and well-separated clusters, while H3 tends to increase global inter-cluster margins. This suggests that structural perturbations primarily enhance intra-class cohesion, whereas degradation simulation contributes to stronger inter-class separation.

Overall, geometric analysis aligns with classification results and reinforces the central observation that augmentation design significantly influences embedding topology.

5.5 Embedding Visualization

Figure 8 presents t-SNE projections of probe embeddings for SwAV (first row), VICReg (second row), and Barlow Twins (third row) under augmentation hypotheses H1–H3 (columns).

Table 2. Geometric structure analysis of the embedding space under different augmentation hypotheses. Higher values indicate better clustering quality for Silhouette (Silh.), Calinski–Harabasz (CH), and Inter/Intra Ratio, while lower values are better for Davies–Bouldin (DB). The top-3 results in each column are highlighted in bold.

Model	H1				H2				H3			
	Silh.↑	DB↓	CH↑	Ratio↑	Silh.↑	DB↓	CH↑	Ratio↑	Silh.↑	DB↓	CH↑	Ratio↑
SimSiam	0.453	0.765	120.050	1.790	0.492	0.715	122.967	1.992	**0.509**	0.701	**124.892**	2.009
SimCLR	0.446	0.775	118.993	1.695	0.480	0.713	121.027	1.981	0.489	0.699	121.982	1.989
SwAV	**0.587**	**0.601**	**133.772**	**2.672**	**0.668**	**0.423**	**144.926**	**3.308**	**0.621**	**0.542**	**138.048**	**2.858**
VICReg	0.466	**0.721**	120.322	1.898	**0.535**	**0.647**	**130.550**	**2.271**	**0.516**	**0.692**	124.640	**2.030**
BYOL	0.459	0.778	120.172	1.845	0.510	0.696	124.967	2.015	0.497	0.705	122.872	2.002
MoCoV2	0.462	0.745	**120.256**	1.799	0.485	0.720	121.569	1.985	0.505	0.712	124.679	2.005
Barlow Twins	**0.477**	**0.713**	**121.050**	**1.997**	**0.512**	**0.690**	**125.180**	**2.022**	**0.527**	**0.684**	**128.522**	**2.183**

Clear structural differences emerge across hypotheses. For instance, in the first row (SwAV), the embeddings under H2 (middle column) form highly compact and well-separated clusters compared to H1 (left column), where partial overlap is visible between several classes. H3 (right column) further preserves cluster cohesion despite the introduction of simulated degradation.

Similarly, VICReg under H3 (second row, right column) exhibits improved inter-class separation relative to H1, supporting its strong linear probing performance reported earlier. These observations suggest that enforcing structural and degradation-oriented priors encourages the model to rely on stable geometric patterns rather than superficial appearance cues.

Although two-dimensional projections do not perfectly preserve global distances, the visual trends are consistent with both the classification results and geometric metrics, reinforcing the impact of augmentation design on embedding topology.

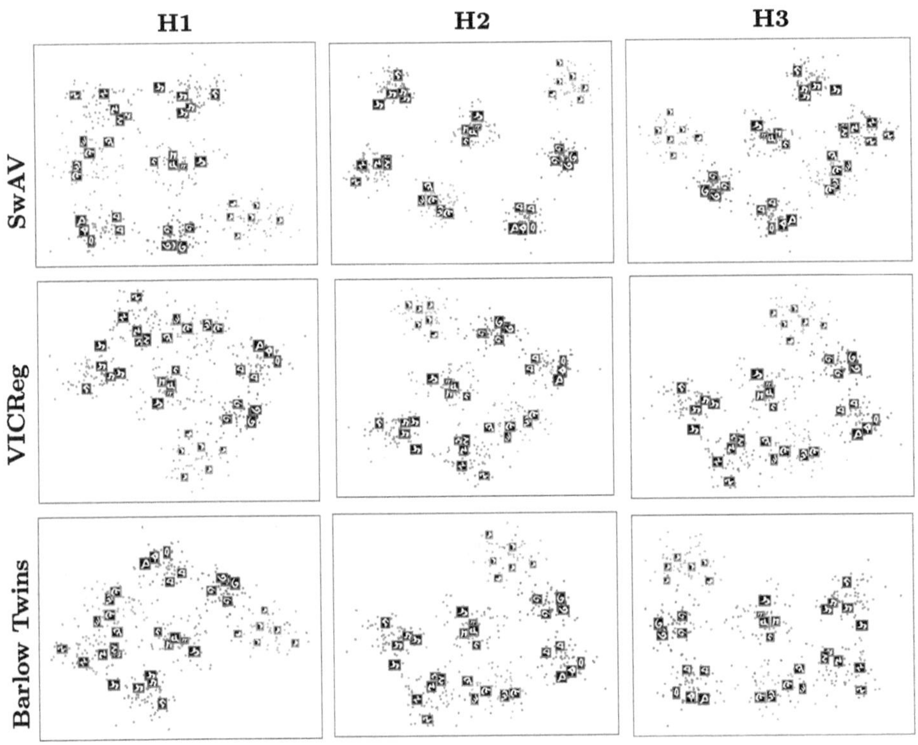

Fig. 8. t-SNE projections of probe embeddings for three representative SSL methods (rows) under augmentation hypotheses H1–H3 (columns). Rows from top to bottom represent SwAV, VICReg, and Barlow Twins, respectively. H2 and H3 consistently yield tighter clusters and clearer inter-class separation compared to H1, visually supporting the quantitative findings.

6 Discussion

The experimental results demonstrate that augmentation design substantially influences the structure of learned representations in historical script analysis. Across all evaluated methods, structural perturbations (H2) and historically and materially motivated degradation (H3) consistently improve both classification accuracy and geometric organization compared to appearance-based transformations (H1). These findings suggest that augmentation acts as a strong inductive bias that defines what the model learns to treat as semantically invariant.

Learning the Structural Core Under Degradation

The performance of H3 deserves particular attention. By introducing simulated erosion, occlusion, and surface irregularities, H3 encourages the model to rely on the persistent structural core of each letter rather than on superficial visual cues. When parts of a glyph are partially removed or corrupted, the model must learn to maintain representation consistency from incomplete visual evidence. This mechanism likely explains the observed increase in inter-class margins within the embedding space.

Importantly, the degradations introduced in H3 are not arbitrary noise injections. They are designed to approximate degradation patterns observed in historical material, thereby aligning the augmentation strategy with the material reality of archaeological artifacts. This alignment appears to enhance both robustness and discriminability.

Structural Invariance as a Paleographic Prior. The consistent improvement under H2 highlights the importance of structural invariance for handwritten historical characters. Unlike natural images, the identity of a letter is defined primarily by stroke configuration rather than color or texture. Encouraging invariance to controlled geometric distortions strengthens intra-class compactness while preserving class boundaries.

From a paleographic perspective, this can be interpreted as embedding a domain-aware prior into the learning process. The augmentation hypothesis effectively encodes assumptions about which variations are stylistic and which are semantically meaningful.

Augmentation Design vs. Algorithm Choice. An important observation is that changing the augmentation hypothesis often produces larger performance shifts than changing the underlying self-supervised framework. In several cases, moving from H1 to H2 within the same model yields greater improvements than switching between SSL algorithms under a fixed hypothesis. This suggests that, in degraded historical scripts, augmentation design may be as critical as algorithmic innovation.

Toward Intra-Object and Writer-Level Analysis. Beyond classification performance, the learned embeddings open promising avenues for historical analysis. One compelling direction is the study of intra-object consistency: do instances of the same letter extracted from a single bowl cluster more tightly than those extracted from different bowls?

Such analysis may provide indirect evidence of scribal idiosyncrasies and contribute to unsupervised writer attribution. Given the sparsity of metadata in many archaeological collections, embedding-based similarity analysis could assist in linking dispersed artifacts or identifying stylistic coherence across objects.
Limitations and Future Directions. This study evaluates representation quality using a probe set of seven letter classes and a single backbone architecture (ResNet-18). While sufficient for controlled geometric analysis, future work should extend evaluation to larger character inventories, additional encoder families such as Vision Transformers, and supervised or semi-supervised baselines for contextualizing absolute performance.

Further exploration of writer-level clustering, cross-collection generalization, and integration with paleographic assessment frameworks represents compelling avenues for future research.

Overall, the findings indicate that carefully formulated invariance hypotheses play a decisive role in how semantic identity is encoded. In the context of historical scripts, augmentation is not merely a training heuristic, but a principled mechanism for embedding material and paleographic knowledge into representation learning.

7 Conclusion

This work investigates how augmentation hypotheses influence self-supervised representation learning for historical Aramaic scripts. Through controlled experiments, we show that structural transformations and historically and materially motivated degradation consistently improve both classification performance and geometric organization compared to appearance-based augmentations.

These results show that the choice of what should remain invariant in degraded historical scripts is not a secondary implementation detail, but a central modeling decision. When aligned with the material structure of the data, augmentation hypotheses can shape representations as strongly as the SSL objective itself.

Beyond improved separability, the learned representations open promising directions for paleographic analysis, including unsupervised writer-level clustering and artifact linkage in settings with sparse metadata.

Overall, this study highlights the importance of aligning augmentation design with the material and historical characteristics of the data, demonstrating that principled invariance modeling is central to representation learning in cultural heritage contexts.

References

1. Atamni, N., Madi, B., Bordman, S., Shapira, D.V., Rabaev, I., El-Sana, J.: Multi-task learning for Hebrew paleography: script classification and date estimation. In: International Conference on Document Analysis and Recognition, pp. 79–97. Springer (2025)
2. Baird, H.S.: The state of the art of document image degradation modelling. In: Digital Document Processing: Major Directions and Recent Advances, pp. 261–279 (2007)
3. Bardes, A., Ponce, J., LeCun, Y.: VICReg: variance-invariance-covariance regularization for self-supervised learning. In: Proceedings of the International Conference on Learning Representations (ICLR) (2022)
4. Boudraa, M., Bennour, A., Nahas, M., Marie, R.R., Al-Sarem, M.: Historical manuscripts analysis: a deep learning system for writer identification using intelligent feature selection with vision transformers. J. Imag. **11**(6) (2025). https://doi.org/10.3390/jimaging11060204
5. Caron, M., Misra, I., Mairal, J., Goyal, P., Bojanowski, P., Joulin, A.: Unsupervised learning of visual features by contrasting cluster assignments. Adv. Neural. Inf. Process. Syst. **33**, 9912–9924 (2020)
6. Chen, M., Xiang, Y., Zheng, Y.: An approach for rejoining ancient Dunhuang manuscripts automatically based on self-supervised learning and encoding with azimuths information. In: 2024 International Conference on Image Processing, pp. 472–481. IEEE (2024). Computer Vision and Machine Learning (ICICML)
7. Chen, M., Xiang, Y., Zheng, Y.: A self-supervised learning approach for pairwise matching of ancient Dunhuang fragments. ACM J. Comput. Cult. Heritage **19**(1), 1–19 (2026)
8. Chen, T., Kornblith, S., Norouzi, M., Hinton, G.: A simple framework for contrastive learning of visual representations. In: International conference on machine learning, pp. 1597–1607. PMLR (2020)
9. Chen, X., Fan, H., Girshick, R., He, K.: Improved baselines with momentum contrastive learning (2020). arXiv:2003.04297 arXiv preprint
10. Chen, X., He, K.: Exploring simple Siamese representation learning. In: Proceedings of the IEEE/CVF Conference on Computer Vision and Pattern Recognition, pp. 15750–15758 (2021)
11. Du, B., Liu, G., Ge, W.: Deep self-supervised learning for oracle bone inscriptions features representation. In: 2021 IEEE 4th International Conference on Information Systems and Computer Aided Education (ICISCAE), pp. 7–11. IEEE (2021)
12. Fleischhacker, D., Kern, R., Göderle, W.: Enhancing OCR in historical documents with complex layouts through machine learning. Int. J. Digit. Libr. **26**(1), 3 (2025)
13. Grill, J.B., et al.: Bootstrap your own latent: a new approach to self-supervised learning. In: Advances in Neural Information Processing Systems (NeurIPS) (2020)
14. He, S., Schomaker, L.: Deep adaptive learning for writer identification based on single handwritten word images. Pattern Recogn. **88**, 64–74 (2019)
15. Kanungo, T., Haralick, R.M., Baird, H.S., Stuezle, W., Madigan, D.: A statistical, nonparametric methodology for document degradation model validation. IEEE Trans. Pattern Anal. Mach. Intell. **22**(11), 1209–1223 (2000)
16. Keglevic, M., Fiel, S., Sablatnig, R.: Learning features for writer retrieval and identification using triplet CNNs. In: 16th International Conference on Frontiers in Handwriting Recognition (ICFHR), pp. 211–216. IEEE (2018)

17. Kirillov, A., et al.: Segment anything. In: Proceedings of the IEEE/CVF International Conference on Computer Vision (ICCV) (2023)
18. Lastilla, L., Ammirati, S., Firmani, D., Komodakis, N., Merialdo, P., Scardapane, S.: Self-supervised learning for medieval handwriting identification: a case study from the Vatican apostolic library. Info. Process. Manag. **59**(3), 102875 (2022)
19. Levene, D.: A Corpus of Magic Bowls: Incantation Texts in Jewish Aramaic from Late Antiquity. Kegan Paul, London (2003)
20. Likforman-Sulem, L., Zahour, A., Taconet, B.: Text line segmentation of historical documents: a survey. IJDAR **9**(2), 123–138 (2007)
21. Naamneh, S., et al.: Classifying the scripts of Aramaic incantation bowls. In: Proceedings of the 7th International Workshop on Historical Document Imaging and Processing (2023). https://doi.org/10.1145/3604951.3605510
22. Nammneh, S., et al.: Detecting spiral text lines in Aramaic incantation bowls. In: Pattern Recognition, pp. 250–264. Springer Nature Switzerland, Cham (2025). https://doi.org/10.1007/978-3-031-78495-8_16
23. Pirrone, A., Beurton-Aimar, M., Journet, N.: Self-supervised deep metric learning for ancient papyrus fragments retrieval. Int. J. Docum. Anal. Recog. (IJDAR) **24**(3), 219–234 (2021)
24. Segal, J.B., Hunter, E.C.D.: Catalogue of the Aramaic and Mandaic Incantation Bowls in the British Museum. British Museum Press, London (2000)
25. Sulaiman, A., Omar, K., Nasrudin, M.F.: Degraded historical document binarization: a review on issues, challenges, techniques, and future directions. J. Imaging **5**(4), 48 (2019)
26. Ultralytics: Ultralytics YOLO. https://github.com/ultralytics/ultralytics (2023), gitHub repository. Accessed 15 June 2026
27. Vidal-Gorène, C., Decours-Perez, A.: Detecting and deciphering damaged medieval Armenian inscriptions using YOLO and vision transformers. In: International Conference on Document Analysis and Recognition, pp. 22–36. Springer (2024)
28. Vlachou-Efstathiou, M., Siglidis, I., Stutzmann, D., Aubry, M.: An interpretable deep learning approach for morphological script type analysis. In: International Conference on Document Analysis and Recognition, pp. 3–21. Springer (2024)
29. Xiong, W., Zhou, L., Yue, L., Li, L., Wang, S.: An enhanced binarization framework for degraded historical document images. EURASIP J. Image Video Process. **2021**(1), 13 (2021)
30. Zbontar, J., Jing, L., Misra, I., LeCun, Y., Deny, S.: Barlow twins: self-supervised learning via redundancy reduction. In: Proceedings of the International Conference on Machine Learning (ICML) (2021)

Decipherment of Oracle Bone Inscription via Component Deconstruction and Alignment

Yuanhui Lin[1,2], Hetao Wu[1,2], Qingju Jiao[3], Yongge Liu[3], and Da-Han Wang[1,2(✉)]

[1] School of Computer and Information Engineering, Xiamen University of Technology, Fujian, China
wangdh@xmut.edu.cn
[2] Fujian Key Laboratory of Pattern Recognition and Image Understanding, Xiamen, China
[3] School of Computer and Information Engineering, Anyang Normal University, Anyang, Henan, China

Abstract. Oracle Bone Inscription (OBI) decipherment is challenging due to severe deformation, noise, and large intra-class variation, making direct recognition unstable. Although diffusion models have shown promise for cross-era glyph translation, existing methods, such as OBSD, often perform poorly under class-disjoint zero-shot decipherment, producing outputs with strong structural noise and semantic drift that hinder accurate mapping to modern characters. To overcome this, we introduce component-aware structural constraints by incorporating radical-like component cues and enforcing component-level alignment during diffusion generation. We propose **DAOD** (*Deconstruction-and-Alignment-based Oracle Decipherment*), a structure-conditioned diffusion framework for OBI-to-modern glyph translation under a class-disjoint (zero-shot) setting. DAOD enhances generation controllability through *Component Deconstruction and Alignment* (CDA), which aligns OBI components with modern components and constructs paired masks and confidence-aware embeddings as diffusion conditions. A pair-level feedback mechanism down-weights unreliable matches using diffusion residuals during training, and *DAOD-OCR* provides auxiliary feature-level supervision with a fixed ResNet-50 classifier (99.26% accuracy on a 9,000-class set). Experiments on EVOBC and HUST-OBC (1,590 classes, 71,698 images) demonstrate consistent improvements over strong baselines.

Keywords: Oracle Bone Inscription · Decipherment · Component Alignment · Diffusion Model · OCR Guidance

1 Introduction

Oracle Bone Inscriptions (OBI), dating back about 3,000 years, constitute one of the earliest mature writing systems in China. To date, among an estimated 4,500

G. A. Fink et al. (Eds.): ICDAR 2026, LNCS 16972, pp. 485–501, 2027.
https://doi.org/10.1007/978-3-032-36023-6_28

discovered oracle bone characters, only approximately 1,500 have been successfully deciphered and mapped to modern Chinese characters. OBI decipherment is challenging because real glyphs often suffer from severe deformation, stroke breakage, background noise, and large intra-class variation, which makes direct recognition unstable and difficult to interpret [16]. Meanwhile, incomplete corpora and the lack of standardized encodings motivate *image-based decipherment*, where a model generates modern-character hypotheses as visual clues. Recent conditional diffusion models provide a strong backbone for such cross-era glyph translation by learning a distribution over plausible modern glyphs conditioned on ancient inputs [7].

A key observation is that expert decipherment is inherently *component-driven*: many OBI glyphs can be explained by a small set of recurring parts and their spatial composition. Importantly, many modern Chinese characters preserve structural cues inherited from ancient glyphs (e.g., component-like parts and relative layouts), even though stroke styles and overall appearances evolve substantially over time. Prior efforts on component-level retrieval and component-level segmentation, as well as recent benchmarks, suggest that fine-grained structural perception remains a major bottleneck in OBI workflows [1,9,10]. As illustrated in Fig. 1, cross-era component correspondences can provide interpretable evidence for decipherment and offer controllable anchors for generation.

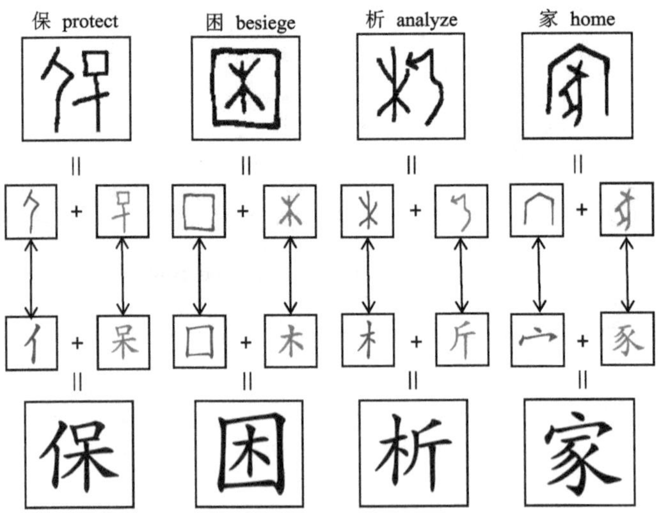

Fig. 1. Motivating examples showing that OBI decipherment can be explained by component correspondences. Top: OBI glyphs; bottom: modern standard characters; middle: decomposed components and the aligned component pairs across eras (arrows), which provide interpretable structural cues for decipherment.

Despite encouraging progress, existing decipherment systems still face two practical challenges in the *class-disjoint (zero-shot)* setting, where test character categories are unseen during training. First, *structural collapse under zero-shot decipherment.* Most pipelines learn at the whole-glyph level, and thus have limited awareness of internal character composition (e.g., components and their spatial organization). When encountering unseen OBI categories with broken strokes and variant layouts, the generator may lose structural constraints and produce corrupted glyphs with spurious artifacts, merged parts, or missing substructures. This issue is particularly pronounced for diffusion-based generation: without explicit, reliable structural anchors, denoising can drift toward visually plausible but structurally invalid shapes. Second, *semantic drift in modern-character mapping.* Purely visual translation tends to overemphasize shape resemblance at the pixel level and may output "shape-correct but meaning-wrong" glyphs: small changes in key strokes or component placement can flip character identity, especially among visually similar rare characters. Therefore, an effective decipherment model should incorporate *explicit component-level structural priors* to prevent structural collapse, and also include a *semantic verification signal* to reduce visually plausible but semantically incorrect generations.

To address these challenges, we revisit OBSD [7], a diffusion-based decipherment approach that formulates OBI decipherment as whole-glyph conditional diffusion generation. OBSD introduces a Local Structure Sampling (LSS) module that learns local correspondences by randomly cropping image patches during training [7]. While LSS encourages learning local patterns, in the class-disjoint zero-shot setting random crops may capture incomplete strokes, background noise, or uninformative fragments, yielding unstable structural cues. As a result, the generation process can still suffer from structural noise (e.g., irregular blobs and broken topology) and semantic drift when mapping to modern characters.

Inspired by the expert habit of "observing shapes and inferring meanings" and "decomposing characters into components", we propose **DAOD** (*Deconstruction-and-Alignment-based Oracle Decipherment*), which improves zero-shot decipherment by replacing random local sampling with *more reliable component-level constraints.* Concretely, DAOD introduces a *Component Deconstruction and Alignment (CDA)* mechanism: we deconstruct an input OBI glyph into compact components and align them to components from a curated modern-character component dictionary, forming paired structural conditions (e.g., masks and confidence-aware embeddings) that are injected into diffusion denoising. This design provides explicit cross-era structural anchors, which better preserves component topology and spatial layout, thereby mitigating structural collapse. Moreover, because component alignment can be noisy under severe degradation, DAOD includes a *training-time pair-level feedback* mechanism that adaptively down-weights unreliable aligned pairs based on diffusion residual signals, preventing erroneous conditions from dominating generation. Finally, to reduce "shape-correct but meaning-wrong" outputs, DAOD incorporates *DAOD-OCR-guided semantic supervision* as an auxiliary constraint during training, encouraging outputs to preserve identity-critical discriminative cues that support correct modern-character mapping.

Our contributions are summarized as follows:

- **DAOD framework.** We propose *DAOD*, a structure-conditioned diffusion framework for OBI-to-modern glyph translation in the class-disjoint (zero-shot) setting, improving robustness and interpretability via explicit component priors.
- **Component-based structural conditions.** We introduce *CDA* to build diffusion conditions from component deconstruction and cross-era alignment, providing paired masks and confidence-aware embeddings that constrain denoising and alleviate structural collapse.
- **Feedback and OCR guidance.** We propose a pair-level feedback mechanism to down-weight unreliable alignments during training, and incorporate DAOD-OCR-guided semantic supervision to reduce semantic drift in modern-character mapping.

2 Related Work

Oracle Bone Inscription (OBI) decipherment is harder than conventional recognition because many glyphs lack standardized encodings and suffer from severe deformation and noise. Early learning-based studies focus on recognition and representation learning under large variations and limited annotations [8,33]. To improve robustness, prior work explores data augmentation and robust training for long-tailed distributions (e.g., CycleGAN-based and dynamic augmentation) [17,27,31], as well as recognition with unsupervised or weakly supervised learning [23,24]. Other directions include alternative representations and few-shot settings [4,12,37], and detecting inscriptions in rubbings with pseudo labels, shape priors, and lightweight detectors [29,34,35].

Structure-aware modeling is particularly relevant since expert interpretation is component-driven. Component-aware recognition and reasoning explicitly model compositional units to improve robustness and generalization [3,18]. Beyond recognition, component-level retrieval and segmentation provide interpretable evidence across variants [9,10], and cross-era character linking builds supervision bridges between ancient glyphs and modern forms [2]. Related efforts also target broader accessibility and visual understanding of oracle bone materials [28].

Recently, decipherment is increasingly formulated as image generation that outputs modern-character hypotheses, including GAN-based translation [5], component "puzzle-piece" reconstruction [25], and diffusion-based controllable generation [7]. These methods connect to document enhancement and restoration using advanced generative models [21,30]. However, visually plausible outputs can still be semantically incorrect under severe degradation, akin to semantic hallucinations in multimodal generation [32,36], motivating explicit structural priors and semantic verification.

Progress is also enabled by datasets and benchmarks. HWOBC is an early handwriting-style dataset [13], while HUST-OBC and EVOBC provide larger-scale standardized resources for recognition and decipherment [6,26], with additional dataset efforts [14,22]. OBI-Bench further evaluates end-to-end workflows and highlights bottlenecks in fine-grained perception and knowledge-aware reasoning [1].

3 Method

3.1 Problem Definition and Overview

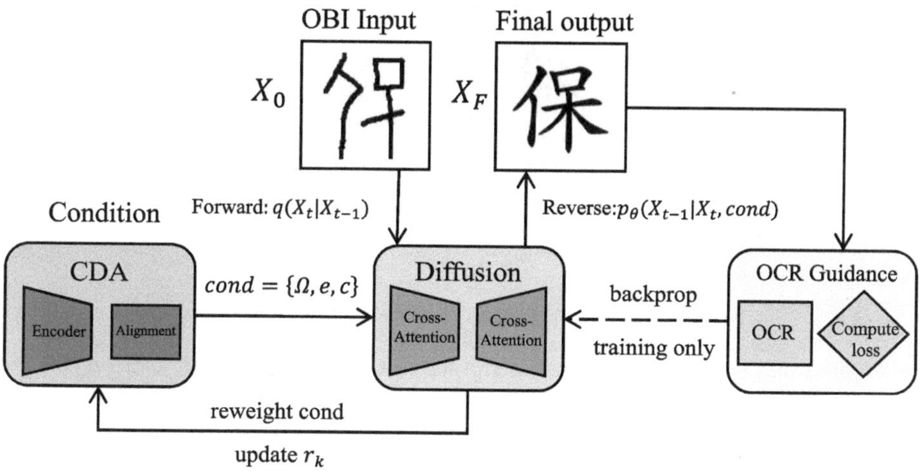

Fig. 2. Overview of DAOD. CDA constructs cond $= \{\Omega, e, c\}$ to condition diffusion denoising $p_\theta(X_{t-1} \mid X_t, \text{cond})$ and generate X_F from X_0. During training, a feedback loop updates pair reliabilities r_k to down-weight unreliable conditions, and OCR guidance adds an auxiliary OCR loss with gradients back-propagated to diffusion (dashed; training only). Inference uses fixed cond without feedback updates or OCR loss.

Given an OBI glyph image X_0, DAOD aims to generate a recognizable modern standard glyph X_F. As shown in Fig. 2, DAOD uses conditional diffusion as the backbone and injects component-level conditions from CDA, where CDA encodes and aligns oracle components to modern components to produce paired masks Ω, embeddings e, and confidences c. To improve robustness, we calibrate condition reliability with a training-time feedback mechanism and further apply OCR-guided semantic supervision; both are disabled at inference, where DAOD performs standard conditional sampling with fixed cond.

3.2 Component Dictionary Construction

DAOD builds component-level priors for diffusion conditioning by constructing a modern component dictionary and extracting stable OBI components.

Modern Component Dictionary. We merge 890 common characters from *Hanzi Writer Data* with 701 manually annotated rare characters, forming 1,590 modern character classes. Each class is associated with a component decomposition and spatial layout, which we use to render unified modern glyph targets for diffusion training.

OBI Component Extraction. Given X_0, we binarize and skeletonize it, then obtain connected components $\mathcal{C} = \{c_i\}_{i=1}^N$ with masks and geometry (area A_i, centroid $\mathbf{p}_i$). To handle fragmentation, we merge nearby small parts using the affinity

$$s_{ij} = \frac{\min(A_i, A_j)}{\max(A_i, A_j)} \cdot \exp\left(-\frac{\|\mathbf{p}_i - \mathbf{p}_j\|_2}{\tau}\right), \tag{1}$$

and keep the dominant components $\mathcal{O} = \{o_k\}_{k=1}^{K_o}$ for CDA alignment. We restrict K_o to a small value (e.g., $K_o \in \{1, 2\}$) because OBI characters typically contain few dominant structures, and excessive decomposition may introduce noise and instability (Sec. 3.4).

3.3 Conditional Diffusion Generation

We adopt a conditional diffusion model to translate an input OBI glyph X_0 into a modern standard character image. For consistency with Fig. 3, we use X_t to denote the image state at timestep t and denote the final reconstructed sample after reverse denoising as $\hat{X}_0$, where we set $X_F = \hat{X}_0$ as the deciphered output.

Following DDPM, the forward process progressively perturbs X_0 with Gaussian noise:

$$q(X_t \mid X_{t-1}) = \mathcal{N}\left(X_t; \sqrt{1-\beta_t}\, X_{t-1}, \beta_t \mathbf{I}\right), \quad t = 1, \ldots, T, \tag{2}$$

where $\{\beta_t\}_{t=1}^T$ is a variance schedule. Equivalently,

$$X_t = \sqrt{\bar{\alpha}_t}\, X_0 + \sqrt{1-\bar{\alpha}_t}\, \epsilon, \quad \epsilon \sim \mathcal{N}(0, \mathbf{I}), \tag{3}$$

with $\alpha_t = 1 - \beta_t$ and $\bar{\alpha}_t = \prod_{s=1}^t \alpha_s$.

As illustrated in Fig. 3, we parameterize the reverse process with a U-Net noise predictor $\epsilon_\theta(\cdot)$ conditioned on timestep t and structural conditions cond:

$$\hat{\epsilon} = \epsilon_\theta(X_t, t \mid \text{cond}). \tag{4}$$

The denoising transition is defined as

$$p_\theta(X_{t-1} \mid X_t, \text{cond}) = \mathcal{N}\left(X_{t-1}; \mu_\theta(X_t, t, \text{cond}), \sigma_t^2 \mathbf{I}\right), \tag{5}$$

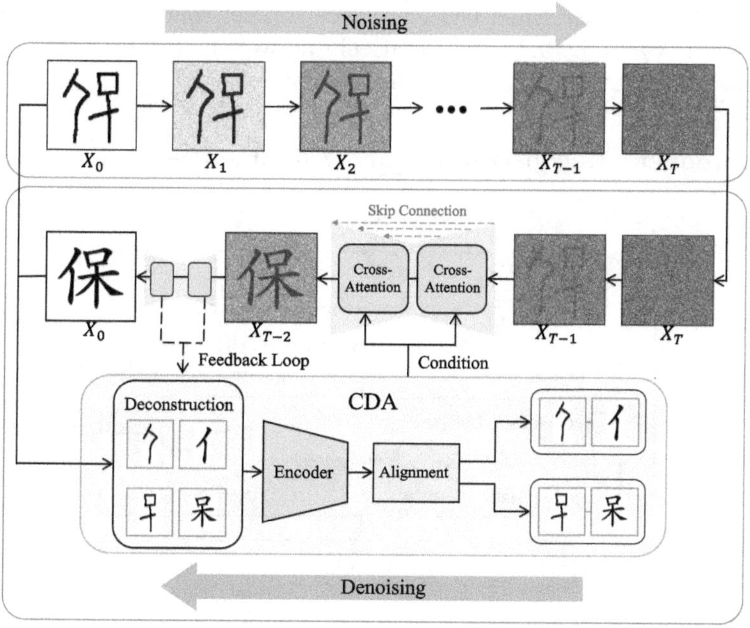

Fig. 3. Diffusion backbone in **DAOD**. Top: forward noising $X_0 \to X_T$. Bottom: reverse denoising with CDA conditions injected into the diffusion U-Net via cross-attention, while skip connections preserve low-level stroke details. A training-time feedback loop calibrates condition reliability (Sec. 3.4).

where $\mu_\theta(\cdot)$ is computed from $\hat{\epsilon}$ using the standard DDPM parameterization, and σ_t^2 follows the chosen sampling schedule.

Condition Injection. As shown in the CDA block of Fig. 3, CDA deconstructs the glyph and aligns components to form structural conditions cond $= \{\Omega, e, c\}$ (Sec. 3.4), where Ω denotes paired spatial masks, e denotes component embeddings, and c denotes confidence scores. We inject e into the diffusion U-Net using cross-attention blocks (Fig. 3), where the noisy feature map serves as the query and component embeddings serve as keys/values. The spatial priors Ω and confidences c modulate attention responses so that reliable component pairs impose stronger constraints. Skip connections are retained to preserve fine stroke details.

Training Objective. We train the diffusion backbone with the noise-prediction loss:

$$\mathcal{L}_{\text{diff}} = \mathbb{E}_{t,\epsilon}\Big[\|\epsilon - \epsilon_\theta(X_t, t \mid \text{cond})\|_2^2\Big], \tag{6}$$

where t is uniformly sampled from $\{1, \ldots, T\}$ and X_t is obtained by Eq. (3). The overall objective additionally includes OCR-guided semantic supervision (Sec. 3.5), while CDA-integrated feedback calibrates condition reliability during training (Sec. 3.4).

Inference. At inference time, CDA provides a fixed cond for the input glyph. Starting from $X_T \sim \mathcal{N}(0, \mathbf{I})$, we iteratively apply Eq. (5) for $t = T, \ldots, 1$ to obtain $\hat{X}_0$, and set $X_F = \hat{X}_0$ as the final output.

3.4 Component Deconstruction and Alignment with Feedback

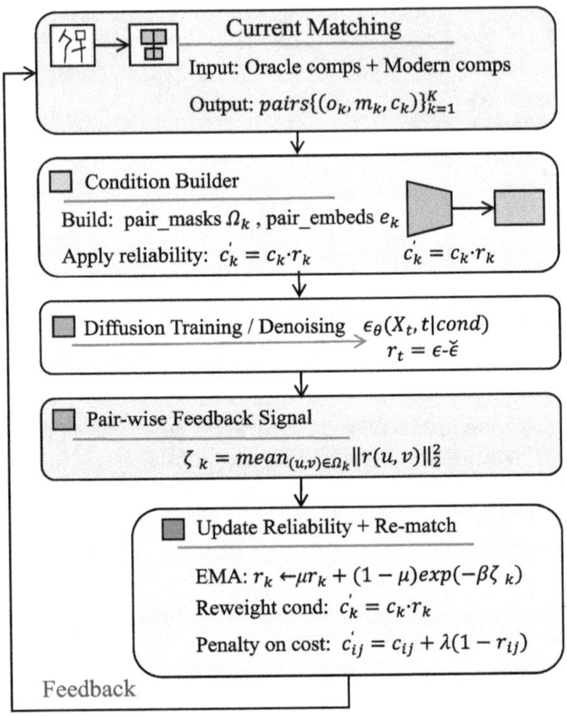

Fig. 4. Training-time feedback inside CDA. Current matching outputs aligned pairs $\{(o_k, m_k, c_k)\}_{k=1}^{K}$. The condition builder constructs paired masks Ω_k and pair embeddings e_k, and applies reliability to reweight confidences $c_k' = c_k r_k$. During diffusion training, the residual $r = \epsilon - \epsilon_\theta(X_t, t \mid \text{cond})$ is pooled to compute pair-wise feedback ζ_k, which updates r_k (EMA) and optionally penalizes matching costs to trigger re-matching.

CDA constructs cross-era component correspondences and converts them into structural conditions for diffusion generation. Given an input glyph X_0 and target label y, we first obtain OBI component candidates $\mathcal{O}$ using the extraction scheme in Sec. 3.2, and retrieve modern components $\mathcal{M}(y)$ from the modern component dictionary (Sec. 3.2). A matching module computes similarity scores and produces a Top-K set of aligned pairs

$$\mathcal{P} = \{(o_k, m_k, c_k)\}_{k=1}^{K}, \quad c_k \in [0, 1], \tag{7}$$

where $o_k \in \mathcal{O}$, $m_k \in \mathcal{M}(y)$, and c_k denotes the initial matching confidence. We empirically use a small K (typically 2 to 4) to retain only the most reliable alignments, thereby filtering low-quality matches and avoiding noisy conditions in diffusion.

Condition Builder. From $\mathcal{P}$, CDA builds structural conditions cond $= \{\Omega, e, c\}$ for the diffusion backbone (Sec. 3.3):

$$\text{cond} = \{\Omega, e, c\}, \quad \Omega = \{\Omega_k\}_{k=1}^{K}, \ e = \{e_k\}_{k=1}^{K}, \ c = \{c_k\}_{k=1}^{K}, \tag{8}$$

where Ω_k is the paired spatial mask and e_k is the pair embedding encoding the matched modern component identity and aligned geometry. During training, CDA further maintains a reliability score r_k for each pair and applies it to reweight the confidence:

$$c'_k = c_k \cdot r_k. \tag{9}$$

Pair-Level Feedback (Training Only). As shown in Fig. 4, feedback is computed from diffusion residuals to calibrate pair reliabilities. For a training step, we sample t and ϵ to obtain X_t by Eq. (3), and predict noise $\hat{\epsilon} = \epsilon_\theta(X_t, t \mid \text{cond})$ (Eq. (4)). We define the residual map as

$$r_t = \epsilon - \hat{\epsilon}. \tag{10}$$

For each aligned pair k, we compute a pair-wise feedback signal by pooling the residual energy within Ω_k:

$$\zeta_k = \text{mean}_{(u,v)\in\Omega_k} \left\| r_t(u,v) \right\|_2^2. \tag{11}$$

We update the reliability score $r_k \in [0,1]$ using an exponential moving average (EMA):

$$r_k \leftarrow \mu r_k + (1-\mu)\exp(-\beta\,\zeta_k), \tag{12}$$

where $\mu \in [0,1)$ controls smoothing and $\beta > 0$ controls sensitivity. The updated r_k is then used in Eq. (9) to reweight conditioning strength for subsequent optimization.

Optional re-matching. When a pair remains unreliable, we penalize its matching cost to encourage re-matching. Let c_{ij} denote the matching cost (or negative similarity) between an OBI component o_i and a modern component m_j, and let r_{ij} be the corresponding reliability. We update the cost as

$$c'_{ij} = c_{ij} + \lambda(1 - r_{ij}), \tag{13}$$

where λ controls the penalty strength. CDA can then re-run matching with $\{c'_{ij}\}$ to refresh $\mathcal{P}$ and rebuild cond.

At inference time, feedback updating and re-matching are disabled. CDA is executed once to obtain a fixed cond, and the diffusion model performs standard sampling to produce X_F (Sec. 3.3).

Algorithm 1. DAOD Training (CDA + Feedback + OCR)

Require: Training sample (X_0, y); component dictionary $\mathcal{D}$; steps T; μ, β; λ_{ocr} (and optional re-match weight λ)

$\mathcal{O} \leftarrow \text{ExtractComponents}(X_0)$, $\mathcal{M}(y) \leftarrow \mathcal{D}[y]$
$\mathcal{P} \leftarrow \text{Match}(\mathcal{O}, \mathcal{M}(y))$, $\text{cond} = \{\Omega, e, c\} \leftarrow \text{BuildCond}(\mathcal{P})$
Initialize $r_k \leftarrow 1$ for $k = 1, \ldots, K$
for each iteration **do**
    Sample $t \sim U\{1, \ldots, T\}$, $\epsilon \sim \mathcal{N}(0, \mathbf{I})$, and form X_t by Eq. (3)
    Reweight $c'_k \leftarrow c_k r_k$ and set $\text{cond} \leftarrow \{\Omega, e, c'\}$
    $\hat{\epsilon} \leftarrow \epsilon_\theta(X_t, t \mid \text{cond})$, $\mathcal{L}_{\text{diff}} \leftarrow \|\epsilon - \hat{\epsilon}\|_2^2$
    $\{\zeta_k\} \leftarrow \text{PoolResidual}(\epsilon - \hat{\epsilon}, \{\Omega_k\})$ ▷ Eq. (11)
    $r_k \leftarrow \mu r_k + (1 - \mu) \exp(-\beta \zeta_k)$ **for all** k ▷ Eq. (12)
    $\mathcal{L}_{\text{ocr}} \leftarrow \text{OCRLoss}(\cdot)$ ▷ Sec. 3.5
    Update θ by minimizing $\mathcal{L}_{\text{diff}} + \lambda_{\text{ocr}} \mathcal{L}_{\text{ocr}}$
    (Optional) Re-run Match with penalized costs using Eq. (13), then rebuild cond
end for

3.5 OCR-Guided Semantic Supervision

Although diffusion generation can produce visually plausible glyphs, appearance similarity alone does not guarantee semantic correctness under heavy degradation and inter-class resemblance. To encourage the reconstructed output to be *recognizable* and semantically aligned with the target, we introduce an OCR-guided semantic supervision term during training.

As illustrated in Fig. 5, we employ a pre-trained OCR network and keep it fixed. During training, the diffusion model reconstructs $\tilde{X}_0$ under CDA condition $\tilde{X}$ via the reverse denoising transition $p_\theta(X_{t-1} \mid X_t, \tilde{X})$. We then feed both the ground-truth modern glyph X_0 and the reconstructed glyph $\tilde{X}_0$ into the OCR network and extract deep features. Let $\varphi(\cdot)$ denote the OCR feature extractor (e.g., the penultimate-layer embedding). The OCR-guided loss is defined as the feature discrepancy:

$$\mathcal{L}_{\text{ocr}} = \left\| \varphi(X_0) - \varphi(\tilde{X}_0) \right\|_2, \tag{14}$$

which matches the supervision signal shown in Fig. 5. This loss encourages $\tilde{X}_0$ to preserve discriminative cues that the OCR model uses for recognition, thus reducing visually plausible but semantically incorrect generations.

The OCR guidance is applied only during training and is disabled at inference. The final objective combines diffusion training (Eq. (6)) and OCR guidance:

$$\mathcal{L} = \mathcal{L}_{\text{diff}} + \lambda_{\text{ocr}} \mathcal{L}_{\text{ocr}}, \tag{15}$$

where λ_{ocr} controls the strength of semantic supervision.

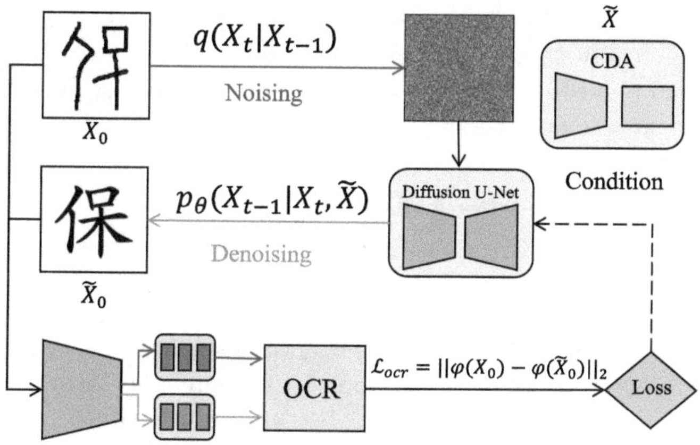

Fig. 5. OCR-guided semantic supervision (training only). CDA provides condition $\tilde{X}$ to the diffusion U-Net. The diffusion model follows forward noising $q(X_t \mid X_{t-1})$ and reverse denoising $p_\theta(X_{t-1} \mid X_t, \tilde{X})$ to reconstruct $\tilde{X}_0$. A fixed OCR network extracts features $\varphi(\cdot)$ from the ground-truth X_0 and the reconstruction $\tilde{X}_0$, and the feature distance is used as an auxiliary loss to provide semantic guidance.

4 Experiments

4.1 Datasets and Evaluation Protocol

We train and evaluate DAOD on two large-scale oracle bone inscription (OBI) datasets, *HUST-OBC* [26] and *EVOBC* [6]. Together, they contain 71,698 OBI glyph images covering 1,590 character classes. For each OBI sample, we associate a target modern standard glyph image as the decipherment goal.

Our task is *decipherment*: given an OBI glyph image X_0, the model generates its modern standard counterpart X_F. To reflect practical decipherment where unseen characters are common and to avoid memorization, we adopt a *class-disjoint* split when partitioning the data. Specifically, we split the datasets into training and test sets with a 9:1 ratio, while ensuring that the character categories in the test set are absent from the training set (i.e., no overlap in classes). This protocol turns evaluation into a *zero-shot* setting at the character-category level and provides a more faithful assessment of decipherment ability.

Since DAOD outputs images, traditional image-generation metrics (e.g., SSIM) are not suitable for evaluating decipherment success: small visual differences may not affect readability, whereas slight structural errors can change the intended character. Following recent practices in OBI decipherment, we employ OCR as a quantifiable proxy metric to approximate human-level evaluation. This measures whether the generated glyph is recognizable and matches the ground-truth label. Concretely, we use two OCR evaluators. (i) *DAOD-OCR* is a lightweight classifier with a ResNet-50 backbone, trained on a 9,000-class Chinese character dataset that includes 1,700 rare modern characters corresponding to OBI categories, to better recognize diffusion-generated outputs and compare them against ground-truth labels. DAOD-OCR achieves 99.26% recognition accuracy on the 9,000-class benchmark. (ii) We additionally report results with the widely-used open-source *PaddleOCR* as an off-the-shelf evaluator to support more objective assessment.

4.2 Comparison with Generative Baselines

We compare the proposed DAOD with representative image-to-image translation and diffusion-based generation baselines, including Pix2Pix [11], Palette [19], CycleGAN [38], BBDM [15], CDE [20], and OBSD [7]. All methods are trained under the same class-disjoint (zero-shot) decipherment split described in Sec. 4.1 and generate modern glyph images from OBI inputs. For fair evaluation, we follow the OCR-based protocol in Sec. 4.1 and report Top-K accuracy using both our self-trained *DAOD-OCR* and the open-source PaddleOCR.

Table 1 summarizes the quantitative results. DAOD achieves the best performance across all Top-K metrics under *DAOD-OCR*, reaching the highest Top-1 accuracy (45.11%) and maintaining clear margins over strong diffusion baselines such as OBSD and CDE across Top-10 to Top-500. Under PaddleOCR, absolute scores are lower but DAOD still performs best (Top-1: 34.55%), mainly because many target characters are rare and thus harder for a general-purpose OCR; this also motivates our dedicated *DAOD-OCR* evaluator.

We attribute these gains to DAOD's explicit structural conditioning and robustness designs. In contrast, appearance-driven translation models (e.g., Pix2Pix and CycleGAN) are more sensitive to stroke breakage and layout variations, leading to distorted structures and low recognizability in the class-disjoint setting. Diffusion-based baselines improve stability, but still lack explicit cross-era structural priors. DAOD injects component-level conditions to constrain the denoising trajectory and further suppresses unreliable correspondences via feedback, resulting in more structurally consistent and semantically recognizable decipherment outputs.

Table 1. Comparison of the proposed DAOD and generative baselines under the class-disjoint decipherment protocol.

Setting		Methods						
Evaluation Tool	Rank	Pix2Pix	Palette	CycleGAN	BBDM	CDE	OBSD	**DAOD (ours)**
DAOD-OCR	Top-1@Acc	0.00%	0.00%	0.00%	19.14%	30.55%	41.88%	**45.11%**
DAOD-OCR	Top-10@Acc	0.00%	0.00%	0.00%	30.20%	45.25%	46.27%	**50.46%**
DAOD-OCR	Top-20@Acc	0.00%	0.00%	0.00%	34.87%	46.24%	48.14%	**51.87%**
DAOD-OCR	Top-50@Acc	0.00%	0.00%	8.50%	38.55%	49.35%	51.46%	**55.26%**
DAOD-OCR	Top-100@Acc	0.00%	2.85%	19.25%	41.76%	53.88%	54.47%	**58.15%**
DAOD-OCR	Top-200@Acc	14.35%	8.10%	37.97%	47.82%	55.45%	57.23%	**61.84%**
DAOD-OCR	Top-500@Acc	17.11%	19.95%	60.10%	59.15%	60.28%	62.28%	**65.95%**
PaddleOCR	Top-1@Acc	0.00%	0.00%	0.00%	7.25%	18.91%	30.10%	**34.55%**

4.3 Ablation Study

We conduct ablation experiments to quantify the contribution of key designs in DAOD. Starting from a plain conditional diffusion backbone (*Diffusion*), we progressively add (i) component deconstruction-and-alignment conditioning (*+CDA*), and (ii) the OCR-guided semantic supervision (*+OCR Guidance*). All variants follow the same class-disjoint (zero-shot) decipherment protocol in Sec. 4.1. We report Top-K accuracy using both our self-trained *DAOD-OCR* evaluator and the external PaddleOCR.

Table 2 shows that CDA-style structural conditioning provides the dominant performance gain across all Top-K metrics. The plain diffusion baseline yields almost no semantic recognizability (e.g., 0.45% Top-1 under DAOD-OCR), indicating that appearance-driven generation without explicit structural priors is insufficient under severe degradation. Adding CDA dramatically improves accuracy (e.g., from 0.45% to 43.95% at Top-1 and from 16.19% to 64.78% at Top-500), validating that component-level conditions effectively constrain generation and prevent structural collapse. This demonstrates that the primary source of performance gain is the structural reformulation from CDA, while hyperparameters like K_o and Top-K serve mainly to suppress noisy alignments and ensure stability. Further introducing OCR guidance yields consistent additional gains (e.g., 45.11% Top-1 and 65.95% Top-500), suggesting that semantic supervision helps reduce visually plausible but incorrect outputs. Similar trends are observed under PaddleOCR (Top-1 improves from 32.37% to 34.55%), demonstrating that the benefit generalizes beyond a single evaluator.

Table 2. Ablation study of DAOD.

Setting		Variants		
Evaluation Tool	Rank	Diffusion	**+CDA**	**+OCR Guidance**
DAOD-OCR	Top-1	0.45%	**43.95%**	**45.11%**
DAOD-OCR	Top-10	2.13%	**48.49%**	**50.46%**
DAOD-OCR	Top-20	4.12%	**50.42%**	**51.87%**
DAOD-OCR	Top-50	6.89%	**53.84%**	**55.26%**
DAOD-OCR	Top-100	9.05%	**56.89%**	**58.15%**
DAOD-OCR	Top-200	10.72%	**59.67%**	**61.84%**
DAOD-OCR	Top-500	16.19%	**64.78%**	**65.95%**
PaddleOCR	Top-1	0.00%	**32.37%**	**34.55%**

4.4 Qualitative Results

We provide qualitative comparisons to assess structural consistency and semantic plausibility of the generated modern glyphs. As shown in Fig. 6, each row corresponds to one test sample, and the columns report the *Input* OBI glyph,

GT modern glyph, representative baselines (*CycleGAN*, *BBDM*, and *OBSD*), two ablations (*w/o CDA*, *w/o OCR*), and *Ours*.

Overall, DAOD produces more legible modern glyphs with cleaner component geometry and more faithful spatial composition. Removing CDA leads to the most unstable results: *w/o CDA* often yields ambiguous structures with misplaced, merged, or collapsed parts, confirming that component-level conditions are crucial for constraining diffusion generation under severe degradation. Among representative baselines, *CycleGAN* frequently exhibits style-transfer artifacts and broken strokes, while diffusion baselines (*BBDM*/*OBSD*) generate visually cleaner outputs but still suffer from component-level mistakes such as missing or shifted strokes, which may harm recognizability. Comparing *w/o OCR* with the full model further shows that OCR guidance improves semantic correctness: without OCR supervision, outputs can appear plausible yet miss subtle discriminative strokes, whereas the full DAOD better preserves identity-critical details and yields more stable and semantically plausible decipherment results.

Fig. 6. Qualitative results on OBI decipherment. Left: input OBI glyphs and ground-truth modern glyphs (GT). Right: outputs from different methods, including baselines and ablated variants of DAOD.

5 Conclusion and Limitations

In this paper, we addressed oracle bone inscription (OBI) decipherment from an image-generation perspective and proposed DAOD, a structure-conditioned diffusion framework for translating OBI glyphs into modern standard Chinese characters. DAOD builds component-level conditions via *Component Deconstruction and Alignment* (CDA), and improves robustness with a training-time pair-level feedback mechanism that down-weights unreliable alignments. We further introduce DAOD-OCR-guided semantic supervision for more recognizable outputs and adopt an OCR-based evaluation under a class-disjoint (zero-shot) protocol. Experiments on *HUST-OBC* and *EVOBC* demonstrate consistent gains over representative translation and diffusion baselines.

Our main limitation is that the OBI component dictionary and decomposition scheme are not yet standardized. The current skeleton-based extraction with heuristic fragment merging can produce coarse or imperfect component partitions under severe stroke breakage and large layout variations, which may propagate errors to component alignment and subsequent conditioning. In future work, we plan to develop a more principled and standardized component definition pipeline, potentially supported by expert annotations and consistent decomposition rules, and to explore stronger cross-era structural priors and more reliable semantic evaluators for improved generalization to highly ambiguous inscriptions.

Acknowledgments. This work was supported by the National Natural Science Foundation of China (No. 62576301), and the Open Project of the State Key Laboratory of Multimodal Artificial Intelligence Systems (No. MAIS2024101).

Disclosure of Interests. The authors have no competing interests to declare that are relevant to the content of this article.

References

1. Chen, Z., Chen, T., Zhang, W., Zhai, G.: OBI-Bench: can LMMs aid in study of ancient script on oracle bones? arXiv preprint arXiv:2412.01175 (2024)
2. Chi, Y., Giunchiglia, F., Shi, D., Diao, X., Li, C., Xu, H.: ZiNet: linking Chinese characters spanning three thousand years. In: Findings of the Association for Computational Linguistics: ACL 2022, pp. 3061–3070 (2022)
3. Diao, X., et al.: RZCR: zero-shot character recognition via radical-based reasoning (2022). arXiv:2207.05842 arXiv preprint
4. Fujikawa, Y., Li, H., Yue, X., Aravinda, C., Prabhu, G.A., Meng, L.: Recognition of oracle bone inscriptions by using two deep learning models. Int. J. Digit. Human. **5**(2), 65–79 (2023)
5. Gao, F., Zhang, J., Liu, Y., Han, Y.: Image translation for oracle bone character interpretation. Symmetry **14**(4), 743 (2022)
6. Guan, H., et al.: An open dataset for the evolution of oracle bone characters: Evobc. arXiv preprint arXiv:2401.12467 (2024)
7. Guan, H., et al.: Deciphering oracle bone language with diffusion models. arXiv preprint arXiv:2406.00684 (2024)
8. Guo, J., Wang, C., Roman-Rangel, E., Chao, H., Rui, Y.: Building hierarchical representations for oracle character and sketch recognition. IEEE Trans. Image Process. **25**(1), 104–118 (2015)
9. Hu, Z., Cheung, Y.M., Zhang, Y., Peiying, Z., Ling, T.P.: Component-level segmentation for oracle bone inscription decipherment. In: Proceedings of the AAAI Conference on Artificial Intelligence, vol. 39, pp. 28116–28124 (2025)

10. Hu, Z., Cheung, Y.M., Zhang, Y., Zhang, P., Tang, P.L.: Component-level oracle bone inscription retrieval. In: Proceedings of the 2024 International Conference on Multimedia Retrieval, pp. 647–656 (2024)
11. Isola, P., Zhu, J.Y., Zhou, T., Efros, A.A.: Image-to-image translation with conditional adversarial networks. In: Proceedings of the IEEE conference on computer vision and pattern recognition, pp. 1125–1134. (2017)
12. Jiang, R., Liu, Y., Zhang, B., Chen, X., Li, D., Han, Y.: OraclePoints: a hybrid neural representation for oracle character. In: Proceedings of the 31st ACM International Conference on Multimedia, pp. 7901–7911 (2023)
13. Li, B., Dai, Q., Gao, F., Zhu, W., Li, Q., Liu, Y.: HWOBC-a handwriting oracle bone character recognition database. In: Journal of Physics: Conference Series, vol. 1651, p. 012050. IOP Publishing (2020)
14. Li, B., et al.: Oracle bone inscriptions multi-modal dataset. arXiv preprint arXiv:2407.03900 (2024)
15. Li, B., Xue, K., Liu, B., Lai, Y.K.: BBDM: image-to-image translation with Brownian bridge diffusion models. In: Proceedings of the IEEE/CVF Conference on Computer Vision and Pattern Recognition. pp. 1952–1961 (2023)
16. Li, J., et al.: A comprehensive survey of oracle character recognition: challenges, datasets, methodology, and beyond. Pattern Recogn. **169**, 111824 (2026). https://doi.org/10.1016/j.patcog.2025.111824
17. Li, J., Wang, Q.F., Huang, K., Yang, X., Zhang, R., Goulermas, J.Y.: Towards better long-tailed oracle character recognition with adversarial data augmentation. Pattern Recogn. **140**, 109534 (2023)
18. Lin, X., Chen, S., Zhao, F., Qiu, X.: Radical-based extract and recognition networks for oracle character recognition. Int. J. Docum. Anal. Recogn. (IJDAR) **25**(3), 219–235 (2022)
19. Saharia, C., et al.: Palette: image-to-image diffusion models. In: ACM SIGGRAPH 2022 Conference Proceedings, pp. 1–10 (2022)
20. Saharia, C., Ho, J., Chan, W., Salimans, T., Fleet, D.J., Norouzi, M.: Image super-resolution via iterative refinement. IEEE Trans. Pattern Anal. Mach. Intell. **45**(4), 4713–4726 (2022)
21. Souibgui, M.A., Kessentini, Y.: De-GAN: a conditional generative adversarial network for document enhancement. IEEE Trans. Pattern Anal. Mach. Intell. **44**(3), 1180–1191 (2020)
22. Wang, M., Deng, W.: A dataset of oracle characters for benchmarking machine learning algorithms. Sci. Data **11**(1), 87 (2024)
23. Wang, M., Deng, W., Liu, C.L.: Unsupervised structure-texture separation network for oracle character recognition. IEEE Trans. Image Process. **31**, 3137–3150 (2022)
24. Wang, M., Deng, W., Su, S.: Oracle character recognition using unsupervised discriminative consistency network. Pattern Recogn. **148**, 110180 (2024)
25. Wang, P., et al.: Puzzle pieces picker: deciphering ancient Chinese characters with radical reconstruction. In: International Conference on Document Analysis and Recognition, pp. 169–187. Springer (2024)
26. Wang, P., et al.: An open dataset for oracle bone character recognition and decipherment. Sci. Data **11**(1), 976 (2024)
27. Wang, W., Zhang, T., Zhao, Y., Jin, X., Mouchere, H., Yu, X.: Improving oracle bone characters recognition via a cycleGAN-based data augmentation method. In: International Conference on Neural Information Processing, pp. 88–100. Springer (2022)

28. Wu, H., Li, K., Zhang, X.Y., Wang, Q., Wang, D.H.: OracleGCD: generalized category discovery for oracle bone scripts. In: International Conference on Document Analysis and Recognition, pp. 459–475. Springer (2025)
29. Yang, Z., Fu, T.: Oracle detection and recognition based on improved tiny-YOLOV4. In: Proceedings of the 2020 4th International Conference on Video and Image Processing, pp. 128–133 (2020)
30. Yang, Z., et al.: DocDiff: document enhancement via residual diffusion models. In: Proceedings of the 31st ACM International Conference on Multimedia, pp. 2795–2806 (2023)
31. Yue, X., Li, H., Fujikawa, Y., Meng, L.: Dynamic dataset augmentation for deep learning-based oracle bone inscriptions recognition. ACM J. Comput. Cult. Herit. **15**(4), 1–20 (2022)
32. Zhang, X., et al.: Shallow focus, deep fixes: enhancing shallow layers vision attention sinks to alleviate hallucination in LVLMs. In: Proceedings of the 2025 Conference on Empirical Methods in Natural Language Processing, pp. 3512–3534 (2025)
33. Zhang, Y.K., Zhang, H., Liu, Y.G., Yang, Q., Liu, C.L.: Oracle character recognition by nearest neighbor classification with deep metric learning. In: 2019 International Conference on Document Analysis and Recognition (ICDAR), pp. 309–314. IEEE (2019)
34. Zhao, Q., et al.: Few-shot cross-modal text detection via clip:. Visual Comput. **41**(12), 9471–9485 (2025)
35. Zhao, Q., Yan, Y., Wang, D.H.: TextMamba: scene text detector with mamba. In: 2025 International Joint Conference on Neural Networks (IJCNN), pp. 1–8. IEEE (2025)
36. Zhao, Q., et al.: MCA-LLaVA: manhattan causal attention for reducing hallucination in large vision-language models. In: Proceedings of the 33rd ACM International Conference on Multimedia, pp. 3981–3990 (2025)
37. Zhao, X., Liu, S., Wang, Y., Fu, Y.: FFD Augmentor: towards few-shot oracle character recognition from scratch. In: Proceedings of the Asian Conference on Computer Vision, pp. 1622–1639 (2022)
38. Zhu, J.Y., Park, T., Isola, P., Efros, A.A.: Unpaired image-to-image translation using cycle-consistent adversarial networks. In: Proceedings of the IEEE International Conference on Computer Vision, pp. 2223–2232 (2017)

Layout Analysis

REPLICA: An Agentic Framework for Visually Faithful Document Reconstruction

R. Raghuveer[1], Anirudh Srinivasan[1], Venkata Kesav Venna[1](✉), S. Sreevatsa[2], Aryan Jain[2], Sahithi Kukkala[2], and Ravi Kiran Sarvadevabhatla[1,2]

[1] BharatGen, Mumbai, India
{raghuveer.r,anirudh.srinivasan,venkat.kesav}@bharatgen.com
[2] IIIT Hyderabad, Hyderabad, India
{tallapragada.s,aryan.j,sahithi.kukkala}@research.iiit.ac.in,
ravi.kiran@iiit.ac.in
https://replica-agents.github.io/

Abstract. PDF documents integrate poorly with the web. To improve accessibility, they are frequently converted into HTML. Existing approaches for document-to-HTML conversion largely focus on retaining text and logical structure. However, the document's physical structure is not preserved. Visual cues such as color, font style are overlooked and design aspects such as content positioning are ignored. To address these shortcomings, we introduce REPLICA, an agentic document-to-HTML framework that generates visually faithful, layout-aware, and semantically enriched HTML which we refer as FID-HTML. Our approach generalizes across diverse document types and multiple languages. For systematic evaluation, we introduce VFDR-Bench spanning 15 languages, and 15+ document types. REPLICA achieves 93% visual fidelity, outperforming strong baselines including GPT-5 and Gemini 3 Pro. Notably, REPLICA with Qwen3-VL-8B achieves a 39% gain overall as compared to Qwen3-VL-30B, despite being 3.75× smaller and operating at comparable inference time, highlighting the effectiveness of structured agentic design over brute-force scale. By preserving visual integrity and semantics in multilingual documents, REPLICA unlocks new possibilities for design-preserving translation, scalable access, and structured search.

Keywords: Document Reconstruction · Physical and Logical Layout Analysis · Document Agents

1 Introduction

PDFs preserve documents with precise control over layout, typography, and visual styling. Yet this fidelity often vanishes when converted for the web. Traditional OCR engines [9,18,50] process page images to recover plain text, discarding layout, typography, and styling. Other workflows convert PDFs to alternate

R. R. A. Srinivasan and V. K. Venna—Equal Contribution.

G. A. Fink et al. (Eds.): ICDAR 2026, LNCS 16972, pp. 505–522, 2027.
https://doi.org/10.1007/978-3-032-36023-6_29

formats such as Markdown [10]. PDF parsers [1,41,46] extract text or lightweight markup directly from the PDF's internal structure but still lose structural hierarchy (e.g., multi-column layouts), fine-grained formatting, and much of the visual and spatial identity.

PDF-to-HTML conversion is a more common alternative in web publishing, digital archiving, and accessibility workflows. Plain HTML with minimal CSS [11] preserves basic layout elements but often simplifies typography, color, and positioning.

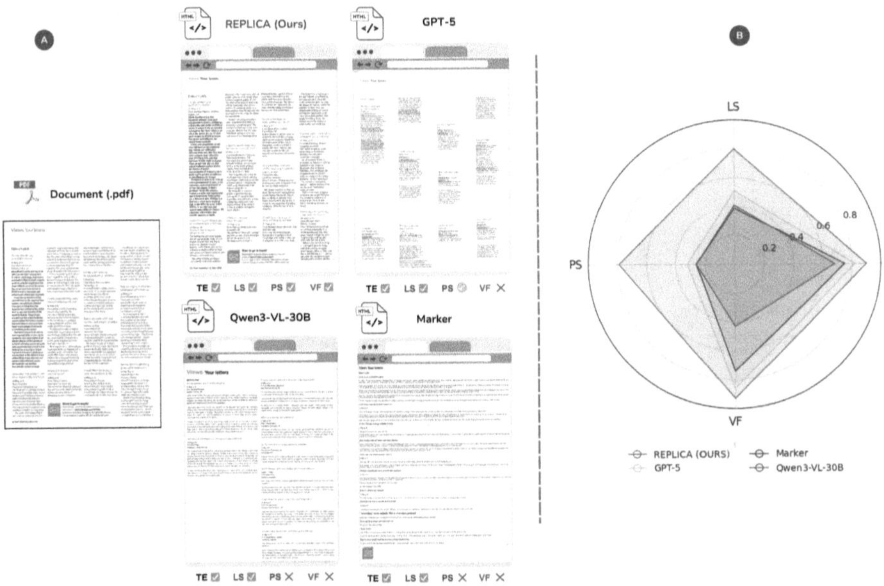

Fig. 1. Ⓐ **Comparison of visual fidelity across four approaches**: our method REPLICA, an open-source Vision-Language Model (Qwen3-VL-30B [64]), a commercial Vision-Language Model (GPT-5 [38]), and a pipeline-based method (Marker [43]). Note that REPLICA faithfully preserves fine-grained layout cues - such as text color, paragraph indentation, font attributes (e.g., bold and italics), font size, and even global structural properties like the correct number of columns. Ⓑ**Radar plot of quantitative performance**: showing the performance of REPLICA relative to baselines. Unlike other baselines, REPLICA reliably preserves all key document aspects after conversion - TE (Text Extraction), PS (Physical Structure), LS (Logical Structure), and VF (Visual Fidelity). Additional qualitative results can be found in Fig. 2.

Existing HTML converters [32,43] often linearize multi-column layouts, distort tables, and misplace elements, compromising both structure and visual fidelity. Tools like LaTeXML [34] work well for specific sources (e.g., LaTeX) but lack cross-format generality. Recent document-focused VLMs [12,28,36,45] generate structured outputs from images, yet struggle with multilingual scripts, degraded scans, complex hierarchies, and fine-grained style preservation.

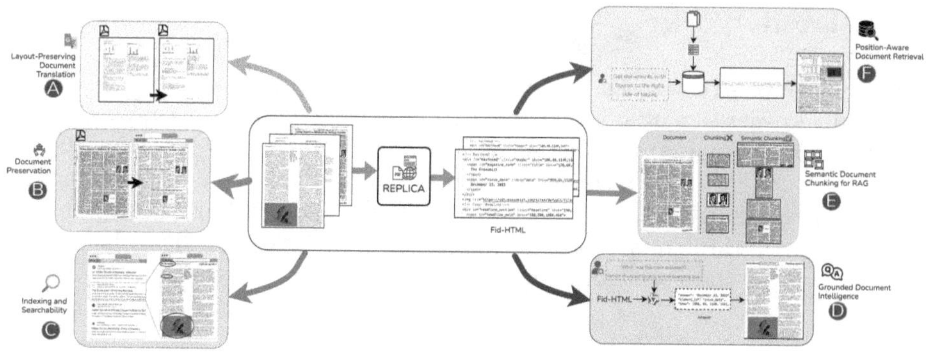

Fig. 2. Downstream Applications of REPLICA enabled by FID-HTML: (A) Layout-Preserving Document Translation, (B) Document Preservation, (C) Indexing & Searchability, (D) Grounded Document Intelligence, (E) Semantic Document Chunking for RAG, (F) Position-Aware Document Retrieval.

To overcome these limitations, we introduce REPLICA, an agentic framework that generates FID-HTML- semantically enriched, layout-aware HTML preserving text, logical structure, physical layout, and visual fidelity (see fig. 1). Beyond plain rendering, it embeds semantic layout classes and spatial cues, directly addressing the layout, typography, and styling losses of prior methods. Our contributions are:

1. FID-HTML, a new web-native format that represents documents in high-fidelity HTML, which is both semantically enriched and preserves visual and spatial fidelity.
2. REPLICA, an agentic framework for end-to-end page reconstruction that segments, localizes, assembles, and refines content to produce FID-HTML.
3. VFDR-Bench, a benchmark for evaluating fidelity and robustness in document-to-web conversion.
4. A suite of new evaluation metrics: Global Position Score (GPS), Local Position Score (LPS), and Visual Fidelity Score (VFS) - introduced alongside the benchmark for evaluating physical structure and visual fidelity in document-to-web conversion.

More broadly, REPLICA expands the utility of enriched HTML outputs beyond basic readability, potentially enabling applications like layout-preserving document translation, grounded document intelligence, Retrieval Augmented Generation (RAG), etc., (see Fig. 2). The presence of semantic tags, preserved reading order, and enriched figure alt-text in FID-HTML ensures seamless compatibility with screen readers on the web-rendered document, further enhancing accessibility.

2 Related Work

Document reconstruction spans two complementary directions. Logical reconstruction recovers semantic structure - reading order, hierarchy (sections, lists, tables), and structured outputs such as HTML or LaTeX - often sacrificing precise layout. Physical reconstruction targets pixel-level fidelity, preserving page geometry, multi-column structure, exact placement, styling, font attributes, spacing, and faithful reinsertion of figures and backgrounds. Together, these paradigms span the spectrum from semantic parsing to layout-preserving recreation.

2.1 Logical Document Reconstruction

Pipeline Parsing: Modular systems split document understanding into layout detection, reading order, OCR, and table/math parsing, then merge outputs via rules or learned modules [15,17,18,23,42,56,59,72]. Tools like Marker [43] and MinerU [55] produce solid Markdown/JSON/HTML but typically linearize layouts and under-specify styling, limiting visual fidelity. [32,43,55]. **End-to-End VLMs:** General-purpose VLMs [3,6,8,64,66,68] support zero-shot document understanding and can generate descriptions or lightweight HTML/JSON , but often miss fine layout and styling details. Specialist models [4,16,17,20,27,31,60] improve hierarchy, tables, and formulas via document-centric training . Recent variants like Qwen3-VL and olmOCR generate HTML-like outputs [45,64], yet most still fall short on visual fidelity, including fonts, spacing, and multi-column placement.

2.2 Physical Document Reconstruction

Approaches for Document Images: Systems such as PDFMathTranslate and image-to-LaTeX regeneration target scientific documents, preserving formulas, tables, and two-column layouts [19,40,48,57]. More recent work like DREAM focuses on high-fidelity digital reconstruction, autoregressively modeling both content and layout to recover precise element positions [25].

Approaches for UI Screenshots: Screenshot-to-HTML methods target UI-centric, text-light layouts [13,14,30,53,61,70], often using iterative rendering and reflection [21,49], along with GUI detection and language-driven refinement [62,63,71]. However, they generalize poorly to complex, text-dense document layouts due to distribution shift.

Our approach (REPLICA) is a layout-aware method that explicitly targets *both* axes: (i) logical reconstruction (semantic tags, reading order, tables/math) and (ii) physical reconstruction (absolute positioning, fonts, spacing, multi-column layout). This yields web-native HTML/CSS that is semantically enriched and faithful to source appearance.

3 FID-HTML: A High-Fidelity Document Representation.

FID-HTML is a high-fidelity, semantically enriched HTML representation generated by, capturing both semantic structure and visual presentation to support faithful reconstruction and machine-readable downstream tasks.

Table 1. Comparison of document representation formats. T.E. denotes Text Extraction; L.S. denotes Logical Structure; P.S. denotes Physical Structure; V.F. denotes Visual Fidelity. R.O. denotes Reading Order and L.H. denotes Logical Hierarchy. 'P' denotes partial support.

Format Type	Subtype	T.E.	L.S.						P.S.		V.F.
		Content	R.O.	L.H.	Tables	Lists	Figures	Formula	Hierarchy	BBoxes	
Markdown	md	✓	P	✓	×	×	P	P	P	×	×
	mmd	✓	P	✓	P	×	P	P	P	×	×
LaTeX	LaTeX	✓	P	✓	✓	✓	✓	✓	✓	P	✓
DocTags	DocTags [36]	✓	P	✓	✓	✓	✓	✓	✓	✓	P
HTML	HTML	✓	P	✓	✓	✓	P	P	P	×	✓
	hOCR [50]	✓	P	✓	✓	✓	P	P	✓	✓	✓
	Qwen-HTML [64]	✓	P	✓	✓	✓	P	P	P	×	✓
	Unstructured HTML [52]	✓	P	✓	✓	✓	P	P	×	×	✓
	FID-HTML (Ours)	✓	✓	✓	✓	✓	✓	✓	✓	✓	✓

Fig. 3. FID-HTML: High-Fidelity Document Representation. FID-HTML is the enriched output format of REPLICA, designed to preserve both the semantic structure and visual presentation of a document. The representation encodes: ❶ *Textual content* ; ❷ *Hierarchical structure* ; ❸ *Semantic tags* ; ❹ *Positional information* ; ❺ *Styling metadata* ; and ❻ *Figures and backgrounds* .

Core Representation. FID-HTML encodes a document along six complementary dimensions to jointly preserve semantics and visual fidelity: ❶ *Textual content* , where complete text is retained at both word and block levels; ❷ *Hierarchical structure* , captured through nested `<div>` containers reflecting sections, subsections, and logical groupings; ❸ *Semantic tags* , including structured elements such as `<li>`, `<table>`, `<tr>`, and `<td>`, along with meaningful class names for semantic grounding; ❹ *Positional information* , encoding global placement via absolute positioning and local alignment via relative containers and <p> wrappers; ❺ *Styling* and metadata , preserving font size, color, and text attributes (bold, italics, underline, strikethrough) through inline CSS; and ❻ *Figures and backgrounds* , where images and page-level backgrounds are layered into the DOM with accessibility-oriented alt-text. Together, these components enable faithful reconstruction while retaining machine-readable semantic structure (see Fig. 3).

Advantages: Unlike traditional parsers that discard layout and styling, FID-HTML preserves the full spectrum of document signals. Semantic tags and logical grouping support downstream LLM/VLM tasks (e.g., understanding, summarization, retrieval) (see Table 2); absolute and relative positioning maintain layout fidelity; styling metadata ensures faithful rendering; and VLM-generated alt-text enhances accessibility - capabilities typically absent in conventional systems (see Table 1).

4 Approach

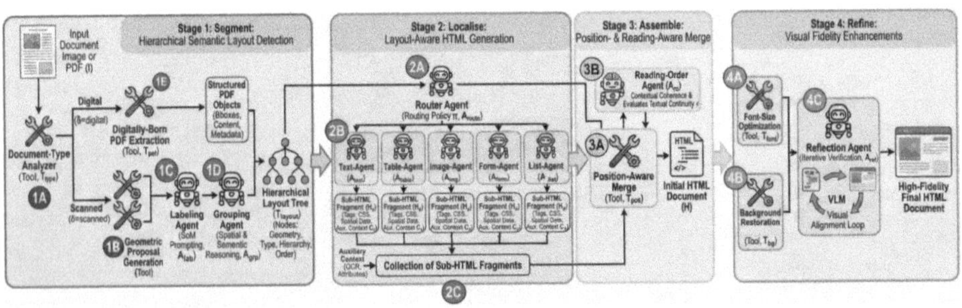

Fig. 4. Overview of REPLICA: An agentic framework for Visually Faithful Document Reconstruction. The agentic framework constructs a hierarchical semantic layout (❶ Segment), generates sub-HTML fragments via routed specialist-agents (❷ Localize), composes them with position- and reading-aware assembly (❸ Assemble), and iteratively refines visual fidelity through font optimization, background restoration, and VLM-based reflection (❹ Refine), producing a high-fidelity final HTML document.

REPLICA is an agentic document-to-web framework that treats documents as structured visual entities, reconstructing them in HTML with structural and visual fidelity.

REPLICA adopts an agentic divide-and-conquer strategy centered on four questions (Fig. 4):

1. *How to divide the document into meaningful components?* (**Segment**) 1
2. *How to convert each component into equivalent HTML?* (**Localize**) 2
3. *How to assemble components into a coherent structure?* (**Assemble**) 3
4. *How to refine structure for visual and structural fidelity?* (**Refine**) 4

We formalize the system as a coordination between reasoning agents and deterministic tool calls. Let $\mathcal{A}$ denote the set of agents and $\mathcal{T}$ the set of tools. Agents perform semantic reasoning, routing, grouping, and generation, while tools execute structured and reproducible operations (e.g., layout detection, reading-order computation, font optimization).

Given a document page image or PDF I, the objective is to produce HTML H whose rendering $R(H)$ aligns with I across TE, LS, PS, and VF dimensions.

4.1 Segment: Hierarchical Semantic Layout Detection

The segment stage builds a semantically grounded layout tree $\mathcal{T}_{\text{layout}}$ encoding geometry, types, and hierarchy. While deterministic detectors localize boundaries well, they lack semantic reasoning and can fail on borderless tables, multi-column wraps, or nested structures. We therefore introduce reasoning agents to validate, refine, and hierarchically organize proposals before generation, preventing downstream error propagation.

Document-Type Analysis (Tool). 1A A document classifier tool $T_{\text{type}} \in \mathcal{T}$ determines whether the input I is digitally-born or scanned:

$$\delta = T_{\text{type}}(I), \quad \delta \in \{\texttt{digital}, \texttt{scanned}\}.$$

This decision governs whether structure is extracted directly from PDF primitives or inferred from visual geometry.

Geometric Proposal Generation (Tool). 1B If $\delta = \texttt{scanned}$, we invoke a layout detection ensemble $\mathcal{T}_{\text{det}} \subset \mathcal{T}$ to generate candidate regions:

$$\mathcal{B} = \{\, b_i \,\}_{i=1}^{N} \quad \text{where} \quad b_i = m(I),\ m \in \mathcal{T}_{\text{det}}.$$

Each region b_i is defined by a bounding box and pixel support. These proposals are geometry-driven and may over-segment, merge incompatible content, or miss hierarchical relationships.

Labeling Agent. 1C A labeling agent $A_{\text{lab}} \in \mathcal{A}$ assigns semantic attributes to each region:

$$(\ell_i, \tau_i) = A_{\text{lab}}(I, b_i),$$

where ℓ_i denotes the structural label (e.g., paragraph, table, figure) and τ_i captures auxiliary attributes (e.g., style or subtype). Using Set-of-Mark prompting [65], the agent jointly reasons over spatial grounding and textual content to also identify mis-segmented regions, detect mixed-content blocks, and reinterpret ambiguous proposals.

Grouping Agent. 1D Given labeled regions, a grouping agent $A_{\text{grp}} \in \mathcal{A}$ constructs a directed layout graph

$$\mathcal{G} = (V, E), \quad V = \{v_i \mid v_i \leftrightarrow b_i\},$$

where nodes correspond to semantic regions. Edges E encode parent–child relations inferred from spatial containment, reading order, and semantic compatibility. During this stage, inconsistent or fragmented proposals can be merged, reattached, or discarded to enforce coherent structure. The final layout tree is obtained by enforcing acyclicity and single-parent constraints:

$$\mathcal{T}_{\text{layout}} = \text{Tree}(\mathcal{G}).$$

This step resolves caption–figure bindings, list nesting, footnote attachment, and other hierarchical ambiguities beyond pure geometric detection.

Digitally-Born PDF Extraction (Tool). 1E If $\delta = \texttt{digital}$, we bypass geometric detection and instead invoke a structured PDF parser $T_{\text{pdf}} \in \mathcal{T}$:

$$\mathcal{B} = T_{\text{pdf}}(I),$$

which directly extracts bounding boxes, textual spans, and font metadata from PDF primitives. In this case, visual grounding and proposal aggregation are unnecessary.

4.2 Localize: Layout-Aware HTML Generation

Given the hierarchical layout tree $\mathcal{T}_{\text{layout}}$, the localize stage converts each node $v \in \mathcal{T}_{\text{layout}}$ into structured sub-HTML. Single-shot VLM generation per region is brittle and hallucination-prone, so we introduce a router agent for hypothesis validation before generation. The generation is carried out by specialist agents.

Router Agent. 2A Each node v is associated with:

$$(I_v, b_v, \ell_v, \phi_v),$$

where b_v its bounding box, I_v is the visual crop of the input document restricted to b_v, ℓ_v the upstream semantic label from the Segment stage, and ϕ_v encodes structural constraints (e.g., parent type, ordering).

A routing agent $A_{\text{route}} \in \mathcal{A}$ selects an appropriate specialist agent:

$$a_v = A_{\text{route}}(I_v, \ell_v, \phi_v), \quad a_v \in \mathcal{A}_{\text{spec}}.$$

Rather than relying solely on the upstream label ℓ_v, the router jointly evaluates visual and contextual cues to detect modality mismatches. If the crop contradicts the predicted type (e.g., an equation labeled as text), it overrides the label and reroutes the node to the correct specialist, preventing cascading hallucinations during generation.

Specialist Agents. 2B Different layout elements require distinct generation strategies. We therefore define a set of specialist agents:

$$\mathcal{A}_{\text{spec}} = \{A_{\text{text}}, A_{\text{table}}, A_{\text{img}}, A_{\text{form}}, A_{\text{list}}, A_{\text{eqn}}\} \subset \mathcal{A}.$$

After routing, each node v is assigned a specialist $a_v \in \mathcal{A}_{\text{spec}}$, which generates localized HTML:

$$H_v = a_v(I_v, b_v, \phi_v),$$

Conditional Tool Invocation. Specialist agents invoke auxiliary tools $\mathcal{T}_{\text{aux}} \subset \mathcal{T}$ only when necessary. For instance, the Text Agent calls OCR for low-resolution or script-complex crops but bypasses it for clean digital text, while the Table Agent queries a structure recognizer when detecting merged headers or irregular grids to obtain explicit rowâĂŞcolumn constraints. This selective orchestration improves fidelity without redundant computation, and since nodes are conditionally independent, sub-HTML generation is parallelizable.

4.3 Assemble: Position and Reading-Aware Merge

The assemble stage merges independently generated sub-HTMLs $\{H_v\}_{v \in \mathcal{T}_{\text{layout}}}$ into a coherent document. While localization is node-level, reconstruction must preserve both hierarchical structure and logical reading flow. We therefore separate structural placement from semantic ordering.

Position-Aware Merge (Tool). 3A A composition tool $T_{\text{pos}} \in \mathcal{T}$ performs a bottom-up merge over $\mathcal{T}_{\text{layout}}$, preserving containment, nesting, and spatial alignment to produce a structurally valid document.

Reading-Order Agent. 3B Spatial heuristics (top-to-bottom and left-to-right sorting) alone fails on complex layouts like multi-column newspapers (B in Fig. 2) or floating elements. A reading-order agent $A_{\text{ro}} \in \mathcal{A}$ therefore computes a semantically grounded ordering relation by evaluating textual continuity and contextual coherence across nodes. The final HTML is linearized under $\prec$, ensuring a natural and logically consistent reading flow.

4.4 Refine: Visual Fidelity Enhancements

Even when segmentation, localization, and assembly are structurally correct, the rendered HTML may visually diverge from the source due to differences between PDF rendering and the browser's CSS box model. To address this, we introduce a refinement stage that operates on the rendered output $R(H)$ and compares it against the original document image I.

Font-Size Optimization (Tool). 4A A deterministic tool $T_{\text{font}} \in \mathcal{T}$ selects maximal non-overflow font scales via bounded search to preserve typographic fidelity.

Background Restoration (Tool). 4B A tool $T_{bg} \in \mathcal{T}$ extracts page-level visual cues (e.g., background colors, shading, overlays) and restores them using CSS layering and absolute positioning to preserve global visual context.

Reflection Agent. 4C A reflection agent $A_{ref} \in \mathcal{A}$ performs closed-loop visual comparison between $R(H)$ and I, identifying perceptual artifacts such as subtle overlaps, wrapping inconsistencies, or misalignment that emerge only after rendering. When discrepancies are detected, the agent injects targeted CSS corrections (e.g., layout adjustments or `shape-outside`) until visual alignment stabilizes. This refinement stage ensures that fidelity is evaluated at the level of appearance rather than structural compliance alone.

4.5 Implementation Details

Segment. For scanned documents, layout regions are detected using three deterministic tools: Doclayout-YOLO-v10 [73] for flat layouts, IndicDLP-YOLO-v10 [37] for hierarchical structures, and Hi-SAM [69] for fine-grained text segmentation. Their predictions are fused using IoU-based merging. Semantic labeling and hierarchical grouping are performed using Qwen3-VL-8B-Instruct, which assigns region types and constructs the nested layout tree.

Localize. Each layout node is routed to a specialized agent (text, table, image, form, list, or equation), all implemented using Qwen3-VL-8B [64]. Additional context is provided through deterministic tools: OCR via Gemma-3-4B-it [51], word localization via DocTR [35], and fine-grained style extraction via TexTAR [22]. For tables, Surya's Table Structure Recognition is used to explicitly recover row–column grids. Structured output validation and bounded retries ensure well-formed HTML.

Assemble. Fragments are merged using geometry and hierarchy signals from segmentation. Reading order is computed using coordinate-based heuristics (top-to-bottom, left-to-right with column handling).

Refine. Font sizes are optimized using `textFit.js` [47] with bounding box constraints from Hi-SAM. Backgrounds are reconstructed using OpenCV-based inpainting guided by region masks. Finally, a reflection agent (Qwen3-VL-8B) performs up to five iterative corrections (`max_iterations=5`) to visually align the rendered HTML with the source document.

5 Experimental Setup

We conduct comprehensive experiments to assess the performance of competing approaches on four key dimensions of document-to-web conversion. Beyond final output quality, we adopt an agentic evaluation protocol that measures intermediate agent decisions (segmentation, routing, refinement) and tool reliability (reading order, font fitting, background restoration), enabling component-level failure attribution in addition to system-level comparison.

VFDR-Bench. We introduce VFDR-Bench, a benchmark for visually faithful document-to-web conversion. Existing efforts target sub-tasks such as text extraction (TE) via OCR benchmarks [45,67] or address logical structure (LS) only limitedly through table-to-HTML [24] and document-to-Markdown [7,29, 39], but none capture all dimensions of the problem. VFDR-Bench fills this gap with 5,000 documents spanning single- and multi-column layouts, scanned and born-digital sources, 15+ languages, and 20+ document categories. All ground-truth annotations are fully expert human annotated to ensure structural, spatial, and visual correctness, providing a rigorous basis for both system-level fidelity evaluation and agent-level analysis.

5.1 System-Level Fidelity Metrics

Text Extraction (TE): Character Recognition Rate (CRR) and Word Recognition Rate (WRR).

Logical Structure (LS): Normalized Tree Edit Distance (NTED). We use TEDS [44] for tables and CDM [54] for formulas.

Physical Structure (PS). Global Position Score (GPS) and Local Position Score (LPS) measure layout- and line-level spatial fidelity. For layout blocks ($\mathcal{B}$) and lines ($\mathcal{L}$), we perform one-to-one matching between ground-truth and predicted elements using maximum IoU. We compute area-normalized spatial precision and recall based on overlap ratios, and report their harmonic mean (F1) as the final score. This formulation penalizes both missed elements (recall) and hallucinated bounding boxes (precision), yielding a balanced measure of physical structure fidelity.

Visual Fidelity (VF): Visual Fidelity Score (VFS) is computed via a VLM-as-a-judge comparing rendered HTML to human-annotated ground truth.

Overall Score: Overall score is calculated as an average across all the axes (TE, LS, PS. VF). To prevent pathological OCR failures from dominating multilingual aggregates, CRR and WRR are clipped at -1 per sample before averaging.

$$\text{Overall} = \frac{\text{TE} + \text{LS} + \text{PS} + \text{VF}}{4}, \quad \text{TE} = \frac{\text{CRR} + \text{WRR}}{2}, \quad \text{PS} = \frac{\text{GPS} + \text{LPS}}{2}.$$

5.2 Agent-Level Evaluation

For the *Segment* stage, we introduce Agnostic Text F1 (AT-F1) and Agnostic Visual F1 (AV-F1). Since downstream Set-of-Mark (SoM) prompting resolves semantic labels, evaluation isolates geometric quality only. We compute area-normalized spatial precision and recall for text and visual regions, and report their harmonic mean (F1). Unlike pure coverage metrics, this penalizes both missed regions (recall) and over-segmentation or bloated boxes (precision), ensuring tight and complete proposals that do not clutter SoM inputs. We additionally

report Hierarchy Consistency (HC), the fraction of pages whose predicted layout trees satisfy structural constraints.
For the *Localize* stage, we introduce the List Structure Penalty (LSP) to assess proper list formatting, while established metrics such as Tree Edit Distance Similarity (TEDS) [44] for tables and Character Detection Matching (CDM) [54] for formulas are also employed.

We further report Routing Accuracy (RA) and Sub-HTML Validity Rate (SVR):

$$RA = \frac{\#\text{Correct Routes}}{\#\text{Total Regions}}, \qquad SVR = \frac{\#\text{Valid DOM Fragments}}{\#\text{Generated Fragments}}.$$

Reflection behavior is quantified via Reflection Gain (RG):

$$RG = \frac{1}{K}\sum_{k=1}^{K}\left(S^{(k)} - S^{(k-1)}\right).$$

5.3 Tool-Level Reliability

$$TCSR = \frac{\#\text{Successful Tool Calls}}{\#\text{Total Tool Calls}}, \qquad FFS = \frac{\#\text{No-Overflow Paragraphs}}{\#\text{Total Paragraphs}},$$

Table 2. Quantitative comparison of document-to-HTML methods across four evaluation dimensions. Text Extraction (TE: WRR, CRR), Logical Structure (LS: NTED), Physical Structure (PS: GPS, LPS), and Visual Fidelity (VF: VFS via VLM-as-a-judge) are reported. The overall score (OS), computed as the mean of TE, LS, PS, and VF (Sect. 5.1), is reported as OS_{all}(all languages), OS_{en}(English), and OS_{zh}(Chinese). Baselines grouped into four categories: Pipeline Tools , Expert VLMs , General VLMs , and Screenshot2HTML . REPLICA represents our agentic method.

Method	T.E.		L.S.	P.S.		V.F.	Overall		
	WRR (↑)	CRR (↑)	NTED (↓)	LPS (↑)	GPS (↑)	VFS (↑)	OS_{all} (↑)	OS_{en} (↑)	OS_{zh} (↑)
Marker [43]	<u>0.68</u>	<u>0.82</u>	0.61	0.19	0.34	0.74	0.53	0.60	0.45
Docling [32]	0.40	0.49	0.62	0.17	0.26	0.55	0.38	0.45	0.12
rolmOCR [2]	0.54	0.67	0.95	0.15	0.07	0.56	0.34	0.42	0.35
smolDocling [36]	0.37	0.54	0.89	0.16	0.23	0.54	0.28	0.35	0.02
olmOCR [45]	0.49	0.64	0.93	0.21	0.09	0.33	0.31	0.38	0.34
Nanonets-OCR-s [33]	0.25	0.42	0.94	0.18	0.10	0.56	0.26	0.34	0.31
GOT-OCR-2.0 [60]	0.36	0.39	0.93	0.15	0.25	0.70	0.32	0.49	0.10
ChatDoc-OCRFlux [5]	0.30	0.38	0.88	0.15	0.24	0.44	0.27	0.39	0.34
Qwen3-VL [64]	0.66	0.79	0.59	<u>0.22</u>	0.30	0.43	0.47	0.48	0.31
Gemma3 [51]	0.58	0.59	0.85	0.19	0.25	0.38	0.36	0.42	0.37
InternVL3 [6]	0.06	0.30	0.82	0.19	0.29	0.31	0.22	0.29	0.27
Gemini-2.5-Pro [8]	0.67	0.42	0.56	0.14	0.31	0.82	0.47	0.47	0.30
GPT-5 [38]	0.65	0.79	<u>0.49</u>	0.21	<u>0.43</u>	<u>0.86</u>	<u>0.58</u>	<u>0.62</u>	<u>0.50</u>
Webcode2m [14]	0.33	0.45	0.82	0.15	0.29	0.39	0.30	0.42	0.13
Waffle VLM Websight [30]	0.14	0.42	0.84	0.11	0.29	0.19	0.22	0.27	0.11
REPLICA (Ours)	**0.88**	**0.93**	**0.20**	**0.73**	**0.86**	**0.93**	**0.86**	**0.93**	**0.89**

$$BSS = \frac{\#\text{Successful Background Restorations}}{\#\text{Eligible Pages}},$$

Reading Order Accuracy (ROA) is measured via Kendall-τ against ground-truth reading order.

Human Validation. Pearson correlation between seven expert annotators and metric-induced rankings exceeds 0.8 across 100 documents, confirming alignment with expert judgment.

6 Results

Overall Performance: REPLICA achieves the best overall results (Table 2), with near-perfect text extraction, strong logical structure preservation, accurate physical structure recovery (over 85%), and the highest visual fidelity (93%). In contrast, all baselines trade off across axes - parsers sacrifice layout, VLMs sacrifice structure or positioning, and Screenshot2HTML models fail to generalize to document settings - whereas REPLICA excels across all simultaneously. Overall, REPLICA sets a new state of the art by excelling across all axes simultaneously.

Robustness across languages and document types. REPLICA sustains high performance across 15 languages and 10+ document types, with minimal variance in text accuracy, structural fidelity, and visual alignment across scripts and layouts. In contrast, existing approaches degrade sharply beyond English and Chinese and remain confined to narrow domains or simple layouts, while REPLICA generalizes reliably across diverse formats.

Qualitative Examples. Figures 1 , 2 present qualitative examples of REPLICA's faithful visual reconstruction across a diverse range of document types - including historic newspapers, multi-column magazines, and financial documents. REPLICA consistently reproduces the original document's appearance in FID-HTML, accurately capturing column boundaries, typography, and spatial layout.

6.1 Ablations

REPLICA Framework Ablations. As shown in Table 3, performance improves consistently with each modular component, validating the agentic decomposition. Strengthening the Segment stage yields large gains in spatial alignment and overall score; introducing the agentic Router and Specialist formulation further produces a substantial jump over monolithic localization; and refinement steps (font correction, background stitching, reflection) deliver steady improvements, culminating in the strongest overall fidelity. The results demonstrate that high-quality reconstruction emerges from coordinated modular reasoning rather than any single stage.

Table 3. REPLICA Framework Ablations. Incremental impact of Segment, Agentic Localize (Router + Specialist Agents), and Refine components on Overall Score (OS), Visual Fidelity (VFS), and Global Position Score (GPS). Each row adds the specified component to the previous configuration.

Configuration	OS (↑)	VFS (↑)	GPS (↑)
Flat Layout (DLN)	0.53	0.74	0.34
+ Hierarchical Layout (IDLP-DLN)	0.60	0.79	0.48
+ Hi-SAM (Full Segment)	0.67	0.82	0.63
+ SoM in Segment	0.71	0.85	0.69
+ Monolithic Localize (w/o Router & Specialists)	0.76	0.88	0.72
+ Agentic Localize (Router + Specialists)	0.86	0.90	0.77
+ Font-Size Fix (fs)	0.89	0.92	0.79
+ Background Restore (bg)	0.90	0.93	0.80
+ Reflection (final)	**0.93**	**0.97**	**0.83**

Table 4. Agentic and Tool Reliability Metrics. Trace-based evaluation of intermediate agents and tool calls.

Category	Metric	Score
Segment Agent	Agnostic Text F1 (AT-F1)	0.98
	Agnostic Visual F1 (AV-F1)	0.99
	Hierarchy Consistency (HC)	0.98
Router Agent	Routing Accuracy (RA)	0.97
Localize Agent	Sub-HTML Validity Rate (SVR)	0.96
Reflection Agent	Reflection Gain (RG)	+0.06
Tools	Tool Call Success Rate (TCSR)	0.995
	Reading Order Accuracy (ROA)	0.98
	Font Fit Success (FFS)	0.97
	Background Stitch Success (BSS)	0.96

Agentic and Tool Reliability. Trace-based evaluation in Table 4 confirms high reliability across coverage, routing, sub-HTML generation, and tool execution, with reflection delivering consistent gains - demonstrating that performance is driven by dependable structured reasoning rather than any single component.

6.2 Limitations

While REPLICA delivers high-fidelity single-page reconstruction, it does not yet support multi-page consistency, potentially fragmenting cross-page structures (e.g., spanning tables or figures). Text attributes (bold, italics, underline, color) are reliably recovered, but precise font identification remains unresolved. Existing methods such as DeepFont [58] and SwoRNet [26] offer limited, mostly

English/Chinese coverage, and large-scale multilingual font recognition remains an open challenge.

7 Conclusion

We introduce REPLICA, an agentic framework for converting raw document image into high-quality document representation (HTML) while preserving both structure and visual appearance. Instead of relying on a single model to do everything, our approach breaks the problem into coordinated stages that employs various agents; detecting layout, generating content with specialized agents, assembling the document in the correct order, and refining the final output based on how it looks after rendering. By combining structured tools with reasoning agents, REPLICA can correct layout mistakes, maintain proper hierarchy, and resolve visual inconsistencies that arise during rendering. The resulting HTML is not only machine-readable but also visually faithful to the original document, enabling reliable use in accessibility, publishing, and document understanding applications.

Acknowledgments. Our sincere gratitude goes to Amruth Sagar K N and late B.V. Khadiravana for their invaluable support. The project was supported by BharatGen at IIIT Hyderabad.

References

1. A. S. Inc.: Pymupdf. https://github.com/pymupdf/PyMuPDF (2024)
2. AI, R.: RolmOCR: A Faster, Lighter Open Source OCR Model (2025)
3. Anthropic: Claude 3.5 sonnet. https://www.anthropic.com/news/claude-3-5-sonnet (2024)
4. Blecher, L., et al.: Nougat: Neural optical understanding for academic documents. In: ICLR (2023)
5. ChatDOC-com: OCRFlux: a multimodal toolkit for converting pdfs and images into markdown. https://github.com/chatdoc-com/OCRFlux (2025)
6. Chen, Z., et al.: InternVL: scaling up vision foundation models and aligning for generic visual-linguistic tasks. In: CVPR (2024)
7. Chen, Z., et al.: MDEval: evaluating and enhancing markdown awareness in large language models. In: WWW (2025)
8. Comanici, G., et al.: Gemini 2.5: pushing the frontier with advanced reasoning, multimodality, long context, and next-generation agentic capabilities. arXiv **2507.06261** (2025)
9. Cui, C., et al.: PaddleOCR 3.0 technical report. arXiv preprint arXiv:2507.05595 (2025)
10. Developers, P.: Pandoc User's Guide. Pandoc (2025). https://pandoc.org/MANUAL.html
11. Documentation, U.: pdf2htmlEX Manual (Ubuntu Xenial manpage) (2016). https://manpages.ubuntu.com/manpages/xenial/man1/pdf2htmlEX.1.html

12. Feng, H., et al.: Dolphin: document image parsing via heterogeneous anchor prompting. arXiv preprint arXiv:2505.14059 (2025)
13. Gui, Y., et al.: UICopilot: automating UI synthesis via hierarchical code generation from webpage designs. In: WWW (2025)
14. Gui, Y., et al.: WebCode2M: a real-world dataset for code generation from webpage designs. In: WWW (2025)
15. He, Y., et al.: PingAn-VCGroup's solution for ICDAR 2021 competition on scientific table image recognition to latex. arXiv preprint arXiv:2105.01846 (2021)
16. Hu, A., et al.: mPLUG-DocOWL2: high-resolution compressing for OCR-free multi-page document understanding. arXiv preprint arXiv:2409.03420 (2024)
17. Huang, Y., et al.: LayoutLMv3: pre-training for document ai with unified text and image masking. In: WWW (2022)
18. Jaided AI: EasyOCR: Ready-to-use OCR with 80+ supported languages. https://github.com/JaidedAI/EasyOCR (2024)
19. Kayal, P., et al.: Tables to latex: structure and content extraction from scientific tables. IJDAR **26** (2022)
20. Kim, G., et al.: DONUT: OCR-free document understanding transformer. In: ECCV (2022)
21. Kolthoff, K., et al.: Zero-shot prompting approaches for LLM-based graphical user interface generation. arXiv 2412.11328 (2024)
22. Kumar, R., et al.: TexTAR: textual attribute recognition in multi-domain and multi-lingual document images. In: ICDAR (2025)
23. Li, C., et al.: PP-OCRv3: more attempts for the improvement of ultra-lightweight OCR system. arXiv preprint arXiv:2206.03001 (2022)
24. Li, M., et al.: TableBank: table benchmark for image-based table detection and recognition. In: LREC (2020)
25. Li, X., et al.: DREAM: document reconstruction via end-to-end autoregressive model. arXiv preprint arXiv:2507.05805 (2025)
26. Li, X., et al.: SwordNet: Chinese character font style recognition network. IEEE Access **10** (2022)
27. Li, Z., et al.: Monkey: image resolution and text labels are important for large multimodal models. In: CVPR (2024)
28. Li, Z., et al.: MonkeyOCR: document parsing with a structure-recognition-relation triplet paradigm. arXiv preprint arXiv:2506.05218 (2025)
29. Li, Z., et al.: ReadOC: a unified benchmark for realistic document structured extraction. arXiv preprint arXiv:2409.05137 (2024)
30. Liang, S., et al.: WAFFLE: multi-modal model for automated front-end development. arXiv preprint arXiv:2410.18362 (2024)
31. Liu, Y., et al.: TextMonkey: an OCR-free large multimodal model for document understanding. arXiv preprint arXiv:2403.04473 (2024)
32. Livathinos, N., et al.: Docling: an efficient open-source toolkit for AI-driven document conversion. arXiv preprint arXiv:2501.17887 (2025)
33. Mandal, S., et al.: Nanonets-OCR-S: a model for transforming documents into structured markdown with intelligent content recognition and semantic tagging. arXiv (2025)
34. Miller, B.: Latexml the manual. Web document (2011)
35. MindEE: DOCTR: Document text recognition. https://github.com/mindee/doctr (2021)
36. Nassar, K., et al.: SmoldOcling: lightweight document understanding via efficient vision-language modeling. arXiv preprint arXiv:2501.04567 (2025)

37. Nath, O., et al.: IndicDLP: a foundational dataset for multi-lingual and multi-domain document layout parsing. In: ICDAR (2025)
38. OpenAI: GPT-5 system card. Tech. rep., OpenAI (2025). https://cdn.openai.com/gpt-5-system-card.pdf
39. Ouyang, L., et al.: OmniDocBench: benchmarking diverse PDF document parsing with comprehensive annotations. In: CVPR (2025)
40. Ouyang, R., et al.: PDFMathTranslate: scientific document translation preserving layouts. arXiv 2507.03009 (2025)
41. P. Team: Pypdfium2: Python bindings for PDFium. https://github.com/pypdfium2-team/pypdfium2 (2024)
42. Padhi, I., et al.: TabFormer: tabular transformers for modeling multivariate time series. In: ICASSP (2021)
43. Paruchuri, V.: Marker. https://github.com/datalab-to/marker (2025)
44. Pawlik, M., Augsten, N.: Tree edit distance: robust and memory-efficient. Inf. Syst. **56** (2016)
45. Poznanski, A., et al.: OLMOCR: open layout modeling for robust OCR-free document understanding. arXiv preprint arXiv:2502.01983 (2025)
46. PYPDF Maintainers: PYPDF: A Pure-Python PDF Library. https://github.com/py-pdf/pypdf (2024)
47. Reed, S.: textfit. https://github.com/STRML/textFit (2014)
48. Safnuk, B., Hu, G.: Reconstructing latex source files from generated PDFs: a neural network approach. In: INDIN, pp. 890–895 (2018)
49. Si, C., et al.: Design2Code: benchmarking multimodal code generation for automated front-end engineering. arXiv preprint arXiv:2403.03163 (2024)
50. Smith, R.: An overview of the Tesseract OCR engine. In: ICDAR (2007)
51. Team, G., et al.: Gemma 3 technical report. arXiv preprint arXiv:2503.19786 (2025)
52. Unstructured-IO: Unstructured: Open-source tools for unstructured data processing. https://github.com/Unstructured-IO (2025)
53. Wan, Y., et al.: Divide-and-conquer: generating UI code from screenshots. arXiv preprint arXiv:2406.16386 (2025)
54. Wang, B., et al.: CDM: a reliable metric for fair and accurate formula recognition evaluation. arXiv preprint arXiv:2409 (2024)
55. Wang, B., et al.: MinerU: an open-source solution for precise document content extraction. arXiv preprint arXiv:2409.18839 (2024)
56. Wang, B., et al.: UniMERNet: a universal network for real-world mathematical expression recognition. arXiv preprint arXiv:2404.15254 (2024)
57. Wang, Z., Liu, J.C.: Translating math formula images to latex sequences using deep neural networks with sequence-level training. IJDAR **24** (2021)
58. Wang, Z., et al.: DeepFont: identify your font from an image. In: ACM MM (2015)
59. Wang, Z., et al.: LayoutReader: pre-training of text and layout for reading order detection. In: EMNLP (2021)
60. Wei, H., et al.: General OCR theory: towards OCR-2.0 via a unified end-to-end model. arXiv preprint arXiv:2409.01704 (2024)
61. Wu, F., et al.: MLLM-based UI2Code automation guided by UI layout information. Proc. ACM Softw. Eng. **2** (2025)
62. Xiao, S., et al.: Ui semantic component group detection: grouping UI elements with similar semantics in mobile graphical user interface. Displays **83** (2024)
63. Xie, M., et al.: UIED: a hybrid tool for GUI element detection. In: ESEC/FSE (2020)
64. Yang, A., et al.: Qwen3 technical report. arXiv preprint arXiv:2505.09388 (2025)

65. Yang, J., et al.: Set-of-mark prompting unleashes extraordinary visual grounding in GPT-4V. arXiv preprint arXiv:2310.11441 (2023)
66. Yang, Z., et al.: The dawn of LMMs: preliminary explorations with GPT-4v. arXiv preprint arXiv:2309.17421 (2023)
67. Yang, Z., et al.: CC-OCR: a comprehensive and challenging OCR benchmark for evaluating large multimodal models in literacy. arXiv preprint arXiv:2412.02210 (2024)
68. Yao, Y., et al.: MiniCPM-V: a GPT-4V-level MLLM on your phone. arXiv preprint arXiv:2408.01800 (2024)
69. Ye, M., et al.: Hi-SAM: marrying segment anything model for hierarchical text segmentation. In: TPAMI (2024)
70. Yun, S., et al.: Web2Code: a large-scale webpage-to-code dataset and evaluation framework for multimodal LLMs. NeurIPS **37** (2024)
71. Zhang, T., et al.: NLDesign: a UI design tool for natural language interfaces. In: TACC-China (2024)
72. Zhao, Y., et al.: DETRs beat YOLOs on real-time object detection. In: CVPR (2024)
73. Zhao, Z., et al.: DocLayout-YOLO: enhancing document layout analysis through diverse synthetic data and global-to-local adaptive perception. arXiv preprint arXiv:2410.12628 (2024)

HKGC: A Hierarchical Knowledge Graph Construction Framework for Structure-Aware RAG

Yingxin Guan, Jian Xing(✉), Zhaohua Zheng, Zhaofu Zeng, Bai Lei, Fanchen Meng, and Haitao Guo

Hainan University, Haikou, China
{24220854120007,xingjian}@hainanu.edu.cn

Abstract. Retrieval-Augmented Generation (RAG) systems often ignore the hierarchical structure of documents, leading to structural–semantic decoupling that undermines coherent reasoning. Standard approaches employ fixed-size sliding windows or heuristic delimiters to partition the text into independent chunks, which causes the retriever to miss relevant context, often leading to incorrect answers during the reasoning process. To address this, we propose HKGC, a novel framework that resolves structural–semantic decoupling through a specific and efficient architectural pipeline. Rather than merely fusing two modalities, HKGC uniquely combines deterministic Finite State Machine (FSM) parsing—enabling zero-latency extraction of the logical document hierarchy—with a rigorous dual-stage generate–discriminate paradigm for precise semantic graph construction. By applying a hybrid entity canonicalization strategy, HKGC aligns these semantic units with layout structures to reconstruct global context, explicitly constraining semantic navigation to valid structural pathways. Experimental results on three domain-specific benchmarks show that HKGC achieves strong performance, outperforming the majority of baselines with an average F1 improvement of 19.5%, reaching up to 34.5% in technical domains. Moreover, the results indicate that HKGC's explicit hierarchical modeling optimizes efficiency, maintaining an average context consumption of only 2,533 tokens and reducing end-to-end latency by up to 10.1%.

Keywords: Retrieval-Augmented Generation · Hierarchical Knowledge Graph · Document Structure Analysis · Structure-Aware Multi-hop QA

1 Introduction

Retrieval-Augmented Generation (RAG) effectively mitigates the inherent limitations of Large Language Models (LLMs), such as factual hallucinations and temporal knowledge decay, by integrating dynamic external knowledge [1,10,20]. The standard pipeline employs a retrieve-then-read architecture: long documents

G. A. Fink et al. (Eds.): ICDAR 2026, LNCS 16972, pp. 523–539, 2027.
https://doi.org/10.1007/978-3-032-36023-6_30

are partitioned into fixed-size, independent text chunks, which are then retrieved via dense vector search and prepended to the input query as contextual evidence [9,14,27].

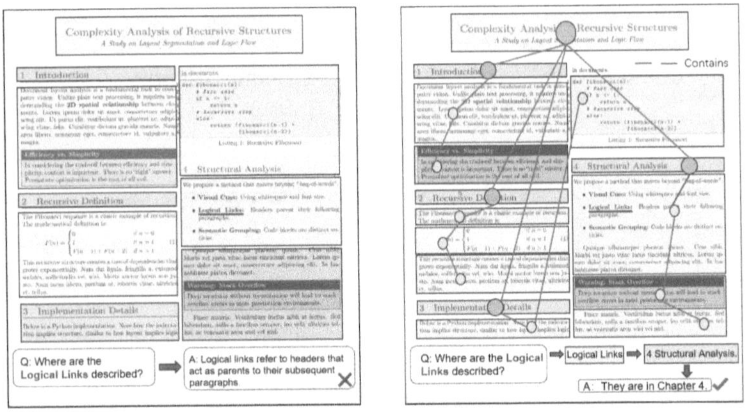

(a) Flat Segmentation (b) Structure-Aware Parsing

Fig. 1. Comparison of Document Representation Paradigms.

However, this flat segmentation ignores the inherent organizational structure of documents, leading to a *structural–semantic decoupling* [29]. Because the retriever only observes isolated fragments, it loses the global context necessary for coherent reasoning. As illustrated in Fig. 1(a), the system may match locally relevant text while missing its structural parent, yielding incorrect answers. While recent studies explore KG-guided paradigms [22,32] and structural priors [23], they typically treat structure as an auxiliary signal rather than first-class knowledge. Consequently, when tackling complex multi-hop questions in professional domains (e.g., technical manuals or financial reports), the fragmented, deeply nested information acts as noise, causing LLMs to generate hallucinations [18,34,44].

To overcome this limitation, we introduce HKGC, a framework that explicitly reintroduces document hierarchy as an intrinsic constraint (Fig. 1(b)). HKGC represents documents as a unified heterogeneous knowledge graph, integrating layout-induced logical hierarchies with textual semantic relations. Specifically, HKGC first constructs a logical Graph (G_{log}) using a deterministic finite-state machine. Operating directly on digitally native text, this parser is driven by explicit layout cues (e.g., font sizes and indentations) to map documents into tree-structured representations (Document → Section → Paragraph). By encoding *parent–child* and *sibling* relations, G_{log} provides a global structural context that constrains local semantics. Consider the following example from a database deployment guide:

Chapter 3: Production Deployment → 3.2 Security Config
"Administrators should set the `allow_anonymous` parameter to False..."

Appendix A: Local Dev Environment → A.1 Quick Start
"For debugging purposes, the `allow_anonymous` parameter can be set to True."

Under flat chunk-based representations, the instruction "set parameter to True" becomes semantically ambiguous. In HKGC, each statement is explicitly grounded in its corresponding section node, enabling the model to distinguish configuration knowledge applicable to production versus local development. While G_{log} provides global structural constraints, capturing fine-grained knowledge requires relational semantics. Therefore, HKGC parallelly constructs a Semantic Graph (G_{sem}) of entity–relation triplets [5]. To minimize hallucinated relations, we adopt a dual-stage Generator–Discriminator paradigm [1] rather than relying on single-pass LLM extraction. Finally, a hybrid entity canonicalization strategy fuses G_{log} and G_{sem} into the unified heterogeneous graph $\mathcal{G}$, establishing cross-index linkages. To bridge offline indexing and online reasoning, HKGC leverages a layout-enriched dual-index space. During inference, dense vector search locates initial evidence anchors, and the unified graph explicitly traces their hierarchical lineage and semantic neighbors via subgraph expansion. This approach inherently filters out structurally irrelevant noise, providing the LLM with a cohesive context that ensures logical consistency and minimizes redundant token consumption [17].

To validate the effectiveness of our framework, we conduct extensive experiments on three domain-specific multi-hop QA benchmarks. In conclusion, our contributions are three-fold:

- HKGC is proposed as a novel framework to resolve structural-semantic decoupling. Its core novelty lies in its specific architectural pipeline: the unique combination of deterministic FSM logical parsing and a rigorous dual-stage semantic graph construction.
- A layout-enriched dual-index space is designed to bridge dense vector retrieval with graph topology. By explicitly injecting hierarchical ancestry into text representations, this mechanism ensures that the retrieval process is inherently structure-aware, effectively resolving semantic ambiguity and minimizing redundant context.
- Extensive experiments show that HKGC achieves a superior accuracy-efficiency trade-off. Compared to the average performance of all baselines, it delivers an overall F1 improvement of 19.5% (up to 34.5% in technical domains) while significantly reducing computational overhead, maintaining a minimal context size of 2,533 tokens, and lowering latency by up to 10.1%.

2 Related Work

Document Layout as Knowledge Representation. Traditionally, Document Layout Analysis (DLA) treats structure as a perceptual signal via vision models [11] or auxiliary visual features in multimodal pre-training [13,36,42]. While recent graph-based methods [4,32] and rule-based parsers [25] introduce relational structures, they primarily capture spatial adjacency or treat hierarchy as a preprocessing artifact. Although recent advancements have begun to bridge this gap by exploring semantic structure parsing for specific elements like tables [21], existing generalized pipelines largely fail to model the global document hierarchy as a first-class, constraint-bearing knowledge prior [12,16,23]. Structure-agnostic chunking discards hierarchical provenance, obscuring the scope of retrieved content [18,41]. HKGC departs from this paradigm by explicitly modeling layout as a logical structure graph (G_{log}), encoding structural constraints directly from native layout cues to preserve document-level dependencies for complex reasoning.

Structure-Aware RAG Models. Standard RAG couples language models with external corpora using a retriever–reader architecture over independent text chunks [19,20]. To support multi-hop reasoning, recent work introduces structural indices over retrieved content [33]. RAPTOR [29] builds tree-based indices via semantic clustering, while HippoRAG [18] and StructRAG [16] explore associative memory. However, these representations often leave document-intrinsic hierarchy implicit or lose fine-grained logical provenance [12,16]. Furthermore, relying solely on semantic association risks reasoning drift in long-tail settings [7,26], and fetching dense subgraphs often introduces substantial redundant context. Recently, integrating structural layouts with semantic graphs has also been explored for complex document understanding [30,38]. While acknowledging these dual-modality efforts, HKGC distinguishes itself through its specific architectural execution: we uniquely combine deterministic FSM logical parsing—enabling zero-latency structural extraction—with a rigorous dual-stage generate-discriminate paradigm to minimize hallucinated relations. This creates a layout-enriched dual-index space that explicitly constrains semantic navigation to valid structural pathways, reducing the redundant context common in existing approaches.

KG-Enhanced Reasoning in RAG. A prominent paradigm for enabling multi-hop reasoning involves dynamically constructing Knowledge Graphs (KGs) directly from the retrieved textual context [2,8,35,40]. Methods like Think-on-Graph (ToG) [31] and ReSearch [3] use LLM agents or reinforcement learning for graph exploration, while DP [23] exploits structural priors for introspection. A critical limitation of these methods, including graph-based summarization models like GraphRAG [5], is their reliance on single-pass extraction guided by in-context learning [15,24,45]. They do not explicitly enforce evidence-level consistency, leaving them susceptible to hallucinated relations. To overcome this and ensure structural validity, HKGC introduces a dual-stage *generate–discriminate* paradigm. By explicitly anchoring extracted semantics back to the document's logical layout hierarchies, our framework enables faithful, evidence-driven multi-hop reasoning without relying on extensive labeled supervision.

3 Method

Figure 2 illustrates the overall architecture of HKGC. To enable structure-aware reasoning, our framework operates through a specific four-stage pipeline—

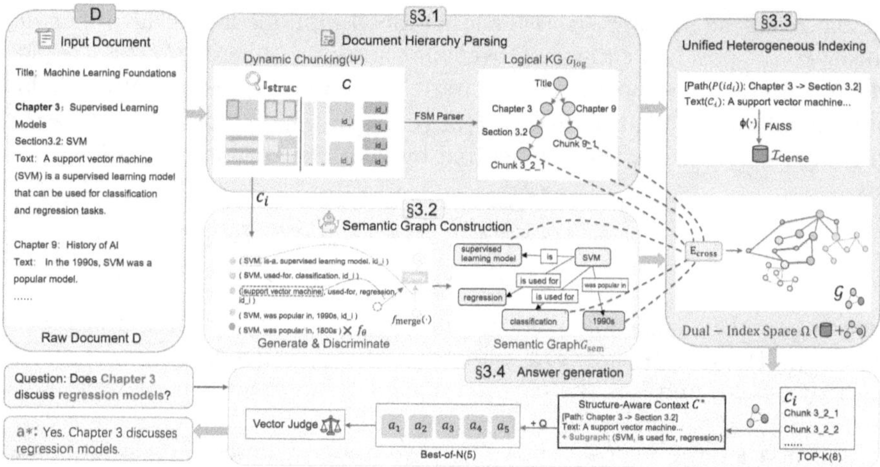

Fig. 2. Overview of the HKGC framework in four stages: (1) Document Hierarchy Parsing (Sect. 3.1) constructs a logical graph G_{log} encoding document topology; (2) Semantic Graph Construction (Sect. 3.2) extracts quadruplets to build the semantic graph G_{sem}; (3) Unified Heterogeneous Indexing (Sect. 3.3) aligns graphs via provenance edges E_{cross}, forming a dual-index space Ω ($\mathcal{G}$ and FAISS index $\mathcal{I}_{\text{dense}}$); (4) Answer Generation (Sect. 3.4) assembles context $\mathcal{C}^*$ and applies a *Vector Judge* to select the final answer a^* from Best-of-N candidates.

uniquely combining deterministic FSM logical parsing with a rigorous dual-stage semantic graph construction to transform raw documents into a highly grounded heterogeneous graph for evidence-guided QA.

3.1 Document Hierarchy Parsing

HKGC employs a structure-aware hierarchical parser to transform raw documents into a logical structure graph G_{log} and a standardized chunk sequence C. This module explicitly preserves layout-induced hierarchies to serve as topological constraints for downstream semantic modeling.

For native digital documents, we first preprocess the paragraph stream into a purified sequence $\mathcal{S} = \{s_1, \ldots, s_n\}$ by removing non-semantic artifacts (e.g., headers, footers). To preserve the document hierarchy, we introduce a dynamic chunking algorithm Ψ that sequentially processes $\mathcal{S}$. Guided by a structural indicator $\mathbb{I}_{\text{struc}}(s_m)$ detecting layout cues (e.g., font sizes, numbering), Ψ adaptively switches between hierarchical aggregation for structured regions and linear merging for free-form text.

Each resulting chunk c_i is assigned a globally unique logical identifier $id_i = \text{Concat}(H(D), \text{pos}_i)$ to support provenance tracking, where $H(D)$ is the document metadata hash and pos_i is the positional index. The standardized chunk sequence is formally defined as $C = \{(id_i, \mathcal{C}_i)\}_{i=1}^{|C|}$. The textual content $\mathcal{C}_i$ of the i-th chunk is deterministically generated by Ψ at the current paragraph pointer m as:

$$\mathcal{C}_i = \Psi(s_m) = \begin{cases} \Phi_{\text{hier}}(\{s_j \mid s_j \in \mathcal{N}_{\text{tree}}(s_m)\}), & \mathbb{I}_{\text{struc}}(s_m) = 1 \\ \Phi_{\text{naive}}(\{s_m, \ldots, s_{m+k}\}, \tau_{\text{len}}), & \text{otherwise} \end{cases} \tag{1}$$

where Φ_{hier} and Φ_{naive} denote hierarchical aggregation and linear merging respectively; $\mathcal{N}_{\text{tree}}(s_m)$ represents subordinate elements under node s_m, τ_{len} is the length threshold, and k is the maximum span of subsequent paragraphs within τ_{len}.

To construct the logical graph G_{log}, we apply a deterministic finite-state-machine (FSM) parser driven by the layout cues. The FSM maintains active states representing hierarchical depth. Upon detecting a structural cue, it transitions states, creating a new structural node and establishing explicit *parent–child* edges. Crucially, this parsing process ensures that leaf chunks are accurately attached to their governing structural nodes, retaining their full hierarchical path as metadata. Let V_{struc} denote the set of higher-level structural nodes, V_{chunk} denote the leaf chunk nodes, and E_{log} represent the hierarchical edges. The logical structure graph is formally constructed as:

$$\begin{aligned} G_{\text{log}} &= \langle V_{\text{log}}, E_{\text{log}} \rangle, \quad \text{where} \quad V_{\text{log}} = V_{\text{struc}} \cup V_{\text{chunk}} \\ E_{\text{log}} &= \{(v_p, \text{has\_child}, v_c) \mid v_p \in V_{\text{struc}}, v_c \in \text{Children}_{\text{FSM}}(v_p)\} \end{aligned} \tag{2}$$

where $\text{Children}_{\text{FSM}}(v_p)$ denotes the child nodes explicitly assigned to v_p by the FSM. This graph acts as the structural foundation for the subsequent heterogeneous indexing phase.For instance, as illustrated in the running example of Fig. 2 (Sect. 3.1), the FSM correctly parses "Chapter 3" and "Section 3.2" as hierarchical ancestors for the chunk describing SVM.

3.2 Semantic Graph Construction

HKGC constructs a semantic graph G_{sem} from the chunk sequence C via a *Generate–Discriminate* paradigm. To extract reliable semantics while minimizing LLM hallucinations, we prompt an LLM $\mathcal{M}$ to generate candidate entity-relation triplets for each chunk $c_i \in C$ and bind them to their logical identifier id_i. Constrained by an instruction template $\mathcal{I}_{\text{prompt}}$ that maps relations r to a predefined ontology, a fault-tolerant parser $\mathcal{P}_{\text{parse}}$ extracts a local candidate quadruplet set $T_{\text{cand}}^{(i)}$:

$$T_{\text{cand}}^{(i)} = \{(h, r, t, id_i) \mid (h, r, t) \in \mathcal{P}_{\text{parse}}\left(\arg\max_Y P(Y \mid \mathcal{C}_i, \mathcal{I}_{\text{prompt}})\right)\} \tag{3}$$

where Y is the generated token sequence and P denotes its conditional probability under $\mathcal{M}$. Each tuple (h, r, t, id_i) represents the head, relation, tail, and provenance ID, respectively.

To suppress spurious relations, a cross-encoder discriminator $f_\theta(h, r, t, \mathcal{C}_i)$ evaluates the textual entailment between each triplet in $T^{(i)}_{\text{cand}}$ and its source text $\mathcal{C}_i$. Retaining only candidates exceeding a confidence threshold τ_{score} and pruning verbose relations (e.g., via a string length limit) yields a validated local set T^*_i. We aggregate these to form the global validated quadruplet set $T^* = \bigcup_{i=1}^{|C|} T^*_i$.

To resolve synonymous entities across chunks, an embedding-based merging function $f_{\text{merge}}(\cdot)$ clusters coreferent nodes via cosine similarity. This canonicalization maps raw entities to a unified vertex set V_{sem} connected by relational edges E_{sem} that explicitly retain provenance metadata. The final semantic property graph is formally defined as:

$$\begin{aligned} G_{\text{sem}} &= \langle V_{\text{sem}}, E_{\text{sem}} \rangle, \quad \text{where} \quad V_{\text{sem}} = \bigcup_{(h,r,t,id_i)\in T^*} \{f_{\text{merge}}(h), f_{\text{merge}}(t)\} \\ E_{\text{sem}} &= \left\{ \left(f_{\text{merge}}(h), r, f_{\text{merge}}(t), id_i\right) \mid (h, r, t, id_i) \in T^* \right\} \end{aligned} \tag{4}$$

This semantically cohesive and traceable graph serves as the knowledge foundation for the subsequent heterogeneous indexing phase.Following our running example (Fig. 2, Sect. 3.2), this stage successfully extracts and validates the (SVM, is used for, regression) triplet from the raw chunk.

3.3 Unified Heterogeneous Indexing

The core innovation of HKGC lies in unifying logical layout and semantic knowledge. As illustrated in Fig. 2 (Sect. 3.3), the primary objective of this module is to fuse the logical graph G_{log} and the semantic graph G_{sem}.

To formally construct the cross-layer edges E_{cross}, we leverage the provenance metadata preserved within the validated quadruplets T^* from Sect. 3.2. Since each quadruplet $(h, r, t, id_i) \in T^*$ inherently binds the extracted relation to its originating chunk, we define E_{cross} by linking the canonicalized semantic entities directly to their corresponding logical chunk nodes in V_{chunk}. By formally representing each logical chunk node $v_c \in V_{\text{chunk}}$ by its unique logical identifier id_i, these provenance edges are mathematically constructed as:

$$E_{\text{cross}} = \left\{ (v, \textit{extracted\_from}, id_i) \mid (h, r, t, id_i) \in T^*, v \in \{f_{\text{merge}}(h), f_{\text{merge}}(t)\} \right\} \tag{5}$$

As shown by the *purple dashed lines* in Fig. 2, E_{cross} allows the purely semantic vertices V_{sem} to inherit the global hierarchical context from G_{log}. The resulting unified heterogeneous graph is formally defined as:

$$\mathcal{G} = G_{\text{log}} \cup G_{\text{sem}} \cup E_{\text{cross}} \tag{6}$$

Functionally, $\mathcal{G}$ acts as the topological backbone of our framework. By explicitly modeling E_{cross}, it ensures logical traceability where any semantic relation can be traced back to its exact layout scope, serving as a hard constraint for downstream multi-hop reasoning.

Furthermore, while $\mathcal{G}$ provides rich semantic context, navigating a large-scale graph for initial evidence retrieval is computationally expensive. To enable rapid recall, we construct a parallel dense vector index. Let $P(id_i)$ denote the explicit hierarchical path string of the chunk identified by id_i, dynamically derived by traversing upward through the ancestry nodes in G_{log} (e.g., [Path: Chapter 3 $\rightarrow$ Sect. 3.2]). Without G_{log}, individual text chunks would remain isolated semantic islands. By concatenating this structural metadata with its corresponding pure chunk text $\mathcal{C}_i$, we inject layout priors into the text. Using a dense encoder $\phi(\cdot)$, these path-enriched representations are vectorized. To ensure that the retrieved vectors can be mapped back to the graph topology, the FAISS index is constructed as a set of key-value pairs associating each logical identifier id_i with its corresponding dense vector:

$$\mathcal{I}_{\text{dense}} = \text{FAISS}\Big(\{(id_i, \phi(\text{Concat}(P(id_i), \mathcal{C}_i))) \mid (id_i, \mathcal{C}_i) \in C\}\Big) \quad (7)$$

Ultimately, the output of this indexing module is a dual-index search space, formally denoted as $\Omega = \{\mathcal{G}, \mathcal{I}_{\text{dense}}\}$. In this paradigm, $\mathcal{I}_{\text{dense}}$ serves as a high-speed entry point to locate initial evidence anchors, while $\mathcal{G}$ provides the topological pathways to constrain and guide subsequent reasoning.

3.4 Answer Generation

During inference, HKGC performs joint retrieval and structure-aware generation to answer a user query Q. As illustrated in Fig. 2 (Sect. refsec:generation), this process leverages the dual-index space Ω to prevent hallucination and enforce structural constraints.

Given a query Q, we employ the dense encoder $\phi(\cdot)$ to retrieve the top-K (e.g., $K = 8$) most relevant logical nodes from the FAISS index $\mathcal{I}_{\text{dense}}$ via standard cosine similarity. Rather than directly concatenating these disjoint chunks, HKGC uses them as initial entry anchors in the unified graph $\mathcal{G}$. Through the provenance edges E_{cross}, we activate the associated semantic subgraphs in G_{sem} and trace their hierarchical lineage in G_{log}. Specifically, we expand these anchors by retrieving their explicitly connected structural ancestors and direct semantic neighbors within $\mathcal{G}$, deterministically linearizing this cohesive, structure-aware context into a prompt $\mathcal{C}^*$. For our running example, the retrieved text about SVM is thus explicitly bounded by its [Chapter 3] metadata and semantically linked to the (SVM, is used for, regression) triplet, matching the user's intent.

With the assembled context $\mathcal{C}^*$ and the original query Q, we prompt the LLM $\mathcal{M}$ to synthesize the final answer. To further improve robustness and mitigate generative drift, HKGC augments standard greedy decoding with a Best-of-N strategy. As depicted in Fig. 2, the LLM generates multiple candidate answers in parallel to form a candidate pool $A = \{a_1, \ldots, a_N\}$ (e.g., $N = 5$). Finally, a semantic re-ranking module, denoted as the *Vector Judge*, evaluates these candidates. Instead of merely measuring query-answer similarity, the Vector Judge utilizes a dedicated evaluation encoder $\psi(\cdot)$ to assess the factual alignment between each candidate a_i and the evidence-grounded prompt. It selects the optimal answer a^* that maximizes the cosine similarity with the concatenated context and query:

$$a^* = \arg\max_{a_i \in A} \frac{\psi(\text{Concat}(Q, \mathcal{C}^*)) \cdot \psi(a_i)}{\|\psi(\text{Concat}(Q, \mathcal{C}^*))\| \cdot \|\psi(a_i)\|} \tag{8}$$

This Vector Judge mechanism favors answers that are not only semantically responsive to the query but structurally grounded in the retrieved layout and knowledge evidence, ensuring reliable and faithful multi-hop reasoning.

Table 1. Statistics of the Structure-Intensive QA Benchmarks. Each benchmark induces a high-density heterogeneous graph with tens of thousands of nodes and relations, forming a long-range multi-hop reasoning space over long-form domain texts.

Dataset	QA Pairs	Chunks	Nodes	Triples	Edges
Tech-QA	100	8,299	53,387	56,288	123,265
Finance-QA	100	4,737	17,102	24,569	54,380
Law-QA	100	3,640	21,494	19,654	43,111

4 Experiments

4.1 Experimental Setup

Datasets. We evaluate HKGC on three new domain-specific QA datasets—Tech-QA, Finance-QA, and Law-QA—sourced from long-form documents. As shown in Table 1, each benchmark features a large-scale heterogeneous graph comprising tens of thousands of nodes and relations. To evaluate structural complexity and long-range dependency modeling, our core evaluation focuses on a diagnostic set of 100 multi-hop questions per domain. Additionally, to facilitate fine-grained hop-level analysis, we curate a parallel subset of 100 single-hop questions per domain to evaluate baseline retrieval capabilities. These questions are synthesized from raw text using adapted protocols from HotpotQA [43] and RAGAS [6], followed by expert verification. Notably, for the core multi-hop instances, we adopt a global random sampling strategy to select source chunks, resulting in a $> 99\%$ probability that paired evidence chunks are spatially separated, thereby explicitly enforcing cross-segment hierarchical reasoning.
Baselines. We compare HKGC against representative baselines across three categories. All methods utilize the same 4-bit quantized Qwen-2.5-32B-Instruct as the reader model: (1) *naive baselines*: *w.o documents* (closed-book) and *w. all documents* (feeding the entire unabridged text); (2) *standard retrievers*: BM25 [28], E5 [37], and BGE-Large-en-v1.5 [39]; (3) *reasoning-enhanced models*: IRCoT [34], GraphRAG [5], ToG [31], and DP [23]. Unless specified, retrieval-based baselines select the top-K ($K = 8$) chunks based on relevance scores.
Evaluation. We report Source Recall (SR) and F1 score for retrieval coverage and lexical accuracy. For generation quality, we introduce the Generation

Quality Score (GQS), where an LLM judge (Qwen-2.5-72B) evaluates predicted answers on a 15 scale based on entity coverage and factual correctness. To assess computational efficiency, we measure token consumption (average context tokens fed to the LLM) and inference latency.

Implementation Details. We deploy the 4-bit quantized Qwen-2.5-32B-Instruct via Ollama for zero-shot inference across all generation tasks. The cross-encoder discriminator f_θ is implemented using BGE-Reranker-Large, with the entailment threshold $\tau_{\text{score}} = 0.8$ and a relation length limit of 15 tokens. Dense vectors are built using BGE-Large-en-v1.5 and FAISS. For answer generation, we apply Best-of-N decoding ($N = 5$).

Table 2. Performance of HKGC and baselines on the core multi-hop test sets across three domain-specific QA benchmarks. We report Source Recall (SR), F1, Generation Quality Score (GQS), and average input Tokens (#Tok). * indicates that SR is the highest among all methods except the *w. all docs* setting. The best performance per dataset per metric is marked in boldface.

Method	Tech-QA				Fin-QA				Law-QA			
	SR	F1	GQS	#Tok	SR	F1	GQS	#Tok	SR	F1	GQS	#Tok
Qwen-2.5-32B Reader with Context from Simple Heuristics												
w/o docs	-	30.79	3.37	54	-	18.29	3.29	38	-	30.03	3.45	45
w. all docs	93.67	14.63	2.97	28.7k	93.68	7.06	2.20	27.2k	80.43	17.68	2.64	26.4k
Qwen-2.5-32B Reader with Context from Standard Retrievers												
Naive (BM25)	49.37	35.67	2.96	2.5k	71.58	32.12	2.39	2.3k	59.78	39.94	2.80	2.2k
Naive (E5)	54.43	17.84	3.10	2.5k	68.42	32.18	2.32	2.3k	43.48	39.47	2.92	2.2k
Std RAG (BGE)	56.96	19.92	2.96	2.5k	70.53	31.50	2.59	2.3k	39.13	26.29	2.93	2.2k
Qwen-2.5-32B Reader with Context from Reasoning and Graph												
IRCoT(2023)	56.96	38.83	3.51	3.7k	70.53	27.00	2.87	3.1k	39.13	38.32	2.84	3.3k
GraphRAG(2024)	41.77	31.01	3.75	7.7k	53.68	22.52	3.12	5.1k	59.78	34.58	3.57	7.4k
ToG(2024)	41.14	30.08	3.39	5.4k	40.22	31.66	3.32	6.5k	44.74	19.41	3.23	6.3k
DP(2025)	35.99	40.57	3.42	5.2k	54.23	**45.26**	**3.53**	6.4k	41.51	37.59	3.32	6.5k
HKGC (Ours)	89.87*	**41.09**	**3.88**	2.7k	88.42*	32.75	3.34	2.5k	78.26*	**40.71**	**3.70**	2.4k

4.2 Results and Analysis

We provide our main experimental results in this section.

(RQ1): How Does HKGC Perform Compared to the Baselines?. Table 2 reports the core performance on three domain-specific multi-hop QA benchmarks. Across all datasets, HKGC achieves state-of-the-art or competitive results, demonstrating a robust balance between evidence coverage, answer accuracy, and end-to-end efficiency.

On Tech-QA, HKGC outperforms standard retrieval pipelines, more than doubling the F1 score of Std RAG-BGE (41.09% vs. 19.92%) with only a marginal token overhead. Furthermore, although explicitly omitted from the main table due to space constraints, our empirical measurements confirm that this token efficiency directly translates to fast response times. For instance, HKGC maintains an average inference latency of 7.88 s, comparable to naive dense retrieval but faster than the *w. all docs* setting (8.63 s). The full-document setting, despite achieving the highest Source Recall, consistently suffers from noise accumulation, leading to high token costs (e.g., 28.7k) and substantially lower F1.

When compared against reasoning-intensive baselines, HKGC consistently outperforms GraphRAG and ToG in both F1 and GQS. Crucially, by effectively filtering structurally irrelevant subgraphs, HKGC accelerates inference by up to 10.1% compared to heavy graph baselines on Tech-QA, while requiring significantly fewer context tokens. On Fin-QA, although DP attains higher peak accuracy and faster response—likely due to its specialized deliberation on dense numerical data—it consumes more than 2.5× the context length of HKGC. This resource overhead renders DP less viable for edge devices or large-scale inference scenarios, where memory constraints and token efficiency are critical. In contrast, HKGC reverses this trend on Tech-QA and Law-QA, significantly outperforming DP in accuracy while maintaining competitive latency. Overall, these results validate that our unified framework successfully balances high retrieval accuracy with strict token economy, providing a robust, fast, and cost-effective solution for complex long-document QA.

Table 3. F1 scores (%) of HKGC and baselines on single-hop (Sgl.) and multi-hop (Mul.) evaluation subsets across Tech-QA, Fin-QA, and Law-QA. The best performance per column is marked in boldface.

Method	Tech-QA		Fin-QA		Law-QA	
	Sgl.	Mul.	Sgl.	Mul.	Sgl.	Mul.
BM25	38.35	35.67	33.05	32.12	43.78	39.94
Standard RAG (BGE)	31.36	19.92	34.49	31.50	23.78	26.29
IRCoT	**46.79**	38.83	26.41	27.00	**44.82**	38.32
GraphRAG	38.26	31.01	28.28	22.52	34.82	34.58
ToG	37.47	30.08	32.79	31.66	26.60	19.41
HKGC (Ours)	44.71	**41.09**	**35.06**	**32.75**	43.63	**40.71**

(RQ2): How Does HKGC Perform Under Increasing Reasoning Complexity?. Table 3 breaks down the F1 performance on single-hop and multi-hop questions. The results reveal a clear trend: while advanced iterative retrievers (e.g., IRCoT) occasionally achieve peak performance on simpler single-hop queries, HKGC demonstrates superior robustness and establishes state-of-the-art accuracy across all multi-hop scenarios.

For single-hop questions, locating a localized, isolated piece of evidence is often sufficient. Consequently, models like IRCoT excel through direct, chunk-level semantic matching. However, as reasoning complexity increases, these baselines experience significant performance degradation. Standard dense retrievers and iterative methods suffer notable performance drops because they intrinsically treat long documents as flat collections of text blocks. Without explicit structural awareness, they are vulnerable to semantic gaps and distractor chunks. Furthermore, even advanced graph-based and exploratory baselines like ToG and GraphRAG struggle significantly with multi-hop queries (e.g., ToG's performance drops to 19.41 on Law-QA). This vulnerability arises because their graph construction and reasoning pathways often discard the vital hierarchical layout context, reducing them to flat semantic networks unable to reliably chain together scattered facts.

In contrast, HKGC consistently maintains high stability, experiencing only minimal degradation when transitioning to complex multi-hop queries. This resilience directly stems from our dual-index design: explicitly modeling document hierarchy and semantic provenance via the unified graph $\mathcal{G}$ preserves the global layout context. This unified representation acts as a crucial structural anchor, preventing reasoning chains from drifting. It effectively restricts the search space and guides the LLM to aggregate multi-hop evidence along accurate hierarchical pathways. Ultimately, these results validate that unifying logical topology with semantic knowledge is essential for robust multi-step inference.

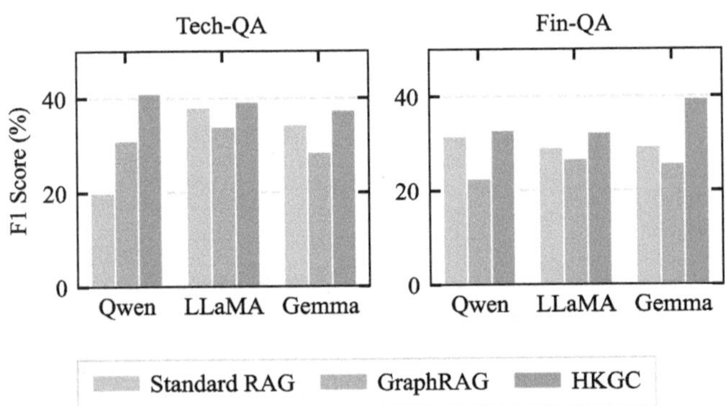

Fig. 3. Performance of HKGC with different readers on the test sets of Tech-QA and Fin-QA.

(RQ3): Does HKGC Generalize Across Different Reader Models?. Figure 3 compares performance under different LLM backbones (Qwen-2.5, LLaMA-3, and Gemma). HKGC consistently outperforms Standard RAG across all readers, demonstrating strong model-agnostic generalization. Compared to GraphRAG, whose performance heavily fluctuates across backbones, HKGC exhibits stable gains, achieving up to a 106% relative F1 improvement over Standard RAG on Tech-QA equipped with Qwen-2.5. Moreover, HKGC effectively mitigates the severe performance variance seen in baselines. For instance, on Tech-QA, while Standard RAG exhibits a ∼18% performance gap between reader models (e.g., LLaMA vs. Qwen), HKGC narrows this variance to under 4%. Even factoring in the inherent domain sensitivities of different LLMs on Fin-QA (where Gemma typically excels), HKGC consistently elevates the overall performance floor, indicating that structure-aware filtering and hierarchical grounding deliver robust reasoning independent of the underlying LLM's raw capabilities.

Table 4. Ablation results on accuracy (F1) and efficiency (Avg. Tokens) across Tech-QA and Fin-QA. ↓ denotes the reduction in token cost of HKGC relative to the *w/o Hierarchy* baseline, highlighting the filtering effect of hierarchical constraints.

Method	Tech-QA		Fin-QA	
	F1	Avg. Tokens	F1	Avg. Tokens
w/o Documents	30.79	54	18.29	38
w/o Graph	19.92	2,503	31.50	2,282
w/o Hierarchy	40.98	3,698	32.69	3,904
HKGC (Ours)	**41.09**	**2,689** (↓27.3%)	**32.75**	**2,527** (↓35.3%)

(RQ4): What is the Cost–performance Trade-off of HKGC?. Table 4 reports the ablation results, revealing how HKGC independently balances accuracy and efficiency. First, removing the semantic graph (*w/o Graph*, which degrades to standard dense retrieval) causes a substantial drop in F1 (e.g., dropping from 41.09% to 19.92% on Tech-QA). This confirms that explicitly modeling relational links is crucial for resolving complex multi-hop queries. However, incorporating this semantic graph without logical structural boundaries (*w/o Hierarchy*) causes the reasoning space to expand significantly. While this variant preserves high accuracy (40.98% on Tech-QA), it expands the context window, substantially increasing token consumption to 3,698 tokens. By unifying both dimensions, the full HKGC framework resolves this dilemma. It achieves comparable or even slightly higher accuracy than the *w/o Hierarchy* setting, while significantly reducing average token usage by 27.3% on Tech-QA and 35.3% on Fin-QA. This demonstrates that hierarchical document modeling acts as an optimal structural filter—eliminating structurally irrelevant evidence and condensing reasoning paths. Ultimately, HKGC delivers a superior accuracy–efficiency trade-off, proving its practical viability for scalable long-document QA.

(RQ5): How Does the Number of Retrieved Text Chunks (K) Affect Reasoning Performance?. Figure 4 illustrates the impact of varying K on reasoning accuracy. As K increases from 1 to 8, F1 consistently improves, indicating that enhanced evidence coverage directly benefits multi-hop reasoning. However, beyond this peak, further enlarging the context (e.g., at $K = 10$) triggers a decline in F1 despite consistently higher Recall. This performance drop is attributed to increased contextual noise and the LLM's vulnerability to distractor chunks. Overall, $K = 8$ effectively avoids this noise accumulation while capturing sufficient structural evidence, achieving the best trade-off. Therefore, it is adopted as our default setting.

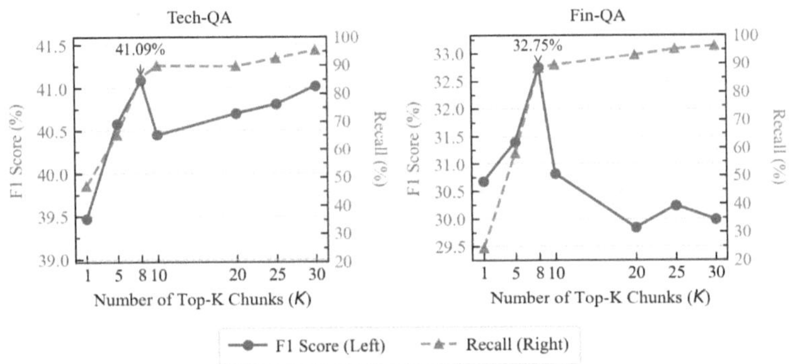

Fig. 4. Impact of the number of retrieved chunks (K) on HKGC performance. F1 scores peak around $K = 8$ on both Tech-QA and Fin-QA, achieving an optimal trade-off between evidence coverage (Recall) and reasoning precision.

5 Conclusion

In this paper, we introduce HKGC, a novel framework that resolves the structural-semantic decoupling problem in complex RAG systems through a specific architectural pipeline. Rather than merely fusing two modalities, HKGC uniquely combines deterministic Finite State Machine (FSM) parsing with a rigorous dual-stage generate-discriminate paradigm. Specifically, HKGC transforms implicit document layouts into explicit graph constraints and accurately grounds semantic units in their logical provenance. Given this unified representation, HKGC employs a structure-aware context assembly to filter structurally irrelevant noise and prevent reasoning drift. Experimental results on three domain-specific QA benchmarks show that HKGC achieves strong performance, outperforming baseline methods with an average F1 improvement of 19.5% (reaching up to 34.5% in technical domains). Moreover, explicit hierarchical modeling optimizes efficiency, condensing average context consumption to only 2,533 tokens and reducing end-to-end latency by up to 10.1% compared to graph-heavy baselines. Ultimately, these findings demonstrate that explicit hierarchical constraints are a crucial component for scalable, precision-sensitive long-document QA. Admittedly, because HKGC relies on layout cues, its applicability to flat, unstructured texts (e.g., plain web articles) remains a limitation regarding broader generalization. While our current deterministic parsing processes digitally native text to ensure zero-latency execution without optical character recognition (OCR) overhead, we acknowledge the richness of visual grammars in document analysis. Moving forward, integrating state-of-the-art vision-based document image segmentation to construct multimodal logical graphs represents a promising direction for our future work.

Acknowledgments. This work is supported by Hainan Province Science and Technology Special Fund under Grant ZDYF2024GXJS008, supported by Hainan Provincial Natural Science Foundation of China under Grant 825RC742, and supported by Hainan University Scientific Research Fund under Grant KYQD(ZR)23182.

Disclosure of Interests. The authors have no competing interests to declare that are relevant to the content of this article.

References

1. Asai, A., Wu, Z., Wang, Y., Sil, A., Hajishirzi, H.: Self-RAG: learning to retrieve, generate, and critique through self-reflection. In: International Conference on Learning Representations (2024)
2. Bian, H.: LLM-empowered knowledge graph construction: a survey (2025). arXiv:2510.20345 arXiv preprint
3. Chen, M., et al.: ReSearch: learning to reason with search for LLMs via reinforcement learning. In: Advances in Neural Information Processing Systems (2025)
4. Chen, Y., et al.: Graph-based document structure analysis. In: International Conference on Learning Representations (2025)
5. Edge, D., et al.: From local to global: a graph RAG approach to query-focused summarization (2024). arXiv:2404.16130 arXiv preprint
6. Es, S., James, J., Anke, L.E., Schockaert, S.: RaGas: automated evaluation of retrieval augmented generation. In: Proceedings of the Conference of the European Chapter of the Association for Computational Linguistics, pp. 150–158 (2024)
7. Ezzabady, M.K., Benamara, F.: Entity quality enhancement in knowledge graphs through LLM-based question answering. In: Proceedings of the Workshop on Generative AI and Knowledge Graphs, pp. 136–145 (2025)
8. Fang, J., Meng, Z., MacDonald, C.: TRACE the evidence: constructing knowledge-grounded reasoning chains for retrieval-augmented generation. In: Findings of the Association for Computational Linguistics: EMNLP, pp. 8472–8494 (2024)
9. Gao, Y., et al.: Retrieval-augmented generation for large language models: a survey. arXiv preprint arXiv:2312.10997 (2023)
10. Guu, K., Lee, K., Tung, Z., Pasupat, P., Chang, M.: Retrieval augmented language model pre-training. In: International Conference on Machine Learning, pp. 3929–3938 (2020)
11. He, K., Gkioxari, G., Dollár, P., Girshick, R.: Mask R-CNN. In: Proceedings of the IEEE International Conference on Computer Vision, pp. 2961–2969 (2017)
12. Huang, H., et al.: Retrieval-augmented generation with hierarchical knowledge. In: Findings of the Association for Computational Linguistics: EMNLP, pp. 6044–6060 (2025)
13. Huang, Y., Lv, T., Cui, L., Lu, Y., Wei, F.: LayoutLMV3: pre-training for document AI with unified text and image masking. In: Proceedings of the ACM International Conference on Multimedia, pp. 4083–4091 (2022)
14. Izacard, G., et al.: Atlas: few-shot learning with retrieval augmented language models. J. Mach. Learn. Res. **24** (2023)

15. Jacovi, A., et al.: A chain-of-thought is as strong as its weakest link: a benchmark for verifiers of reasoning chains. In: Proceedings of the Annual Meeting of the Association for Computational Linguistics, pp. 4615–4634 (2024)
16. Jia, R., Zhang, B., Méndez, S.J.R., Omran, P.G.: StructRAG: structure-aware RAG framework with scholarly knowledge graph. In: Companion Proceedings of the ACM Web Conference (2025)
17. Jiang, H., et al.: LongLLMLingua: accelerating and enhancing LLMs in long context scenarios via prompt compression. In: Proceedings of the Annual Meeting of the Association for Computational Linguistics, pp. 1658–1677 (2024)
18. Gutierrez, J., Shu, B., Gu, Y., Yasunaga, Y., Su, M., Y.: HippoRAG: neurobiologically inspired long-term memory for large language models. In: Advances in Neural Information Processing Systems, pp. 59532–59569 (2024)
19. Karpukhin, V., et al.: Dense passage retrieval for open-domain question answering. In: Proceedings of the Conference on Empirical Methods in Natural Language Processing, pp. 6769–6781 (2020)
20. Lewis, P., et al.: Retrieval-augmented generation for knowledge-intensive NLP tasks. In: Advances in Neural Information Processing Systems (2020)
21. Li, X., Dong, J., Wong, R.: From surface to semantics: semantic structure parsing for table-centric document analysis (2025). arXiv:2508.10311 arXiv preprint
22. Ma, C., Chen, Y., Wu, T., Khan, A., Wang, H.: Large language models meet knowledge graphs for question answering: Synthesis and opportunities. In: Proceedings of the Conference on Empirical Methods in Natural Language Processing, pp. 24589–24608 (2025)
23. Ma, J., et al.: Deliberation on priors: trustworthy reasoning of large language models on knowledge graphs. Adv. Neural. Inf. Process. Syst., (2025)
24. Nguyen, M.V., et al.: Direct evaluation of chain-of-thought in multi-hop reasoning with knowledge graphs. In: Findings of the Association for Computational Linguistics: ACL, pp. 2862–2883 (2024)
25. Niu, S., et al.: Tree-KG: an expandable knowledge graph construction framework for knowledge-intensive domains. In: Proceedings of the Annual Meeting of the Association for Computational Linguistics, pp. 18516–18529 (2025)
26. Qian, H., Zhang, P., Liu, Z., Mao, K., Dou, Z.: MemoRAG: moving towards next-gen RAG via memory-inspired knowledge discovery (2024). arXiv:2409.05591 arXiv preprint
27. Ram, O., et al.: In-context retrieval-augmented language models. Trans. Assoc. Comput. Ling. **11**, 1316–1331 (2023)
28. Robertson, S.E., Zaragoza, H.: The probabilistic relevance framework: BM25 and beyond. Found. Trends Inf. Retr. **3**(4), 333–389 (2009)
29. Sarthi, P., Abdullah, S., Tuli, A., Khanna, S., Goldie, A., Manning, C.D.: RAPTOR: recursive abstractive processing for tree-organized retrieval. In: International Conference on Learning Representations (2024)
30. Sourati, Z., et al.: LAD-RAG: layout-aware dynamic rag for visually-rich document understanding (2025). arXiv:2510.07233 arXiv preprint
31. Sun, J., et al.: Think-on-graph: deep and responsible reasoning of large language model on knowledge graph. In: International Conference on Learning Representations (2024)
32. Tan, X., Wang, X., Liu, Q., Xu, X., Yuan, X., Zhang, W.: Paths-over-Graph: knowledge graph empowered large language model reasoning (2025). arXiv preprint
33. Trivedi, H., Balasubramanian, N., Khot, T., Sabharwal, A.: MuSiQue: multihop questions via single-hop question composition. In: Transactions of the Association for Computational Linguistics (2022)

34. Trivedi, H., Balasubramanian, N., Khot, T., Sabharwal, A.: Interleaving retrieval with chain-of-thought reasoning for knowledge-intensive multi-step questions. In: Proceedings of the Annual Meeting of the Association for Computational Linguistics, pp. 10014–10037 (2023)
35. Vrandečić, D., Krötzsch, M.: WikiData: a free collaborative knowledgebase. Commun. ACM **57**(10), 78–85 (2014)
36. Wang, D., et al.: DocLLM: a layout-aware generative language model for multimodal document understanding. In: Proceedings of the Annual Meeting of the Association for Computational Linguistics, pp. 8529–8548 (2024)
37. Wang, L., et al.: Text embeddings by weakly-supervised contrastive pre-training. arXiv preprint arXiv:2212.03533 (2022)
38. Wang, S., Zhou, Y., Fang, Y.: BookRAG: a hierarchical structure-aware index-based approach for retrieval-augmented generation on complex documents (2025). arXiv:2512.03413 arXiv preprint
39. Xiao, S., Liu, Z., Zhang, P., Muennighoff, N., Lian, D., Nie, J.Y.: C-Pack: packed resources for general Chinese embeddings. In: Proceedings of the International ACM SIGIR Conference on Research and Development in Information Retrieval, pp. 641–649 (2024)
40. Xiong, S., Yang, Y., Fekri, F., Kerce, J.C.: TILP: differentiable learning of temporal logical rules on knowledge graphs (2023). arXiv preprint
41. Xu, P., et al.: Retrieval meets long context large language models. In: International Conference on Learning Representations (2024)
42. Xu, Y., Li, M., Cui, L., Huang, S., Wei, F., Zhou, M.: LayoutLM: pre-training of text and layout for document image understanding. In: Proceedings of the ACM SIGKDD International Conference on Knowledge Discovery and Data Mining (2020)
43. Yang, Z., et al.: HotpotQA: a dataset for diverse, explainable multi-hop question answering. In: Proceedings of the Conference on Empirical Methods in Natural Language Processing, pp. 2369–2380 (2018)
44. Yoran, O., Wolfson, T., Ram, O., Berant, J.: Making retrieval-augmented language models robust to irrelevant context. In: International Conference on Learning Representations (2024)
45. Zhao, R., Li, X., Joty, S., Qin, C., Bing, L.: Verify-and-edit: a knowledge-enhanced chain-of-thought framework. In: Proceedings of the Annual Meeting of the Association for Computational Linguistics, pp. 5823–5840 (2023)

Complex Layout Classification in the Wild: A Low-Resource Approach with Layout-Preserving Augmentations

Sharva Gogawale(✉), Iddo Hakim, Gal Grudka, Mohammad Suliman, Omer Ventura, Daria Vasyutinsky-Shapira, Berat Kurar-Barakat, and Nachum Dershowitz

School of Computer Science and AI, Tel Aviv University, Ramat Aviv, Tel Aviv, Israel
{sharvag,iddoh,galgrudka,suliman,omerventura}@mail.tau.ac.il,
{dariashap,berat,nachumd}@tauex.tau.ac.il

Abstract. Many digitized corpora suffer from low resources because annotations may be scarce, page scans are noisy and of poor resolution, or layouts are structurally complex in ways that negatively affect the quality of automatic transcription. Developing robust classification models for low-resource languages is inhibited by the lack of large-scale annotated data and by the frequent semantic complexity of page layouts. To this end, we have curated a complex-layout dataset, manually classified into eight distinct layout types based on their separator regions. To overcome data scarcity, we propose a novel training strategy in the form of a CNN-based classifier that employs strong, domain-aware augmentations to improve generalization. We utilize narrow anisotropic Gaussian masking to suppress incidental textual details while preserving essential separations, compelling the model to learn global geometric arrangements. Additionally, we implement reflection-induced label transformations to enrich the training distribution while maintaining label consistency across asymmetric categories. The results demonstrate that layout-specific augmentations can substantially improve page-level layout classification under severe annotation scarcity.

Keywords: Document layout analysis · complex layout classification · historical documents · data augmentation · low-resource learning

1 Introduction

Document layout analysis is a fundamental task in document understanding, which serves as a critical prerequisite for downstream pipelines, such as optical character recognition (OCR), automated transcription [16], entity extraction, and digital reconstruction of reading orders. Within layout analysis, layout classification refers to the task of assigning the layout of a page to a specific category. This is especially useful for historical documents, as it allows each page to be directed to the best processing pipeline. Then, later steps (e.g. segmentation, OCR, and reading-order reconstruction) can be adapted to the specific layout.

G. A. Fink et al. (Eds.): ICDAR 2026, LNCS 16972, pp. 540–557, 2027.
https://doi.org/10.1007/978-3-032-36023-6_31

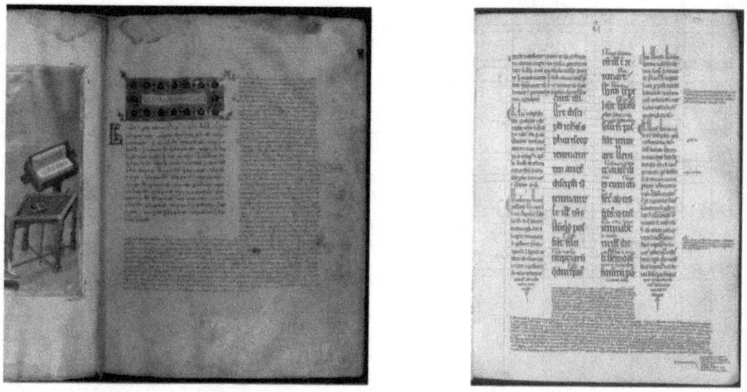

Fig. 1. Sample parchment pages with complicated layouts. Left: Greek Gospels with catena (10th c. Byzantium), National Library of Greece. Right: Gospels in Latin with glosses (England, 1300–1350), Cambridge, Queens' College, MS 26.

Modern documents such as digital PDFs, invoices, or scientific papers usually follow clean and well-organized grid structures, often called "Manhattan" layouts. In contrast, historical documents are much more irregular and complex, and often contain a high level of visual noise. This makes layout classification not merely a visual segmentation task but an important preliminary step in separating the logical structures from physical attributes. Correspondingly, layout analysis can be broadly categorized into visual-based physical layout analysis and semantics-based logical analysis [7]. Visual-based analysis relies on appearance and geometric cues to differentiate regions with different visual attributes, whereas logical analysis builds upon this segmentation by organizing regions according to the semantic role and relationships expressed in the text [15].

Complex layouts (Fig. 1) are often found in medieval manuscripts before the era of print (until the mid-16th century), and are present in recently printed books, too. Developing robust models for classifying such layouts can support layout-aware segmentation and OCR pipelines for both manuscripts and printed books. Because of the uniformity of printed scripts, it is easier to first solve this problem for printed books, and then proceed to manuscripts.

Two natural layouts for a page containing a main text along with a translation or commentary are two columns or top and bottom. We refer to these layouts, as well as one region per page, as "simple." But more complex options suggest themselves, especially when the main part is considerably shorter than the supplementary parts. Rather than two vertical columns, the supplement may be wrapped around the main text in several ways: L-shaped and its mirror image (frequently on facing pages); C-shaped and its mirror image (already attested in the 8th century Codex Zacynthius, containing the Gospel of Luke with a running catena commentary); O-shaped with the main text in the center. These arrangements culminate in the Glossa Ordinaria or "Talmudic" layout (included in our catch-all Y layout), with three (or more) main components: a central text, with one commentary on the outside (L or C) and another on the inside (L or C), with

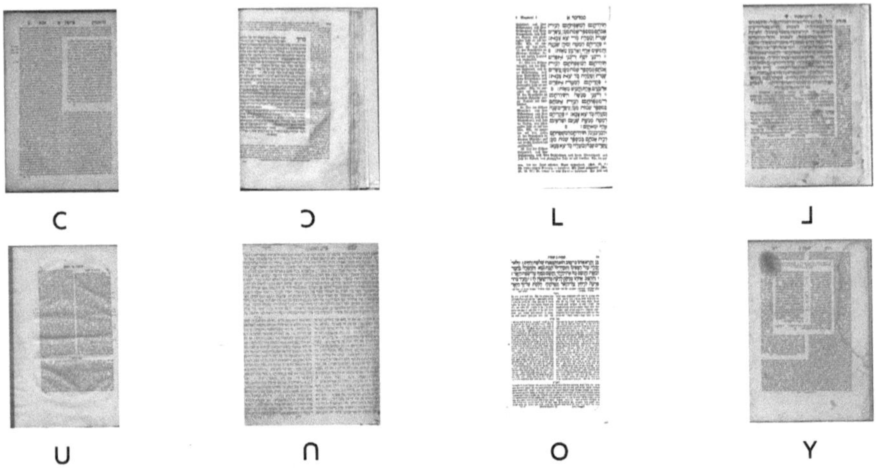

Fig. 2. Samples of complex layout classes.

facing pages usually having a mirrored layout. Also, instead of a full-width main text on the top of the page with upper and lower texts, the upper alone may be split into two columns in what we call the U format, or likewise, the lower text may be split in an inverted U format. When there are three components, top, bottom, and middle, one or more may be split into two columns. This variety of layouts continued to be used when books with such texts were printed. Figure 2 presents illustrative examples of (Hebrew) books for each class.

Of special interest for manuscripts is the layout of the colophon page. These are unique for every manuscript, but their classification, as well as that of the many other decorative formats (e.g. *tashqil* in Samaritan manuscripts or micrography in Hebrew ones), are beyond the scope of this paper.

Despite strong progress in document layout analysis, most benchmark datasets focus on layout analysis for English-language articles and similarly standardized document types. The recent advancements in deep learning-based layout analysis have been driven by large-scale annotated benchmarks like PubLayNet [17] and DocLayNet [9], which predominantly focus on English-language scientific articles or modern document types. For instance, in [11], document page classification is addressed using deep neural networks that leverage visual layout cues to distinguish broad structural categories. However, these improvements do not easily apply to low-resource historical collections with complex layouts. Fully supervised methods need large amounts of annotated data and often fail to handle unseen layout structures. Digitizing such material faces many challenges, and supervised methods trained on available datasets struggle with the unseen, non-rectangular layouts of medieval manuscripts.

It has been shown that pixel-based segmentation approaches often perform poorly when dealing with complex document layouts, especially when simple structural assumptions fail to capture the diverse organization of the page [3].

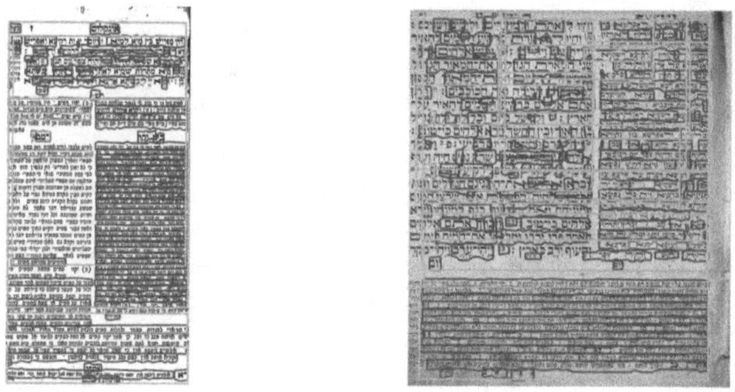

Fig. 3. Complex layout samples with faulty segmentation.

Consequently, performance degradation is commonly observed in datasets containing intricate layout configurations. Even popular open-source text recognition environments like kraken and eScriptorium [14], which are designed for historical and non-Latin scripts, rely mostly on pixel-level segmentation. This makes it hard for them to reliably separate meaningful regions in documents with complex layouts. As illustrated in Fig. 3, these systems often fail to distinguish between semantically distinct text blocks in complex C- or L-shaped configurations. Thus, there is a critical need for robust classification models capable of learning from minimal samples without requiring massive-scale annotation efforts. These limitations motivate methods that can learn from tiny annotated datasets while focusing on stable geometric cues such as separator structure rather than overfitting to incidental textual content or style.

To address these challenges, we study complex-layout classification in a low-resource historical setting by curating a compact dataset and proposing a training strategy that emphasizes layout-defining separator geometry. The main contributions are as follows: (1) We curate the CLC dataset, a specialized corpus of 155 high-resolution Hebrew pages manually annotated with eight distinct, semantically complex layout classes. (2) We propose a *Narrow Anisotropic Gaussian Augmentation* strategy that suppresses incidental textual details while preserving essential separator structures, compelling the model to learn global geometric arrangements rather than overfitting to font or content. (3) We provide a systematic evaluation under severe data scarcity, including backbone comparison, augmentation ablation, external blind-corpus evaluation and a preliminary label-efficiency analysis.

2 Related Work

Layout Analysis Benchmark Datasets. Layout analysis has benefited substantially from the release of large-scale annotated benchmarks. PubLayNet [17] is a widely used public dataset built primarily from modern scientific articles, enabling strong supervised training for document segmentation and structural understanding. DocLayNet [9] expands this direction by covering more diverse and visually complex layouts drawn from a variety of public document sources. The PRImA [1] layout analysis dataset is comprised of heterogeneous document types (e.g. magazines and technical articles), comprising 305 pages grouped into six categories, namely, Financial Reports, Manuals, Scientific Articles, Laws & Regulations, Patents, and Government Tenders; the majority of documents are in Latin scripts. More recently, M^6Doc [3] explicitly targets diversity in both layout and language, including multi-layout settings (rectangular, Manhattan, non-Manhattan, and multi-column Manhattan) and multilingual content (Chinese and English). Collectively, these datasets have shaped the dominant evaluation protocols and model designs for layout analysis under modern, well-resourced conditions.

Historical Layout Datasets. Several datasets have been compiled to address historical documents with unique challenges due to noise, degradation, and layout complexity. NewsEye [5] comprises Austrian newspaper pages from the 19th and early 20th century with carefully corrected text, including 148 training pages and 13 validation pages, with ground truth created using Transkribus by the Austrian National Library. DIVA-HisDB [13] includes 150 pages from three medieval Latin manuscripts from the Carolingian period. A Brazilian historical newspaper dataset contains 140 page images sourced from the Brazilian National Digital Library (BND) [11]. For non-Latin scripts, SCUT-CAB [2] addresses the complex layouts of ancient Chinese books with 4000 manually annotated images, while the HJDataset [12] provides a large-scale resource for historical Japanese documents, including hierarchical structures and reading-order information. These datasets highlight the growing recognition that historical layout analysis requires specialized resources; they also underscore how fragmented the landscape remains across scripts and traditions. U-DIADS-Bib [19] introduced a few-shot benchmark for ancient manuscript layout segmentation. The ICDAR 2025 FEST competition [18] targeted few-shot text-line segmentation of handwritten documents. Our focus is on page-level separator-driven layout classification as an upstream routing step before segmentation and OCR.

Page-Level Classification. Approaches to page-level layout classification have evolved from feature engineering to deep learning. Early work often relied on manually engineered features. A study on page layout classification for legal documents categorized pages into structural types such as single column, marginal single column, and two column formats. The results showed that feature-engineered methods (e.g. SVM and random forest) leveraging geometric layout cues achieved accuracies around 94%, emphasizing the value of explicit structural features [6]. With the advent of deep learning, convolutional neural networks (CNNs) became the dominant strategy. A prior study demonstrated that CNNs could effectively discriminate between diverse categories such as letters, memos, invoices, etc., by analyzing whole-page visual cues [1]. Additionally, frameworks for graphical object detection [10] have leveraged transfer learning and domain adaptation to identify graphical elements like tables and figures, mitigating the scarcity of annotated training data.

Hebrew Layout Resources. In the specific context of Hebrew document layout analysis, resources are notably scarce. The primary existing benchmark is NetLay [4], which contains approximately 1,300 pages from printed Hebrew (or Hebrew-alphabet) books. NetLay utilizes a multi-label encoding approach with VGG16 to classify pages into four categories, namely no-text, single-column, double-column, and complex. However, the label space remains relatively coarse with respect to separator-defined, non-rectangular historical layouts (e.g. C- or L-shaped configurations), which motivates finer-grained datasets and modeling assumptions that explicitly target separator geometry. Motivated by these gaps, our work focuses on separator-defined complex layout classification for low-resource historical Hebrew pages, complementing existing Hebrew resources (e.g. NetLay) with a finer-grained label space tailored to non-rectangular and multi-region structures. Our goal is to provide a targeted, publicly usable dataset and baseline that isolates the layout-classification problem under severe annotation constraints and emphasizes global geometric structure.

3 Complex Layout Classification (CLC) Dataset

We curate a compact, task-specific corpus to address complex layout classification with the high-level goal of improving downstream text transcription systems by routing pages to layout-aware pipelines. The dataset comprises 155 high-resolution document images sourced from the National Library of Israel's digital collections and printed books from various periods. To evaluate the efficacy of our model in a low-resource setting, we work with a small but representative dataset, a common constraint in computational humanities where large-scale annotated data are scarce.

The images are manually categorized into our eight classes. The complexity of these layouts is not merely structural but also semantic; pages often contain separate but related works by different authors, thereby necessitating models that can capture complex, multi-block hierarchical relationships. They use two main script types: square (block letters) and "Rashi" (curvilinear). To support robust learning from small data, all images are retained at their native scan resolution. Table 1 summarizes the distribution across the eight layout classes, the script types present in each, and the annotation source. Fragments in non-Hebrew characters are a common phenomenon in complex layouts, as such texts typically include commentaries or translations into local vernaculars or appear in scholarly editions.

Ground truth annotations were established through a rigorous protocol. To maximize structural diversity, candidate pages were restricted to a single representative page per printed book. Initial categorizations were performed by trained annotators, and all labels were subsequently reviewed and verified by an expert paleographer, resolving the disagreements to produce a single ground-truth label per image. The complete dataset, including the images, book identifiers, and the ground truth, is available at https://github.com/TAU-CH/midrash_clc.

Table 1. CLC dataset statistics. Sq = square (block letters); Ra = Rashi (curvilinear). Pages sourced from the National Library of Israel digital collections.

Class	Count	% Total	Script Types
C	21	13.5	Sq, Ra
Ↄ	19	12.3	Sq, Ra
L	28	18.1	Sq, Ra
⅃	22	14.2	Sq, Ra
O	22	14.2	Sq, Ra
U	16	10.3	Sq, Ra
∩	13	8.4	Sq, Ra
Y	14	9.0	Sq, Ra
Total	**155**	**100.0**	—

4 Method

We address the task of complex-layout classification, where document images are to be assigned to one of eight predefined layout types. Let $\mathcal{C}$ denote the set of layout classes, $\mathcal{C}$ = {C, Ↄ, L, ⅃, U, ∩, O, Y}. These classes correspond to page layouts containing text regions shaped as C, horizontally reflected C, L, horizontally reflected L, U, vertically reflected U, or O, with irregular region forms grouped under the Y class.

The primary challenge in classifying document images into these eight classes is the severe scarcity of annotated data, with only a handful of samples available per class. To address this limitation, we propose a CNN-based framework that employs strong augmentation strategies to improve generalization.

In layout classification, the defining feature of each class lies in its separator regions, which partition the page into characteristic geometric layouts defined in $\mathcal{C}$. In contrast, the textual content, including font type, size, or semantic information does not determine the layout category. Therefore, any augmentation procedure that preserves separator regions encourages the model to focus on the global arrangement of separators rather than incidental textual details.

In particular, we adopt two complementary families of augmentations: (1) Narrow anisotropic Gaussian based masking, which applies multi-oriented anisotropic Gaussian filters with a narrow standard deviation to suppress text regions while preserving separator regions; and (2) horizontal and vertical reflections, which enrich the diversity of training samples while ensuring label consistency in asymmetric categories. Together, these augmentations significantly expand the effective training distribution and enable robust CNN learning despite the limited availability of annotated data.

Narrow Anisotropic Gaussian Based Masking. We augment the document images using a multi-oriented anisotropic Gaussian filter that preserves sepa-

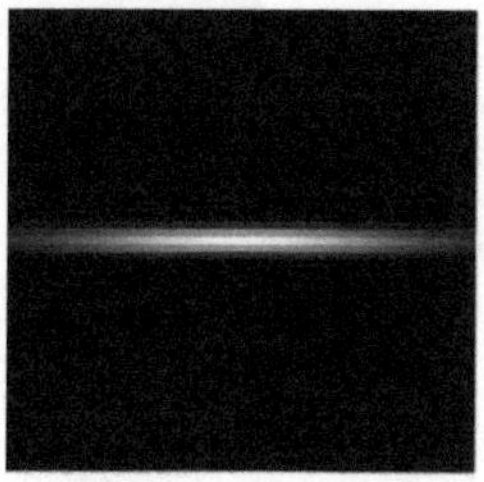

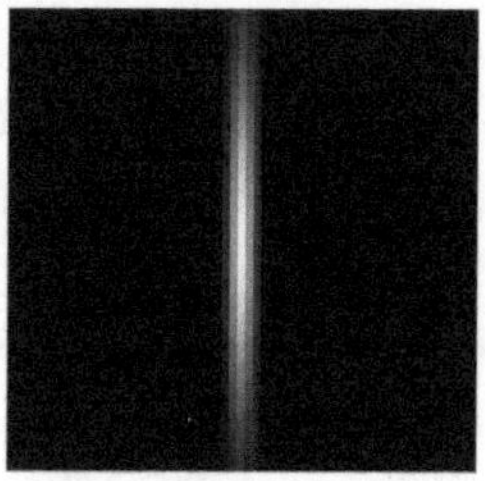

Fig. 4. Narrow anisotropic Gaussians used for masking ($\sigma_x = 1$; for visualization, $\sigma_y = 15$ is shown). Left: horizontal ($\theta = 0$). Right: vertical ($\theta = \pi/2$).

rator regions. For orientation θ and standard deviations σ_x, σ_y, the kernel is

$$G_{\sigma_x,\sigma_y,\theta}(x,y) = \frac{1}{2\pi\sigma_x\sigma_y}\exp\left(-\frac{(x\cos\theta + y\sin\theta)^2}{2\sigma_x^2} - \frac{(-x\sin\theta + y\cos\theta)^2}{2\sigma_y^2}\right)$$

Given a document image $I(x,y)$, where (x,y) are pixel coordinates, containing low-contrast text on a high-contrast background, we fix $\sigma_x = 1$ and filter $I(x,y)$ with $G_{1,\sigma_y,\theta}$ at orientations $\theta \in \{0, \pi/2\}$ (Fig. 4). For each pixel, the maximum response across the two orientations is computed to form the response map $R_{\sigma_y}(x,y)$. The directional response and the subsequent binary mask are defined as: $R_{\sigma_y}(x,y) = \max_{\theta\in\{0,\pi/2\}}(I * G_{1,\sigma_y,\theta})(x,y)$; $M_{\sigma_y}(x,y) = \mathbf{1}[\tilde{R}_{\sigma_y}(x,y) \geq t]$, where $\tilde{R}_{\sigma_y}$ denotes the min-max normalized response map scaled to $[0,1]$, and $\mathbf{1}[\cdot]$ is the indicator function. We then form the masked augmentation by preserving pixels in separator regions and suppressing the remaining content: $I'(x,y) = 255 \cdot M_{\sigma_y}(x,y) + \big(1 - M_{\sigma_y}(x,y)\big)\big(255 - I(x,y)\big)$. Pixels outside the separator mask are inverted to suppress fine-grained text texture while retaining global page structure and high-contrast separator geometry.

The use of $\sigma_x = 1$ together with multi-orientation filtering ensures that separator region pixels consistently yield high responses. In our case, we set $\theta \in \{0, \pi/2\}$, since separator regions in the dataset are oriented only vertically or horizontally. At vertical separator regions, the maximum response is produced by the vertically oriented narrow Gaussian, whereas at horizontal separator regions it is produced by the horizontally oriented narrow Gaussian. Because the filters are narrow, their placement at separator regions near text region boundaries still yields high responses without blurring, despite the presence of neighboring low-contrast pixels within the text regions (Fig. 5).

Horizontal and Vertical Reflections. To further enrich the training distribution, we apply horizontal and vertical reflections, which preserve separator regions while modifying the overall page structure. In contrast, rotations and shearings are deliberately excluded, as they alter separator orientations and thereby disrupt the structural integrity of the layout. Translations are avoided, since they may shift crucial separator regions outside the field of view, invalidating the

רבינו עובדי' : ברטנורא 34

מבדיל: ו הבדיל בחטאת. שמלק שני סימני'
כחטאת: או לא הבדיל בעולה. שמלק כס
סי אחד: פסול. דכל שנוי שסוף לפני עבודת
הדם פסול: ולא מיצה דם הגוף פסול.
דעיקר דמים בגוף סוף דמכיחי: ז חטאת
העוף שמלקה שלא לשמה. מליקה בעוף
במקום שחיטה בבהמה. ומילוי בעוף במקום
זריקת הדם בבהמה. וסממנים ספוסלת
בשחיטה וזריקת דם בבהמה. פוסלת במליקה
ובמילוי הדם בעוף אבל קבלה והלוך אין
בעוף: עולת העוף כשרה. כדתנן בריש

זבחים פרק ו

הרשן. שסע ולא הבדיל. ואם הבדיל
כשר. וספגו במלח וזרקו על גבי
האשים: ז לא הסיר לא את המוראה
ולא את הנוצה ולא את בני מעים
היוצאין עמה ולא ספגו במלח. כל
ששינה בה מאחר שמיצה את דמה
כשרה. הבדיל בחטאת ולא הבדיל
בעולה פסולה. מיצה דם הראש ולא
מיצה דם הגוף פסולה. מיצה דם הגוף
ולא מיצה דם הראש כשרה: ז חטאת
העוף שמלקה שלא לשמה. מיצה דמה

שלא לשמה. או לשמה ושלא לשמה. או שלא לשמה ולשמה. פסולה. עולת

Fig. 5. Illustration of separator region preservation in a document with an L-shaped layout. On vertical separator regions, no matter how narrow they are, a narrow vertical Gaussian (red) will lead to a high response. Similarly, on horizontal separator regions, no matter how narrow they are, a narrow horizontal Gaussian (green) will lead to a high response. (Color figure online)

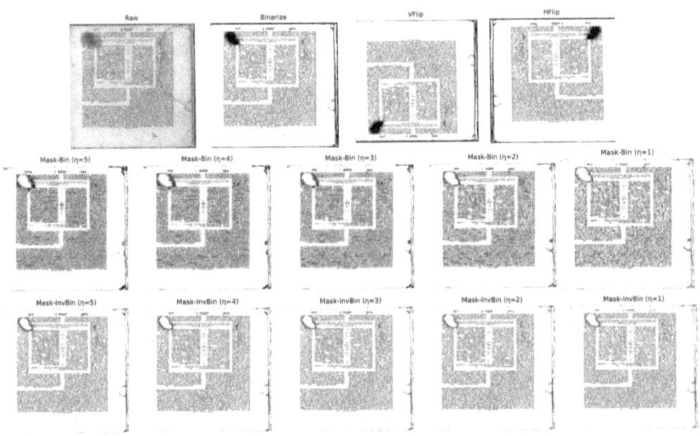

Fig. 6. Comprehensive view of the augmentation suite. A single training image is expanded into eleven distinct intensity representations (binary, masked, and inverse-masked), alongside their valid spatial reflections.

class-defining structure. And scalings are omitted because all input images are rescaled to a fixed size prior to training, rendering additional scaling redundant.

Reflections do require relabeling; however, the label set is closed under reflection, ensuring that the new labels remain within the $\mathcal{C}$. Horizontal reflection exchanges C ↔ Ɔ and L ↔ ⅃. Vertical reflection exchanges U ↔ ∩ and turns L and ⅃ into Y. Two classes, O and Y, are (usually) retained under reflection.

Augmentations. To overcome severe data scarcity, we implement a combinatorial but task-constrained augmentation suite that preserves class-defining separator topology, in contrast to generic appearance perturbations which perturb arbitrary image regions without structural awareness. For each training image, we first binarize, while also extracting an inverse-binarized variant to minimize polarity sensitivity. We then generate masking-based variants by applying our narrow anisotropic Gaussian operator across five strength settings, $\eta \in \{1, 2, 3, 4, 5\}$. This gives both binarized-masked and inverse-binarized-masked modalities, resulting in a transformed pool of $|\mathcal{T}| = 11$ distinct intensity representations per image (1 plain binary, 5 masked binary, and 5 inverse-masked binary) as illustrated in Fig. 6. Subsequently, each of the 11 modalities is subjected to a spatial reflection protocol using deterministic label mappings. Horizontal reflection is universally applied, exchanging asymmetric pairs, while preserving symmetric classes. Vertical reflection is applied exclusively to the subset of classes where structural semantics remain well-defined (U, ∩, O, C, Ɔ, Y). By enumerating all valid spatial and intensity combinations, this pipeline effectively acts as a dataset multiplier, yielding 22 to 33 training instances per original image depending on class validity under vertical reflection. Finally, to mitigate the residual class imbalance within this expanded set, we employ weighted random sampling during optimization.

ConvNeXt-Based Classifier. ConvNeXt [8] is a modernized Convolutional Network which retains inductive biases that matter for layout: locality, translation equivariance, and hierarchical composition—properties well aligned with separator-driven global geometry under scarce labels. A ConvNeXt block replaces classic ResNet bottlenecks with: (i) large-kernel depthwise convolution for long-range spatial aggregation, and (ii) pointwise "MLP-style" mixing for channel interactions, while using LayerNorm/GELU for stable optimization.

Let $x \in \mathbb{R}^{H \times W \times C}$ be an input feature map. A depthwise convolution with kernel K operates per-channel:

$$\mathrm{DWConv}(x)_{i,j,c} = \sum\nolimits_{u,v \in K} w_{u,v,c}\, x_{i+u,\, j+v,\, c}.$$

Channel mixing is then performed by pointwise projections (equivalent to 1×1 convolutions): For a channel vector $\mathbf{z}_{i,j} \in \mathbb{R}^C$ at spatial location (i, j), this operates as a linear transformation $\mathrm{PW}(\mathbf{z})_{i,j} = \mathbf{W}\mathbf{z}_{i,j}$. A simplified ConvNeXt block can be written as:

$$y = x + \mathcal{F}(x), \qquad \mathcal{F}(x) = \mathrm{PW}_2(\sigma(\mathrm{LN}(\mathrm{PW}_1(\mathrm{DWConv}(x))))),$$

where LN is LayerNorm, σ is the GELU activation, PW_1 is an expansion layer that increases the channel dimension by a factor of 4, and PW_2 is a projection layer that reduces it back to the original dimension.

For classification, the network outputs logits $g(x) \in \mathbb{R}^8$, and is trained with standard cross-entropy loss.

We fine-tune an ImageNet-pretrained ConvNeXt-Tiny, adapting the stem to 1-channel inputs by averaging RGB pretrained weights, and replacing the final classifier for 8 classes. We use *staged fine-tuning*: we train the classifier head first and then unfreeze the backbone with a smaller learning rate, which reduces overfitting early and improves stability under scarce supervision. We first establish a strong *supervised baseline* by fine-tuning standard image backbones for 8-way page-level layout classification (C, Ɔ, L, ⅃, U, ∩, O, Y).

Input images were resized to 224×224 pixels and binarized using Otsu's thresholding prior to ingestion. We utilized a stratified data split, allocating 60% to training, 20% to validation, and 20% to testing. To prevent destabilizing the pretrained weights, we employed a two-stage fine-tuning strategy for a maximum of 100 epochs using a batch size of 16. For the first 10 epochs, we froze all non-classifier parameters and trained the head using the AdamW optimizer (initial learning rate 10^{-3}, weight decay 10^{-4}) with a cosine annealing schedule ($T_{\text{max}} = 70$, $\text{lr}_{\text{min}} = 10^{-5}$). Subsequently, the entire network was unfrozen, and AdamW was reinitialized with a lower learning rate of 10^{-4} (weight decay 10^{-4}) and a new cosine schedule ($T_{\text{max}} = 90$, $\text{lr}_{\text{min}} = 10^{-6}$). Training was regulated by early stopping with a patience of 10 epochs based on validation loss.

Active Learning Protocol. We also study an active learning protocol to evaluate label efficiency in our low-resource setting, which reflects historical document analysis, where structurally complex layouts often require domain-expert annotation and labels are expensive to obtain. Therefore, even modest reductions in labeling effort are practically useful.

Let $\mathcal{D}_{\text{train}}$ be the training split. At round t, we maintain a labeled set $\mathcal{L}_t \subseteq \mathcal{D}_{\text{train}}$ and an unlabeled pool $\mathcal{U}_t = \mathcal{D}_{\text{train}} \setminus \mathcal{L}_t$. We initialize with $\rho_0 \approx 0.2$ and add $\Delta\rho \approx 0.2$ each round until the pool is exhausted. At each round, we train ConvNeXt-Tiny with our **Full** pipeline (Binary + Reflections + Anisotropic Masking) on the labeled set, then select a batch $\mathcal{S}_t \subset \mathcal{U}_t$ of size B according to an acquisition function $a_t(x)$: $\mathcal{S}_t = \arg\max_{S\subseteq\mathcal{U}_t,\,|S|=B} \sum_{x\in S} a_t(x)$, followed by $\mathcal{L}_{t+1} \leftarrow \mathcal{L}_t \cup \mathcal{S}_t$ and $\mathcal{U}_{t+1} \leftarrow \mathcal{U}_t \setminus \mathcal{S}_t$. We compare three simple acquisition rules: (i) *Random* ($a_t(x)$ uniform); (ii) *Entropy*, using predictive entropy $a_t(x) = -\sum_{c=1}^{8} p(c \mid x) \log\bigl(p(c \mid x)+\epsilon\bigr)$; and (iii) *Margin*, using the negative gap between the two highest class probabilities, $a_t(x) = -\bigl(p_{(1)}(x) - p_{(2)}(x)\bigr)$, where $p_{(1)}(x)$ and $p_{(2)}(x)$ denote the largest and second-largest entries of $p(\cdot \mid x)$. For the uncertainty rules we select the B samples with the highest $a_t(x)$.

Learning Curve Summary. Across labeled budgets $\rho_t = |\mathcal{L}_t|/|\mathcal{D}_{\text{train}}|$, we evaluate accuracy and macro-F_1 on a fixed held-out split and summarize label-efficiency using the area under the learning curve (AULC).

5 Results

Backbone Architecture Comparison. To determine the optimal architecture for complex layout classification under severe data constraints, we evaluated a diverse set of deep learning models: classical CNNs (VGG16, ResNet-18), a highly optimized mobile architecture (EfficientNet-B0), a pure vision transformer (ViT-B/16), and a modernized CNN (ConvNeXt-Tiny). All architectures were trained using our proposed Full layout-preserving pipeline (Binary+Flips+Masking) to ensure a fair comparison. Because the test set is small (31 pages), we report macro-averaged precision, recall, and F_1-score alongside accuracy to account for class imbalance.

Table 2. Left: Backbone comparison under our full layout-preserving pipeline on the CLC dataset. Right: NetLay-adapted multi-label baseline models *without* layout-preserving augmentations. Test set: 31 pages. Best in bold.

Model	Params	Acc.	P	R	F_1	NetLay	Acc.	P	R	F_1
ConvNeXt-Tiny	28.6M	**0.90**	**0.93**	**0.88**	**0.88**	EfficientNet-				
EfficientNet-B0	5.3M	0.45	0.39	0.44	0.41	V2-S	**0.52**	**0.60**	**0.48**	**0.49**
VGG16	138M	0.45	0.39	0.44	0.41	VGG16	0.45	0.42	0.42	0.40
ResNet-18	11.7M	0.23	0.16	0.20	0.17	ViT-B/16	0.23	0.30	0.19	0.17
ViT-B/16	86.6M	0.23	0.19	0.20	0.15					

Table 3. External blind evaluation on the test corpus (20 samples per class).

Class	Wrong	Right	Accuracy (%)	Class	Wrong	Right	Accuracy (%)
C	1	19	95	O	13	7	35
Ↄ	8	12	60	U	0	20	100
L	0	20	100	∩	4	16	80
⅃	0	20	100	Y	0	20	100
				Overall	26	134	83.75

Compared to vision transformers, which typically require extensive supervision to learn strong spatial priors, ConvNeXt relies less on learning global structure from scratch. The large receptive field from depthwise large-kernel convolutions allows the network to capture the global geometric arrangement of separator regions (e.g. distinguishing a U layout from an L layout), and parameter-efficient depthwise operations reduce overfitting risk and improve efficiency. This dynamic is visually confirmed in Fig. 7, where the network's attention is predominantly localized along separator regions, block boundaries, and inter-column gaps rather than dense textual content. Such qualitative evidence demonstrates that our proposed preprocessing and augmentations successfully compel the classifier to anchor its predictions on global separator geometry.

Table 2 summarizes the performance of the evaluated architectures. We additionally adapt the NetLay multi-label encoding [4] to our 8-class task (Table 2, right); the best result reaches only 52% accuracy and 0.49 macro-F_1 without layout-preserving augmentations, confirming that the separator-aware training strategy is the decisive contribution. ConvNeXt-Tiny performs substantially better than the others, achieving an accuracy of 0.90 and a macro-F_1 of 0.88, substantially outperforming EfficientNet-B0 and VGG16 (acc. 0.45, macro-F_1 0.41), as well as ResNet-18 and ViT-B/16 (macro-F_1 $\leq$ 0.17). On a per-class basis, ConvNeXt achieves perfect or near-perfect performance on most layouts (e.g. Ↄ/⅃/U), with remaining errors concentrated in the rarest and most ambiguous classes (Y and ∩), suggesting confusion among separator patterns when examples are extremely limited. In low-resource regimes, ViTs often underperform

Fig. 7. GradCAM visualizations of the ConvNeXt-Tiny classifier. Warmer colors indicate regions with higher contribution to the predicted layout class.

unless heavily regularized or trained on large datasets. ResNet-18 suffers a similar fate, likely due to its smaller receptive field and aggressive spatial downsampling, which can weaken the thin, delicate separator lines crucial for identifying medieval layout classes. The strong performance of ConvNeXt-Tiny supports our data-scarcity hypothesis: with limited samples, the model cannot rely on memorizing textures. Instead, CNN inductive biases encourage learning contiguous edges, while the ViT-inspired design helps integrate these cues into coherent global layout structures. The only notable struggle for ConvNeXt-Tiny is the Y class (recall of 0.33, $F_1 = 0.50$), which represents the most intricate layout with central text flanked by multiple distinct elements. This slight degradation highlights that, while ConvNeXt effectively models global structure from minimal data, highly fragmented layouts with inner and outer separators remain challenging for zero-to-few-shot historical document analysis.

Evaluation on a Blind Corpus. To examine how the classifier behaves outside our CLC dataset, we conducted a blind-corpus sanity check on a large corpus drawn from the book collections of the National Library of Israel (NLI). These were not used for training, validation, or hyperparameter tuning. We emphasize that this is not a fully labeled out-of-domain benchmark; rather, it is a precision-oriented stress test on high-confidence predictions in a large unlabeled archive. Inference was performed over a directory containing 36,215 book folders. For each folder, up to 10 page images were randomly sampled (or fewer when unavailable), yielding a maximum of 362,150 candidate pages. The trained classifier model (Full configuration: Binary+Flips+Masking) was applied without any fine-tuning. To focus on highly confident predictions, we retained only outputs whose maximum softmax probability exceeded 0.99. This procedure identified: 3,932 book folders containing at least one page classified as complex; and 9,339 page images satisfying the confidence threshold.

Because the corpus lacks ground truth layout annotations, we conducted a manual visual evaluation on a stratified sample of the high-confidence predictions. For each of the eight classes, 20 images were randomly selected. Each image was visually inspected and assigned its true layout class, which was then compared to the model prediction. See Table 3. Out of 160 inspected samples, 134 were correct, giving an overall accuracy of 83.75% on this verified subset. Performance was lower for Ↄ and especially for O layouts.

Table 4. Ablation results on CLC using ConvNeXt-Tiny (31 pages), showing **Acc**uracy and macro-averaged **P**recision, **R**ecall, and F_1. Best in **bold**.

Config	Acc.	P	R	F_1
Binary only	0.32	0.34	0.31	0.29
Binary+Flips	0.45	0.44	0.47	0.42
Binary+Masking	0.68	0.69	0.61	0.59
Binary+Std. Aug.	0.68	0.76	0.62	0.63
Full (Masking+Flips)	**0.90**	**0.93**	**0.88**	**0.88**

η	Acc.	P	R	F_1
η=1	0.55	0.56	0.53	0.52
η=3	0.71	0.82	0.70	0.69
η=5	0.81	0.85	0.79	0.79

Ablation. To systematically quantify the contributions of our proposed layout-preserving augmentations, we conducted an extensive ablation study using the best-performing ConvNeXt-Tiny architecture. All experiments are evaluated on the same 8-way test set (31 pages).

Ablation Settings. We consider (i) **Binary only**, which uses only binarized inputs without geometric augmentations; (ii) **Binary+Flips**, which additionally applies horizontal/vertical reflections with label-consistent remapping for asymmetric layouts; (iii) **Binary+Masking**, which uses only the proposed narrow anisotropic Gaussian masking to suppress incidental text while preserving separator structure; (iv) **Full**, which combines Binary+Flips+Masking; and (v) **Mask strength** sweeps, where we vary a masking hyperparameter $\eta \in \{1, 3, 5\}$ controlling the strength of anisotropic suppression.

Quantitative Results. Table 4 shows that Binary preprocessing alone performs poorly (acc. 0.32, macro-F_1 0.29), indicating that simple foreground/background normalization is insufficient to learn separator-defined global geometry from tiny supervision. Notably, classes with nested semantic regions (⅃ and Y) completely fail to converge (F_1: 0.00) in this low-resource regime. Reflection-based augmentation in the **Binary+Flips** strategy, together with a fixed label-transformation mapping, improves accuracy to 0.45. While this successfully expands the dataset and enforces orientation invariance (boosting C's F_1 from 0.44 to 0.62), the model remains heavily distracted by local textual textures, resulting in marginal gains for complex asymmetric layouts. We also evaluate a **Binary+Standard Augmentations** baseline using common transforms (random rotation, affine, perspective, and Gaussian blur), which achieves 0.68 accuracy and 0.63 macro-F_1. While this exceeds Binary+Flips, the full pipeline with separator-preserving masking and label-consistent reflections yields substantially stronger performance, confirming that task-aware augmentation design matters beyond generic regularization.

In contrast, **Binary+Masking** produces a substantial jump (acc. 0.68, macro-F_1 0.59), supporting the hypothesis that the task is fundamentally *separator-driven*. By stochastically suppressing high-frequency textual components while preserving the contiguous geometric paths of separators, the masking

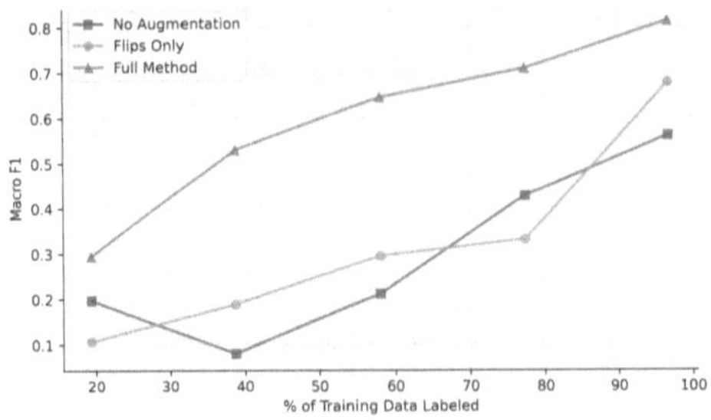

Fig. 8. Active learning under label scarcity. Macro-F_1 learning curves are shown for three augmentation settings, demonstrating that the full layout-preserving pipeline improves performance throughout the low-label regime.

forces the CNN's inductive bias to focus strictly on layout-defining topological boundaries. Hence, the model is compelled to base its predictions on global geometric structures, such as long separator strokes, block boundaries, and their spatial arrangement, rather than local content- or style-specific patterns.

Finally, combining all components (**Full**) achieves the best overall performance (acc. 0.90, macro-F_1 0.88), indicating that flips and masking are complementary: flips increase effective sample diversity and promote invariance, while masking acts as a structure-preserving regularizer that improves generalization.

Effect of Masking Strength. Varying the masking hyperparameter highlights a clear sensitivity: weaker masking (η=1) underperforms (acc. 0.55, macro-F_1 0.52), while moderate-to-strong masking improves robustness (η=3: acc. 0.71, macro-F_1 0.69; η=5: acc. 0.81, macro-F_1 0.79). This trend suggests that sufficiently strong suppression is required to consistently remove text cues; otherwise, the model partially relies on residual texture signals that do not transfer across pages. This directional filtering acts as a line-tracking mechanism: it aggressively blurs orthogonal high-frequency noise (text) while constructively reinforcing continuous low-frequency paths (separators). Nevertheless, the best performance is obtained by the **Full** configuration, implying that masking alone is not a substitute for geometric augmentation and label-consistent symmetry handling.

Minimizing Cost via Active Learning. Annotation remains a bottleneck in historical document analysis, since complex layouts often require domain-aware inspection. We therefore evaluate whether active learning can improve label efficiency when expanding CLC.

Figure 8 shows that the dominant factor in label-efficient learning is the augmentation strategy. Under entropy acquisition, the full layout-preserving pipeline achieves a macro-F_1 AULC of 0.61, compared with 0.30 for Binary+Flips and 0.28 for Binary only. This large shift indicates that separator-preserving masking and label-consistent reflections improve learning consistently across all labeling budgets, rather than only at the final training phase.

Evaluating the acquisition rules reveals a secondary effect relative to the augmentation strategy. On accuracy, the AULC is identical for Random, Entropy, and Margin acquisition (0.65), with all three converging to the same final accuracy of 0.87. However, macro-F_1 is more sensitive to the acquisition choice: Margin yields the highest AULC (0.62), compared with 0.61 for both Entropy and Random. Its advantage appears mainly in the mid-budget regime: at 60% labeled data, Margin reaches a macro-F_1 of 0.69 (compared to 0.59 for Random), and at 80% it reaches 0.76 (compared to 0.73 for Random). At the earliest post-initial budget, however, Margin slightly trails Random, suggesting that acquisition-rule differences are less reliable when the labeled set is extremely small. Overall, these results suggest that the primary label-efficiency gain comes from the separator-aware augmentation pipeline, while margin-based acquisition provides a lightweight mechanism for prioritizing structurally informative pages during dataset expansion.

6 Conclusion

We addressed the challenging task of complex-layout classification in low-resource historical documents, motivated by the practical need to route heterogeneous documents to layout-aware transcription and segmentation pipelines. To support controlled evaluation under annotation scarcity, we curated the CLC dataset, consisting of 155 high-resolution pages spanning eight separator-defined layout categories. We introduced layout-preserving augmentations that suppress incidental textual detail while retaining separator structure, and combined them with label-consistent reflection transformations to expand the effective training distribution without introducing label noise. Masking, combined with label-consistent spatial reflections, achieved 90% accuracy on our dataset. Furthermore, an external blind evaluation on the massive NLI corpus demonstrated strong generalization behavior yielding an 83.75% accuracy on a manually verified high-confidence subset, suggesting that the classifier can generalize beyond the curated dataset. Finally, our preliminary active-learning study shows that the augmentation pipeline is the primary driver of label efficiency, while margin-based acquisition provides a lightweight strategy for prioritizing structurally informative pages during dataset expansion. Future work will extend the approach to handwritten manuscript pages and integrate script-type recognition for layout-aware OCR routing.

Acknowledgments. This research was funded in part by the European Union as part of "MiDRASH" [https://www.midrash.eu] (ERC project no. 101071829, with principal investigators: Nachum Dershowitz, Tel Aviv University; Judith Olszowy-Schlanger, EPHE-PSL; Avi Shmidman, Bar-Ilan University; and Daniel Stökl Ben Ezra, EPHE-PSL). Views and opinions expressed are, however, those of the authors only and do not necessarily reflect those of the European Union or the European Research Council Executive Agency. Neither the European Union nor the granting authority can be held responsible for them.

References

1. Antonacopoulos, A., Bridson, D., Papadopoulos, C., Pletschacher, S.: A realistic dataset for performance evaluation of document layout analysis. In: ICDAR, pp. 296–300 (2009)
2. Cheng, H., Jian, C., Wu, S., Jin, L.: SCUT-CAB: a new benchmark dataset of ancient Chinese books with complex layouts for document layout analysis. In: ICFHR (2022)
3. Cheng, H., et al.: M^6Doc: a large-scale multi-format, multi-type, multi-layout, multi-language, multi-annotation category dataset for modern document layout analysis. In: CVPR, pp. 15138–15147 (2023)
4. Gogawale, S., et al.: NetLay: layout classification dataset for enhancing layout analysis. Magazén: Int. J. Digital Public Human. **5**, 1–14 (2024)
5. Hamdi, A., et al.: A multilingual dataset for named entity recognition, entity linking and stance detection in historical newspapers. In: SIGIR 2021, pp. 2328–2334 (2021)
6. Josi, F., Wartena, C., Heid, U.: Preparing legal documents for NLP analysis: improving the classification of text elements by using page features. In: 8th International Conference on Natural Language Processing (NATP), vol. 12 (2022)
7. Lee, J., Hayashi, H., Ohyama, W., Uchida, S.: Page segmentation using a convolutional neural network with trainable co-occurrence features. In: ICDAR, pp. 1023–1028 (2019)
8. Liu, Z., Mao, H., Wu, C.Y., Feichtenhofer, C., Darrell, T., Xie, S.: A ConvNet for the 2020s. In: CVPR (2022)
9. Pfitzmann, B., Auer, C., Dolfi, M., Nassar, A.S., Staar, P.W.J.: DocLayNet: a large human-annotated dataset for document-layout segmentation. In: KDD 2022, pp. 3743–3751 (2022)
10. Saha, R., Mondal, A., Jawahar, C.V.: Graphical object detection in document images. In: ICDAR, pp. 51–58 (2019)
11. Júnior, S., Paixão, E.S.D., Alvarez, T., A.B.: Comparative performance of YOLOv8, YOLOv9, YOLOv10, and YOLOv11 for layout analysis of historical documents images. Appl. Sci. **15**(6) (2025)
12. Shen, Z., Zhang, K., Dell, M.: A large dataset of historical Japanese documents with complex layouts. In: CVPR Workshops, pp. 2336–2343 (2020)
13. Simistira, F., Seuret, M., Eichenberger, N., Garz, A., Liwicki, M., Ingold, R.: DIVA-HisDB: a precisely annotated large dataset of challenging medieval manuscripts. In: ICFHR, pp. 471–476 (2016)
14. Stokes, P., Kiessling, B., Stökl Ben Ezra, D., Tissot, R., Gargem, E.H.: The eScriptorium VRE for manuscript cultures. Classics@ J. **18** (2021)
15. Yang, X., Yumer, E., Asente, P., Kraley, M., Kifer, D., Giles, C.L.: Learning to extract semantic structure from documents using multimodal fully convolutional neural networks. In: CVPR, pp. 4342–4351 (2017)
16. Zhang, C., Ibrayim, M., Hamdulla, A.: A methodological study of document layout analysis. In: VRHCIAI, pp. 12–17 (2022)
17. Zhong, X., Tang, J., Yepes, A.J.: PubLayNet: largest dataset ever for document layout analysis. In: ICDAR, pp. 1015–1022 (2019)

18. Zottin, S., De Nardin, A., Branca, G., Piciarelli, C., Foresti, G.L.: ICDAR 2025 competition on few-shot text line segmentation of ancient handwritten documents (FEST). In: ICDAR, pp. 586–602. Springer, Cham (2026)
19. Zottin, S., De Nardin, A., Colombi, E., Piciarelli, C., Pavan, F., Foresti, G.L.: U-DIADS-Bib: a full and few-shot pixel-precise dataset for document layout analysis of ancient manuscripts. Neural Comput. Appl. **36**(20), 11777–11789 (2024)

GRaF-Net: a Multi-branch Gated Residual Architecture for Floor Plan Semantic Segmentation

Axel De Nardin(✉), Silvia Zottin, Claudio Piciarelli, and Gian Luca Foresti

Artificial Vision and Machine Learning Lab, University of Udine, Udine, Italy
{axel.nardin,silvia.zottin,claudio.piciarelli,gianluca.foresti}@uniud.it

Abstract. Semantic segmentation of floor plan documents is a critical task in document image analysis, enabling applications such as indoor modeling, spatial reasoning, and 3D reconstruction. However, accurate segmentation remains challenging due to thin structural boundaries, symbol interference, high inter-class similarity, and substantial variability in drawing styles. To address these challenges, we propose GRaF-Net, a gated residual architecture that introduces a multi-branch decoding strategy for enhanced structural and semantic feature recovery.

In GRaF-Net, a deep encoder first extracts hierarchical feature representations from the input document image. The decoder adopts a multi-branch design to capture complementary spatial information at different receptive fields, facilitating precise boundary delineation and reliable room-type classification. Within each decoding branch, gated convolutional blocks dynamically regulate feature propagation, selectively enhancing informative activations while suppressing noise and ambiguities common in floor plan documents. Gated convolutions are integrated throughout the network to dynamically regulate information flow, enabling selective feature amplification and effective refinement of high-resolution segmentation outputs. Extensive experiments on benchmark datasets demonstrate that GRaF-Net consistently outperforms strong baseline segmentation models, particularly in complex layouts with ambiguous boundaries. The proposed architecture provides a robust and efficient framework for high-fidelity floor plan document segmentation.

Keywords: Floor Plan Segmentation · Document Analysis · Floor Plan Analysis

1 Introduction

Architectural floor plan documents provide structured spatial information but remain challenging to analyze due to thin structural boundaries, ambiguous room layouts, and stylistic variability. Semantic segmentation is a fundamental step

G. A. Fink et al. (Eds.): ICDAR 2026, LNCS 16972, pp. 558–574, 2027.
https://doi.org/10.1007/978-3-032-36023-6_32

in floor plan understanding, enabling the extraction of both structural elements (e.g., walls, doors and windows) and semantic regions. Accurate segmentation supports downstream applications such as 3D reconstruction, spatial reasoning, and digital environment modeling [17].

Floor plan segmentation presents distinct challenges compared to natural image segmentation. Structural information is conveyed primarily through thin line primitives, which are easily degraded during downsampling. At the same time, room regions often exhibit similar visual appearances and differ mainly through spatial configuration. Variations in drawing style, scale, and quality further complicate reliable segmentation.

Existing methods typically operate either on vector-based [27] or raster-based representations [7]. Vector-based approaches leverage explicit geometric primitives, enabling structured reasoning but requiring access to vector data or intermediate reconstruction steps. Raster-based methods, in contrast, directly process document images and offer broader applicability, but must infer geometric structure implicitly from pixel-level features. This places stronger demands on the model's ability to maintain structural information across the feature hierarchy.

A key challenge in floor plan segmentation is balancing structural precision and semantic abstraction. Structural features such as walls require high spatial fidelity, whereas room classification depends on broader contextual reasoning. In this work, we address this challenge through controlled cross-scale gated refinement, enabling selective interaction between structural and semantic representations. Specifically:

- We propose GRaF-Net, a gated residual architecture for joint room and boundary segmentation in floor plan documents.
- We introduce a gated convolutional encoder that improves the preservation of thin structural elements while maintaining semantic abstraction.
- We propose a Gated Residual Learning (GRL) module that performs controlled cross-scale refinement of semantic features using structural cues.
- We validate the proposed framework on CubiCasa5K, R2V, and R3D datasets, demonstrating consistent improvements over strong raster-based baselines.

The remainder of the paper is organized as follows. In Sect. 2 we perform a comprehensive review of the literature on floor plan analysis and segmentation. In Sect. 3 we provide a detailed description of the proposed framework. In Sect. 4 we define the experimental setup and present a thorough comparison with the current state-of-the-art approaches. Furthermore, we show the results of the ablation study we conducted to better understand the effects of the key components of our framework. Finally, in Sect. 5 we provide our final remarks regarding the presented work.

2 Related Work

Automatic understanding of architectural floor plan documents requires robust semantic interpretation of structured graphical elements under highly heterogeneous drawing conventions. Floor plan analysis has therefore become a central topic in document image understanding and computer vision, with applications ranging from property management to automated design assessment [9,19,20].

Several studies concentrated on structural element extraction, particularly boundary detection and the identification of walls and openings [1,4,10,23]. These approaches evolved from rule-based geometric pipelines to deep neural networks capable of learning more discriminative visual representations. Beyond structural parsing, recent studies have explored automated layout synthesis using generative [16,32] and diffusion-based models [21,30], as well as 3D indoor reconstruction from panoramic RGB images [3,14,22,28] or RGB-D sequences [13]. These directions further emphasize the importance of reliable semantic segmentation as a foundational component.

Floor plan segmentation, a core subtask, is generally addressed through vector-based or raster-based strategies. Vector-based methods convert raster drawings into structured geometric representations. For instance, Raster-to-Vector (R2V) [12] detects junctions and reconstructs wall layouts via integer programming, while VectorFloorSeg (VFS) [27] leverages graph neural networks to explicitly model spatial relationships between architectural components. Although naturally suited for downstream geometric modeling, such approaches often depend on heuristics and multi-stage post-processing, which can limit robustness under irregular layouts or varying drafting styles.

Beyond boundary modeling, textual cues have also been exploited to enhance semantic understanding. RC-Net [25] integrates room-type text information through a dedicated feature branch and introduces row and column consistency constraints across feature maps. This strategy encourages uniform semantic labeling within each room and produces cleaner segmentation boundaries. A related hybrid strategy is adopted by Lu et al. [15] in the DDFP framework, which targets rural residential floor plans characterized by high stylistic variability. The approach combines a joint neural network for simultaneous prediction of geometric elements and textual information with a post-processing room segmentation stage to reconstruct structured room layouts. More recently, De Nardin et al. [7] proposed the M&M architecture, which operates entirely in the raster domain and explicitly disentangles room segmentation from boundary recognition through a dual-branch design. A cross-intra-branch attention module enables adaptive information exchange between branches while mitigating feature interference.

In parallel, the emergence of large foundation models has introduced a complementary paradigm. The Segment Anything Model (SAM) [11] has been adapted to architectural documents by Piciarelli et al. [18], demonstrating that prompt-based segmentation can achieve accurate room extraction across heterogeneous datasets with minimal interaction. Zhang and Zhang [33] further combine SAM with LLMs in a few-shot framework for accessibility-oriented analysis, integrating segmentation with higher-level semantic reasoning. While these approaches exhibit strong generalization capabilities, their effectiveness remains dependent on prompt design and external semantic modules.

Although raster-based segmentation methods achieve strong performance, they typically rely on standard convolutional encoders and uniform multi-scale fusion without explicitly regulating structural–semantic interactions. This can lead to loss of fine structural detail or propagation of low-level noise into semantic predictions. The proposed approach addresses this limitation through gated feature modulation and controlled cross-scale residual refinement.

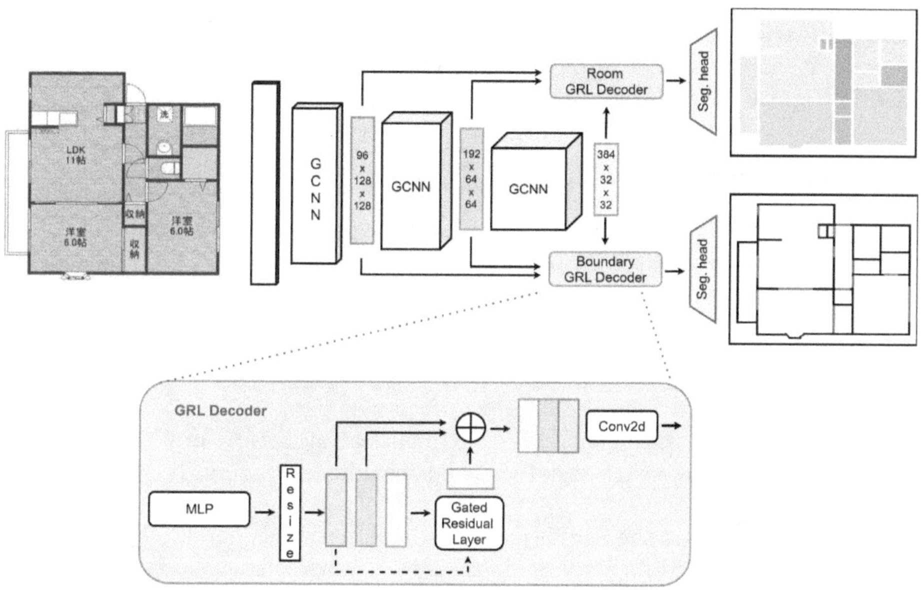

Fig. 1. High-level overview of the proposed dual-branch architecture for floor plan segmentation. The pipeline leverages a Gated CNN-based encoder (GCNN) to extract features at multiple scales from the input images. The feature maps extracted by the last three blocks of the encoder are then fed to two symmetrical MLP decoders, leveraging Gated Residual Learning (GRL) modules for cross-scale refinement of the feature maps, which are then fused at a channel level and the resulting combined map is fed into the respective segmentation heads to produce the final output.

3 Method

In this paper, we address floor plan parsing as a dual-task dense prediction problem, where semantic room regions and structural boundaries must be inferred jointly from a rasterized architectural drawing. Floor plans combine thin, high-frequency structural elements (walls, separators) with large homogeneous regions (rooms), requiring representations that preserve fine geometry while maintaining semantic abstraction across scales.

To this end, we propose GRaF-Net, a gated dual-decoder architecture built on a convolutional encoder and a SegFormer-style multi-scale decoder augmented with a Gated Residual Learning (GRL) module. The core design principle is controlled cross-scale refinement: high-level semantic features are modulated by low-level structural cues through multiplicative gating within a residual formulation. Unlike standard skip connections, the proposed decoder performs gated residual refinement prior to fusion, enabling selective structural modulation of semantic features.

The resulting model preserves structural sharpness in boundary prediction while maintaining semantic coherence in room segmentation, within a unified dual-task framework. A high-level overview of the proposed architecture is provided in Fig. 1.

3.1 Architecture Overview

Given an input image $I \in R^{H \times W \times 3}$, the encoder produces a hierarchy of feature maps at multiple spatial resolutions. These features are passed to a dual decoder that predicts:

- a room segmentation map: $S_r \in R^{H \times W \times C_r}$
- a boundary segmentation map: $S_b \in R^{H \times W \times C_b}$

The two decoder branches share an identical structure but do not share parameters. This allows each task to specialize while leveraging the same encoder backbone.

The decoder follows the SegFormer paradigm: each encoder stage is projected into a common embedding space, resized to a unified spatial resolution, and fused via channel concatenation. We extend this design with a gated residual refinement between the deepest and shallowest decoder features.

3.2 Gated Convolution Encoder

Floor plan segmentation requires simultaneously preserving thin, high-frequency structural elements (e.g., walls, separators, openings) and capturing global semantic context for room classification. To address this dual requirement, we adopt a MambaOut-like [29] hierarchical convolutional encoder architecture, in which token mixing is performed via gated depthwise convolutions rather than self-attention. This choice provides strong spatial inductive bias, computational efficiency, and improved sensitivity to fine geometric patterns that are critical in architectural drawings.

The encoder is organized into four stages. Each stage begins with a spatial downsampling operation followed by multiple Gated CNN (GCNN) blocks. The first stage uses a two-layer 3×3 convolutional stem (stride 2) to gradually reduce spatial resolution while preserving low-level structural detail. The remaining stages apply a 3×3 stride-2 convolution to progressively enlarge the receptive field and increase channel dimensionality. This design produces a multi-scale feature hierarchy, where early stages emphasize geometric precision and deeper stages encode broader layout semantics.

Gated CNN Block. Given an input feature map $X \in R^{B \times H \times W \times C}$, each GCNN block first applies layer normalization and channel expansion via a linear projection. The expanded representation is partitioned into three branches: a gating branch g, an identity branch i, and a convolutional branch c.

The convolutional branch undergoes depthwise convolution (kernel size k) to model localized spatial interactions, which are essential for capturing thin wall boundaries and small structural elements. For efficiency, convolution may be applied only to a subset of channels. The filtered feature $c^{'}$ is concatenated with the identity branch and multiplicatively modulated by the activated gating signal:

$$v = GELU(g) \cdot Concat(i, c^{'}) \tag{1}$$

This multiplicative gating mechanism enables adaptive feature selection: structurally meaningful responses (e.g., consistent wall edges) are amplified, while noisy textures, annotations, or stylistic variations common in floor plans are suppressed.

The gated representation is projected back to the original channel dimension and added to the input via a residual connection:

$$Y = Linear(v) + X \tag{2}$$

Overall, the encoder produces structurally aware multi-scale representations that retain boundary sharpness in early layers while progressively integrating global spatial context in deeper layers. This balance between geometric fidelity and semantic abstraction is fundamental for accurate room and boundary segmentation in complex floor plan documents.

3.3 Gated Residual Learning Decoder

While the encoder extracts hierarchical representations that progressively abstract structural and semantic information, accurate floor plan segmentation requires reinjecting fine geometric detail into high-level semantic features. Standard SegFormer-style decoders align and concatenate multi-scale features uniformly, implicitly assuming that shallow and deep representations contribute equally. However, in floor plan imagery, shallow features predominantly encode thin structural cues, whereas deeper features capture room-level semantics and global layout context. Treating them symmetrically may blur boundaries or propagate low-level noise.

To address this, we introduce a Gated Residual Learning (GRL) module that performs controlled cross-scale refinement prior to multi-scale fusion. Let $\hat{F}_d$

denote the deepest projected feature and $\hat{F}_s$ the shallowest aligned feature. Instead of directly concatenating them, the GRL module adaptively refines the deep semantic representation using structural cues before aggregation. Concretely, a gated transformation, built upon the same gated convolutional principle adopted in the encoder, is applied to the deep feature. Through depthwise spatial filtering and multiplicative gating, the transformation selectively amplifies geometrically meaningful responses (e.g., wall alignments and separators) while suppressing spurious activations such as annotations or stylistic artifacts common in architectural drawings. The refined representation is then combined with the shallow feature through a residual formulation:

$$\hat{F}_{ref} = G(\hat{F}_d) + \hat{F}_s \tag{3}$$

where $G(\cdot)$ denotes the gated refinement operator. Unlike the encoder blocks, which operate within a single scale to enhance feature selectivity, GRL explicitly performs cross-scale modulation, ensuring that structural detail influences semantic abstraction in a controlled and adaptive manner. The residual formulation preserves optimization stability while preventing shallow features from overwhelming high-level context. The GRL module is applied independently in both decoder branches (room and boundary), enabling task-specific refinement strategies.

3.4 Multi-scale Fusion

After GRL-based refinement of the deepest feature, the decoder performs scale-aligned aggregation of all projected encoder representations. Each encoder stage is first mapped to a common embedding dimension through an MLP and spatially resized to a shared resolution. This produces a set of feature maps that are semantically comparable and spatially aligned, enabling structured multi-scale integration.

Rather than merging features additively, we concatenate them along the channel dimension, preserving the distinct information carried by each scale. The concatenated tensor is then compressed through a 1×1 convolution followed by Batch Normalization and ReLU activation. This projection performs learned channel mixing across scales, allowing the network to infer task-dependent weighting of fine structural cues and broader contextual information within a unified embedding space.

The room and boundary branches implement this fusion independently. Although architecturally symmetric, their parameters are not shared, allowing each task to learn a distinct multi-scale integration strategy. Room segmentation emphasizes global spatial coherence, while boundary prediction relies more heavily on high-frequency structural signals; separate fusion layers permit these different inductive biases to emerge naturally during training.

3.5 Datasets

For our experiments, we used three widely adopted datasets for raster-based floor plan segmentation, each offering distinct characteristics in terms of scale, annotation richness, and structural diversity.

CubiCasa5K [8] is one of the largest publicly available floor plan image datasets, comprises 5,000 annotated floor plans collected from an initial set of 15,000 mostly Finnish images. The dataset provides ground-truth labels for 12 room types (e.g., living room, bedroom, kitchen) and 11 furniture or structural elements (e.g., doors, windows, fireplaces). Images vary substantially in resolution and style, with subsets categorized as architectural, high-quality black-and-white, and more colorful marketing-style plans. The dataset includes predefined training, validation, and test splits to support consistent evaluation.

R2V (Raster-to-Vector) [12] consists of 815 annotated floor plan images with pixel-wise labels for structural elements and room types. Originally derived from earlier collections such as the LIFULL HOME's dataset, this set predominantly includes plans with rectangular layouts and uniform wall thickness, making it suitable for methods focused on regular architectural designs. The standard split uses 715 images for training and 100 for testing, and annotations were manually labeled for walls, doors, windows, and room-type regions.

R3D [31] is a smaller dataset of 232 floor plans that complements R2V by including images with more varied room shapes and nonuniform wall thickness, reflecting a broader range of real-world architectural complexity. Each image has a fixed resolution of 512×512 pixels, and the dataset split consists of 170 images for training and 53 for testing. The inclusion of irregular layouts in R3D provides a more challenging benchmark for segmentation models when compared with predominantly regular layouts in R2V.

4 Experiments

4.1 Implementation Details

All experiments were conducted on a Linux-based system equipped with an NVIDIA A100 GPU with 40 GB of VRAM. Training was performed for a maximum of 200 epochs with a batch size of 8 and an initial learning rate of $1e-4$. Optimization was carried out using the Adam optimizer with a weight decay of $1e-5$. An early stopping strategy was employed to prevent overfitting, terminating training if the validation loss did not improve for 10 consecutive epochs, after an initial warm-up period of 10 epochs. All input floor plan images were resized to a fixed spatial resolution of 512×512 pixels to ensure consistent processing across datasets.

4.2 Metrics

For quantitative evaluation, following the previous literature on floor plan segmentation [12,26], we selected two standard metrics for floor plan segmentation performance evaluation: overall accuracy (Eq. 4), and Mean Intersection over Union (Mean IoU) (Eq. 5).

$$\text{Overall accuracy} = \frac{1}{C}\sum_i \frac{N_{ii}}{\sum_j N_{ij}}, \tag{4}$$

$$\text{Mean IoU} = \frac{1}{C}\sum_i \frac{N_{ii}}{\sum_j N_{ij} + \sum_j N_{ji} - N_{ii}}, \tag{5}$$

where C is the total number of classes, N_{ii} is the number of correctly-predicted pixels for class i, N_{ij} is the number of pixels of ground-truth class i that are predicted as class j and, N_{ji} is the number of pixels predicted as class i that belong to ground-truth class j.

Table 1. Comparison between the proposed approach and the current SOTA on the CubiCasa5k, R2V and R3D datasets. All results have been obtained by applying the proposed model an end-to-end fashion without relying on any additional post-processing steps. The best-performing method for each dataset and metrics is reported in bold. While the "*" denotes the models taking vectorial images as input instead of the rasterized ones.

		CubiCasa5K		**R2V**		**R3D**	
	Params(M)	mAcc	mIoU	mAcc	mIoU	mAcc	mIoU
PSPNet [34]	19.97	0.70	0.34	0.88	0.70	0.84	0.50
DLV3+ [2]	20.67	0.72	0.58	0.88	0.69	0.85	0.50
DeepFP [31]	28.91	0.59	0.47	0.89	0.74	0.89	0.63
FPNet [24]	-	-	-	0.88	0.64	0.94	0.73
OffsetGA [26]	-	-	-	0.93	0.77	0.91	0.74
RC-Net [25]	-	0.76	0.64	0.87	0.73	0.84	0.75
VFS* [27]	51.39	0.75	0.62	0.90	0.80	-	-
M&M [7]	26.65	0.84	0.61	**0.94**	0.83	0.94	0.74
Ours	36.61	**0.87**	**0.65**	**0.94**	**0.85**	**0.95**	**0.78**

4.3 Quantitative Results

Table 1 reports a comparison between GRaF-Net and prior floor plan segmentation approaches on the three most prominent datasets available in the literature: CubiCasa5K, R2V, and R3D. Methods marked with (*) operate on vectorial floor plan representations, while the remaining approaches, including ours, work on rasterized images. This distinction is important, as vector-based models benefit from explicit geometric primitives and structured layouts, whereas raster-based methods must infer such structure directly from raw pixels.

On CubiCasa5K, GRaF-Net achieves 0.87 mAcc and 0.65 mIoU, improving upon the previous best-reported results, achieved by VFS(*), by 12% and 3%, respectively. In particular, the model surpasses the strongest raster-based baselines and also exceeds the vector-based VFS approach. The improvement in mIoU is especially significant in this dataset, which exhibits substantial stylistic variation and drafting inconsistencies. These gains are consistent with the intended effect of gated structural refinement, which aims to preserve boundary precision while improving semantic consistency.

On R2V, GRaF-Net reaches 0.94 mAcc and 0.85 mIoU, establishing once again a new State-Of-The-Art (SOTA). Compared to the strongest prior method, GRaF-Net achieved a non-trivial improvement in mIoU of 2%. Notably, this dataset has historically favoured vector-based methods such as OffsetGA, due to its regular layouts and consistent wall geometry. Achieving superior performance with a purely raster-based pipeline indicates that the proposed gated residual refinement effectively captures geometric alignment and structural relationships that were previously modelled explicitly through vector representations.

The robustness of the approach is further reflected on R3D, where GRaF-Net attains 0.95 mAcc and 0.78 mIoU, again surpassing the previous best results by 1% and 4% respectively. The key characteristic of the R3D dataset is that it introduces irregular room shapes and non-uniform wall thickness, making precise boundary delineation more challenging. The consistent improvement over both raster and vector-based baselines demonstrates that the proposed architecture generalizes robustly across datasets with markedly different structural properties, even without relying on any additional post-processing strategies for segmentation refinement.

Overall, GRaF-Net consistently improves upon the previous state of the art across all evaluated datasets, including those where vector-based representations were previously considered advantageous, on both the considered metrics. In particular, the consistent improvement on the mIoU metric, which penalizes boundary misalignment and fragmented predictions more strongly, indicates that the improvements stem primarily from enhanced structural precision rather than marginal increases in pixel-wise classification confidence. From a document analysis perspective, this is particularly relevant: accurate region overlap and clean boundary separation directly impact downstream applications such as vectorization, layout reasoning, and CAD reconstruction. These results support the effectiveness of controlled cross-scale gated refinement as a principled mechanism for structural reasoning in raster-based floor plan parsing.

4.4 Qualitative Results

Figure 2 presents a qualitative comparison between the segmentation maps produced by GRaF-Net and those obtained by representative competing methods across the three datasets considered in this study, namely CubiCasa5K, R2V, and R3D. For each example, we show the input floor plan image, the corresponding ground truth annotation, and the predictions generated by the proposed approach, domain-specific floor plan segmentation models, and general-purpose

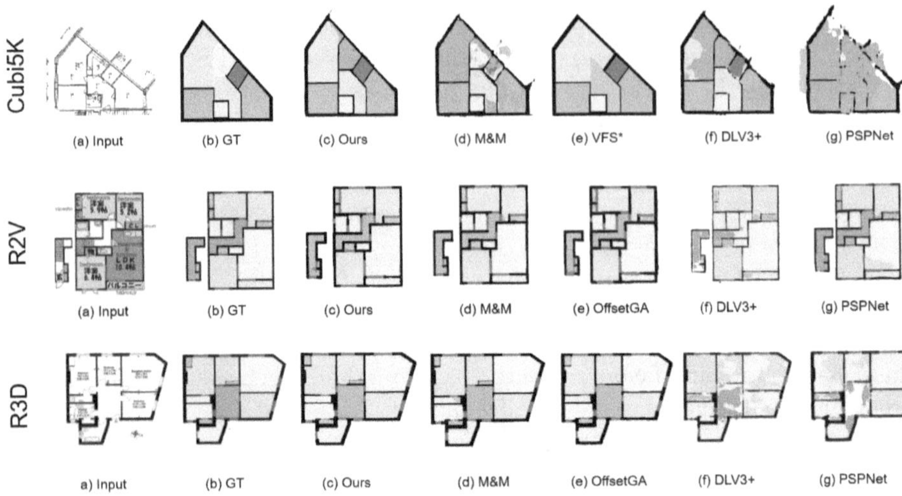

Fig. 2. Qualitative comparison between the proposed framework and a selection of competitor floor plan segmentation models on the 3 datasets selected from this study, namely CubiCasa5K, R2V, and 3RD. All the segmentation outputs have been obtained by reimplementing the respective models, with the exception of the VFS* and OffsetGA ones.

semantic segmentation architectures. All results were obtained using our own reimplementations, except for VFS and OffsetGA, whose official implementations were not publicly available.

On the R2V dataset, which is characterized by regular layouts, uniform wall thickness, and relatively consistent drawing conventions, most methods achieve visually accurate segmentation results. General-purpose models such as DeepLabV3+ and PSPNet already produce structurally coherent predictions, reflecting the lower structural ambiguity of this dataset. Consequently, the qualitative improvements achieved by GRaF-Net are more subtle in this scenario. Nevertheless, upon closer inspection, the proposed model produces more consistent boundary delineation, particularly in narrow wall regions and junction areas. These improvements are reflected in reduced boundary discontinuities and fewer instances of semantic leakage between adjacent rooms. This behavior is consistent with the design of the gated convolutional encoder, which selectively reinforces structurally meaningful features while suppressing irrelevant activations.

The qualitative differences become more pronounced on the R3D dataset, which introduces irregular room geometries and non-uniform wall thickness, more closely resembling real-world architectural drawings. In this setting, general-purpose segmentation models exhibit increased structural errors, including fragmented wall predictions and misclassification of small or irregularly shaped regions. Domain-specific approaches achieve improved robustness but still exhibit localized noise and boundary inconsistencies. In contrast, GRaF-Net produces more coherent segmentation maps, preserving structural continuity and reducing spurious predictions. The gated residual learning module plays a key role in this behavior by enabling controlled cross-scale integration of structural cues into high-level semantic representations, improving both boundary precision and semantic region consistency.

The most significant qualitative differences are observed on the CubiCasa5K dataset, which represents the most challenging scenario due to its larger scale, stylistic diversity, and higher structural complexity. In this dataset, both general-purpose and domain-specific methods exhibit substantial degradation in segmentation quality, including boundary fragmentation, incorrect merging of adjacent rooms, and increased prediction noise. The VFS method produces visually clean segmentation maps but incorrectly merges certain adjacent regions, resulting in clear semantic inconsistencies. GRaF-Net, on the other hand, successfully preserves the separation between most semantic regions and produces structurally consistent boundaries across the layout. This demonstrates the effectiveness of the proposed gated refinement mechanism in handling heterogeneous document conditions and preserving structural detail even in challenging scenarios. However, similarly to competing approaches, the proposed method still exhibits occasional errors in highly ambiguous regions where structural cues are weak or visually degraded.

Overall, the qualitative results confirm that GRaF-Net improves structural coherence and semantic consistency across floor plan segmentation tasks. The improvements are particularly evident in challenging scenarios characterized by irregular geometries and stylistic variability, where preserving thin structural boundaries and preventing semantic leakage is critical. These findings support the effectiveness of the gated convolutional encoding and gated residual refinement mechanisms in enhancing structural sensitivity and improving segmentation robustness in document image analysis applications.

Visualization of the Gated Attention Mechanism. Figure 3 presents activation maps from the gate attention module of the boundary and room decoder branches across four representative floor plan examples. Rather than specializing independently in their respective tasks, the two branches exhibit a complementary division of labor driven by the cross-scale gated refinement introduced by the GRL module. The boundary branch shows broad, region-filling activations, with warm responses spanning room interiors and suppressions emerging at wall locations. This behavior encodes boundaries implicitly through regional contrast rather than direct edge detection, providing spatial context for inferring precise boundaries from transitions between adjacent regions. Conversely, the room branch exhibits sharply localized activations along wall lines and structural edges, while suppressing room interiors. This edge-sensitive behavior reflects the GRL module's controlled injection of low-level geometric details into high-level semantic representations, preserving structural precision across decoder scales. Together, the two branches provide complementary signals, regional context from the boundary branch and geometric precision from the room branch, which are

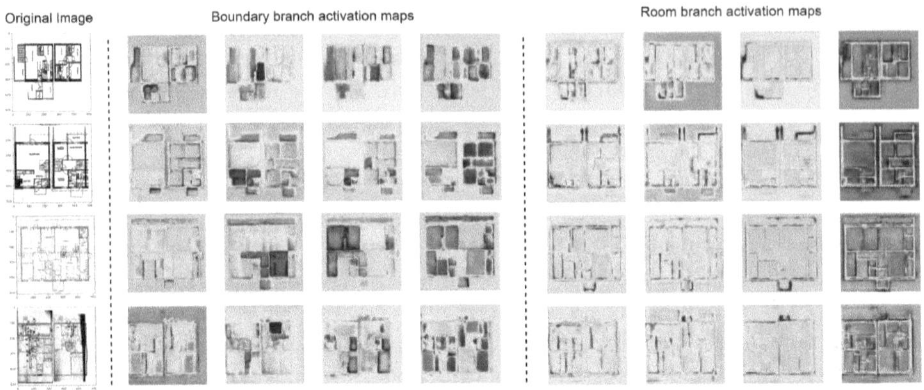

Fig. 3. Activation maps extracted from the boundary (left) and room (right) decoder branches across four representative floor plan examples. The boundary branch exhibits broad, region-filling responses that encode boundaries implicitly through contrast between adjacent room areas, while the room branch displays sharply localized activations along wall lines and structural edges, reflecting the GRL module's injection of low-level geometric detail into high-level semantic representations.

fused through gated residual refinement into coherent joint predictions. This specialization helps explain the consistent performance gains in Table 1, particularly in mIoU, which is highly sensitive to boundary misalignment.

4.5 Ablation Study

To validate the contribution of each individual component introduced in the proposed architecture, we report the results of an ablation study conducted on the CubiCasa5k dataset. Specifically, we test and compare 5 different scenarios: the first one is represented by the baseline model, which is represented by an architecture identical to the proposed one but with the gated connections replaced by the corresponding regular layers, both in the encoder and the decoder portion of the model. Next, we test the impact of including the gated (GCNN) and gated residual (GRL) connections individually in the encoder and decoder, respectively, as well as the performance gain achieved by relying on both simultaneously. Finally, we also compare the performance of our model with a baseline relying on regular gated MLP layers (GMLP) in the decoder portion of the pipeline, instead of the proposed Gated Residual Learning module. The results of the ablation study are reported in Table 2, with the worst results across the different configurations being reported in red and the best performance being reported in bold. As we can see, the baseline model, without gated modules, achieves the lowest performance, demonstrating how the introduction of each of the modules of the proposed architectures leads to a substantial improvement in performance across both metrics and all the segmentation categories. Interestingly, while the introduction of the encoder module leads to a very substantial improvement in performance on its own, the decoder module provides the largest gains when paired with the GCNN based Encoder, especially when it comes to the segmentation of the boundary categories as opposed to when used on its own. When both modules are combined, they lead to an increase in the IoU metric of +6% and +17% for the boundary and room categories, respectively, with an overall improvement of 17%. Finally, we can also observe how replacing the proposed GRL module with regular gated MLPs in the decoder part of the architecture consistently leads to a decrease in performance, with the overall IoU dropping from 0.65 to 0.62 (-3%). Overall, these results support the hypothesis that gated feature modulation improves segmentation performance by enabling controlled interaction between structural and semantic representations.

Table 2. Ablation study showing the effect of each component on the final performance of the framework in terms of overall accuracy and mIoU on both the boundary and room-types classes individually, as well as combined. Each column represents a different model configuration, with the ✓ symbol indicating the modules included.

Metrics		Accuracy/mIoU				
Modules	Encoder GNN	–	✓	–	✓	✓
	Decoder GMLP	–	–	–	✓	–
	Decoder GRL	–	–	✓	–	✓
Cubi5K	Boundaries	0.96 / 0.64	**0.97** / 0.66	0.96/ 0.65	**0.97** / 0.69	**0.97** / **0.70**
	Rooms	0.78 / 0.49	0.87 / 0.65	0.80 / 0.52	0.87 / 0.64	**0.88** / **0.66**
	Overall	0.76 / 0.48	0.86 / 0.63	0.78 / 0.51	0.85 / 0.62	**0.87** / **0.65**

4.6 Limitations

While GRaF-Net improves structural and semantic segmentation performance, several limitations remain. First, the model operates purely in the raster domain and does not explicitly enforce topological constraints, which may lead to inconsistencies in highly ambiguous regions. Second, the approach does not incorporate textual or symbolic cues, which may provide additional semantic information. Finally, the computational cost of gated refinement is higher than that of standard convolutional decoders, which may impact deployment in resource-constrained environments.

5 Conclusions

In this work, We introduced GRaF-Net, a gated residual architecture for semantic segmentation of floor plan documents. The proposed framework improves structural and semantic feature interaction through gated convolutional encoding and gated residual refinement. A key component of the proposed architecture is the Gated Residual Learning module, which enables adaptive interaction between structural and semantic features and contributes to improved segmentation performance. This mechanism improves boundary precision and region consistency without relying on vector-level inputs or post-processing. The dual-branch decoder further allows task-specific refinement for room and boundary segmentation, ensuring that both semantic and structural aspects of the document are accurately captured. Extensive experiments on CubiCasa5K, R2V, and R3D demonstrate that GRaF-Net consistently improves over previous state-of-the-art methods, achieving margins of +2% to +4% mIoU. Notably, these gains are obtained using a purely raster-based pipeline, outperforming approaches that operate on vector representations. This highlights the effectiveness of controlled gating mechanisms for recovering structural information directly from document images, making the approach broadly applicable to both digital and scanned floor plans. Overall, the proposed architecture offers a robust and efficient framework for floor plan segmentation. Future work will explore instance-level reasoning, vector reconstruction modules, and the extension of gated residual refinement to other structured document tasks such as technical drawings. We will also investigate in-domain transfer learning techniques [5,6] to assess adaptability in low-annotation settings.

Acknowledgments. The paper was partially supported by MUR FISA "I-TERAPI" Project - FISA-2023-00297 - G23C25000440006.

References

1. Ahmed, S., Weber, M., Liwicki, M., Langenhan, C., Dengel, A., Petzold, F.: Automatic analysis and sketch-based retrieval of architectural floor plans. Pattern Recogn. Lett. **35**, 91–100 (2014). https://doi.org/10.1016/j.patrec.2013.04.005, frontiers in Handwriting Processing
2. Chen, L.C., Zhu, Y., Papandreou, G., Schroff, F., Adam, H.: Encoder-decoder with atrous separable convolution for semantic image segmentation. In: Proceedings of the European Conference on Computer Vision (ECCV), pp. 801–818 (2018)
3. Cruz, S., Hutchcroft, W., Li, Y., Khosravan, N., Boyadzhiev, I., Kang, S.B.: Zillow indoor dataset: Annotated floor plans with 360° panoramas and 3d room layouts. In: 2021 IEEE/CVF Conference on Computer Vision and Pattern Recognition (CVPR), pp. 2133–2143 (2021). https://doi.org/10.1109/CVPR46437.2021.00217
4. De Las Heras, L.P., Terrades, O.R., Robles, S., Sánchez, G.: Cvc-fp and sgt: a new database for structural floor plan analysis and its groundtruthing tool. Int. J. Document Anal. Recogn. (IJDAR) **18**(1), 15–30 (2015). https://doi.org/10.1007/s10032-014-0236-5
5. De Nardin, A., Zottin, S., Colombi, E., Piciarelli, C., Foresti, G.L.: Is imagenet always the best option? an overview on transfer learning strategies for document layout analysis. In: Foresti, G.L., Fusiello, A., Hancock, E. (eds.) Image Analysis and Processing - ICIAP 2023 Workshops, pp. 489–499. Springer Nature Switzerland, Cham (2024)
6. De Nardin, A., Zottin, S., Piciarelli, C., Foresti, G.L., Colombi, E.: In-domain versus out-of-domain transfer learning for document layout analysis. Int. J. Document Anal. Recogn. (IJDAR) **28**(2), 161–175 (2025)

7. De Nardin, A., Zottin, S., Toma, A., Piciarelli, C., Foresti, G.L.: A multibranch and multiattention framework for floor plan segmentation. Comput.-Aided Civil Infrastruct. Eng. **40**(23), 3674–3687 (2025). https://doi.org/10.1111/mice.70030
8. Kalervo, A., Ylioinas, J., Häikiö, M., Karhu, A., Kannala, J.: Cubicasa5k: A dataset and an improved multi-task model for floorplan image analysis. In: Felsberg, M., Forssén, P.E., Sintorn, I.M., Unger, J. (eds.) Image Analysis, pp. 28–40. Springer International Publishing, Cham (2019)
9. Khade, R., Jariwala, K., Chattopadhyay, C.: A comprehensive survey of floor plan image analysis and related applications. Int. J. Document Anal. Recogn. (IJDAR) (2025). https://doi.org/10.1007/s10032-025-00528-8
10. Kim, S., Park, S., Kim, H., Yu, K.: Deep floor plan analysis for complicated drawings based on style transfer. J. Comput. Civ. Eng. **35**(2), 04020066 (2021)
11. Kirillov, A., et al.: Segment anything. In: Proceedings of the IEEE/CVF International Conference on Computer Vision (ICCV), pp. 4015–4026 (2023)
12. Liu, C., Wu, J., Kohli, P., Furukawa, Y.: Raster-to-vector: Revisiting floorplan transformation. In: 2017 IEEE International Conference on Computer Vision (ICCV), pp. 2214–2222 (2017). https://doi.org/10.1109/ICCV.2017.241
13. Liu, C., Wu, J., Furukawa, Y.: FloorNet: A Unified Framework for Floorplan Reconstruction from 3D Scans. In: Ferrari, V., Hebert, M., Sminchisescu, C., Weiss, Y. (eds.) ECCV 2018. LNCS, vol. 11210, pp. 203–219. Springer, Cham (2018). https://doi.org/10.1007/978-3-030-01231-1_13
14. Liu, C., Schwing, A.G., Kundu, K., Urtasun, R., Fidler, S.: Rent3d: Floor-plan priors for monocular layout estimation. In: 2015 IEEE Conference on Computer Vision and Pattern Recognition (CVPR), pp. 3413–3421 (2015). https://doi.org/10.1109/CVPR.2015.7298963
15. Lu, Z., et al.: Data-driven floor plan understanding in rural residential buildings via deep recognition. Inf. Sci. **567**, 58–74 (2021). https://doi.org/10.1016/j.ins.2021.03.032
16. Nauata, N., Hosseini, S., Chang, K.H., Chu, H., Cheng, C.Y., Furukawa, Y.: House-gan++: Generative adversarial layout refinement network towards intelligent computational agent for professional architects. In: 2021 IEEE/CVF Conference on Computer Vision and Pattern Recognition (CVPR), pp. 13627–13636 (2021). https://doi.org/10.1109/CVPR46437.2021.01342
17. Noichl, F., Collins, F.C., Braun, A., Borrmann, A.: Enhancing point cloud semantic segmentation in the data-scarce domain of industrial plants through synthetic data. Comput.-Aided Civil Infrastruct. Eng. **39**(10), 1530–1549 (2024)
18. Piciarelli, C., Zottin, S., De Nardin, A., Foresti, G.L.: Segmenting anything in architecture: A tailored approach to segmenting floor plan documents. In: Rodolà, E., Galasso, F., Masi, I. (eds.) Image Analysis and Processing - ICIAP 2025, pp. 416–428. Springer Nature Switzerland, Cham (2026)
19. Pizarro, P.N., Hitschfeld, N., Sipiran, I., Saavedra, J.M.: Automatic floor plan analysis and recognition. Autom. Constr. **140**, 104348 (2022). https://doi.org/10.1016/j.autcon.2022.104348
20. Pokuciński, S., Mrozek, D.: Methods and applications of space understanding in indoor environment—a decade survey. Appl. Sci. **14**(10) (2024). https://doi.org/10.3390/app14103974
21. Shabani, M.A., Hosseini, S., Furukawa, Y.: Housediffusion: Vector floorplan generation via a diffusion model with discrete and continuous denoising. In: 2023 IEEE/CVF Conference on Computer Vision and Pattern Recognition (CVPR), pp. 5466–5475 (2023). https://doi.org/10.1109/CVPR52729.2023.00529

22. Sun, C., Hsiao, C.W., Sun, M., Chen, H.T.: Horizonnet: Learning room layout with 1d representation and pano stretch data augmentation. In: 2019 IEEE/CVF Conference on Computer Vision and Pattern Recognition (CVPR), pp. 1047–1056 (2019). https://doi.org/10.1109/CVPR.2019.00114
23. Sun, Y., Zhou, M., Cao, Z.: Gated-dual-attention-based full-resolution floor plan segmentation network for topological semantic mapping. IEEE Trans. Syst. Man Cybern. Syst. 1–14 (2026). https://doi.org/10.1109/TSMC.2025.3648921
24. Upadhyay, A., Dubey, A., Kuriakose, S.M.: Fpnet: Deep attention network for automated floor plan analysis. In: Coustaty, M., Fornés, A. (eds.) International Conference on Document Analysis and Recognition, pp. 163–176. Springer, Cham (2023). https://doi.org/10.1007/978-3-031-41498-5_12
25. Wang, T., Meng, W.L., Lu, Z.D., Guo, J.W., Xiao, J., Zhang, X.P.: Rc-net: Row and column network with text feature for parsing floor plan images. J. Comput. Sci. Technol. **38**(3), 526–539 (2023)
26. Wang, Z., Sun, N.: Offset-guided attention network for room-level aware floor plan segmentation. IEEE Access **11**, 63667–63677 (2023). https://doi.org/10.1109/ACCESS.2023.3288598
27. Yang, B., Jiang, H., Pan, H., Xiao, J.: Vectorfloorseg: Two-stream graph attention network for vectorized roughcast floorplan segmentation. In: Proceedings of the IEEE/CVF Conference on Computer Vision and Pattern Recognition (CVPR), pp. 1358–1367 (2023)
28. Yang, S.T., Wang, F.E., Peng, C.H., Wonka, P., Sun, M., Chu, H.K.: Dula-net: A dual-projection network for estimating room layouts from a single rgb panorama. In: 2019 IEEE/CVF Conference on Computer Vision and Pattern Recognition (CVPR), pp. 3358–3367 (2019). https://doi.org/10.1109/CVPR.2019.00348
29. Yu, W., Wang, X.: Mambaout: Do we really need mamba for vision? In: Proceedings of the IEEE/CVF Conference on Computer Vision and Pattern Recognition, pp. 4484–4496 (2025)
30. Zeng, P., Gao, W., Yin, J., Xu, P., Lu, S.: Residential floor plans: Multi-conditional automatic generation using diffusion models. Autom. Constr. **162**, 105374 (2024). https://doi.org/10.1016/j.autcon.2024.105374
31. Zeng, Z., Li, X., Yu, Y.K., Fu, C.W.: Deep floor plan recognition using a multi-task network with room-boundary-guided attention. In: Proceedings of the IEEE/CVF International Conference on Computer Vision, pp. 9096–9104 (2019)
32. Zhang, H., Savov, A., Dillenburger, B.: Maskplan: Masked generative layout planning from partial input. In: 2024 IEEE/CVF Conference on Computer Vision and Pattern Recognition (CVPR), pp. 8964–8973 (2024). https://doi.org/10.1109/CVPR52733.2024.00856
33. Zhang, H., Zhang, R.: Few-shot learning with large foundation models for automated segmentation and accessibility analysis in architectural floor plans. J. Infrastruct. Intell. Resil. **4**(2), 100137 (2025)
34. Zhao, H., Shi, J., Qi, X., Wang, X., Jia, J.: Pyramid scene parsing network. In: Proceedings of the IEEE Conference on Computer Vision and Pattern Recognition, pp. 2881–2890 (2017)

Recognition of Tables and Formulas

Beyond Rows to Reasoning: Agentic Retrieval for Multimodal Spreadsheet Understanding

Anmol Gulati(✉), Sahil Sen, Waqar Sarguroh, and Kevin Paul

Commercial Technology and Innovation Office, PricewaterhouseCoopers, New York, USA
{anmol.gulati,sahil.sen,waqar.sarguroh,kevin.paul}@pwc.com

Abstract. Recent advances in multimodal Retrieval-Augmented Generation (RAG) enable Large Language Models (LLMs) to analyze enterprise spreadsheet workbooks with millions of cells, cross-sheet dependencies, and embedded visuals. However, state-of-the-art approaches exclude critical context through single-pass retrieval, lose data resolution through compression, and exceed LLM context windows through naive full-context injection, preventing reliable multi-step reasoning over complex workbooks. We introduce Beyond Rows to Reasoning (BRTR), a multimodal agentic framework for spreadsheet understanding that replaces single-pass retrieval with an iterative tool-calling loop, supporting end-to-end Excel workflows from complex analysis to structured editing. With over 200 h of expert human evaluation, BRTR achieves state-of-the-art performance across three frontier spreadsheet understanding benchmarks, surpassing prior methods by 25% points on FRTR-Bench, 7 points on SpreadsheetLLM, and 32 points on FINCH. We evaluate five multimodal embedding models, identifying NVIDIA NeMo Retriever 1B as the top performer for mixed tabular and visual data, across nine LLMs. Ablation experiments confirm that the planner, retrieval, and iterative reasoning each contribute highly, and cost analysis shows GPT-5.2 achieves the best efficiency–accuracy trade-off. Throughout all evaluations, BRTR maintains full auditability through explicit tool-call traces.

Keywords: Multimodal RAG · Spreadsheet understanding · Agentic retrieval · Enterprise QA · Tool-calling agents · Document analysis

1 Introduction

Recent advances in multimodal Retrieval-Augmented Generation (RAG) [14] have enabled Large Language Models (LLMs) to analyze enterprise spreadsheet workbooks containing complex tabular data, cross-sheet dependencies, and embedded visual artifacts [8,10]. Current approaches follow two main strategies:

A. Gulati—Now at Patronus AI.

G. A. Fink et al. (Eds.): ICDAR 2026, LNCS 16972, pp. 577–593, 2027.
https://doi.org/10.1007/978-3-032-36023-6_33

compression-based methods encode worksheets into condensed representations that fit within LLM context windows [8], while retrieval-based methods chunk and embed spreadsheet content for semantic search at query time [10]. However, both commit to a single pass with no mechanism for iterative refinement (Fig. 2). Real analysts cross-reference across sheets, follow cell dependencies and gather evidence incrementally before drawing conclusions; current pipelines cannot mimic this behavior because the LLM lacks the ability to re-query, explore adjacent regions, or refine its search based on intermediate results. Additionally, existing work has not comprehensively evaluated multimodal embedding models for spreadsheet content, leaving practitioners without guidance on which embeddings best capture the semantic structure of mixed tabular and visual data.

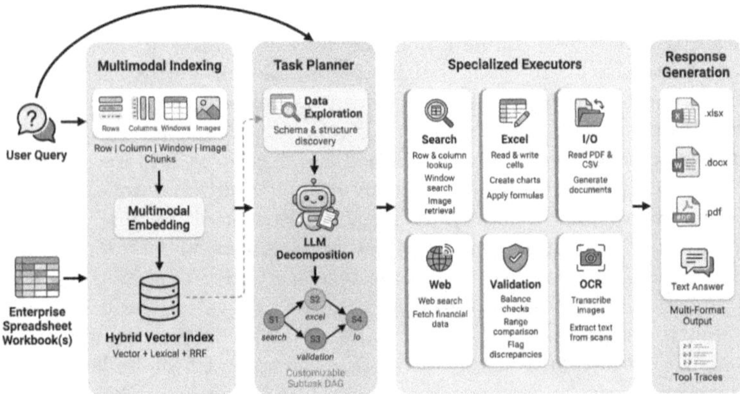

Fig. 1. BRTR pipeline: multimodal indexing, agentic task planning, specialized tool execution, and multi-format response generation with end-to-end tool-trace auditability.

In this paper, we introduce Beyond Rows to Reasoning (BRTR), a multimodal agentic framework for spreadsheet understanding that replaces single-pass retrieval with an iterative tool-calling loop, enabling LLMs to invoke structured search tools, inspect results, and refine queries across multiple turns for end-to-end Excel analysis and editing (Fig. 1). We make two contributions:

- **Multimodal embedding model evaluation:** We present a comprehensive comparison of five multimodal embedding models for spreadsheet content retrieval on FRTR-Bench, providing practitioners with the first empirical guidance on embedding selection for mixed tabular and visual data (Sect. 4.1).
- **BRTR agentic framework:** BRTR equips LLMs with structured search tools and a context budgeting mechanism that controls token growth. Supported by +200 h of human evaluation, BRTR achieves state-of-the-art accuracy on FRTR-Bench and the SpreadsheetLLM benchmark, outperforming single-pass retrieval and compression baselines. We extend BRTR with

spreadsheet manipulation and multimodal interaction tools, achieving state-of-the-art performance on the FINCH finance and accounting benchmark demonstrating generalization to composite, multi-step enterprise workflows spanning calculation, formatting, visualization, and cross-file validation across heterogeneous artifacts including spreadsheets, PDFs, and images.

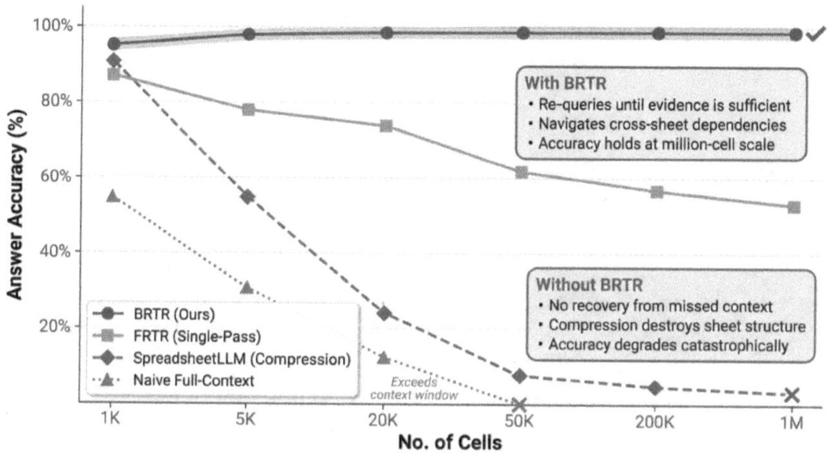

Fig. 2. Conceptual illustration of answer accuracy vs. spreadsheet size. BRTR maintains near-perfect accuracy by iteratively re-querying, while single-pass and compression methods degrade with complexity. Naïve full-context exceeds context windows beyond 50K cells. Tables 2 and 3 provide empirical support for these trends.

2 Related Work

2.1 Retrieval-Based Spreadsheet Understanding

Spreadsheet understanding poses unique challenges due to two-dimensional layouts, flexible formatting, and mixed content types. Early table understanding methods adapted transformer architectures for structured data. TaPas [11] extended BERT to encode tables by flattening rows and adding positional embeddings for columns and rows, enabling weakly supervised question answering without generating logical forms. TUTA [27] introduced tree-based transformers with a bi-dimensional coordinate tree to capture both spatial and hierarchical structure, achieving state-of-the-art performance on cell type and table type classification tasks through structure-aware attention mechanisms.

For enterprise spreadsheets, compression and retrieval approaches have emerged to handle workbooks that exceed LLM context limits. SpreadsheetLLM [8] introduced SheetCompressor, an encoding framework with structural-anchor-based compression, inverse index translation, and data-format-aware aggregation, achieving 25× compression while improving table detection by 25.6%.

From Rows to Reasoning (FRTR) [10] proposed multi-granular chunking at row, column, and window levels with hybrid lexical-dense retrieval, enabling question answering over workbooks with millions of cells. Upstream of retrieval, LlamaSheets [17] addresses the pre-processing stage, using hierarchical header extraction and adaptive table segmentation to transform messy spreadsheets into structured data for downstream consumption. While these methods advance scalability, they commit to a single pass: the system retrieves or extracts a fixed context before generation, with no mechanism for iterative refinement if retrieval misses critical evidence.

2.2 Agentic and Iterative Retrieval for Structured Data

Recent work explores multi-step interaction and tool-based exploration for table understanding, building on the ReAct paradigm [30] and broader advances in agentic RAG and tool selection [18,26]. One family of approaches uses code execution as the primary action space: SheetAgent [4] generates Python with SQL-based subview narrowing, TableZoomer [29] combines query-aware sub-table generation with Program-of-Thoughts execution, and SheetBrain [28] pairs a Python sandbox with a validation module that triggers re-execution on failure. A second family replaces free-form code with structured commands or learned policies: SheetMind [31] uses a multi-agent decompose–act–reflect loop with a Backus–Naur Form grammar, while TableMind [12] trains an agent via supervised fine-tuning and reinforcement learning for multi-turn tool invocation. However, these methods either depend on code execution, which can be unreliable when agents enter unproductive loops, or require model fine-tuning with specialized supervision. BRTR avoids both limitations by equipping general-purpose LLMs with structured search tools through native function-calling, requiring no fine-tuning and constraining tool calls to prevent runaway loops. A separate line of work develops specialized pre-training for table reasoning [3,5,16], but requires substantial training resources, limiting enterprise accessibility.

2.3 LLM-Powered Spreadsheet Products

Commercial products increasingly embed LLMs into spreadsheet applications, including Claude for Excel [1], Microsoft Copilot Agent Mode [24], Gemini in Google Sheets [9], and Claude Cowork [2], but all remain scoped to the active workbook without semantic indexing. Benchmarks such as SpreadsheetBench [23], InstructExcel [25], FINCH [7], and SpreadsheetArena [13] have standardized evaluation of these systems. BRTR differs by indexing content into a searchable vector store, enabling agentic cross-file retrieval with full tool-trace auditability.

3 Methodology

BRTR utilizes a retrieval pipeline with an agentic loop that enables iterative evidence acquisition. The framework consists of three core components: (1) multimodal indexing and retrieval, which chunks spreadsheet content and embeds it for semantic search (Sect. 3.1); (2) an agentic tool-calling loop, where the LLM invokes structured search tools and refines queries across multiple turns (Sect. 3.2); and (3) a planner-executor architecture that decomposes composite workflows into dependency graphs of sub-tasks dispatched to specialized executors (Sect. 3.3). Figure 1 illustrates the complete pipeline.

3.1 Multimodal Indexing and Retrieval

BRTR builds on the FRTR chunking and retrieval infrastructure [10] (Fig. 1, left). Given a workbook, the system extracts four chunk types (row, column, rectangular window, and embedded image), each preserving metadata such as sheet name, cell coordinates, and headers. All chunks are embedded using a multimodal embedding model; based on the evaluation in Sect. 4.1, we adopt NVIDIA NeMo Retriever 1B.

At query time, BRTR performs hybrid retrieval combining dense cosine-similarity search with BM25 lexical matching, the latter being important for exact numerical values and cell references [10]. Results across all chunk types are fused via Reciprocal Rank Fusion (RRF) [6]:

$$\text{score}_{\text{RRF}}(c, q) = \sum_{r \in \mathcal{R}} \frac{1}{k + \text{rank}_r(c, q)} \tag{1}$$

where $\mathcal{R}$ is the set of ranked lists (dense, lexical $\times$ chunk type), $\text{rank}_r(c, q)$ is the position of chunk c in list r, and $k = 60$. The top-K chunks ($K = 10$) are passed to the agent as initial context before iterative refinement begins.

3.2 Agentic Tool-Calling Loop

The core innovation of BRTR is replacing single-pass retrieval with an iterative tool-calling loop where the LLM invokes search tools, inspects results, and refines queries until sufficient evidence is gathered.

Search Tools. The agent has access to five search tools: `search_rows` retrieves row chunks, `search_columns` retrieves column chunks, `search_windows` retrieves rectangular cell regions, `search_images` retrieves embedded visual content, and `search_all` performs a unified search across all content types. Each tool accepts a query and returns the top-K matching chunks with their metadata. All tools use the same hybrid dense-lexical retrieval pipeline described in Sect. 3.1. Row and column searches additionally accept optional coordinate parameters (row number, column number) that filter results to chunks at specific positions, enabling the agent to combine semantic relevance with precise cell-level targeting.

Algorithm 1. BRTR Planner-Executor Pipeline

Require: Task q, corpus $\mathcal{C}$, LLM $\mathcal{M}$, K=10, T=50
Ensure: Answer a, tool trace τ
Phase 1: Decompose
$e \leftarrow$ ExploreData($q, \mathcal{C}, K$) {Vector-search quoted terms for column locations}
$P, o \leftarrow \mathcal{M}(q, e)$ {Subtask DAG $\{(s_i, type_i, deps_i)\}_{i=1}^{n}$ and output type o}
Phase 2: Execute subtask DAG
$done \leftarrow \{\}$; $\tau \leftarrow []$
while $|done| < n$ **do**
  $ready \leftarrow \{s_i \in P : deps_i \subseteq done\}$
  for all $s_i \in ready$ **in parallel do**
    $tools_i \leftarrow$ GetToolSet($type_i$) {Type-specific tools}
    $ctx_i \leftarrow \{s_i.desc,\ q,\ \text{Summarize}(done[deps_i])\}$
    for $t = 1, \ldots, T$ **do**
      $resp \leftarrow \mathcal{M}(ctx_i,\ tools_i)$
      for all tool call f in $resp$ **do**
        $r \leftarrow$ Execute(f)
        $\tau \leftarrow \tau \cup \{(f, r)\}$; $ctx_i \leftarrow ctx_i \cup \{f, r\}$
      end for
    end for
    $done[s_i] \leftarrow resp$
  end for
end while
Phase 3: Synthesize
$a \leftarrow \mathcal{M}(q,\ \text{Summarize}(done),\ o)$ {Merge all executor outputs}
return $(a,\ \tau)$

Iterative Refinement Process. Given a user query, the LLM analyzes it to determine which content types are relevant and invokes one or more search tools. After inspecting the returned chunks, the model decides whether additional evidence is needed. If initial results are insufficient or ambiguous, the LLM refines its search query, requests different content types, or applies coordinate filtering to examine specific rows or columns identified in prior results. Each agent invocation is bounded by $T = 50$ tool-call iterations, a limit that prevents unproductive search loops; the loop terminates early when the model determines it has gathered sufficient evidence. In the planner configuration (Sect. 3.3), this budget applies independently to each executor sub-task. In practice, frontier models average 3–6 tool calls per query, well below this ceiling (Fig. 3). Across all frontier-model runs on FRTR-Bench, over 90% of queries terminate within 8 tool calls and 99% within 15; on FINCH, over 90% of executor sub-tasks complete within 20 calls.

Context Management. As the agent accumulates tool-call results across iterations, the conversation context grows. Image chunks, embedded as base64-encoded content, are the primary driver of token consumption. To prevent context overflow, BRTR applies an always-prune strategy: whenever new images

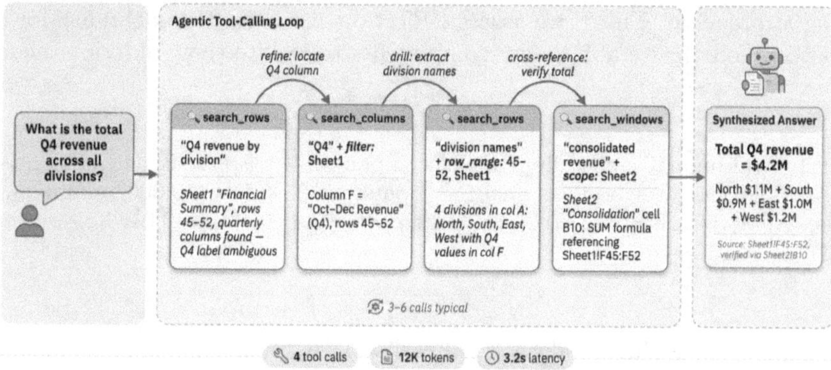

Fig. 3. Representative BRTR tool-call trace.

are retrieved, all prior image content is stripped from earlier tool-call responses, retaining their textual metadata (sheet name, coordinates, alt-text) while removing the encoded image data. The agent's reasoning about prior images is preserved in the conversation history, so retaining only the most recent batch avoids overflow without sacrificing accumulated understanding.

3.3 Planner-Executor Architecture

To generalize to composite, multi-step enterprise workflows, we extend BRTR with a planner and six specialized executors (Fig. 1; Algorithm 1). The *planner* decomposes workflow instructions into a dependency graph of sub-tasks; independent branches execute concurrently. Six executors handle distinct capabilities: *Excel* (read/write cells, formulas, formatting, charts), *IO* (PDFs, CSVs, DOCX), *web* (search, URL fetching, financial data), *validation* (accounting invariants such as balance-sheet identity and debit-credit equality), *vision* (image transcription and text extraction from scanned documents via the LLM's native multimodal capabilities, not a dedicated OCR stack), and *search* (the base BRTR retrieval tools described above). Each executor operates within the agentic loop described above. The planner dispatches ready sub-tasks to the appropriate executor and feeds completed results into downstream dependencies. This decomposition directly addresses the long-horizon composition bottleneck identified in prior work [7], where monolithic agents suffer error accumulation across multi-task workflows.

4 Experiments

We evaluate BRTR through four experiments. First, we compare multimodal embedding models to identify the best retrieval backbone for spreadsheets. Second, we evaluate the full BRTR agentic framework on two benchmarks against existing approaches. Third, we assess BRTR on the FINCH benchmark for generalization. Fourth, we ablate key components to isolate their individual contributions.

Table 1. Multimodal embedding model comparison on FRTR-Bench (146 queries, 28 files). Bold indicates best per metric.[1](Bootstrap 95% CIs (10,000 resamples, 146 queries): NeMo Recall@10 Δ=0.09 vs. Cohere, CI [0.03, 0.15], p<0.01; NeMo vs. Jina Δ=0.27, p<0.001.)

	$K = 5$			$K = 10$		
Embedding Model	nDCG	Recall	MAP	nDCG	Recall	MAP
Amazon Titan	0.26	0.39	0.27	0.33	0.47	0.27
Cohere v4	**0.33**	0.45	**0.35**	0.40	0.51	0.34
Amazon Nova	0.30	0.45	0.32	0.40	0.58	0.33
NVIDIA NeMo	0.33	**0.47**	0.33	**0.42**	**0.60**	**0.34**
Jina v4	0.19	0.26	0.19	0.23	0.33	0.19

4.1 Experiment 1: Multimodal Embedding Model Comparison

Experimental Setup. Five multimodal embedding models are evaluated on a subset of FRTR-Bench comprising 146 queries across 28 files. Each query targets specific spreadsheet content requiring retrieval of relevant chunks containing rows, columns, cell windows, or embedded images. The models span proprietary and open-source architectures across multiple providers: Amazon Titan Multimodal, Cohere Embed v4, Amazon Nova Multimodal, NVIDIA NeMo Retriever 1B, and Jina Embeddings v4. Because spreadsheet data admits many valid retrieval paths to a correct answer, computing exhaustive relevance judgments for each query is infeasible. Instead, we retrieve the top-10 chunks by cosine similarity and evaluate relevance using GPT-5 as an LLM judge, comparing retrieved content against ground-truth annotations. We validate this automated assessment through the downstream human evaluation in Sect. 4.2: retrieval quality is ultimately reflected in end-to-end answer accuracy, which is judged by human evaluators across all benchmarks. Furthermore, better retrieval can lead to fewer total searches, reducing the LLM's overall token usage.

Metrics. We report standard retrieval metrics at cutoffs $K \in \{5, 10\}$: nDCG@K (normalized Discounted Cumulative Gain) evaluates ranking quality with position-weighted relevance; Recall@K measures the fraction of relevant chunks appearing within the top-K results; and MAP@K (Mean Average Precision) captures precision across all recall levels.

Results. Table 1 presents retrieval performance across all five embedding models at K=5 and K=10. NVIDIA NeMo Retriever achieves the highest Recall@10 (0.60) and also leads in nDCG@10 (0.42) and MAP@10 (0.34), demonstrating the strongest overall retrieval quality. Cohere v4 excels at early-stage ranking, achieving the best nDCG@5 (0.33) and MAP@5 (0.35), suggesting it places relevant results higher, a property that is valuable when retrieving fewer chunks to minimize token consumption. Amazon Nova achieves competitive Recall@10 (0.58) and nDCG@10 (0.40), trailing NVIDIA NeMo by only 2% points on Recall@10. Jina v4 shows substantially lower performance across all metrics, with Recall@10 of only 0.33, a gap of 27% points compared to NVIDIA NeMo, indicating that not all multimodal embedding models generalize effectively to structured spreadsheet content with mixed text and visual elements.

Discussion. For agentic systems like BRTR that iteratively refine searches, higher Recall@10 is preferable, making breadth of initial retrieval more valuable than precision at shallow cutoffs. Based on these findings, we adopt NVIDIA NeMo Retriever as the default embedding model for all subsequent experiments.

Table 2. Cross-technique performance on FRTR-Bench (159 queries, 30 files). SSLLM = SpreadsheetLLM; CiE = Claude in Excel. BRTR achieves up to 99% accuracy, 25 pp above single-pass FRTR. Operational cost columns report means. Bold indicates highest accuracy per technique.

	Accuracy				Cost (BRTR)		
Model	BRTR	FRTR	SSLLM	CiE	Calls	Tokens	Latency
GPT-5	**0.99**	0.73	0.18	–	4.01	68,061	166.34
Gemini 3 Pro	**0.99**	0.68	0.29	–	5.40	48,462	217.77
Claude Opus 4.6	0.98	**0.74**	**0.34**	**0.84**	5.20	462,614	143.23
GPT-5.2	0.98	0.72	0.32	–	3.05	20,127	90.81
Claude Sonnet 4.5	0.96	**0.74**	0.13	0.69	6.20	92,202	163.83
Gemini 2.5 Pro	0.85	0.67	0.24	–	2.23	15,729	111.04
GPT-4o	0.67	0.49	0.06	–	3.25	50,745	140.20
LLaMA 4 Maverick 17B	0.67	0.56	0.23	–	1.99	38,483	92.32

4.2 Experiment 2: BRTR Agentic Framework Evaluation

Experimental Setup. BRTR is evaluated across nine LLMs spanning frontier and open-weight models, using NVIDIA NeMo Retriever embeddings based on Experiment 1 findings.[1] Two baselines are compared on both benchmarks: (1)

[1] All proprietary models were accessed via their respective APIs using the latest available versions as of January 2026, with temperature set to 0 and default sampling parameters throughout.

FRTR [10], single-pass hybrid retrieval followed by generation; and (2) SpreadsheetLLM [8], SheetCompressor encoding of entire workbooks. On FRTR-Bench we additionally report Claude in Excel [1] as a commercial product baseline; on the SpreadsheetLLM benchmark we additionally include naïve full-context, which passes uncompressed spreadsheet content directly to the LLM. Claude in Excel is evaluated only on FRTR-Bench: the SpreadsheetLLM benchmark is already well-saturated by existing methods, so we prioritized the more challenging FRTR-Bench. It is excluded from FINCH because it operates within a single workbook and cannot ingest multiple spreadsheets to produce new output files, disqualifying it from the majority of FINCH workflows.

The evaluation spans two benchmarks. FRTR-Bench consists of 159 queries across 30 enterprise workbooks spanning 155 sheets, nearly four million cells, 656K rows, and 53 embedded images [10], where each query requires locating specific cells, cross-referencing multiple sheets, or interpreting embedded charts. The SpreadsheetLLM benchmark comprises 298 single-sheet, text-only questions requiring numerical reasoning, formula synthesis, cell lookup, and format interpretation across diverse individual worksheets. All BRTR configurations use top-K=10 initial retrieval with a maximum of 50 tool calls per query. We exclude agentic systems (SheetAgent [4], SheetBrain [28], TableMind [12]) that target different task formulations or lack public implementations.

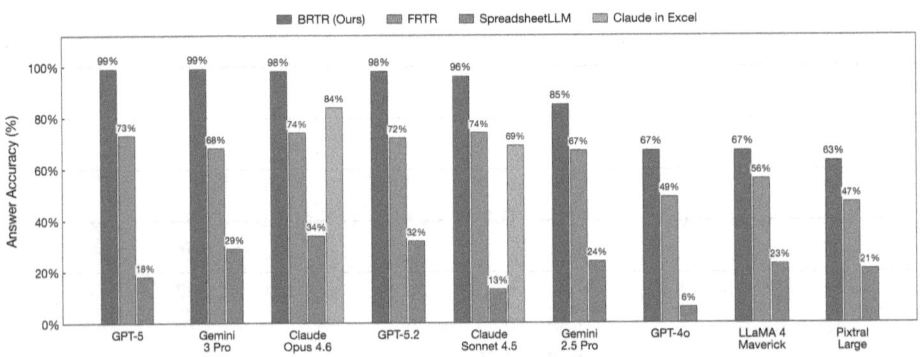

Fig. 4. Answer accuracy on FRTR-Bench across nine LLMs and four techniques. BRTR (teal) consistently dominates, achieving up to 99% accuracy with frontier models, 25% points above the best single-pass FRTR baseline. SpreadsheetLLM compression (lavender) performs poorly throughout, confirming that cross-sheet references require retrieval rather than compression alone.

Metrics. Answer Accuracy measures the fraction of queries answered correctly, with correctness determined by semantic equivalence to ground truth, judged by four evaluators with hands-on experience in enterprise finance and accounting data. A deterministic pre-screening step applies exact and fuzzy string matching on numbers, cell references, and formulas, achieving 100% agreement with human judgments on these matchable items by construction; all semantic and partial-match judgments are made solely by human reviewers. Evaluators compared each model response against ground-truth annotations across all model and technique configurations on FRTR-Bench (159 queries), the SpreadsheetLLM benchmark (298 queries), and the FINCH benchmark (172 workflows), totaling over 200 person-hours. Mean Tool Calls captures the average number of search tool invocations per query. Mean Tokens reports total token consumption across input context, tool results, and generated responses. Mean Latency measures end-to-end inference time from query submission to final answer.

Results on FRTR-Bench. Table 2 and Fig. 4 present performance across all approaches on FRTR-Bench. BRTR with frontier models achieves exceptional accuracy: GPT-5 and Gemini 3 Pro both reach 99% (157/159 queries correct), an improvement of 25% points over the best single-pass FRTR baseline (Claude Opus 4.6 and Claude Sonnet 4.5, both at 74%). Claude Opus 4.6 and GPT-5.2 each achieve 98% with BRTR; among these, GPT-5.2 demonstrates notable efficiency with only 20,127 mean tokens and 90.81 s mean latency, the lowest resource consumption among high-accuracy configurations. Claude Sonnet 4.5 reaches 96% accuracy but requires 6.20 tool calls on average compared to 3.05 for GPT-5.2, suggesting a more exploratory search strategy. In addition, while Claude Opus 4.6 approaches state-of-the-art performance, it consumes nearly 10× the amount of tokens of Gemini 3 Pro, illustrating a tradeoff between token consumption and accuracy.

Iterative retrieval benefits models across capability levels: GPT-4o improves from 49% (FRTR) to 67%, and Gemini 2.5 Pro reaches 85% with only 2.23 tool calls. However, smaller models such as LLaMA 4 Maverick 17B show limited gains, suggesting insufficient reasoning capacity for effective multi-step search. Among baselines, SpreadsheetLLM compression performs poorly because intense compression excludes key information [10]. Claude in Excel reaches 84% but trails BRTR by 15 points, partly because it cannot read embedded images. Of 23 image-dependent queries, BRTR's only 2 errors are LLM vision misreads (overlapping bar-chart labels and a low-resolution scan), demonstrating that iterative image search with context pruning substantially closes this gap.

Results on SpreadsheetLLM Benchmark. Table 3 shows BRTR achieves state-of-the-art performance on the SpreadsheetLLM benchmark, with GPT-5 reaching 98% accuracy and Claude Sonnet 4.5 achieving 97%. These results surpass SpreadsheetLLM's own compression approach (89–91% for frontier models), demonstrating that iterative retrieval generalizes effectively beyond enterprise workbooks to diverse table understanding tasks. FRTR achieves competitive performance on this benchmark (84–87% for frontier models), approaching SpreadsheetLLM compression while using comparable token budgets, which suggests single-pass retrieval may suffice for simpler table understanding tasks. Nonetheless, BRTR's 7–11% point improvement over FRTR demonstrates clear value from iterative refinement. The naïve full-context baseline significantly underper-

Table 3. Cross-technique performance on the SpreadsheetLLM (SSLLM) benchmark (298 queries). Naïve = naïve full-context injection. BRTR achieves 97–98% accuracy with frontier models, while SSLLM compression reaches 89–91% and naïve approaches yield only 55–68%. FRTR, SSLLM, and Naïve use mean context sizes of 6,920, 5,811, and 13,631 tokens respectively. Operational cost columns report means. Bold indicates highest accuracy per technique. Models sorted by BRTR accuracy.

	Accuracy				BRTR Cost	
Model	BRTR	FRTR	SSLLM	Naïve	Tokens	Latency (s)
GPT-5	**0.98**	**0.87**	0.90	**0.68**	156,306	273.36
Claude Sonnet 4.5	0.97	0.84	**0.91**	0.55	166,117	280.59
Gemini 2.5 Pro	0.79	0.85	0.89	0.67	27,675	223.79
GPT-4o	0.56	0.77	0.78	0.35	256,523	315.93
LLaMA 4 Maverick 17B	0.30	0.53	0.73	0.41	18,608	408.73

forms all methods (35–68% accuracy), confirming that unstructured spreadsheet content overwhelms LLMs without compression or targeted retrieval [15].

Discussion. Across both benchmarks, BRTR's gains are larger on FRTR-Bench (25 percentage points over single-pass retrieval) than on the SpreadsheetLLM benchmark (7–11 percentage points), reflecting that iterative search most directly addresses the cross-sheet reasoning demands of complex enterprise workbooks. A capability threshold also emerges: frontier models consistently exceed 95% accuracy with BRTR, while smaller models show a striking reversal on the SpreadsheetLLM benchmark, where they perform worse with BRTR than with SpreadsheetLLM compression. This suggests that effective agentic reasoning requires not only function-calling capabilities but also the capacity to plan multi-step search strategies and refine queries based on partial evidence.

Table 4. Cross-technique performance on the FINCH benchmark (172 workflows). Naïve denotes single-call API-based agents; results for Claude Sonnet 4.5 and Gemini 3 Pro are reported by [7].

	Accuracy		BRTR Cost (Median)		
Model	BRTR	Naïve	Calls	Tok. (K)	Lat. (min)
Claude Opus 4.6	**0.95**	**0.63**	37	358	4.8
GPT-5.2	**0.95**	0.46	33	223	4.3
Claude Sonnet 4.5	0.87	0.20	25	305	2.7
Gemini 3 Pro	0.73	0.27	28	380	7.9
LLaMA 4 Maverick 17B	0.12	0.01	8	113	0.5

4.3 Experiment 3: FINCH Finance and Accounting Benchmark

To evaluate generalization to composite, multi-step enterprise workflows, we assess BRTR on FINCH [7], a benchmark of 172 finance and accounting workflows over 1,710 spreadsheets (27M cells) plus PDFs and images. 78.5% of workflows involve multiple interdependent tasks; the best prior accuracy is 41.9% [7].

Experimental Setup. Using the BRTR planner-executor architecture described in Sect. 3.3, we evaluate five models: Claude Opus 4.6, GPT-5.2, Claude Sonnet 4.5, Gemini 3 Pro, and LLaMA 4 Maverick 17B. Correctness is determined by expert human reviewers following the same evaluation protocol as Experiment 2, with LLM-based pre-screening and final pass/fail decisions made by human evaluators. We compare against a naïve baseline of single-call API-based agents evaluated on the same 172 workflows.

Results. Table 4 presents accuracy on FINCH. BRTR with Claude Opus 4.6 achieves 95.3% (164/172 workflows); GPT-5.2 follows at 94.8% (163/172). Improvements over naïve single-call agents are substantial: +48.9 percentage points for GPT-5.2, +66.9 for Claude Sonnet 4.5, and +46.0 for Gemini 3 Pro. Operational costs are heavily right-skewed, so we report medians; GPT-5.2 achieves the best efficiency trade-off at 223K tokens per workflow compared to 358K for Opus, while trailing by only 0.5 percentage points. LLaMA 4 Maverick 17B achieves only 11.6%, consistent with the capability threshold observed in Experiment 2.

Discussion. BRTR's planner-driven decomposition directly addresses the long-horizon composition bottleneck identified in prior work [7], where accuracy drops sharply on workflows with more than two tasks. By converting composite workflows into a dependency graph with concurrent execution, the planner prevents error accumulation that degrades monolithic approaches. Of the 8 tasks BRTR fails on, failures cluster into formula evaluation errors, instruction-following omissions, and incomplete multi-sheet generation, suggesting that execution fidelity rather than retrieval or planning is the primary bottleneck.

4.4 Ablation Study

To isolate the contribution of each BRTR component, we evaluate four ablated configurations on a representative 50-task subset of FINCH using Claude Opus 4.6. We ablate on FINCH rather than FRTR-Bench or SpreadsheetLLM because BRTR already saturates those benchmarks at 99% and 98%, leaving insufficient headroom for meaningful component-level comparison. The subset preserves the full benchmark's distribution across business types and task categories (all within ±4%), covers all artifact modalities (spreadsheets, PDFs, images, web search), and intentionally includes all eight tasks on which BRTR fails on the full benchmark, biasing the subset toward harder cases and yielding a conservative accuracy estimate.

Configurations. No Planner removes the planner-executor architecture: a single monolithic agent accesses the full tool set across all executors without task decomposition or specialized routing, retaining the iterative agent loop and hybrid BRTR retrieval. *Naïve* is the Claude Opus 4.6 API without any BRTR retrieval, as performed in the FINCH evaluation [7]. The LLM accesses excel files by reading ranges directly. *Full Context* replaces retrieval with direct file injection, serializing source files into the LLM prompt; the iterative loop and write tools are retained but all search tools are removed. *No Agent Loop* removes the iterative agent loop: a single LLM call with tools bound, where tool calls are executed but outputs are not fed back for further reasoning.

Table 5. Ablation study on a representative 50-task FINCH subset (Claude Opus 4.6). Each row removes one or more BRTR components. Tool Calls, Tokens, and Latency report medians. Bold indicates highest accuracy.

Configuration	Accuracy	Tool Calls	Tokens (K)	Latency (s)
BRTR (Full)	**0.88**	33	312	248
No Planner	0.68	25	521	175
Naïve	0.46	–	438	190
Full Context	0.52	16	583	114
No Agent Loop	0.00	1	11	5

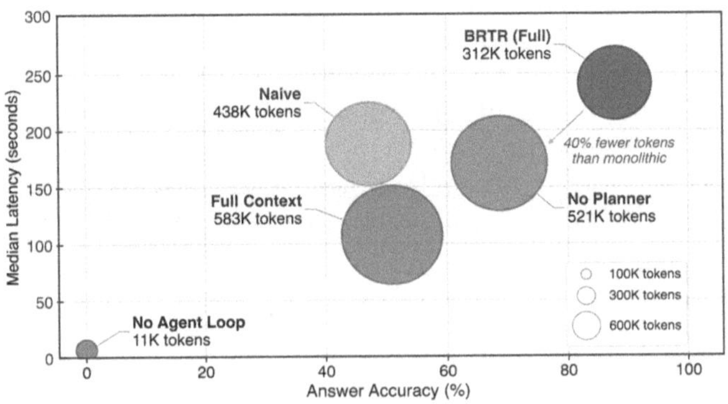

Fig. 5. Ablation on a 50-task FINCH subset (Claude Opus 4.6). Position encodes accuracy (x) and latency (y); bubble area encodes median token consumption. BRTR (Full) achieves the highest accuracy with 40% fewer tokens than No Planner.

Results. Table 5 and Fig. 5 present the ablation results. BRTR achieves 88% accuracy on the subset, compared to 68% for No Planner (−20 percentage points), 46% for Naïve (−42), 52% for Full Context (−36), and 0% for No Agent Loop. Each component contributes meaningfully, but the planner delivers the largest single improvement, particularly on cross-sheet retrieval tasks (86% vs. 43% without the planner). Retrieval is critical for web search (67% vs. 0% under Full Context) and tasks without source files to inject. The planner also improves token efficiency: despite more tool calls (median 33 vs. 25), BRTR consumes 40% fewer tokens (312K vs. 521K) than No Planner by decomposing tasks into focused sub-tasks with smaller context windows. Full Context consumes the most tokens (583K) yet achieves only 52%, and 20% of tasks are unreachable without retrieval (7 exceed the 200K context window; 3 require web search). Removing the agent loop entirely yields 0% accuracy, confirming that multi-turn reasoning is essential.

4.5 Cost Analysis

BRTR's iterative approach increases token consumption relative to single-pass baselines. On FRTR-Bench, frontier models consume 20K–92K tokens per query (Table 2), compared to 7.7K for FRTR and 12.7K for SpreadsheetLLM. On FINCH, the most capable configurations consume a median of 223K–358K tokens per workflow (Table 4), reflecting the multi-step nature of composite enterprise tasks. GPT-5.2 consistently achieves the best efficiency–accuracy trade-off across all benchmarks, matching frontier accuracy with lower token consumption.

4.6 Limitations

BRTR's iterative tool-calling loop consumes 3–12× more tokens than single-pass baselines; prompt caching [21], semantic caching, and efficient tool selection [19,20,22] can mitigate this in production. Smaller open-weight models do not benefit from agentic retrieval and can even underperform compression baselines, currently restricting BRTR to frontier-class LLMs; however, the framework is model-agnostic and requires no fine-tuning, so it will benefit as open-weight function-calling improves.

5 Conclusion

We presented Beyond Rows to Reasoning (BRTR), a multimodal agentic framework for spreadsheet understanding that replaces single-pass retrieval with an iterative tool-calling loop, supporting end-to-end Excel workflows from complex analysis to structured editing. A comprehensive evaluation of five multimodal embedding models on FRTR-Bench identified NVIDIA NeMo Retriever 1B as the top performer for mixed tabular and visual data. Across nine LLMs, supported by over 200 h of expert human evaluation, BRTR achieved state-of-the-art performance across three frontier spreadsheet understanding benchmarks, surpassing prior methods by 25 percentage points on FRTR-Bench, 7 points on SpreadsheetLLM, and 32 points on FINCH. Ablation experiments confirmed that the planner, retrieval, and iterative reasoning each contribute substantially, and cost analysis showed GPT-5.2 achieves the best efficiency–accuracy trade-off. Throughout all evaluations, BRTR maintained full auditability through explicit tool-call traces. Future work will explore extending BRTR to additional structured document types beyond spreadsheets, incorporating learned retrieval strategies that adapt to query complexity, and evaluating the framework in production enterprise deployments with real-time user feedback.

References

1. Anthropic: Claude in Excel: AI-powered spreadsheet assistant. https://support.anthropic.com/en/articles/12650343-claude-in-excel (2025)
2. Anthropic: Claude Cowork: Agentic desktop AI for knowledge work. https://claude.com/blog/cowork-research-preview (2026)
3. Cao, L., et al.: Fortune: Formula-driven reinforcement learning for symbolic table reasoning in language models (2025). https://arxiv.org/abs/2505.23667
4. Chen, Y., et al.: SheetAgent: towards a generalist agent for spreadsheet reasoning and manipulation via large language models. In: Proceedings of the ACM Web Conference (WWW) (2025). https://doi.org/10.1145/3696410.3714962
5. Cheng, Z., et al.: FORTAP: using formulas for numerical-reasoning-aware table pretraining. In: Proceedings of the 60th Annual Meeting of the Association for Computational Linguistics (ACL), Volume 1: Long Papers, pp. 1150–1166 (2022). https://doi.org/10.18653/v1/2022.acl-long.82
6. Cormack, G.V., Clarke, C.L.A., Buettcher, S.: Reciprocal rank fusion outperforms condorcet and individual rank learning methods. In: Proceedings of the 32nd International ACM SIGIR Conference on Research and Development in Information Retrieval, pp. 758–759. SIGIR '09, Association for Computing Machinery, New York, NY, USA (2009). https://doi.org/10.1145/1571941.1572114
7. Dong, H., et al.: Finch: Benchmarking finance & accounting across spreadsheet-centric enterprise workflows (2026). https://arxiv.org/abs/2512.13168
8. Dong, H., et al.: SpreadsheetLLM: encoding spreadsheets for large language models. In: Proceedings of the 2024 Conference on Empirical Methods in Natural Language Processing, EMNLP (2024)
9. Google: Collaborate with Gemini in Google Sheets. https://support.google.com/docs/answer/14356410 (2025)
10. Gulati, A., Sen, S., Sarguroh, W., Paul, K.: From rows to reasoning: a retrieval-augmented multimodal framework for spreadsheet understanding. In: Proceedings of the 32nd ACM SIGKDD Conference on Knowledge Discovery and Data Mining (KDD) (2026). https://doi.org/10.1145/3770855.3817962
11. Herzig, J., Nowak, P.K., Müller, T., Piccinno, F., Eisenschlos, J.: TAPAS: weakly supervised table parsing via pre-training. In: Proceedings of the 58th Annual Meeting of the Association for Computational Linguistics, pp. 4320–4333. Association for Computational Linguistics (2020). https://doi.org/10.18653/v1/2020.acl-main.398
12. Jiang, C., Cheng, M., Tao, X., Mao, Q., Ouyang, J., Liu, Q.: TableMind: An autonomous programmatic agent for tool-augmented table reasoning (2025). https://arxiv.org/abs/2509.06278

13. Kundurthy, S., et al.: SpreadsheetArena: Decomposing preference in LLM generation of spreadsheet workbooks (2026). https://www.meridian.ai/research/SpreadsheetArena.pdf, preprint, February 2026
14. Lewis, P., et al.: Retrieval-augmented generation for knowledge-intensive NLP tasks (2021). https://arxiv.org/abs/2005.11401
15. Liu, N.F., et al.: Lost in the middle: how language models use long contexts. Trans. Assoc. Comput. Linguist. **12**, 157–173 (2024). https://doi.org/10.1162/tacl_a_00638
16. Liu, Q., et al.: TAPEX: table pre-training via learning a neural SQL executor. In: International Conference on Learning Representations (ICLR) (2022). https://openreview.net/forum?id=O50443AsCP
17. LlamaIndex: LlamaSheets: Turn messy spreadsheets into AI-ready data. https://www.llamaindex.ai/ (2025)
18. Lumer, E., et al.: Tool and agent selection for large language model agents in production: A survey. Preprints (2025). https://doi.org/10.20944/preprints202512.1050.v1
19. Lumer, E., Gulati, A., Subbiah, V.K., Basavaraju, P.H., Burke, J.A.: ScaleMCP: dynamic and auto-synchronizing model context protocol tools for LLM agents. In: Proceedings of the International Joint Conference on Computational Intelligence (IJCCI) (2025)
20. Lumer, E., Gulati, A., Subbiah, V.K., Basavaraju, P.H., Burke, J.A.: MemTool: optimizing short-term memory management for dynamic tool calling in LLM agent multi-turn conversations. In: Proceedings of the 48th European Conference on Information Retrieval, ECIR (2026)
21. Lumer, E., et al.: Don't break the cache: An evaluation of prompt caching for long-horizon agentic tasks (2026). https://arxiv.org/abs/2601.06007
22. Lumer, E., Subbiah, V.K., Burke, J.A., Basavaraju, P.H., Huber, A.: Toolshed: Scale tool-equipped agents with advanced rag-tool fusion and tool knowledge bases (2024). https://arxiv.org/abs/2410.14594
23. Ma, Z., et al.: SpreadsheetBench: Towards challenging real world spreadsheet manipulation (2024). https://arxiv.org/abs/2406.14991
24. Microsoft: Building agent mode in Excel. https://techcommunity.microsoft.com/blog/excelblog/building-agent-mode-in-excel/4457320 (2025)
25. Payan, J., et al.: InstructExcel: A benchmark for natural language instruction in excel (2023). https://arxiv.org/abs/2310.14495
26. Singh, A., Ehtesham, A., Kumar, S., Khoei, T.T.: Agentic retrieval-augmented generation: A survey on agentic rag (2025). https://arxiv.org/abs/2501.09136
27. Wang, Z., et al.: TUTA: tree-based transformers for generally structured table pre-training. In: Proceedings of the 27th ACM SIGKDD Conference on Knowledge Discovery & Data Mining, pp. 1780–1790. KDD '21, ACM (2021). https://doi.org/10.1145/3447548.3467434
28. Wang, Z., et al.: SheetBrain: A neuro-symbolic agent for accurate reasoning over complex and large spreadsheets (2025). https://arxiv.org/abs/2510.19247
29. Xiong, S., et al.: TableZoomer: A collaborative agent framework for large-scale table question answering (2025). https://arxiv.org/abs/2509.01312
30. Yao, S., et al.: ReAct: synergizing reasoning and acting in language models. In: International Conference on Learning Representations (ICLR) (2023). https://openreview.net/forum?id=WE_vluYUL-X
31. Zhu, R., et al.: SheetMind: An end-to-end LLM-powered multi-agent framework for spreadsheet automation (2025). https://arxiv.org/abs/2506.12339

Beyond the Page Break: An LLM-Based Solution for Cross-Page Table Reconstruction

Yu Tang[1,2], Hongwei Li[3], Yixuan Cao[1,2], and Ping Luo[1,2](✉)

[1] State Key Laboratory of AI Safety, Institute of Computing Technology, Chinese Academy of Sciences (CAS), Beijing 100190, China
[2] University of Chinese Academy of Sciences, CAS, Beijing 100049, China
luop@ict.ac.cn
[3] Beijing Paoding Technology Co., Ltd., Beijing, China

Abstract. In multi-page documents, a table may span more than one page, known as a **cross-page table**. Since many downstream tasks rely on the structural integrity of such tables, their reconstruction has become a critical subtask in multi-page document parsing. However, existing multi-page document parsing methods either ignore cross-page table reconstruction entirely or implement it via heuristic rules, leading to erroneous fragmentation or poor generalization. In particular, real-world cross-page tables often exhibit challenging patterns—such as **vertical splitting**, **header repetition**, **cross-page cells**, and **dimension mismatch**—which rule-based methods struggle to resolve robustly. To overcome these limitations, we propose **TabStitcher**, an LLM-based, two-stage framework for cross-page table reconstruction that can be seamlessly integrated with existing single-page parsers. It first identifies table fragments across consecutive pages and then merges them into a coherent, unified table structure. Furthermore, to address the scarcity of costly manually annotated data for training, we propose a data synthesis framework that can generate large-scale cross-page table instances with challenging patterns, enabling the model to learn to handle such cases effectively. Experiments on real-world benchmarks demonstrate TabStitcher's effectiveness in both detection (98.3 F1-score) and merging tasks (96.6 TEDS). Besides, its integration with diverse off-the-shelf single-page parsers yields substantial improvements (at least 11.9 CPT-EDS higher than the best baseline) in end-to-end cross-page table reconstruction. To the best of our knowledge, this work presents the first systematic study of cross-page table reconstruction in multi-page document parsing and offers a practical solution to this challenging task.

Keywords: Cross-page table reconstruction · Multi-page document parsing · Challenging cross-page table patterns · LLM-based

1 Introduction

Document parsing is a process of converting PDF or image-based documents into structured, machine-readable formats like Markdown or HTML, serving

G. A. Fink et al. (Eds.): ICDAR 2026, LNCS 16972, pp. 594–610, 2027.
https://doi.org/10.1007/978-3-032-36023-6_34

as the cornerstone of many downstream tasks, such as document information extraction and document-based question answering. For multi-page documents, mainstream parsing methods typically consist of two stages. In the first stage, the document is parsed page by page. In the second stage, the single-page parsing results are integrated into a document-level parsing output. While such a strategy offers substantial efficiency advantages over fully end-to-end parsing, it also faces significant challenges due to the presence of cross-page elements—where a single coherent structure, such as a paragraph or a table, is split across consecutive pages. Since many downstream tasks require their structural integrity in the final parsing output, it is necessary to reconstruct these cross-page elements from single-page parsing results.

Among all cross-page elements, the reconstruction of cross-page tables is particularly critical. This is first because of their high prevalence, especially in financial documents such as annual reports. In an analysis of approximately one million pages from collected financial documents, we find that up to 15% of tables span multiple pages. Second, tables often encode structured quantitative information, where even minor structural biases (e.g., missing rows or misaligned columns) can lead to significant inaccuracies or faulty conclusions in downstream tasks. For instance, in tabular numerical calculations within documents [8]—such as computing the sum of all rows in each column—leaving cross-page table fragments unmerged or merging them with column misalignment can both lead to erroneous results, which is unacceptable for applications in the financial domain.

However, when combining single-page parsing results, existing methods [1, 5, 11–13, 15] either neglect the cross-page table reconstruction or rely on heuristic rules, which often result in erroneous fragmentation or incorrect merging. For example, if the table fragment on the first page is not the last element on that page (as shown in Fig. 1), or if the fragment on the second page is not the first element (as shown in Fig. 2), rule-based approaches may erroneously assume they belong to different tables. In particular, as shown in Fig. 1 to 4, real-world cross-page tables may exhibit challenging patterns, such as **vertical splitting**, **header repetition**, **cross-page cells**, and **dimension mismatch**, which further complicate their reconstruction. For instance, for the cross-page table shown in Fig. 3, a single cell's text "*Indirect energy*" is split into "*Indirect*" and "*energy*" across pages. Simple row-wise concatenation of table fragments treats the two parts as separate cells, producing a biased table reconstruction result.

To overcome these limitations, we introduce **TabStitcher**, an LLM-based pipeline for accurate cross-page table reconstruction in multi-page document parsing. The pipeline consists of two stages—the detection stage and the merging stage. In the detection stage, the content of each pair of consecutive pages is input into the LLM to detect cross-page table fragments. In the subsequent merging stage, these detected fragments are fed again into the LLM to produce semantically coherent reconstructed tables. For minimizing inference overhead and simplifying deployment, TabStitcher employs a shared lightweight LLM with 0.6 billion parameters to perform both detection and merging tasks, guided by task-specific prompts. Besides, because the detection stage operates on independent page pairs and most merging operations can be processed in parallel, the pipeline can efficiently handle long documents with hundreds of pages.

Furthermore, since publicly available cross-page table datasets are scarce and manual annotation is costly, we develop a cross-page table data synthesis framework. It takes a single table as input and splits it into two parts to simulate cross-page table fragments. It can offer two key advantages. First, it enables the large-scale generation of cross-page table data for model training. Second, it can synthesize cross-page tables exhibiting diverse challenging patterns like real-world scenarios, which allows our model to learn robust reconstruction strategies for such cases.

Given the absence of publicly available benchmarks, to evaluate TabStitcher's performance in real-world scenarios, we construct a benchmark curated from real-world English and Chinese documents. Extensive experiments on the benchmark demonstrate TabStitcher's effectiveness. In the detection task, TabStitcher achieves an average F1-score of 98.3, significantly outperforming the state-of-the-art baseline by 6.4 points. In the merging task, it achieves an average RTEDS score that is 2.5 points higher than the best baseline. Further studies show that TabStitcher can maintain consistent performance under challenging patterns, while existing methods suffer from significant performance degradation. Moreover, in evaluating end-to-end cross-page table reconstruction performance, diverse off-the-shelf single-page parsers integrated with TabStitcher consistently exhibit substantial improvements (at least 11.9 CPTEDS higher than the best baseline), demonstrating TabStitcher's excellent compatibility and adaptability. All these results confirm that TabStitcher delivers both high accuracy and strong generalization for cross-page table reconstruction in real-world scenarios.

We summarize our contributions as follows:

- We propose TabStitcher, a two-stage LLM-based pipeline to accurately reconstruct cross-page tables for multi-page document parsing. Compared to existing approaches, our method achieves more semantically coherent and structurally accurate results. Moreover, owing to its plug-and-play design, our method can be seamlessly integrated into various off-the-shelf single-page parsers, endowing them with the ability to parse cross-page tables.
- We first conduct a systematic analysis of cross-page table patterns and identify four challenging patterns. Based on that, we design a cross-page table data synthesis framework for large-scale training without costly manual annotations, enabling our model to learn how to handle those challenging cases effectively.
- We build a real-world benchmark and conduct extensive experiments to validate TabStitcher's effectiveness. On both the detection and merging tasks, TabStitcher significantly outperforms the baselines. Further experimental results highlight the robustness of TabStitcher in handling cross-page tables with challenging patterns. Additionally, successful integration with diverse single-page parsers demonstrates TabStitcher's excellent compatibility and adaptability.

2 Related Work

For multi-page document parsing, contemporary approaches can be broadly categorized into two streams: end-to-end parsing and page-wise parsing.

2.1 End-to-End Parsing

End-to-end approaches are typically based on large vision-language models, taking all page images as input and performing sequence-to-sequence generation. Depending on whether they are specifically designed and trained for document tasks, these models can be further categorized into two branches:

General-Purpose Models. Notable general-purpose models include GPT-4V [7], the Gemini series [4,14], the Qwen-VL series [2,16], and the InternVL series [17,23]. These models benefit from large-scale pre-training on diverse vision-language data and exhibit strong zero-shot performance on the document parsing task.

Document-Specific Models. For document-specific models, optimized architectures tailored to document structural features enable their superior performance in capturing layout information, reading order, and fine-grained text localization. Representative works include TextMonkey [10], GOT [18], and the mPLUG-DocOwl series [6,20].

However, despite various optimizations, end-to-end methods still suffer from low inference speed in parsing multi-page documents due to long input contexts and output sequences. Moreover, attention dilution and the difficulty of modeling long-range dependencies can also lead to degraded model performance in such scenarios.

2.2 Page-Wise Parsing

Page-wise parsing methods decompose the multi-page document parsing workflow into two stages: page-level parsing and post-processing. In the page-level parsing stage, all pages are parsed independently. In the following post-processing stage, the individual page-level parsing results are aggregated to produce a document-level output. Since page parsing can be executed in parallel, page-wise parsing approaches are typically more efficient than end-to-end approaches and have thus become the dominant paradigm in practice.

Unfortunately, most existing approaches either overlook this reconstruction of the cross-page tables entirely or implement it using manually designed rules. For example, Nougat [3], Dolphin [5], Nanonets-OCR-S [11], and olmOCR [13] simply concatenate the page contents without considering continuity. Docling [1] and MonkeyOCR [9] introduce simple heuristics to identify and merge cross-page paragraphs, but can not deal with cross-page tables. To the best of our knowledge, MinerU [15] and Marker [12] are the only two existing methods that support cross-page table reconstruction. MinerU employs a sophisticated rule set to detect and merge cross-page table fragments. Marker invokes an external multimodal large language model to determine whether two table fragments in

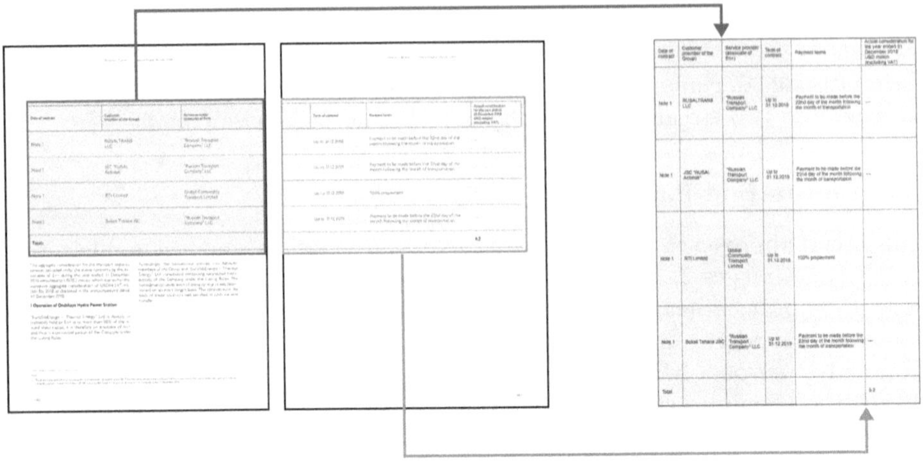

Fig. 1. A cross-page table example with vertical splitting. Left: two consecutive pages with a cross-page table. Right: the reconstructed table. The table is vertically split into left (highlighted in purple) and right (highlighted in orange) fragments. (Color figure online)

adjacent pages should be merged and the merging direction (vertical or horizontal), but the final merging is performed via simple row-wise or column-wise concatenation. Nevertheless, both MinerU and Marker are inherently limited in generalizability. For example, MinerU cannot detect vertically split tables, while Marker fails to handle repeated headers.

3 Challenging Patterns in Cross-Page Tables

In real-world documents, various factors—arising from document design features or bias in the prior table structure recognition stage—always introduce challenging patterns that make accurate reconstruction of cross-page tables non-trivial. Based on an analysis of a diverse set of real-world instances, we identify four key challenging patterns in cross-page tables below. All of them cannot be easily resolved by rule-based strategies, highlighting the need for a merging approach that incorporates contextual reasoning and layout understanding—capabilities that motivate our use of large language models.

3.1 Vertical Splitting

In most cases, cross-page tables are split horizontally along rows. However, some wide tables may be split vertically instead. For example, in Fig. 1, the cross-page table is split along columns, producing left (highlighted in purple) and right (highlighted in yellow) fragments across two consecutive pages.

Before integrating table fragments, it is important to determine the merging direction. However, this cannot be achieved by simply comparing the row or column counts of fragments. First, fragments may coincidentally share the same row and column counts. Second, and more critically, as discussed in Sect. 3.4, the extracted table fragments may suffer from dimension mismatch problems, leading to inaccurate estimates of row and column counts.

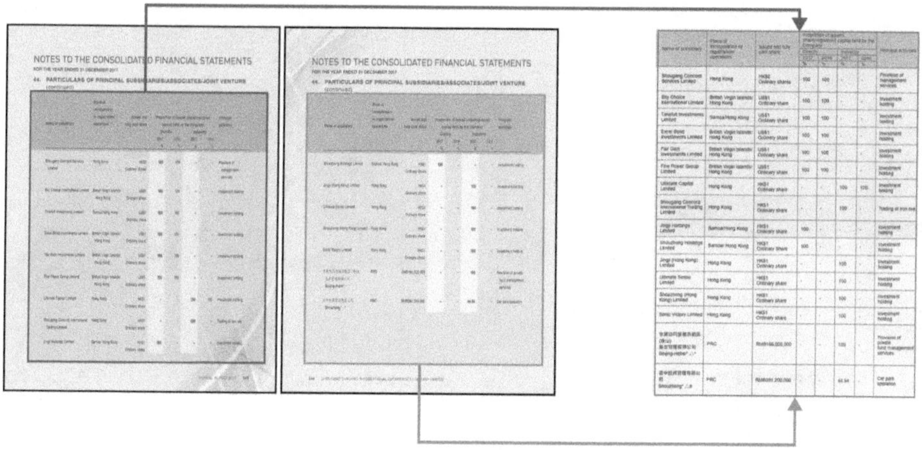

Fig. 2. A cross-page table example with header repetition. Left: two consecutive pages with a cross-page table. Right: the reconstructed table. Both fragments share the identical table header (highlighted in red). (Color figure online)

3.2 Header Repetition

When a table spans multiple pages, it is common practice to repeat the column headers (for horizontal splits) or row headers (for vertical splits) on each page to improve readability. For example, table fragments shown in Fig. 2 share the same column header (highlighted in red).

When merging fragments with duplicated headers, models should retain only the first header and discard the others. A normal heuristic for repeated header identification is to compute the longest prefix of matching rows (for column headers) or columns (for row headers). However, this approach has two major limitations. First, identical leading rows or columns are not necessarily headers—they may represent repeated data entries. Blindly treating them as headers can lead to incorrect suppression of valid content. Second, real-world repeated headers may exhibit minor variations across pages due to rendering inconsistencies or parsing errors.

3.3 Cross-Page Cells

Another challenging pattern occurs when a table is split in the middle of cells. In such cases, the text content within the cell may be fragmented across the page

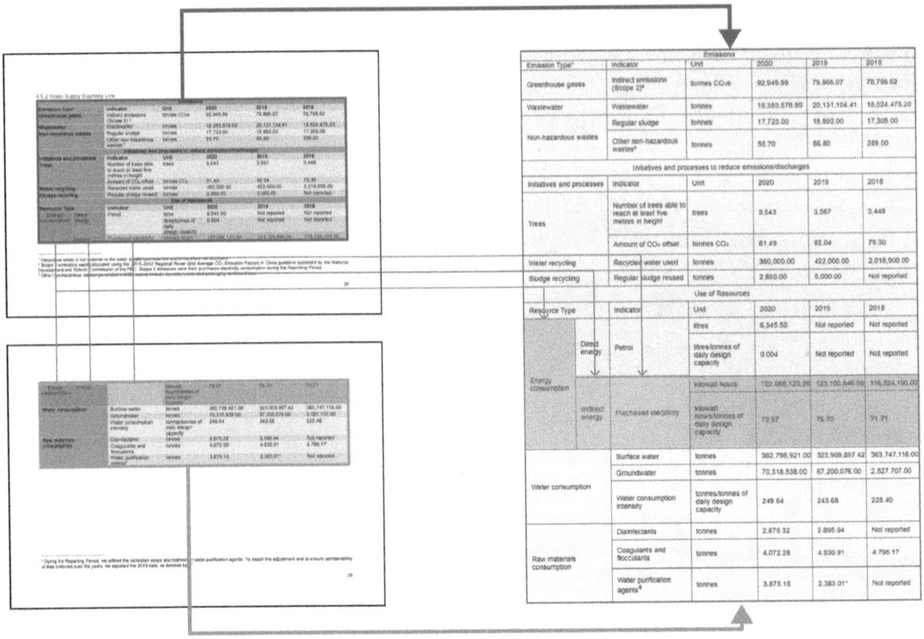

Fig. 3. A cross-page table example with cross-page cells. Left: two consecutive pages with a cross-page table. Right: the reconstructed table. Though similarly located at the boundary, the cells highlighted in red are cross-page cells, while the cells highlighted in green are not. (Color figure online)

break. For instance, in the cross-page table shown in Fig. 3, three cells are split across two pages (highlighted in orange): (1) The first cell's text "Energy consumption" is repeated in both cell fragments; (2) The second cell's text "Indirect energy" is cleanly split into "Indirect" and "energy"; (3) The third cell, though split, retains the full text "Purchased electricity" in its first fragment, with no content in the second.

This introduces ambiguity in merging table fragments, as the model must identify cross-page cells and merge them. This is non-trivial: even cells within the same row or column may differ in their cross-page status. For example, in Fig. 3, the cells highlighted in red and green, although similarly located in the last row of the first fragment and the first row of the second fragment, are independent cells in fact.

3.4 Dimension Mismatch

Due to various factors in the upstream table structure recognition stage, the detected number of columns or rows in a table fragment may deviate from the ground truth, leading to a dimension mismatch among fragments of the same cross-page table.

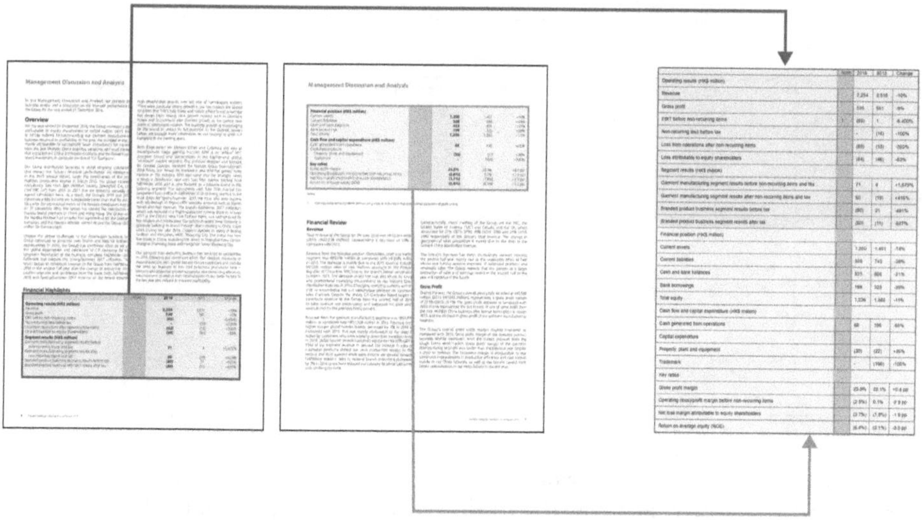

Fig. 4. A cross-page table example with dimension mismatch. Left: two consecutive pages with a cross-page table. Right: the reconstructed table. The column "Note" (highlighted in red) is empty on the second page. Since the table is borderless, the parsing of the second page may miss this column.

This phenomenon frequently occurs in borderless tables. For example, in Fig. 4, the column "Note" (highlighted in red) is empty on the second page. Due to the lack of explicit delimiters, the parser may fail to detect this column, leading to a mismatch in column count between the two fragments.

A more subtle source stems from the parser's bias in recognizing spanning attributes of cells. For example, in Fig. 5, the last three cells in the first table fragment each correspond to three columns in the second fragment (i.e., colspan = 3), but the parser may fail to infer this spanning behavior without seeing the continuation on the next page.

4 TabStitcher

Figure 6 illustrates an overview of TabStitcher, our LLM-based pipeline to accurately reconstruct cross-page tables in multi-page document parsing. The pipeline consists of two sequential stages: detection and merging. A shared lightweight LLM is employed in both stages, with task-specific prompts guiding the model toward distinct behaviors. This unified architecture reduces deployment complexity, minimizes inference overhead, and enables implicit knowledge transfer across stages while preserving functional specialization.

Notably, we choose an LLM-based architecture rather than an MLLM-based architecture for two main reasons. First, while integrated with an off-the-shelf single-page parser, the model can directly leverage the parsing results to access both semantic and layout information of tables without extra visual modules.

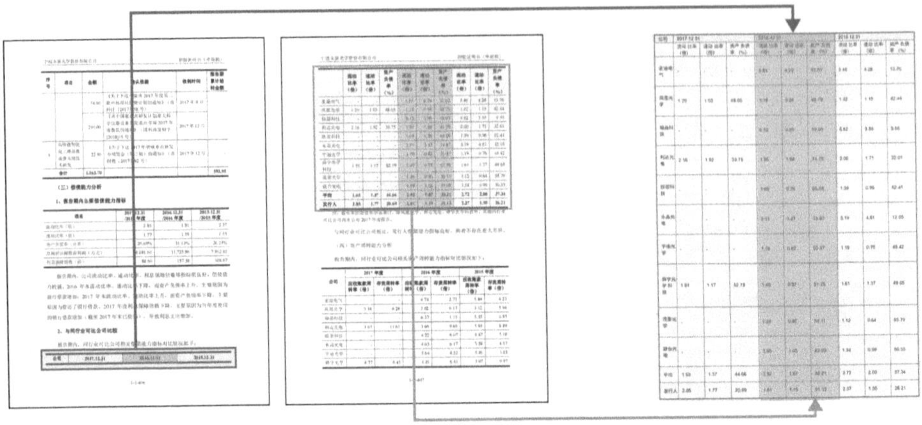

Fig. 5. Another cross-page table example with dimension mismatch. Left: two consecutive pages with a cross-page table. Right: the reconstructed table. In the first fragment, the last three cells each have `colspan=3`, corresponding to three columns in the second fragment. One such cell and its matching columns in the second fragment are highlighted in red for illustration.

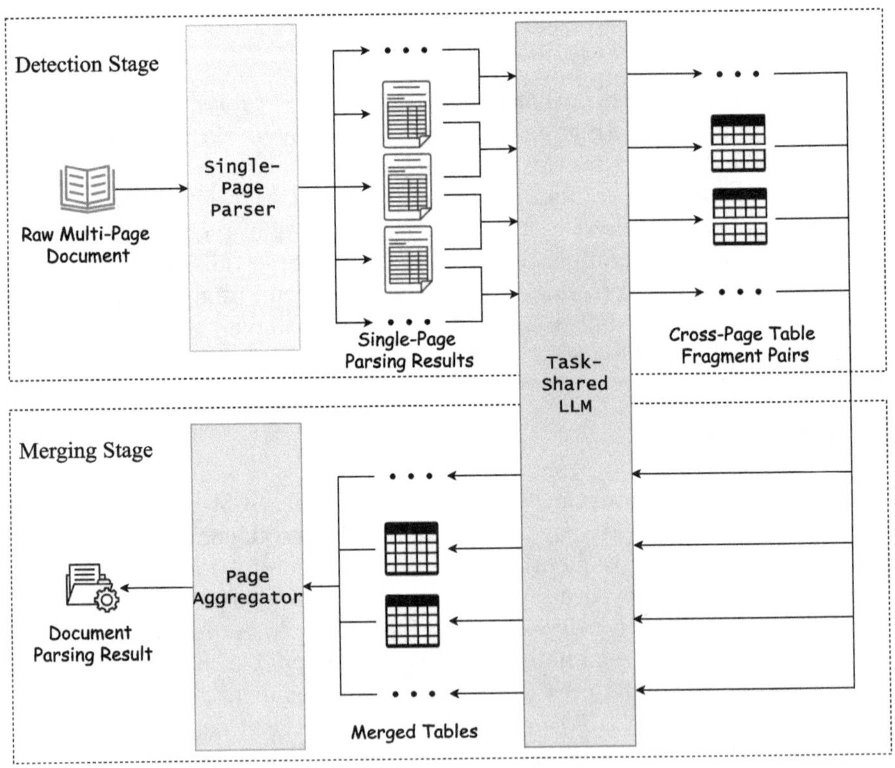

Fig. 6. The overview of the TabStitcher pipeline.

Second, and more importantly, as shown in Sect. 5, our data synthesis framework enables large-scale generation of high-quality synthetic textual training data tailored for LLMs. In contrast, generating realistic, high-fidelity images of cross-page tables for MLLM training remains prohibitively challenging.

4.1 Detection Stage

The goal of the detection stage is to identify table fragments that logically belong to the same cross-page table across consecutive pages. In this stage, for every two consecutive pages, the prompt consists of their single-page parsing results produced by an off-the-shelf parser. Within the parsing results, tables are represented in HTML to support complex structural features such as cell spanning, while all other content is formatted in Markdown. Notably, since parsing results in other formats like LaTeX can be readily converted to such a format, our method is compatible with a wide range of off-the-shelf document parsers without requiring architectural modifications.

In the detection prompt, the elements (e.g., paragraphs, tables) within a single-page parsing result are sequentially indexed in reading order. The LLM is instructed to determine whether any tables are split across the page boundary. If so, it returns a set of tuples, each containing the corresponding indexes of a cross-page table fragment pair. This enables direct mapping for the subsequent merging stage.

Formally, for a document with N pages, denoted as $\mathcal{P} = \langle P_1, P_2, \ldots, P_N \rangle$, let $\mathcal{E}_i = \{e_i^1, e_i^2, \ldots\}$ be the single-page parsing result of page P_i, which consists of block-level elements. The detection stage performs $N-1$ independent LLM invocations:

$$\mathcal{D}_i = \mathit{Detect}_\theta(\mathcal{E}_i, \mathcal{E}_{i+1}), \quad i = 1, \ldots, N-1, \tag{1}$$

where Detect_θ denotes the LLM parameterized by θ, guided by a detection-specific prompt. The output $\mathcal{D}_i$ is a set of tuples (e_i^k, e_{i+1}^l), each indicating that table fragment $e_i^k \in \mathcal{E}_i$ on page P_i and $e_{i+1}^l \in \mathcal{E}_{i+1}$ on page P_{i+1} logically belong to the same cross-page table. The full set of detected cross-page table fragment pairs is aggregated as:

$$\mathcal{D} = \bigcup_{i=1}^{N-1} \mathcal{D}_i. \tag{2}$$

This formulation results in $\mathcal{O}(N)$ computational complexity and supports full parallelization across all page pairs, enabling scalable processing of long documents.

4.2 Merging Stage

In the merging stage, all detected cross-page table fragment pairs in $\mathcal{D}$ are reintegrated into full tables. Each merging prompt to the LLM consists of a pair of cross-page table fragments identified in the detection stage, which are represented in HTML format. The LLM is prompted to generate a single, semantically and structurally consistent HTML table that integrates both fragments.

Formally, the merging stage takes $\mathcal{D}$ as input and produces a set of merged elements $\mathcal{M}$. For each pair $(e_i^k, e_{i+1}^l) \in \mathcal{D}$, the merged result $m_{i,i+1}^{k,l} \in \mathcal{M}$ is determined as follows:

$$m_{i,i+1}^{k,l} = Merge_\theta(e_i^k, e_{i+1}^l) \tag{3}$$

where Merge$_\theta(\cdot)$ represents the same LLM parameterized by θ under a table merging prompt that generates the unified table.

Finally, the document-level parsing output $\mathcal{R}$ is constructed by replacing all detected cross-page table fragments with their corresponding merged table in the original single-page results $\{\mathcal{E}_i\}_{i=1}^N$, and then concatenating these single-page parsing results in page order:

$$\mathcal{R} = Aggregate\left(\{\mathcal{E}_i\}_{i=1}^N; \mathcal{D}, \mathcal{M}\right). \tag{4}$$

While most merging operations are independent and can be executed in parallel, for these tables that span three or more pages, their fragments are merged sequentially to maintain global structural consistency. This hybrid execution strategy ensures both efficiency and accuracy in complex scenarios.

5 Synthesis of Cross-Page Table Data

To address the scarcity of cross-page table data and avoid costly manual annotation, we propose a synthetic data generation framework. It takes a complete HTML table as input and transforms it into a pair of table fragments that simulate realistic cross-page instances.

The synthesis process consists of the following stages:

1. **Split Location Sampling.** We randomly select a row index at which to split the table into top and bottom fragments. This simulates the common case of horizontally split tables in multi-page documents.
2. **Simulation of Cross-Page Cells.** We select cells in the last row of the first fragment and split them across the page boundary. The text content of each selected cell is either randomly split between the two resulting fragments or duplicated in both.
3. **Simulation of Header Repetition.** If the split occurs below the header region, we duplicate the header rows at the beginning of the second fragment.
4. **Simulation of Dimension Mismatch.** We optionally perform the following operations: (1) Empty Column Removal: We remove entirely empty columns from fragments, mimicking the behavior of parsers on borderless tables, where such columns are often not detected. (2) Spanning Attribute Reduction: We modify `colspan` attributes in cases where spanning behavior cannot be reliably inferred from local layout cues. Specifically, in a table fragment, if all cells in a given column are merged with cells in the preceding column, we merge them and reduce the `colspan` values of these cells.
5. **Simulation of Vertical Splitting**: To cover both vertical and horizontal splitting patterns, we transpose the original table and its fragments to generate synthetic vertical splitting examples.

Notably, each operation is applied probabilistically and may be skipped in practice. The pipeline above can directly generate data for the merging task. For the detection task, we similarly apply the pipeline in splitting the content of a page with tables to simulate two consecutive pages.

6 Experimental Setup

6.1 Training Setup

Training Dataset. We build a document collection, which consists of publicly available documents crawled from the internet, as well as our private internal documents in both English and Chinese. For each page in the collection, we apply a high-precision closed-source single-page parser to get its single-page parsing result. Then, we use the method described in Sect. 5 to synthesize training samples. This training dataset includes about 400,000 samples for the detection task and 100,000 samples for the merging task.

Training Settings. The LLM used in TabStitcher is fine-tuned based on Qwen3-0.6B [19]. Training is conducted using the LLaMA-Factory library [21] with the following hyperparameters: a batch size of 8, a learning rate of 1×10^{-5}, the AdamW optimizer, and a cosine learning rate schedule. The model is trained for one epoch on 8 NVIDIA GeForce RTX 4090 GPUs.

6.2 Evaluation Setup

Datasets. The dataset consists of 1200 real-world consecutive page pairs randomly sampled from our document collection that were not used in training. To avoid trivial cases in the detection task, each page includes at least one table. In the dataset, half of the page pairs contain cross-page tables (positive subset), while the other half do not (negative subset). For positive samples, each page pair is manually annotated to localize potential cross-page table fragments, along with the fully reconstructed cross-page table as ground truth. Both positive and negative subsets contain approximately equal numbers of English and Chinese samples. In evaluating the detection task, both the positive and negative subsets are used. In evaluating the merge task and the end-to-end reconstruction task, only the positive subset is used.

Baselines. We compare our TabStitcher with MinerU [15], Marker [12] and GPT-4o [7]. MinerU employs heuristic rules to detect and merge cross-page table fragments. Marker calls an external MLLM (`gemini-2.0-flash` by default) to detect cross-page table fragments and decide the merging direction, but the merging itself is performed via simple row-wise or column-wise concatenation. GPT-4o leverages the parsing results and page images as the context during both the detection and merging stages. We also substitute the fine-tuned LLM in TabStitcher with pretrained Qwen3-0.6B and Qwen3-32B models [19] as additional baselines, while holding all other components fixed. Furthermore, in the merging task, a naive merging strategy (denoted as "Naive") is also compared, which directly conducts row-wise concatenation for table fragments.

Metrics. In the detection task, we calculate the precision, recall, and F1-score for each model. For positive samples, a prediction is considered true when the model correctly identifies all cross-page table fragment pairs. For negative samples, a prediction is deemed true only if the model correctly determines that no merging is required.

In the merging task, we evaluate merging quality using the Tree-Edit-Distance-based Similarity (TEDS) score [22], which measures the structural similarity between the reconstructed table and the ground-truth. The higher scores indicate better structural fidelity. Moreover, to better quantify the improvement of each model over the naive merging strategy (e.g., direct row-wise concatenation), we also introduce a Relative TEDS (RTEDS) metric, defined as:

$$RTEDS_{model} = \frac{TEDS_{model} - TEDS_{Naive}}{100 - TEDS_{Naive}} \times 100, \tag{5}$$

Notably, in both the detection and merging tasks, the quality of the upstream single-page parser may affect the final results. To ensure a fair comparison, we uniformly adopt the parsing output from the high-precision closed-source single-page parser.

For evaluating the end-to-end cross-page table reconstruction performance, we proposed the Cross-Page TEDS (CPTEDS) as the metric, defined as:

$$CPTEDS = \begin{cases} TEDS, & \text{if the cross-page table fragments is correctly detected} \\ 0, & \text{otherwise} \end{cases} \tag{6}$$

7 Experimental Results

7.1 Detection Task

As shown in Table 1, overall, TabStitcher achieves the highest recall scores (97.3), demonstrating its strong capability to identify potential cross-page table fragments. And it also attains nearly perfect precision (99.3), as well as MinerU (99.0), indicating that they are highly effective at filtering out spurious merges. This is expected for MinerU, as its rule-based criteria require a strict combination of layout and textual conditions to trigger a merge decision.

Overall, TabStitcher achieves an F1-score of 98.3, far surpassing the best baseline (+6.4). These results confirm that our LLM-based approach enables robust detection of cross-page table fragments across diverse languages and complex layouts.

Table 1. The evaluation results of the detection task. The best entries are marked in **bold**. Note that for Qwen3-0.6B, since it produces negative predictions for all samples, its precision and F1-score are undefined.

Method	Type	English			Chinese			Overall		
		P.	R.	F1.	P.	R.	F1.	P.	R.	F1.
MinerU	Rule-based	98.9	60.7	75.3	99.2	76.1	86.1	99.0	68.7	81.1
Marker	MLLM-based	96.3	72.4	82.7	91.2	40.0	55.6	94.4	55.7	70.1
GPT-4o	MLLM-based	94.9	87.7	91.2	94.5	91.0	92.7	94.7	89.3	91.9
Qwen3-0.6B	LLM-based	-	0.0	-	-	0.0	-	-	0.0	-
Qwen3-32B	LLM-based	69.0	55.2	61.3	73.3	76.1	74.7	71.5	66.0	68.6
TabStitcher	LLM-based	**99.3**	**95.2**	**97.2**	**99.4**	**99.4**	**99.4**	99.3	**97.3**	**98.3**

Notably, although similarly applying the LLM in the detection task, TabStitcher significantly outperforms the same-sized pretrained Qwen3-0.6B and even far surpasses the much larger Qwen3-32B and GPT-4o, highlighting the effectiveness of the training. Furthermore, the comparison with the MLLM-based Marker and GPT-4o also validates the rationale behind our choice of an LLM-based architecture instead of an MLLM-based one.

7.2 Mering Task

Table 2. The evaluation results of the merging task. The best entries are marked in **bold**.

Method	Type	TEDS			RTEDS		
		EN	ZH	Avg.	EN	ZH	Avg.
Naive	Rule-based	82.4	77.1	79.8	0.0	0.0	0.0
MinerU	Rule-based	86.5	80.2	83.4	23.3	13.5	17.8
Marker	Rule-based	87.0	81.8	84.4	26.1	20.5	22.8
GPT-4o	MLLM-based	96.7	95.5	96.1	81.3	80.3	80.7
Qwen3-0.6B	LLM-based	72.9	67.8	70.4	-54.0	-40.6	-46.5
Qwen3-32B	LLM-based	91.2	88.8	90.0	50.0	51.1	50.5
TabStitcher	LLM-based	**97.2**	**96.0**	**96.6**	**84.1**	**82.5**	**83.2**

As shown in Table 2, in the merging task, TabStitcher substantially outperforms the best baseline by 0.5 TEDS, or 2.5 RTEDS. The result demonstrates TabStitcher's superior ability to reconstruct coherent cross-page table structures.

Furthermore, we conduct additional experiments to investigate the impact of challenging cross-page table patterns on model performance. Among the positive subset, about 52% of cross-page tables exhibit at least one of the challenging patterns, and some instances even display multiple challenging patterns simultaneously. Notably, since these samples are randomly selected from the document collection, this also reflects the distribution of challenging patterns in real-world cross-page tables.

The results in Table 3 show that the performance of all baselines degrades significantly under challenging patterns. For instance, on vertical splitting cases, the TEDS scores of Naive, MinerU, and Qwen3-0.6B drop below 55. In contrast, TabStitcher maintains consistently strong performance (nearly 90 TEDS or higher) across all challenging pattern types, demonstrating its robustness and generalizability.

Table 3. Distribution and influence of challenging patterns—cross-page cells (CC), header repetition (HR), dimension mismatch (DM), and vertical splitting (VS)—on model performance in the merging task, measured by the TEDS score.

Pattern	Ratio	Naive	MinerU	Marker	Qwen3-0.6B	Qwen3-32B	GPT-4o	TabStitcher
CC	19.7%	74.9	78.2	77.3	66.6	84.5	86.2	**89.6**
HR	32.0%	82.1	87.3	80.5	67.2	92.6	96.1	**98.3**
DM	11.3%	75.1	73.8	76.5	71.4	83.2	88.2	**94.5**
VS	1.6%	50.3	52.0	68.7	54.9	85.2	89.0	**91.8**

7.3 End-to-End Reconstruction Task

Table 4. The evaluation results of the end-to-end cross-page table reconstruction task, measured by the CPTEDS score.

Method	Page Parser	English	Chinese	Overall
Mineru	Mineru	57.8	68.8	63.3
Qwen3-32B	Mineru	54.7	69.3	62.0
GPT-4o	Mineru	70.9	76.2	73.6
TabStitcher	Mineru	**87.0**	**89.6**	**88.3**
Marker	Marker	63.1	32.9	48.0
Qwen3-32B	Marker	51.7	59.5	55.6
GPT-4o	Marker	64.8	72.1	68.5
TabStitcher	Marker	**79.7**	**81.1**	**80.4**

To verify that TabStitcher can be seamlessly integrated with diverse off-the-shelf single-page parsers and enable them to effectively reconstruct cross-page tables, we evaluate the end-to-end cross-page table reconstruction performance of TabStitcher and baselines with MinerU and Marker as the single-page parsers.

As shown in Table 4, TabStitcher achieves substantial improvements over all baselines—including the native cross-page merging strategies of MinerU and Marker as well as the much larger Qwen3-32B model. When integrated with MinerU, TabStitcher attains a CPTEDS score of 88.3, outperforming the best baseline by 25.0 points; when integrated with Marker, it achieves 80.4 CPTEDS, surpassing the best baseline by 11.9 points. All the results demonstrate TabStitcher's excellent compatibility and adaptability.

8 Conclusion

We present TabStitcher, a lightweight LLM-based approach for cross-page table reconstruction. By precisely detecting and merging cross-page table fragments, TabStitcher produces accurate and coherent reconstructed tables. To enable scalable training, we further introduce a data synthesis framework that can generate cross-page table instances exhibiting diverse challenging patterns, allowing the model to learn robust handling strategies. Experiments on real-world benchmarks demonstrated the effectiveness and robustness of TabStitcher. We believe this work can help to establish a solid foundation for future research in the field.

Acknowledgments. This work was supported by the National Natural Science Foundation of China (No.62206265, 62076231).

Disclosure of Interests. The authors have no competing interests to declare that are relevant to the content of this article.

References

1. Auer, C., et al.: Docling technical report. arXiv preprint arXiv:2408.09869 (2024)
2. Bai, S., et al.: Qwen2. 5-VL technical report. arXiv preprint arXiv:2502.13923 (2025)
3. Blecher, L., Cucurull, G., Scialom, T., Stojnic, R.: Nougat: neural optical understanding for academic documents. arXiv preprint arXiv:2308.13418 (2023)
4. Comanici, G., et al.: Gemini 2.5: pushing the frontier with advanced reasoning, multimodality, long context, and next generation agentic capabilities. arXiv preprint arXiv:2507.06261 (2025)
5. Feng, H., et al.: Dolphin: document image parsing via heterogeneous anchor prompting. arXiv preprint arXiv:2505.14059 (2025)
6. Hu, A., et al.: MPLUG-docowl2: high-resolution compressing for OCR-free multi-page document understanding. arXiv preprint arXiv:2409.03420 (2024)
7. Hurst, A., et al.: GPT-4o system card. arXiv preprint arXiv:2410.21276 (2024)
8. Li, H., Yang, Q., Cao, Y., Yao, J., Luo, P.: Cracking tabular presentation diversity for automatic cross-checking over numerical facts. In: Proceedings of the 26th ACM SIGKDD International Conference on Knowledge Discovery & Data Mining, pp. 2599–2607 (2020)

9. Li, Z., et al.: MonkeyOCR: document parsing with a structure-recognition-relation triplet paradigm. arXiv preprint arXiv:2506.05218 (2025)
10. Liu, Y., et al.: TextMonkey: an OCR-free large multimodal model for understanding document. arXiv preprint arXiv:2403.04473 (2024)
11. Mandal, S., Talewar, A., Ahuja, P., Juvatkar, P.: Nanonets-OCR-S: a model for transforming documents into structured markdown with intelligent content recognition and semantic tagging (2025)
12. Paruchuri, V.: Marker (2024). https://github.com/VikParuchuri/marker
13. Poznanski, J., et al.: OLMOCR: unlocking trillions of tokens in pdfs with vision language models. arXiv preprint arXiv:2502.18443 (2025)
14. Team, G., et al.: Gemini 1.5: unlocking multimodal understanding across millions of tokens of context. arXiv preprint arXiv:2403.05530 (2024)
15. Wang, B., et al.: MINERU: an open-source solution for precise document content extraction. arXiv preprint arXiv:2409.18839 (2024)
16. Wang, P., et al.: Qwen2-VL: enhancing vision-language model's perception of the world at any resolution. arXiv preprint arXiv:2409.12191 (2024)
17. Wang, W., et al.: InternVL3.5: advancing open-source multimodal models in versatility, reasoning, and efficiency. arXiv preprint arXiv:2508.18265 (2025)
18. Wei, H., et al.: General OCR theory: towards OCR-2.0 via a unified end-to-end model arXiv preprint. arXiv:2409.01704 (2024)
19. Yang, A., et al.: Qwen3 technical report. arXiv preprint arXiv:2505.09388 (2025)
20. Ye, J., et al.: MPLUG-DOCOWL: modularized multimodal large language model for document understanding. arXiv preprint arXiv:2307.02499 (2023)
21. Zheng, Y., Zhang, R., Zhang, J., Ye, Y., Luo, Z.: LlamaFactory: unified efficient fine-tuning of 100+ language models. In: Proceedings of the 62nd Annual Meeting of the Association for Computational Linguistics (volume 3: System Demonstrations), pp. 400–410 (2024)
22. Zhong, X., ShafieiBavani, E., Jimeno Yepes, A.: Image-based table recognition: data, model, and evaluation. In: In: Vedaldi, A., Bischof, H., Brox, T., Frahm, JM. (eds.) European Conference on Computer Vision, pp. 564–580. Springer (2020). https://doi.org/10.1007/978-3-030-58589-1_34
23. Zhu, J., et al.: InternVL3: exploring advanced training and test-time recipes for open-source multimodal models. arXiv preprint arXiv:2504.10479 (2025)

PatentME: A Dataset and Reference-Free Post-OCR Verification Task for Printed Mathematical Expression Recognition

François Wieckowiak[1,2](✉), Véronique Eglin[1], Tony Bonnet[2], Stéphane Bres[1], and Laëtitia Rousseau[2]

[1] INSA Lyon, CNRS, Ecole Centrale de Lyon, Universite Claude Bernard Lyon 1, Université Lumière Lyon 2, LIRIS, UMR5205, 69621 Villeurbanne, France
[2] Luminess, Paris, France
francois.wieckowiak@liris.cnrs.fr
https://liris.cnrs.fr,https://www.luminess.eu

Abstract. Mathematical Expression Recognition (MER) is used for the digitization of technical documents in scientific databases, but erroneous recognition can lead to indexing errors. Most MER systems are trained on clean, digitally born images with LaTeX markup, and few are evaluated on scanned documents with noise and multiple font styles. There is also a lack of reference-free metrics for evaluation when the ground truth markup is unknown. To address these issues, we propose a new task of post-OCR visual verification supported by two datasets. **PatentME-OCR** consists of 41k images of mathematical expressions with varying fonts and noise, along with their MathML markup strings. Zero-shot evaluation of four different MER architectures pretrained on digitally born clean images shows poor out-of-distribution generalization, with an average visual exact match of only 15.7%. Using their erroneous predictions, we designed **PatentME-Siamese** for a new style-invariant post-OCR verification task, in which a judge model must determine if the prediction image is semantically equivalent to the reference image, despite uncontrolled visual variations. Our CNN-Siamese baseline achieved a recall of 78% at 94% precision, outperforming a zero-shot GPT-5-mini baseline which achieves 39% recall at 81% precision. Our Siamese distance correlates with markup-based scores ($\rho = 0.77$) and can serve as a trustworthy reference-free evaluation metric. The datasets and models are available at https://github.com/fwieckowiak/PatentME.

Keywords: Printed Mathematical Expression Recognition · Siamese Metric Learning · Reference-free Evaluation

1 Introduction

Mathematical Expression Recognition (MER) is a core component of digitization of scientific and technical documents. Accurate conversion of mathematical

G. A. Fink et al. (Eds.): ICDAR 2026, LNCS 16972, pp. 611–628, 2027.
https://doi.org/10.1007/978-3-032-36023-6_35

expressions into structured markup such as LaTeX or MathML enables reliable indexing, search, and retrieval of knowledge in large-scale databases. Incorrect markup leads to indexing errors and potentially failed retrieval, so making sure recognition systems make no mistakes is critical. Current MER research evaluates their models with metrics such as BLEU, edit distance, or exact match. These metrics require the reference ground-truth markup to be computed, so they cannot be used in real world digitization scenarios of never-seen-before documents where the only reference available is the raster image of the expression. Pixel-based or layout-based similarity metrics can fail: two expressions may appear nearly identical while they visually differ semantically, and conversely, the same formula rendered in different fonts, resolutions, or noise levels may appear visually distinct. Previous approaches in Mathematical Information Retrieval (MIR [31]) or image-based similarity have not fully addressed this issue, as they either rely on textual ground truth (MathML or LaTeX) or are sensitive to uncontrolled visual variations. This gap motivates the need for a reference-free verification framework able to recognize semantic equivalence under visual domain shift. To address this limitation, we investigate an underexplored problem: reference-free verification of MER outputs under strong visual variability. In this paper, we propose a novel approach to validate if a prediction matches the input image, while remaining robust to variations in style, scale and image quality. The contributions of our work are threefold: (i) we introduce **PatentME-OCR**, a dataset of 41k printed mathematical expressions from European patents showing strong style variability, (ii) we propose a reference-free post-OCR verification task and construct **PatentME-Siamese** to enable learning style-invariant similarity metrics without textual ground truth and (iii) we validate our approach with baseline methods, showing that it reliably detects OCR errors and supports high-precision evaluation in real-world scenarios. The paper is organized as follows. Section 2 reviews related work and highlights the gaps our approach addresses. Section 3 presents our datasets and reference-free verification methodology. Section 4 reports experimental results, and Sect. 5 discusses the findings. Finally, Sect. 6 concludes and outlines future directions.

2 Literature Review

2.1 Diversity of Mathematical Expression Datasets

The literature on mathematical expression datasets is divided between handwritten and printed expressions. Our work focuses on printed mathematical expressions (PME), even though handwritten datasets have had a lot of importance in the ME recognition domain, such as The CROHME competition [29], organized since 2011, or more recently, the MathWriting dataset [14].

PME datasets are quite diverse and heterogeneous. Early works such as [12] extracted 5 500 display and inline LaTeX expressions from both scanned (75 to 600 dpi) and digitally born documents. [5] and [26] evaluated MER on 600 dpi scans of scientific articles and reported noise and artifacts from scanning, symbol touching due to binarization, and high variability in character sets, fonts, italicization, and layouts. Recently, PME datasets are usually synthetically generated by rendering LaTeX markups extracted from LaTeX documents, and do not include any noise or deformation introduced by real-world scanning processes. A foundational contribution is im2latex100k [10]. It is built from the KDD Cup 2003 arXiv dataset [13], from which the authors extracted 100k display LaTeX mathematical expressions and rendered them as images. It is the primary benchmark for MER today. Several datasets use a similar acquisition pipeline. Im2latex90k and 140k [25] filtered erroneous expressions from the 100k version, the IBEM dataset [4] explicitly includes both inline and displayed expressions, while ME20K [11] gather LaTeX expressions from Zhixue.com. In [23], the MathNet MER model was trained on im2latexv2. This dataset differs from previous ones by re-rendering the im2latex100k dataset in 59 different fonts at 600 dpi to increase visual variability and model robustness. Recently, both printed and handwritten datasets have been merged into large scale datasets, such as [18], which was used to train to the TexTeller model [19], designed for improved generalization thanks to its large-scale and diverse training dataset. There is no ME dataset taking its source from patent documents. DeepPatent2 [2], PatFig [6], and PatentLMM [24] focus on technical drawings and figure captioning of patents downloaded from the USTPO, EPO, or Google Patent directly. Overall, the current large scale ME datasets, consisting of digitally born clean images, lack the noise and diversity of scanned and patent documents.

2.2 Metrics for Visual Mathematical Expression Similarity

For MER evaluation, standard OCR evaluation metrics such as Exact Match, BLEU, or edit distance are commonly used to compare the predicted markup to ground-truth markup, but they are poorly suited to mathematical expressions because small syntactic differences can change visual or structural meaning [15]. Metrics based on rendered images or symbol layout graphs partially address this limitation [28,32], but they still rely on a textual ground truth, which would not be available in real-world scenarios where only reference images exist. Purely visual MER evaluation metrics such as IMEGE [3] assume that both the reference and predicted images are generated under controlled and identical conditions (font, resolution, layout, noise). These metrics become unreliable when reference images originate from real-world scans or photographs. Style-agnostic visual metrics, on the other hand, may rely on recursive X–Y decomposition [33], connected component analysis [34], or tree-based structural representations [27,30]. In another line of work, [1] employs a multimodal embedding framework that combines a ResNet-based visual encoder with a graph representation of the ME. These approaches are more robust to rendering variations and noise, but most still rely on the reference MathML to construct the layout tree or enrich the connected components analysis, making them reference-dependent metrics. This limitation motivates the design of visual similarity metrics that operate solely at the image level, without access to textual annotations. To date, no prior work has introduced a fully style-agnostic metric adapted to conditions in which font, rendering engine, resolution, and noise are uncontrolled, and where no ground-truth textual representation of the expression is provided.

2.3 Siamese Networks for Document Analysis and Recognition

Our work focuses on image-only verification under strong style variations between the reference and query inputs. It is similar to face or signature verification, which are often solved with Siamese Neural Networks (SNNs) [9]. They map inputs into an embedding space and compare them with a distance metric. They learn to keep similar samples closer together and different samples apart. They have been applied to document recognition and OCR-related problems such as word spotting in historical manuscripts [7] and Urdu character recognition [17]. Siamese architectures also prove effective for recognition of low-quality inputs, such as text from medical imaging documents affected by low resolution and noise [20]. For mathematical expressions, [21] uses SNNs to encode images of formulas from arXiv and learn similarities between expressions. The learned encoder is then used for mathematical search. The authors emphasize the importance of negative sampling strategies, i.e., selecting hard negatives during training to improve performance. Siamese contrastive learning and supervised learning are combined in [16] for handwritten mathematical expression recognition. The encoder is trained on paired handwritten and rendered printed expressions to minimize both a contrastive loss and a cross-entropy OCR loss simultaneously. These studies demonstrate that Siamese Neural Networks prove useful in document and symbol recognition tasks, including mathematical expressions, by learning robust similarity embeddings that can generalize to unseen or visually variable inputs.

3 Methodology

3.1 Problem Formulation

Let a reference image of a mathematical expression X be associated with a correct MathML representation y^+ and an incorrect MathML representation y^-, and their corresponding visual rendering Y^+ and Y^- (Fig. 1).

Our objective is to teach a model that, given the reference image X and a candidate image $\hat{Y}$, can determine whether the candidate image corresponds to a correct or incorrect MathML markup string. The task is not trivial because the reference image X can be of any font, quality and style, so we cannot directly compare X and $\hat{Y}$ using simple image similarity metrics. To do so, we will employ a Siamese neural network S (illustrated in Fig. 2) that can embed both images into a shared embedding space, and compute the distance between the resulting vectors. With $\mathbf{e}_X$ and $\mathbf{e}_{\hat{Y}}$ the embedded expressions, semantic similarity is measured using their angular distance. For very close vectors, a cosine distance would compress small differences, whereas the angular distance is linear in θ,

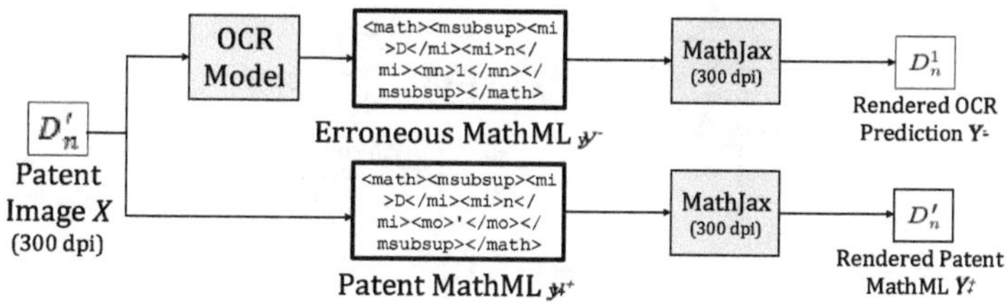

Fig. 1. Generation pipeline of the reference image X, positive image Y^+ and negative image Y^-.

providing finer discrimination between highly similar embeddings. Early experiments also suggested that the angular distance is more effective for our task. It is defined as:

$$d(\mathbf{e}_X, \mathbf{e}_{\hat{Y}}) = \frac{1}{\pi} \arccos\left(\frac{\langle \mathbf{e}_X, \mathbf{e}_{\hat{Y}} \rangle}{\|\mathbf{e}_X\| \, \|\mathbf{e}_{\hat{Y}}\|} \right), \qquad d \in [0, 1].$$

Using a Contrastive loss, we can teach the model to keep the embedding of X and Y^+ close and put the ones of X and Y^- apart. We choose a Contrastive rather than a Triplet loss because it directly optimizes the distance between positive and negative pairs, providing an interpretable similarity score. Triplet loss enforces relative ordering between an anchor, positive, and negative sample, which is less aligned with our verification task. Our Contrastive loss is defined as:

$$\mathcal{L}(X, \hat{Y}, \, z) = z \, d(\mathbf{e}_X, \mathbf{e}_{\hat{Y}})^2 + (1 - z) \max\big(0, m - d(\mathbf{e}_X, \mathbf{e}_{\hat{Y}})\big)^2,$$

with $z = \begin{cases} 1, & \text{if } \hat{Y} = Y^+ \\ 0, & \text{if } \hat{Y} = Y^- \end{cases}$, and $m > 0$ is a margin hyperparameter.

The classification decision $\hat{z}$ is obtained by thresholding the distance:

$$\hat{z} = \begin{cases} 1, & \text{if } d(\mathbf{e}_X, \mathbf{e}_{\hat{Y}}) \le \tau \\ 0, & \text{otherwise} \end{cases}, \text{ with } \tau \text{ the selected similarity threshold.}$$

In high-precision scenarios, the decision threshold τ is selected on a validation set using a Precision–Recall curve to enforce a very low number of false positives. In our setting, positive pairs correspond to images representing the same formula, while negative pairs correspond to prediction errors. We therefore choose τ to achieve at least 95% precision on the positive class, ensuring that pairs predicted as identical are almost always correct and avoiding incorrectly classifying negative pairs as positive. After finding this threshold, model performance is evaluated using recall on the positive class, which measures how many true identical pairs are successfully retrieved. Missing some positive pairs is less critical, since these false negatives can be rechecked by downstream models or manual verification.

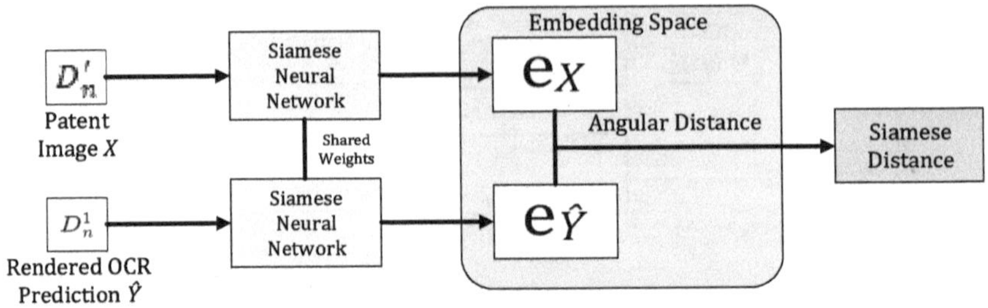

Fig. 2. The SNN architecture for the verification task. The reference and predicted images are encoded by a shared network and compared in the embedding space.

3.2 PatentME-OCR

We extracted 40 955 mathematical expression images from 448 European patents and their MathML representations from the patent XMLs[1]. MathML simplification and normalization was applied to ensure reliable conversion and rendering using MathJax, without affecting visual rendering and meaning. The image dimensions and MathML lengths vary considerably. Figure 3 show histograms of these features, while Fig. 4 illustrates the variability in image quality, style, and font. We call this version of this testing dataset **PatentME-OCR**.

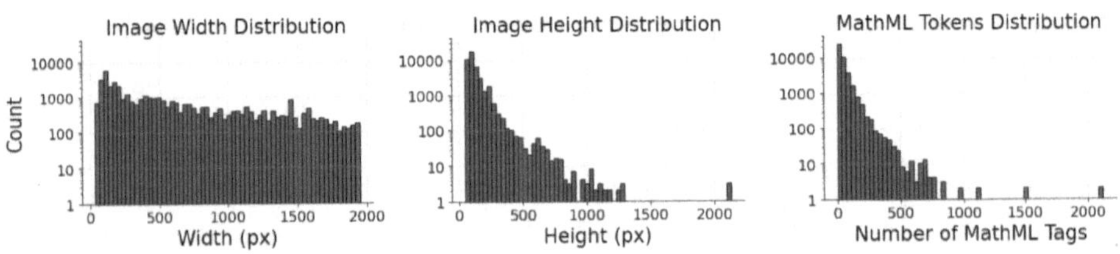

Fig. 3. Log-scale distribution of image width, image height, and number of MathML tokens across the 40,955 expressions of the **PatentME-OCR** dataset.

Using the PiE-MER library [28] and without any fine-tuning, we evaluated four MER systems on the reference images X, obtained their predictions $\hat{y}$, and rendered images $\hat{Y}$. All of these models generate LaTeX, which is then converted into MathML using LaTeXML. The selected models are Pix2tex [8], a simple CNN+ViT encoder-decoder architecture, Qwen2-VL-OCR2-2B-Instruct [22], a multi-modal VLM fine-tuned for OCR, English understanding, and LaTeX-based math problem solving (prompt in Fig. 5), MathNet [23], trained on the multi-font i2lv2 dataset using a Convolutional Vision Transformer architecture (with input images rescaled to 600 dpi to match the input requirements), and Texteller2.0 [19], trained on 7.5M diverse formulas, which claims strong performance on rare symbols, complex multi-line expressions, and matrices.

[1] https://data.epo.org/expert-services/.

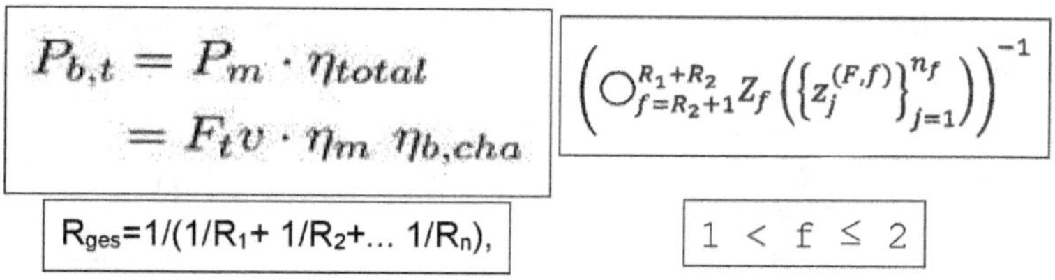

Fig. 4. A few mathematical expressions from **PatentME-OCR** with different sizes, qualities and fonts.

System prompt
You are a Mathematical Expression Recognition Assistant. You will provide the LaTeX of the expression given the image. Do not transcribe the equation number. Provide ONLY the LaTeX string between $$ delimiters.

Fig. 5. Prompt used with Qwen2-VL-OCR2-2B-Instruct.

The evaluation metrics are selected to align with our target applications: structural correctness in MathML for MIR as well as perceptual similarity between rendered formulas for image-based post-OCR validation. We therefore select metrics that focus on these two modalities: Levenshtein distance and exact match ratio, applied to both the MathML strings and the rendered ground truth and prediction images. To compute Image Levenshtein distance, columns of pixels are treated as characters [10].

3.3 PatentME-Siamese

The **PatentME-Siamese** dataset consists of the positive and negative pairs (X, Y^+) and (X, Y^-) defined in Sect. 3.1. We generate the negative samples Y^- by rendering the incorrect MathML predictions of the Texteller model, which achieved the best performance on **PatentME-OCR** (see Sect. 4.1). This ensures that the erroneous predictions are the most subtle, compared to errors from a lower-performing model that would be more obvious. Figure 6 illustrates subtle and obvious errors. This hard negative mining strategy encourages the SNN to learn fine-grained visual differences between correct and incorrect expressions.

Corresponding positive samples Y^+ are generated by rendering the ground-truth MathML from the patent XML (See Fig. 1).

This procedure resulted in the **PatentME-Siamese** dataset, consisting of 67,680 pairs evenly split between positive (33,840) and negative (33,840) samples. Each sample includes the paths to the reference patent image X and the query image $\hat{Y}$, a binary label, and the metadata associated with the texteller evaluation and original patent. The dataset is split as follows: 60,972 training pairs, 3,346 validation pairs, and 3,362 test pairs.

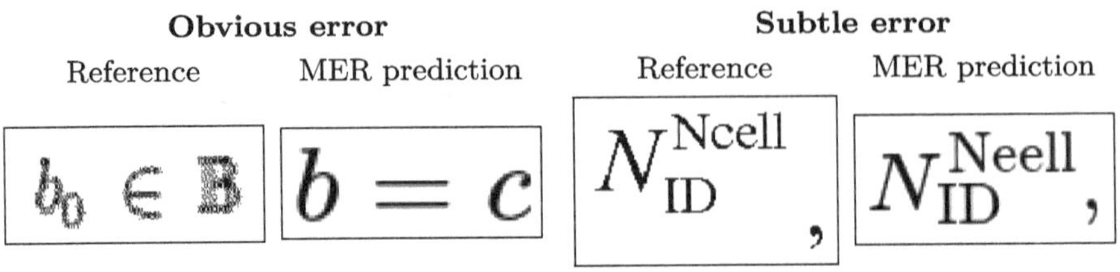

Fig. 6. Obvious vs. subtle negative samples in **PatentME-Siamese**.

3.4 Siamese Network Architecture

Our Siamese pipeline is illustrated in Fig. 2. We select a lightweight ResNet-18 CNN as the encoder, which has proven effective for mathematical expression embedding [1] and enables rapid experimentation. A fast and compact architecture is also desirable for real-time post-OCR verification. The network is trained from scratch, as our task differs from ImageNet classification and the dataset size is sufficient to learn a task-specific representation. Our loss and distance function are the contrastive loss and angular distance described in Sect. 3.1. The contrastive margin m is treated as a hyperparameter to be optimized. An optimal threshold τ is calculated to reach a minimum precision of 95% on the validation set. Positive or negative labels are obtained using the threshold on the test set, and compared to the ground truth to produce the final confusion matrix. We then report the corresponding recall, which measures the proportion of positive pairs correctly retrieved. We also perform a study on the impact of input image preprocessing. For the Resnet18 SNN, we set all of our input images to be of a fixed square of size either 224^2 or 512^2 pixels, and we compare two rescaling strategies: resizing while preserving the aspect ratio with top/bottom or left/right padding (Padding), and rescaling to fully fill the square input size (Scaling). These preprocessing methods are illustrated in Fig. 7.

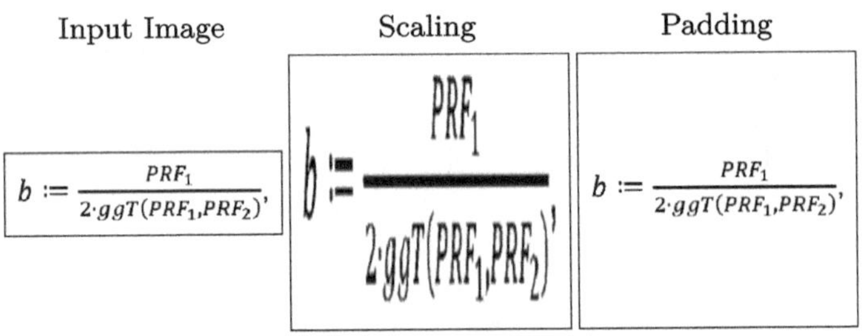

Fig. 7. The two preprocessing methods evaluated in our study.

We optimize the margin m using a Bayesian search trained 14 times over 20 epochs, with the objective of maximizing test recall @95% precision. The final model, with the best input preprocessing and margin, is then trained 7 times for 50 epochs. We use a batch size of 128 and the Adam optimizer with a cosine-annealing learning rate scheduler to improve convergence. We report the losses, average precision, and test recall over time, as well as the validation Precision-Recall curve and test confusion matrix of the best model on its best epoch.

3.5 GPT-5-Mini Baseline

We use GPT-5-mini as a non-learning VLM-based baseline to compare against our Siamese approach. The prompt in Fig. 8 instructs the model to determine whether the reference and query images represent the same or different expression.

System prompt
You are an expert AI system for rigorous mathematical expression comparison. Task: Given two images, determine whether the represented mathematical expressions are semantically identical. Ignore differences in font, size, and equation numbering. Bold and italic carry semantic meaning and must be treated as differences. Layout differences are significant: a stacked (vertical) fraction and an inline fraction (e.g., 1/2) must be considered different. Output strictly a JSON object with the following field: {distance: a float in [0,1]}. The distance value must satisfy: 0 indicates identical expressions, 1 indicates completely different expressions.

Fig. 8. Prompt used with GPT-5-mini

To limit the token usage, we scale down the images such that their longest side does not exceed 512 pixels while preserving aspect ratio. Like the Siamese model, GPT-5-mini returns a distance value that represents how similar the two expressions X and $\hat{Y}$ are. We use the same pipeline to find τ on the validation set and evaluate on the test set.

4 Results

4.1 PatentME-OCR

Four MER systems were evaluated with no fine-tuning on the 40 955 samples of the **PatentME-OCR** dataset: Pix2tex, Qwen, MathNet, and Texteller. Table 1 summarizes the evaluation results across the chosen metrics. Levenshtein represents the average Levenshtein distance over the corresponding modality (MathML or Image). Exact Match indicates the proportion of expressions that exactly match the reference ground truth in each modality.

Overall, Texteller and MathNet achieved the highest scores on **PatentME-OCR**, followed by Qwen and Pix2tex. However, all systems perform substantially worse than their reported results on their original datasets, illustrating the challenge of OCR on highly variable patent data.

Table 1. Evaluation of 4 MER systems on the **PatentME-OCR** dataset. The metrics are the Levenshtein distance and exact match ratio, for the MathML and Image modalities (in %). Best score per metric in bold.

Model	MathML		Image	
	Levenshtein ↓	Exact Match ↑	Levenshtein ↓	Exact Match ↑
Pix2tex	41.99	6.14	67.23	10.64
Qwen	35.29	6.12	60.09	14.82
MathNet	32.63	10.19	56.12	18.70
Texteller	**32.57**	**12.07**	**55.34**	**18.76**
Average	35.62	8.63	59.70	15.73

4.2 PatentME-Siamese

Siamese Model: Table 2 reports the impact of input resolution (224^2 vs. 512^2) and preprocessing method (Padding vs. Scaling) on the Siamese network described in Sect. 3.4, trained for 50 epochs with margin $m = 0.5$. Padding consistently yields higher test recall than scaling at both resolutions, with the best performance obtained using 512^2 inputs with padding with a best recall of 72.99% at epoch 41. Models trained with 224^2 inputs reach peak recall earlier (epochs 27–28), indicating early convergence and limited gains from additional training. Bayesian optimization of the margin parameter m was conducted over 14 trials using 512^2 inputs with padding. An optimal margin of $m = 0.33$ was found for our architecture.

Table 2. Impact of input preprocessing on the test recall at 95% validation precision. Training over 50 epochs.

Input Processing	Final Recall (%)	Best Recall (%)	Epoch of Best
Padding 224^2	59.49	62.64	28
Scaling 224^2	43.19	55.68	27
Padding 512^2	**68.71**	**72.99**	41
Scaling 512^2	27.36	31.23	36

Figure 9 reports the performance of the final model (margin $m = 0.33$, 512^2 inputs with padding) over 7 runs. It shows the average training and validation losses, validation average precision, and test recall across 50 epochs.

Figure 10 reports the evaluation of the best run on the test set, including the validation precision-recall curve, test confusion matrix, and predicted test distance distributions. With a decision threshold $\tau = 0.069$ selected to ensure at least 95% validation precision, the model achieves a true negative rate of 95.2% and a recall of 78% on positive pairs in the test set. The distribution of distances for positive and negative samples shows a clear separation with minimal overlap. The decision threshold $\tau = 0.069$ effectively separates this space between the two groups while limiting the number of false positives. The observed angular distances lie in [0, 0.5] rather than [0,1] due to the embedding optimization: vectors for positive pairs are close and negatives are separated just enough to satisfy the contrastive margin. Manual verification of the false positives (FP) and false negatives (FN) produced by the Siamese model was conducted to confirm the ground-truth labels. Some predictions initially labeled as errors were found to be correct upon human review, indicating minor inconsistencies in the dataset. Out of 369 false negatives, 29 were corrected to true negatives after manual verification. Similarly, 12 out of 80 false positives were corrected to true positives. Overall, the corrected samples represent 1.22% of the entire test set.

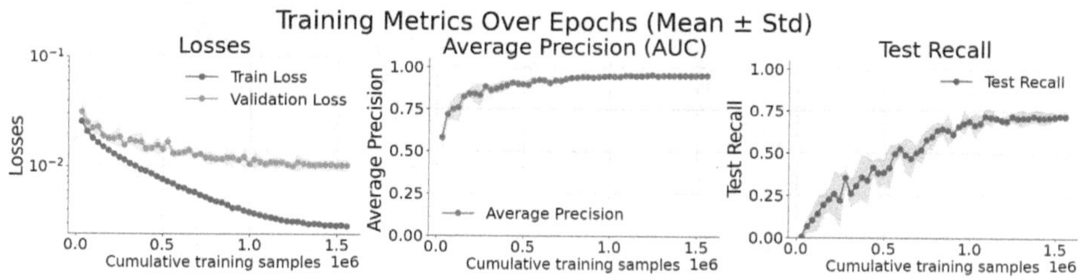

Fig. 9. Evolution of the training metrics for the final SNN model (margin $m = 0.33$, 512^2 input with padding) averaged over 7 runs.

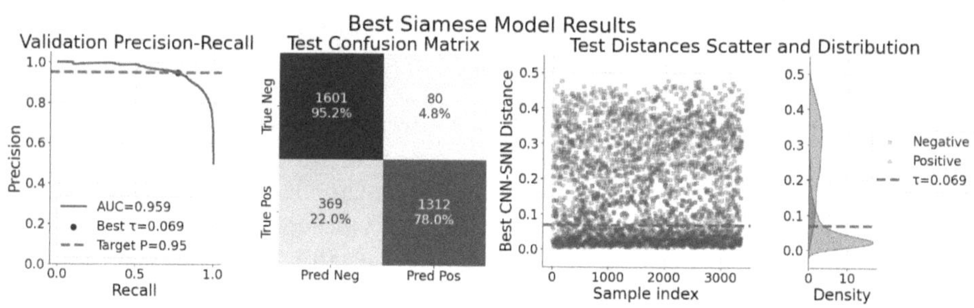

Fig. 10. Validation precision-recall curve, test confusion matrix, and distribution of predicted distances on the test set for the best SNN model.

GPT-5-Mini Approach: The same protocol was applied to the GPT-5-mini baseline with a threshold selected on the validation set to ensure at least 95% precision. In practice, only $\tau = 0$ satisfied this requirement, labeling all samples as negative. To circumvent this, the test threshold was set to $\tau = 0.00^+$, meaning only samples with a distance exactly equal to 0 are labeled as positive. At this operating point on the test set, the model achieves a true negative rate of 91.0% and a recall of 39.0% on positive pairs. Figure 11 shows the PR curve, test confusion matrix, and distance distributions for this baseline. Compared to the Siamese model, GPT-5-mini shows lower retrieval performance: its recall is much lower (39% vs 78%) even with a slightly lower true negative rate (91% vs 95.2%). The distribution plot shows distances clustered near 0 and 1, reflecting a sharp separation between inputs perceived as similar versus different. However, many negative samples appear near 0, where ideally only positives should be located. This overlap corresponds to potential false positives and explains why a very strict test threshold of $\tau = 0.00^{+}$ was required to maintain the desired precision.

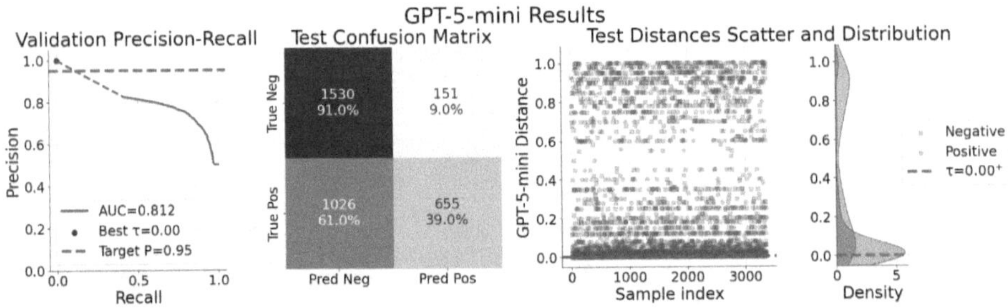

Fig. 11. Validation precision-recall curve, test confusion matrix, and visualization of predicted distances between samples on the test set for the GPT-5-mini baseline. Only $\tau = 0$ satisfies a 95% minimum validation precision.

4.3 Correlation of Siamese Distance with Patent-ME OCR Metrics

Table 3 reports the Spearman correlation between model distances and OCR-based metrics across all test samples. The Siamese distance is computed directly between pairs of image inputs X and $\hat{Y}$, whereas the OCR-based metrics compare the ground-truth markup y^{+} with the OCR prediction $\hat{y}$. By comparing these distances, we can confirm whether the Siamese metric reflects real differences in the OCR output and can be trusted as a reference-free metric for MER performance in real-world scenarios. The reference-free Siamese distance shows a stronger correlation with both token-level (MathML) and image-level Levenshtein metrics compared to the GPT distance, indicating better alignment with markup-based metrics.

Figure 12 shows the relationship between Siamese distances and the two MER metrics from Sect. 4.1. The scatter of MathML Levenshtein versus Siamese distances appears roughly linear, reflecting a more uniform distribution of the MathML metric. In contrast, the scatter of Image Levenshtein versus Siamese distances shows a cluster of points in the top-right of the figure, reflecting the strongly uneven distribution of the image-level metric, with many samples above

Table 3. Spearman correlation between model distances and OCR-based metrics. Higher coefficients indicate a stronger monotonic relationship.

Model Distance	MathML Levenshtein	Image Levenshtein
Siamese Distance	0.770	0.765
GPT Distance	0.634	0.604

0.7. Overall, the plots illustrate the different distributions of the two metrics while remaining consistent with the Spearman correlations reported in Table 3.

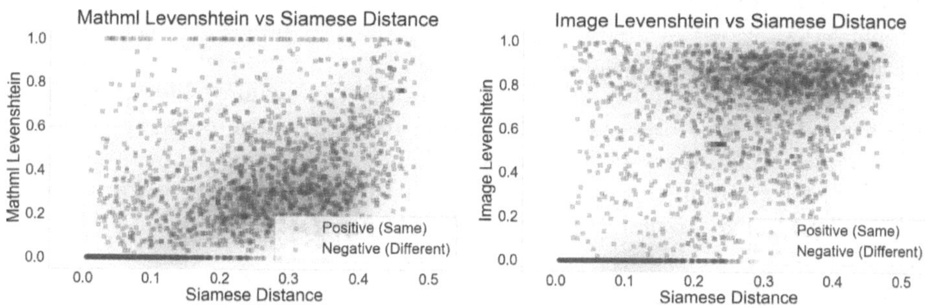

Fig. 12. Scatter plot of Siamese distances versus MathML and image-level Levenshtein distances. The MathML distances show a roughly linear trend, while the image-level distances are clustered at high values.

5 Discussion

5.1 Evaluation on PatentME-OCR

The evaluation of the different models on the **PatentME-OCR** dataset shows that current MER systems struggle with heterogeneous and noisy scanned images and the requirement to produce correct MathML representations. Errors range from ambiguous symbol confusion to complete failure on very low-quality images (see Fig. 6). Models such as Texteller and Mathnet, trained on more diverse datasets, perform slightly better. This confirms that improving dataset diversity, with multiple sources (digitally born, scanned, handwritten) and variations in fonts and styles, is essential for advancing robustness in MER. The models generate LaTeX expressions, but the evaluation is performed on the MathML and image modalities, since LaTeX ground truth is not available in **PatentME-OCR**. Conversion from LaTeX to MathML is performed using the well-established *LaTeXML*, ensuring a consistent conversion process across all models. Despite being tailored for OCR and LaTeX-based math problem solving, the evaluated Qwen variant (2B parameters) performs far below models specifically designed for MER. It also remains relatively small compared to recent Qwen models of up to 72B parameters. Our model was prompted to output LaTeX for fair comparison with other MER systems, though using a better prompt or fine-tuning to generate MathML directly could further enhance performance.

5.2 Evaluation on PatentME-Siamese

Impact of Input Processing: The input resolution analysis revealed that using a small input size leads to a clear drop in performance. Equations in **PatentME-OCR** vary uniformly between 0 and 2000 pixels in width, and most of them are under 512 pixels in height (Fig. 3). Reducing them to 224^2 compresses fine structural details and removes critical visual information. In contrast, 512^2 preserves enough details for accurate recognition at the cost of a higher GPU usage and training time. For the rescaling strategies, padding consistently outperformed scaling. The model is sensitive to symbol shapes and relative positioning. Scaling distorts these properties, while padding preserves them even if it introduces unused white regions. Also, some patent expressions include equation numbers in the images that are to be ignored in the recognition. They can artificially increase the image width and force aggressive down-scaling of the patent image, thus degrading performance.

GPT vs Siamese: Our ResNet18 SNN baseline achieves strong performance despite its simplicity and limited training set of 61k samples. Even under these constraints, it outperforms the GPT baseline on the **PatentME-Siamese** comparison task. Vision Transformer-based encoders could likely capture even finer visual differences and further improve performances. The naive GPT-5-mini approach performs worse, likely because it is not optimized for the specific comparison task. The main source of errors appears to be data quality rather than model limitations. Qualitative observations show that both simple and complex equations can be correctly matched, suggesting robustness to expression complexity. Many mistakes come from extremely low-quality patent scans, ambiguous symbols and subtle differences such as a missing period or comma at the end of the formula. Manual verification of the False Positive and False Negative of the test set revealed 1.22% of mislabeled samples in test set, mostly due to errors from the patent XML themselves. Removing corrupted samples from the training set would likely reduce noise and increase the reliability of the learned metric.

Correlation Analysis: A moderate correlation is observed between the Siamese embedding distance and standard MER metrics. This suggests that the learned metric is meaningful and can be trusted in real-world verification scenarios. Though lightweight, our baseline architecture proves useful for visual verification in tasks like database proofreading or archive digitization. Using larger architectures such as Vision Transformers with an expanded dataset could likely improve performance further, especially in terms of recall.

6 Conclusion And Future Work

In this study, we identified a lack in reference-free evaluation metrics for MER when the original image may be subject to uncontrolled variations in font, scale and quality. We proposed a new task of post-OCR visual verification supported by two datasets. **PatentME-OCR** is a dataset of 41k printed mathematical expression images extracted from European patents with strong variability in fonts, sizes and scanning noise. It provides a new benchmark for MER systems that are usually trained on clean and digitally born data. Using erroneous predictions on **PatentME-OCR**, we constructed **PatentME-Siamese**. It consists of positive and hard negative image pairs designed to learn a style-invariant similarity metric between images without access to textual ground truth. We evaluated a lightweight ResNet-based SNN and a zero-shot vision–language model on this task. The Siamese metric performed better in terms of precision and recall, and correlation analyses demonstrated that it is suitable for high-precision evaluation in real-world scenarios. The datasets, models, and instructions on how to gather more data are available in our GitHub repository[2].

Besides implementing larger encoders such as Vision Transformers and increasing the size of the training datasets, future work should focus on preprocessing steps such as automatic removal of equation numbers to reduce noise, and Curriculum Learning strategies with specific hard negative mining to improve performance on the verification task. For MER, developing new tokenizers and robust architectures to generate MathML instead of LaTeX is now made possible thanks to our patent data acquisition pipeline.

Disclosure of Interests. This work was carried out as part of a CIFRE PhD in collaboration between LIRIS and Luminess (CIFRE grant no. 2023/0356). It uses data sourced from EPO databases. © European Patent Organisation.

References

1. Ahmed, S., Davila, K., Setlur, S., Govindaraju, V.: Equation attention relationship network (EARN): a geometric deep metric framework for learning similar math expression embedding. In: 2020 25th International Conference on Pattern Recognition (ICPR), pp. 6282–6289 (2021). https://doi.org/10.1109/ICPR48806.2021.9412619
2. Ajayi, K., et al.: DeepPatent2: a large-scale benchmarking corpus for technical drawing understanding. Sci. Data **10**(1), 772 (2023). https://doi.org/10.1038/s41597-023-02653-7

[2] https://github.com/fwieckowiak/PatentME.

3. Álvaro, F., Sánchez, J.-A., Benedí, J.-M.: An image-based measure for evaluation of mathematical expression recognition. In: Sanches, J.M., Micó, L., Cardoso, J.S. (eds.) IbPRIA 2013. LNCS, vol. 7887, pp. 682–690. Springer, Heidelberg (2013). https://doi.org/10.1007/978-3-642-38628-2_81
4. Anitei, D., Sánchez, J.A., Benedí, J.M., Ñoya, E.: The IBEM dataset: a large printed scientific image dataset for indexing and searching mathematical expressions. Pattern Recogn. Lett. **172**, 29–36 (2023). https://doi.org/10.1016/j.patrec.2023.05.033
5. Ashida, K., Okamoto, M., Imai, H., Nakatsuka, T.: Performance evaluation of a mathematical formula recognition system with a large scale of printed formula images. In: Second International Conference on Document Image Analysis for Libraries (DIAL 2006), pp. 12 pp.–331 (2006). https://doi.org/10.1109/DIAL.2006.30
6. Aubakirova, D., Gerdes, K., Liu, L.: PatFig: generating short and long captions for patent figures. In: 2023 IEEE/CVF International Conference on Computer Vision Workshops (ICCVW), pp. 2835–2841 (2023). https://doi.org/10.1109/ICCVW60793.2023.00305
7. Barakat, B.K., Alasam, R., El-Sana, J.: Word spotting using convolutional Siamese network. In: 2018 13th IAPR International Workshop on Document Analysis Systems (DAS), pp. 229–234 (2018). https://doi.org/10.1109/das.2018.67
8. Blecher, L.: lukas-blecher/LaTeX-OCR (2021). https://github.com/lukas-blecher/LaTeX-OCR. Accessed 17 Feb 2026
9. Bromley, J., Guyon, I., LeCun, Y., Säckinger, E., Shah, R.: Signature verification using a "Siamese" time delay neural network. In: Proceedings of the 7th International Conference on Neural Information Processing Systems, NIPS 1993, pp. 737–744 (1993). https://dl.acm.org/doi/10.5555/2987189.2987282
10. Deng, Y., Kanervisto, A., Ling, J., Rush, A.M.: Image-to-markup generation with coarse-to-fine attention (2016). http://arxiv.org/abs/1609.04938, preprint, arXiv:1609.04938
11. Fu, Y., Liu, T., Gao, M., Zhou, A.: EDSL: an encoder-decoder architecture with symbol-level features for printed mathematical expression recognition. In: Document Analysis and Recognition - ICDAR 2023, pp. 134–151. Springer, Cham (2023). https://doi.org/10.1007/978-3-031-41676-7_8
12. Garain, U., Chaudhuri, B.B.: A corpus for OCR research on mathematical expressions. IJDAR **7**(4), 241–259 (2005). https://doi.org/10.1007/s10032-004-0140-5
13. Gehrke, J., Ginsparg, P., Kleinberg, J.: Overview of the 2003 KDD cup. ACM SIGKDD Explor. Newsl. **5**(2), 149–151 (2003). https://doi.org/10.1145/980972.980992
14. Gervais, P., Fadeeva, A., Maksai, A.: Mathwriting: a dataset for handwritten mathematical expression recognition. In: Proceedings of the 31st ACM SIGKDD Conference on Knowledge Discovery and Data Mining V.2, pp. 5459–5469. Association for Computing Machinery (2025). https://doi.org/10.1145/3711896.3737436
15. Lapointe, A., Blostein, D.: Issues in performance evaluation: a case study of math recognition. In: 2009 10th International Conference on Document Analysis and Recognition, pp. 1355–1359 (2009). https://doi.org/10.1109/ICDAR.2009.247
16. Lin, Q., Huang, X., Bi, N., Suen, C.Y., Tan, J.: CCLSL: combination of contrastive learning and supervised learning for handwritten mathematical expression recognition. In: Computer Vision - ACCV 2022, pp. 577–592. Springer, Cham (2023)
17. Naseer, A., Zafar, K.: Meta-feature based few-shot Siamese learning for Urdu optical character recognition. Comput. Intell. **38**(5), 1707–1727 (2022). https://doi.org/10.1111/coin.12530

18. OleehyO: OleehyO/latex-formulas-80M (2024). https://huggingface.co/datasets/OleehyO/latex-formulas-80M. Accessed 17 Feb 2026
19. OleehyO: OleehyO/TexTeller (2024). https://huggingface.co/OleehyO/TexTeller. Accessed 17 Feb 2026
20. Osagie, E., Ji, W., Helian, N.: Medical image character recognition using attention-based Siamese networks for visually similar characters with low resolution. In: Proceedings of the Third International Conference on Innovations in Computing Research (ICR 2024), pp. 110–119. Springer, Cham (2024). https://doi.org/10.1007/978-3-031-65522-7_10
21. Pfahler, L., Schill, J., Morik, K.: The search for equations – learning to identify similarities between mathematical expressions. In: Brefeld, U., Fromont, E., Hotho, A., Knobbe, A., Maathuis, M., Robardet, C. (eds.) ECML PKDD 2019. LNCS (LNAI), vol. 11908, pp. 704–718. Springer, Cham (2020). https://doi.org/10.1007/978-3-030-46133-1_42
22. Sakthi, P.: prithivMLmods/qwen2-VL-OCR2-2b-instruct (2025). https://huggingface.co/prithivMLmods/Qwen2-VL-OCR2-2B-Instruct. Accessed 17 Feb 2026
23. Schmitt-Koopmann, F.M., Huang, E.M., Hutter, H.P., Stadelmann, T., Darvishy, A.: Mathnet: a data-centric approach for printed mathematical expression recognition. IEEE Access **12**, 76963–76974 (2024). https://doi.org/10.1109/ACCESS.2024.3404834
24. Shukla, S., Sharma, N., Gupta, M., Mishra, A.: Patentlmm: large multimodal model for generating descriptions for patent figures. In: Proceedings of the Thirty-Ninth AAAI Conference on Artificial Intelligence and Thirty-Seventh Conference on Innovative Applications of Artificial Intelligence and Fifteenth Symposium on Educational Advances in Artificial Intelligence. AAAI Press (2025). https://doi.org/10.1609/aaai.v39i19.34257
25. Singh, S.S.: Teaching machines to code: neural markup generation with visual attention (2018). http://arxiv.org/abs/1802.05415, preprint, arXiv:1802.05415
26. Uchida, S., Nomura, A., Suzuki, M.: Quantitative analysis of mathematical documents. IJDAR **7**(4), 211–218 (2005). https://doi.org/10.1007/s10032-005-0142-y
27. Wang, Z., Zhang, M., Baraniuk, R.G., Lan, A.S.: Scientific formula retrieval via tree embeddings. In: 2021 IEEE International Conference on Big Data (Big Data), pp. 1493–1503 (2021). https://doi.org/10.1109/BigData52589.2021.9671942
28. Wieckowiak, F., Eglin, V., Bonnet, T., Bres, S., Rousseau, L.: A multimodal evaluation pipeline for mathematical expression recognition: comparisons of datasets, metrics, and models. In: Document Analysis and Recognition – ICDAR 2025, pp. 120–136 (2026). https://doi.org/10.1007/978-3-032-04627-7_7
29. Xie, Y., et al.: ICDAR 2023 crohme: competition on recognition of handwritten mathematical expressions. In: Document Analysis and Recognition - ICDAR 2023, pp. 553–565 (2023). https://doi.org/10.1007/978-3-031-41679-8_33
30. Zanibbi, R., Davila, K., Kane, A., Tompa, F.W.: Multi-stage math formula search: Using appearance-based similarity metrics at scale. In: Proceedings of the 39th International ACM SIGIR Conference on Research and Development in Information Retrieval, pp. 145–154 (2016). https://doi.org/10.1145/2911451.2911512
31. Zanibbi, R., Mansouri, B., Agarwal, A.: Mathematical information retrieval: search and question answering. Found. Trends® Inf. Retrieval (2025). https://doi.org/10.1561/1500000095

32. Zanibbi, R., Mouchère, H., Viard-Gaudin, C.: Evaluating structural pattern recognition for handwritten math via primitive label graphs. In: SPIE Proceedings, vol. 8658, p. 865817 (2013). https://doi.org/10.1117/12.2008409
33. Zanibbi, R., Yu, L.: Math spotting: retrieving math in technical documents using handwritten query images. In: 2011 International Conference on Document Analysis and Recognition, pp. 446–451 (2011). https://doi.org/10.1109/ICDAR.2011.96
34. Zanibbi, R., Yuan, B.: Keyword and image-based retrieval of mathematical expressions. In: Document Recognition and Retrieval XVIII, vol. 7874, pp. 141–149 (2011). https://doi.org/10.1117/12.873312

Multilingual Table Recognition: A Benchmark Dataset and A Local-Global Hybrid Model

Nam Tuan Ly[1,2](✉), Atsuhiro Takasu[3,4], and Masaki Nakagawa[2]

[1] Tokyo Metropolitan University, Tokyo 192-0397, Japan
namly@tmu.ac.jp
[2] Tokyo University of Agriculture and Technology, Tokyo 184-8588, Japan
nakagawa@cc.tuat.ac.jp
[3] National Institute of Informatics, Tokyo 101-8430, Japan
takasu@nii.ac.jp
[4] The Graduate University for Advanced Studies (SOKENDAI), Kanagawa 240-0193, Japan

Abstract. The tabular format is widely used to present important information in communication, research, and data analysis across many languages. However, most prior works on table recognition have focused on English tables. To facilitate multilingual table recognition with deep learning, we create and release the largest publicly available multilingual table recognition dataset, WikiMTR, containing more than 14 million tables with corresponding structured HTML annotations across over 20 languages. We also propose a novel local-global hybrid model for recognizing table images and converting them into HTML code. The proposed model comprises a local-global encoder that extracts both local and global features, and three decoders for the three subtasks of table recognition. Extensive experiments across three large-scale table recognition datasets, including WikiMTR, confirm that our model performs well for multilingual table recognition and achieves comparable results that favorably compare with prior works. The ablation study also confirms the effectiveness of the local-global hybrid encoder.

Keywords: Local-Global Feature · Multilingual Table Recognition · Multilingual Table Dataset · End-to-End

1 Introduction

Tables are an information-rich format that is widely used in research papers, books, handwritten notes, invoices, financial documents, and many other areas. Thus, table analysis is an important technique in document image understanding. Table image recognition (*table recognition* hereinafter) is a key step in table analysis; it aims to recognize the table structure and the text content of each table cell from an input table image and then represent them in a machine-readable format, such as HTML [10, 16, 40] and LaTeX [2, 13]. Table recognition is challenging due to the diversity of table styles and the complexity of table structures; it has thus been attracting the attention of researchers worldwide.

G. A. Fink et al. (Eds.): ICDAR 2026, LNCS 16972, pp. 629–644, 2027.
https://doi.org/10.1007/978-3-032-36023-6_36

In recent years, due to the rapid development of deep neural networks (DNNs) and the availability of large-scale table datasets, many effective DNN-based table recognition methods [28–30, 33] have been proposed. These methods can be divided into two categories: two-stage and end-to-end approaches. Most existing techniques [9, 14, 27–30, 33, 36] are two-stage approaches. They divide the main problem (table recognition) into two subproblems: table structure recognition and table cell content recognition. Then, they solve each subproblem independently using two separate systems. Two-stage approaches have achieved competitive results on large-scale datasets of table images. However, these methods rely on training datasets containing rich annotation information; they are also difficult to train and infer because they consist of multiple separate components. Conversely, end-to-end approaches aim to solve the main problem using a single system and are usually easy to train and to infer, achieving excellent results across various tasks, including machine translation [35], speech recognition [1, 4], and text recognition [22, 23]. Many researchers have been working on end-to-end approaches and achieved competitive results compared to the two-stage approaches [12, 21, 24]. However, most of these methods employed a CNN backbone that primarily focuses on local information to extract visual features from table images, making it difficult to recognize large table images. In this paper, building upon the local attention model [24], we introduce a local-global hybrid encoder that leverages both local and global information, improving the model's performance in recognizing large table images.

While tabular data are widely used across many languages, most large-scale tabular datasets [39, 40] and prior table recognition research [32] focus only on English tables. To facilitate the research on multilingual table recognition, we create and release WikiMTR, the largest publicly available image-based multilingual table recognition dataset built from Wikipedia, because table structures and styles are different in different languages. We also conduct the very first experiments on multilingual table recognition. The source code and dataset will be publicly available at https://github.com/namtuanly/WikiMTR.

The main contributions of this paper are as follows:

- We propose a local-global hybrid model for large table recognition.
- We create and release the first large-scale multilingual table recognition dataset.
- Extensive experiments across three large-scale table recognition datasets, including a multilingual dataset, demonstrate the superiority of the proposed model and its effectiveness for multilingual table recognition.

The rest of this paper is organized as follows. In Sect. 2, we briefly review related works. In Sect. 3, we provide an overview of the proposed model. Section 4 describes the overview of the WikiMTR dataset. Section 5 presents the experimental results for table recognition and their analysis. Finally, we discuss the limitations of our work in Sect. 6 and draw conclusions in Sect. 7.

2 Related Work

Table understanding in document images can be divided into two main steps: (1) table detection: detecting the bounding boxes (bboxes) of tables in documents [6]; (2) table recognition: recognizing table structure information and the content of each table cell [6]. Here, we provide a brief survey of table recognition methods.

Table recognition methods can be divided into two categories: two-stage and end-to-end. A large body of prior work [9, 14, 27–30, 33, 36] adopts a two-stage formulation, treating structure recognition and cell-content recognition as separate problems handled by independent modules. Because cell-content recognition can often be delegated to off-the-shelf OCR engines, much of this line of research concentrates on the structure-recognition stage alone [9, 14, 28, 30, 33, 36]. From 1990 to 2004, table structure recognition methods [9, 14, 36] primarily relied on handcrafted features and heuristic rules and were primarily applied to simple table structures or predefined table formats. Since the 2010s, driven by the success of deep learning, many deep learning–based methods for table structure recognition [28, 30, 33] have been proposed, achieving significant improvements. S. Schreiber et al. [33] presented a twofold system, DeepDeSRT, that uses Faster R-CNN [31] and FCN [18] for table detection and table structure recognition. S. Raja et al. [30] proposed a table structure recognition model, TabStruct-Net, that jointly predicts aligned cell regions and localized cell relations.

Some recent works [27, 29, 37, 38] focus on table structure recognition and table cell content recognition to build complete table recognition systems. J. Ye et al. [37] introduced TableMASTER, a model with two decoders for predicting table structures and table cell locations. They further integrated a text line detector to identify text lines in each table cell and employed a text line recognizer based on MASTER [19] to recognize them, thereby constructing a complete table recognition system. TableMASTER won third place in the ICDAR 2021 competition [10]. Similar to [37], A. Nassar et al. [27] proposed TableFormer, which recognizes table structures and table cell locations from an input table in a PDF file and then uses the text content extracted from the PDF to build a table recognition system. L. Qiao et al. [29] proposed a table structure recognition system, LGPMA, built on Mask R-CNN. Then, they combined LGPMA with single-line text detection and recognition models to obtain OCR information. This model won first place in the ICDAR 2021 competition [10].

In recent years, due to rapid advances in deep learning and the availability of large-scale datasets of table images, researchers have focused on end-to-end approaches [2, 40]. Y. Deng et al. [2] cast table recognition as image-to-LaTeX translation and applied the IM2TEX architecture [3] directly to table images. X. Zhong et al. [40] introduced the encoder-dual-decoder (EDD) framework, pairing one decoder for table structure with another for per-cell text. However, these methods do not leverage multi-task learning [2] or utilize cell detection signals [40], resulting in mediocre performance compared to current SOTA approaches. Furthermore, they do not provide individual cell locations—information that is crucial for downstream tasks such as hierarchical recognition [11].

Most recently, Ly et al. [24] proposed an end-to-end model with a CNN encoder and three decoders equipped with a local attention mechanism for table recognition, demonstrating the effectiveness of local attention for recognizing large tables. However, the CNN encoder in [24] primarily captures local features, which limits its performance on large table images, and its evaluation is restricted to English tables. In this paper, we address these limitations by introducing a local-global hybrid encoder that extracts both local and global visual features, and by releasing WikiMTR, the first large-scale multilingual table recognition dataset, on which we conduct the first experiments on multilingual table recognition.

3 The Proposed Method

In this study, we propose a local-global hybrid model to address the limitations of local-visual-features-based methods for recognizing large-sized table images. The proposed model is composed of four main components: a local-global hybrid encoder for extracting both local and global features from an input image and three decoders (structure decoder, cell-bbox decoder, and cell-content decoder) for predicting the three subtasks of table recognition, as shown in Fig. 1. For an input table image, the hybrid encoder first produces a unified sequence that embeds both local and global visual cues. The structure decoder consumes this sequence and autoregressively emits the structure tokens describing the table layout. Each time a token corresponding to a new cell is emitted, the cell-bbox and cell-content decoders are activated; conditioned on the structure decoder's hidden state at that position, they regress the cell's bounding box and decode its textual content, respectively. We will describe the four components of our model in the following subsections.

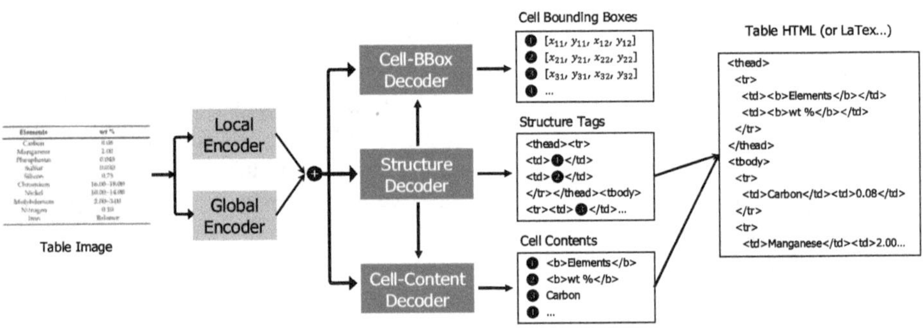

Fig. 1. The overview of the proposed model.

3.1 Local-Global Hybrid Encoder

At the top of the proposed model, the local-global hybrid encoder extracts both local and global visual features from an input image and encodes them as a sequence of fixed-size features. The proposed local-global hybrid encoder comprises three modules: (i) a CNN backbone for local feature extraction, (ii) a Swin Transformer [17] encoder for global context modeling, and (iii) a fusion module that combines the two feature streams.

Given an input table image $I \in R^{h \times w \times c}$, the CNN backbone extracts a local feature grid $F^{loc} \in R^{h' \times w' \times d_{loc}}$, where h' and w' depend on the input resolution and the down-sampling operations of the CNN, d_{loc} denotes the channel dimensions (feature/embedding dimensions) of the CNN-based local feature map. In our implementation, we adopt a ResNet-style backbone for its strong ability to capture fine-grained patterns, such as edges, strokes, and cell boundaries.

In parallel, we employ a Swin Transformer encoder to capture broader layout and long-range dependencies. Swin Transformer builds hierarchical feature maps via patch partition and patch merging, and uses shifted-window self-attention to efficiently exchange information across neighboring windows. Specifically, the image is partitioned into non-overlapping patches, projected into token embeddings, and processed through multiple Swin stages, producing a global feature grid $F^{glob} \in R^{h'' \times w'' \times d_{glob}}$, where d_{glob} denotes the channel dimensions (feature/embedding dimensions) of the Swin-based global feature map. Compared with purely convolutional features, F^{glob} provides stronger global structural cues, such as row/column alignment, header regions, and multi-cell spanning patterns.

To combine the complementary information, we design a fusion module that aligns and merges F^{loc} and F^{glob}. First, we use 1×1 convolutions (or linear projections) to align channel dimensions. Then, we fuse them by concatenation followed by a projection layer:

$$F^{fuse} = Proj\left(\left[F^{loc}; F^{glob}\right]\right) \tag{1}$$

where $[\cdot; \cdot]$ denotes channel-wise concatenation and $\text{Proj}(\cdot)$ is implemented as a 1×1 convolution (or linear layer) with normalization and nonlinearity. Finally, the fused feature grid is flattened into a fixed-length sequence of features $F \in \mathbb{R}^{N \times d}$ with $N = h\prime \times w\prime + h'' \times w''$ and d is the fused channel dimension. This sequence of features is used as the encoder output and fed into the subsequent decoder to predict table structure and content.

3.2 Three Decoders

Structure Decoder

At the top of the shared encoder, the structure decoder uses the encoded sequence of fixed-size features to predict a sequence of HTML tags, tokenized at the HTML tag level, similar to [40]. Following the local attention model [24], we use n stacked local decoder layers followed by a linear layer and a softmax layer to build the structure decoder. The local attention mechanism enables the structure decoder to focus more effectively on salient local features of neighboring tokens when predicting a structure token. The stack of local decoder layers uses the encoded sequence of fixed-size features as both value and key vectors. Its input query vector is the right-shifted sequence of structure tokens (the table structure target in training and the output of the structure decoder in inference) after being fed through the embedding layer and the positional encoding layer. Finally, the output of the stack of local decoder layers is fed into the linear layer. Then, the softmax layer generates a sequence of structure tokens that represent the table structure.

Cell-Bbox Decoder

The cell-bbox decoder is responsible for localizing each detected cell. As illustrated in Fig. 1, it comprises a single local decoder layer, a linear projection, and a sigmoid activation. Whenever the structure decoder emits a new-cell token, this branch is activated: the corresponding structure decoder hidden state serves as the query, while the encoded feature sequence provides the keys and values. The local decoder layer's output is then mapped through the linear and sigmoid layers to regress the four normalized coordinates of the cell's bounding box.

Cell-Content Decoder
Similar to the cell-bbox decoder, the cell-content decoder retrieves the hidden states from the structure decoder that correspond to tokens representing a new cell. The decoder uses these hidden states to predict the text content of the table cells. In this work, the text in table cells is tokenized at the character level, and the cell-content decoder serves as a text recognizer. As shown in Fig. 1, the cell-content decoder consists of an embedding layer, a positional encoding layer, *m* stacked local decoder layers, and is followed by a linear layer and a softmax layer. The following is the decoding mechanism of the cell-content decoder.

When the structure decoder produces a new-cell token, the cell-content decoder is activated. The encoded feature sequence from the encoder serves as both the keys and the values. On the query side, the right-shifted cell content target (or, at inference time, the decoder's own previous outputs) passes through the embedding and positional-encoding layers and is then summed with the structure decoder hidden state associated with that cell. The resulting query is processed by the stacked local decoder layers, whose output is projected by a linear layer and normalized by a softmax to produce the cell's text content.

3.3 Network Training

The whole model is trained end-to-end using stochastic gradient descent on pairs of table images and their annotations of the table structure, as well as each cell box and its text content. The overall loss of the proposed model is defined as follows:

$$\mathcal{L} = \lambda_1 \mathcal{L}_{\text{struc.}} + \lambda_2 \mathcal{L}_{\text{cont.}} + \lambda_3 \mathcal{L}_{\text{bbox}} \tag{2}$$

where $\mathcal{L}_{\text{struc.}}$ and $\mathcal{L}_{\text{cont.}}$ are the table structure recognition loss and the cell content prediction loss, respectively, that are implemented in cross-entropy loss; $\mathcal{L}_{\text{bbox}}$ is the cell bbox regression loss, which is optimized using the L1 loss; and λ_1, λ_2, and λ_3 are weight hyperparameters.

4 WikiMTR

Recently, several large-scale table (structure) recognition datasets have been released to the research community, including PubTabNet [40], FinTabNet [39], and PubTables-1M [34]. However, to the best of our knowledge, these datasets only consist of English tables, and there is currently a lack of table recognition datasets in other languages. This limits the development and evaluation of models designed for multilingual or non-English table recognition, which is increasingly important in global applications.

To address this gap, Ly et al. [20] created and released WikiTableSet, a dataset comprising approximately 1 million tables in three languages: English, French, and Japanese. In this paper, we introduce WikiMTR, a multilingual table dataset spanning over 20 widely used languages to advance research on multilingual table recognition. As a result, WikiMTR comprises more than 14 million tables across over 20 languages, as summarized in Table 1, compared with previous large-scale datasets, and in Table 2,

Table 1. Comparison of datasets for table structure recognition (TSR) and table recognition (TR).

Datasets	Samples	TSR	TR	Languages
PubTabNet [40]	510K	✔	✔	English
FinTabNet [39]	113K	✔	✔	English
PubTables-1M [34]	948K	✔	✔	English
WikiTableSet [20]	1M	✔	✔	English, French, Japanese
WikiMTR (ours)	> 14M	✔	✔	> 20 languages

which shows the number of samples for each of the top 12 most widely used languages. To increase the diversity of table styles in WikiMTR, we randomly vary cell padding, border width, and border styles (including fully bordered, borderless, and no-border tables) during synthetic table image generation. Each language-specific subset is then divided into training, validation, and testing sets. We also plan to extend WikiMTR to include over 50 widely used languages and release it to the community. WikiMTR is generated by rendering Wikipedia tables, with structure and cell annotations obtained directly from the source markup. It naturally includes cells spanning multiple rows and columns. However, it does not cover real-world document degradations (e.g., noise, skew, blur, or low resolution), handwritten tables, or multi-page tables. These limitations are discussed in Sect. 6.

Table 2. Number of samples per language.

Subset	Samples
English	4,291,914
German	1,656,004
French	1,309,113
Spanish	789,926
Italian	774,355
Japanese	690,941
Chinese	613,191
Dutch	592,042
Polish	525,861
Portuguese	487,228
Korean	252,390
Vietnamese	143,292

5 Experiments

We conduct experiments on standard benchmarks for table recognition to evaluate the performance of the proposed model and the effectiveness of the local-global hybrid approach. The datasets are discussed in Sect. 5.1; the implementation details and evaluation metrics are described in Sect. 5.2; the ablation studies are explained in Sect. 5.3; a comparison between the proposed model and previous methods is presented in Sect. 5.4; the experiments of multilingual table recognition are presented in Sect. 5.5; and Sect. 5.6 describes the visualization results.

5.1 Datasets

PubTabNet

X. Zhong et al. [40] published a large-scale table image dataset called PubTabNet, created by collecting scientific articles from the PubMed Central Open Access Subset. It consists of over 568,000 table images with corresponding annotations on table structures, text content, and the bounding boxes of all non-empty table cells. Depending on its complexity, a table is "simple" if it does not contain row spans or column spans; otherwise, it is "complex". This dataset comprises 500,777 training samples, 9,115 validation samples, and 9,064 final evaluation samples, used in the training, development, and final evaluation phases of the ICDAR 2021 competition, respectively.

PubTabNet250

PubTabNet250 is a subset we curate by retaining only PubTabNet tables whose number of structure tokens exceeds 250, yielding 114,111 training and 2,161 validation images. We use it to assess the effectiveness of the local-global hybrid encoder on large tables.

FinTabNet

X. Zheng et al. [39] introduced FinTabNet, a large-scale corpus drawn from the annual reports of S&P 500 companies that, as in PubTabNet, provides HTML structure and cell-bounding-box annotations. It contains roughly 112K table images split into training, testing, and validation partitions at an 81%:9.5%:9.5% ratio.

5.2 Implementation Details and Evaluation Metrics

Implementation Details

In the local-global hybrid encoder, inspired by [19], we use a ResNet-31 backbone [7] with multi-aspect global context attention (GCAttention) applied after each residual block to build the CNN backbone. All input images are resized to 520 × 520 pixels before being fed into the local-global hybrid encoder.

In the three decoders, all local decoder layers share the same architecture: eight attention heads, an input feature size of 512, and a feed-forward layer with 2048 nodes. The number of decoder layers in the structure and cell-content decoders is set to $n = 3$ and $m = 1$, respectively. The structure and cell tokens (in the decoding process) have maximum sequence lengths of 600 and 150, respectively. The weight hyperparameters are set to $\lambda_1 = \lambda_2 = \lambda_3 = 1$ (same contribution of each decoder).

The proposed model is implemented in the MMCV library [26] and trained on four NVIDIA A100 40G GPUs with a batch size of eight. Optimization starts with a learning rate of 1e-3 for the first 12 epochs, after which it is decayed to 1e-4 and kept until convergence

Evaluation Metrics
To evaluate the table recognition performance of the proposed model, we use the tree-edit-distance-based similarity (TEDS) metric [40]. Representing the prediction and the ground truth as a tree structure of HTML tags, it is calculated as follows:

$$\mathrm{TEDS}(T_a, T_b) = 1 - \frac{\mathrm{EditDist}(T_a, T_b)}{\max(|T_a|, |T_b|)} \tag{3}$$

where T_a and T_b represent tables in a tree-structured HTML format, *EditDist* denotes the tree-edit distance, and $|T|$ represents the number of nodes in T. We also denote TEDS-struc. as the TEDS score between two tables, computed only from table structure information.

To evaluate the performance of the cell-content decoder, we use the cell content accuracy (CCA) metric, which is defined as follows:

$$\mathrm{CCA}(S, S') = \frac{100}{|S'|} \sum_{i=0}^{|S'|-1} \begin{cases} 1 \; \mathit{if} \; S[i] = S'[i] \\ 0 \; \mathit{otherwise} \end{cases} \tag{4}$$

where *S* and *S'* are the lists of predicted cell content and the ground-truth cell content, respectively.

5.3 Ablation Studies

In this section, we conducted ablation experiments on the PubTabNet250 dataset to evaluate the effectiveness of the local-global hybrid visual features compared with only local or global visual features in the encoder. To do that, we prepared two variations that are identical to the proposed model except for using a local (ResNet) or a global (Swin Transformer) encoder. Table 3 shows their table recognition results on the validation set of the PubTabNet250 dataset in different sequence lengths of the table structure tokens.

As shown in Table 3, Global outperforms Local on all PubTabNet250 validation sets, indicating that global vision features improve local vision features for table recognition. Furthermore, Local-Global outperforms both Local and Global on the PubTabNet250 validation set, and it is even more dominant on tables with a large number of table structure tokens. The results imply that combining local and global features improves table recognition performance, especially for large tables, demonstrating the effectiveness of both local and global vision information.

Table 3. Table recognition results (TEDS) with different visual encoders.

Encoders	TEDS (%)			
	All	len > 500	len > 600	len > 700
Local Features (ResNet)	94.95	92.99	91.80	90.21
Global Features (Swin)	95.17	93.40	92.38	91.17
Local-Global Features (ResNet + Swin)	96.05	95.28	94.78	94.01

len: number of table structure tokens.

5.4 Comparison with SOTA

FinTabNet

This experiment evaluates the performance of the proposed model on the FinTabNet dataset in terms of TEDS (Table recognition) and TEDS-struc. (Table structure recognition). To make a fair comparison with previous methods, we do not use any data augmentation or ensemble techniques. Table 4 compares the TEDS and TEDS-struc. Scores between the proposed model and the previous methods on the test set of the FinTabNet dataset. With TEDS of 97.19% and TEDS-struc. of 98.96%, the proposed model achieves state-of-the-art results in table structure recognition and comparable results to MuTabNet [12] in table recognition.

Table 4. Table recognition results (%) on FinTabNet test set.

Models	TEDS-struc	TEDS
EDD [40]	90.60	–
GTE [39]	87.14	–
GTE$^{(PT)}$ [39]	91.02	–
TableFormer [27]	96.80	–
SPRINT [15]	98.03	–
GridFormer [25]	98.63	–
WSTabNet [20]	98.72	95.32
Local Attention Model [24]	98.85	95.74
MuTabNet [12]	98.87	**97.69**
BGTR [8]	98.89	–
The Proposed Method	**98.96**	97.19

PubTabNet

In this experiment, we evaluate the proposed model's performance for table recognition on the PubTabNet dataset. To fairly compare with the previous methods, we also do not employ any data augmentation or ensemble techniques. Table 5 shows the table recognition results of the proposed model compared with previous methods on the validation set of the PubTabNet dataset. As shown in Table 5, the proposed model outperforms all previous methods that do not employ data augmentation or ensemble techniques. Furthermore, with TEDS of 98.29% on simple tables and TEDS of 95.78% on complex tables, the proposed model also achieves state-of-the-art results on both simple and complex table images.

Table 5. Table recognition results on PubTabNet validation set.

Models	TEDS		
	Simple	Complex	All
EDD [40]	91.20	85.40	88.30
TabStruct-Net [30]	–	–	90.10
GTE [39]	–	–	93.00
TableFormer [27]	95.40	90.10	93.60
SEM (3) [38]	94.80	92.50	93.70
LGPMA + OCR (1) [29]	–	–	94.60
GridFormer [25]	–	–	95.84
TRUST [5]	–	–	96.20
VCGoup (2) [37]	–	–	96.26
WSTabNet [20]	97.89	95.02	96.48
BGTR [8]	–	–	96.57
Multi-Tasks Model [21]	97.92	95.36	96.67
Local Attention Model [24]	98.07	95.42	96.77
MuTabNet [12]	98.16	95.53	96.87
The Proposed Model	**98.29**	**95.78**	**97.07**
VCGoup + ME (2) [37]	–	–	96.84

Although we do not use any data augmentation as well as ensemble techniques, the proposed method achieves better results compared to VCGroup + ME [15], which requires additional annotations of text-line bounding boxes of the text content in each table cell and employs three model ensembles in both table structure recognition and text line recognition models.

5.5 Multilingual Table Recognition

In this section, we conduct experiments on WikiMTR, a dataset comprising tables in multiple languages, to evaluate the proposed model's performance in multilingual table recognition. To the best of our knowledge, this is the first study addressing this task, and it can serve as a baseline for future research. Due to computational resource constraints, we conduct experiments on table images from three widely used languages: English, French, and Chinese. To ensure a balanced comparison across languages, we randomly select an equal number of tables (450K) from each language to construct both the monolingual training sets and the multilingual training set. The monolingual datasets are referred to as **EN450K**, **FR450K**, and **ZH450K**, and the multilingual dataset combining all three is denoted as **EN + FR + ZH450K**. To ensure a fair comparison in terms of training sample size between the multilingual and monolingual datasets, we also create a second multilingual dataset, **EN + FR + ZH150K**, by sampling 150K tables from each monolingual dataset. We then train our model independently on each of the two multilingual datasets and compare their performance with that of models trained individually on the monolingual subsets. The performance of the trained models is evaluated on the test sets for each language using table structure recognition (TEDS-Struct), cell content recognition (CCA), and overall table recognition (TEDS) to examine the impact of the training language on model performance, as summarized in Table 6.

Table 6. Multilingual table recognition results.

Training Set	Testing Set (%)								
	English			French			Chinese		
	TEDS -struc	CCA	TEDS	TEDS -struc	CCA	TEDS	TEDS -struc	CCA	TEDS
EN + FR + ZH450K	99.06	68.75	96.49	99.49	82.73	96.20	98.44	73.72	96.04
EN + FR + ZH150K	98.89	71.11	95.99	99.45	81.05	95.16	98.36	69.98	95.06
EN450K	98.72	71.31	96.32	98.16	n/a	n/a	96.87	n/a	n/a
FR450K	95.39	n/a	n/a	99.39	83.28	94.89	97.39	n/a	n/a
ZH450K	95.31	n/a	n/a	97.83	n/a	n/a	98.27	66.78	95.09

As shown in Table 6, the models trained on individual monolingual subsets achieve good TEDS-Struct scores on the test sets of the other languages, indicating that the proposed model generalizes well to table structure recognition across languages. Moreover, we observe that the model trained on the combined multilingual dataset (**EN + FR + ZH450K**) outperforms the monolingual models across all evaluation metrics. This indicates that training on the combined multilingual data improves performance not only in table structure recognition but also in cell content recognition and overall table recognition, compared to training on monolingual subsets alone. We also observe that the model trained on the multilingual dataset (**EN + FR + ZH150K**) achieves comparable performance to each monolingual model on its respective test set, despite having fewer training samples per language.

5.6 Visualization Results

In this section, we show the visualization results of the proposed model on PubTabNet. As shown in Fig. 2, the proposed model predicts complex table structures, including the bboxes and contents of all table cells, including empty cells and cells spanning multiple rows/columns.

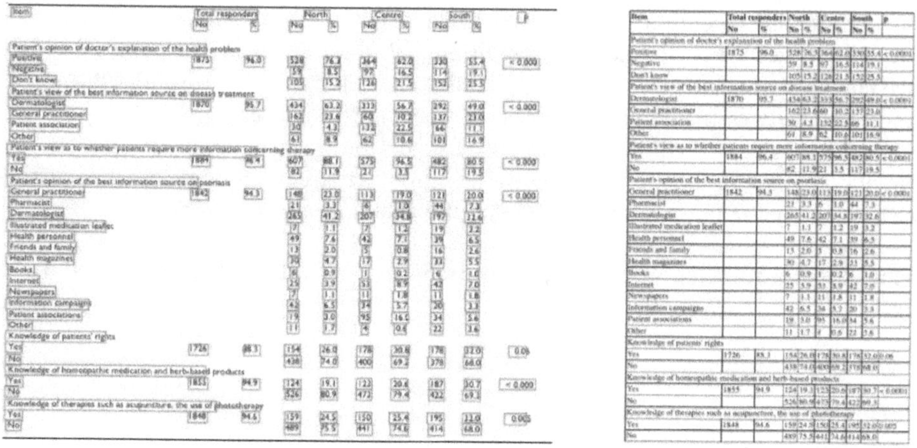

Fig. 2. Visualization results of the proposed model on PubTabNet.

6 Limitations

In this section, we discuss the limitations of our study and directions for future work. First, the three benchmark datasets do not include real-world document degradations such as noise, skew, blur, low resolution, handwritten content, or multi-page tables. Therefore, the generalization of the proposed model to real-world scanned documents remains to be validated. Extending the evaluation to such challenging real-world scenarios is an important direction for future work.

Second, although WikiMTR covers more than 20 languages, our experiments are limited to English, French, and Chinese due to computational constraints. We also do not evaluate low-resource languages or complex scripts such as Arabic and Indic scripts. Evaluating a broader range of languages and scripts remains an important direction for future work.

Third, our multilingual experiments are limited to comparisons among variants of the proposed model trained on different subsets of WikiMTR. We do not evaluate external state-of-the-art methods on WikiMTR; thus, the proposed model currently serves as the primary baseline for this dataset.

Finally, we evaluate cell content using only cell content accuracy (CCA), which measures exact cell-level matches. Additional metrics such as character error rate (CER) and word error rate (WER), together with a more detailed analysis of multilingual text and failure cases, would provide a more comprehensive evaluation and are left for future work.

7 Conclusions

In this study, we propose a local-global hybrid model to address the limitations of local-visual-features-based methods for recognizing large-sized table images. Extensive experiments on three large-scale table image datasets demonstrate that the proposed model outperforms existing methods; it can process not only simple tables well but also complex tables. Ablation studies demonstrate the effectiveness of the local-global hybrid encoder for table recognition, particularly for large tables. To advance multilingual table recognition, we introduce WikiMTR, a multilingual table dataset covering over 20 widely used languages, and present the first empirical study in this domain to examine the impact of the languages of the training samples on model performance.

In the future, we will test the proposed model on other table image datasets. We also plan to explore the model's effectiveness in other structural domains of document analysis and recognition, such as diagrams and charts.

Acknowledgments. This work was partially supported by MEXT/JSPS KAKENHI Grant Numbers JP24K20789 and JP24H00738.

References

1. Bahdanau, D., et al.: End-to-end attention-based large vocabulary speech recognition. In: Proceedings of the International Conference on Acoustics, Speech and Signal Processing (ICASSP), pp. 4945–4949 (2016)
2. Deng, Y., et al.: Challenges in end-to-end neural scientific table recognition. In: Proceedings of the International Conference on Document Analysis and Recognition, ICDAR, pp. 894–901 (2019). https://doi.org/10.1109/ICDAR.2019.00148
3. Deng, Y., et al.: Image-to-markup generation with coarse-to-fine attention. In: 34th International Conference on Machine Learning, ICML 2017, vol. 3, pp. 1631–1640 (2016). https://doi.org/10.48550/arxiv.1609.04938
4. Dong, L., et al.: Speech-transformer: a no-recurrence sequence-to-sequence model for speech recognition. In: Proceedings of the IEEE International Conference on Acoustics, Speech and Signal Processing - Proceedings, pp. 5884–5888 (2018)
5. Guo, Z., et al.: TRUST: an accurate and end-to-end table structure recognizer using splitting-based transformers (2022). https://doi.org/10.48550/arXiv.2208.14687
6. Hashmi, K.A., et al.: Current status and performance analysis of table recognition in document images with deep neural networks. IEEE Access. **9**(2021), 87663–87685 (2021). https://doi.org/10.1109/ACCESS.2021.3087865
7. He, K., et al.: Deep residual learning for image recognition. In: 2016 IEEE Conference on Computer Vision and Pattern Recognition (CVPR), pp. 770–778 (2016)
8. Hu, L., Huang, S.: Enhancing table structure recognition via bounding box guidance. In: Pattern Recognition. ICPR 2024. Lecture Notes in Computer Science. LNCS, vol. 15320, pp. 209–225 (2025). https://doi.org/10.1007/978-3-031-78498-9_15
9. Itonori, K.: Table structure recognition based on textblock arrangement and ruled line position. In: Proceedings of 2nd International Conference on Document Analysis and Recognition (ICDAR '93), pp. 765–768 (1993). https://doi.org/10.1109/ICDAR.1993.395625

10. Jimeno Yepes, A., et al.: ICDAR 2021 competition on scientific literature parsing. document analysis and recognition – ICDAR 2021. In: ICDAR 2021. Lecture Notes in Computer Science. LNCS, vol. 12824, pp.605–617 (2021). https://doi.org/10.1007/978-3-030-86337-1_40
11. Jirachanchaisiri, P., et al.: TRH2TQA: table recognition with hierarchical relationships to table question-answering on business table images. In: 2025 IEEE/CVF Winter Conference on Applications of Computer Vision (WACV), pp. 8844–8852 (2025). https://doi.org/10.1109/WACV61041.2025.00857
12. Kawakatsu, T.: Multi-cell decoder and mutual learning for table structure and character recognition. In: Document Analysis and Recognition - ICDAR 2024. ICDAR 2024. LNCS, vol. 14804, pp. 389–405 (2024). https://doi.org/10.1007/978-3-031-70533-5_23
13. Kayal, P., et al.: ICDAR 2021 competition on scientific table image recognition to LaTeX. In: Document Analysis and Recognition – ICDAR 2021. ICDAR 2021. LNCS, vol. 12824, pp. 754–766 (2021) https://doi.org/10.1007/978-3-030-86337-1_50
14. Kieninger, T.G.: Table structure recognition based on robust block segmentation. **3305**, 22–32 (1998). https://doi.org/10.1117/12.304642
15. Kudale, D., et al.: SPRINT: script-agnostic structure recognition in tables. In: Document Analysis and Recognition - ICDAR 2024. ICDAR 2024. LNCS, vol. 14808, pp. 350–367 (2024). https://doi.org/10.1007/978-3-031-70549-6_21
16. Li, M., et al.: TableBank: a benchmark dataset for table detection and recognition (2019). https://doi.org/10.48550/arxiv.1903.01949
17. Liu, Z., et al.: Swin transformer: hierarchical vision transformer using shifted windows. In: IEEE International Conference on Computer Vision, pp. 9992–10002 (2021). https://doi.org/10.1109/ICCV48922.2021.00986
18. Long, J., et al.: Fully convolutional networks for semantic segmentation. In: 2015 IEEE Conference on Computer Vision and Pattern Recognition (CVPR). 07–12 June 2015, pp. 3431–3440 (2015). https://doi.org/10.1109/CVPR.2015.7298965
19. Lu, N., et al.: MASTER: multi-aspect non-local network for scene text recognition. Pattern Recogn. **117**, 107980 (2021). https://doi.org/10.1016/J.PATCOG.2021.107980
20. Tuan, L.N., et al.: Rethinking image-based table recognition using weakly supervised methods. In: Proceedings of the 12th International Conference on Pattern Recognition Applications and Methods (ICPRAM 2023), Lisbon, Portugal (2023)
21. Tuan, L.N., Atsuhiro, T.: An end-to-end multi-task learning model for image-based table recognition. In: Proceedings of the 18th International Joint Conference on Computer Vision, Imaging and Computer Graphics Theory and Applications (VISIGRAPP 2023) (2023)
22. Ly, N.T., et al.: An attention-based end-to-end model for multiple text lines recognition in Japanese historical documents. In: Proceedings of the International Conference on Document Analysis and Recognition, ICDAR, pp. 629–634 (2019)
23. Ly, N.T., et al.: Training an end-to-end model for offline handwritten Japanese text recognition by generated synthetic patterns. In: Proceedings of International Conference on Frontiers in Handwriting Recognition, ICFHR (2018)
24. Ly, N.T., Takasu, A.: An end-to-end local attention based model for table recognition. In: Document Analysis and Recognition - ICDAR 2023. ICDAR 2023. LNCS, vol. 14188, pp. 20–36 (2023). https://doi.org/10.1007/978-3-031-41679-8_2
25. Lyu, P., et al.: GridFormer: towards accurate table structure recognition via grid prediction. In: MM 2023 - Proceedings of the 31st ACM International Conference on Multimedia, pp. 7747–7757 (2023). https://doi.org/10.1145/3581783.3611961
26. MMCV Contributors 2018. {MMCV: OpenMMLab} Computer Vision Foundation (2018)
27. Nassar, A., et al.: TableFormer: table structure understanding with transformers (2022). https://doi.org/10.48550/arxiv.2203.01017

28. Prasad, D., et al.: CascadeTabNet: an approach for end to end table detection and structure recognition from image-based documents. In: IEEE Computer Society Conference on Computer Vision and Pattern Recognition Workshops. 2020-June, pp. 2439–2447 (2020). https://doi.org/10.48550/arxiv.2004.12629
29. Qiao, L., et al.: LGPMA: complicated table structure recognition with local and global pyramid mask alignment. In: Document Analysis and Recognition – ICDAR 2021. ICDAR 2021. LNCS, vol. 12821, pp. 99–114 (2021). https://doi.org/10.1007/978-3-030-86549-8_7/TABLES/4
30. Raja, S., et al.: Table structure recognition using top-down and bottom-up cues. In: Computer Vision – ECCV 2020. ECCV 2020. LNCS, vol. 12373, pp. 70–86 (2020). https://doi.org/10.1007/978-3-030-58604-1_5/FIGURES/8
31. Ren, S., et al.: Faster R-CNN: towards real-time object detection with region proposal networks. IEEE Trans. Pattern Anal. Mach. Intell. **39**(6), 1137–1149 (2015). https://doi.org/10.48550/arxiv.1506.01497
32. Salaheldin Kasem, M., et al.: Deep learning for table detection and structure recognition: a survey. ACM Comput. Surv. **56**(12), 41 (2024). https://doi.org/10.1145/3657281
33. Schreiber, S., et al.: DeepDeSRT: deep learning for detection and structure recognition of tables in document images. In: Proceedings of the International Conference on Document Analysis and Recognition, ICDAR, vol. 1, pp. 1162–1167 (2017). https://doi.org/10.1109/ICDAR.2017.192
34. Brandon, S., et al.: PubTables-1M: towards comprehensive table extraction from unstructured documents. In: Proceedings of the IEEE/CVF Conference on Computer Vision and Pattern Recognition (CVPR), pp. 4634–4642 (2022)
35. Vaswani, A., et al.: Attention is all you need. In: Advances in Neural Information Processing Systems, pp. 5999–6009 (2017)
36. Wang, Y., et al.: Table structure understanding and its performance evaluation. Pattern Recogn. **37**(7), 1479–1497 (2004). https://doi.org/10.1016/J.PATCOG.2004.01.012
37. Ye, J., et al.: PingAn-VCGroup's Solution for ICDAR 2021 competition on scientific literature parsing task B: table recognition to HTML (2021). https://doi.org/10.48550/arxiv.2105.01848
38. Zhang, Z., et al.: Split, embed and merge: an accurate table structure recognizer. Pattern Recogn. **126**, 108565 (2022). https://doi.org/10.1016/J.PATCOG.2022.108565
39. Zheng, X., et al.: Global table extractor (GTE): a framework for joint table identification and cell structure recognition using visual context. In: 2021 IEEE Winter Conference on Applications of Computer Vision (WACV), pp. 697–706 (2021). https://doi.org/10.1109/WACV48630.2021.00074
40. Zhong, X., et al.: Image-based table recognition: data, model, and evaluation. In: Computer Vision – ECCV 2020. ECCV 2020. LNCS, vol. 12366, pp. 564–580 (2020). https://doi.org/10.1007/978-3-030-58589-1_34/TABLES/3

Text and Symbol Recognition

What Can Low Resource Languages Learn From Each Other?

P. Achyuth, Kahaan Shah(✉), and Chetan Arora

IIT Delhi, New Delhi, India
{achyuth.cstaff,kahaan.cstaff,chetan}@cse.iitd.ac.in

Abstract. Despite the rapid advancement of Vision-Language Models (VLMs), their linguistic reach remains largely confined to high-resource languages, leaving the majority of the world's 7,000+ living languages on the wrong side of a growing digital divide. This disparity is especially pronounced in Optical Character Recognition (OCR), where low-resource scripts lack the massive datasets required for traditional scaling laws. We investigate OCR adaptation in extreme data-scarce regimes (<10K real and <250K synthetic images), demonstrating that conventional fine-tuning strategies often reach a performance ceiling. Our key finding reveals a structural inefficiency in language-specific adaptation: while higher layers of specialized models diverge to capture unique script nuances, the lower layers learn redundant, highly similar features. Motivated by this observation, we propose PSMC (Pre-train, Specialize, Merge, and Co-train), a data-efficient framework that capitalizes on a cross-script "transfer effect". Our approach first derives language-specific experts from a high-resource base model, then employs task arithmetic to fuse these experts into a unified, high-performance multilingual backbone. Extensive evaluation across 10 Indian scripts (supporting 20+ languages) shows that PSMC achieves on average ∼13% relative reduction in Word Error Rate (WER) over individual specialist models without increasing the parameter count. Our results indicate that joint training in the merged latent space facilitates a constructive knowledge transfer that benefits all constituent scripts, providing a scalable pathway for inclusive VLM development. Project Page: https://vision-iitd.github.io/projects/PSMC.

Keywords: Scene Text Recognition · Printed Text Recognition · Multilingual OCR

1 Introduction

The recent surge in Vision-Language Models (VLMs) has enabled unprecedented capabilities in cross-modal understanding. However, these advancements have largely been asymmetric. While high-resource languages enjoy robust support,

P. Achyuth and K. Shah—These authors contributed equally to this work.

G. A. Fink et al. (Eds.): ICDAR 2026, LNCS 16972, pp. 647–662, 2027.
https://doi.org/10.1007/978-3-032-36023-6_37

a significant majority of the world's 7,000+ living languages remain sidelined, resulting in a widening digital divide. This disparity is particularly critical in Optical Character Recognition (OCR), which serves as the fundamental gateway for digitizing and preserving the cultural and administrative heritage of low-resource communities.

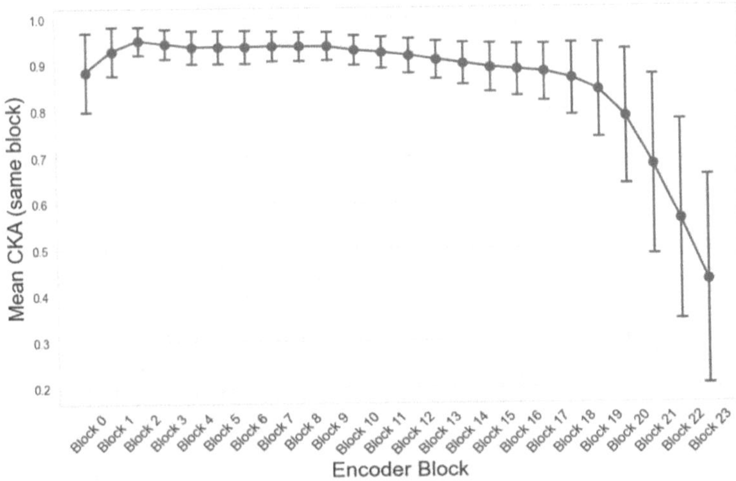

Fig. 1. To quantitatively assess the similarity of learned representations across language-specific models, we employ Centered Kernel Alignment (CKA) [11], a representation similarity metric that is invariant to orthogonal transformations and isotropic scaling. We consider a base model trained on the full English dataset and further fine-tuned on 250K samples (10K real, 240K synthetic) of each script. We choose 25 scene and 25 printed text images per language and pass them through each model, extracting features at each encoder layer. We then perform a CKA analysis at each layer (more details in the experiments section). The plot shows the average similarity between layers at the same index, averaged across all images and model pairs. Observe that the representations in the lower layers are very similar, while language specific features are learned in the deeper layers of each encoder. This motivates us to co-train across languages, allowing the encoder to jointly learn the early script-agnostic robust representations from varied samples while accommodating script-specific features in the final encoder layers.

Current paradigms for extending OCR to new scripts rely on massive data scaling; a luxury not afforded by the "long-tail" of low-resource languages. In extreme data-scarce regimes, where real training samples are often capped at fewer than 10,000 images, standard adaptation strategies begin to falter. Language-specific fine-tuning or delta training, while effective in high-resource contexts, often results in overfitting or catastrophic forgetting of the general visual features learned during the model's base pre-training. Consequently, these models reach a performance ceiling that fails to capture the nuances of diverse scripts.

Through an architectural analysis of language-specific "specialists", we identify a crucial structural redundancy: while the higher layers of these models diverge to master script-specific characters, the lower layers learn remarkably similar, domain-agnostic visual features (see Fig. 1). This observation suggests that maintaining separate, isolated models for each language is not only parameter-inefficient but also ignores a latent opportunity for cross-linguistic synergy.

In this work, we propose PSMC (Pre-train, Specialize, Merge, and Co-train), a novel, data-efficient framework that leverages joint training across diverse scripts to overcome the limitations of data scarcity. Our approach moves beyond simple fine-tuning by first deriving language-specific experts ('S' stage) from a common pretrained base model ('P' stage), and then employing task arithmetic to fuse their weight offsets into a single, unified multilingual backbone ('M' stage). This merging process effectively "pools" the commonalities of different scripts while preserving their unique diagnostic features. Finally, we perform a joint co-training ('C' stage) with all the languages to disentangle representations across languages.

By evaluating PSMC across 10 Indian scripts, covering more than 20 languages, we demonstrate that our unified model consistently outperforms individual specialist models. Our findings reveal a constructive transfer effect: by training in a merged latent space, the model leverages shared structural patterns between scripts (e.g., the shared phonetic roots and stroke styles across scripts) to reduce the Word Recognition Rate (WER) by an average of $\sim$13%.

Table 1. Indic Languages and Scripts

Script	Languages
Devanagari	Hindi, Marathi, Sindhi, Kashmiri, Dogri, Nepali, Bodo, Maithili, Sanskrit, Konkani
Bengali-Assamese	Bengali, Assamese, Santali, Manipuri
Odia	Odia
Gujarati	Gujarati
Kannada	Kannada
Malayalam	Malayalam
Meitei	Manipuri
Gurmukhi	Punjabi
Tamil	Tamil
Telugu	Telugu

Table 2. Public Datasets Used

Dataset	Languages	Train Size	Test Size
IndicSTR-Roadside (Real, Scene) [17]	Bengali, Gujarati *, Hindi, Kannada, Malayalam, Marathi Oriya, Punjabi, Tamil, Telugu	175K	50K
IndicSTR12 (Real, Scene) [14]	Assamese, Bengali, Odia Marathi, Hindi, Kannada Urdu, Telugu, Malayalam Gujarati, Punjabi, Tamil	20K	7K
Mozhi (Real, Printed) [16]	Assamese, Bengali, Odia, Manipuri Marathi, Hindi, Kannada Urdu, Telugu, Malayalam Gujarati, Punjabi, Tamil	960K	120K
IndicSTR12-Synth (Synthetic, Scene) [14]	Assamese, Bengali, Odia, Manipuri Marathi, Hindi, Kannada Urdu, Telugu, Malayalam Gujarati, Punjabi, Tamil	24M	6M
IIIT-Synthetic-R (Synthetic, Printed) [20]	Assamese, Bengali, Odia, Manipuri Marathi, Hindi, Kannada Telugu, Malayalam Gujarati, Punjabi, Tamil	6M	4.5M

* We found that the Gujarati set had several Hindi samples, mislabelled GT and samples with no visible text. We present results after omitting these.

Contributions. Our contributions are summarized as follows:

- We identify a hierarchical redundancy in specialized OCR models, providing a motivation for model merging in data-constrained environments.
- We introduce PSMC, a four-stage framework that utilizes task arithmetic to create a single high-performance multilingual OCR model.
- We provide a systematic evaluation across 10 low-resource Indian scripts, proving that joint refinement in a merged space yields superior results to isolated specialization.
- We release the code and datasets used in this study to facilitate the expansion of OCR support for the global long-tail of languages.

2 Related Work

Multimodal OCR and Transfer Learning. Recent advancements in Optical Character Recognition (OCR) have been dominated by large-scale Vision-Language Models (VLMs) from industry, such as Qwen-VL [2], InternVL [23], and DeepSeek-OCR [24]. While these models achieve impressive results on holistic page-level understanding through massive data scaling, they diverge from the traditional OCR paradigm that prioritizes high-fidelity word- and line-level recognition - a granular focus still essential for accurate archival digitization. Parallel to these industrial scales, research has increasingly turned toward transfer learning to address the data scarcity inherent in many scripts. In the non-Indic context, Laurent and Lauar [12] demonstrated the efficacy of adapting a pre-trained TrOCR [13] backbone to create specialized single-language models for low-resource Spanish. Within the Indic domain, Baek et al. [1] investigated multilingual training on the MLT-19 dataset using CNN-based architectures; however, while they noted broad benefits to multilingualism, their approach failed to surpass established benchmarks for Bangla and Hindi. Similarly, Gunna et al. [8] explored transfer learning trajectories from high-resource English and Hindi to other Indic scripts using CNN and Bi-LSTM backbones on the MLT-17 and MLT-19 datasets. We build upon these foundational efforts but depart from the traditional "one-to-one" transfer paradigm. Rather than merely adapting a model to a single target, we investigate how co-training and task arithmetic can synthesize knowledge from multiple language-specific experts. Our work demonstrates that by strategically merging these representations, we can develop a unified multilingual model that not only matches but exceeds the performance of isolated specialist models.

Indic Languages OCR Datasets. The development of robust OCR for Indic scripts has been supported by several key datasets, varying significantly in scale and domain. At the word level, IndicSTR-Roadside [17] and Mozhi [16] represent the most substantial resources; the former provides 17,500 scene-text samples across 10 languages, while the latter offers 80,000 printed-text samples across 12 languages. Additionally, the IndicSTR12 dataset [14], an extension of IIIT-ILST [15], contributes 27,000 scene-text images (for 12 languages) along with a large-scale synthetic corpus of 2M training and 500K test samples per language. However, data scarcity remains a critical bottleneck for multilingual research. Established multilingual benchmarks such as MLT-17 [19] and MLT-19 [18] contain only approximately 4,000 training samples and, notably, lack publicly available annotated test sets for prominent languages like Hindi and Bengali. Although recent efforts have attempted to address the "long-tail" of scripts, such as Mozhi-LR [22] for low-resource printed text and IndicVisionBench [6] for page-level VLM evaluation, these resources are currently not accessible to the public.

Indic Language OCR Models. The current landscape of word-level Indic OCR is primarily defined by language-specific "specialists" developed alongside major data releases. The authors of IndicSTR-Roadside [17] and the Mozhi dataset [16] established the prevailing state-of-the-art (SOTA) by training individual expert models using both CNN-based and Transformer-based architectures. Although these models provide high-fidelity recognition for their respective target scripts, they are typically optimized in isolation, lacking a unified multilingual representation. Earlier efforts in the field focused on benchmarking using the IIIT-ILST [15] and IndicSTR12 [14] datasets. For instance, Saluja et al. [21] utilized the IIIT-ILST dataset to benchmark their OCR system. These earlier works largely relied on traditional deep learning pipelines, often combining CNNs for feature extraction with RNNs for sequence modeling, before the recent shift toward the Vision-Language and Transformer architectures that characterize modern SOTA baselines. Building upon these established baselines, we demonstrate that rather than maintaining separate experts, a unified model can leverage the commonalities between these architectures to achieve superior performance through joint training and model merging.

Model Merging and Weight Editing. Weight editing has emerged as a computationally efficient paradigm for synthesizing a unified, multi-task model from a collection of single-task specialists. The field is motivated by the ambitious goal of absorbing the capabilities of multiple expert models into a single parameter space without requiring the prohibitive resources of full-scale joint training. Early approaches relied on straightforward Weight Averaging [10,25]. However, these methods often suffered from significant performance degradation due to parameter interference and the lack of a consistent coordinate system across models. This challenge was largely addressed by the introduction of Task Arithmetic and Task Vectors [9], which define model specialization as a directional vector in weight space relative to a shared initialization. By grounding experts to a common base, task arithmetic allows for the additive combination of capabilities. To further mitigate interference between competing task vectors, recent advancements have focused on resolving sign conflicts and redundant parameters. Techniques such as TIES-Merging [26] introduce sign-consensus mechanisms to prevent conflicting weight updates from neutralizing one another. Similarly, sparsification methods like Breadcrumbs [5] isolate the most influential parameters within each task vector, reducing the noise introduced during the fusion process. We build upon these principles in our PSMC framework, utilizing task arithmetic to unify script-specific expertise into a singular, high-performance multilingual backbone.

3 Datasets and Backbone Architecture

Modality and Language Coverage. We focus our investigation on word-level recognition for two primary reasons. First, publicly available page-level datasets for Indic scripts remain extremely sparse. Second, the word-level modality offers superior data efficiency; a single digitized page or scene image typically yields hundreds of distinct word-level samples. In extreme data-scarce regimes, this granularity is essential for maximizing the supervisory signal from limited raw data. Our experiments span 10 Indic scripts (supporting 20+ languages) as detailed in Table 1. Notably, for languages like Manipuri, which utilize both Meitei and Bengali-Assamese scripts, we include representations for both.

Real-World Benchmarks. We curate our training and evaluation sets from prominent public resources accessible via the NLTM OCR initiative [20]. Specifically, we utilize IndicSTR-Roadside [17] and IndicSTR12 [14] to represent scene-text environments, and the Mozhi dataset [16] for printed-text settings. These datasets constitute the most comprehensive and recent benchmarks for Indic OCR. A detailed statistical summary of these resources is provided in Table 2.

Synthetic Data Generation. To augment our low-resource training, we utilize SynthTiger [27] and TRDG [4] to generate a diverse corpus of synthetic word-level images. Our generation pipeline mimics both scene and printed text by varying typography, chromaticity, background textures, and kerning. In our constrained regime, we generate 240K synthetic samples per language, split equally between scene and printed modalities. We further supplement this with pre-existing synthetic data from IndicSTR12 [14] and IIIT-Synthetic-R [20].

Backbone Architecture: Parseq. We adopt Parseq [3] as our architectural backbone, as it has consistently demonstrated state-of-the-art performance in both Indic scene and printed text recognition [14,17].

- Vision Encoder: We utilize a ViT-L architecture comprising 24 self-attention blocks with a hidden dimensionality of 1024.
- Decoupled Decoder: The decoder consists of 3 Transformer blocks that perform cross-attention with the visual embeddings to generate character-level predictions.
- Multilingual Charset: To support a unified multilingual model, we define a comprehensive charset of 987 characters spanning all 10 target scripts, enabling the model to learn a shared representation across scripts (Fig. 2).

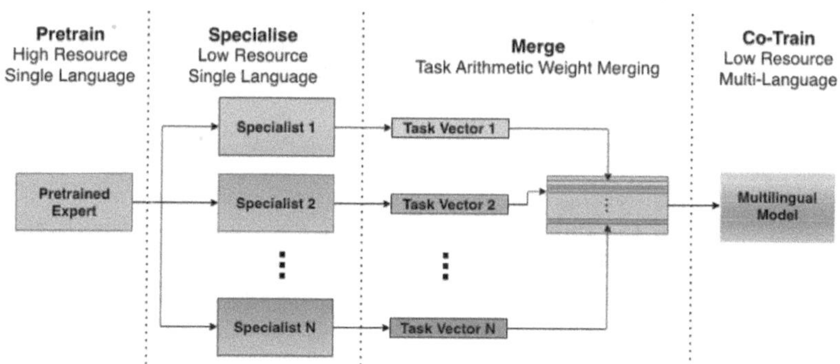

Fig. 2. PSMC Framework

4 Methodology: The PSMC Framework

We introduce Pre-train, Specialize, Merge, and Co-train (PSMC), a structured fine-tuning paradigm designed to optimize multilingual Indic OCR under extreme low-resource constraints. Our approach is motivated by empirical evidence of cross-script feature alignment in the early stages of visual processing (see Fig. 1), allowing the model to explicitly leverage transferable representations across diverse scripts.

4.1 Transfer Learning from a Pre-trained Anchor: Stages P and S of the PSMC Framework

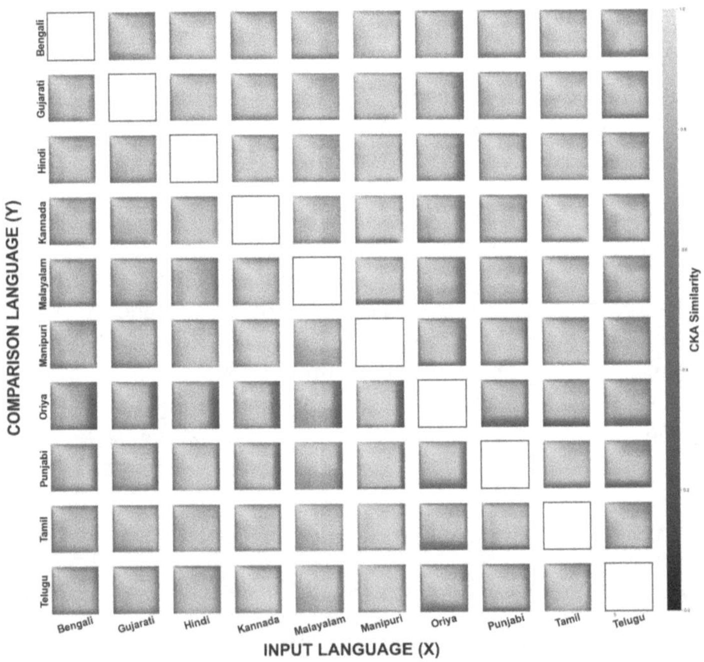

Fig. 3. A comprehensive visualization to highlight the feature similarity of layer-wise encoder features across the individual specialist models: Following the same experimental set up as Fig. 1 Heatmap (X, Y) depicts the averaged pairwise CKA feature similarity across all the 24 encoder layers between Model X and Y for input images from language X. Note the high similarity spreads in the upper left corners of all the heatmaps demonstrating the similar high-level, script-agnostic visual cues learnt in the early layers across all specialists.

Transfer learning forms the bedrock of our strategy. Rather than training from a random initialization, which leads to unstable optimization and poor generalization in data-scarce regimes, we initialize our models from a pre-trained anchor model, θ_{En}. We select English as the anchor language due to the abundance of high-quality, diverse annotated data covering various lighting conditions, domains, and typographies. We train θ_{En} on a robust mixture of synthetic and real data, following the original Parseq training protocol [3]. This provides a mature representation space that captures fundamental visual primitives (edges, curves, and strokes) essential for character recognition. As empirically observed in Fig. 1, these low-level primitives are largely script-agnostic, facilitating an effective transfer to Indic scripts. For each target language $\ell \in \mathcal{L}$, we derive a specialized model θ_ℓ by fine-tuning the anchor model on language-specific data:

$$\mathcal{L} = \{\text{Hindi}, \text{Malayalam}, \text{Tamil}, \text{Punjabi}, \dots\}. \tag{1}$$

This specialization allows the model to adapt its high-level semantic layers to specific character geometries while retaining the robust low-level features of the anchor.

4.2 Multi-script Co-training: Stage C of the PSMC Framework

Our analysis of the encoder blocks via Centered Kernel Alignment (CKA) (see Fig. 1 and Fig. 3) reveals high structural similarity in the initial layers of all expert encoders. This redundancy suggests that separate models for each language independently relearn the same stroke and curve features. To exploit this, we employ multi-script co-training, where the model is jointly optimized across the entire set of target languages. Formally, we construct a joint training distribution over the union of $\mathcal{L}$ to produce a single shared model, $\mathcal{M}_{\text{unified}}$. This exposure to a richer, heterogeneous mixture of character shapes within a single optimization trajectory prompts the network to learn more generalized and robust coarse-level representations. As demonstrated in Tables 3 and 4, this multilingual co-training consistently outperforms isolated fine-tuning under identical per-language data constraints. This confirms that in strictly low-resource regimes, the "transfer effect" from multi-script co-training provides measurable gains in both robustness and out-of-distribution generalization.

4.3 Task Arithmetic and Joint Refinement: Stage M of the PSMC Framework

Weight merging strategies, such as Task Arithmetic, have demonstrated significant success in zero-shot generalization for CLIP-based architectures in image classification tasks [5,9,26]. While these classification frameworks typically employ dynamically generated task-specific heads, a unified multilingual OCR model requires a singular, shared architecture capable of simultaneous multi-script inference without head-switching. We therefore leverage Task Arithmetic to synthesize an intermediate merged model, θ_{Merged}. This stage is designed to preserve the shared visual primitives observed in the lower encoder layers while aggregating the divergent, language-specific features captured by individual experts. While θ_{Merged} encapsulates latent multilingual knowledge, it often

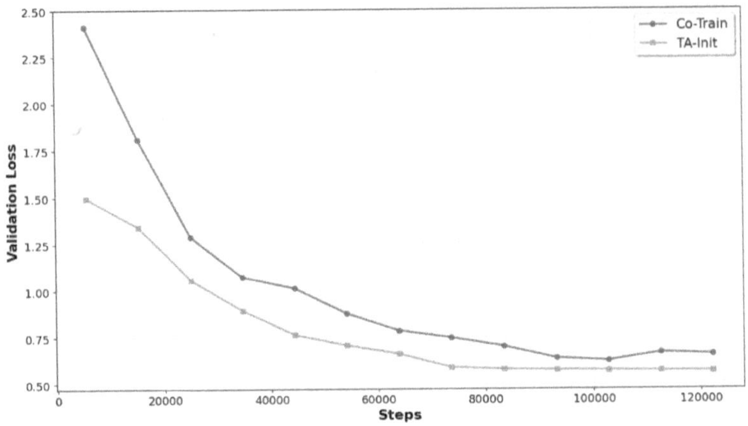

Fig. 4. Validation loss for the task arithmetic initialization vs the simple pretrained initialization

Table 3. WER and NED scores on the Mozhi Dataset (lower is better). The Baseline, Skyline, Ind-FT models are unilingual and Co-FT, PSMC are unified multilingual models. The Baseline and Skyline models were trained on full scale data, while the Ind-FT, Co-FT and PSMC were trained under low resource constraints.

Language	Baseline [16]		Skyline		Ind-FT		Co-FT		PSMC	
	WER	NED	WER	NED	WER	NED	WER	NED	WER	NED
Assamese	3.50	1.00	**1.62**	**0.40**	2.84	0.79	2.48	0.63	2.06	0.48
Bengali	2.10	0.60	1.55	0.45	1.73	0.51	1.45	0.37	**1.29**	**0.36**
Gujarati	6.10	3.50	**2.88**	**1.00**	5.32	1.85	3.90	1.39	3.71	1.28
Hindi	3.70	1.80	**1.49**	**0.42**	2.18	0.71	1.85	0.57	1.76	0.54
Kannada	9.30	2.30	**2.31**	**0.57**	4.88	1.38	2.99	0.70	2.86	0.62
Malayalam	2.30	0.30	**1.14**	0.34	2.16	0.68	1.28	0.23	1.10	**0.21**
Manipuri	3.10	1.00	1.86	0.67	3.81	1.26	1.95	**0.66**	**1.81**	0.68
Marathi	3.10	0.80	3.57	2.21	4.17	1.56	**1.84**	**0.60**	1.94	0.62
Oriya	5.20	2.80	**4.49**	**1.81**	6.19	2.43	4.75	1.82	4.92	1.87
Punjabi	1.30	0.50	**0.73**	**0.25**	2.11	0.74	0.97	0.30	1.18	0.45
Tamil	8.20	2.00	**6.68**	**1.55**	8.78	2.42	7.33	1.67	6.94	1.61
Telugu	6.40	3.20	1.69	0.54	3.17	0.74	1.65	**0.31**	**1.63**	0.32
Average	4.53	1.65	**2.50**	**0.85**	3.95	1.26	2.70	0.77	2.60	0.75

suffers from functional interference, a conflict between weight updates from different scripts. To resolve this, we introduce a final joint refinement stage that aligns the merged weights across the entire architecture. We find that utilizing a merged initialization rather than a simple pretrained anchor leads to faster convergence and superior final accuracy (see Fig. 4).

Table 4. WER and NED Scores on the IndicSTR-Roadside Dataset (lower is better). The Baseline, Skyline, Ind-FT models are unilingual and Co-FT, PSMC are unified multilingual models. The Baseline and Skyline models were trained on full scale data, while the Ind-FT, Co-FT and PSMC were trained under low resource constraints.

Language	Baseline [17]		Skyline		Ind-FT		Co-FT		PSMC	
	WER	NED	WER	NED	WER	NED	WER	NED	WER	NED
Bengali	14.66	7.25	**9.10**	3.45	10.92	3.85	9.78	3.61	9.22	**3.27**
Gujarati	18.09	11.88	**15.35**	**5.08**	19.51	6.75	18.97	6.65	17.72	6.18
Hindi	12.76	4.99	11.12	**4.17**	12.50	4.25	15.74	10.69	**11.08**	5.41
Kannada	20.73	12.36	**6.82**	2.24	8.08	2.50	6.96	2.31	7.12	**2.16**
Malayalam	19.69	10.58	8.42	2.18	9.18	2.52	8.12	1.95	**7.56**	**1.83**
Marathi	14.50	5.53	12.26	3.86	11.14	**3.49**	14.74	9.70	**10.20**	4.88
Oriya	**13.47**	**4.87**	14.02	5.16	18.72	6.86	17.24	5.80	16.96	5.59
Punjabi	15.73	8.54	**13.42**	**5.54**	18.20	8.04	15.28	6.17	13.68	5.94
Tamil	**13.65**	**4.37**	16.50	4.84	20.18	5.79	18.04	5.13	17.22	4.75
Telugu	15.06	7.82	14.64	5.58	15.50	5.65	13.98	**4.75**	**14.58**	4.98
Average	15.83	7.82	**12.16**	**4.21**	14.39	4.97	13.88	5.68	12.53	4.50

Mathematical Formulation. Following Ilharco et al. [9], we define a task vector as the displacement in parameter space between a pretrained base model and its fine-tuned counterpart. Let θ_{En} denote the parameters of the pretrained English anchor model. For each target language ℓ, we obtain a specialized expert θ_ℓ by fine-tuning θ_{En} on language-specific data. The task vector for language ℓ is defined as:

$$\tau_\ell = \theta_\ell - \theta_{\text{En}} \tag{2}$$

This vector represents the specific direction in weight space required to adapt the English anchor toward the unique orthographic features of language ℓ. Given a set of L language experts, we linearly combine their respective task vectors to produce the intermediate merged model:

$$\theta_{\text{Merged}} = \theta_{\text{En}} + \frac{1}{L}\sum_{\ell=1}^{L} \tau_\ell. \tag{3}$$

By grounding the merging process in θ_{En}, we ensure that the unified model retains a consistent structural foundation while benefiting from the collective expertise of all specialized language vectors.

5 Experiments and Results

Evaluation Metrics. To rigorously evaluate model performance, we employ Normalized Edit Distance (NED), as defined in ICDAR19-ReCTS [28]. This metric accounts for the relative sequence length and is expressed as:

$$\text{NED} = \frac{1}{N} \sum_{i=1}^{N} \frac{D(s_i, \hat{s}_i)}{\max(|s_i|, |\hat{s}_i|)} \tag{4}$$

where N is the total number of samples, D is the Levenshtein distance, s_i denotes the ground truth, and $\hat{s}_i$ is the predicted string. We report the NED as a percentage. We also report the Word Error Rate (WER), representing the percentage of word-level predictions that contain an error.

Data Composition and "Skyline" Baselines. Our "Skyline" models represent the upper bound of performance, trained on the full aggregate of available real and synthetic data. The training corpus for each language comprises approximately 80K real printed words (Mozhi), 18.5K real scene-text samples (IndicSTR12, IndicSTR-Roadside), and a massive synthetic set of 16M images (split equally between printed and scene modalities). We establish these baselines using two NVIDIA A100 GPUs with a batch size of 128 and a learning rate of 7×10^{-4}. As shown in Tables 3 and 4, our individual "Skyline" experts consistently outperform existing state-of-the-art baselines across all target scripts.

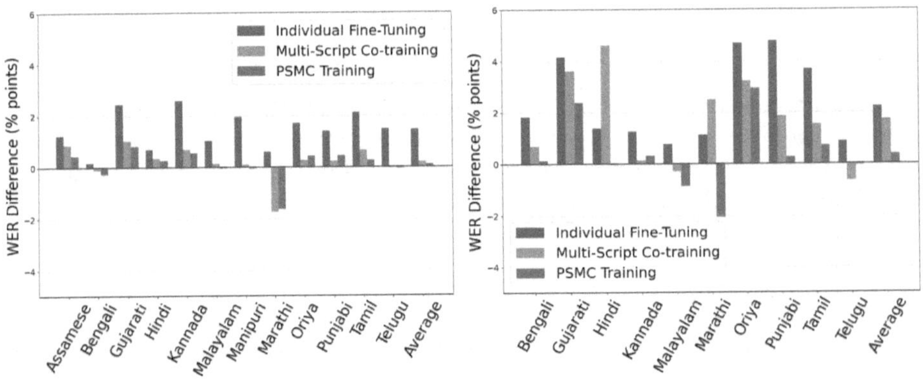

Fig. 5. Difference in WER (lower is better) with respect to the Skyline WER metrics for (left) Mozhi, and (right) IndicSTR-Roadside.

Representation Similarity Analysis. To investigate the structural commonalities across specialized experts, we utilize Centered Kernel Alignment (CKA) [11]. CKA provides a robust measure of similarity between neural representations that is invariant to orthogonal transformations and isotropic scaling. For two feature matrices $\mathbf{X}$ and $\mathbf{Y}$ extracted from n samples, CKA is computed as:

$$\text{CKA}(\mathbf{X}, \mathbf{Y}) = \frac{\text{HSIC}(\mathbf{X}\mathbf{X}^\top, \mathbf{Y}\mathbf{Y}^\top)}{\sqrt{\text{HSIC}(\mathbf{X}\mathbf{X}^\top, \mathbf{X}\mathbf{X}^\top) \cdot \text{HSIC}(\mathbf{Y}\mathbf{Y}^\top, \mathbf{Y}\mathbf{Y}^\top)}}, \tag{5}$$

where HSIC represents the Hilbert-Schmidt Independence Criterion [7]. Our cross-model CKA analysis (Fig. 3) reveals that while deeper layers diverge to capture script-specific nuances, the early layers across all experts remain highly similar. This finding confirms that models independently relearn the same visual primitives, providing a strong empirical justification for our merging and co-training strategy.

(a) Errors in Kannada vattak-akshara strokes

(b) Malayalam letters confused with numbers and punctuation

(c) Errors recognising Marathi letter ळ

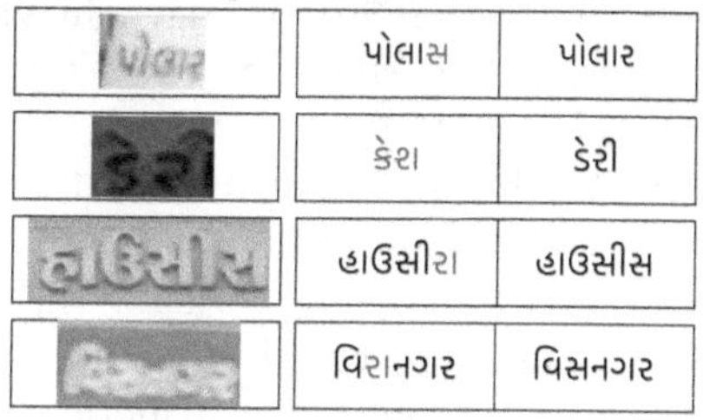

(d) Gujarati letter ૨ confused with શ and સ

Fig. 6. Mistakes in individually finetuned model vs final model. Columns of sub-figures show (left) input image, (center) individually finetuned model output and (right) the PSMC model output.

Low-Resource Regimes: From Fine-Tuning to PSMC. In our low-resource experiments, we restrict each language to only 10K real samples (split equally between scene and printed modalities) and 240K synthetic images.

- Individual Fine-Tuning: Fine-tuning the English anchor model separately for each script leads to localized specialists but fails to capture cross-script synergies.
- Naive Co-Training: Jointly training a single model on the union of all low-resource samples yields a relative WER reduction of 32% on Mozhi and 3.5% on IndicSTR-Roadside compared to the individual fine-tuning.
- PSMC (Task Arithmetic + Co-Training): By initializing the joint training from the merged weight space θ_{Merged} (derived via task arithmetic), we achieve the best performance.

As illustrated in Fig. 5, PSMC significantly minimizes the performance gap relative to the "Skyline" models, often staying within 0.4% points of the full-data upper bound. Furthermore, the validation loss curves in Fig. 4 demonstrate that PSMC provides a superior initialization, leading to faster convergence and a deeper local minimum compared to naive co-training. Qualitative improvements are detailed in Fig. 6, showcasing PSMC's ability to resolve subtle character ambiguities that individual fine-tuned models frequently misidentify.

6 Conclusion

In this work, we addressed the critical data bottleneck that hinders the extension of robust Vision-Language Models to low-resource Indic scripts. By investigating the structural evolution of specialized OCR experts, we identified a significant hierarchical redundancy: while higher layers diverge to capture script-specific nuances, lower-level visual primitives remain largely script-agnostic. This observation provided the empirical foundation for our PSMC (Pre-train, Specialize, Merge, and Co-train) framework. By utilizing task arithmetic to unify language-specific experts into a singular backbone, followed by joint refinement, we demonstrated that cross-script "transfer effects" can compensate for extreme data scarcity. Our results across 10 Indian scripts show that PSMC consistently outperforms individual fine-tuning, achieving an average WRR improvement of $\sim$2% and nearing the performance of "skyline" models trained on significantly larger datasets. Ultimately, our framework offers a scalable and parameter-efficient pathway for bridging the digital divide in OCR, ensuring that the benefits of modern multimodal AI are accessible to the global long-tail of languages.

Future Work. Future work on this topic could focus on expanding to low-resource languages outside the Indian context, such as languages from Africa, the Middle East and South America. In particular, more challenging languages would be those that have different writing paradigms from the languages they are being co-trained with. For example, co-training Arabic, which is written from right to left, with English. Further work to quantify the optimal balance between real and synthetic data would also help to efficiently train multilingual models.

Acknowledgments. This work is supported by MeitY, Government of India, through the NLTM Bhashini project and TATA 1mg.

References

1. Baek, J., Matsui, Y., Aizawa, K.: Cross-lingual learning in multilingual scene text recognition. In: ICASSP 2024 - 2024 IEEE International Conference on Acoustics, Speech and Signal Processing (ICASSP), pp. 6770–6774 (2024). https://doi.org/10.1109/ICASSP48485.2024.10445946
2. Bai, S., et al.: Qwen3-Vl technical report (2025). https://arxiv.org/abs/2511.21631
3. Bautista, D., Atienza, R.: Scene text recognition with permuted autoregressive sequence models. In: Avidan, S., Brostow, G., Cissé, M., Farinella, G.M., Hassner, T. (eds.) Computer Vision - ECCV 2022, pp. 178–196. Springer Nature Switzerland, Cham (2022)
4. Belval, E.: Text recognition data generator (2019). https://github.com/Belval/TextRecognitionDataGenerator?tab=readme-ov-file. Accessed 25 Feb 2026
5. Davari, M., Belilovsky, E.: Model breadcrumbs: scaling multi-task model merging with sparse masks. In: Leonardis, A., Ricci, E., Roth, S., Russakovsky, O., Sattler, T., Varol, G. (eds.) Computer Vision - ECCV 2024, pp. 270–287. Springer Nature Switzerland, Cham (2025)

6. Faraz, A., et al.: IndicVisionBench: benchmarking cultural and multilingual understanding in VLMs. In: The Fourteenth International Conference on Learning Representations (2026). https://openreview.net/forum?id=LmJoLn04iL
7. Gretton, A., Bousquet, O., Smola, A., Schölkopf, B.: Measuring statistical dependence with Hilbert-Schmidt norms. In: Jain, S., Simon, H.U., Tomita, E. (eds.) Algorithmic Learning Theory, pp. 63–77. Springer, Heidelberg (2005)
8. Gunna, S., Saluja, R., Jawahar, C.V.: Transfer learning for scene text recognition in Indian languages. In: Barney Smith, E.H., Pal, U. (eds.) ICDAR 2021. LNCS, vol. 12916, pp. 182–197. Springer, Cham (2021). https://doi.org/10.1007/978-3-030-86198-8_14
9. Ilharco, G., Ribeiro, M.T., Wortsman, M., Schmidt, L., Hajishirzi, H., Farhadi, A.: Editing models with task arithmetic. In: The Eleventh International Conference on Learning Representations (2023). https://openreview.net/forum?id=6t0Kwf8-jrj
10. Jin, X., Ren, X., Preotiuc-Pietro, D., Cheng, P.: Dataless knowledge fusion by merging weights of language models. In: The Eleventh International Conference on Learning Representations (2023). https://openreview.net/forum?id=FCnohuR6AnM
11. Kornblith, S., Norouzi, M., Lee, H., Hinton, G.: Similarity of neural network representations revisited. In: Chaudhuri, K., Salakhutdinov, R. (eds.) Proceedings of the 36th International Conference on Machine Learning. Proceedings of Machine Learning Research, vol. 97, pp. 3519–3529. PMLR, 09–15 June 2019. https://proceedings.mlr.press/v97/kornblith19a.html
12. Lauar, F., Laurent, V.: Spanish TrOCR: leveraging transfer learning for language adaptation (2024). https://arxiv.org/abs/2407.06950
13. Li, M., et al.: TrOCR: transformer-based optical character recognition with pre-trained models. In: Proceedings of the Thirty-Seventh AAAI Conference on Artificial Intelligence and Thirty-Fifth Conference on Innovative Applications of Artificial Intelligence and Thirteenth Symposium on Educational Advances in Artificial Intelligence, AAAI'23/IAAI'23/EAAI'23. AAAI Press (2023). https://doi.org/10.1609/aaai.v37i11.26538
14. Lunia, H., Mondal, A., Jawahar, C.V.: IndicSTR12: a dataset for Indic scene text recognition. In: Coustaty, M., Fornés, A. (eds.) Document Analysis and Recognition - ICDAR 2023 Workshops, pp. 233–250. Springer Nature Switzerland, Cham (2023)
15. Mathew, M., Jain, M., Jawahar, C.V.: Benchmarking scene text recognition in Devanagari, Telugu and Malayalam. In: ICDAR MOCR Workshop (2017)
16. Mathew, M., Mondal, A., Jawahar, C.V.: Towards deployable OCR models for Indic languages. In: Antonacopoulos, A., Chaudhuri, S., Chellappa, R., Liu, C.L., Bhattacharya, S., Pal, U. (eds.) Pattern Recognition, pp. 167–182. Springer Nature Switzerland, Cham (2025)
17. Mondal, A., Tulsyan, K., Jawahar, C.V.: Indic scene text on the roadside. In: Barney Smith, E.H., Liwicki, M., Peng, L. (eds.) Document Analysis and Recognition - ICDAR 2024, pp. 263–278. Springer Nature Switzerland, Cham (2024)
18. Nayef, N., et al.: ICDAR2019 robust reading challenge on multi-lingual scene text detection and recognition — RRC-MLT-2019. In: 2019 International Conference on Document Analysis and Recognition (ICDAR), pp. 1582–1587 (2019). https://doi.org/10.1109/ICDAR.2019.00254
19. Nayef, N., et al.: ICDAR2017 robust reading challenge on multi-lingual scene text detection and script identification - RRC-MLT. In: 2017 14th IAPR International Conference on Document Analysis and Recognition (ICDAR), vol. 01, pp. 1454–1459 (2017). https://doi.org/10.1109/ICDAR.2017.237

20. NLTM OCR: NLTM OCR datasets (2025). https://ilocr.iiit.ac.in/dataset/. Accessed 25 Feb 2026
21. Saluja, R., Maheshwari, A., Ramakrishnan, G., Chaudhuri, P., Carman, M.: OCR on-the-Go: robust end-to-end systems for reading license plates and street signs. In: 2019 International Conference on Document Analysis and Recognition (ICDAR), pp. 154–159 (2019). https://doi.org/10.1109/ICDAR.2019.00033
22. Sarkar, A., Mondal, A., Lehal, G.S., Jawahar, C.V.: Printed OCR for extremely low-resource Indic languages. In: Kakarla, J., Balasubramanian, R., Murala, S., Vipparthi, S.K., Gupta, D. (eds.) Computer Vision and Image Processing, pp. 108–122. Springer Nature Switzerland, Cham (2026)
23. Wang, W., et al.: InternVL3.5: advancing open-source multimodal models in versatility, reasoning, and efficiency (2025). https://arxiv.org/abs/2508.18265
24. Wei, H., Sun, Y., Li, Y.: DeepSeek-OCR: contexts optical compression (2025). https://arxiv.org/abs/2510.18234
25. Wortsman, M., et al.: Model soups: averaging weights of multiple fine-tuned models improves accuracy without increasing inference time. In: Chaudhuri, K., Jegelka, S., Song, L., Szepesvari, C., Niu, G., Sabato, S. (eds.) Proceedings of the 39th International Conference on Machine Learning. Proceedings of Machine Learning Research, vol. 162, pp. 23965–23998. PMLR, 17–23 July 2022. https://proceedings.mlr.press/v162/wortsman22a.html
26. Yadav, P., Tam, D., Choshen, L., Raffel, C., Bansal, M.: TIES-merging: resolving interference when merging models. In: Thirty-Seventh Conference on Neural Information Processing Systems (2023). https://openreview.net/forum?id=xtaX3WyCj1
27. Yim, M., Kim, Y., Cho, H.-C., Park, S.: SynthTIGER: synthetic text image GEneratoR towards better text recognition models. In: Lladós, J., Lopresti, D., Uchida, S. (eds.) ICDAR 2021. LNCS, vol. 12824, pp. 109–124. Springer, Cham (2021). https://doi.org/10.1007/978-3-030-86337-1_8
28. Zhang, R., et al.: ICDAR 2019 robust reading challenge on reading Chinese text on signboard. In: 2019 International Conference on Document Analysis and Recognition (ICDAR), pp. 1577–1581 (2019). https://doi.org/10.1109/ICDAR.2019.00253

A Benchmark of State-Space Models vs. Transformers and BiLSTM-Based Models for Historical Newspaper OCR

Merveilles Agbeti-Messan[1](✉), Thierry Paquet[1], Clément Chatelain[2], Pierrick Tranouez[1], and Stéphane Nicolas[1]

[1] LITIS EA4108, University of Rouen Normandy, Rouen, France
{komlan-epe-nsin.agbeti-messan,thierry.paquet,pierrick.tranouez, stephane.nicolas}@univ-rouen.fr

[2] LITIS EA4108, INSA of Rouen Normandy, Rouen, France
clement.chatelain@insa-rouen.fr

Abstract. End-to-end OCR for historical newspapers remains challenging, as models must handle long text sequences, degraded print quality, and complex layouts. While Transformer-based recognizers dominate current research, their quadratic complexity limits efficient paragraph-level transcription and large-scale deployment. We investigate linear-time State-Space Models (SSMs), specifically Mamba, as a scalable alternative to Transformer-based sequence modeling for OCR.

We present to our knowledge, the first OCR architecture based on SSMs, combining a CNN visual encoder with bi-directional and autoregressive Mamba sequence modeling, and conduct a large-scale benchmark comparing SSMs with Transformer- and BiLSTM-based recognizers. Multiple decoding strategies (CTC, autoregressive, and non-autoregressive) are evaluated under identical training conditions alongside strong neural baselines (VAN, DAN, DANIEL) and widely used off-the-shelf OCR engines (PERO-OCR, Tesseract OCR, TrOCR, Gemini).

Experiments on historical newspapers from the Bibliothèque nationale du Luxembourg, with newly released $> 99\%$ verified gold-standard annotations, and cross-dataset tests on Fraktur and Antiqua lines, show that all neural models achieve low error rates ($\sim$2% CER), making computational efficiency the main differentiator. Mamba-based models maintain competitive accuracy while halving inference time and exhibiting superior memory scaling ($1.26\times$ vs $2.30\times$ growth at 1000 chars), reaching 6.07% CER at the severely degraded paragraph level compared to 5.24% for DAN, while remaining $2.05\times$ faster.

We release code, trained models, and standardized evaluation protocols to enable reproducible research and guide practitioners in large-scale cultural heritage OCR available at https://github.com/MarcoPerson/ssm-ocr-benchmark.

Keywords: State-Space Models · Historical Document OCR · Transformer Architectures · Paragraph-Level Recognition · Benchmarking

G. A. Fink et al. (Eds.): ICDAR 2026, LNCS 16972, pp. 663–680, 2027.
https://doi.org/10.1007/978-3-032-36023-6_38

1 Introduction

Cultural heritage institutions worldwide face the monumental task of digitizing millions of historical documents to ensure their preservation and accessibility. Among these materials, historical newspapers represent a particularly challenging category due to their physical degradation, non-standardized typography, complex multi-column layouts, and linguistic diversity [16,30]. The Bibliothèque nationale du Luxembourg (BnL), like its counterparts such as the Bibliothèque nationale de France (BnF) and the Library of Congress, maintains extensive newspaper collections spanning over a century, written in multiple languages (Luxembourgish, French, German), and exhibiting varying degrees of preservation quality. Accurate automatic transcription of these documents through Optical Character Recognition (OCR) is essential not only for full-text search and digital humanities research but also for reducing the prohibitive cost of manual transcription [23].

Recent advances in deep learning have dramatically improved OCR performance on contemporary documents, with Transformer or Attention based architectures achieving near-human accuracy on clean printed text. Models such as TrOCR [21], DAN (Document Attention Network) [6], and VAN [7] have established new benchmarks by leveraging self-attention mechanisms to capture long-range dependencies in document images. However, these architectures face practical deployment challenges when processing longer text regions. While autoregressive Transformers can handle paragraphs or full pages, they incur substantial computational costs: the $O(n^2)$ complexity of self-attention results in prohibitive memory requirements due to large key-value caches during autoregressive decoding, and sequential token generation leads to high inference latency [14]. These limitations are particularly problematic for heritage institutions digitizing millions of newspaper pages, where batch processing throughput and memory efficiency are critical operational constraints alongside recognition accuracy.

Concurrently, a new family of sequence models has emerged: State-Space Models (SSMs), exemplified by the Mamba architecture [14]. Unlike Transformers, SSMs achieve $O(n)$ computational complexity through selective state-space mechanisms inspired by control theory, enabling efficient processing of long sequences while maintaining a constant memory footprint during inference. Mamba has demonstrated impressive results in natural language processing and computer vision tasks [22,36], positioning itself as a viable alternative to Transformers for sequence modeling. Despite this promise, SSMs remain unexplored in the context of text recognition from document images, and their comparative performance against established Transformer and recurrent neural network baselines has not been systematically evaluated on historical documents.

Furthermore, the OCR literature exhibits a persistent divide between autoregressive (AR) and non-autoregressive (non-AR) decoding paradigms. AR models, such as DAN [6], generate characters sequentially by conditioning on previously predicted tokens, enabling rich contextual modeling at the cost of inference speed. Non-autoregressive models trained with Connectionist Temporal Classification (CTC) [11] assume conditional independence between time steps, enabling a single forward pass followed by sequential decoding. This offers faster inference than autoregressive models, but may sacrifice accuracy due to the lack of explicit language modeling. This dichotomy has been explored in isolated studies, but a unified evaluation comparing AR and non-AR variants of SSMs, Transformers, and BiLSTM-based architectures on the same dataset and same task is absent from the literature. Additionally, the role of tokenization strategies, character-level versus subword-level (BPE), has not been rigorously assessed for historical document OCR [4], where archaic spellings and out-of-vocabulary words are prevalent.

To address these gaps, this paper presents a comprehensive benchmark of State-Space Models versus Transformers and BiLSTM-based models for historical newspaper OCR, evaluated at both line and paragraph granularities on the BnL corpus. Our study makes the following contributions:

- **First application of State-Space Models to document OCR:** We introduce a Mamba-based architecture with three decoding variants (CTC, autoregressive, and non-autoregressive) for text recognition in historical documents, demonstrating that SSMs constitute a viable alternative to Transformers for this task and opening a new research direction in heritage document digitization.
- **Systematic comparison across decoding paradigms:** We evaluate six neural architectures spanning three decoding strategies (CTC, autoregressive, and non-autoregressive cross-entropy) and three model families (SSMs, Transformers, and BiLSTMs), providing the first controlled benchmark of these paradigms on historical documents under identical experimental conditions. We additionally compare them against four widely-used production OCR systems (PERO-OCR [18], Tesseract [29], TrOCR [21], Gemini [10]).
- **Dual-granularity evaluation:** We assess the models at line and paragraph levels, providing insights into how architectures scale when processing longer text regions, a critical consideration for practical deployment in mass digitization workflows.
- **Comprehensive efficiency and robustness analysis:** Beyond recognition accuracy, we report computational metrics (parameters, memory, inference latency, throughput) and investigate model behavior across varying sequence lengths and document degradation levels, enabling practitioners to make informed architectural choices based on operational constraints.
- **High-quality reproducible benchmark:** We release our standardized train/validation/test splits derived from gold-standard human annotations, with test splits manually corrected to $> 99\%$ accuracy, along with trained model checkpoints and evaluation code, establishing a reliable foundation for future research in historical newspaper OCR.

The remainder of this paper is organized as follows: Sect. 2 reviews related work in historical OCR, Transformer, BiLSTM, and SSM architectures, as well as paragraph-level recognition. Section 3 describes our methodology, detailing the architectures evaluated and training protocols. Section 4 presents our experimental setup and the BnL datasets. Section 5 reports results and in-depth analyses. Section 6 concludes with directions for future work.

2 Related Work

2.1 Historical Document OCR and Newspaper Challenges

Historical newspapers present unique challenges: image degradation (blur, bleed-through), complex layouts (multi-column, advertisements), and typographic variation across space and time (e.g., Antiqua vs Fraktur typefaces). These factors make recognition accuracy sensitive to preprocessing, segmentation, and the chosen decoding paradigm.

Large-scale digitization often relies on OCR systems mixing layout analysis, line extraction, and line-level recognisers. Transkribus has operationalised trainable handwriting recognition in cultural heritage [23]. Open-source recognisers like PERO-OCR [18] illustrate the importance of robust CTC-style line recognisers in production. Evaluations show modern neural approaches significantly outperform legacy systems like Tesseract [29] on degraded documents, but performance remains sensitive to domain-specific fine-tuning [30].

Benchmarking has been shaped by datasets and competitions: READ-BAD [12] for baseline detection, DIVA-HisDB [28] for annotated historical pages, and the IMPACT dataset [16] for historical print. ICDAR 2019 included historical document competitions [31], though evaluations still focus on line-level rather than paragraph-level settings.

2.2 Sequence Modeling: CTC, Attention, and Transformers

Most recognizers decompose into a visual encoder and sequence decoder. Three paradigms have emerged for sequence modeling in OCR, differing primarily in how they handle the alignment between visual features and output tokens, and whether they model dependencies between output symbols.

CTC-based approaches. Connectionist Temporal Classification (CTC) [11] sidesteps explicit alignment by marginalizing over all monotonic paths between encoder frames and output symbols. This enables parallel decoding with a conditional independence assumption between output characters. The CRNN [26] combining CNN features with recurrent networks and CTC remains a strong line-level baseline. Recent work explores bidirectional GRUs [2] or convolutional architectures while retaining CTC. VAN [7] demonstrates competitive results on handwritten paragraphs in CTC mode. PyLaia [24] is a widely-adopted CTC framework for historical documents. While computationally efficient, CTC's independence assumption prevents explicit language modeling, motivating attention-based alternatives.

Autoregressive attention and Transformers. In contrast to CTC's parallel decoding, autoregressive (AR) approaches treat OCR as conditional sequence generation, where each output token is predicted sequentially conditioned on previously generated symbols. This enables rich modeling of linguistic dependencies at the cost of inference latency proportional to sequence length. The Decoupled Attention Network [32] decouples alignment from decoding states, SAR [20] employs attention for visual alignment, and ASTER [27] adds rectification for irregular text. Transformer OCR strengthened AR decoding with pre-trained components. TrOCR [21] combines pre-trained image and text Transformers. Vision Transformer encoders with Transformer decoders [1,17] demonstrate pure-attention effectiveness. For historical newspapers, quadratic attention complexity and sequential AR decoding make long sequences expensive to process at inference time. DAN [6] introduced segmentation-free document attention for full-page handwritten recognition.

Hybrid and non-autoregressive Transformers. Seeking to bridge the speed of CTC with the supervision quality of autoregressive models, recent work explores non-autoregressive (NAR) Transformer decoders that predict all positions in parallel using learned queries. Non-autoregressive Transformer decoders predict tokens in parallel. PARSeq [1] operates in both AR and non-AR modes with competitive accuracy on scene text. DANIEL [4] used BPE (Byte-Pair Encoding) subword tokenization, improving efficiency for morphologically rich languages, and speeding up the decoding process at inference time.

2.3 Paragraph-Level Recognition

Paragraph-level recognition aims to recognise text without explicit line segmentation, reducing pipeline brittleness but increasing sequence length and complexity, particularly for historical newspapers where line detection may fail.

Paragraph models. VAN [7] introduces vertical attention to progressively read lines from full paragraph images. The "Scan, Attend and Read" (FPHR) system [3] uses a CNN encoder with an attention-based recurrent decoder for end-to-end handwritten paragraph recognition, demonstrating competitive results against prior MDLSTM-CTC baselines. SPAN [5] performs paragraph recognition without explicit line break annotations. The "Start, Follow, Read" [33] learns to read full pages in reading order without requiring line segmentation annotations.

Challenges for newspapers. Paragraph settings amplify issues: longer sequences exacerbate AR decoding latency; CTC must compress long visual streams while maintaining coherence; layout variability increases encoder burden; memory for full-page attention can be prohibitive. Recent architectures handle 100+ character paragraphs [4,7], though inference speed remains a bottleneck.

2.4 State-Space Models as Attention Alternatives

Structured state-space models (SSMs) offer efficient long-range modeling with favorable scaling versus Transformers. S4 [13] introduced efficient parameterisation based on HiPPO, showing strong performance on long-range benchmarks. S4D extended this with improved training stability.

Selective state-space models. Mamba [14] introduced input-dependent selective SSMs, enabling content-adaptive propagation while maintaining linear-time processing. Vision Mamba (ViM) [36] and VMamba [22] adapted SSMs to 2D vision via spatial flattening or directional scanning.

SSMs for document understanding. VL-Mamba [25] explores SSMs for multimodal document tasks. DocMamba [15] introduces efficient document pre-training, competing with LayoutLM on information extraction at lower cost. For text recognition, some recent works explore SSMs: MambaSTR [35] for scene text with masked state space modeling; TextMamba [34] for detection; and a TinyML model for Korean handwriting [8] demonstrating compact Mamba variants for resource-constrained devices.

Limitations and open challenges in historical document OCR. These works focus on scene text or modern handwriting with short sequences (<30 characters) and clean images. No prior work systematically evaluates SSMs on historical newspaper OCR, where sequences are longer (50–500+ characters at paragraph level), images degraded, and typographic variation high. The interaction between SSM backbones and decoding strategies (CTC, AR, non-AR) remains unexplored in OCR.

Positioning. Existing literature provides strong baselines across CTC, attention, and Transformer paradigms, plus paragraph-level models. However, a controlled comparison of modern SSM backbones against Transformers on historical newspapers at both line and paragraph granularities is absent. Our study targets this gap, benchmarking decoding paradigms and model families under a unified protocol on the BnL dataset.

3 Mamba-Based OCR

This section describes our Mamba-based architectures for historical newspaper OCR. We first formalize the sequence recognition problem and three decoding paradigms, then present our model designs with the core Mamba formulation.

3.1 Preliminaries

We formulate OCR as a sequence prediction task: given an image $\mathbf{I}$ containing text, we predict a character sequence $\mathbf{y} = (y_1, \dots, y_T)$ with $y_t \in \mathcal{V}$, where $\mathcal{V}$ is the character vocabulary. Three decoding paradigms dominate current approaches, differing in how they factorize $P(\mathbf{y}|\mathbf{I})$.

Autoregressive (AR) decoding factorizes $P(\mathbf{y}|\mathbf{I}) = \prod_{t=1}^{T} P(y_t|y_{<t}, \mathbf{I})$, where $y_{<t}$ denotes all previously generated tokens, enabling rich language modeling at the cost of T sequential forward passes.

Connectionist Temporal Classification (CTC) [11] uses a blank symbol to marginalize over monotonic alignments of length $L \geq T$: $P(\mathbf{y}|\mathbf{I}) = \sum_{\mathbf{a} \in \mathcal{A}(\mathbf{y})} \prod_{\ell=1}^{L} P(a_\ell|\mathbf{I})$. CTC assumes conditional independence given the image, allowing efficient parallel frame-wise prediction with greedy or beam search decoding.

Non-autoregressive cross-entropy (NAR-CE) predicts all positions simultaneously using learned position queries: $P(\mathbf{y}|\mathbf{I}) = \prod_{t=1}^{T_{\max}} P(y_t|\mathbf{I})$, where $T_{\max}$ is a fixed maximum length. This combines parallel decoding with direct cross-entropy supervision, avoiding CTC's alignment search.

3.2 Mamba Formulation

State Space Models (SSMs) originate from control theory, where they describe dynamic systems through a latent state evolving over time. For an input $x(t) \in \mathbb{R}$ and state $h(t) \in \mathbb{R}^N$, the continuous dynamics are:

$$h'(t) = \mathbf{A}h(t) + \mathbf{B}x(t), \quad y(t) = \mathbf{C}h(t) \tag{1}$$

with $\mathbf{A} \in \mathbb{R}^{N\times N}$ (state transition), $\mathbf{B} \in \mathbb{R}^{N\times 1}$ (input projection), and $\mathbf{C} \in \mathbb{R}^{1\times N}$ (output mapping). For discrete sequences like text, these are discretized with timescale Δ:

$$h_t = \bar{\mathbf{A}}h_{t-1} + \bar{\mathbf{B}}x_t, \quad y_t = \mathbf{C}h_t \tag{2}$$

where $\bar{\mathbf{A}} = \exp(\Delta\mathbf{A})$ and $\bar{\mathbf{B}} = (\Delta\mathbf{A})^{-1}(\exp(\Delta\mathbf{A}) - \mathbf{I}) \cdot \Delta\mathbf{B}$.
In practice, $\mathbf{A}$, $\mathbf{B}$, $\mathbf{C}$, and Δ are *learned parameters* optimized during training.

Mamba [14] introduces *selective* SSMs by making $\mathbf{B}$, $\mathbf{C}$, and Δ input-dependent functions rather than fixed matrices:

$$\mathbf{B} = \text{Linear}_B(\mathbf{x}), \quad \mathbf{C} = \text{Linear}_C(\mathbf{x}), \quad \Delta = \text{softplus}(\text{Linear}_\Delta(\mathbf{x})) \tag{3}$$

for input $\mathbf{x} \in \mathbb{R}^{L\times D}$. This allows the model to dynamically filter or retain information based on content, analogous to attention mechanisms, while maintaining $O(n)$ linear complexity via hardware-aware parallel scanning. All linear projections (Linear_B, Linear_C, Linear_Δ) are trainable neural layers that adapt to the OCR task during backpropagation.

3.3 Mamba-Based Architectures

We propose three Mamba-based models sharing a common visual encoder and connector but employing different decoding strategies. Figure 1 provides an overview of all variants.

Shared CNN Encoder. To ensure fair comparison, we adopt the CNN encoder from DAN [6]: five convolutional layers with batch normalization and max-pooling produce a 2D feature map of dimension $H' \times W' \times 256$, flattened to a sequence $\mathbf{x} \in \mathbb{R}^{L\times 256}$ where $L = H' \times W'$.

Bidirectional Mamba Context Modeling. A single bidirectional Mamba connector processes the visual sequence to capture global context in both directions. This layer employs a residual architecture with normalization and feed-forward components. Given the input sequence $\mathbf{x} \in \mathbb{R}^{L\times 256}$, we first apply layer normalization and a GELU-activated linear transformation:

$$\tilde{\mathbf{x}} = \text{GELU}(\text{Linear}_1(\text{LayerNorm}_1(\mathbf{x}))) \tag{4}$$

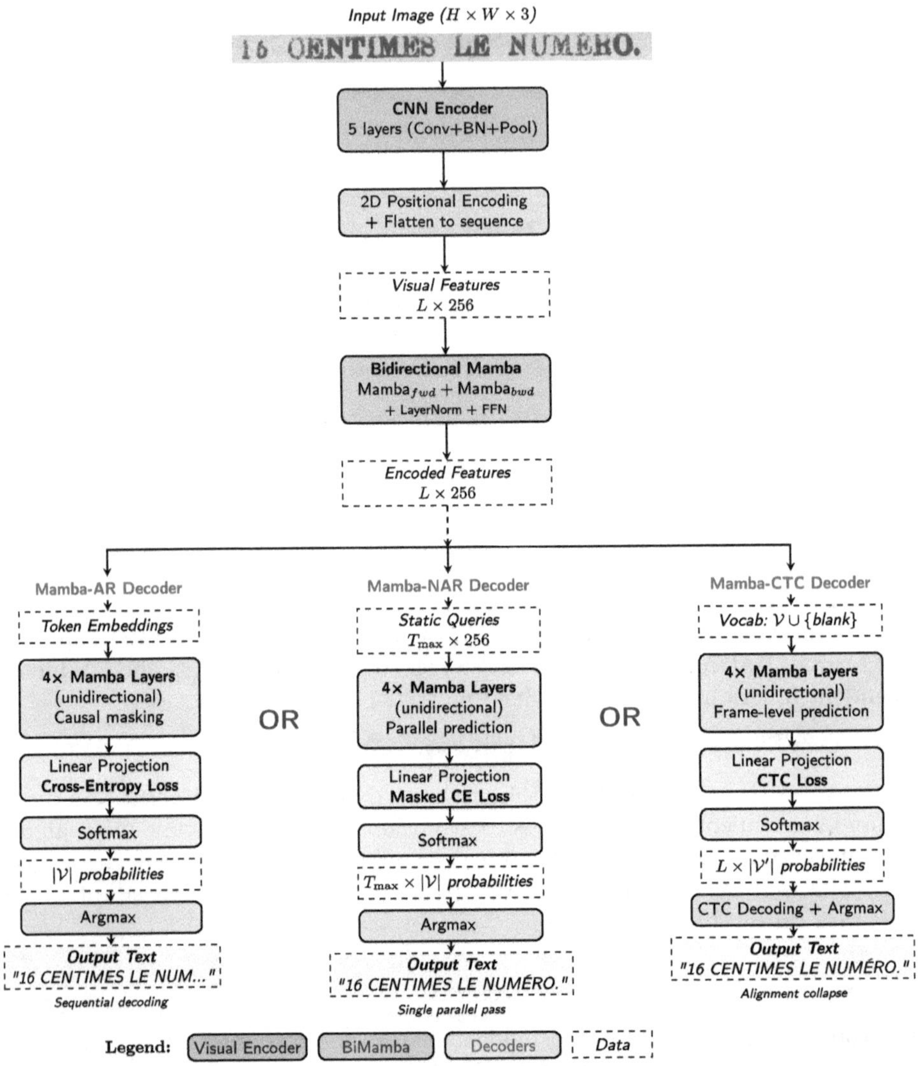

Fig. 1. Mamba-based OCR architecture. Shared visual encoder (blue) extracts features, bidirectional Mamba connector (purple) models global context, and three decoder variants (orange) implement different decoding strategies: autoregressive (AR), non-autoregressive with static queries (NAR), and CTC alignment.

The transformed sequence is then processed bidirectionally using the same Mamba block. Forward processing propagates information left-to-right, while backward processing reverses the sequence, applies Mamba, and reverses the output:

$$\mathbf{h}^{\text{fwd}} = \text{Mamba}(\tilde{\mathbf{x}}), \quad \mathbf{h}^{\text{bwd}} = \text{flip}(\text{Mamba}(\text{flip}(\tilde{\mathbf{x}}))) \tag{5}$$

where flip(·) reverses the temporal dimension. The forward and backward outputs are combined via addition (not concatenation), normalized, and passed through a feedforward layer with residual connections.

Table 1. Comparison of evaluated architectures. Mamba variants differ only in decoding strategy.

Model	Backbone	Decoding	Tokenization	Params (M)	Pre-trained?
Mamba-based (ours)					
Mamba-CTC	CNN + SSM	CTC	Character	~14.1	No
Mamba-AR	CNN + SSM	Autoregressive	Character	~14.1	No
Mamba-NAR	CNN + SSM	NAR (CE)	Character	~14.1	No
Transformer baselines					
VAN [7]	CNN + Transformer + LSTM	CTC	Character	~2.4	No
DAN [6]	CNN + Transformer	Autoregressive	Character	~7.1	No
DANIEL [4]	CNN + Transformer (BART)	Autoregressive	BPE (XLM-R)	~146.5	Partially
Off-the-shelf systems					
PERO-OCR [18]	CNN + BiLSTM	CTC	Character	–	Yes (EU newspapers)
Tesseract v5 [29]	LSTM	–	–	–	Yes (generic)
TrOCR (zero-shot) [21]	ViT + Transformer	Autoregressive	BPE	~334	Yes (LAION, etc.)
TrOCR (fine-tuned) [21]	ViT + Transformer	Autoregressive	BPE	~334	Yes + BnL fine-tuning
Gemini [10]	VLM (proprietary)	Prompted	–	–	Yes (proprietary)

Decoding Variants. All variants use four unidirectional Mamba layers with an expansion factor of 6 (projecting the 256-dimensional input to an inner dimension of 1536) and a state dimension of 256 (the recurrent state size) before their respective output heads.

Mamba-AR. At each step t, visual context $\mathbf{h} \in \mathbb{R}^{L\times 256}$ is concatenated with token embeddings $[\mathbf{e}_1, \ldots, \mathbf{e}_t] \in \mathbb{R}^{t\times 256}$ along the sequence dimension. The 4 Mamba layers process this with causal masking, fusing modalities at input level rather than via explicit cross-attention. Training uses teacher forcing with cross-entropy; inference is sequential.

$$\mathbf{d}_t = \text{Mamba}_{\text{dec}}([\mathbf{h}; \mathbf{e}_{1:t}]), \quad P(y_t|y_{<t}) = \text{softmax}(\mathbf{W}_{\text{AR}}\mathbf{d}_t) \tag{6}$$

Mamba-CTC. Projects Mamba outputs to vocabulary $|\mathcal{V}'|$ with CTC loss and greedy decoding.

$$\mathbf{z} = \text{Mamba}_{\text{dec}}(\mathbf{h}), \quad P(a_\ell|\mathbf{I}) = \text{softmax}(\mathbf{W}_{\text{CTC}}\mathbf{z}_\ell) \tag{7}$$

Mamba-NAR. Uses $T_{\max} = 500$ learned queries $\mathbf{Q} \in \mathbb{R}^{T_{\max}\times 256}$ for parallel prediction with cross-entropy. The decoder emits $T_{\max}$ logits in a single pass.

$$\mathbf{d} = \text{Mamba}_{\text{dec}}([\mathbf{h}; \mathbf{Q}]), \quad P(y_t|\mathbf{I}) = \text{softmax}(\mathbf{W}_{\text{NAR}}\mathbf{d}_t) \tag{8}$$

Table 1 summarizes all models with parameter counts.

Table 2. Line-level dataset statistics.

Split	Fraktur	Antiqua	Total
Train	9,817	13,497	23,314
Validation	701	964	1,665
Test	3,507	4,821	8,328

Table 3. Paragraph-level dataset statistics (1–135 lines/paragraph).

Metric	Train	Valid	Test
Total Paragraphs	20,604	1,795	1,540
Total Lines	308,692	26,844	22,668
Avg Lines/Para	14.98	14.95	14.72
Avg Chars/Para	617.86	616.59	605.68
Avg Chars/Line	40.31	40.30	40.22

4 Experimental Setup

4.1 Datasets

BnL Historical Newspaper Corpus. We evaluate on two datasets (one at line-level, the other at paragraph-level) from the Bibliothèque nationale du Luxembourg's public domain historical newspaper collection[1], covering pre-1878 publications in German, French, and Luxembourgish.

Line-Level Dataset. The line-level dataset contains 33,000 manually transcribed text lines with double-keying verification (>99.95% accuracy), split by font type: 14,000 Fraktur and 19,000 Antiqua. We use the following train/validation/test splits: 23,314/1,665/8,328 lines. The character vocabulary size is 129, including 19th-century specific glyphs (accents, ligatures, punctuation) (Table 2).

Paragraph-Level Dataset. Derived from *L'Indépendance Luxembourgeoise* (1877), this dataset comprises 23,939 text regions extracted from 304 newspaper issues. Table 3 provides full corpus statistics.

For our main experiments, we filter to severely degraded paragraphs with 1–10 lines, which covers 86% of all paragraphs while keeping sequence length under 1000 characters. This yields 13,013 training, 1,156 validation, and 976 test paragraphs. The character vocabulary for paragraph data is 121. We additionally train DANIEL on the full range (1–135 lines) to assess extreme-length handling.

Ground Truth Creation. While METS/ALTO files provide layout coordinates, their embedded transcriptions are unreliable. For training/validation sets, we generate pseudo-ground truth using ROVER [9] to combine outputs from Gemini, PERO-OCR, and Tesseract with weighted voting (Gemini: 5, PERO: 3, Tesseract: 3). This weighting reflects Gemini's superior accuracy while allowing PERO and Tesseract to override it only when they agree. We acknowledge potential circularity from evaluating Gemini on ROVER-derived labels; however, our manually corrected gold-standard test set (976 paragraphs) eliminates this bias for final metrics.

Image Preprocessing. All images are normalized to 300 DPI. Line and paragraph images are padded to minima of 100 pixels (height) and 1000 pixels (width) respectively, while preserving aspect ratio.

[1] https://data.bnl.lu/data/historical-newspapers/.

4.2 Baseline Systems

We compare against strong baselines representing different architectural families and usage scenarios.

Transformer-Based Models

DAN [6] is our primary autoregressive Transformer baseline. It shares the same CNN encoder as our Mamba models, followed by an 8-layer Transformer decoder with 4 attention heads, ensuring direct architectural comparability. **DANIEL** [4] extends DAN with Byte-Pair Encoding (BPE) subword tokenization using XLM-RoBERTa and a 4-layer BART decoder [19], testing whether subword tokens better handle historical spelling variations than character-level models. **VAN** [7] is our CTC-based Transformer baseline. It augments the shared CNN encoder with an implicit line segmentation attention mechanism followed by a unidirectional LSTM before CTC projection, demonstrating competitive accuracy on long sequences without autoregressive decoding.

Off-the-Shelf OCR Systems

PERO-OCR [18] is a production-grade CNN-BiLSTM-CTC system pre-trained on European historical newspapers. We evaluate it zero-shot using the `pero_eu_cz_print_newspapers_2022-09-26` checkpoint, representing out-of-the-box performance. **Tesseract v5** [29] provides a lower-bound baseline, illustrating the gap between classical image processing with LSTM-based recognition and modern neural approaches, on degraded historical documents. **TrOCR** [21] combines a Vision Transformer encoder with a Transformer decoder, pre-trained on large-scale document data. We evaluate two configurations: zero-shot inference and fine-tuning on our BnL training set. **Gemini** [10] represents the frontier of foundation model OCR. We query the Gemini API with structured prompts requesting line-by-line text extraction in XML format. Given its proprietary nature, these results indicate current VLM capabilities rather than providing a reproducible benchmark.

4.3 Training Configuration

Pre-training Strategy. To mitigate overfitting on limited BnL data, we pre-train DAN and all Mamba variants on 1 million synthetic lines generated from French Wikipedia articles. For paragraph-level Mamba-AR and DAN, we adopt curriculum learning: pre-training proceeds on synthetic paragraphs of gradually increasing length (1–10 lines). Even after transitioning to real BnL data, we maintain 10% synthetic paragraphs in each batch to preserve domain robustness. DANIEL and TrOCR leverage their existing pre-training, while VAN requires none for line-level CTC [7]. All CTC models (Mamba-CTC, VAN, PERO-OCR) use greedy decoding without external language models. Beam search with LM fusion could improve accuracy but would complicate comparison across paradigms.

Optimization and Hyperparameters. All models use AdamW with learning rates model-specific: 10^{-4} for DAN, VAN, and Mamba variants; 10^{-5} for DANIEL (fine-tuning BART); 10^{-5} with cosine annealing to 10^{-10} for TrOCR (500-step warmup). Batch size is 4 for line-level and 1 for paragraph-level recognition due to memory constraints.

Data Augmentation. We apply augmentations (probability 0.5) to simulate historical document variability: DPI rescaling, perspective distortion, elastic deformation, morphological operations, color jitter, Gaussian blur/noise, and sharpening, all applied in random order.

4.4 Evaluation Metrics

Recognition Quality. We report Character Error Rate (CER) and Word Error Rate (WER) as normalized Levenshtein distance: CER $= \frac{S+D+I}{N} \times 100\%$, WER $= \frac{S_w+D_w+I_w}{N_w} \times 100\%$. CER is our primary metric as it directly measures transcription accuracy independent of tokenization.

Computational Efficiency. We measure inference latency (ms/image, on A100), throughput (images/sec), and peak GPU memory (GB) during inference, reflecting deployment constraints for large-scale digitization.

5 Results

5.1 Line-Level Recognition

Tables 4 and 5 present line-level recognition results on Antiqua and Fraktur test sets respectively.

Accuracy. Neural models trained on BnL data achieve competitive recognition quality on Antiqua (1.83–2.66% CER), with Mamba-AR and DAN tied for best accuracy. VAN achieves lowest WER (3.39%). On Fraktur, performance degrades slightly (3.01–3.20% CER), with VAN and Mamba-NAR achieving best results. DANIEL's substantial drop on Fraktur (6.18% vs 2.37% CER on Antiqua) likely reflects BPE tokenizer vocabulary mismatch. Off-the-shelf systems perform poorly: zero-shot TrOCR fails catastrophically (15.46–35.98% CER), while fine-tuned TrOCR achieves 3.62–4.71% CER at prohibitive cost. Gemini (7.23–10.50% CER) lags specialized models despite massive scale.

Efficiency. VAN dominates throughput (161.5 img/s on Antiqua) with lowest latency (6.19 ms), followed by Mamba-CTC (129.2 img/s). Autoregressive models incur substantial overhead: DAN requires 156.5 ms per line (25× slower than VAN), while Mamba-AR achieves 53.3 ms (2.9× faster than DAN). Figure 2 visualizes the accuracy-latency trade-off for both scripts.

5.2 Paragraph-Level Recognition

Table 6 presents results on severely degraded paragraph images. DAN achieves best accuracy (5.24% CER), but VAN delivers the most practical profile: 6.42%

Table 4. Line-level recognition performance on Antiqua test set (Best in **bold**).

Model	CER (%, ↓)	WER (%, ↓)	Latency (ms, ↓)	Throughput (img/s, ↑)	Inf Mem (MB)	Train Mem (GB)
Mamba-based (ours)						
Mamba-CTC	2.18	5.16	7.74	129.2	543	7.17
Mamba-AR	**1.83**	**3.84**	53.3	18.7	2898	8.06
Mamba-NAR	2.66	8.26	13.6	73.5	960	6.31
Transformer baselines						
VAN	1.85	**3.39**	**6.19**	**161.5**	1379	14.14
DAN	**1.83**	3.61	156.5	6.4	706	6.74
DANIEL	2.37	4.99	23.8	41.6	2986	10.80
Off-the-shelf systems						
PERO-OCR	13.50	26.19	30.9	32.4	–	–
Tesseract v5	4.79	16.06	97.9	10.2	–	–
TrOCR (zero-shot)	15.46	60.97	69.5	14.4	1389	–
TrOCR (fine-tuned)	3.62	8.95	284.3	3.5	1472	27.70
Gemini	7.23	19.11	~1000	~1	–	–

CER with 17.2 ms latency (23× faster than DAN). Mamba-AR maintains competitive accuracy (6.07% CER) while outpacing DAN by 2.05× (195.6 ms vs 401.2 ms). DANIEL's efficiency advantage over DAN (112.1 ms vs 401.2 ms) demonstrates BPE tokenization benefits for reducing autoregressive sequence length. We report only Mamba-AR at paragraph level. Mamba-CTC and Mamba-NAR were designed for single-line recognition and lack multi-line layout modeling mechanisms, making them unsuitable for paragraph images containing stacked text lines.

5.3 Memory Scaling Analysis

We used synthetic data at controlled lengths (100/300/600/1000 chars) specifically to isolate the architectural scaling behavior from data-quality effects. On this synthetic data, all evaluated models achieve ~1% CER across all length bins, so the scaling story is cleanly about efficiency rather than accuracy degradation. Table 7 and Fig. 3 examine peak inference memory as sequence length increases from 100 to 1000 characters. Mamba-AR exhibits near-linear growth (1.26× increase), while DAN scales quadratically (2.30× increase) due to expanding key-value caches. VAN shows favorable scaling until 600 characters, after which memory usage doubles (2.00×). DANIEL's growth (1.90×) reflects BPE compression reducing token count while still following Transformer scaling laws.

Table 5. Line-level recognition performance on Fraktur test set (Best in **bold**).

Model	CER (%, ↓)	WER (%, ↓)	Latency (ms, ↓)	Throughput (img/s, ↑)	Inf Mem (MB)	Train Mem (GB)
Mamba-based (ours)						
Mamba-CTC	3.20	6.84	8.26	121.0	564	7.33
Mamba-AR	3.06	5.73	56.0	17.8	2915	8.01
Mamba-NAR	3.02	7.63	13.7	73.0	981	7.07
Transformer baselines						
VAN	**3.01**	5.21	**6.53**	**153.3**	1473	16.27
DAN	3.03	**4.88**	166.0	6.0	703	**6.10**
DANIEL	6.18	14.11	23.7	41.9	3109	12.79
Off-the-shelf systems						
PERO-OCR	15.73	32.44	33.2	30.1	–	–
Tesseract v5	21.42	65.13	100.1	10.0	–	–
TrOCR (zero-shot)	35.98	98.40	72.0	13.9	1389	–
TrOCR (fine-tuned)	4.71	10.79	414.5	2.4	1472	19.36
Gemini	10.50	29.59	∼1000	∼1	–	–

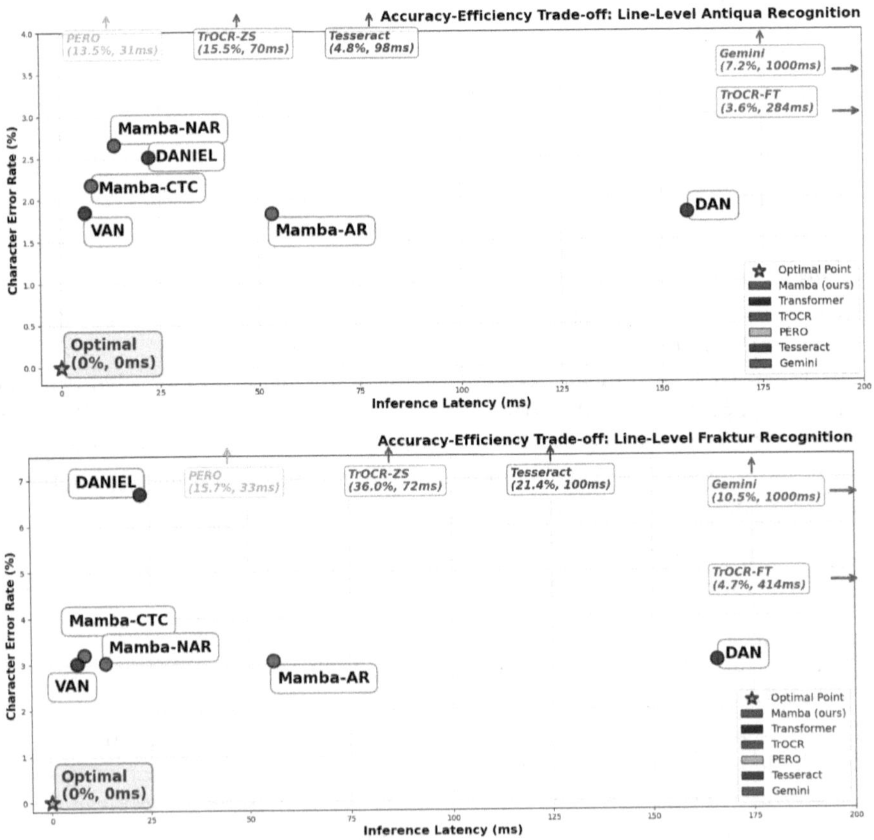

Fig. 2. Accuracy vs latency trade-off on line-level recognition. **Left:** Antiqua test set, Mamba-AR achieves DAN-level accuracy with 2.9× lower latency. **Right:** Fraktur test set, VAN and Mamba-NAR provide optimal speed-accuracy balance for Gothic script.

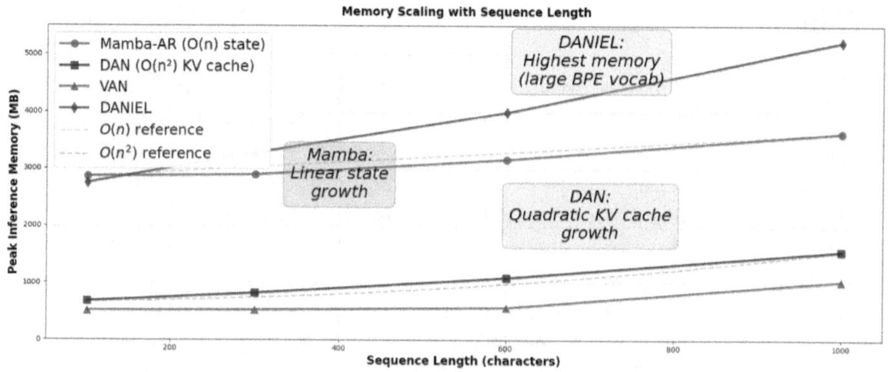

Fig. 3. Peak inference memory vs sequence length. Mamba-AR scales linearly ($O(n)$) while DAN quadratically ($O(n^2)$), confirming theoretical predictions.

Table 6. Severely degraded paragraph-level recognition performance on BnL test set (976 paragraphs, 1–10 lines). Best results in **bold**.

Model	CER (%, ↓)	WER (%, ↓)	Latency (ms, ↓)	Throughput (img/s, ↑)	Inf Mem (MB)	Train Mem (GB)
Neural models						
Mamba-AR	6.07	8.81	195.6	4.4	3021	14.56
VAN	6.42	**7.64**	**17.2**	**58.1**	620	8.80
DAN	**5.24**	9.14	401.2	2.5	948	19.17
DANIEL	6.18	9.53	112.1	8.9	3723	29.40
Off-the-shelf systems						
PERO-OCR	12.58	19.34	35.8	27.9	–	–
Tesseract v5	12.12	24.37	105.7	9.5	–	–
Gemini	6.06	9.65	~1000	~1	–	–

Table 7. Peak inference memory scaling with sequence length. Mamba's higher baseline (2865 MB vs 676 MB) reflects larger parameter count (14.1M vs 7.1M), but exhibits linear $O(n)$ growth (1.26×) vs DAN's quadratic $O(n^2)$ (2.30×), enabling better scalability for long sequences.

Model	Peak Memory (MB)				Growth Factor (vs 100 chars)			
	100	300	600	1000	100	300	600	1000
Mamba-AR	2,865	2,892	3,160	3,622	1.00×	1.01×	1.10×	1.26×
DAN	676	819	1,087	1,551	1.00×	1.21×	1.61×	2.30×
VAN	515	524	567	1,031	1.00×	1.02×	1.10×	2.00×
DANIEL	2,751	3,274	3,987	5,219	1.00×	1.19×	1.45×	1.90×

These results empirically validate SSM's theoretical $O(n)$ complexity advantage: for 1000-character sequences.

6 Conclusion

This paper presents the first application of State-Space Models to historical document OCR, introducing three Mamba-based architectures with distinct decoding strategies (CTC, autoregressive, and non-autoregressive). Through systematic evaluation on the BnL newspaper corpus (8,328 lines, 976 annotated paragraphs, Antiqua and Fraktur scripts), we demonstrate that Mamba offers a compelling alternative to Transformers, particularly for paragraph-level recognition where computational efficiency is critical.

Mamba-AR achieves line-level accuracy comparable to DAN (1.83% CER both) with 2.9× faster inference (53.3 ms vs 156.5 ms). At paragraph level, Mamba-AR delivers 2.05× speedup over DAN (195.6 ms vs 401.2 ms) while maintaining CER within 0.83% (6.07% vs 5.24%). Memory scaling experiments empirically confirm Mamba's $O(n)$ complexity advantage over Transformers' $O(n^2)$ at 1000-character sequences, memory grows 1.26× for Mamba versus 2.30× for DAN, validating theoretical predictions and enabling larger batch sizes for long-sequence processing.

Our findings reveal that while all neural models achieve competitive line-level accuracy (~2% CER), computational efficiency differentiates them at paragraph level. VAN emerges as the optimal production baseline, delivering 58.1 img/s throughput with 6.42% CER—23× faster than DAN with only 1.18% CER penalty. Character-level tokenization suffices for historical text. BPE-based DANIEL's Fraktur collapse (6.18% vs 2.37% CER on Antiqua) demonstrates vocabulary mismatch risks. Foundation models (TrOCR, Gemini) underperform specialized systems despite massive pre-training, revealing persistent domain adaptation gaps for degraded historical documents.

We release our code, trained checkpoints, and BnL benchmark splits upon publication to establish a reproducible foundation for future research. As cultural heritage institutions digitize billion-page archives, this work guides practitioners toward architectures balancing quality with deployment feasibility, and invites further exploration of efficient sequence modeling for document understanding.

References

1. Bautista, D.: Scene text recognition with permuted autoregressive sequence models. In: Proceedings of the 17th ECCV (2022)
2. Bluche, T.: Gated convolutional recurrent neural networks for multilingual handwriting recognition. In: ICDAR (2017)
3. Bluche, T., et al.: Scan, attend and read: End-to-end handwritten paragraph recognition with MDLSTM attention. In: ICDAR (2017)
4. Constum, T.: Daniel: a fast document attention network for information extraction and labelling of handwritten documents. In: IJDAR (2025)
5. Coquenet, D., et al.: SPAN: A simple predict & align network for handwritten paragraph recognition. In: 16th ICDAR (2021)
6. Coquenet, D.: Dan: A segmentation-free document attention network for handwritten document recognition. IEEE TPAMI (2023)

7. Coquenet, D.: End-to-end handwritten paragraph text recognition using a vertical attention network. IEEE TPAMI (2023)
8. Cui, S.: Context-aware TinyML model for Korean handwriting recognition under online assisted learning scenes. Internet Technol. Lette. (2025)
9. Fiscus, J.G.: A post-processing system to yield reduced word error rates: Recognizer output voting error reduction (rover). In: IEEE Workshop on Automatic Speech Recognition and Understanding Proceedings (1997)
10. Google: Gemini (gemini-2.0-flash-001 version), [large language model] (2025)
11. Graves, A.: Connectionist temporal classification: labelling unsegmented sequence data with recurrent neural networks. In: Proceedings of the 23rd ICML (2006)
12. Grüning, T.: READ-BAD: A new dataset and evaluation scheme for baseline detection in archival documents (2018)
13. Gu, A.: Efficiently modeling long sequences with structured state spaces. In: ICLR (2022)
14. Gu, A.: Mamba: Linear-time sequence modeling with selective state spaces. In: COLM (2024)
15. Hu, P.: Docmamba: Efficient document pre-training with state space model. In: AAAI (2025)
16. IMPACT Centre of Competence: IMPACT dataset of historical newspapers (2011)
17. Kim, G.: OCR-free document understanding transformer. In: ECCV (2022)
18. Kol, O.: PERO OCR: Towards fully trainable OCR system. In: Document Analysis and Recognition - ICDAR Workshops (2021)
19. Lewis, M.: Bart: Denoising sequence-to-sequence pre-training for natural language generation, translation, and comprehension. In: Proceedings of the 58th ACL (2020)
20. Li, H.: Show, attend and read: A simple and strong baseline for irregular text recognition. In: AAAI (2019)
21. Li, M.: TrOCR: Transformer-based optical character recognition with pre-trained models. In: AAAI (2023)
22. Liu, Y.: Vmamba: Visual state space model. In: NeurIPS (2024)
23. Muehlberger, G.: Transforming scholarship in the archives through handwritten text recognition: Transkribus as a case study. J. Documentation (2019)
24. Puigcerver, J.: Are multidimensional recurrent layers really necessary for handwritten text recognition? In: ICDAR (2017)
25. Qiao, Y.: Vl-mamba: Exploring state space models for multimodal learning, arXiv preprint (2024) (2024).
26. Shi, B., et al.: An end-to-end trainable neural network for image-based sequence recognition and its application to scene text recognition. IEEE TPAMI (2017)
27. Shi, B.: ASTER: An attentional scene text recognizer with flexible rectification. IEEE TPAMI (2019)
28. Simistira, F., et al.: DIVA-HisDB: A precisely annotated large dataset of challenging medieval manuscripts. In: 15th ICFHR (2016)
29. Smith, R.: An overview of the tesseract ocr engine. In: ICDAR (2007)
30. Springmann, U.: OCR of historical printings with an application to building diachronic corpora: A case study using the RIDGES herbal corpus. In: Digital Humanities Quarterly (2017)
31. Sánchez, J.A.: ICDAR 2019 competition on historical document recognition. In: ICDAR (2019)
32. Wang, T.: Decoupled attention network for text recognition. In: AAAI (2020)
33. Wigington, C.: Start, follow, read: End-to-end full-page handwriting recognition. In: ECCV (2018)

34. Zhao, Q.: Textmamba: Scene text detector with mamba. In: IJCNN (2025)
35. Zhou, B.: Mambastr: Scene text recognition with masked state space model. In: ICIC (2025)
36. Zhu, L.: Vision mamba: Efficient visual representation learning with bidirectional state space model. In: Proceedings of the 41st ICML (2024)

Evolution-Guided Diffusion for Oracle Bone Script Decipherment

Haotian Chen[1,2], Hetao Wu[1,2], Qingju Jiao[3], Yongge Liu[3], and Da-Han Wang[1,2](✉)

[1] School of Computer and Information Engineering, Xiamen University of Technology, Xiamen, Fujian, China
[2] Fujian Key Laboratory of Pattern Recognition and Image Understanding, Xiamen, China
wangdh@xmut.edu.cn
[3] School of Computer and Information Engineering, Anyang Normal University, Anyang, Henan, China

Abstract. Oracle Bone Script (OBS), originating from the late Shang Dynasty over three millennia ago, represents the earliest known mature writing system in China and the direct ancestor of modern Chinese characters. OBS characters span multiple historical stages and undergo significant and nonlinear structural transformations during their long-term evolution, making direct mapping from OBS to modern characters a challenging task. Although diffusion-based methods have shown promise for this problem, most existing approaches treat it as a single-stage glyph translation task and fail to fully exploit the structural continuity inherent in Chinese character evolution. To address this limitation, we propose a multi-stage conditional diffusion framework that integrates stage embeddings and evolutionary curriculum learning under a shared-parameter design. By explicitly modeling multiple historical stages, the model captures stable structural patterns across evolution. Furthermore, an intermediate-stage replay mechanism preserves key evolutionary cues during training and improves generalization. Experimental results demonstrate consistent improvements over diffusion baselines, indicating the effectiveness of leveraging character evolution for assisting ancient script decipherment.

Keywords: Oracle Bone Script · Historical Script Evolution · Diffusion Models · Multi-stage Modeling · Curriculum Learning

1 Introduction

Oracle Bone Script (OBS), originating from the late Shang Dynasty over three millennia ago, is the earliest known mature writing system in China and the direct ancestor of modern Chinese characters [2]. As an important carrier of historical records, OBS preserves rich linguistic, cultural, and social information. More than 4,500 distinct OBS characters have been identified, yet only about

G. A. Fink et al. (Eds.): ICDAR 2026, LNCS 16972, pp. 681–697, 2027.
https://doi.org/10.1007/978-3-032-36023-6_39

1,600 have been reliably deciphered [15,21,22]. A large number of characters still lack confirmed correspondences to modern Chinese forms, making computationally assisted decipherment an important research direction in document analysis and ancient script studies [6,8,11,23,24,28,29].

Recent advances in generative modeling have opened new possibilities for OBS decipherment. OBSD [6] first introduced conditional diffusion models [9] to this task by directly learning the mapping from OBS images to modern Chinese characters, demonstrating promising performance through image generation [19]. Subsequent studies further explored generation frameworks incorporating semantic modeling and recognition-oriented feature learning [1,10,14,16,18,26,30,34]. Despite these advances, most existing methods still follow a *single-stage cross-era mapping* paradigm that directly transforms OBS into modern characters. Such long-range cross-era transformation introduces substantial structural discrepancies, often leading to glyph misalignment and unstable generation, as illustrated in Fig. 1.

In fact, Chinese character evolution is gradual rather than abrupt. Although stroke styles vary across historical periods, higher-level structural configurations (e.g., left–right or top–bottom layouts) often remain relatively stable across stages, as shown in Fig. 2. This suggests that intermediate historical stages can serve as structural bridges, alleviating the large structural gap in direct cross-era mapping.

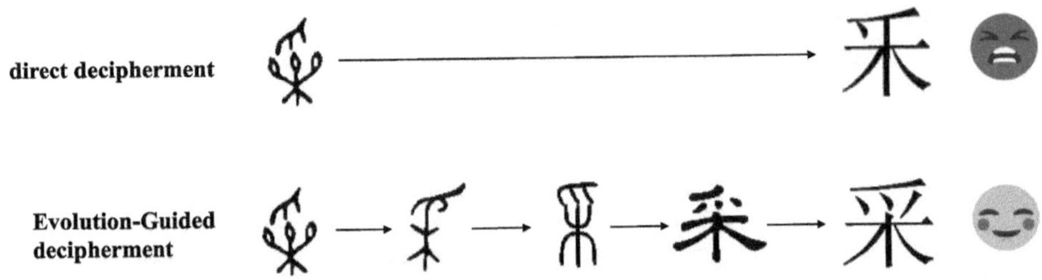

Fig. 1. Comparison between single-stage and multi-stage decipherment paradigms. Top: Single-stage direct decipherment performs one-step OBS-to-modern mapping, requiring large cross-era structural transformation, which may cause structural misalignment and unstable generation. Bottom: Evolution-guided multi-stage decipherment introduces intermediate historical stages as transitional bridges, progressively aligning character structures and improving generation stability.

Motivated by these observations, we formulate OBS decipherment as a multi-stage evolutionary learning problem with modern Traditional Chinese as the final target. To this end, we propose a parameter-shared stage-conditional diffusion framework that integrates data from multiple historical stages within a unified model, transforming conventional single-stage mapping into a progressive multi-stage generation process. By introducing stage-aware conditioning, the model learns cross-era structural continuity within a shared representation space, improving structural alignment and generation stability.

Furthermore, we introduce an evolutionary curriculum learning strategy to better exploit the bridging role of intermediate stages. The model progressively learns transitional structural representations following the historical evolution order, while an intermediate-stage replay mechanism preserves previously acquired cross-stage knowledge during later training, improving both stability and generalization.

The main contributions of this work are summarized as follows:

- Inspired by the evolutionary nature of Chinese characters, we model Oracle Bone Script decipherment from a multi-stage evolutionary perspective, where intermediate historical scripts are introduced as auxiliary supervision to explicitly capture cross-era structural continuity.
- We propose a parameter-shared stage-conditional diffusion framework with learnable stage embeddings, enabling a unified model to integrate multiple historical stages and improve structural alignment in cross-era glyph generation.
- We design an evolutionary curriculum training strategy with intermediate-stage replay, which progressively guides the model from intermediate historical structural learning toward modern character generation, improving training stability and generalization.

2 Related Work

2.1 Oracle Bone Script Recognition

Early research on Oracle Bone Script (OBS) primarily focused on character recognition, evolving from handcrafted feature engineering to deep learning-based approaches [7,11,17,31–33]. The release of large-scale benchmark datasets further accelerated progress in closed-set recognition [5,12,22]. However, recognition-oriented methods rely on predefined character categories and therefore remain limited when dealing with characters whose correspondences are unknown [16].

2.2 Oracle Bone Script Decipherment

To address this limitation, prior studies have explored both reasoning-based and generative approaches. Zhang et al. [29] proposed a case-based reasoning framework that infers unknown characters by analogy with previously interpreted ones. Generative modeling has recently emerged as a promising direction. Sundial-GAN [1] uses cascaded GANs to simulate cross-era glyph transformation, while OBSD [6] introduced conditional diffusion models [9] to learn a direct mapping from OBS to modern Chinese characters. Related generative studies have also explored ancient character modeling and data augmentation [4,10,27].

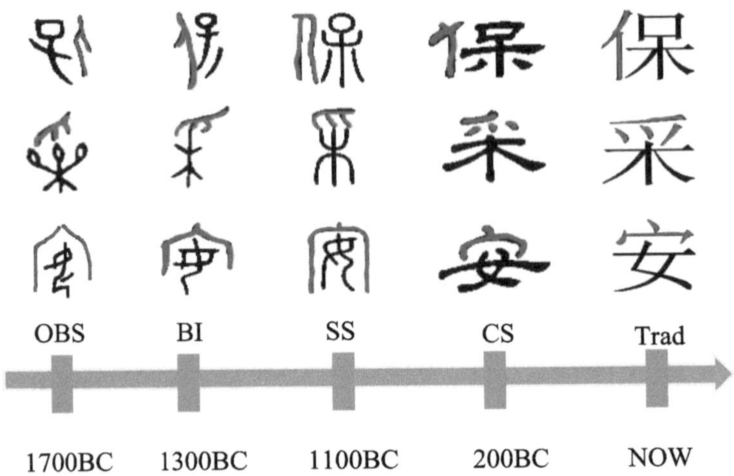

Fig. 2. Illustration of structural continuity in Chinese character evolution. From Oracle Bone Script (OBS), Bronze Inscriptions (BI), Seal Script (SS), and Clerical Script (CS) to modern Traditional Chinese (Trad), character forms undergo gradual morphological transformation while preserving overall structural configurations. Red annotations highlight corresponding structural components shared across historical stages, demonstrating cross-era continuity. (Color figure online)

Recent work further incorporates semantic information through vision–language models. Wu et al. [26] proposed a text–image dual-conditional Stable Diffusion model guided by textual semantics, while Peng et al. [18] leveraged large vision–language models to link glyph evidence with semantic reasoning for interpretable decipherment.

Despite these advances, most generative approaches adopt a *single-stage cross-era mapping* paradigm that directly translates OBS into modern characters, largely overlooking the continuous evolution of Chinese character structures. Motivated by this observation, we model OBS decipherment as a multi-stage evolutionary learning problem and integrate multiple historical stages within a parameter-shared diffusion framework to capture cross-era structural continuity.

3 Method

3.1 Overview

The overall framework consists of two hierarchical components: the proposed **Evolution-Guided Multi-Stage Modeling (EGMM)** module and a refinement module adapted from OBSD [6]. EGMM formulates Oracle Bone Script (OBS) decipherment as a stage-aware multi-stage diffusion generation problem, while the refinement module performs structural enhancement and stroke-level correction on the generated results.

Specifically, OBS decipherment is modeled as the conditional distribution:

$$p(x_{\text{target}} \mid x_{\text{OBS}}, S), \tag{1}$$

where S denotes the target historical stage. Stage information is injected into the diffusion network through learnable stage embeddings, enabling controllable one-to-many generation across different historical stages.

To better exploit structural continuity in Chinese character evolution, we further introduce an evolutionary curriculum learning strategy with intermediate-stage replay. The model first learns structural transitions through intermediate historical scripts before shifting toward modern Traditional Chinese as the final decipherment target. The overall framework is illustrated in Fig. 3.

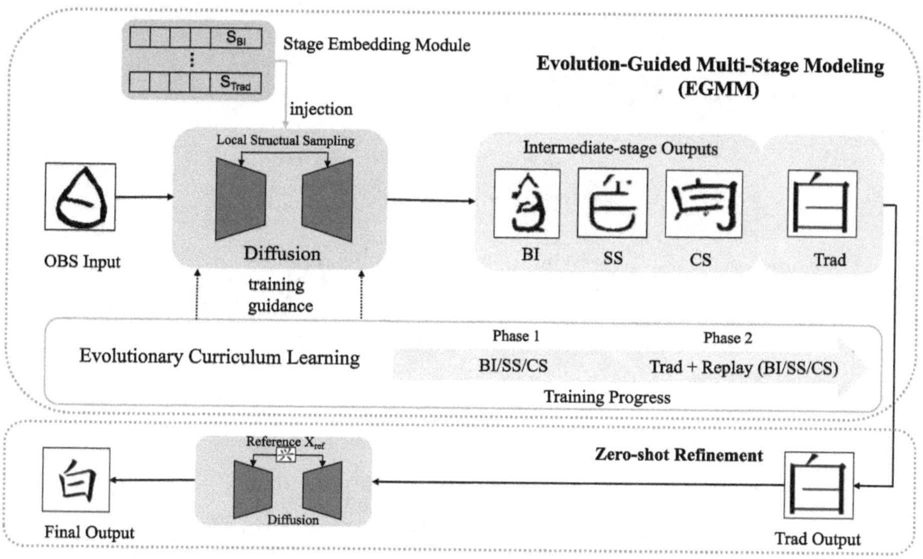

Fig. 3. Overview of the EGD-OBS decipherment pipeline. The EGMM module injects stage embeddings into a diffusion model to generate characters across multiple historical stages. Training is guided by Evolutionary Curriculum Learning with intermediate replay. A zero-shot refinement module further improves the generated Traditional Chinese characters.

3.2 Multi-stage Evolution Modeling

We formulate OBS decipherment as a multi-stage conditional diffusion generation task to explicitly model the continuous structural evolution of Chinese characters, as illustrated in Fig. 4. Instead of directly learning a one-step mapping from OBS to modern Chinese characters, we introduce intermediate historical stages to reduce cross-era structural discrepancy and improve generalization.

The training dataset is defined as:

$$\mathcal{S} = \{(\tilde{X}_i, X_i^{(s)}, s_i)\}_{i=1}^{N}, \tag{2}$$

where $\tilde{X}_i$ denotes the OBS input image, and $X_i^{(s)}$ denotes the corresponding target image at stage $s_i \in \{0, 1, 2, 3\}$, representing Bronze Inscription (BI), Seal Script (SS), Clerical Script (CS), and Traditional Chinese, respectively.

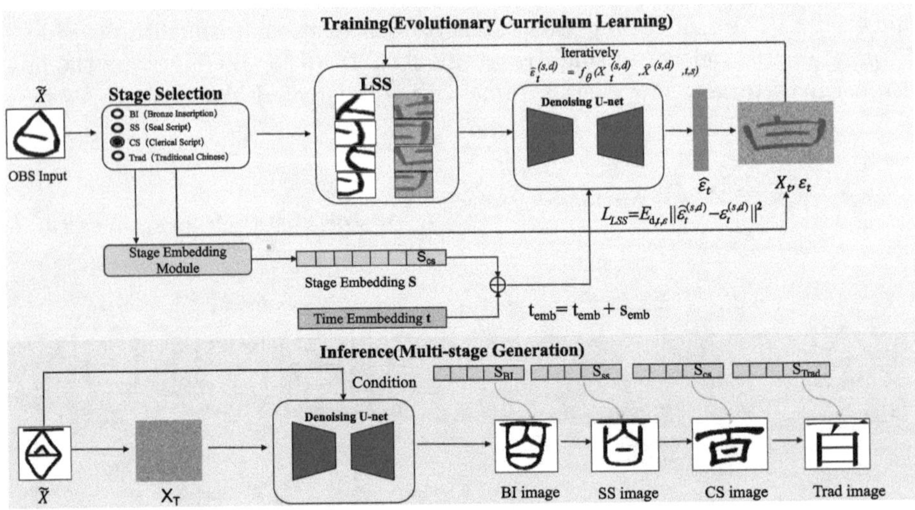

Fig. 4. Overview of the EGMM decipherment pipeline.

Forward Diffusion Process. We adopt the standard Denoising Diffusion Probabilistic Model (DDPM) [9] as the generative backbone. The forward diffusion process progressively adds Gaussian noise to the target-stage image:

$$q(X_1, \ldots, X_T \mid X_0^{(s)}) = \prod_{t=1}^{T} q(X_t \mid X_{t-1}), \tag{3}$$

where

$$q(X_t \mid X_{t-1}) = \mathcal{N}\Big(X_t; \sqrt{1-\beta_t} X_{t-1}, \beta_t I\Big), \tag{4}$$

where β_t denotes a predefined noise variance schedule.

Using the reparameterization trick, the above process can be equivalently written as

$$q(X_t \mid X_0^{(s)}) = \mathcal{N}\Big(X_t; \sqrt{\bar{\alpha}_t} X_0^{(s)}, (1-\bar{\alpha}_t) I\Big), \tag{5}$$

$$X_t = \sqrt{\bar{\alpha}_t} X_0^{(s)} + \sqrt{1-\bar{\alpha}_t}\, \epsilon, \tag{6}$$

where $\epsilon \sim \mathcal{N}(0, I)$ and $\bar{\alpha}_t = \prod_{k=1}^{t}(1-\beta_k)$.

Reverse Diffusion Formulation. Given OBS input $\tilde{X}$ and target stage s, the model learns:

$$p_\theta(X_{t-1} \mid X_t, \tilde{X}, s), \tag{7}$$

parameterized as:

$$p_\theta(X_{t-1} \mid X_t, \tilde{X}, s) = \mathcal{N}\left(X_{t-1}; f_\theta(X_t, \tilde{X}, t, s), \sigma_t^2 I\right), \tag{8}$$

where $f_\theta(\cdot)$ denotes the conditional denoising network.

Training follows the standard noise prediction objective:

$$\mathcal{L}_{\text{simple}} = \mathbb{E}_{X_0^{(s)}, t, \epsilon}\left[\left\|\epsilon - f_\theta\left(\sqrt{\bar{\alpha}_t} X_0^{(s)} + \sqrt{1-\bar{\alpha}_t}\epsilon,\ \tilde{X},\ t,\ s\right)\right\|_2^2\right], \tag{9}$$

where $\epsilon \sim \mathcal{N}(0, I)$.

Patch-Based Diffusion with Localized Structural Sampling. Due to the substantial structural discrepancy between OBS glyphs and modern Chinese characters, whole-character diffusion often leads to unstable generation. Following OBSD [6], we adopt Localized Structural Sampling (LSS) and reformulate diffusion at the patch level.

Given OBS image $\tilde{X} \in \mathbb{R}^{H \times W \times 3}$ and target image $X_0^{(s)} \in \mathbb{R}^{H \times W \times 3}$, we randomly sample D spatially corresponding patch pairs:

$$\tilde{X}^{(d)},\ X_0^{(s,d)} \in \mathbb{R}^{\tilde{p} \times \tilde{p} \times 3}, \quad d = 1, 2, \ldots, D, \tag{10}$$

with $\tilde{p} = 64$.

Patch-level diffusion is defined as:

$$X_t^{(s,d)} = \sqrt{\bar{\alpha}_t} X_0^{(s,d)} + \sqrt{1-\bar{\alpha}_t}\, \epsilon_t^{(d)}, \tag{11}$$

where $\epsilon_t^{(d)} \sim \mathcal{N}(0, I)$.

The reverse process becomes:

$$p_\theta\left(X_{0:T}^{(s,d)} \mid \tilde{X}^{(d)}\right) = p\left(X_T^{(s,d)}\right) \prod_{t=1}^{T} p_\theta\left(X_{t-1}^{(s,d)} \mid X_t^{(s,d)}, \tilde{X}^{(d)}, t, s\right). \tag{12}$$

The patch-level objective is:

$$\mathcal{L}_{\text{LSS}} = \mathbb{E}_{d,t,\epsilon}\left[\left\|\hat{\epsilon}_t^{(d)} - \epsilon_t^{(d)}\right\|_2^2\right], \quad \hat{\epsilon}_t^{(d)} = f_\theta\left(X_t^{(s,d)}, \tilde{X}^{(d)}, t, s\right). \tag{13}$$

This multi-stage formulation enables the shared diffusion model to learn consistent structural correspondences across historical writing systems.

3.3 Historical Stage Injection

Although multi-stage supervision provides multiple training targets, the model still lacks explicit awareness of stage-specific structural distributions. To address this, we introduce a historical stage injection mechanism.

Stage Embedding. Given a discrete historical stage label $s \in \{0, 1, 2, 3\}$ corresponding to BI, SS, CS, and Traditional Chinese, we define a learnable stage embedding:

$$\mathrm{Emb}_s \in \mathbb{R}^d, \tag{14}$$

where d denotes the embedding dimension.

Timestep Embedding. Following standard diffusion models, timestep t is encoded as:

$$\mathrm{Emb}_t^{(2i)} = \sin\left(\frac{t}{10000^{2i/d}}\right), \quad \mathrm{Emb}_t^{(2i+1)} = \cos\left(\frac{t}{10000^{2i/d}}\right). \tag{15}$$

Joint Time–Stage Conditioning. We fuse timestep and stage embeddings:

$$\mathrm{Emb}_{t,s} = \mathrm{Emb}_t + \lambda \cdot \mathrm{Emb}_s, \tag{16}$$

where λ is learnable.

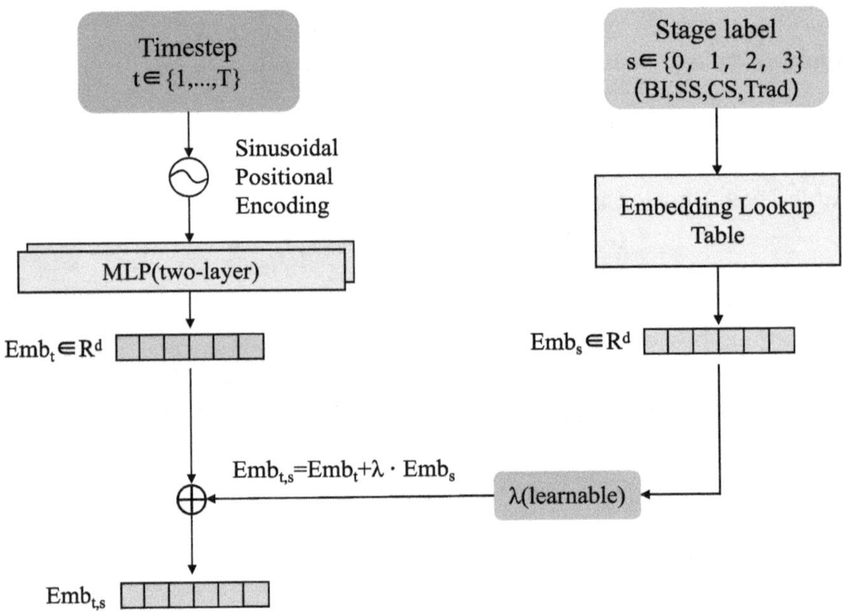

Fig. 5. Joint time–stage embedding mechanism.

As shown in Fig. 5, the fused embedding $Emb_{t,s}$ is injected into U-Net residual blocks via additive conditioning, enabling the parameter-shared diffusion model to adapt generation behavior according to the specified historical stage.

3.4 Evolutionary Curriculum Training with Intermediate-Stage Replay

Directly optimizing all target stages jointly often leads to unstable convergence. We therefore introduce an evolutionary curriculum learning strategy with intermediate-stage replay.

Two-Phase Curriculum Strategy. Training is divided into two phases. In the first phase, the model learns mappings from OBS to intermediate historical scripts (BI, SS, CS), encouraging the learning of evolution-invariant structural patterns. In the second phase, training shifts to modern Traditional Chinese as the final decipherment target, allowing the model to transfer previously learned structural knowledge.

Algorithm 1. Evolutionary Curriculum Training

Require: Training set $\mathcal{S} = \{(\tilde{X}_i, X_i^{(s_i)}, s_i)\}_{i=1}^N$; conditional denoiser $f_\theta(\cdot)$; total epochs E; diffusion steps T; replay ratio ρ.

Initialize parameters θ
Set curriculum switch epoch $E_{\text{sw}} \leftarrow E/2$
for *epoch* $= 1, \ldots, E$ **do**
  for each mini-batch $(\tilde{X}, X^{(s)}, s) \sim \mathcal{S}$ **do**
    Sample timestep $t \sim \text{Uniform}(\{1, \ldots, T\})$, noise $\epsilon \sim \mathcal{N}(0, I)$
    Stage selection (curriculum + replay):
    if *epoch* $\leq E_{\text{sw}}$ **then**
      choose $s \leftarrow s_{\text{mid}}$ {intermediate stage (BI/SS/CS)}
    else
      sample $u \sim \text{Uniform}(0, 1)$
      $s \leftarrow \begin{cases} s_{\text{trad}}, & u \geq \rho, \\ s_{\text{mid}}, & u < \rho \end{cases}$
    end if
    Localized Structural Sampling: $(\tilde{X}^{\text{LSS}}, X_0^{(s),\text{LSS}}) \leftarrow \text{LSS}(\tilde{X}, X^{(s)})$
    $X_t^{\text{LSS}} \leftarrow \sqrt{\bar{\alpha}_t}\, X_0^{(s),\text{LSS}} + \sqrt{1 - \bar{\alpha}_t}\, \epsilon$
    $\hat{\epsilon} \leftarrow f_\theta(X_t^{\text{LSS}}, \tilde{X}^{\text{LSS}}, t, s)$
    $\mathcal{L} \leftarrow \|\hat{\epsilon} - \epsilon\|_2^2$
    Update θ by minimizing $\mathcal{L}$
  end for
end for
return θ

Intermediate-Stage Replay. To prevent forgetting intermediate-stage knowledge, we periodically replay intermediate targets during the second phase:

$$s = \begin{cases} s_{\text{trad}}, & \text{with probability } p, \\ s_{\text{mid}}, & \text{with probability } 1 - p, \end{cases} \tag{17}$$

where s_{trad} denotes Traditional Chinese and s_{mid} denotes an intermediate stage.

Effect on Generalization. Intermediate-stage replay acts as an evolution-consistent regularizer, preserving shared structural representations while optimizing toward the final decipherment objective. This improves training stability and enhances zero-shot generalization to structurally complex character categories.The overall evolutionary curriculum training procedure is summarized in Algorithm 1.

4 Experiments

4.1 Implementation Details

The diffusion model is optimized using Adam with $\beta_1 = 0.9$, $\beta_2 = 0.999$, weight decay 10^{-4}, and a learning rate of 2×10^{-5}. Training uses a batch size of 8, each containing eight 64×64 local patches. The model is trained for 450 epochs, with the first half focusing on intermediate historical stages. All experiments are conducted on a single NVIDIA A800 GPU.

4.2 Dataset and Evaluation Metric

To train and evaluate the proposed method, we select 1,588 character categories from the HUST-OBS [25] and EVOBC datasets [5], comprising 72,152 images in total. The test set contains only characters with known decipherment outcomes to ensure objective evaluation. The dataset is split at the character-category level into training and test subsets with a 9:1 ratio, with no category overlap between them, establishing a strict zero-shot evaluation setting.

We adopt the OCR recognition accuracy as the primary quantitative evaluation metric. Two OCR systems are employed: a self-trained classifier based on a ResNet-101 backbone trained on a large-scale dataset containing 82,312 Chinese character categories, and the open-source OCR system PaddleOCR [3]. For each generated image, the OCR prediction is compared with the ground-truth label, and overall performance is reported using Top-k accuracy.

4.3 Quantitative Results

We comprehensively evaluate the proposed method under two evaluation protocols, namely single-round and multi-round decipherment, using both EGD-OBS-OCR and PaddleOCR [3] as automatic recognition tools and reporting Top-k accuracy. The single-round evaluation measures the model's direct decipherment capability from a single generation, while the multi-round evaluation allows multiple sampling attempts for the same input image, simulating the iterative process of repeated reasoning and trial in real-world decipherment.

As shown in Table 1, under the EGD-OBS-OCR evaluation the proposed EGD-OBS achieves a Top-1 accuracy of 44.8%, while under PaddleOCR evaluation the Top-1 accuracy reaches 33.4%, both surpassing OBSD [6] (41.7%/30.2%). Compared with image-to-image translation baselines including Palette [19], CycleGAN [35], BBDM [13], CDE [26], and Dual-approx Bridge [20],

EGD-OBS maintains advantages at the Top-500 level. This indicates that multi-stage structural modeling not only improves first-choice accuracy but also enhances candidate coverage within the generated set.

Table 2 reports the cumulative success rates under the multi-round evaluation protocol. Using EGD-OBS-OCR, the success rate increases from 44.8% in the first trial to 83.1% after ten trials. A similar trend is observed with PaddleOCR, where performance improves from 33.4% to 61.3%. These results indicate that repeated sampling effectively leverages the stochastic generation process of diffusion models, substantially improving decipherment success through candidate accumulation.

Table 1. Comparison of single-round decipherment success rate between EGD-OBS and different methods.

Evaluation Tool	Rank	Palette	Cycle GAN	BBDM	CDE	Dual-approx Bridge	OBSD	EGD-OBS (ours)
EGD-OBS -OCR	Top-1	0.0%	0.0%	19.2%	32.0%	31.5%	41.7%	**44.8%**
	Top-10	0.0%	0.0%	29.7%	48.0%	47.7%	51.0%	**54.7%**
	Top-20	0.0%	0.0%	35.1%	51.2%	49.7%	55.1%	**58.6%**
	Top-50	0.0%	9.1%	40.2%	53.2%	53.0%	58.0%	**63.7%**
	Top-100	4.1%	20.3%	43.1%	57.5%	56.8%	62.3%	**65.1%**
	Top-200	9.0%	38.2%	46.7%	60.3%	60.0%	63.6%	**66.3%**
	Top-500	20.1%	61.2%	59.2%	65.2%	64.3%	65.6%	**68.3%**
PaddleOCR	Top-1	0.0%	0.0%	7.5%	19.3%	18.4%	30.2%	**33.4%**

Table 2. Top-1 accuracy under multi-round decipherment of EGD-OBS.

Number of Trial	1	2	3	4	5
EGD-OBS-OCR	44.8%	59.3%	70.3%	78.6%	79.5%
PaddleOCR	33.4%	43.2%	49.5%	53.8%	56.5%
Number of Trial	6	7	8	9	10
EGD-OBS-OCR	80.3%	81.2%	82.6%	83.1%	83.1%
PaddleOCR	58.3%	60.2%	60.4%	61.3%	61.3%

4.4 Qualitative Results

Figure 6 presents qualitative comparisons of different image-to-image generation models on the oracle bone decipherment task. Traditional GAN-based

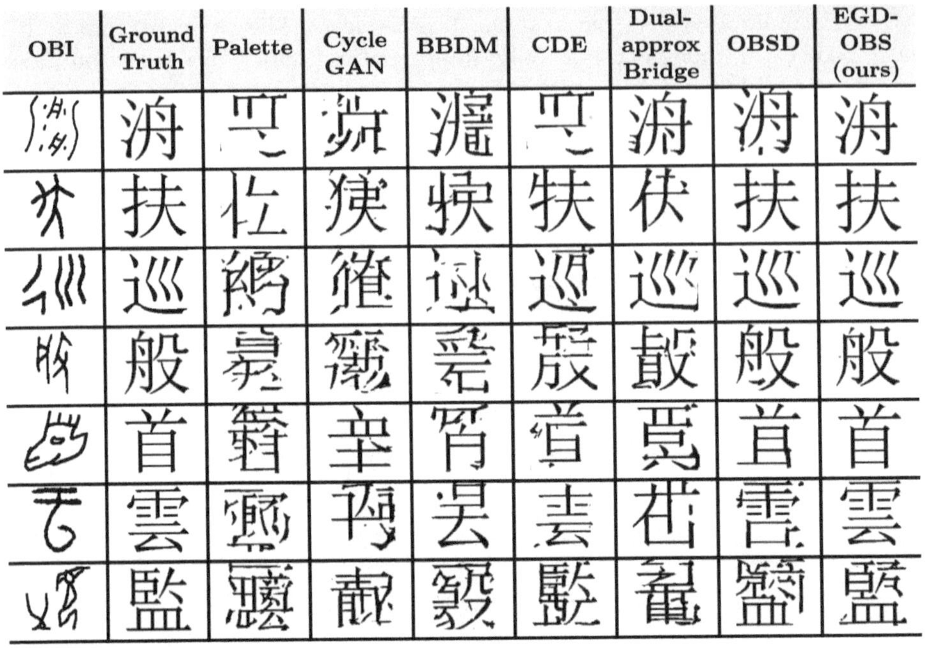

Fig. 6. Qualitative comparison of decipherment results across different methods.

approaches, such as Palette [19] and CycleGAN [35], often struggle to preserve global structural integrity when reconstructing complex characters. Their outputs typically appear as disordered stroke combinations. As a result, it becomes difficult to form meaningful and recognizable character structures.

Diffusion-based models, including BBDM [13], CDE [20], and Dual-approx Bridge [26], produce clearer contours and more coherent overall structures. However, when confronted with large cross-era morphological discrepancies, these methods may still generate local structural shifts. In contrast, guided by the proposed multi-stage evolutionary modeling mechanism, EGD-OBS better preserves high-level structural configurations while progressively refining stroke details, resulting in characters that more closely resemble the ground-truth modern forms in both global layout and local morphology.

Notably, the last row presents a failure case. When encountering character categories that are structurally complex and sparsely represented in the training data, EGD-OBS can still preserve the overall structural framework but fails to fully recover the target character. This limitation arises because such rare characters exhibit weak structural coupling with other character categories, making it difficult to learn reliable structural correspondences from image-level supervision alone.

4.5 Ablation Study

To evaluate the contribution of each component in EGD-OBS, we conduct ablation experiments by progressively introducing Stage Embedding, Curriculum Learning, and Replay. Quantitative results are reported in Table 3.

(a) OBSD-style Baseline. To fairly evaluate the contribution of each component, we adopt the original OBSD framework as the baseline and directly apply it to our multi-stage dataset. Although intermediate-stage data are introduced, the model still lacks explicit stage-aware conditioning, making it difficult to establish stable structural correspondences between oracle bone script and different historical stages. The results indicate that simply introducing intermediate-stage data is insufficient to achieve reliable cross-stage structural alignment.

(b) + Stage Embedding. We introduce learnable stage embeddings to explicitly encode stage-specific structural characteristics within a shared-parameter framework. This enables the model to better distinguish different historical writing stages. As shown in Table 3, this modification significantly improves all evaluation metrics, highlighting the importance of explicit stage conditioning.

(c) + Curriculum Learning. Building on the stage-embedded model, we further introduce a two-stage curriculum learning strategy. The model first focuses on intermediate-stage mappings and then gradually shifts toward the final target stage. Compared with uniform multi-stage training, curriculum learning further improves performance, indicating that a structured optimization path facilitates more stable cross-stage mapping and better evolutionary modeling.

(d) + Replay (Full Model). Finally, we incorporate a replay mechanism into the curriculum framework. During later training stages, intermediate samples are periodically revisited to preserve structural constraints and mitigate stage forgetting. The complete model achieves the best performance across all evaluation metrics, demonstrating that replay effectively reinforces evolutionary continuity and stabilizes training.

In conclusion, the ablation results demonstrate the effectiveness of the proposed EGD-OBS framework for multi-stage oracle bone script decipherment.

4.6 Decipherment on Genuine Undeciphered OBS

To evaluate the practical applicability of the proposed model in real archaeological scenarios, we conducted inference on a small set of genuine Oracle Bone Script (OBS) characters that have not yet been formally deciphered by experts. As shown in Fig. 7, due to the absence of authoritative ground-truth annotations, the evaluation in this section is primarily qualitative. For each input character, multiple samplings with different random seeds were performed to generate diverse candidate interpretations, allowing us to examine the stability and variability of the model outputs. In several cases, consistent structural components or radicals appear in different samplings, potentially providing useful visual cues for future scholarly investigation. It should be emphasised that this study does

Table 3. Quantitative results of the ablation experiment.

Metric	Rank	OBSD-style (Baseline)	+ Stage Embedding	+ Curriculum Learning	+ Repay
EGD-OBS OCR	Top-1	3.7%	32.1%	39.0%	44.8%
	Top-10	5.4%	46.0%	50.2%	54.7%
	Top-20	7.5%	48.2%	54.5%	58.6%
	Top-50	9.5%	51.3%	58.2%	63.7%
	Top-100	12.8%	54.7%	61.0%	65.1%
	Top-200	14.7%	57.2%	63.2%	66.3%
	Top-500	21.0%	61.3%	67.1%	68.3%
PaddleOCR	Top-1	3.0%	18.6%	28.7%	33.4%

Fig. 7. Qualitative results on genuine undeciphered OBS. The first column shows the input glyph, followed by ten independent sampling results.

not aim to provide definitive decipherment results; any interpretation of undeciphered characters ultimately requires careful validation by domain experts. We hope that the candidate results generated by our model can serve as auxiliary references for ongoing research in Oracle Bone Script studies.

5 Conclusion

In this work, we propose EGD-OBS, an evolution-guided conditional diffusion framework for Oracle Bone Script (OBS) decipherment that models the continuous and multi-stage nature of Chinese character evolution. The proposed framework integrates data from multiple historical scripts within a parameter-shared diffusion architecture and introduces learnable stage embeddings to capture cross-era structural continuity. By combining an evolutionary curriculum learning strategy with intermediate-stage replay, EGD-OBS progressively learns structural transitions across historical stages and ultimately focuses on modern Traditional Chinese as the final decipherment target. Experimental results show that EGD-OBS achieves consistent performance improvements over existing generative baselines. These results demonstrate that leveraging historical evolutionary information for structural alignment improves both generation stability and decipherment accuracy. Moreover, the proposed evolutionary modeling paradigm provides a general framework for other pictographic or logographic writing systems characterized by gradual historical evolution.

Acknowledgements. This work was supported by National Natural Science Foundation of China (No. 62576301) and the Open Project of the State Key Laboratory of Multimodal Artificial Intelligence Systems (No. MAIS2024101).

References

1. Chang, X., Chao, F., Shang, C., Shen, Q.: Sundial-GAN: a cascade generative adversarial networks framework for deciphering oracle bone inscriptions. In: Proceedings of the ACM International Conference on Multimedia, pp. 1195–1203. ACM MM (2022)
2. Christina, C.: The Chinese history that is written in bone. Am. Sci. **106**(3), 133 (2018)
3. Cui, C., et al.: PaddleOCR 3.0 technical report. arXiv preprint arXiv:2507.05595 (2025)
4. Gao, J., Giunchiglia, F., Zhao, T., Zhao, X., Li, C., Xu, H.: Oracle bone heritage data augmentation based on two-stage decomposition GANs. NPJ Herit. Sci. **13**, 230 (2025). https://doi.org/10.1038/s40494-025-02049-3
5. Guan, H., et al.: An open dataset for the evolution of oracle bone characters: EVOBC. arXiv preprint arXiv:2401.12467 (2024)
6. Guan, H., et al.: Deciphering oracle bone language with diffusion models. In: Proceedings of the Annual Meeting of the Association for Computational Linguistics (ACL) (2024)
7. Guo, J., Wang, C., Roman-Rangel, E., Chao, H., Rui, Y.: Building hierarchical representations for oracle character and sketch recognition. IEEE Trans. Image Process. **25**(1), 104–118 (2016). https://doi.org/10.1109/TIP.2015.2500019

8. Han, X., Bai, Y., Qiu, K., Liu, Z., Sun, M.: ISOBS: an information system for oracle bone script. In: Proceedings of the 2020 Conference on Empirical Methods in Natural Language Processing (EMNLP), pp. 227–233 (2020)
9. Ho, J., Jain, A., Abbeel, P.: Denoising diffusion probabilistic models. In: Advances in Neural Information Processing Systems (NeurIPS), vol. 33, pp. 6840–6851 (2020)
10. Huang, H., et al.: AGTGAN: Unpaired image translation for photographic ancient character generation. arXiv preprint arXiv:2303.07012 (2023)
11. Huang, S., Wang, H., Liu, Y., Shi, X., Jin, L.: OBC306: a large-scale oracle bone character recognition dataset. In: Proceedings of the International Conference on Document Analysis and Recognition (ICDAR), pp. 681–688 (2019)
12. Jiang, R., Liu, Y., Zhang, B., Chen, X., Li, D., Han, Y.: OraclePoints: a hybrid neural representation for oracle character. In: Proceedings of the ACM International Conference on Multimedia, pp. 7901–7911. ACM MM (2023)
13. Li, B., Xue, K., Liu, B., Lai, Y.K.: BBDM: Image-to-image translation with Brownian bridge diffusion models. In: Proceedings of the IEEE/CVF Conference on Computer Vision and Pattern Recognition (CVPR), pp. 1952–1961 (2023)
14. Li, C., et al.: OracleFusion: assisting the decipherment of oracle bone script with structurally constrained semantic typography. In: Proceedings of the IEEE/CVF International Conference on Computer Vision (ICCV), pp. 19893–19902 (2025)
15. Li, J., et al.: A comprehensive survey of oracle character recognition: challenges, datasets, methodology, and beyond. Pattern Recogn. **169**, 111824 (2026)
16. Liu, J., Wei, J., Hu, T., Li, X., Liu, G., Liu, D.: OBI-CMF: self-supervised learning with contrastive masked frequency modeling for oracle bone inscription recognition. NPJ Herit. Sci. **13**, 55 (2025). https://doi.org/10.1038/s40494-025-01644-8
17. Meng, L., Kamitoku, N., Yamazaki, K.: Recognition of oracle bone inscriptions using deep learning based on data augmentation. In: Proceedings of Metrology for Archaeology and Cultural Heritage, pp. 33–38 (2018)
18. Peng, K., Zhao, M., Yu, H., Fu, T., Li, B.: Interpretable oracle bone script decipherment through radical and pictographic analysis with LVLMs. arXiv preprint arXiv:2508.10113 (2025). https://doi.org/10.48550/arXiv.2508.10113
19. Saharia, C., et al.: Palette: image-to-image diffusion models. In: ACM SIGGRAPH 2022 Conference Proceedings, pp. 1–10 (2022)
20. Saharia, C., Ho, J., Chan, W., Salimans, T., Fleet, D.J., Norouzi, M.: Image super-resolution via iterative refinement. IEEE Trans. Pattern Anal. Mach. Intell. **45**(4), 4713–4726 (2022)
21. Sommerschield, T., et al.: Machine learning for ancient languages: a survey. Comput. Linguist. **49**(3), 703–747 (2023)
22. Wang, M., Deng, W.: A dataset of oracle characters for benchmarking machine learning algorithms. Sci. Data **11**(1), 87 (2024)
23. Wang, M., Deng, W., Liu, C.L.: Unsupervised structure-texture separation network for oracle character recognition. IEEE Trans. Image Process. **31**, 3137–3150 (2022)
24. Wang, P., et al.: Puzzle pieces picker: deciphering ancient Chinese characters with radical reconstruction. In: Proceedings of the International Conference on Document Analysis and Recognition (ICDAR), pp. 169–187. Springer (2024)
25. Wang, P., et al.: An open dataset for oracle bone character recognition and decipherment. Sci. Data **11**, 976 (2024). https://doi.org/10.1038/s41597-024-03807-x
26. Wu, R., Wang, S., Li, X., Liu, D., et al.: A text image dual conditional stable diffusion model for oracle bone inscription decipherment. NPJ Herit. Sci. **13**, 453 (2025). https://doi.org/10.1038/s40494-025-02019-9

27. Yang, Z., Peng, D., Kong, Y., Zhang, Y., Yao, C., Jin, L.: FontDiffuser: one-shot font generation via denoising diffusion with multi-scale content aggregation and style contrastive learning. In: Proceedings of the AAAI Conference on Artificial Intelligence, vol. 38, no. 7, pp. 6603–6611 (2024). https://doi.org/10.1609/aaai.v38i7.28482
28. Zhang, C., et al.: Data-driven oracle bone rejoining: a dataset and practical self-supervised learning scheme. In: Proceedings of the ACM SIGKDD International Conference on Knowledge Discovery and Data Mining (KDD), pp. 4482–4492 (2022)
29. Zhang, G., Liu, D., Smyth, B., Dong, R.: Deciphering ancient Chinese oracle bone inscriptions using case-based reasoning. In: International Conference on Case-Based Reasoning (ICCBR), pp. 309–324. Springer (2021)
30. Zhang, X., et al.: Shallow focus, deep fixes: enhancing shallow layers vision attention sinks to alleviate hallucination in LVLMs. In: Proceedings of the 2025 Conference on Empirical Methods in Natural Language Processing, pp. 3512–3534 (2025)
31. Zhang, Y.K., Zhang, H., Liu, Y.G., Yang, Q., Liu, C.L.: Oracle character recognition by nearest neighbor classification with deep metric learning. In: Proceedings of the International Conference on Document Analysis and Recognition (ICDAR), pp. 309–314 (2019)
32. Zhao, Q., Liang, L., Zhang, X., Zhang, T., Li, J., Liu, Y.: Few-shot cross-modal text detection via CLIP. Vis. Comput. **41**(12), 9471–9485 (2025)
33. Zhao, Q., Yan, Y., Wang, D.H.: TextMamba: scene text detector with mamba. In: 2025 International Joint Conference on Neural Networks (IJCNN), pp. 1–8. IEEE (2025)
34. Zhao, Q., et al.: MCA-LLaVA: manhattan causal attention for reducing hallucination in large vision-language models. In: Proceedings of the 33rd ACM International Conference on Multimedia, pp. 3981–3990 (2025)
35. Zhu, J.Y., Park, T., Isola, P., Efros, A.A.: Unpaired image-to-image translation using cycle-consistent adversarial networks. In: Proceedings of the IEEE International Conference on Computer Vision (ICCV), pp. 2223–2232 (2017)

Towards Universal Khmer Text Recognition

Marry Kong[1], Rina Buoy[1,2(✉)], Sovisal Chenda[1], Nguonly Taing[1], Masakazu Iwamura[2], and Koichi Kise[2]

[1] Techo Startup Center, Ministry of Economy and Finance, Phnom Penh, Cambodia
rina.buoy@tsc.gov.kh
[2] Osaka Metropolitan University, Osaka, Japan
https://github.com/rinabuoy/Towards-Universal-Khmer-Text-Recognition

Abstract. Khmer is a low-resource language characterized by a complex script, presenting significant challenges for optical character recognition (OCR). While printed document text recognition has advanced because of available datasets, performance on other modalities, such as handwritten and scene text, remains limited by data scarcity. Training modality-specific models for each modality does not allow cross-modality transfer learning, from which modalities with limited data could otherwise benefit. Moreover, deploying many modality-specific models results in significant memory overhead and requires error-prone routing of each input image to the appropriate model. On the other hand, simply training on a combined dataset with a non-uniform data distribution across different modalities often leads to degraded performance on underrepresented modalities. To address these challenges, we propose a universal Khmer text recognition (UKTR) framework capable of handling diverse text modalities. Central to our method is a novel modality-aware adaptive feature selection (MAFS) technique designed to adapt visual features according to a particular input image modality and enhance recognition robustness across modalities. Extensive experiments demonstrate that our model achieves state-of-the-art (SoTA) performance. Furthermore, we introduce the first comprehensive benchmark for universal Khmer text recognition, which we release to the community to facilitate future research. Our datasets and models can be accessed via this repository(https://github.com/rinabuoy/Towards-Universal-Khmer-Text-Recognition).

Keywords: low-resource language · text recognition · multi-modalities · non-uniform distribution. modality-adaptation

1 Introduction

The development of Khmer text recognition has received significant attention over the past decade. However, many previous methods [1,3–5] have primarily been designed to operate only on printed document modalities. This is mainly

G. A. Fink et al. (Eds.): ICDAR 2026, LNCS 16972, pp. 698–715, 2027.
https://doi.org/10.1007/978-3-032-36023-6_40

(a) Document text

(b) Scene text

(c) Handwritten text

Fig. 1. Sample images from our new datasets.

because it is relatively straightforward to generate synthetic printed document training data in sufficient quantity and quality. In contrast, generating high-fidelity synthetic data for other modalities, such as handwritten and scene documents as shown in Fig. 1, remains technically challenging. Consequently, progress in Khmer scene text recognition, and particularly in handwritten text recognition, has been severely limited by the lack of high-quality, real-world training and benchmark datasets.

Due to the non-uniform distribution of Khmer text recognition data across different modalities (more printed document datasets and fewer scene and handwritten datasets), most previous methods [6–9] have been specifically designed and trained to recognize a single text modality. As a result, cross-modality learning is either not exploited at all or, at best, limited to pre-trained weight initialization [7]. Moreover, deploying multiple modality-specific models results in significant memory overhead and requires routing of each input image to the appropriate model, making it unsuitable or error-prone for an end-to-end OCR pipeline. Conversely, training a unified Khmer text recognition model on a combined dataset spanning all modalities with unequal proportions often leads to degraded performance on underrepresented modalities (i.e., handwritten and scene texts).

Fig. 2. Sample Khmer text layout (with permission [5].) Blue: consonant subscript. Green: base consonant. Orange: dependent vowel. Purple: diacritic. Best viewed in color. (Color figure online)

To achieve robust recognition performance across all text modalities, this paper proposes a universal Khmer text recognition framework (UKTR) capable of handling diverse text modalities. This is achieved through a novel modality-aware adaptive feature selection (MAFS) technique that dynamically selects the most relevant visual features for accurate text recognition across modalities. In addition, the proposed UKTR framework employs a joint connectionist temporal classification (CTC) [10] and Transformers-based decoder [11] to generate text in a non-autoregressive (i.e., generating all tokens in parallel) and autoregressive manner (i.e., generating one token at a time), respectively. The two decoders present a latency-accuracy trade-off within the same model, with the CTC decoder being faster but less accurate and the Transformer-based decoder being slower but more accurate.

To facilitate robust training and evaluation of the proposed UKTR framework, we construct the first joint general Khmer scene and handwritten text datasets and benchmarks, which will serve as a valuable resource for the research community beyond this work. Experimental results demonstrate that the proposed UKTR model achieves state-of-the-art (SoTA) performance on these new multi-modality and existing benchmarks.

Our contributions can be summarized as follows:

1. We propose the UKTR framework that recognizes Khmer text across printed, scene, and handwritten modalities, robustly. The proposed framework is based on a novel MAFS technique that adaptively selects the most relevant visual features required for accurate text recognition across different modalities.
2. Our UKTR model supports both non-autoregressive and autoregressive text generation, enabling a flexible trade-off between latency and accuracy during inference.
3. We construct the first joint general Khmer scene and handwritten text datasets and benchmarks.
4. Our proposed UKTR model achieves SoTA performance on the new multi-modality and existing benchmarks.

2 Related Work

Khmer is a complex abugida writing system with a large inventory of characters, including consonants (base and subscript forms), dependent vowels, independent vowels, and diacritics [1,2], as illustrated in Fig. 2. Khmer text is written from left to right with optional spaces. Characters are often stacked by attaching subscripts to a base consonant, forming character clusters with ligatures [1]. Such complex stacking structures pose significant challenges for accurate character recognition.

Compared with Latin-based languages, Khmer is a low-resource and low-attention language, with limited research and publicly available datasets. Nonetheless, over the past decade, deep learning-based approaches have been proposed to advance Khmer text recognition, particularly for the printed text modality. Prior to the deep learning era, Khmer OCR methods primarily focused on isolated character recognition, employing wavelet-based feature descriptors [12] and support vector machine (SVM) classifiers [13]. The performance and generalization capabilities of these classical machine learning approaches were significantly limited.

For printed documents and synthetic scene text modalities, Buoy et al. [3] introduced a sequence-to-sequence (seq2seq) Khmer textline recognition framework with an attention mechanism, consisting of a convolutional visual feature extractor and an autoregressive decoder based on gated recurrent units (GRUs) for character decoding. Buoy et al. [1] extended this work by employing a 2D visual feature extractor to enhance feature representation and a Transformers-based decoder [11] to improve character decoding performance. Their models were trained on large-scale synthetically generated document and scene text datasets. Similarly, Keo et al. [14] trained both CTC-based and attention-based Khmer text recognition models using synthetic word-level training images. Since Khmer textlines can be very long and the decoding process is prone to attention misalignment, Buoy et al. [5] proposed a new recognition approach based on a partially autoregressive (PAR) decoder. While previous methods treat individual characters as atomic units, the authors introduced a non-autoregressive framework that generates text at the character cluster level, significantly reducing decoding time while maintaining recognition accuracy. In addition, off-the-shelf OCR systems such as Surya-OCR [15], Tesseract-OCR [16], and emerging multimodal models, including DeepSeek-OCR (v1 and v2) [17,18], are also capable of recognizing Khmer text for certain fonts. Among these, Tesseract-OCR has served as a strong baseline for Khmer OCR.

For the real scene text modality, Nom et al. [6] published the first real-world Khmer scene text dataset (KhmerST) comprising images captured using smartphone cameras. The dataset was designed to support both text detection and recognition tasks. For text recognition, they fine-tuned a Transformers-based OCR (TrOCR) [19] model, achieving superior performance compared to Tesseract-OCR on Khmer text. Furthermore, Nom et al. [8] introduced an expanded version of the dataset, called WildKhmerST, which contains more diverse and challenging real-world samples. Nom et al. [7] subsequently investigated cross-lingual learning strategies for Khmer text recognition by exploring various pre-trained visual encoders and language decoders, demonstrating significant improvements in both recognition accuracy and generalization capability.

For the handwritten text modality, previous research has been largely limited to isolated character recognition [20] and short-word recognition [9], primarily focusing on historical palm-leaf manuscripts. Recently, Ngin et al. [21] constructed a Khmer online handwriting dataset that is used for Khmer writer verification purposes. Consequently, Khmer handwritten text recognition, particularly for modern Khmer, remains relatively underexplored.

In summary, current Khmer OCR research primarily focuses on developing modality-specific models. However, training separate models for each modality prevents cross-modality transfer learning, which could otherwise benefit modalities with limited data. In an end-to-end OCR pipeline, deploying separate models for each input modality results in significant memory overhead and increases the likelihood of errors, as each input must be routed to the appropriate model.

Therefore, in this paper, we propose a holistic approach to Khmer text recognition, aiming to accurately recognize Khmer text across all modalities. We address these challenges from both the dataset and model architecture perspectives.

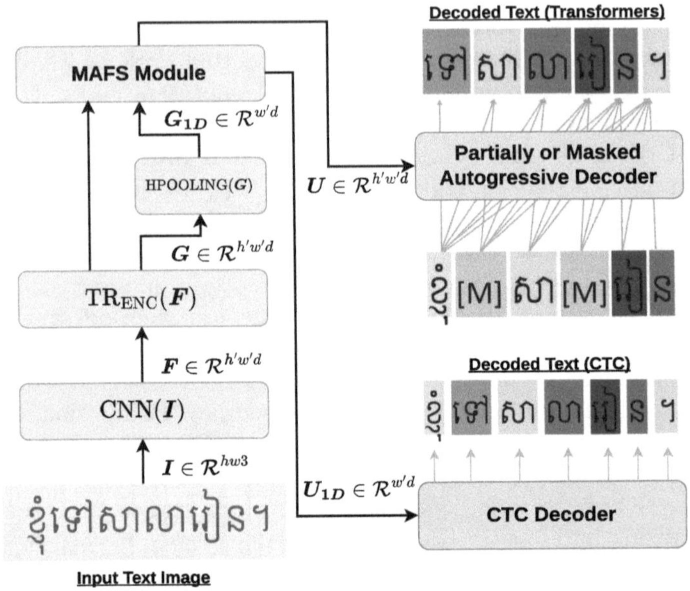

Fig. 3. The proposed UKTR framework.

3 Methodology

3.1 The Proposed Universal Khmer Text Recognition Framework (UKTR)

The proposed UKTR framework, as shown in Fig. 3, is designed to robustly recognize Khmer text across multiple modalities. It supports text generation in both non-autoregressive (NAR) and autoregressive (AR) modes. The architecture comprises a base convolutional visual encoder, a modality-aware adaptive feature selector, a CTC-based NAR decoder, and a Transformers-based AR decoder.

3.2 Visual Encoder

The visual encoder is designed to extract the essential visual features required for accurate character recognition. Following [5], our visual encoder consists of a base convolutional network for visual feature extraction, followed by a Transformers-based encoder to capture sequential dependencies.

Mathematically, the visual encoder can be expressed as

$$\boldsymbol{F} = \text{CNN}(\boldsymbol{I}) \tag{1}$$
$$\boldsymbol{G} = \text{TR}_{ENC}(\boldsymbol{F}) \tag{2}$$
$$\boldsymbol{G}_{1D} = \text{HPOOLING}(\boldsymbol{G}), \tag{3}$$

where CNN and TR_{ENC} are the base convolutional feature extractor and the Transformers encoder network, respectively. $\boldsymbol{I}$ is an input RGB image, $\boldsymbol{F}$ and $\boldsymbol{G}$ are 2D feature maps with a depth of d. Since the CTC decoder expects 1D feature maps, HPOOLING is used to average the 2D feature maps over the height dimension.

The base CNN feature extractor consists of six sequential ResNet blocks with increasing channel dimensions from 32 to 512, where each block is formed by 1×1 and/or 3×3 convolutions repeated multiple times. Spatial downsampling by a factor of $(2, 2)$ is applied in ResNet-Blocks 1 and 4, while the remaining blocks preserve the spatial resolution. The TR_{ENC} network uses an embedding dimension (d) of 512, with three stacked encoder layers, eight attention heads, a dropout rate of 0.1, and a feed-forward network dimension of 2048 ($4 \times d$). The visual encoder's model size and complexity are given in Table 1.

Table 1. The visual encoder's complexity and model size. Params: model parameters. Flops: floating operations on an input of 32×116 pixels.

Name	Params.	Flops
ResNet backbone (visual features)	13.0M	3.2G
Transformers encoder (sequential features)	9.5M	2.2G
Total	22.5M	5.4G

3.3 Modality-Aware Adaptive Feature Selector (MAFS)

To enable the proposed framework to robustly recognize Khmer text across different modalities, we introduce a modality-aware adaptive feature selector, which consists of three components: modality router, modality adapter, and multi-modality aggregator. Mathematically, the modality-aware adaptive feature selector can be expressed as

$$\boldsymbol{z} = \text{GPOOLING}(\boldsymbol{G}) \tag{4}$$

$$\boldsymbol{r} = \text{ROUTER}(\boldsymbol{z}) \tag{5}$$

$$\boldsymbol{H} = \text{ADAPTER}(\boldsymbol{G}) \tag{6}$$

$$\boldsymbol{H_{1D}} = \text{ADAPTER}(\boldsymbol{G_{1D}}) \tag{7}$$

$$\boldsymbol{U} = H \cdot \boldsymbol{r} \tag{8}$$

$$\boldsymbol{U_{1D}} = H_{1D} \cdot \boldsymbol{r}, \tag{9}$$

where GPOOLING is an average global pooling operator, which takes $\boldsymbol{G}$ and returns an average vector $\boldsymbol{z} \in R^d$. ROUTER is the router network, whick takes $\boldsymbol{z}$ and returns a probability vector $\boldsymbol{r} \in R^n$ over n text modalities (default: 5). ADAPTER is a feature projection network with a default project dimension (p) of 128, which takes $\boldsymbol{G}$ and $\boldsymbol{G_{1D}}$ and returns $\boldsymbol{H}$ and $\boldsymbol{H_{1D}}$, respectively. $\boldsymbol{U_{1D}}$ and $\boldsymbol{U}$ are the modality-aware adaptive features, which are marginalized over all modality sources, for the CTC and Transformers decoders, respectively.

As shown in Fig. 4, the role of the ADAPTER is to adapt visual features to different modalities (five adapters are used by default). Since the input modality is not known in advance and is not assumed to be available, the ROUTER estimates the probability distribution over all modalities. Based on this probability distribution, the adapted features are combined accordingly. This design eliminates the need for prior knowledge of the image modality and is therefore well suited for real-world deployment.

Although there are three discrete modalities, document, scene, and handwritten, real images rarely belong to only one modality. Instead, they can be represented within a simplex spanning these three modalities. For example, a text image may be both handwritten and scene simultaneously. By default, we set this hyperparameter (n) to five, providing additional capacity to model hybrid or transitional cases.

3.4 Text Decoders

To support both non-autoregressive (NAR) and autoregressive (AR) decoding within the same UKTR model, we incorporate both CTC and Transformers-based decoders. The CTC decoder aligns the $\boldsymbol{U_{1D}}$ feature maps with the target sequence $\boldsymbol{Y}$ by minimizing a CTC loss ($\boldsymbol{l_{CTC}}$) between the predicted token distributions $\hat{\boldsymbol{Y}}_{CTC}$, where $y_i \in \mathbb{R}^c$ (c is the number of unique tokens), and $\boldsymbol{Y}$.

The Transformers decoder converts the $\boldsymbol{H}$ feature maps along with the context vector (i.e., $\boldsymbol{Y}$ shifted one token to the right) into $\hat{\boldsymbol{Y}}_{TR}$, where $y_i \in \mathbb{R}^c$. It is trained by minimizing the cross-entropy loss ($\boldsymbol{l_{TR}}$) between $\hat{\boldsymbol{Y}}_{TR}$ and $\boldsymbol{Y}$. The Transformers decoder consists of three stacked layers with an embedding dimension (d) of 512, 8 attention heads, a feed-forward network dimension of 2048, and a dropout rate of 0.1. Following Buoy et al. [5], a masking ratio of 30% is applied. The Transformers decoder's complexity and model size are given in Table 2.

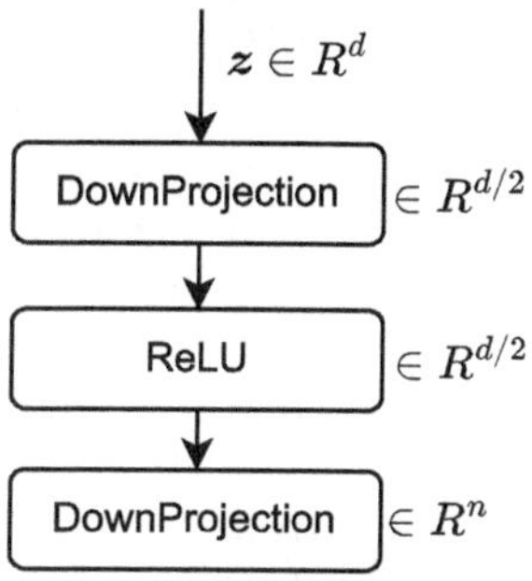

(a) The ROUTER network.

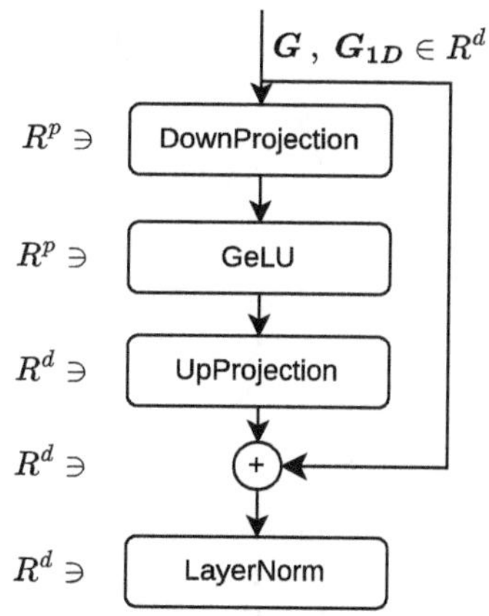

(b) The ADAPTER network.

Fig. 4. The designs of the ROUTER and ADAPTER networks.

As for the tokenizer, we extended the same Khmer character cluster tokenizer [5] by adding case-sensitive English characters, numbers, and symbols. The number of unique tokens (c) is 11,899. Mathematically, the total loss can be expressed as

$$l_{Total} = l_{CTC} + l_{TR}. \tag{10}$$

Table 2. The Transformers decoder's complexity and model size. Flops: floating operations on an input of 32 × 116 pixels.

Name	Params.	Flops
Transformers decoder (AR decoding)	24.08M	0.82G

4 Datasets

Previous methods [1,3–5] have primarily relied on synthetic training data, as it is technically convenient to generate large-scale datasets. In contrast, real training data for scene and handwritten text modalities remain scarce for the Khmer script. In the following subsections, we first describe existing public datasets, followed by a description of our newly developed datasets designed to address the gaps in real Khmer scene and handwritten text data.

Existing Datasets: Here are the existing available Khmer text recognition datasets:

- **Buoy et al.** [1]**'s Synthetic Document and Scene Textline Data**: Buoy et al. [1] synthetically generated approximately 2.8 million document and scene textline images, comprising 1.5 million document images and 1.3 million scene images, using 11 commonly used Khmer fonts.
- **KHOB** [22]: KHOB is a collection of Khmer textline images manually cropped from scanned low-resolution Khmer PDF documents. The processed dataset [5] contains 1,318 textline samples.
- **SynthText** [23]: 70,000 textline images were extracted from 10,000 synthetically generated Khmer identity card images.
- **Long et al.** [24]**'s HierText Data**: This dataset includes training (8,281 images), validation (1,724 images), and test (1,634 images) sets, with annotations for both Latin printed and handwritten textlines and words. After cropping into textlines and removing vertical texts from the training and validation sets, the dataset contains approximately 518,726 images.
- **Nom et al.** [6]**'s KhmerST**: This dataset contains real Khmer scene text images, mostly logos and billboards, captured in natural scenes (both indoor and outdoor). It includes a mixture of short words and long textlines. The dataset originally comprised 1,544 manually annotated images, yielding a total of 3,237 cropped text images. The processed KhmerST dataset [5] contains 3,022 text images. The dataset is particularly challenging due to complex imaging conditions and diverse artistic fonts.
- **Nom et al.** [8]**'s WildKhmerST**: This dataset represents an expanded version of KhmerST, containing 29,601 annotated textlines from 10,000 images captured in diverse public locations across Cambodia, such as streets, signboards, supermarkets, and commercial areas. It includes challenging text samples, including artistic, blurred, low-light, curved, and occluded text, as well as text embedded in complex backgrounds.
- **Soy** [25]**'s Khmer Handwritten Data (KH)**: This dataset contains approximately 4.2k handwritten and printed Khmer and Latin textline images. It is divided into training (3,991 images) and evaluation (211 images) sets.

The summary of the existing datasets and sample preview images are provided in Table 3 and Fig. 5, respectively.

New Datasets in This Study: To address the challenge of limited availability of real Khmer scene and handwritten text images, we introduce two new datasets for these modalities, as described below:

- **General Khmer Scene Text Data (GKST):** We manually annotated 4,221 Khmer scene text images. Similar to the KhmerST and WildKhmerST datasets, these images were captured using smartphone cameras under general indoor and outdoor conditions, exhibiting natural distortions and perspective

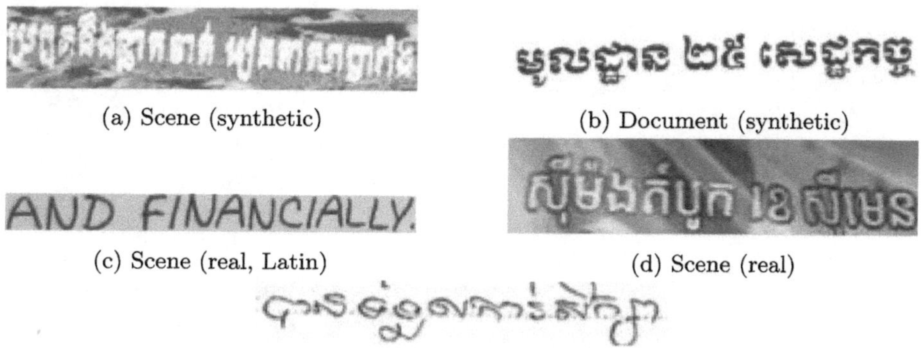

(a) Scene (synthetic) (b) Document (synthetic)

(c) Scene (real, Latin) (d) Scene (real)

(e) Handwritten

Fig. 5. Sample cropped images from the existing datasets. (a) Buoy et al. [1]. (b) KHOB. (c) HierText. (d) KhmerST. (e) Soy [25].

transformations. However, a key distinction is that we started with general scene images and annotated the text within them. In contrast, the KhmerST and WildKhmerST datasets contain focused, close-up text images, as shown in Fig. 6. The dataset is divided into train (4,009 images) and evaluation (212 images) sets.

- **General Khmer Handwritten Text Data (KHT):** Similarly, we manually annotated 14,168 Khmer handwritten text images from diverse sources, including handwritten birth certificates, exam papers, and notes, as shown in Fig. 7. The dataset is divided into training (13,457 images) and evaluation (711 images) sets.

(a) WildKhmerST (b) GKST (our)

Fig. 6. Sample images from the WildKhmerST dataset and our GKST dataset.

The summary of the new datasets and sample preview images are provided in Table 3 and Fig. 8, respectively.

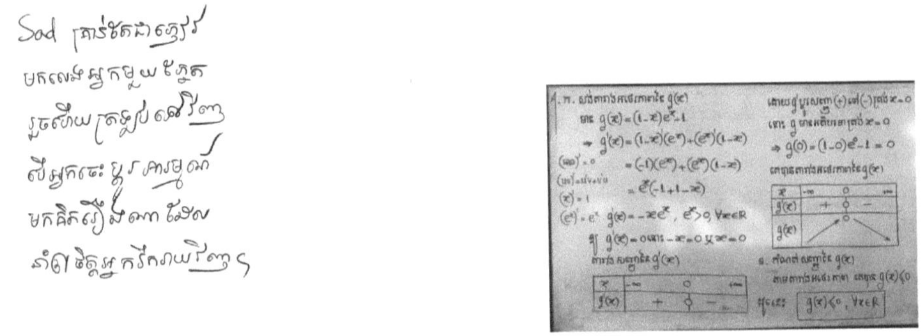

(a) Handwritten text on paper

(b) Handwritten text on whiteboard

Fig. 7. Sample images from our KHT dataset.

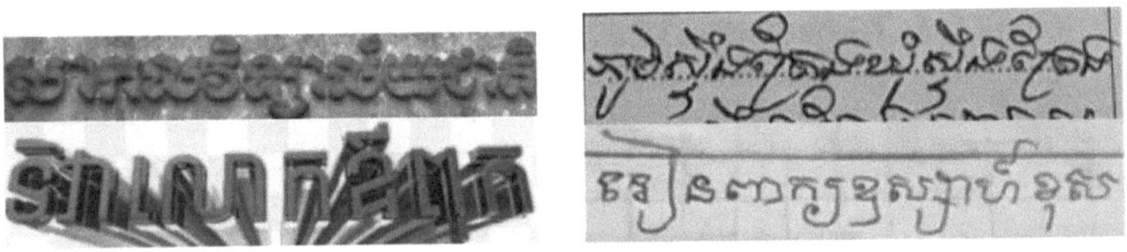

(a) Scene (real)

(b) Handwritten (real)

Fig. 8. Sample cropped images from our new datasets.

Document, Scene & Handwritten, and Evaluation Datasets: We categorize the above datasets into three groups according to their primary modalities as follows:

- **Document (D)**: The datasets in this group are primarily synthetic but large-scale, with the exception of the Latin HierText dataset. This group includes Buoy et al. [1], SynthText [23], and HierText [24]. The objective of this group is to learn visual representations of both Khmer and Latin scripts from large-scale printed document text data.
- **Scene & Handwritten (S&H)**: The datasets in this group consist of real data sources, primarily comprising scene and handwritten text. The group includes WildKhmerST [8], the training sets of Soy [25], and our newly collected datasets.
- **Evaluation**: To maintain consistency with the previous methods, the evaluation group includes the KhmerST [6] dataset (S), the KHOB [22] dataset (D), the evaluation sets of KH [25] (D&H), and our newly introduced GKST and KHT datasets (S&H).

Table 3. Summary of datasets, sizes, usage, and modalities. Eval.: evaluation. Doc.: document.

Dataset	Size		Purpose			Modality		
	Train	Eval.	Training (D)	Adapting (S&H)	Eval.	Doc.	Scene	Handwritten
Buoy et al. [1]	2.8M	–	✓	–	–	✓	✓	–
KHOB [22]	–	1,318	–	✓	✓	✓	–	–
SynthText [23]	70k	–	✓	–	–	✓	–	–
HierText [24]	518,726	–	✓	–	–	–	✓	✓
KhmerST [6]	–	3,022	–	✓	✓	–	✓	–
WildKhmerST [8]	29,601	–	–	✓	–	–	✓	–
KH [25]	3,991	211	–	✓	✓	✓	–	✓
GKST (ours)	4,009	212	–	✓	✓	–	✓	–
KHT (ours)	13,457	711	–	✓	✓	–	–	✓

5 Experiments, Results, and Discussion

5.1 Experimental Setup

The UKTR models were first trained on the large-scale document (D group) datasets to learn robust visual representations of Khmer and Latin characters. During this phase, a cyclic learning rate schedule was employed, with a minimum value of 10^{-5} and a maximum of 10^{-4}. This training phase lasted five full epochs with a batch size of 32 images.

The trained UKTR models then underwent a modality-adapting training phase on the real scene and handwritten (S&H group) datasets for 50 epochs. During this phase, a lower cyclic learning rate schedule was employed, with a minimum value of 10^{-6} and a maximum of 10^{-5}. In both phases, the Adam optimizer was used, with a gradient clipping value of 50.

To prevent the models from underperforming on printed document text images during the modality-adapting training phase, we sampled an equal number of document images and mixed them with the S&H images. This mixing strategy, together with the proposed MAFS module, enables the models to maintain robustness across all text modalities.

5.2 Results and Discussion

In terms of evaluation metrics, only character error rate (CER) is reported for evaluating Khmer text recognition performance in the subsequent analyses, as Khmer is an unsegmented script (i.e., it has no explicit word boundaries).

Recognition Performance of the UKTR Models: Table 4 presents the recognition accuracy comparisons of our UKTR models against existing methods across evaluation datasets of different modalities. The results show that our models (**UKTR (D + S&H)**) outperform the previous methods by a significant margin on all evaluation datasets except KHOB. Specifically, our best model using the Transformers decoder achieved CERs of 2.37%, 2.19%, 4.11%, 3.34%, and 6.10% on the KHOB, KhmerST, KH, GKST, and KHT datasets, respectively.

On the KHOB dataset (D), our best-performing model marginally underperformed compared to Buoy et al. [5], which achieved a CER of 2.13%. This is because their model, with its dedicated Transformers decoder, was specifically optimized and trained for recognizing printed document texts. In contrast, our UKTR models, employing both joint CTC and Transformers decoders, were trained and optimized to recognize Khmer text across multiple modalities.

For our UKTR models, comparing the Transformers decoder with the CTC decoder shows that the former moderately outperforms the latter. Specifically, the CER improvements achieved by the Transformers decoder are 0.90%, 0.83%, 1.78%, 1.07%, and 3.42% on the KHOB, KhmerST, KH, GKST, and KHT datasets, respectively. These results highlight the important role of language modeling enabled by the Transformers decoder in Khmer character recognition. However, these gains come at the cost of increased recognition latency, since the CTC decoder generates all tokens simultaneously, whereas the Transformers decoder generates tokens sequentially, one at a time (see Table 5).

For our UKTR models, comparing the general training phase (i.e.,**UKTR (D Only)**) using the synthetic datasets only with the modality-adapting training phase (**UKTR (D + S&H)**) using both real scene and handwritten datasets shows that the modality-adapted models achieve significantly lower CERs across all evaluation datasets. This demonstrates that our modality-adapting strategy successfully retains general training capabilities (for printed document recognition) while simultaneously acquiring new capabilities for additional modalities (handwritten and scene texts). This is facilitated by the proposed MAFS module, which is modality-aware and can adaptively select visual features based on the specific text modality. The ablation study of the proposed MAFS module will be discussed in the next subsection.

In summary, our UKTR models achieve SoTA recognition performance across all modalities, including document, scene, and handwritten texts. In addition, the same UKTR models can be configured to perform both autoregressive and non-autoregressive decoding.

Ablation Study of the MAFS Module: To evaluate the robustness of the proposed MAFS module, we established a baseline model in which the module was removed. Table 5 compares the CERs of models with and without the MAFS module. The results show that removing the MAFS module leads to a significant degradation in modality-adapting performance across all text modalities, for both decoder types. This occurs because, during modality-adapting training, the model without the MAFS module struggles to learn the relevant visual features required to accurately recognize Khmer text for a given input modality. Consequently, the model fails to retain its general training capabilities (for

Table 4. Character error rates (CER) on the evaluation datasets of different modalities (D,S,&H). * : model-specific tokenizer. Char.: character. KCC: Khmer character cluster. **Bold**: best. *Italic*: second best.

	Dec.	Tok.	KHOB(D)	KhmerST(S)	KH(D,H)	GKST(S)	KHT(H)
Tesseract-OCR [16]	CTC	Char.	9.19	40.96	–	–	–
Surya-OCR [15]	Tr.	*	17.69	43.21	–	–	–
Buoy et al. [1]	Tr.	Char.	3.03	–	–	–	–
Buoy et al. [4]	CTC	KCC	2.33	–	–	–	–
Nom et al. [6]	Tr.	*	–	17.00	–	–	–
Soy et al. [25]	Tr.	*	–	–	17.00	–	–
Buoy et al. [5]	Tr.	KCC	**2.13**	7.01	–	–	–
UKTR (D Only)	CTC	KCC	2.70	10.06	32.96	10.50	45.90
UKTR (D Only)	Tr.	KCC	2.55	14.34	36.30	15.26	52.52
UKTR (D + S&H)	CTC	KCC	2.46	*3.02*	*5.89*	*4.41*	*9.52*
UKTR (D + S&H)	Tr.	KCC	*2.37*	**2.19**	**4.11**	**3.34**	**6.10**

printed document text recognition) and is also unable to acquire new capabilities for scene and handwritten text modalities.

Table 5. Character error rates (CER) w/ and w/o the MAFS module (**D + S&H**). *: model-specific tokenizer. Char.: character. KCC: Khmer character cluster. **Bold**: best. *Italic*: second best.

	Decoder	Tokenizer	Time(ms)	KHOB	KhmerST	KH	GKST	KHT
UKTR w/o MAFS	CTC	KCC	23	3.85	4.30	8.17	6.15	11.89
UKTR w/o MAFS	Tr.	KCC	775	3.93	5.48	5.84	5.39	*7.66*
UKTR w/ MAFS	CTC	KCC	27	*2.46*	*3.02*	*5.89*	*4.41*	9.52
UKTR w/ MAFS	Tr.	KCC	777	**2.37**	**2.19**	**4.11**	**3.34**	**6.10**

Effect of the Number of Modality Sources: The ROUTER network estimates a probability distribution over modality sources ($n = 5$). To evaluate the impact of this choice, we also considered two baseline cases with $n = 3$ and $n = 8$.

Table 6 indicates that the impact of the number of modality sources (n) on performance is subtle and dataset-dependent. Setting $n = 3$ slightly improves recognition on KHOB, KhmerST, GKST, and KHT, but slightly reduces accuracy on KH. Increasing n to eight yields minor gains on KHOB at the expense of other datasets. Overall, $n = 3$ and $n = 5$ provide essentially equivalent performance. While $n = 3$ already captures the three principal modalities effectively, $n = 5$ provides additional capacity to model hybrid or transitional cases with marginal but consistent improvements on several benchmarks. By contrast, $n = 8$ may introduce redundancy and optimization difficulty, which can explain the lack of further gains.

Table 6. Character error rates (CER) for $n = 3$, $n = 5$, and $n = 8$. * : model-specific tokenizer. Char.: character. KCC: Khmer character cluster. **Bold**: best. *Italic*: second best.

	Decoder	Tokenizer	KHOB	KhmerST	KH	GKST	KHT
UKTR ($n = 3$, **D + S&H**)	CTC	KCC	2.43	3.03	7.12	4.57	9.87
UKTR ($n = 3$, **D + S&H**)	Tr.	KCC	*2.29*	**2.12**	5.17	**2.99**	**5.82**
ine **UKTR** ($n = 5$, **D + S&H**)	CTC	KCC	2.46	3.02	5.89	4.41	*9.52*
UKTR ($n = 5$, **D + S&H**)	Tr.	KCC	2.37	*2.19*	**4.11**	*3.34*	*6.10*
ine **UKTR** ($n = 8$, **D + S&H**)	CTC	KCC	2.45	3.09	6.50	4.79	9.71
UKTR ($n = 8$, **D + S&H**)	Tr.	KCC	**2.07**	2.31	*4.67*	3.78	6.44

Qualitative Assessment: Figure 9 presents qualitative comparisons between the ground-truth texts and the model predictions obtained using the CTC and Transformer decoders for several randomly selected input images. Consistent with the quantitative results, predictions from the Transformer decoder are generally more accurate than those from the CTC decoder. Except for a highly blurry input image (fourth row), the Transformer decoder successfully extracts the correct texts across all modalities. In contrast, the CTC decoder tends to struggle when the input images are either severely blurred (fourth row) or contain complex backgrounds (sixth row) or with complex handwritten text (second row).

6 Limitations and Future Work

We identify the following limitations associated with this study and future directions as follows:

1. The ROUTER network is trained by minimizing the overall recognition loss (l_{Total}) rather than any modality-specific supervision loss. Thus, future research should introduce an explicit modality supervision loss to enable the ROUTER network to learn the labelled modality distributions. In addition, further ablation studies examining alternative MAFS designs should be conducted, including hard-routing mechanisms, adapter-only variants (no router), and general mixture-of-experts (MoE) architectures.
2. Although the UKTR model is trained by jointly optimizing two decoders, CTC and Transformers, the choice of decoder during inference is independent of the input text image. Future research could explore a decoder router network that dynamically selects the most suitable decoder for a given image, optimizing both latency and accuracy.

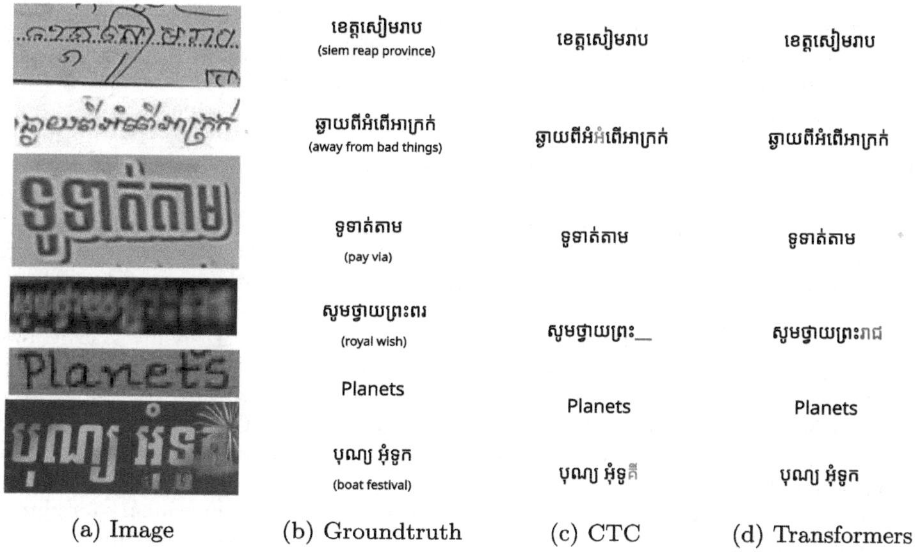

Fig. 9. Sample images, groundtruth texts, and model predictions (CTC vs. Transformers).

3. Further comparisons with general Latin-based scene text recognition models are required to further strengthen the robustness of the proposed method. Future work will adapt and benchmark the SoTA CTC-based and Transformer-based methods on Khmer datasets to provide a more comprehensive evaluation.

7 Conclusions

This paper presented a UKTR framework capable of recognizing Khmer text across multiple modalities, including document, scene, and handwritten texts. To address the challenge of limited availability of real Khmer scene and handwritten training data and benchmarks, we introduced new datasets for these modalities. Experimental results demonstrate that the proposed UKTR models achieved SoTA recognition performance on all benchmark datasets.

Acknowledgments. This study was supported by JSPS Kakenhi Grant Number 26K14965 and JST CRONOS Grant Number JPMJCS24K4.

References

1. Buoy, R., Iwamura, M., Srun, S., Kise, K.: Toward a low-resource non-Latin-complete baseline: an exploration of khmer optical character recognition. IEEE Access. **11**, 128044–128060 (2023)

2. Kaing, H., et al.: Towards tokenization and part-of-speech tagging for Khmer: Data and discussion. Trans. Asian Low-Resour. Lang. Inf. Process. **20**, 1–16 (2021)
3. Buoy, R., Taing, N., Chenda, S., Kor, S.: Khmer printed character recognition using attention-based Seq2Seq network. Ho Chi Minh City Open Univ. J. Sci.-Eng. Technol. 3–16 (2022)
4. Buoy, R., Iwamura, M., Srun, S., Kise, K.: Language-aware non-autoregressive khmer textline recognition. In: International Conference on Pattern Recognition and Artificial Intelligence, pp. 339–353 (2024)
5. Buoy, R., Chenda, S., Taing, N., Kong, M., Iwamura, M., Kise, K.: Addressing the attention drift problem for khmer long textline recognition. Int. J. Doc. Anal. Recogn. (IJDAR) 1–26 (2025)
6. Nom, V., Bakkali, S., Luqman, M., Coustaty, M., Ogier, J.: KhmerST: a low-resource khmer scene text detection and recognition benchmark. Inm: Proceedings of the Asian Conference on Computer Vision, pp. 1777–1792 (2024)
7. Nom, V., Keo, S., Bakkali, S., Luqman, M., Coustaty, M., Ogier, J.: Cross-Lingual Learning for Low-Resource Khmer Scene Text Detection and Recognition. In: International Conference On Document Analysis And Recognition, pp. 347–365 (2025)
8. Nom, V., et al.: WildKhmerST: a comprehensive dataset and benchmark for khmer scene text detection and recognition in the wild. In: International Conference On Document Analysis And Recognition, pp. 351–368 (2025)
9. Valy, D., Verleysen, M., Chhun, S., Burie, J.: Character and text recognition of khmer historical palm leaf manuscripts. In: 2018 16th International Conference On Frontiers In Handwriting Recognition (ICFHR), pp. 13–18 (2018)
10. Graves, A., Fernández, S., Gomez, F., Schmidhuber, J.: Connectionist temporal classification: labelling unsegmented sequence data with recurrent neural networks. In: Proceedings Of The 23rd International Conference On Machine Learning, pp. 369–376 (2006)
11. Vaswani, A., et al.: Attention is all you need. Adv. Neural Inf. Process. Syst. **30** (2017)
12. Chey, C., Kumhom, P., Chamnongthai, K.: Khmer printed character recognition by using wavelet descriptors. Int. J. Uncertainty Fuzziness Knowl.-Based Syst. **14**, 337–350 (2006)
13. Sok, P., Taing, N.: Support vector machine (SVM) based classifier for khmer printed character-set recognition. In: Signal and Information Processing Association Annual Summit and Conference (APSIPA), 2014 Asia-Pacific, pp. 1–9 (2014)
14. Keo, S., Coustaty, M., Bakkali, S., Rossinyol, M.: State-of-the-Art Khmer Text Recognition Using Deep Learning Models. In: ASEAN Conference on Emerging Technologies 2024 (2024)
15. Paruchuri, V., Team, D., Surya: A lightweight document OCR and analysis toolkit (2025). https://github.com/VikParuchuri/surya, GitHub repository
16. Google, Tesseract OCR Engine. https://github.com/tesseract-ocr/tesseract, Accessed 16 Feb 2026
17. Wei, H., Sun, Y., Li, Y.: Deepseek-ocr: Contexts optical compression. ArXiv Preprint ArXiv:2510.18234 (2025)
18. Wei, H., Sun, Y., Li, Y.: DeepSeek-OCR 2: Visual Causal Flow. ArXiv Preprint ArXiv:2601.20552 (2026)
19. Li, M., et al.: Trocr: Transformers-based optical character recognition with pre-trained models. In: Proceedings Of The AAAI Conference On Artificial Intelligence, Vol. 37, 13094–13102 (2023)
20. Annanurov, B., Noor, N.: A compact deep learning model for Khmer handwritten text recognition. IAES Int. J. Artif. Intell. **10**, 8938–8938 (2021)

21. Ngin, K., Valy, D., Phauk, S., Pin, V., Yin, B., Ang, S.: KhmerWriterID: toward robust khmer writer verification using deep learning: CNN, RNN, and hybrid approaches. IEEE Access (2026)
22. EKYCSolutions Ekycsolutions/Khmer-OCR-benchmark-dataset: A standardized benchmark dataset for Khmer Optical Character Recognition (OCR) engine. GitHub, https://github.com/EKYCSolutions/khmer-ocr-benchmark-dataset
23. Seanghay SynthKhmer-10k. (Hugging Face, 2024). https://huggingface.co/datasets/seanghay/SynthKhmer-10k, Accessed 17 Feb 2026
24. Long, S., Qin, S., Panteleev, D., Bissacco, A., Fujii, Y., Raptis, M.: Towards End-to-End Unified Scene Text Detection and Layout Analysis. In: Proceedings of the IEEE/CVF Conference on Computer Vision and Pattern Recognition (2022)
25. Soy, V., Kimly, Y., Malis, L., Chean, B.: Khmer Handwritten OCR using Pretrained TrOCR Architecture (2024). https://www.researchgate.net/publication/385417010_Khmer_Handwritten_OCR_using_Pretrained_TrOCR_Architecture, Poster presentation

Author Index

G. A. Fink et al. (Eds.): ICDAR 2026, LNCS 16972, pp. 717–719, 2027.
https://doi.org/10.1007/978-3-032-36023-6

FSC
www.fsc.org
MIX
Papier | Fördert
gute Waldnutzung
FSC® C083411

Kolibri 360 GmbH
Ferdinand-Jühlke-Straße 7
99095 Erfurt, Deutschland
produktsicherheit@kolibri360.de